Nanomaterials and Nanotechnology in Medicine

Nanomaterials and Nanotechnology in Medicine

Edited by

Dr. P.M. Visakh
Assistant Professor, Department of Physical Electronics
TUSUR University,
Tomsk, Russia

WILEY

This edition first published 2023

Registered Offices
John Wiley & Sons, Inc., 111 River Street, Hoboken, NJ 07030, USA
John Wiley & Sons Ltd, The Atrium, Southern Gate, Chichester, West Sussex, PO19 8SQ, UK

Editorial Office
The Atrium, Southern Gate, Chichester, West Sussex, PO19 8SQ, UK

For details of our global editorial offices, customer services, and more information about Wiley products visit us at www.wiley.com.

Wiley also publishes its books in a variety of electronic formats and by print-on-demand. Some content that appears in standard print versions of this book may not be available in other formats.

Library of Congress Cataloging-in-Publication Data

Names: P. M., Visakh, editor.
Title: Nanomaterials and nanotechnology in medicine / edited by Dr. P.M. Visakh, Assistant Professor, Department of Physical Electronics, TUSUR University, Tomsk, Russia.
Description: Hoboken, NJ : Wiley, 2022. | Includes bibliographical references and index.
Identifiers: LCCN 2022027360 (print) | LCCN 2022027361 (ebook) | ISBN 9781119558033 (hardback) | ISBN 9781119558040 (adobe pdf) | ISBN 9781119558095 (epub)
Subjects: LCSH: Nanomedicine. | Nanotechnology. | Nanostructured materials–Therapeutic use. | Medical technology.
Classification: LCC R857.N34 N25643 2022 (print) | LCC R857.N34 (ebook) | DDC 610.285–dc23/eng/20220802
LC record available at https://lccn.loc.gov/2022027360
LC ebook record available at https://lccn.loc.gov/2022027361

Cover Design: Wiley
Cover Image: © Volodymyr Horbovyy/Shutterstock

Set in 9.5/12.5pt STIXTwoText by Straive, Pondicherry, India
Printed and bound by CPI Group (UK) Ltd, Croydon, CR0 4YY

C9781119558033_030922

Contents

List of Contributors

Elham Abohamzeh
Department of Energy, Materials, and Energy Research Center (MERC)
Karaj, Iran

Chander Amgoth
Department of Chemistry
Zhejiang University
Hangzhou, China

António Francisco Ambrósio
University of Coimbra, Coimbra Institute for Clinical and Biomedical Research (iCBR), Faculty of Medicine
Coimbra, Portugal

and

University of Coimbra
Center for Innovative Biomedicine and Biotechnology (CIBB), Coimbra, Portugal

and

Clinical Academic Center of Coimbra (CACC)
Coimbra, Portugal

and

Association for Innovation and Biomedical Research on Light and Image (AIBILI)
Coimbra, Portugal

V.K. Ameena Shirin
College of Pharmaceutical Sciences
Government Medical College
Kozhikode, Kerala, India

Saritha Appukuttan
Department of Chemistry
Amrita Vishwa Vidyapeetham
Kollam, Kerala, India

Liliana Aranda Lara
Facultad de Medicina
Universidad Autónoma del Estado de México
Toluca, Estado de México, Mexico

Ehsan Aryan
Antimicrobial Resistance Research Center
Department of Medical Bacteriology and Virology, Qaem University Hospital
School of Medicine, Mashhad University of Medical Sciences
Mashhad, Iran

Hongzhen Bai
Department of Chemistry
Zhejiang University
Hangzhou, China

Francesco Baino
Institute of Materials Physics and Engineering
Department of Applied Science and Technology, Politecnico di Torino
Torino, Italy

Ajit Behera
Department of Metallurgical & Materials Engineering
National Institute of Technology
Rourkela, Odisha, India

Asit Behera
School of Mechanical Engineering, KIIT
Bhubaneswar, India

Rupali S. Bhadale
Shobhaben Pratapbhai Patel School of Pharmacy and Technology Management
SVKM's NMIMS
Mumbai, Maharashtra, India

Elisa J. Campos
University of Coimbra, Coimbra Institute for Clinical and Biomedical Research (iCBR), Faculty of Medicine
Coimbra, Portugal

and

University of Coimbra
Center for Innovative Biomedicine and Biotechnology (CIBB)
Coimbra, Portugal

and

Clinical Academic Center of Coimbra (CACC), Coimbra, Portugal

António Campos
University of Coimbra, Coimbra Institute for Clinical and Biomedical Research (iCBR), Faculty of Medicine
Coimbra, Portugal

and

University of Coimbra
Center for Innovative Biomedicine and Biotechnology (CIBB), Coimbra, Portugal

and

Clinical Academic Center of Coimbra (CACC)
Coimbra, Portugal

and

Department of Ophthalmology, Centro Hospitalar Leiria E.P.E., Leiria, Portugal

and

ciTechCare, Center for Innovative Care and Health Technology, Polytechnic Institute of Leiria, Leiria, Portugal

Shuai Chen
Department of Chemistry
Zhejiang University
Hangzhou, China

D. Dharun Daniel Raj
Department of Biotechnology
Maulana Abul Kalam Azad University of Technology
Haringhata, West Bengal, India

Moataz A. Elsawy
Polymer Laboratory, Petrochemical Department
Egyptian Petroleum Research Institute
Nasr City, Cairo, Egypt

Anila Fariq
Department of Biotechnology
University of Kotli
Azad and Jammu Kashmir, Pakistan

and

Department of Biotechnology
Fatima Jinnah Women University
Rawalpindi, Pakistan

Guillermina Ferro-Flores
Departamento de Materiales Radiactivos
Instituto Nacional de Investigaciones Nucleares
Ocoyoacac, Estado de México Mexico

Jayalatha Gopalakrishnan
Department of Polymer Science and Rubber Technology
Cochin University of Science and Technology
Kochi, Kerala, India

Eggehard Holler
Nanomedicine Research Center in the Department of Neurosurgery
Cedars Sinai Medical Center
Los Angeles, CA, USA

Keila Isaac-Olivé
Facultad de Medicina
Universidad Autónoma del Estado de México
Toluca, Estado de México, Mexico

Liron L. Israel
Nanomedicine Research Center in the Department of Neurosurgery
Cedars Sinai Medical Center
Los Angeles, CA, USA

Jitha S. Jayan
Department of Chemistry
Amrita Vishwa Vidyapeetham
Kollam, Kerala, India

Kuruvilla Joseph
Department of Chemistry
Indian Institute of Space Science and Technology
Valiyamala, Kerala, India

Saeid Kargozar
Tissue Engineering Research Group (TERG), Department of Anatomy and Cell Biology
School of Medicine
Mashhad
University of Medical Sciences
Mashhad, Iran

V. Karthik
Department of Biotechnology
K.S. Rangasamy College of Technology
Tiruchengode, Tamil Nadu, India

Farzad Kermani
Department of Materials Engineering
Faculty of Engineering
Ferdowsi University of Mashhad (FUM)
Mashhad, Iran

Gerardo Leyva-Gómez
Facultad de Química
Universidad Nacional Autónoma de México
Ciudad de México, Mexico

Vaishali Y. Londhe
Shobhaben Pratapbhai Patel School of Pharmacy and Technology Management
SVKM's NMIMS, Mumbai
Maharashtra, India

Nancy Evelyn Magaña Vergara
Facultad de Ciencias Químicas
Universidad de Colima
Colima, Mexico

João Martins
University of Coimbra, Coimbra Institute for Clinical and Biomedical Research (iCBR), Faculty of Medicine
Coimbra, Portugal

and

University of Coimbra
Center for Innovative Biomedicine and Biotechnology (CIBB), Coimbra, Portugal

and

Clinical Academic Center of Coimbra (CACC)
Coimbra, Portugal

and

University of Coimbra, Coimbra Institute for Biomedical Imaging and Translational Research (CIBIT), Institute for Nuclear Sciences Applied to Health (ICNAS)
Coimbra, Portugal

Néstor Mendoza Muñoz
Facultad de Ciencias Químicas,
Universidad de Colima
Colima, Mexico

Zahra Meshkat
Antimicrobial Resistance Research Center
Department of Medical Bacteriology and Virology, Qaem University Hospital
School of Medicine
Mashhad
University of Medical Sciences
Mashhad, Iran

Majid G. Mobarhan
Metabolic Syndrome Research Center
School of Medicine
Mashhad
University of Medical Sciences
Mashhad, Iran

Mohamed H. Mostafa
Polymer Laboratory, Petrochemical Department
Egyptian Petroleum Research Institute
Nasr City, Cairo, Egypt

S.S. Mohapatra
Department of Mechanical Engineering
NIT Rourkela, Odisha, India

Ranjan K. Mohapatra
Department of Chemistry
Government College of Enginnering
Keonjhar, Odisha, India

Enrique Morales-Avila
Facultad de Química
Universidad Autónoma del Estado de México
Toluca, Estado de México, Mexico

T.G.N. Nagarjun
Department of Food Technology
K.S. Rangasamy College of Technology
Tiruchengode, Tamil Nadu, India

Simin Nazarnezhad
Tissue Engineering Research Group (TERG)
Department of Anatomy and Cell Biology
School of Medicine
Mashhad
University of Medical Sciences
Mashhad, Iran

Blanca Ocampo-García
Departamento de Materiales Radiactivos
Instituto Nacional de Investigaciones Nucleares
Ocoyoacac, Estado de México, Mexico

S. Poornima
Department of Biotechnology
K.S. Rangasamy College of Technology
Tiruchengode, Tamil Nadu, India

K. Pramod
College of Pharmaceutical Sciences
Government Medical College
Kozhikode, Kerala, India

David Quintanar-Guerrero
Facultad de Estudios Superiores Cuautitlán
Universidad Nacional Autónoma de México
Mexico, Mexico

Ramya Rajan
Department of Chemistry
Amrita Vishwa Vidyapeetham
Kollam, Kerala, India

Majid Rezayi
Medical Toxicology Research Center
Mashhad
University of Medical Sciences
Mashhad, Iran

and

Department of Medical Biotechnology and Nanotechnology, School of Medicine
Mashhad University of Medical Sciences
Mashhad, Iran

Kobra S. Rizi
Antimicrobial Resistance Research Center
Department of Medical Bacteriology and Virology, Qaem University Hospital
School of Medicine
Mashhad
University of Medical Sciences
Mashhad, Iran

and

Department of Medical Biotechnology and Nanotechnology, School of Medicine
Mashhad
University of Medical Sciences
Mashhad, Iran

Chinnu Sabu
College of Pharmaceutical Sciences
Government Medical College
Kozhikode, Kerala, India

A.K. Sahoo
School of Mechanical Engineering
KIIT
Bhubaneswar, India

Renu Sankar
College of Pharmaceutical Sciences
Government Medical College
Kozhikode, Kerala, India

M. Saravanan
AMR and Nanotherapeutics Laboratory
Department of Pharmacology
Saveetha Dental College
Saveetha Institute of Medical and Technical Sciences
Chennai, India

Ayesha Selhaba
Department of Biotechnology
University of Kotli
Azad and Jammu Kashmir, Pakistan

Ahmad Shafee
Public Authority of Applied Education & Training, College of Technological Studie
Applied Science Department
Shuwaikh, Kuwait

M. Sheikholeslami
Department of Mechanical Engineering
Babol Noshirvani University of Technology
Babol, Iran

and

Renewable energy systems and nanofluid applications in heat transfer Laboratory
Babol Noshirvani University of Technology
Babol, Iran

R. Subbaiya
Department of Biological Sciences
School of Mathematics and Natural Sciences
The Copperbelt University
Riverside, Kitwe, Zambia

Guping Tang
Department of Chemistry
Zhejiang University
Hangzhou, China

P. M. Visakh
Department of Physical Electronics
TUSUR University
Tomsk, Russia

A. Vigneshwaran
School of Biosciences and Technology
Vellore Institute of Technology
Vellore, Tamil Nadu, India

Azra Yasmin
Department of Biotechnology
Fatima Jinnah Women University
Rawalpindi, Pakistan

Kaxi Yu
Department of Chemistry
Zhejiang University
Hangzhou, China

Anum Zulfiqar
Department of Biotechnology
Fatima Jinnah Women University
Rawalpindi, Pakistan

Preface

This book, *Nanomaterials and Nanotechnology in Medicine*, summarizes recent research accomplishments in the area of nanomedicine. It discusses topics such as: challenges and opportunities, nanoneurology, nanomolecular diagnostics, nanopharmaceuticals, the role of nanotechnology in biological therapies, nanomaterials for gene therapy, nanomaterials in ophthalmology/ocular diseases, nanonutrition, rubber-based nanocomposites for biomedical applications etc. The book is intended to provide a valuable reference source for university and college faculties, professionals, post-doctoral research fellows, senior graduate students, researchers, and medical doctors. The various chapters in this book are written by prominent researchers from industry, hospitals, academia, and government and private research laboratories across the globe. It provides an up-to-date record of the major findings and observations in the field of nanomedicine.

Chapter 1 provides an overview of the main points of the various chapters by focusing on scope, state-of-the-art, applications, new challenges and opportunities of nanomedicine. The second chapter describes important applications of nanoneurology. In this chapter the author introduces the topic and discusses recent advances in nanomaterials and their application in: neurodegenerative diseases; nanomaterial application for stroke; nanomaterial application for spinal cord injury; nanomaterial application for brain tumor; the potential adverse effects of nanomaterials in nanoneurology; and regulatory issues.

Nanodiagnostics, nanoparticles for molecular diagnostics, the comparison between nanomaterials and other materials in molecular diagnostics, and the prospects of nanodiagnostics are described in Chapter 3. The fourth chapter focuses on nanopharmaceuticals. In this chapter, after the introduction and identification of recent advances, the authors discuss the following topics: liposomes in nanopharmaceuticals, polymeric nanoparticles in nanopharmaceuticals, solid lipid nanoparticles in nanopharmaceuticals, dendrimers in nanopharmaceuticals, and quantum dots in nanopharmaceuticals. The role of nanotechnology in biological therapies is the emphasis in the fifth chapter. In this chapter, the authors address different topics such as biological therapies, nanoparticles in biological therapies, application of nanotechnology in biological therapies, and the advantages and disadvantages of nanoparticles in biological therapies.

The sixth chapter's focus is gene therapy and includes nanomaterials and their physicochemical properties, target organ biocompatibility/toxicity, and gene delivery. The following chapter concentrates on nanomaterials in ophthalmology and ocular diseases. In this chapter, the author explains ocular anatomy, physiological barriers in the eye, methods of ocular disease treatment, and the use of nanomedicine in ocular therapy. Nanonutrition, the focus of

Chapter 8, covers topics such as nanostructured delivery systems, nanoparticles based on inorganic materials, and metal nanoparticles. Rubber-based nanocomposites provide the subject content for Chapter 9 and includes discussion of biomedical applications for rubber-based nanoparticles including their applications and their use in nanocomposites.

Chapter 10 provides an overview of the materials development of nanoparticles and the challenges and opportunities encountered including the scope and state-of-the-art of nanomedicine. The next chapter, 11, describes the past, present, and future aspects of nanomedicine, that is, research and education in nanomedicine. In discussing tissue engineering and regenerative medicine the authors cover the general concepts, implantable nanomaterials to regenerate living tissues, nanomaterials as carriers for therapeutic agents, nanofibrous scaffolds, and nanotopography techniques for tissue-engineered scaffolds. The authors of Chapter 12 address a wide range of topics under the umbrella of nanorobotics. This includes nanorobots in: nanomedicine; medical imaging; targeted drug delivery; enzymatic nanolithography; a biomimetic approach; cell biochips; precision surgery; and detoxification. They further discuss the fabrication, toxicity, administration and retrieval of nanorobots as well as the clinical presence of nanorobots and their reproducibility and standardization. Chapter 13's focus is on nanosensors and their application where the authors discuss the classification and fabrication of nanosensors, and inorganic and biopolymer-derived nanosensors.

Inorganic nanoparticles for drug-delivery applications are addressed in Chapter 14. This chapter contains explanations and discussions of the synthesis of inorganic nanoparticles, as well as inorganic nanoparticles' properties and functionalization, and drug delivery by: quantum dots; mesoporous silica nanoparticles; silver, gold, and iron oxide nanoparticles. Also contained in this chapter is a consideration of hybrid systems and the prospects for inorganic nanoparticle applications. Chapter 15 introduces intelligent nanomaterials, including: their design and function; the range for medicines; and the type of stimuli required in medicine as well as their clinical applications and potential risk factors in use.

Polymer-based nanocomposites, Chapter 16, addresses topics such as: polymers for biomedical fields, the synthesis of nanocomposites; characterization tools for nanocomposites; the size, shape, and morphology of nanocomposites, and polymer nanocomposites for various applications including use in molecular diagnosis, biopharmaceutics. Chapter 17 considers the effect of the toxicity and physiochemical properties of nanoparticle applications and includes discussion of the biodistribution, mechanism, and excretion of nanomaterials. The following chapter, Chapter 18, focuses on multifunctional nanoparticles in medical applications. The authors discuss diagnostic and therapeutic application, and external-stimuli responsive nanoparticles as part of their account. The authors in the final chapter provide information on and discussion of the antimicrobial properties and applications of nanoparticles.

Finally, I would like to express my sincere gratitude to all the contributors of this book, who provided excellent support in the successful completion of this venture. I am grateful to them for the commitment and the sincerity they have shown toward their contributions in the book. Without their enthusiasm and support, the compilation of book could have not been possible. I would like to thank all the reviewers who have taken their valuable time to make critical comments on each chapter. I also thank John Wiley & Sons for recognizing the demand for such a book, and for realizing the increasing importance of *Nanomaterials and Nanotechnology in Medicine* and for starting such a new project area.

Dr. Visakh P. M.

1

Nanomaterials and Nanotechnology in Medicine

Medical Applications: Challenges and Opportunities

P. M. Visakh

Department of Physical Electronics, TUSUR University, Tomsk, Russia

1.1 Nanoneurology

In principle, nanosystems with surface modifications can provide blood-brain barrier (BBB)-crossing, imaging, and therapeutic treatment especially in the case of surface-modified, encapsulating nanosystems. They can also reduce off-target effects. Frequently, nanosystems are designed to combine ample targeting, imaging, and drug loads designed for high functional capacity. In many cases, disease pathology can cause the impairment of the BBB. For example, in brain tumors or metastasis, rapid angiogenesis in the growing tumor results in leaky vessels. The phenomenon that distinguishes tumor from healthy vessels is known as the enhanced permeability and retention effect (EPR) [1, 2].

The main medicinal applications for nanomaterials are imaging and therapy that require targeted delivery. Nanomaterials can be applied in separate or combined "theranostic" fashion since nanomaterials can often carry multiple functionalities. More frequently, targeting moieties are attached to a nanosystem for tracking the administered drug and its safety to unfold its pharmaceutical activity only at the intended site and avoid side ("off-target") effects.

Therapeutic agents that can be delivered by nanotechnology or that are under investigation vary from low molecular weight molecules such as synthetic pharmaceuticals and peptides and high molecular weight proteins such as antibodies, enzymes, and, most challenging, nucleic acids. The mechanisms underlying treatments are versatile as well and may include immunotherapy, photodynamic therapy, thermotherapy, and gene delivery. Nanoparticles, which include micelles, dendrimers, and spheres, are often being studied for central nervous system (CNS) delivery. Micelles [3] are spherical structures of amphiphilic molecules that are organized in accordance with their hydrophobic tail forming the core and their hydrophilic head contacting the aqueous solvent. In contrast with polymers of linear platforms, dendrimers are repetitively branched nanomolecules.

Nanomaterials and Nanotechnology in Medicine, First Edition. Edited by P.M. Visakh.

Another type of nanomaterial includes carbon nanotubes (CNTs), consisting of a crosslinked carbon–carbon structure. Members of the fullerene structural family are allotropes of carbon with a cylindrical nanostructure, either single- or multi-walled involving entirely sp2-hybrid carbon atoms. Since they have unique properties, in term of strength as well as electrical (semiconductors), optical, and thermal properties, they are also studied for BBB delivery, as well as fullerenes of spherical structure. The biocompatibility is disputed. In recent years, cell-derived delivery systems have been studied for CNS delivery. One example is the "nanoghost" (NG) system [4, 5].

Using mostly but not exclusively mesenchymal stem cells [45–48], these derived vesicles are prepared by physical (e.g., sonication) and chemical treatment of cells to remove cytoplasm and nuclei leaving a small (~200 nm) membrane compartment that has inherited targeting capabilities from the "mother" cell. Current NGs are derived from cancer stem cells to treat tumors. Alzheimer's disease (AD) and Parkinson's disease (PD) are the most common types, and some other examples include Huntington's disease, amyotrophic lateral sclerosis (ALS), prion disease, tauopathies, and several types of non-AD dementia, which is a collective term used to describe various symptoms of cognitive decline. Unlike other neurodegenerative diseases such as PD, where there are treatments available, the complexity of AD leads to the fact that to date, only the symptoms are treatable and there is no cure for the disease. As the age of the population is on the rise, so is the number of people who will be affected by the disease in older age [6]. Therefore, AD is one of the most studied neurodegenerative diseases.

Studies suggest that passive diffusion across the BBB may be reduced due to a thickened basement membrane and that the BBB paracellular route remains unaffected, potentially lowering the ability of nanomedicines to cross AD BBB; therefore, the extent of BBB disruption in AD remains controversial [7]. In another study, magnetic Fe_3O_4 nanoparticles coated with oleic acid molecules were used as a nanocarrier loaded with α-synuclein RNAi plasmid [8].

N-isopropylacrylamide derivative (NIPAm-AA) was photo-immobilized onto the oleic acid molecules, and short hairpin RNA (shRNA) was absorbed. The same method was used to absorb nerve growth factor (NGF) to NIPAm-AA to specifically promote neuronal uptake via NGF receptor-mediated endocytosis.

PD, is another neurodegenerative disease characterized by a progressive loss of dopaminergic neurons and an elevated level of intracytoplasmic Lewy bodies (abnormal aggregates of α-synuclein protein) in the remaining neurons located in the substantia nigra (SN) pars compacta, a region of the midbrain. This loss of neurons is accompanied by the death of astrocytes and a significant increase in the number of microglia in the SN. In order to study both the mechanism and drug delivery for PD, 1-methyl-4-phenyl-1,2,3,6-tetrahydropyridine (MPTP) is often used to facilitate PD in mice. In a recent study, graphene quantum dots (GQDs) were used to prevent α-synucleinopathy in PD by inhibiting fibrillization of α-syn and interacting directly with mature fibrils [9].

For that purpose, GQDs were synthesized by placing carbon fibers in a mixture of strong acid. Initial tests showed that in the absence of GQDs, α-syn monomers assembled into mature fibrils, while the GQDs predominantly inhibited α-syn fibrillization. GQDs also induced the dissociation of α-syn fibrils into short fragments. In a study, a functionalized nanogel-based nanovector was selectively internalized in activated mouse or human astrocytes [10]. The nanogel was based on two main polymers, polyethylene glycol 8000 and linear polyethylenimine 2500. Rolipram, an anti-inflammatory drug, when administered by these nanovectors, limited the inflammatory response in A1 astrocytes, reducing iNOS and Lcn2, which in turn reverses the toxic effect of proinflammatory astrocytes on motor neurons *in vitro*, showing advantages over conventionally administered anti-inflammatory therapy. Nanoscale immunoconjugates (NIC) treatment of mice bearing intracranial GL261 glioblastoma (GBM) results in an increase of CD8+ T cells, NK cells, and macrophages with a decrease of regulatory T cells (Tregs) in the brain tumor area. Survival of GBM-bearing mice treated with NIC combination is significantly longer compared to animals treated with single checkpoint inhibitor-bearing NICs or free a-CTLA-4 and a-PD-1. The same polymeric backbone was also used to deliver antisense oligo nucleotides (AONs) against extracellular matrix (ECM) laminins [11].

One example are iron oxide NPs, which may be used for heat therapy in brain tumor systems [12]. The increased temperature in the tumor can reach a certain temperature which can cause cell death, but also recruit the local immune system to fight the tumor or influence gene expression [13, 14]. However, not all iron oxide particles will reach the target. Some will be cleared mainly to the liver by macrophages via the reticuloendothelial system (RES).

1.2 Nanomolecular Diagnostics

Molecular diagnostics is defined as the detection of genomic variations for the diagnosis, prediction, and monitoring feedback of therapy. The integration of genomics, molecular biology techniques, and laboratory medicine has led to the foundation of molecular diagnostics. Together these elements provide accurate characterization of inherited diseases essential for an accurate diagnosis. High-throughput next-generation sequencing methods and genome-wide association studies provide instrumental discernments into the disease mechanisms and allow physicians to assess disease predilection, devise precise diagnostic methods, and personalize a cure [15]. In recent years, the advancement of molecular technologies has remarkably accelerated the growth of molecular diagnostics. Diagnostic molecular biology has been extensively used in immunology, hematology, microbiology, and clinical biochemistry [16].

Nanodiagnostics is defined as the use of nanotechnology for state-of-the-art clinical diagnostic purposes to meet the demands of enhanced sensitivity and prognosis of disease in clinical diagnostics. The increased need for accuracy can be met only through a diagnostically significant interaction between analyte molecules and signal-generating particles to detect a single analyte molecule. Nanotechnology has empowered one-to-one interaction between analytes and signal-generating particles in the size range of proteins (1–20 nm) and other biomolecules [17]. Nanotechnology features matter with dimensions between 1 and 100 nm. This field is playing a substantial role in molecular diagnostics by incorporating new methods based on the inimitable properties of nanometer-scale materials. They include biobarcodes, dendrimers, nanowires, nanocantilevers, nanopores, and nanofluidic devices [18].

Nanowires are categorized on the basis of metals used—that is, silver nanowires, gold nanowires, and magnetic nanowires. One of the nanowires of great interest is gold nanoparticles-decorated silicon nanowires (AuNPs@SiNWs). These nanowires are highly efficient near-infrared (NIR) hyperthermia agents for the destruction of cancerous cells. Three different types of cancerous cells are destructed in 3 mins when these cells are treated with AuNPs@SiNWs by producing efficient heat [19]. CNTs were first reported by Iijima and Ebbese and Ajayan isolated CNTs in large quantities [20]. CNTs are defined as tubular structures with a diameter typically measured in nanometers. The most promising feature of nanotubes is the target delivery of drug toward the tumor [21].

Quantum dots (QDs) are zero-dimensional semiconductor nanocrystals with very unique properties [22]. Carbon-based QDs that consist of graphite QDs and carbon QDs were first observed during the purification of single walled carbon nanotubes (SWCNTs) in 2004 and in 2006 from laser ablation of graphite and cement [23]. They are also known as carbon nanolights because of their good solubility and strong luminescence [24]. In the field of biomedicines, it has the potential application in biosensing, drug delivery, and bioimaging because of less toxicity and good biocompatibility [25]. Carbon QDs have low toxicity, and good biocompatibility provides promising features for fluorescent bioimaging ad multimodal bioimaging of cells and tissues. Carbon nanodots are also present in our daily-life foods such as coffee and breads, proving that they are more biocompatible than the inorganic nanodots. Due to high solubility in water, nontoxicity, excellent biocompatibility, good cell absorbency, multicolor emission, flexibility in surface modification, and high photostability, nanodots are also used as biosensors. Nanorobots, also known as computerized surgical nanorobots or nanobots, can function like a surgeon inside the body under the control of experts [26]. These nanobots are extensions of injectables which go into the body and gather information through images. Properties like chemical sensing; receiving information from, to, and outside the body; locomotion; and computation play very crucial roles in giving output. Picowatts are important for propelling a nanobot at a speed of 1 mm s^{-1} [27].

Nanocantilevers are being introduced in nanorobots. The main significant property of nanocantilevers is real-time detection. Simply put, they can be used to control objects at the single-molecule level. Nanocantilevers were fabricated in arrays and selected according to biomolecules. The nanocantilevers vary with the variation in biomolecules. Nano-based cardiovascular imaging is used for monitoring the live physiological system with high sensitivity and no pain. This live imaging is used for the understanding of pathological conditions and proper diagnosis. The majority of the nano-based cardiovascular imaging modules include thrombus imaging, theranostic imaging, stem cell imaging, and graft imaging [28].

1.3 Nanopharmaceuticals

The first conceptualization of nanopharmaceuticals could be fairly attributed to Paul Ehrlich when, at the beginning of the twentieth century, he introduced the concept of "magic bullets," a term used to refer to "ideal therapeutic agents" that act specifically against a particular pathogen without causing damage to host cells. One of the specific applications of nanotechnology has been in the creation of advanced drug-delivery systems, which have given rise to the term "nanopharmaceuticals." In order to be classified as a "nanopharmaceutical," Weissig et al. [29] suggest that the drug product has to meet two major criteria: (i) nanoengineering has to be employed during the manufacturing process, and (ii) the nanomaterial used has to be either essential for the therapeutic activity or has to confer additional and unique properties to the active drug entity.

There are two approaches used in the construction of nanopharmaceuticals: (i) the top-down approach, consisting of the step-wise size reduction of large pieces by mechanical or physicochemical methods to create smaller structures, and (ii) the bottom-up approach, involving the construction of nanometric structures through the self-assembly of atoms or molecules induced by physical or chemical forces.

During the last decade, the diversification of nanoplatforms has allowed other types of nanopharmaceuticals that are already in clinical trials or approved; we refer here to protein-based, micelle-based, and inorganic NPs including pure metals (Au, Ag, etc.) and metal oxides. The majority of nanopharmaceuticals have been employed directly for therapeutic use. The majority of the drugs included among nanopharmaceuticals are pre-approved drugs; the discovery of novel drugs is not mandatory. Nanoencapsulation solely improves the pharmacokinetic and/or pharmacodynamic characteristics of the drug, increasing the efficacy of the treatments. Nanopharmaceuticals face a series of challenges, including ensuring low toxicity in patients as well as in the environment, the designing of manufacturing methods that represent a good cost–benefit ratio, and, finally, generating a regulatory framework that provides a rapid arrival on the market to products proven to be safe and effective. Nanopharmaceuticals must prove that they can be a revolutionary technological advance to improve the performance of already available or of novel therapeutic agents. Novel applications are now in the pipeline and will soon appear on the market; gene therapy, stimuli–response nanopharmaceutics, and theranostic agents are examples of the next generation of nanodrugs. Liposomes are a submicron and vesicular system for the transport and release of drugs; they are composed of a bilayer or multiple layers with a central aqueous compartment. The development of formulations with liposomes dates from a considerable time back: 1965 [30]. This period has made it possible for these to be one of the drug-delivery systems possessing a broad impact in the clinic.

Nanopharmaceuticals comprise a new generation of therapeutic-containing nanomaterials with unique properties and high potential that are capable of solving the constant and ongoing challenge for physicians in proposing novel therapeutic treatments to improve their patients' quality of life.

One of the main advantages of using polymeric nanoparticles (PNPs) with respect to other nanosystems is the possibility of utilizing biocompatible and biodegradable polymers such as poly(alpha-hydroxy acids), poly(anhydrides), poly(ortho esters), poly(amino acids), etc., avoiding the requirement to remove the residual polymer within the patient's body. PNPs show several advantages in relation to other colloidal systems (e.g. liposomes): (i) better stability in biological fluids and during storage, (ii) easy preparation and diversity in preparation techniques, (iii) reproducibility and easy large-scale manufacturing, and (iv) batch-to-batch reproducibility. Another recent, innovative proposal in nanodrug-delivery systems comprises polymeric lipid hybrid NPs or lipomers, which are nanocontainers composed of a polymeric core and lipid shell that imparts physicochemical stability and biocompatibility to the NPs. Lipomers have been reported as a blend of the positive attributes of both liposomes and PNPs wherein their individual innate flaws are negated. These lipomers have demonstrated good results as drug-delivery systems to treat cancer and infectious diseases, as well as for theranostic purposes [31, 32]. The application of nanotechnology in nanopharmaceutical products is an emerging field that provides novel approaches to revolutionizing conventional medical treatments. These nanosystems are designed to modify the properties of drug molecules such as solubility, half-life, hall biocompatibility, and release characteristics. They can also increase the possibility of crossing the cell membranes and decreasing unwanted adverse effects [33].

Dendrimers offer several mechanisms of interactions with different drugs or biomolecules. The interior is well suited for host-guest interaction and for the incorporation of hydrophobic/hydrophilic molecules inside their empty cavities (nanoscale containers), which present around the core through hydrophobic interactions and hydrogen bonding. Thus, these can be used as dendritic boxes and unimolecular micelles [34].

In addition, terminal functionalities provide a platform for conjugation by covalent bonding (the prodrug approach) of the drug and targeting moieties. These peripheral functional groups can be employed to tailor-make the properties of dendrimers, enhancing their versatility. Dendrimers have also been employed to track the drug in specific sited by magnetic resonance imaging (MRI); for instance, Gonawala et al. developed water-insoluble Combretastatin conjugated to a water-soluble G3-succinamic acid Poly(AMidoAMine) (PAMAM) dendrimer, obtaining nanomolecules ranging in size from 3 to 5 nm to monitor cerebral blood flow across the blood-tumor barrier, observing that the intratumor blood vessels collapsed, leading to necrosis at the core of the tumor [35].

1.4 Role of Nanotechnology in Biological Therapies

Biological therapy is defined as a type of treatment which employs our own organic capabilities (biomolecules) as a source of drugs to treat various diseases. For example, the therapies based on biomolecules for the treatment of infections, cancer, immune system suppression or stimulation, as well as biosystems to directly or indirectly inhibit the proliferation of cancer cells. Biomolecule-coated CNTs (fullerenes) also release heat under radiofrequency fields [8–10] The thermal properties of nanotubes conjugated to different biomolecules could be useful in the therapy of deep-tissue tumors, where heat can induce necrosis or apoptosis, since radiofrequency easily penetrates the tissues [36]. AuNPs attach to target-specific biomolecules (peptides, antibodies, siRNA, aptamers, etc.), which can be luminescent and plasmonic [37].

Their luminescent properties are suitable for acquiring biomedical images, while the photothermal effect in plasmonic AuNPs produces localized heating under irradiation or exposition to laser light or to radiofrequency fields. The combination of polymeric and magnetic nanoparticles (MNPs) functionalized with biomolecules produces smart nanosystems with a response to pH changes, redox fluctuations, or thermal stimuli for drug release, as well as thermal therapy for metastatic cancer, if they are subjected to alternating magnetic fields (AMFs) [38–40]. Therefore, they can be used against tumors by merging hyperthermia with stimuli-dependent controlled drug delivery [41]. Biological therapies with humanized, chimeric, or human antibodies and in general fusion proteins modulate the immunosystem response to treat diseases [42]. Antibodies and protein fusion have been developed for B or T cell–directed therapies. B cell–directed treatments are based on the depletion and inhibition of costimulation, survival, or differentiation of B cells, while T cell–directed therapies are based on T cell activation [43]. Due to the large variety of NPs, a simple classification is difficult. It is well known that literature lacks a standardized nomenclature [44]. For example, although organic NPs are made from organic matter, the classification excludes carbon-coated NPs. When designing NPs, size is the most important factor to be considered. For drug delivery, organic NPs are in the size range from 5 to 200 nm and include liposomes, PNPs, dendrimers, micelles, and nanocapsules. Liposomes, lipid-based NPs, and PNPs are the most studied nanosystems [45–47].

PNPs are submicron-sized colloidal particles, formed by biodegradable or nondegradable polymers [48, 49]. This type of nanoparticle has promising properties that make it useful as a delivery vehicle. The encapsulation is relatively easy, with high capacity and low toxicity [50]. It is well known that encapsulation of hydrophobic drugs improves their bioavailability and tolerability. Also, PNPs are functionalized with specific ligands to make them able to reach a specific receptor site, thus improving the tumor-targeting properties with practically no toxic effects. The main properties of polymeric nanosystems are biocompatibility, biodegradability, and the advantage to encapsulate a great number of therapeutic drugs (alone or combined). Exosomes are a bilayer membrane containing several types of phospholipids and proteins that are involved in several biological and pathological processes. They provide specialized cell-to-cell communications useful for drug delivery systems in targeting cells. Exosomes originate from endosomes and have a size between 40 and 100 nm. An exhaustive investigation has been carried out to use exosomes as drug delivery approaches [51, 52]. Therapeutic applications of functionalized AuNPs arc widely used because of their multifunctional properties as well as their simple synthesis, controllable size, strength dispersity, easy ligand exchange, biocompatibility, and absence of toxicity. Due to the "gentle" chemical nature of gold, functionalization with target-specific ligands is relatively easy, mainly via gold-thiolate bonding.

QDs are inorganic nanomaterials (e.g. CdSe, CdS, CdTe/CdS, ZnS, PbS, and InP) with optical properties for different bioapplications due to their emissions in the NIR region. Although QDs have been widely used for diagnosis, therapeutic and theranostic alternatives are possible. The conjugation of QDs to biomolecules such as antibodies has also been explored for biocompatible systems such as anti-VEGFR2-QD655 (QDs with emission at 655 nm) [53], bevacizumab-QD [54] and Anti-HER2-QD [55]. In addition to conventional targeting and delivery, QDs have also been used to deliver peptides and proteins (cell-penetrating peptide and His6-tagged beta-D-galactosidase: JB577-Qds-GALC) [56], nucleic acids (siRNA) [57], drugs, polymers [58], polysaccharides [59], etc. Conventional NPs are used for the transport of biomolecules such as proteins and drugs based on nucleic acids [60–68], interleukines [69–71], monoclonal antibodies [72–74], enzymes [75, 76], peptides [77, 78] and substances of natural origin [78–80]. The *in vivo* stability of these biological substances—in particular, proteins (of easy renal clearance and enzymatic degradation)—is significantly improved by being protected by the nanoparticle [81]. Cellular internalization is vital for the therapy to be effective. The primary role is played by the cell membrane [82, 83], although the size, shape, load, and stiffness of the nanoparticle also play a significant role. The main mechanism of internalization of nanoparticle-based drug delivery systems (NDDS) is endocytosis (phagocytosis and pinocytosis); particles larger than 500 nm are internalized by phagocytosis, while smaller ones are internalized by pinocytosis (caltrin-independent or not) that may or not be mediated by receptors [84].

The type of endocytosis determines the degree of internalization. In caltrin-mediated endocytosis, the particle reaches the lysosomal compartment, but if mediated by caveolin, the particle does not reach the compartment. When the surface of the NPs is functionalized with ligands for targeted therapy, the fundamental mechanism of internalization is receptor-mediated endocytosis [85]. The changes that occur as the NPs penetrate the tumor tissue influence the release. Tumor tissue is characterized by low oxygen levels and low perfusion, which cause hypoxia. The pH is more acidic as a result of the production of carbon dioxide and lactic acid [86]; this difference in pH causes variation in the concentrations of several enzymes that contribute to particle degradation and drug release [87].

1.5 Nanomaterials for Gene Therapy

Nanotechnology has found increased applications in biomedical therapies due to their nanosize and the ability to mimic the native molecules of the biological system of the human body. Current understanding of the potential toxicity of nanomaterials is very limited because of their extremely small size. They could be retained longer within the biological system. Not only in therapies but also in day-to-day applications, nanotechnology could lead to exposure to NPs, which can get accumulated in the biological system by ingestion, inhalation, and penetration. *In vivo* studies on rodents and humans [88] focused on the clearance of nanomaterials from organs of the human body after exposure by inhalation. From this study it was evident that lungs (as the port for entry of the nanomaterial), the liver, and kidneys are the major organs affected by nanomaterials. If the nanomaterial enters the human body via ingestion or inhalation, they can cause tissue injury by accumulating in tissue. But mostly, when they enter the bloodstream, they can be translocated to different parts of the body via blood circulation [89]. In the case of lungs, airway and alveolar macrophages are the first line of defense against nanomaterials. But it was observed in case studies that any nanomaterial aggregates less than 100 nm were not cleared by phagocytosis [90, 91].

If the nanomaterial gets opsonized, then clearance can be assured by opsonin-dependent phagocytosis. As discussed in target organ toxicity, the liver is the major organ in which nanomaterials get deposited. When nanomaterials get deposited in the lysozyme of kupffer cells (KC) cells in the liver, they can activate lysozyme-associated degradation of nanomaterial. The nanomaterial when enters the acidic pH of the KC cells lot of hydrolase enzymes are generated, which degrades the nanomaterial, making nanotherapy unsuccessful [92]. It was observed that haemoxygenase-1, peroxidase and α-glucosidase facilitated in the degradation of nanomaterials internalized into liver cells. In order to carry out successful gene delivery, the delivery vector must be chosen carefully. Vectors are the agents that carry the genetic material intended to cure the disease and transport the material to the target cell or tissue. The success of gene delivery lies in decreasing the protein expressions that can cause harm to the host and down-regulation of genes that cause tumors and other side-effects [93].

Commonly used inorganic NPs for gene delivery are AuNPs, CNTs, fullerenes, MNPs, QDs, silica NPs, and calcium phosphate NPs. The AuNPs are used for the delivery of genes into mesenchymal stem cells. The AuNPs are cationic so that they can carry the genes to the target location without undergoing any changes. They also have antibacterial peptides that enhance the transfection efficiency and provide an antibacterial activity to the same [94]. The CNTs are used as a gene delivery vehicle because they have better biocompatibility, high length-to-diameter ratio, and more surface area. The CNT surface can be modified so that it can deliver small interfering ribonucleic acid (siRNA), pDNA, and micro ribonucleic acid (miRNA) into the target mammalian cells. The CNTs can be single-walled or multi-walled [95]. They are employed to treat cancer, HIV, and diabetes [96]. The process of gene transfer using MNPs is called Magnetofection. The plasmids form complexes with the polymer-coated MNPs. The complex formation is called a Magnetoplex. Gene regulation is a kind of vast mechanism used by all prokaryotic and eukaryotic cells to control the production of specific gene products like proteins and RNAs. It increases the adaptability of the cells by express protein when it needed. In multicellular organisms, the regulation of gene expression in the same genome sequence leads to cell differentiation (creating different cell types) and embryo morphogenesis [97]. This helps to understand how the biological evolution performs at a molecular level.

The carbon-based nanomaterials with high efficiency have many concerns, mainly to avoid immune response while performing cellular metabolic regulations. With the help of nanocarriers we can avoid immune responses by coating the functional component that regulates gene expression over the surface of NPs so it can easily recognize target cells and aid intracellular entry [98]. AuNPs are a perfect example of a gene regulatory nanocarrier. These AuNPs are widely used in medical applications because of their remarkable properties such as high biocompatibility, low cytotoxicity, ability to easily bind with thiol derivatives by covalent bonding as well as without change its complex form tightly anchored with genetic speck [99]. In several studies involving photothermal therapy or chemotherapy, these gold nanocarriers were used as secondary material to increase the efficacy of gene regulation in cancer cells. During the gene regulation process in cancer cells, antisense sequences coated over the AuNPs which complementary to the target mRNAs are incorporated in the cancer cells that degrade mRNAs [100]. The fluorescence emitted by the nanocarrier was helps to monitor and confirm the mRNAs gets reduced.

During the process of protein synthesis from DNA, all the gene expression stages from transcription to post-translational modification were regulated. For example, the modification of histone in chromatin structure by DNA methylation, ncRNA, or DNA binding proteins regulates the gene expression as inheritable, commonly known as epigenetic regulation [101]. In DNA transcription, as a structural modification of DNA segment which wound around by tightly packed histone protein are leads to DNA supercoiling, these processes were temporarily modified by phosphorylation or permanently modified by methylation. Such modifications create a direct impact on the regulation of gene expressions [102, 103]. Though nanomedicine has more promising results in gene therapy, it also imparts a great challenge to public health in terms of regulatory issues. Many research studies state that not all nanomaterial is toxic for a human system in case of therapeutic applications of nanomaterials, but it is important to screen undesirable properties to aid in producing bio-nanomaterial at least with minimum risk. In the development of novel drugs, nanotechnology contributes especially in the area of cancer therapy and the treatment of neurological disorders. Biological vectors like viruses are one of the ways that nanotechnology may facilitate gene therapy. Immunologically, since the nanomaterials are much less reactive than the vectors, the vectors allow for the delivery of the maximum quantity of genetic material.

1.6 Nanotools for the Treatment of Ocular Diseases

The treatment of ocular diseases is quite challenging because of the anatomy, physiology, and biochemistry of the eye. The development of new and effective therapeutic approaches for ocular diseases may be reached with the contribution of nanotechnology. Nanomedicine is already an established market. According to research by the Business Communications Company (BCC), the nanomedicine market is forecast to reach nearly $300 billion by 2022. Several nanomedicine tools offer already efficient solutions for ophthalmological issues, and many novel products are expected to appear in this field soon.

The precorneal tear film is continuously produced and drained by the lacrimal system; the eyelids play an important role in the distribution of the tear fluid as they spread it all over the anterior surface of the eye—that is, the cornea and the conjunctiva. The tear film is made of three layers (lipid, aqueous, and mucus), but there is no clear-cut barrier between these layers. The inner mucus layer is in direct contact with the epithelial cells of the cornea and conjunctiva and is mainly composed of water (95%) and high molecular weight (0.5–20 MDa) glycoproteins (mucins). Mucins reduce the surface tension between the aqueous layer and the cornea, thus allowing the aqueous layer to adhere to the epithelial cells of the cornea [104]. This central aqueous layer represents the bulk of the tear film. Topical administration is the most common and patient-friendly route for ocular disease treatment, particularly for eye diseases that affect the anterior segment. Nevertheless, it is known that less than 5% of administrated drug is able to reach the aqueous humor. The limited contact time with the ocular surface due to elimination mechanisms, such as tear turnover, nasolacrimal drainage, and systemic absorption in the precorneal area, and the simultaneous restricted entry into the eye due to the cornea barrier account for this ineffectiveness.

Intravitreal administration is the most used route for the treatment of posterior segment diseases, since the aforementioned elimination mechanisms and barriers are overcome. However, this approach is invasive, requiring an operating room setting and an experienced clinician. Furthermore, frequent injections are generally needed, which may cause pain, endophthalmitis, sterile inflammation, glaucoma, retinal detachment, and cataracts. Pharmacokinetic insights concerning intravitreal drug administration were deeply discussed elsewhere [105, 106].

Currently, the use of nanotechnology has been investigated for several ophthalmic applications in the treatment of eye diseases [107]. NPs have been receiving attention as one of the novel drug delivery systems that overcome the barriers of the eye, including the cornea, conjunctiva, and blood-retinal barrier (BRB) [108]. Also, the number of intravitreal injections is expected to decrease by using NPs [109], when compared to the use of intravitreal implants and intravitreal injections. NPs may have potential application for several medical purposes, such as gene therapy, drug delivery (treatment and/or prevention of diseases), and imaging [110, 111]. Furthermore, these systems are able to combine both therapy and diagnosis, delivering therapeutic drugs and diagnostic imaging agents simultaneously, and they are remarkably useful tools for theranostics [112]. As drug delivery systems, NPs can provide: (i) sustained delivery; (ii) targeted delivery to specific cells or tissues; (iii) improved delivery of water-insoluble and large molecules; and (iv) reduced side-effects, minimizing toxicological reactions.

Nevertheless, toxicity of NPs on neuronal cells is still questionable. Size, dose, interaction time, chemical composition (core and coating), surface charge (zeta potential), and water solubility are factors that contribute to neuronal toxicity of NPs. Therefore, NPs must be thoroughly characterized before their application. The neuronal toxicity due to NPs is likely to occur via the formation of reactive oxygen species [113] and the alteration of gene expression patterns [114]. Microglial cells are able to phagocyte NPs, thus increasing the production of intracellular reactive oxygen species and reactive nitrogen species [115], as well secretion levels of proinflammatory cytokines such as tumor necrosis factor (TNF), IL-1β, and IL-6 [116].

There are various mechanisms responsible for the light-triggered release of therapeutic agents, including photocleavage, photoisomerization, and photo-cross-linking [117, 118]. Light-responsive NPs also have been used for the intracellular delivery of biomolecules, namely proteins/peptides, such as green fluorescent protein (GFP) and hemoglobin, and siRNA [119]. Light-responsive NPs for ophthalmic application are still in their early stages of development. Almutairi and collaborators described a light-sensitive degradable polymer containing a quinone-methide self-immolative moiety, which was triggered to degrade by UV irradiation through a light-sensitive group (4,5-dimethoxy-2-nitrobenzyl) along the backbone [120]. Later, for the first time, they reported the *in vivo* light-triggered drug release in the eye [121]. The drug-loaded NPs and pure NPs (without drug) can also be dispersed into contact lenses. Silver NPs, whose antimicrobial [122] and toxic [123] activity are well-known, have been incorporated into contact lenses. Rad and co-authors evaluated the antibacterial efficacy of silver NP-laden contact lenses [124].

Keratitis is described as an infectious ophthalmic inflammation, which can develop from bacteria, such as *Pseudomonas Aeruginosa*. The biofilm formation on the contact lens surface is assumed to be an example for hydrogel contact lens–induced infection. Silver can be loaded on the contact lens surface and released as an antimicrobial agent, thus reducing bacterial colonization. New drug delivery nanosystems need to be designed to improve the bioavailability of drugs, to allow better cellular targeting, to better release profile control, to lower side-effects, and to improve patient compliance. Despite the promising results already reported with NPs in preclinical studies, most of them still need further research. The cost of development and manufacture, the ability to scale up, the lack of knowledge about the impact of NPs on living systems, and the need for approval by the regulatory authorities have been hampering their entrance in the pharmaceutical market.

1.7 Nanotechnology Applications in Food and Nutrition Science

Nanotechnology maintains the covenant of improving the quality of human life and advancing knowledge in a magnitude of methods. The science that will determine the investigation of materials that are one-billionth of a meter in size (smaller than 100 nm) has grown remarkably over the past decades [125]. The development, usage, and manipulation of materials, devices, or systems at the nanometer scale were done in the field of nanotechnology [126]. Nanocochleates are a new drug delivery set based upon to be trapped of optimal drug molecules into the multilayered structure including solid-lipid bilayer in the form of sheets rolled up in spiral form [127]. Due to their high surface-to-volume ratio and new physicochemical effects, nanomaterials can influence color, solubility, and thermodynamics of food substances and so improve and modify the texture, taste, and appearance of food [128, 129]. These new effects supply opportunities to rectify the sensory qualities of food such as taste, texture, and color.Nanomaterials also can be employed to improve preservation mechanisms for food. Recruiting nanosensors and nanopackaging materials provide fast, sensitive, and reliable detection of microbial contamination, harmful chemicals, and pesticides. On the other hand, nanoencapsulation systems can increase the quality of food processing by enabling the delivery of bioactive compounds for increasing the bioavailability in foods [130]. Maintenance of the bioactive ingredients in the food industry for protection against the non-standard process and environmental conditions and correctly delivered to the target organs and cells should be formulated under the nanoencapsulation technique [131].

Archaea (Archaeobacteria) are considered to be distinct from eubacteria and eukaryotes, and they include aerobic, anaerobic, thermophilic, extremely thermophilic, thermoacidophilic, and halophilic microorganisms. All lipid extracts from individual species of Archaea consist of polar ether lipids and from 5 to 20% neutral lipids. Biopolymer NPs were first designed using albumin and non-biodegradable synthetic polymers such as polyacrylamide and poly (methyl acrylate). The risks of chronic toxicity due to the intracellular and/or tissue overloading of non-degradable polymers were soon considered as a major limitation for the systemic management of polyacrylamide and poly (methyl acrylate) NPs in humans. The chemical structure of liposomes includes phospholipids or synthetic amphiphiles merged with sterols, such as cholesterol, to impress membrane permeability. The thin-film hydration method (followed by extrusion) is the most widely used preparation for liposomes, in which lipid components with or without a drug are soluble in an organic solvent [132]. The effects of nanofibers can be specified as having a low unit length (diameter), high surface-area-to-volume ratio, high strength value, low basis weight, high porosity, and small pore size. These make them indispensable in numerous usages [133–136].

Examples of the most important applications of cellulosic nanofibers (a type of nanofiber) in food science are (i) immobilization of bioactive substances such as enzymes, vitamins, and antimicrobials; (ii) nutraceutical delivery systems and controlled release of materials; (iii) as biosensors; (iv) filtration; and (v) for reinforcing composites and in films [137]. Various polymers such as keratin, collagen, silk fibroin, cellulose, gelatin, poly(lactic acid) (PLA), poly(lactic-co-glycolic acid) (PLGA), poly(ethylene-co-vinyl acetate) (PEVA), and polysaccharides such as alginate and chitosan can be utilized in the production of nanofibers [138]. Routine methods for the formation of alginate hydrogel microparticles forfood use involves ionotropic gelation via manipulation of laminar flow, spray-drying, microsized liposome entrapment, emulsification, lyophilization, and homogenization. Alginate and chitosan layer-by-layer coating of nanoliposomes and electrospinning provided superior stability of vitamin B9 or vitamin C in an acidic environment, relating to behavior in the gastric environment and natural fruit juices [139–141]. A limitation of the usage of

PNPs in the food industry is that the host is exposed repeatedly to nano-engineered particles for long periods [142]. So must be used only food-grade components as particles for polymeric particle synthesis process planned for food utilization in the form of natural polymers, and biodegradable and biocompatible surfactants [143]. Furthermore, chitosan and many other natural polymers such as gelatin and sodium alginate have many advantages over conventional synthetic pathways [144].

Chitosan, alginate, pectin, cellulose, carrageenan, hyaluronate, gelatin, dextran, zein (a corn protein rich in prolamines), xanthan, albumin, milk proteins (casein), and whey proteins (β-lactoglobulin and α-lactalbumin) as natural polymers are the most widely used biopolymers in this field [145]. Carbon-based nanomaterials have been used in the development of high-performance sensing devices for food safety inspections to produce, identify, and increase the sensing signals [146]. According to their spatial dimensions, carbon-based nanomaterials can be roughly divided into fullerenes (zero-dimensional), CNTs (one-dimensional), graphene (two-dimensional), graphene coil (multidimensional), etc. Activated carbon (AC) has been widely used as a sorbent for conventional wastewater treatment due to its large surface area and high potential to adsorb an extended spectrum of organic and inorganic contaminants [147–150]. The mono-targeted therapies or "smart drugs" have been designed over the past few years for the treatment of chronic diseases such as cardiovascular diseases, metabolic diseases, cancer, and neurological diseases. The main reason for these chronic disorders is a defect in signaling pathways. Targeting a single pathway among many of the pathways involved is not likely to be effective in the prevention and treatment of these diseases [151].

Several studies over the past five decades indicate that the herbal extract curcumin has antioxidant, anti-inflammatory, antiviral, antimicrobial, antihypertensive, and anticancer effects [152–155]. Curcumin shows poor solubility and low absorption rates. Therefore, its silver NPs were formed to increase solubility, absorption, andeffectiveness against tumors [156]. Another effect of curcumin NPs is their protective property against heavy metal-induced liver damage in humans [157]. Toxic results of the heavy-metal ions—for example, arsenic (As), cadmium (Cd), chromium (Cr), copper (Cu), lead (Pb), and mercury (Hg)—are shown primarily as oxidative stress [158]. Nanotitanium particles as the whitest plastic packaging and white ink materials are manufactured with TiO_2 for keep from ultraviolet light or for aesthetic applications [159]. These transparent nanoform particles are UV-resistant and UV-absorbant [160]. They can be used as UV-protected food packaging by incorporating them into plastic (e.g., co polyether-ester) [161–163]. They can be used in filler materials and plastic containers, bottles, and boxes [164]. Peiró et al. (2006) studied the behavior of photocatalytic oxygen consumption in the form of nanocrystalline TiO2 composite films. They exhibited high scavenging activity and fast oxygen reduction in the presence of ethanol due to the photocatalytic redox reaction [165].

1.8 Rubber Nanocomposites for Biomedical Applications

During the past few years, rubber nanocomposites have gained major attention in the area of biomedical applications. Since the 1940s, elastomers have been known to be used for a wide range of biomedical products. They are generally categorized into three: (i) commodity elastomers used in the biomedical field, (ii) medical grade elastomers ideal for short-term physiological contact, and (iii) elastomers suitable for implants or long-term physiological contact. The prerequisites of rubbers/elastomers for biomedical applications include high purity; biocompatibility; desirable physical, chemical, and mechanical properties; ease of fabrication; sterilizability; and high stability. For the manufacture of biomedical elastomers, the catalysts selected for polymerization should be less toxic and the additives for modifying the polymer properties—such as fillers, plasticizers, accelerators, antioxidants, curatives, pigments, and stabilizers—should be non-toxic and non-leaching. It has been observed that the presence of residual monomers, oligomers, catalysts, and polymerization additives can lead to adverse biological reactions.

For applications such as catheters with balloon and wound dressing, the manufacturing process proceeds with the dispersion of silicone in solvents such as Freon or trichloroethane to be dipped or cast resulting in a uniformly smooth and flawless surface. Organosilicones, which contain low molecular weight dimethyl silioxane with silanol end groups, a tetraethoxy silane curing agent, a tin octoate catalyst, and X-ray contrasting substances, are effective for treating embolisms in kidney arteries. Silicone elastomer implants are widely used in plastic surgery that includes reconstruction of breasts, chin, nose, ear armature, etc. The polyurethanes have been used for many years for products including blood bags and dialysis tubing. The flexibility of these materials obviates the need for the addition of plasticizers that can leach during use. They also resist the possibility of shrinking, cracking, and kinking. In addition, they have resistance to hydrolysis and fungus.

The poly(ether-urethane-urea) segments with the trade names Biomer and Mitrathane have been used in cardiovascular applications. Polyurethane-ureas are processed via solution casting since heating of these polymers to a softening temperature may lead to degradation. The factors that decide the quality of the product depends on the concentration of polymer in solution, solvent characteristics, mold surface, humidity, etc. Biomer exhibits excellent blood compatibility and resilience and hence is notable as a coating for cannulas, catheters, left ventricular assist blood pumps, intra-aorta balloon pumps, etc. It has been reported that interactions of polyurethanes with body fluids can result in failure due to calcification, environmental stress cracking, and chain scission. The deposition of calcium hydroxyapatite on the implanted device is referred to as calcification. When in contact with body fluids, polyurethane can undergo environmental stress cracking under stress conditions. In addition, polyetherurethanes can develop cracks on surfaces that are in contact with tissues. This may be due to the auto-oxidative degradation process accelerated by the corrosion products of metallic components in the implanted device. Polyurethanes with varying degrees of hydrophobicity and hydrophilicity are excellent candidates for the replacement of articular cartilage, artificial joint prosthesis, and percutaneous implantable devices.

Styrenic thermoplastic elastomers have been used for medical tubing due to their cleanliness, chemical inertness, dimensional stability, ease of sterilization, and low gas and vapor transmission. It has been reported that tubing for peristaltic pumps made of styrenic thermoplastic elastomers (TPEs) lasts 30 times longer than silicone rubber. Some catheters and tubing made of polyvinyl alcohol (PVC) have been replaced with TPE to prevent plasticizer migration and extraction. TPEs are flexible at low temperatures so that they can replace natural rubber with no chemical irritants/proteins. Reconstructive surgery is mostly performed when there is a deformity or defect in human organs caused by congenital factors or an injury or degradation process such as osteoarthiritis. The prerequisites for a bioimplant material include high strength and flexibility, good biocompatibility, chemical stability, non-magnetic behavior, ease of fabrication, and disinfection. Although silicone rubber is the commonly used biomaterial for reconstructive surgery, it has inferior bioactivity. Hydroxyapatite has outstanding biological performance in that it promotes new bone formation via degradation in the microenvironment of the organism producing a pore-like structure that facilitates adherence and growth of tissues [166].

Silicone catheters are widely used in clinical applications for administering drugs, intravenous fluids, and blood products; for draining urine or fluid from the urethra or gastrointestinal tract; and for parenteral nutrition. However, they are associated with a high risk of nosocomial infections in the bloodstream [167] and urinary tract and with ventilator-related pneumonia [168, 169]. Metallic alloys have been widely accepted for surgical implants. Stainless steel 316L has received great attention owing to its biocompatibility, corrosion resistance, availability, and low cost. For long-term applications in aggressive biological environments, there are possibilities that stainless steel may corrode and get accumulated in tissues surrounding the implants. Several *in vitro* and *in vivo* studies have substantiated that the metal ions (corrosion products) released can cause toxic effects on seminiferous epithelium of mouse testis, alter the cellular population of the spleen, and in humans, alter lymphocyte-surface antigens [170]. The scaffold is one of the essential elements for bone defect regeneration. It controls the survival, colonization, migration of seed cells, cell proliferation, osteogenesis, etc. The scaffold materials should be osteoinductive—that is, be able to stimulate the differentiation of cells toward osteoblasts. For promoting *in vivo* new bone formation, the conventional approach is to add vascular endothelial growth factor, platelet-derived growth factor, or cytokine of bone morphogenetic protein 2 to the scaffolds. The addition of zinc and silicon to scaffolds can promote the formation of osteoblasts and calcification of bone, respectively.

References

1 Carmeliet, P. and Jain, R.K. (2011). Principles and mechanisms of vessel normalization for cancer and other angiogenic diseases. *Nature Reviews Drug Discovery [Perspective]* [Internet] Jun 1;10:417.

2 Anchordoquy, T.J., Barenholz, Y., Boraschi, D. et al. (2017). Mechanisms and barriers in cancer nanomedicine: addressing challenges, looking for solutions. *ACS Nano* 11 (1): 12–18.

3 Wang, X.-J., Peng, C.-H., Zhang, S. et al. (2019). Polysialic-acid-based micelles promote neural regeneration in spinal cord injury therapy. *Nano Letters* 19 (2): 829–838.

4 Toledano Furman, N.E., Lupu-Haber, Y., Bronshtein, T. et al. (2013). Reconstructed stem cell nanoghosts: a natural tumor targeting platform. *Nano Letters* 13 (7): 3248–3255.

5 Kaneti, L., Bronshtein, T., Malkah Dayan, N. et al. (2016). Nanoghosts as a novel natural nonviral gene delivery platform safely targeting multiple cancers. *Nano Letters* 16 (3): 1574–1582.

6 Gaugler, J., James, B., Johnson, T. et al. (2019). 2019 Alzheimer's disease facts and figures. *Alzheimer's Dement* 15 (3): 321–387.

7 Umlauf, B.J. and Shusta, E.V. (2019). Exploiting BBB disruption for the delivery of nanocarriers to the diseased CNS. *Current Opinion in Biotechnology* 60: 146–152.

8 Niu, S., Zhang, L.-K., Zhang, L. et al. (2017). Inhibition by multifunctional magnetic nanoparticles loaded with alpha-synuclein RNAi plasmid in a Parkinson's disease model. *Theranostics* 7 (2): 344–356.

9 Kim, D., Yoo, J.M., Hwang, H. et al. (2018). Graphene quantum dots prevent **α**-synucleinopathy in Parkinson's disease. *Nature Nanotechnology* 13 (9): 812–818.

10 Vismara, I., Papa, S., Veneruso, V. et al. (2020). Selective modulation of A1 astrocytes by drug-loaded nano-structured gel in spinal cord injury. *ACS Nano* 14, 1: 360–371.

11 Sun, T., Patil, R., Galstyan, A. et al. (2019). Blockade of a Laminin-411–notch axis with CRISPR/Cas9 or a nanobioconjugate inhibits glioblastoma growth through tumor-microenvironment cross-talk. *Cancer Research* 79 (6): 1239–1251.

12 Silva, A.C., Oliveira, T.R., Mamani, J.B. et al. (2011). Application of hyperthermia induced by superparamagnetic iron oxide nanoparticles in glioma treatment. *International Journal of Nanomedicine* 6: 591.

13 Frey, B., Weiss, E.-M., Rubner, Y. et al. (2012). Old and new facts about hyperthermia-induced modulations of the immune system. *International Journal of Hyperthermia* 28 (6): 528–542.

14 Lohr, F., Hu, K., Huang, Q. et al. (2000). Enhancement of radiotherapy by hyperthermia-regulated gene therapy. *International Journal of Radiation Oncology, Biology, Physics* 48 (5): 1513–1518.

15 Goebel, E.A., Vidal, A., Matias-Guiu, X., and Gilks, C.B. (2017). The evolution of endometrial carcinoma classification through application of immunohistochemistry and molecular diagnostics: past, present and future. *Virchows Archiv.* 1–12.

16 Beilby, J. (2006). Diagnostic molecular biology. *Clinical Biochemist Reviews.* 27 (1): 3.

17 Azzazy, H.M., Mansour, M.M., and Kazmierczak, S.C. (2006). Nanodiagnostics: a new frontier for clinical laboratory medicine. *Clinical Chemistry.* 52 (7): 1238–1246.

18 Tang, Y.W., Procop, G.W., and Persing, D.H. (1997). Molecular diagnostics of infectious diseases. *Clinical Chemistry.* 43 (11): 2021–2038.

19 Su, Y., Wei, X., Peng, F. et al. (2012). Gold nanoparticles-decorated silicon nanowires as highly efficient near-infrared hyperthermia agents for cancer cells destruction. *Nano Letters.* 12: 1845–1850.

20 Pang, L.S., Saxby, J.D., and Chatfield, S.P. (1993). Thermogravimetric analysis of carbon nanotubes and nanoparticles. *Journal of Physical Chemistry.* 97: 6941–6942.

21 Fabbro, C., Ali-Boucetta, H., Ros, T.D. et al. (2012). Targeting carbon nanotubes against cancer. *Chemical Communications.* 48: 3911–3926.

22 Bera, D., Qian, L., Tseng, T.-K., and Holloway, P.H. (2010). Quantum dots and their multimodal applications: a review. *Materials.* 3: 2260–2345.

23 Xu, X., Ray, R., Gu, Y. et al. (2004). Electrophoretic analysis and purification of fluorescent single-walled carbon nanotube fragments. *Journal of the American Chemical Society* 126: 12736–12737.

24 Baker, S.N. and Baker, G.A. (2010). Luminescent carbon nanodots: emergent nanolights. *Angewandte Chemie International Edition.* 49: 6726–6744.

25 Shen, J., Zhu, Y., Yang, X., and Li, C. (2012). Graphene quantum dots: emergent nanolights for bioimaging, sensors, catalysis and photovoltaic devices. *Chemical Communications* 48: 3686–3699.

26 Krishnamurthy, S. and Vijayasarathy, S. (2016). Role of nanomaterials in clinical dentistry. In: *Nanobiomaterials in Dentistry* (ed. Alexandru Mihai Grumezescu), 211–240. Elsevier.

27 Ramsden, J.J. (2016). The nano/bio interface. In: *Nanotechnology*, 2e (ed. J.J. Ramsden), 61–90. Oxford: William Andrew Publishing.

28 Deb, S., Ghosh, K., and Shetty, S.D. (2015). Nanoimaging in cardiovascular diseases: current state of the art. *Indian Journal of Medical Research* 141 (3): 285.

29 Weissig, V., Pettinger, T., and Murdock, N. (2014). Nanopharmaceuticals (part 1): products on the market. *International Journal of Nanomedicine* 9: 4357. https://doi.org/10.2147/IJN.S46900.

30 Allen, T.M. and Cullis, P.R. (2013). Liposomal drug delivery systems: from concept to clinical applications. *Advanced Drug Delivery Reviews* 65 (1): 36–48. http://doi.org/10.1016/j.addr.2012.09.037.

31 Jain, A., Sharma, G., Kushwah, V. et al. (2017). Methotrexate and beta-carotene loaded-lipid polymer hybrid nanoparticles: a preclinical study for breast cancer. *Nanomedicine* 12 (15): 1851–1872. http://doi.org/10.2217/nnm-2017-0011.

32 Jose, C. et al. (2018). Polymeric lipid hybrid nanoparticles: properties and therapeutic applications. *Critical Reviews TM in Therapeutic Drug Carrier Systems* 35 (6): 555–588. http://doi.org/10.1615/CritRevTherDrugCarrierSyst.2018024751.

33 Bawa, R. (2010). Nanopharmaceuticals: nanopharmaceuticals. *European Journal of Nanomedicine* 3 (1): http://doi.org/10.1515/EJNM.2010.3.1.34.

34 Tomalia, D.A. (2005). Birth of a new macromolecular architecture: dendrimers as quantized building blocks for nanoscale synthetic polymer chemistry. *Progress in Polymer Science* 30 (3–4): 294–324. https://doi.org/10.1016/j.progpolymsci.2005.01.007.

35 Gonawala, S. et al. (2018). MRI monitoring of cerebral blood flow after the delivery of nanocombretastatin across the blood brain tumor barrier. *Journal of Nanomedicine & Nanotechnology* 9 (05): 1–15. http://doi.org/10.4172/2157-7439.1000516.

36 Gannon, C.J., Cherukuri, P., Yakobson, B.I. et al. (2007). Carbon nanotube-enhanced thermal destruction of cancer cells in a noninvasive radiofrequency field. *Cancer: Interdisciplinary International Journal of the American Cancer Society.* 110 (12): 2654–2665.

37 Zhou, C., Yang, S., Liu, J. et al. (2013). Luminescent gold nanoparticles: a new class of nanoprobes for biomedical imaging. *Experimental Biology and Medicine* 238 (11): 1199–1209.

38 Wolski, P., Nieszporek, K., and Panczyk, T. (2018). Multimodal, pH sensitive, and magnetically assisted carrier of doxorubicin designed and analyzed by means of computer simulations. *Langmuir* 34 (7): 2543–2550.

39 DeNardo, S.J., DeNardo, G.L., Miers, L.A. et al. (2005). Development of tumor targeting bioprobes (111In-chimeric L6 monoclonal antibody nanoparticles) for alternating magnetic field cancer therapy. *Clinical Cancer Research* 11 (19): 7087s–7092s.

40 Qiu, L., Hu, Q., Cheng, L. et al. (2016). cRGDyK modified pH responsive nanoparticles for specific intracellular delivery of doxorubicin. *Acta Biomaterialia* 30: 285–298.

41 Mai, B.T., Fernandes, S., Balakrishnan, P.B. et al. (2018). Nanosystems based on magnetic nanoparticles and thermo-or pH-responsive polymers: an update and future perspectives. *Accounts of Chemical Research* 51 (5): 999–1013.

42 Schirrmacher, V. (2019). From chemotherapy to biological therapy: a review of novel concepts to reduce the side effects of systemic cancer treatment. *International Journal of Oncology* 54 (2): 407–419.

43 Smith, R.M., Clatworthy, M.R., and Jayne, D.R. (2010). Biological therapy for lupus nephritis—tribulations and trials. *Nature Reviews Rheumatology* 6 (9): 547.

44 Pelaz, B., Alexiou, C., Alvarez-Puebla, R.A. et al. (2017). Diverse applications of nanomedicine. *ACS Nano* 11 (3): 2313–2381.

45 Kumari, P., Ghosh, B., and Biswas, S. (2016). Nanocarriers for cancer-targeted drug delivery. *Journal of Drug Targeting.* 24 (3): 179–191.

46 Xin, Y., Yin, M., Zhao, L. et al. (2017). Recent progress on nanoparticle-based drug delivery systems for cancer therapy. *Cancer Biology & Medicine.* 14 (3): 228.

47 Jeevanandam, J., Barhoum, A., Chan, Y.S. et al. (2018). Review on nanoparticles and nanostructured materials: history, sources, toxicity and regulations. *Beilstein Journal of Nanotechnology.* 9 (1): 1050–1074.

48 Sur, S., Rathore, A., Dave, V. et al. (2019). Recent developments in functionalized polymer nanoparticles for efficient drug delivery system. *Nano-Structures & Nano-Objects.* 20: 100397.

49 Masood, F. (2016). Polymeric nanoparticles for targeted drug delivery system for cancer therapy. *Materials Science and Engineering: C.* 60: 569–578.

50 Duan, X., He, C., Kron, S.J. et al. (2016). Nanoparticle formulations of cisplatin for cancer therapy. *Wiley Interdisciplinary Reviews: Nanomedicine and Nanobiotechnology.* 8 (5): 776–791.

51 Ha, D., Yang, N., and Nadithe, V. (2016). Exosomes as therapeutic drug carriers and delivery vehicles across biological membranes: current perspectives and future challenges. *Acta Pharmaceutica Sinica B.* 6 (4): 287–296.

52 Batrakova, E.V. and Kim, M.S. (2015). Using exosomes, naturally-equipped nanocarriers, for drug delivery. *Journal of Controlled Release.* 219: 396–405.

53 Carbary-Ganz, J.L., Barton, J.K., and Utzinger, U. (2014). Quantum dots targeted to vascular endothelial growth factor receptor 2 as a contrast agent for the detection of colorectal cancer. *Journal of Biomedical Optics.* 19 (8): 086003.

54 Gazouli, M., Bouziotis, P., Lyberopoulou, A. et al. (2014). Quantum dots-bevacizumab complexes for in vivo imaging of tumors. *In vivo.* 28 (6): 1091–1095.

55 Nifontova, G., Ramos-Gomes, F., Baryshnikova, M. et al. (2019). Cancer cell targeting with functionalized quantum dot-encoded polyelectrolyte microcapsules. *Frontiers in Chemistry.* 7: 34.

56 Dawson, G. (2016). Quantum dots and potential therapy for K rabbe's disease. *Journal of Neuroscience Research.* 94 (11): 1293–1303.

57 Lin, G., Chen, T., Zou, J. et al. (2017). Quantum dots-siRNA nanoplexes for gene silencing in central nervous system tumor cells. *Frontiers in Pharmacology.* 8: 182.

58 Omidi, M., Hashemi, M., and Tayebi, L. (2019). Microfluidic synthesis of PLGA/carbon quantum dot microspheres for vascular endothelial growth factor delivery. *RSC Advances.* 9 (57): 33246–33256.

59 Joshi, P.N., Agawane, S., Athalye, M.C. et al. (2017). Multifunctional inulin tethered silver-graphene quantum dots nanotheranostic module for pancreatic cancer therapy. *Materials Science and Engineering: C.* 78: 1203–1211.

60 Chou, M.-J., Yu, H.-Y., Hsia, J.-C. et al. (2018). Highly efficient intracellular protein delivery by cationic polyethyleneimine-modified gelatin nanoparticles. *Materials.* 11 (2): 301.

61 Das, P. and Jana, N.R. (2016). Length-controlled synthesis of calcium phosphate nanorod and nanowire and application in intracellular protein delivery. *ACS Applied Materials & Interfaces.* 8 (13): 8710–8720.

62 Emadi, F., Amini, A., Gholami, A. et al. (2017). Functionalized graphene oxide with chitosan for protein nanocarriers to protect against enzymatic cleavage and retain collagenase activity. *Scientific Reports.* 7: 42258.

63 Liu, Y. and Cao, A. (2017). Encapsulating proteins in nanoparticles: batch by batch or one by one. In: *Methods in Enzymology*, vol. 590, 1–31. Elsevier.

64 Suner, S.S., Sahiner, M., Sengel, S.B. et al. (2018). Responsive biopolymer-based microgels/nanogels for drug delivery applications. In: *Stimuli Responsive Polymeric Nanocarriers for Drug Delivery Applications*, vol. 1 (ed. Types and Triggers), 453–500. Elsevier.

65 Buck, J., Grossen, P., Cullis, P.R. et al. (2019). Lipid-based DNA therapeutics: hallmarks of non-viral gene delivery. *ACS Nano.* 13 (4): 3754–3782.

66 Barba, A.A., Bochicchio, S., Dalmoro, A. et al. (2019). Lipid delivery systems for nucleic-acid-based-drugs: from production to clinical applications. *Pharmaceutics.* 11 (8): 360.

67 Elsana, H., Olusanya, T.O., Carr-Wilkinson, J. et al. (2019). Evaluation of novel cationic gene based liposomes with cyclodextrin prepared by thin film hydration and microfluidic systems. *Scientific Reports.* 9 (1): 1–17.

68 Chen, E.Y., Liu, W.F., Megido, L. et al. (2018). Understanding and utilizing the biomolecule/nanosystems interface. In: *Nanotechnologies in Preventive and Regenerative Medicine*, 207–297. Elsevier.

69 Rangel-Corona, R. Soto-Vázquez, R., Weiss-Steider, B. et al. (2014). *Propuesta biotecnológica para el tratamiento del cáncer de cérvix uterino, empleando liposomas*. Anales de la Real Academia Nacional de Farmacia.

70 Christian, D.A. and Hunter, C.A. (2012). Particle-mediated delivery of cytokines for immunotherapy. *Immunotherapy.* 4 (4): 425–441.

71 Duncan, S.A., Dixit, S., Sahu, R. et al. (2019). Prolonged release and functionality of Interleukin-10 encapsulated within PLA-PEG nanoparticles. *Nanomaterials.* 9 (8): 1074.

72 Gdowski, A., Ranjan, A., Mukerjee, A. et al. (2015). Development of biodegradable nanocarriers loaded with a monoclonal antibody. *International Journal of Molecular Sciences.* 16 (2): 3990–3995.

73 Sousa, F., Fonte, P., Cruz, A. et al. (2018). Polyester-based nanoparticles for the encapsulation of monoclonal antibodies. In: *Recombinant Glycoprotein Production* (ed. V. Picanço-Castro and K. Swiech), 239–253. Springer.

74 Çalış, S., Atar, K.Ö., Arslan, F.B. et al. (2019). Nanopharmaceuticals as drug-delivery systems: for, against, and current applications. In: *Nanocarriers for Drug Delivery* (ed. S.S. Mohapatra, S. Ranjan, N. Dasgupta, R. Kumar Mishra, and S. Thomas), 133–154. Elsevier.

75 Dean, S.N., Turner, K.B., Medintz, I.L. et al. (2017). Targeting and delivery of therapeutic enzymes. *Therapeutic Delivery.* 8 (7): 577–595.

76 Bamburowicz-Klimkowska, M., Poplawska, M., and Grudzinski, I.P. (2019). Nanocomposites as biomolecules delivery agents in nanomedicine. *Journal of Nanobiotechnology.* 17 (1): 48.

77 Choonara, B.F., Choonara, Y.E., Kumar, P. et al. (2014). A review of advanced oral drug delivery technologies facilitating the protection and absorption of protein and peptide molecules. *Biotechnology Advances.* 32 (7): 1269–1282.

78 Duo, Y., Yang, M., Du, Z. et al. (2018). CX-5461-loaded nucleolus-targeting nanoplatform for cancer therapy through induction of pro-death autophagy. *Acta Biomaterialia.* 79: 317–330.

79 Dhule, S.S., Penfornis, P., He, J. et al. (2014). The combined effect of encapsulating curcumin and C6 ceramide in liposomal nanoparticles against osteosarcoma. *Molecular Pharmaceutics.* 11 (2): 417–427.

80 Muhamad, N., Plengsuriyakarn, T., and Na-Bangchang, K. (2018). Application of active targeting nanoparticle delivery system for chemotherapeutic drugs and traditional/herbal medicines in cancer therapy: a systematic review. *International Journal of Nanomedicine.* 13: 3921.

81 Tu, J., Boyle, A.L., Friedrich, H. et al. (2016). Mesoporous silica nanoparticles with large pores for the encapsulation and release of proteins. *ACS Applied Materials & Interfaces.* 8 (47): 32211–32219.

82 Bareford, L.M. and Swaan, P.W. (2007). Endocytic mechanisms for targeted drug delivery. *Advanced Drug Delivery Reviews.* 59 (8): 748–758.

83 Zhao, J. and Stenzel, M.H. (2018). Entry of nanoparticles into cells: the importance of nanoparticle properties. *Polymer Chemistry.* 9 (3): 259–272.

84 Juárez, B.H. (2011). *Nanoparticulas semiconductoras coloidales y aplicaciones.* Anales de Química.

85 Bahrami, B., Hojjat-Farsangi, M., Mohammadi, H. et al. (2017). Nanoparticles and targeted drug delivery in cancer therapy. *Immunology Letters.* 190: 64–83.

86 Uthaman, S., Huh, K.M., and Park, I.-K. (2018). Tumor microenvironment-responsive nanoparticles for cancer theragnostic applications. *Biomaterials Research.* 22 (1): 22.

87 Yoo, J., Park, C., Yi, G. et al. (2019). Active targeting strategies using biological ligands for nanoparticle drug delivery systems. *Cancers (Basel).* 11 (5): 640.

88 Wang, B., He, X., Zhang, Z. et al. (2013). Metabolism of nanomaterials in vivo: blood circulation and organ clearance. *Accounts of Chemical Research* 46 (3): 761–769. https://doi.org/10.1021/ar2003336.

89 Kermanizadeh, A., Gaiser, B.K., Johnston, H. et al. (2014). Toxicological effect of engineered nanomaterials on the liver. *British Journal of Pharmacology* 171 (17): 3980–3987. http://doi.org/10.1111/bph.12421.

90 He, X., Zhang, H., Ma, Y. et al. (2013). Lung deposition and extrapulmonary translocation of nano-ceria after intratracheal instillation. *Nanotechnology* 21 (28): 285103. https://doi.org/10.1088/0957-4484/21/28/285103.

91 Geiser, M. (2010). Update on macrophage clearance of inhaled micro-and nanoparticles. *Journal of Aerosol Medicine and Pulmonary Drug Delivery.* 23: 207–217.

92 Zhu, M., Feng, W., Wang, B. et al. (2008). Comparative study of pulmonary responses to nano- and submicron-sized ferric oxide in rats. *Toxicology.* 247: 102–111.

93 Katragadda, C.S., Choudhury, P.K., and Murthy, P.N. (2010). Nanoparticles as non-viral gene delivery vectors. *Indian Journal of Pharmaceutical Education and Research.* 44 (2): 109–120.

94 Peng, L.-H., Huang, Y.-F., Niu, J. et al. (2019). Integration of antimicrobial peptides with gold nanoparticles as unique non- viral vectors for gene delivery to mesenchymal stem cells with antibacterial activity biomaterials integration of antimicrobial peptides with gold nanoparticles as unique non-viral vectors for gene delivery to mesenchymal stem cells with antibacterial activity. *Biomaterials.* 103: 137–149. http://doi.org/10.1016/j.biomaterials.2016.06.057.

95 Behnam, B., Shier, W.T., Hashem, A. et al. (2013). Non-covalent functionalization of single-walled carbon nanotubes with modified polyethyleneimines for efficient gene delivery. *International Journal of Pharmaceutics* 454 (1): 204–215. http://doi.org/10.1016/j.ijpharm.2013.06.057.

96 Bates, K. and Kostas, K. (2013). Carbon nanotubes as vectors for gene therapy: past achievements, present challenges and future goals. *Advanced Drug Delivery Reviews.* 65 (15): 2023–2033. https://doi.org/10.1016/j.addr.2013.10.003.

97 Monod, J. and Jacob, F. (1961). Teleonomic mechanisms in cellular metabolism, growth, and differentiation. *Cold Spring Harbor Symposia on Quantitative Biology* 26: 389–401. http://dx.doi.org/10.1101/sqb.1961.026.01.048.

98 Bhat, M., Robichaud, N., Hulea, L. et al. (2015). Targeting the translation machinery in cancer. *Nature Reviews Drug Discovery* 14 (4): 261–278. http://doi.org/10.1038/nrd4505.

99 Ma, X., Wu, Y., Jin, S. et al. (2011). Gold nanoparticles induce autophagosome accumulation through size-dependent nanoparticle uptake and lysosome impairment. *ACS Nano* 5 (11): 8629–8639. http://doi.org/10.1021/nn202155y.

100 Rudnitzki, F., Feineis, S., Rahmanzadeh, R. et al. (2018). siRNA release from gold nanoparticles by nanosecond pulsed laser irradiation and analysis of the involved temperature increase. *Journal of Biophotonics* 11 (9): http://doi.org/10.1002/jbio.201700329.

101 Sierra, M.I., Valdés, A., Fernández, A.F. et al. (2016). The effect of exposure to nanoparticles and nanomaterials on the mammalian *epigenome. International Journal of Nanomedicine.* 11: 6297–6306. https://doi.org/10.2147/IJN.S120104.

102 Moustakas, A., Pardali, K., Gaal, A., and Heldin, C.H. (2002). Mechanisms of TGF-β signaling in regulation of cell growth and differentiation. *Immunology Letters* 82 (1–2): 85–91. http://doi.org/10.1016/S0165-2478(02)00023-8.

103 Miller, D.M., Thomas, S.D., Islam, A. et al. (2012). c-Myc and cancer metabolism. *Clinical Cancer Research 18*(20): 5546–5553. https://doi.org/10.1158/1078-0432.CCR-12-0977.

104 Bansil, R. and Turner, B.S. (2006). Mucin structure, aggregation, physiological functions and biomedical applications. *Current Opinion in Colloid & Interface Science* 11 (2): 164–170.

105 del Amo, E.M. et al. (2017). Pharmacokinetic aspects of retinal drug delivery. *Progress in Retinal and Eye Research* 57: 134–185.

106 del Amo, E.M. and Urtti, A. (2015). Rabbit as an animal model for intravitreal pharmacokinetics: clinical predictability and quality of the published data. *Experimental Eye Research* 137: 111–124.

107 Kompella, U.B. et al. (2013). Nanomedicines for back of the eye drug delivery, gene delivery, and imaging. *Progress in Retinal and Eye Research* 36: 172–198.

108 Jo, D.H. et al. (2011). Nanotechnology and nanotoxicology in retinopathy. *International Journal of Molecular Sciences* 12 (11): 8288–8301.

109 Moshfeghi, A.A. and Peyman, G.A. (2005). Micro- and nanoparticulates. *Advanced Drug Delivery Reviews* 57 (14): 2047–2052.

110 Sahoo, S.K. et al. (2008). Nanotechnology in ocular drug delivery. *Drug Discovery Today.* 13 (3–4): 144–151.

111 Fangueiro, J.F. et al. (2015). Current nanotechnology approaches for the treatment and management of diabetic retinopathy. *European Journal of Pharmaceutics and Biopharmaceutics* 95, Part B: 307–322.

112 Kelkar, S.S. and Reineke, T.M. (2011). Theranostics: combining imaging and therapy. *Bioconjugate Chemistry* 22 (10): 1879–1903.

113 Iavicoli, I. et al. (2012). Toxicological effects of titanium dioxide nanoparticles: a review of in vivo studies. *Journal of Nanomaterials* 2012: 36.

114 Wang, J. et al. (2009). Expression changes of dopaminergic system-related genes in PC12 cells induced by manganese, silver, or copper nanoparticles. *Neurotoxicology* 30 (6): 926–933.

115 Wang, Y. et al. (2011). Microglial activation, recruitment and phagocytosis as linked phenomena in ferric oxide nanoparticle exposure. *Toxicology Letters* 205 (1): 26–37.

116 Xue, Y. et al. (2012). Four types of inorganic nanoparticles stimulate the inflammatory reaction in brain microglia and damage neurons in vitro. *Toxicology Letters* 214 (2): 91–98.

117 Fomina, N. et al. (2012). Photochemical mechanisms of light-triggered release from nanocarriers. *Advanced Drug Delivery Reviews* 64 (11): 1005–1020.

118 Karimi, M. et al. (2017). Smart nanostructures for cargo delivery: uncaging and activating by light. *Journal of the American Chemical Society* 139 (13): 4584–4610.

119 Lino, M.M. and Ferreira, L. (2018). Light-triggerable formulations for the intracellular controlled release of biomolecules. *Drug Discovery Today* 23 (5): 1062–1070.

120 Fomina, N. et al. (2010). UV and near-IR triggered release from polymeric nanoparticles. *Journal of the American Chemical Society* 132 (28): 9540–9542.

121 Huu, V.A.N. et al. (2015). Light-responsive nanoparticle depot to control release of a small molecule angiogenesis inhibitor in the posterior segment of the eye. *Journal of Controlled Release* 200: 71–77.

122 Lemire, J.A. et al. (2013). Antimicrobial activity of metals: mechanisms, molecular targets and applications. *Nature Reviews Microbiology* 11: 371.

123 Vazquez-Muñoz, R. et al. (2017). Toxicity of silver nanoparticles in biological systems: does the complexity of biological systems matter? *Toxicology Letters* 276: 11–20.

124 Shayani, R.M. et al. (2016). Antibacterial activity of silver nanoparticle-loaded soft contact lens materials: the effect of monomer composition. *Current Eye Research* 41 (10): 1286–1293.

125 Nickols-Richardson, S. and Piehowski, K. (2008). Nanotechnology in nutritional sciences. *Minerva Biotecnologica.* 20 (3): 117.

126 Yu, H., Park, J.-Y., Kwon, C.W. et al. (2018). An overview of nanotechnology in food science: preparative methods, practical applications, and safety. *Journal of Chemistry* 2018.

127 Bhosale, R.R., Ghodake, P.P., Mane, A., and Ghadge, A. (2013). Nanocochleates: a novel carrier for drug transfer. *Journal of Scientific and Innnovative Research.* 2: 964–969.

128 Zhu, M., Nie, G., Meng, H. et al. (2012). Physicochemical properties determine nanomaterial cellular uptake, transport, and fate. *Accounts of Chemical Research* 46 (3): 622–631.

129 Singh, T., Shukla, S., Kumar, P. et al. (2017). Application of nanotechnology in food science: perception and overview. *Frontiers in Microbiology* 8: 1501.

130 Chaudhry, Q., Scotter, M., Blackburn, J. et al. (2008). Applications and implications of nanotechnologies for the food sector. *Food Additives and Contaminants* 25 (3): 241–258.

131 Assadpour, E. and Mahdi, J.S. (2019). A systematic review on nanoencapsulation of food bioactive ingredients and nutraceuticals by various nanocarriers. *Critical Reviews in Food Science and Nutrition* 59 (19): 3129–3151.

132 Zhang, H. (2017). Thin-film hydration followed by extrusion method for liposome preparation. In: *Liposomes* (ed. Gerard G.M. D'Souza), 17–22. Springer.

133 Huang, Z.M., Zhang, Y.Z., Kotaki, M., Ramakrishna, S. (2003). *Composites Science and Technology*; 63: 2223.

134 Graham, K., Gogins, M., and Schreuder-Gibson, H. Incorporation of electrospun nanofibers into functional structures. *International Nonwovens Journal* 2004 (2): 1558925004os-1300209.

135 Sen, K. and Babu, K.M. (2004). Studies on Indian silk. I. Macrocharacterization and analysis of amino acid composition. *Journal of Applied Polymer Science* 92 (2): 1080–1097.

136 Huang, C., Wang, S., Zhang, H. et al. (2006). High strength electrospun polymer nanofibers made from BPDA–PDA polyimide. *European Polymer Journal* 42 (5): 1099–1104.

137 Rezaei, A., Nasirpour, A., and Fathi, M. (2015). Application of cellulosic nanofibers in food science using electrospinning and its potential risk. *Comprehensive Reviews in Food Science and Food Safety* 14 (3): 269–284.

138 Smith, G.B. (ed.) (2011). *Green nanotechnology. Nanostructured thin films IV*. International Society for Optics and Photonics.

139 Alborzi, S., Lim, L.T., and Kakuda, Y. (2010). Electrospinning of sodium alginate-pectin ultrafine fibers. *Journal of Food Science* 75 (1): C100–C107.

140 Alborzi, S., Lim, L.-T., and Kakuda, Y. (2013). Encapsulation of folic acid and its stability in sodium alginate-pectin-poly (ethylene oxide) electrospun fibres. *Journal of Microencapsulation* 30 (1): 64–71.

141 Liu, W., Tian, M., Kong, Y. et al. (2017). Multilayered vitamin C nanoliposomes by self-assembly of alginate and chitosan: long-term stability and feasibility application in mandarin juice. *LWT* 75: 608–615.

142 Sabliov, C.M. and Astete, C.E. (2015). Polymeric nanoparticles for food applications. In: *Nanotechnology and Functional Foods: Effective Delivery of Bioactive Ingredients*, vol. 272 (ed. C. Sabliov, H. Chen, and R. Yada).

143 Cristina, M., Sabilo, H.C., and Rickey, Y. (2015). Polymeric nanoparticles for food applications. In: *Nanotechnology and Functional Foods* (ed. C. Sabliov, H. Chen, and R. Yada). Chichester: Wiley.

144 Rana, V. and Sharma, R. (2019). *Recent Advances in Development of Nano Drug Delivery. Applications of Targeted Nano Drugs and Delivery Systems.*, 93–131. Elsevier.

145 Martău, G.A., Mihai, M., and Vodnar, D.C. (2019). The use of chitosan, alginate, and pectin in the biomedical and food sector – biocompatibility, bioadhesiveness, and biodegradability. *Polymers* 11 (11): 1837.

146 Pan, M., Yin, Z., Liu, K. et al. (2019). Carbon-based nanomaterials in sensors for food safety. *Nanomaterials* 9 (9): 1330.

147 Das, R., Ali, M.E., Hamid, S.B.A. et al. (2014). Carbon nanotube membranes for water purification: a bright future in water desalination. *Desalination* 336: 97–109.

148 Qu, X., Alvarez, P.J., and Li, Q. (2013). Applications of nanotechnology in water and wastewater treatment. *Water Research* 47 (12): 3931–3946.

149 Liu, X., Wang, M., Zhang, S., and Pan, B. (2013). Application potential of carbon nanotubes in water treatment: a review. *Journal of Environmental Sciences* 25 (7): 1263–1280.

150 Smith, S.C. and Rodrigues, D.F. (2015). Carbon-based nanomaterials for removal of chemical and biological contaminants from water: a review of mechanisms and applications. *Carbon* 91: 122–143.

151 Bordoloi, D., Roy, N.K., Monisha, J. et al. (2016). Multi-targeted agents in cancer cell chemosensitization: what we learnt from curcumin thus far. *Recent Patents on Anti-Cancer Drug Discovery* 11 (1): 67–97.

152 Tsang, C.K., Kelly, T.L., Sailor, M.J., and Li, Y.Y. (2012). Highly stable porous silicon–carbon composites as label-free optical biosensors. *ACS Nano* 6 (12): 10546–10554.

153 Bhullar, K.S., Jha, A., Youssef, D., and Rupasinghe, H. (2013). Curcumin and its carbocyclic analogs: structure-activity in relation to antioxidant and selected biological properties. *Molecules.* 18 (5): 5389–5404.

154 Pacher, P., Beckman, J.S., and Liaudet, L. (2007). Nitric oxide and peroxynitrite in health and disease. *Physiological Reviews* 87 (1): 315–424.

155 Kukongviriyapan, U., Pannangpetch, P., Kukongviriyapan, V. et al. (2014). Curcumin protects against cadmium-induced vascular dysfunction, hypertension and tissue cadmium accumulation in mice. *Nutrients* 6 (3): 1194–1208.

156 Garg, S. and Garg, A. (2018). Encapsulation of curcumin in silver nanoparticle for enhancement of anticancer drug delivery. *International Journal of Pharmaceutical Sciences and Research* 9: 1160–1166.

157 García-Niño, W.R. and Pedraza-Chaverrí, J. (2014). Protective effect of curcumin against heavy metals-induced liver damage. *Food and Chemical Toxicology.* 69: 182–201.

158 Flora, S., Mittal, M., and Mehta, A. (2008). Heavy metal induced oxidative stress & its possible reversal by chelation therapy. *Indian Journal of Medical Research* 128 (4): 501.

159 Bumbudsanpharoke, N., Choi, J., and Ko, S. (2015). Applications of nanomaterials in food packaging. *Journal of Nanoscience and Nanotechnology* 15 (9): 6357–6372.

160 Blasco, C. and Picó, Y. (2011). Determining nanomaterials in food. *TrAC Trends in Analytical Chemistry* 30 (1): 84–99.

161 Peiro, A.M., Colombo, C., Doyle, G. et al. (2006). Photochemical reduction of oxygen adsorbed to nanocrystalline TiO2 films: a transient absorption and oxygen scavenging study of different TiO2 preparations. *Journal of Physical Chemistry B* 110 (46): 23255–23263.

162 Ghanbarzadeh, B., Oleyaei, S.A., and Almasi, H. (2015). Nanostructured materials utilized in biopolymer-based plastics for food packaging applications. *Critical Reviews in Food Science and Nutrition* 55 (12): 1699–1723.

163 Lawrence, G., Lyons, K., and Wallington, T. (2013). *Food Security, Nutrition and Sustainability*. Earthscan.

164 Sahu, J.K. (2014). *Introduction to Advanced Food Process Engineering*. CRC Press.

165 Rapsomanikis, A., Papoulis, D., Panagiotaras, D. et al. (2014). Nanocrystalline TiO2 and halloysite clay mineral composite films prepared by sol-gel method: synergistic effect and the case of silver modification to the photocatalytic degradation of basic blue-41 azo dye in water. *Global NEST Journal* 16 (3): 485–498.

166 Suzuki, T., Hukkanen, M., Ohashi, R. et al. (2000). Growth and adhesion of osteoblast-like cells derived from neonatal rat calvaria on calcium phosphate ceramics. *Journal of Bioscience and Bioengineering* 89 (1): 18–26.

167 Stoclin, A., Rotolo, F., Hicheri, Y. et al. (2020). Ventilator-associated pneumonia and bloodstream infections in intensive care unit cancer patients: a retrospective 12-year study on 3388 prospectively monitored patients. *Supportive Care in Cancer* 28 (1): 193–200.

168 Filippo, A.D. and Gaudio, A.D. (2003). Device-related infections in critically ill patients. Part II: prevention of ventilator-associated pneumonia and urinary tract infections. *Journal of Chemotherapy* 15 (6): 536–542.

169 Tenke, P., Mezei, T., Bőde, I., and Köves, B. (2017). Catheter-associated urinary tract infections. *European Urology Supplements* 16 (4): 138–143.

170 Tracana, R.B., Sousa, J.P., and Carvalho, G.S. (1994). Mouse inflammatory response to stainless steel corrosion products. *Journal of Materials Science: Materials in Medicine* 5 (9–10): 596–600.

2

Nanoneurology

Liron L. Israel and Eggehard Holler

Nanomedicine Research Center in the Department of Neurosurgery, Cedars Sinai Medical Center, Los Angeles, CA, USA

2.1 Introduction and Recent Advances

Treating and imaging diseases of the central nervous system (CNS) can present a special challenge by the unique protection of the brain and spinal cord by tight barriers, such as the blood-brain barrier (BBB), the blood cerebrospinal fluid (CSF) barrier (BCSFB), and the arachnoid barrier. A "degenerated" blood-tumor barrier (BTB) [1–3], which is derived from the BBB, protects brain tumors. Nevertheless, among present-day available treatment modalities, therapeutic agents are needed that can be applied systemically and efficiently delivered across the BBB for specific treatment. The endothelial cell layer surrounding the brain vasculature comprises the BBB. The molecular outfit of the cells is arranged with tight junctions between cells and endocytic and transcytotic highly controlled transmembrane pathways which spontaneously move certain low molecular weight nutrients, lipids, and lipoproteins from the blood to the parenchyma [4, 5]. This barrier function is critical to protect the brain from both pathogens and potentially neurotoxic molecules, as well as to maintain brain homeostasis. A CNS blood vessel is formed mainly by the endothelial cells connected by tight junctions (Figure 2.1). These cells are surrounded by pericytes and a basement membrane. Astrocytes then connect with both endothelial cells and pericytes, as well as with neurons, while microglia, the brain's resident immune cells, regulate immune responses [4, 5]. Oxygen from the blood and carbon dioxide from the brain can diffuse freely, and small lipophilic molecules (typically with Mw smaller than $400\,g\,mol^{-1}$) can usually cross the BBB [6]. While the BBB is usually maintained in non-pathological conditions, certain diseases can cause the dysfunction of the BBB or certain areas of it. Even though the BBB may be impaired in certain pathological conditions, the delivery of drugs to the brain at therapeutic levels can still be insufficient or completely impossible. A drug that is applied in high levels to enforce high delivery can be prone to initiate off-target effects. Low specific delivery and off-target effects are remaining problems. Since delivery across the BBB is so poor, a nanomedicine approach is being investigated extensively, since it may enable putting multiple moieties on the same molecule. In principle, nanosystems with surface modifications can provide BBB-crossing, imaging, and therapeutic treatment,

Nanomaterials and Nanotechnology in Medicine, First Edition. Edited by P.M. Visakh.

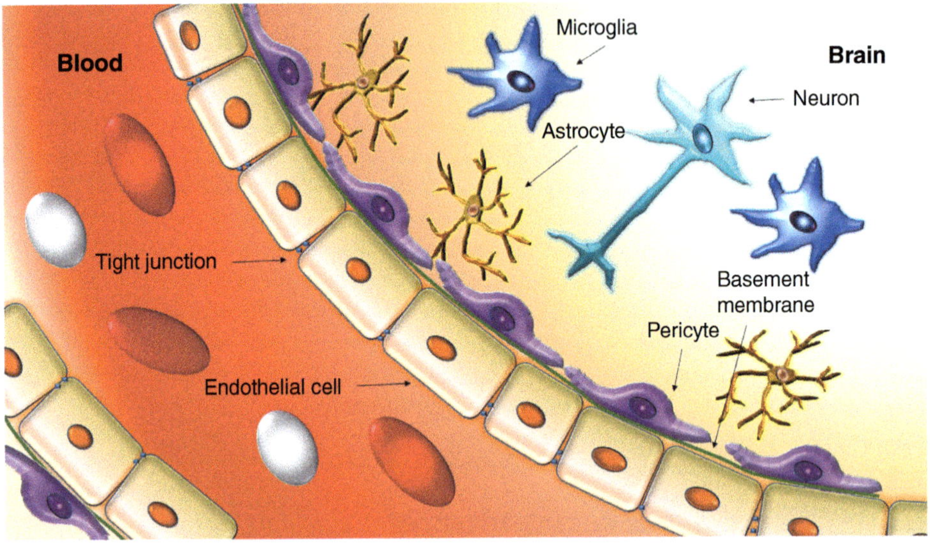

Figure 2.1 Schematic structure of the BBB.

especially in the case of surface-modified, encapsulating nanosystems, and can reduce off-target effects. However, nanosystems are often designed to combine ample targeting, imaging, and drug loads designed for high functional capacity. Nanosystems tend to be designed to be overstuffed with reagents for targeting and killing despite the fact that single cells require only a single encounter. The downside of the design could be shallow penetration into tissue because of high hydrodynamic diameters and initiating undesired side-effects because of the high accumulation of drug.

In many cases, disease pathology can cause the impairment of the BBB. For example, in brain tumors or metastasis, rapid angiogenesis in the growing tumor results in leaky vessels. The phenomenon that distinguishes tumor from healthy vessels is known as the enhanced permeability and retention (EPR) effect [7, 8]. The effect of compromised vasculature structure has been used to facilitate delivery of nanosystems to the tumor. The slow rate of delivery has often been compensated by implementing long cycling in the blood system. To date, attempts of utilizing the EPR for tumor-specific delivery was not successful in clinical trials [9–11]. The cause of the low rate was partially the nature of the EPR effect. The tumor abnormal often variable vasculature generated unfavorably deregulated extracellular matrix (ECM) components and elevated interstitial fluid pressure, which all work against efficient delivery [12, 13]. Following the unsuccessful history of EPR, in the past years a huge effort has been invested to enhance BBB crossing using active targeting, such as receptor-mediated transcytosis. Receptors that are expressed on the luminal membrane of brain endothelial cells recognize vectors, recognize specific antibodies or peptides, and function as ferries to deliver nanosystems across the BBB. Transferrin receptor (TfR), insulin receptor, and low-density lipoprotein receptor-related protein 1 (LRP-1) are three prominent examples of receptors targeted [14–19] transcytosis pathways. The option of drug delivery that bypasses systemic delivery through the BBB is actively being researched, with the nasal aside oral route seeming to be the favored approach.

Several methods to bypass the BBB are invasive methods. One example is intrathecal administration, considered to be one of the least invasive routes, where agents are injected into the subarachnoid space of the spinal cord. A promising non-invasive bypass is intranasal (IN) delivery. In this method, therapeutic agents are sprayed in the nasal cavity to facilitate absorption by the olfactory and respiratory mucosa. After moving either paracellularly or transcellularly through the olfactory and respiratory epithelia, drugs may undergo extracellular transport, before endocytosis into the peripheral trigeminal nerve, or may be taken up into the olfactory sensory neurons by endocytosis. IN delivery is non-invasive, is painless, does not require sterile preparation, and can be often administered by the patients themselves. The downside of this method, however, is that there are wide distribution patterns and penetration mechanisms as with BBB permeation, and results frequently vary from drug to drug, questioning the mechanisms involved in IN. Additionally, the protective barriers of nasal mucosa limit the efficiency of IN therapy, as only 1% or less of the drug reaches the brain after IN administration. Hence, research needs to develop suitable formulations to overcome these problems [20].

The main medicinal applications for nanomaterials are imaging and therapy that require targeted delivery. Nanomaterials can be applied in separate or combined "theranostic" fashion since nanomaterials can often carry multiple functionalities. More frequently, targeting moieties are attached to a nanosystem for tracking the administered drug and its safety to ensure that it unfolds its pharmaceutical activity only at the intended site and avoids side ("off-target") effects. Targeting is even more important in cases of the diseased brain, since the low ratio of roughly 1% of the administered pharmaceutical ingredient (API) across the BBB requires the application of high doses. In brain treatment, many clinical trials fail because of the inability to cross the BBB, which may also lead to adverse effects [21]. Common methods for brain imaging include magnetic resonance imaging (MRI) which is more commonly used for brain tumors, and positron emission tomography (PET), which often employed in Parkinson's disease (PD) and Alzheimer's disease (AD). Besides, computed tomography (CT) is also used, and new preclinical techniques are under development such as magnetic particle imaging (MPI) [22–25].

Therapeutic agents that can be delivered by nanotechnology or are under investigation vary from low molecular weight molecules such as synthetic pharmaceuticals and peptides and high molecular weight proteins such as antibodies, enzymes, and, most challenging, nucleic acids. The mechanisms underlying treatments are versatile as well and may include immunotherapy, photodynamic therapy, thermotherapy, and gene delivery. Some of the mentioned moieties can be delivered to the extracellular matrix, the cell membrane, or intracellularly. Intracellular delivery requires passage through the cellular membrane and frequently the endosomal, mitochondrial, or nuclear membranes to release the "payload" in the pharmaceutically active compartment.

Nanomaterials consist in general of the nanoplatform or incipient which represents the material, in most cases carrying the drug by encapsulation. In some cases, the nanoplatform itself can be used as the therapeutic agent (nanodrug). An example is the iron oxide nanoparticles (IONPs), which in the presence of an alternating external magnetic field can raise the temperature in a tumor, eventually causing its elimination. The use of IONPs is known as "hyperthermic" therapy. Another example is the iron particle induced polarization of macrophages to adopt the "M1" phenotype. In an example, intravenous Ferumoxytol applied to a mouse serving as a model prevented the development of liver metastasis from the intravenous inoculation of tumor cells. Fluorescence-activated cell sorting (FACS) and histopathology studies showed that the observed tumor growth inhibition was accompanied by the increased presence of proinflammatory M1 macrophages in the tumor tissue [26]. Whether peripheral functions also will be active in the brain has to be confirmed.

2.2 Types of Nanomaterials

Several types of nanomaterials are being applied for CNS delivery. Polymeric nanoconjugates (molecular nanoconjugate, mini-nanodrugs, single macromolecules) are being studied extensively mainly due to their extremely small size and plurality of functional ligands, which can be beneficial for cross-BBB delivery and disease treatments [27]. These polymeric chains usually contain multiple repeating units that can be functionalized and they are usually biodegradable and non-toxic, making them ideal for drug delivery. A prominent example is β-poly(L-malic acid), which is biodegradable to carbon dioxide and water, and was recently shown to cross the normal BBB of healthy mice when functionalized with tri-Leucine (LLL) and a BBB-crossing peptide, such as Angiopep-2 (AP2) which targets LRP-1 [19]. Crystalline NPs such as IONPs [28], gold nanoparticles (AuNPs) [29], and even nanodiamonds [30, 31] are also being considered for CNS delivery. Since some of them, such as IONPs and AuNPs, can also be detected my MRI [32] or CT (respectively) [33]. An additional advantage of some crystalline NPs is that some—iron oxide, for example (maghemite: γ-Fe_2O_3 and magnetite: Fe_3O_4)—are magnetic, which may be utilized for additional purposes, such as magnetic targeting [34], magnetic particle imaging (MPI), and treatment using hyperthermia [22]. Quantum dots (QDs) are fluorescent semiconductor nanocrystals, whose absorption energy levels are quantized when excited, achieving properties directly related to the size of the QDs and the emission of photons in a narrow, symmetric energy band [35]. These and more properties make them attractive tools for imaging purposes. However, the toxicity and biocompatibility of QDs (CdSe, for example) are questionable, with associated stabilization issues [36]. Consequently, biological interfacing is often required. When used for CNS delivery, QDs are usually encapsulated or coated [37]. Although inorganic cores of the particles are in the 1–5 nm diameter range, in some cases adhesive coatings for ligand attachment and protection against aggregation and toxicity render these nanoparticles in sizes of 30 nm and higher, which prohibits their deep tissue penetration and kidney clearance, thus promoting deposition-induced toxicities. NPs, which include micelles, dendrimers, and spheres, are often being studied for CNS delivery. Micelles [38] are spherical structures of amphiphilic molecules that are organized in accordance with their hydrophobic tail forming the core and their hydrophilic head contacting the aqueous solvent. In contrast with polymers of linear platforms, dendrimers are repetitively branched nanomolecules. Depending on the lipophilic or hydrophilic nature of their interior, micelles and dendrimers can be considered as "sponges" that dissolve drugs that could be leaky to the solvent to a variable extent. Lipid-based NPs represent biological vesicles of controllable stability and toxicity and are competent for encapsulation of a variety of structurally different drug molecules. Liposomes of single lipid bilayers serving drug encapsulation have been successfully delivered to the CNS.

Another type of nanomaterials includes nanotubes consisting of crosslinked carbon–carbon structure (CNTs). Members of the fullerene structural family are allotropes of carbon with a cylindrical nanostructure, either single- or multi-walled involving entirely sp2-hybrid carbon atoms. Since they have unique properties, in terms of strength as well as electrical (semiconductors), optical, and thermal properties, they are also studied for BBB delivery [39], along with fullerenes of spherical structure. Their biocompatibility is disputed.

When a local and not systemic treatment is an option, hydrogels/nanogels [40] can be used. Nanogels have a 3D-structure forming a network of crosslinked polymers with hydrophilic or amphiphilic properties able to swell and hold a great amount of water (which is why they are preferred for local and not systemic delivery). The network is maintained holding a variety of natural polymers, synthetic polymers, or a combination thereof, covalently or physically crosslinked including non-covalent bonds and hydrogen bonds, electrostatic and hydrophobic interactions [41]. As drug delivery systems, nanogels can be used mostly in disorders such as stroke or spinal cord injury (SCI). The system is also designed to enclose bioactive substances such as nucleic acids, proteins, and cytokines for nasal delivery [41].

In recent years, cell-derived delivery systems have been studied for CNS delivery. One example is the "nanoghost" (NG) system [42, 43]. Using mostly but not exclusively mesenchymal stem cells [44–47], these derived vesicles are prepared by physical (e.g., sonication) and chemical treatment of cells to remove cytoplasm and nuclei, leaving a small (~200 nm) membrane compartment that has inherited targeting capabilities from the "mother" cell [43]. Current NGs are derived from cancer stem cells (CSCs) to treat tumors.

Finally, an additional prominent cell-based method for nanoscale drug delivery to the CNS is based on exosomes. Exosomes are bio-NPs (30–100 nm) that are secreted by many biological cells [48], unlike NGs, which are artificially derived from cells [49–51]. Some reports indicated that exosomes may have an immune privileged status that can efficiently deliver drugs by the pathway of mononuclear phagocyte uptake and migration to the brain [50, 52–54].

An important role in treatment is the targeting of the nanocarrier to the site of delivery and the appropriate release from the carrier, taking into account the specific features as nanopolymers, micelles, dendrimers, liposomes, etc. Active targeting usually involves tethering of a ligand on the nanoplatform that is sterically accessible to bind a receptor at the site of delivery. The ligand is covalently bound and is not required to be cleaved from the carrier platform. For the release of the drug at the site of its activity—often cell cytoplasm or organelles—drugs (prodrugs) must be cleaved from the carrier platform, or the drug must be released by an "opening" of the transporting capsule. Both cleavage and release from encapsulation must be triggered in response to the microenvironment of the pharmacologically selected site. The trigger mechanism can involve acidic pH, disulfide reduction, at-site enzymatic activity, or externally controlled heat or light-induced activation.

A schematic representation of some of the nanosystems mentioned can be seen in Figure 2.2.

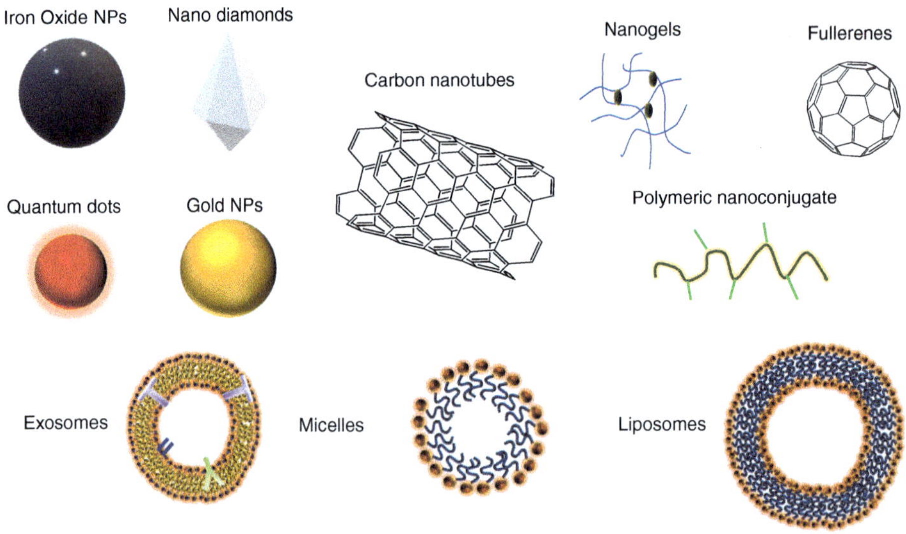

Figure 2.2 Schematic structures of nanomaterials used for CNS delivery.

2.3 Nanomaterial Applications for Neurodegenerative Diseases

Neurodegenerative diseases occur when nerve cells in the brain or peripheral nervous system lose function over time and ultimately die. The risk of being affected by a neurodegenerative disease increases dramatically with age, although there are also cases of "familial" or genetically triggered disease (the opposite of the "sporadic" or non-genetic form), and in that case, it will usually erupt at a younger age. AD and PD are the most common types, and some other examples include Huntington's disease, amyotrophic lateral sclerosis (ALS), prion disease, tauopathies, and several types of non-AD dementia, which is a collective term used to describe various symptoms of cognitive decline. In some brain pathologies, the BBB is impaired or breached, but it is not always to the extent that a large flux of nanomaterials will pass into the brain, and when it does, it is usually very late in the disease.

2.3.1 Alzheimer's Disease

In AD, the main pathologies include chronic neuro-inflammation, aggregates of a protein called amyloid beta (Aβ), which is found normally in the brain in the form of $A\beta_{1\text{-}40}$ but in pathological conditions another version of it, $A\beta_{1\text{-}42}$, is formed in large quantities which causes aggregation. Additionally, NFTs or neurofibrillary tangles are formed, derived from a hyperphosphorylated Tau protein which causes the collapse of a transport system called microtubule structure inside the neurons. Additional characteristics include oxidative stress and mitochondrial damage. Unlike other neurodegenerative diseases such as PD, where there are treatments available, the complexity of AD leads to the fact that, to date, only the symptoms are treatable and there is no cure for the disease. As the age of the population is on the rise, so is the number of people who will be affected by the disease in older age [55,56]. Therefore, it is one of the most studied neurodegenerative diseases. The BBB breakdown mechanism is still being studied [57, 58], and postmortem analysis of AD patients reviles the breakdown of the BBB, starting in the hippocampus area, which is related to memory function [59–63]. Some imaging techniques were also used to evaluate neurovascular dysfunction in AD [64]. However, other studies suggest that passive diffusion across the BBB may be reduced due to a thickened basement membrane and that the BBB paracellular route remains unaffected [65], potentially lowering the ability of nanomedicines to cross AD BBB. Therefore, the extent of BBB disruption in AD remains controversial [66]. Another issue is what pathology to target, since in this case, there are supporters that say the inflammation is the cause of deterioration, while others point to other pathologies such as the Aβ plaques. Yet an additional challenge is the possible change in receptor density in pathological conditions. For example, in AD, one receptor that is used in many cases to facilitate BBB crossing is the LRP-1, which as mentioned earlier, is down-regulated [67]. Considering all these challenges, nanomaterials are being studied extensively to try and address these issues and more. In the field of imaging, small molecules for PET such as amyloid radiotracer ^{11}C-PIB (its structure is based on the molecule Thioflavin-T, a known staining agent for amyloid plaques) are still dominant [68]. MRI was used to evaluate neurovascular dysfunction, but the method is not used for clinical diagnosis [64]. However, some nanomedicine research was done in the field of Aβ detection to try managing the disease more efficiently [69, 70]. One example is the use of a small size polymeric nanoconjugate with a β-poly(L-malic acid) (PMLA) backbone, to deliver gadolinium complexed by 1,4,7,10-Tetraazacyclododecane-1,4, 7,10-tetraacetic acid (Gd-DOTA) as an MRI contrast, to image Aβ plaques *ex vivo* in both human and mice AD brains. Curcumin was conjugated to the polymer, since it can cross the BBB and can also bind Aβ [71]. Another group of researchers used a paramagnetic GF3 NP, decorated with luminescent conjugated oligothiophenes (LCOs) known to selectively bind to Aβ aggregates, for a dual modality imaging, combining MRI and two-photon fluorescent imaging [72]. However, the NP design for *in vivo* imaging showed that a major fraction of the probe was trapped in the liver, hence hampering accessibility for BBB penetration and cerebral amyloid targeting. Different research suggests that the presence of excess metal ions in the brain is most probably one of the main causative factors for the aggregation of $A\beta_{1\text{-}42}$ proteins and, therefore, imaging metal cations may be beneficial for early diagnosis of AD [73]. For that purpose, a nanoprobe (UCHQ) was designed, which combines two main components: up-conversion NPs ($NaYF_4$ or Yb/Er/Tm nanocrystals, UCNPs) used for the detection and up-conversion luminescence (UCL) imaging of Cu^{2+} upon 980 nm exposure and the chelator 8-hydroxyquinoline-2-carboxylic acid (HQC) used for chelating Cu^{2+} and AD therapy. The results show that the emission intensity of UCHQ is highly dependent on the Cu^{2+} concentrations due to the luminescence resonance energy transfer (LRET) from UCNPs to HQC-bonded Cu^{2+}. Additionally, the as-constructed UCHQs could be used for UCL imaging of Aβ both in cells and AD mice *ex vivo*, while *in vivo* imaging was done in a zebrafish embryo model. Per the authors, since the particle can capture Cu^{2+} cations, it can also potentially inhibit plaque formation.

In an approach different from diagnosis by imaging, a design of an artificial peptide nanonetwork biosensor, mimicking the natural pathological peptides' pathogenesis process, was conceived. Periphery platelets can secrete Aβ and induce cross-linking and aggregation to form a surface peptide nanonetwork, resulting in large numbers of poly-tyrosine strands being covalently trapped in the network to serve as an efficient signal amplifier through the electrochemical oxidation of tyrosine. This sensitive and quantitative method can effectively discriminate AD patients based on the detected potential neurodegenerative activity of the platelets. Platelets can be readily collected via peripheral blood sampling, which is considered non-invasive. When it is activated to process and release Aβ, oxidative stress results in self-catalyzed cross-linking. This Aβ assembly and covalent coupling to form a 2D peptide network provides a nanosensing surface for effective signal amplification [74].

Although progress is seen in the field, crossing the BBB in AD mice models efficiently for early diagnosis using nanomaterials remains a challenge, and so a large variety of possible solutions are being investigated.

The same challenge exists for nanomaterials for AD therapy. Unlike AD imaging where there are some efficient PET radiotracers, AD has no known cure, and the current small molecule therapeutics which include Donepezil, Rivastigmine, Galantamine, which are Acetyl Cholinesterase inhibitors, are only addressing the disease symptoms. These small molecules were aimed at the "cholinergic hypothesis", which claims memory loss and deterioration of cognitive functions are due to acetylcholine loss. Another small molecule which addresses the symptoms is Memantine, a low-affinity uncompetitive antagonist at glutamatergic N-methyl-d-aspartate (NMDA) receptors [75]. That fact leads to a massive research effort for AD treatments, many of which are based on nanotechnology [76–79].

One study describes a nanocomposite that can mitigate Aβ-mediated neurotoxicity and eliminate Aβ aggregates in AD mice by inhibiting the formation of oligomeric Aβ [80]. The small (14 ± 4 nm) nanocomposite contains multiple Aβ-binding peptides with the sequence KLVFF (Aβ, 16–20 Lys-Leu-Val-Phe-Phe) integrated on the surface. This structure allows the nanocomposite to co-assemble with Aβ and eventually inhibits the self-aggregation of Aβ. Per the authors, the presence of the nanocomposite leads to the formation of Aβ/nanocomposite co-assembled nanoclusters (50–100 nm) instead of neurotoxic Aβ oligomers and fibrils. Furthermore, the clusters can facilitate the resident microglia. Their hypothesis is that oligomeric Aβ triggers microglia to secrete proinflammatory cytokines, which eventually down-regulate the expression of Aβ-binding scavenger receptors on microglia. Therefore, treatments that convert Aβ oligomers to nonpathogenic clusters might regain microglia's phagocytosis. The nanocomposite with KLVFF (NC-KLVFF) is prepared by wrapping a protein molecule with a cross-linked KLVFF-containing polymer layer, which was synthesized by *in situ* polymerization of acrylamide (AAM), acryloyl-Lys-Leu-Val-Phe-Phe (Acr-KLVFF), and N,N′-methylenebis (acrylamide) (BIS) on the surface of the protein [80].

In some cases, antibody-based treatments can be very efficient, but since many of them can't cross the BBB, they are often attached to a nanosystem that may carry the antibody across the BBB. In an opposite case [81], an $A\beta_{1\text{-}42}$-binding antibody was attached to a polymeric nanosystem. However, it was designed to trap and clear blood circulating $A\beta_{1\text{-}42}$, and by doing so, lowering brain $A\beta_{1\text{-}42}$ levels via a "sink effect." Per this theory, reducing the peripheral $A\beta_{1\text{-}42}$ levels would cause in increased amount of it to be cleared from the brain to the periphery and therefore reduce $A\beta_{1\text{-}42}$ burden inside the brain parenchyma. Therefore, that NP does not have to cross the BBB for effective AD treatment. NPs with high affinity for the $A\beta_{1\text{-}42}$ peptide were produced by coupling an anti-$A\beta_{1\text{-}42}$ monoclonal antibody at the Poly ethylene glycol (PEG) chain ends of co-polymer NPs via biotin-streptavidin (Bio-Sav) ligation [81]. Fourteen-month-old Tg2576 AD-like transgenic male mice and 9 wild type (WT) age-matched littermates were used for anti-A$\beta_{1\text{-}42}$-NP evaluation. Tg2576 mice carry a transgene coding for the 695-amino acid isoform of human Alzheimer Aβ precursor protein carrying the Swedish mutation. After treatment with anti-A$\beta_{1\text{-}42}$-NPs, mice were sacrificed and brain hemispheres were analyzed for the soluble and insoluble Aβ content by ELISA. The results showed that the treatment did not induce any significant reduction of insoluble brain Aβ burden, whereas a significant reduction (−20%) of the Triton-soluble A$\beta_{1\text{-}40}$ and A$\beta_{1\text{-}42}$ levels was detected by comparison with the soluble A$\beta_{1\text{-}40}$ and A$\beta_{1\text{-}42}$ levels in mice treated with PBS [81].

In an interesting example, a phytopharmaceutical called piperine (PIP), which is a main active alkaloid in *Piper nigrum*, which possesses beneficial effects on cognitive function in AD [82], was encapsulated by ionically cross-linked chitosan/tripolyphosphate (TPP) NPs for brain targeting of PIP via the nasal pathway in AD mice. Another study reports a mesoporous nano-selenium (MSe) release delivery system (MSe-Res/Fc-β-CD/Bor) based on borneol (Bor) target, β-cyclodextrin nanovalves (Fc-β-CD) with loaded resveratrol (Res, a stilbenoid produced by several plants in response to injury or when the plant is under attack by pathogens). Previous experiments have shown that MSe-Res/Fc-β-CD/Bor first releases Bor by interacting with blood or intracellular esterases, allowing the nanosystem to pass through the BBB. Subsequently, the Fc-β-CD is opened by the redox (H_2O_2) response to the release of Res at the lesion site. Per this research, MSe-Res/Fc-β-CD/Bor inhibited Aβ aggregation (Aβ), mitigated oxidative stress, and suppressed tau hyperphosphorylation, while successfully improving memory impairment in APP/PS1 mice [83]. In another example, a construct of a methylene blue (MB, a tau aggregation inhibitor) loaded nanocomposite (fabricated by controlled assembly of ultrasmall ceria nanocrystals and MRI enabling iron oxide nanocrystals onto the surface of mesoporous silica NPs), which possesses high binding affinity to hyperphosphorylated tau, was also shown to inhibit multiple key pathways of tau-associated AD pathogenesis including mitochondrial oxidative stress. The nanocomposite, which was also conjugated to Amino-T807, an amino substituent of T807 which is a tau tracer, via macrocyclic chelator 1,4,7-triazacyclononane-1, 4,7-triacetic acid (NOTA), which can be further labeled with ^{68}Ga for PET, demonstrated the prevention of neuronal death both *in vitro* and *in vivo* [84]. Since the *in vivo* experiments were done in rats which were stereotaxically microinjected with Okadiac acid ammonium salt (OA) or saline into the right dorsal hippocampus in order to induce AD, the BBB-crossing capability of this nanocomposite in intact BBB or transgenic AD mice model is unclear. However, this nanocomposite, as it enables targeting, dual modality imaging, and treatment ("theranostic") in one nano system or nano composite, is an example of the possibilities nanomedicine can provide [84].

2.3.2 Parkinson's Disease

PD, is another neurodegenerative disease characterized by a progressive loss of dopaminergic neurons and an elevated level of intracytoplasmic Lewy bodies (abnormal aggregates of α-synuclein protein) in the remaining neurons located in the substantia nigra (SN) pars compacta, a region of the midbrain. This loss of neurons is accompanied by the death of astrocytes and a significant increase in the number of microglia in the SN. In order to study both the mechanism and drug delivery for PD, (1-methyl-4-phenyl-1,2,3,6-tetrahydropyridine) (MPTP) is often used to facilitate PD in mice. It is a prodrug to the neurotoxin

(1-methyl-4-phenylpyridinium) (MPP+), which causes permanent symptoms of PD by destroying dopaminergic neurons in the SN of the brain.

Early in the disease, the most obvious symptoms are shaking, rigidity, slowness of movement, and difficulty with walking since motor neurons are affected. Dementia also becomes common in the advanced stages of the disease. Most conventional therapies to date are symptomatic and usually aim at reconstituting dopaminergic signaling in the surviving neurons. Since the dopamine neurotransmitter levels are dropping, increasing them is the most common treatment. Although some publications support BBB dysfunction [63], dopamine cannot cross the BBB, and therefore, a precursor to dopamine, Levodopa (L-Dopa), is given along with an inhibitor for the enzyme that transforms L-Dopa to dopamine. Since the inhibitor cannot cross the BBB, the transformation from L-Dopa to dopamine will happen only in the brain. One neuroimaging method for PD, PET, is based on the same principle in which positron emitter 18F-DOPA is used along with an inhibitor. 18F-DOPA then transforms to 18F-Dopamine only in the brain, which allows for an efficient neuroimaging of PD. Other neuroimaging methods include MRI, specifically through iron-sensitive T_2^*, and susceptibility weighted imaging (SWI) sequences, both of which can demonstrate the absence of the characteristic "swallow tail" imaging pattern in the dorsolateral SN [23, 84–86]. Nanomedicine, however, has a lot to offer when it comes to sensing. For example, one study describes electrochemical sensor nanocomposite containing graphene oxide/ZnO nanorods (GR/ZnO). The graphite screen printed electrodes (GSPE) described were developed for the simultaneous detection of L-Dopa and tyrosine [87]. The designed nanosensor showed excellent performance such as high sensitivity, good selectivity, and analytical application in real samples. The combination of graphene oxide/ZnO nanorods nanocomposite with the screen-printed electrode was found to be favorable for amplifying electrochemical signals [87].

Among PD cases, approximately 5–10% of all patients suffer from a monogenic form of PD. The vast majority of PD cases, however, are caused by the combined action and interaction of many DNA sequence variants of moderate to weak effect sizes, which are also present in the general population, as well as environmental and lifestyle exposures [88]. While conventional medicine mainly treats disease symptoms, many nanomedicines under investigation are aiming at both reducing the symptoms and addressing the mechanisms that cause cell death. One cell death mechanism suspected to be involved is the abnormal aggregation of α-synuclein mentioned before, while other suspected cell-death mechanisms include proteasomal and lysosomal system dysfunction [89] and reduced mitochondrial activity [88, 90]. Oxidative stress is also believed to be responsible for the dopaminergic neuron loss [63], and so, many nanomedicines target reactive oxygen species (ROS). For example, in one study, ceria NP systems were used for selective scavenging of mitochondrial, intracellular, and extracellular ROS in MPTP-induced PD mice [91].

In a different approach, since dopamine itself cannot cross the BBB, one study described a co-delivery of borneol and dopamine intranasally. Since the Lactoferrin (Lf) receptor is overexpressed on the apical surface of respiratory epithelial cells, borneol and lactoferrin co-modified NPs (Lf-BNPs) were used to encapsulate dopamine [92]. Other strategies to overcome the BBB may include the temporary opening of the BBB by a focused ultrasound (FUS) [93]. For example, a two-pronged treatment strategy was described, composed of MRI-guided FUS and brain-penetrating NPs (BPNs). The treatment provides widespread but targeted glial cell-line derived neurotrophic factor (GDNF) transgene expression in the brain following systemic administration [94]. MRI-guided FUS allows circulating gene vectors to partition into the brain tissue by noninvasive and transient opening of the BBB within the areas where FUS is applied. Once beyond the BBB, BPNs provide widespread and uniform GDNF expression throughout the targeted brain tissue. GDNF plasmids or negative control reporter gene bearing plasmids (pBALs) were condensed with a blend of polyethylenimine (PEI) and PEI-PEG conjugates to yield GDNF-BPNs or pBALs-BPNs, respectively. GDNF-BPNs were injected intravenously and delivered to the lesioned striatum in the PD-induced mice with MRI-guided FUS. Per the authors, animals treated with GDNF-BPNs and FUS exhibited 2.8-fold and 2.2-fold increases in dopamine levels compared to control animals at weeks 6 and 12, respectively [94].

Another approach was described when substantive decreases in neuro inflammation and nigrostrital degeneration were achieved in mouse models of PD when bone marrow–derived macrophages (BMMs) were loaded with the nano-formulated antioxidant enzyme catalase, or "nanozyme" [95]. Per the authors, nanozyme-loaded BMMs targeted diseased sites and improved transport across the BBB. To evaluate whether BMM nanozyme carriage affected transport across the BBB, brain/plasma ratios for cell-incorporated and cell-free nanozymes were determined. MPTP-intoxicated mice were injected into the tail vein with ^{3}H-nanozyme loaded into the macrophage-carriers, or free NPs. Nanozymes loaded into BMMs entered the brain at a faster rate than cell-free NPs. The researchers also showed that inflammation induced in the brain (or peripheral organs) attracted macrophages loaded with nanozymes, resulting in the increased drug accumulation in these tissues [95]. In a different study, researchers used catalase to treat PD; however, it was encapsulated in exosomes [50], naturally occurring nanosized vesicles. Exosomes are comprised of natural lipid bilayers with the abundance of adhesive proteins that readily interact with cellular membranes. Researchers used exosomes secreted by monocytes and macrophages, which may provide an opportunity to avoid entrapment in mononuclear phagocytes of the host immune system, as well as enhance delivery of incorporated drugs to target cells. The size of the obtained catalase-loaded exosomes (exoCAT) was in the range of 100–200 nm. The exosomes were taken up by neuronal cells *in vitro*, while a considerable amount was detected in the PD mouse brain following IN administration. To visualize the ability of exosomes to target and deliver their payload to inflamed brain tissues, confocal imaging studies were conducted in a PD mouse model. The images revealed a wide distribution of exosomes throughout the brain—in particular, the cerebral frontal cortex, central sulcus, and cerebellum. Interestingly, the number of exosomes delivered upon the IN administration was greater than those delivered through intravenous injection. The anti-inflammatory effect of exoCAT was demonstrated by the significant reduction in microgliosis as measured by CD11b expression and the decrease in astrocytosis as demonstrated by astroglia glial fibrillary acidic protein (GFAP) levels [50].

In another study, magnetic Fe_3O_4 NPs coated with oleic acid molecules were used as a nanocarrier loaded with Alpha-Synuclein RNAi Plasmid [96]. N-isopropylacrylamide derivative (NIPAm-AA) was photo-immobilized onto the oleic acid molecules, and short hairpin RNA (shRNA) was absorbed. The same method was used to absorb nerve growth factor (NGF) to NIPAm-AA to specifically promote neuronal uptake via NGF receptor-mediated endocytosis. Per the authors, shRNA plasmid could be the release mechanism in the target cell due to the temperature and pH sensitivity of NIPAm-AA interference with α-syn synthesis. To assess whether the efficacy of the NPs, quantitative real-time PCR (qPCR) was used to measure the mRNA expression of α-syn and tyrosine hydroxylase (TH, which catalyzes the formation of L-Dopa) genes in various treatment groups. A decrease in the mRNA levels of α-syn and enhancement of TH mRNA levels were found for the NPs group compared with the saline group in the MTPT mice model of PD [96].

In a recent study, graphene QDs were used to prevent α-synucleinopathy in PD by inhibiting fibrillization of α-syn and interacting directly with mature fibrils [97]. For that purpose, graphene quantum dots (GQDs) were synthesized by placing carbon fibers in a mixture of strong acid. Initial tests showed that in the absence of GQDs, α-syn monomers assembled into mature fibrils, while the GQDs predominantly inhibited α-syn fibrillization. GQDs also induced the dissociation of α-syn fibrils into short fragments. The *in vivo* permeability of the BBB was studied by using GQD-biotin for immunohistochemical analysis of the brain. After intraperitoneal (i.p.) injection, a considerable amount of GQD-biotin was detected in the CNS olfactory bulb, neocortex, midbrain, and cerebellum. Additionally, compared to WT mice, mice administered with GQDs were protected against α-syn preformed fibrils–induced loss of dopaminergic neurons.

2.4 Nanomaterial Applications for Strokes

A stroke occurs when a blood vessel that carries oxygen and nutrients to the brain is either blocked by a clot or ruptures. When that happens, part of the brain cannot get the blood, which causes hypoxia and consequently the death of brain cells. Clot obstructing is called an ischemic stroke, and a blood vessel rupturing and preventing blood flow to the brain is called a hemorrhagic stroke. A transient ischemic attack (TIA), or "mini stroke," is usually caused by a temporary clot [98]. In pathology and anatomy, ischemic tissue potentially destined for infarction is not irreversible, and the target of any acute therapies is known as "penumbra" (Figure 2.3). The ischemic penumbra then denotes a region that is functionally impaired but potentially salvageable. The ischemic core by contrast refers to the irreversibly damaged tissue distal to an occluded blood vessel. With time, the infarct core expands into the penumbra, and therapeutic opportunity could be lost. Since the brain has high intrinsic metabolic activity and large concentrations of the neurotransmitter excitotoxin glutamate, it is especially vulnerable to ischemic insult. The challenge is to restore

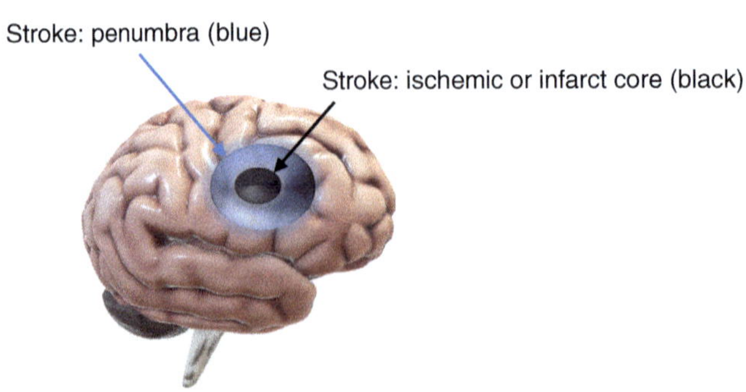

Figure 2.3 Schematic representation of stroke core and penumbra.

adequate blood flow quickly and to protect viable tissue from mechanisms that lead to cell and tissue demise [99]. Within the penumbra, several mechanisms can irreversibly damage brain tissue. Excitotoxicity and calcium overload are major factors involved in the early stages of ischemic cell death. Glutamate, the most common neurotransmitter, accumulates in the extracellular space due to reuptake, energy, and ion pump failure. Oxidative and nitrosative stress are also powerful mediators of ischemic injury. It may tip the balance between pro-death and pro-survival pathways, or ROS and perhaps reactive nitrogen species may act directly as executioners of cell death. Focal ischemia also evokes a robust inflammatory response that begins within a few hours of onset and typifies the secondary or delayed response to ischemia. It involves activation of microglia and astrocytes as well as an influx of hematogenous cells recruited by cytokines, adhesion molecules, and chemokines across the activated blood vessel wall [99].

As for imaging acute stroke, the classic methods such as CT, MRI, or magnetic resonance angiography remain most effective [100, 101]; therefore, in the field of nanomedicine for stroke, the imaging capability will usually be part of a theranostic system. Most efforts in the field are aimed at reducing damage to the tissue and inducing tissue repair [102–105]. In one study, a spontaneous formation of brain-derived neurotrophic factor (BDNF) NPs by simple mixing of the native BDNF and PEG-PGA block copolymer in water was used [106]. Low levels of BDNF are linked to delayed neurological recovery, depression, and cognitive impairment following stroke. However, systemically administered BDNF in its native form has minimal therapeutic effect due to its poor BBB permeability and short serum half-life. In this study, an NP polyion complex formulation of BDNF (nano-BDNF) was administered to male C57BL/6J mice after experimental ischemic stroke. BDNF electrostatically couples with the oppositely charged PGA chain and entraps into a particle core surrounded by a shell of uncharged water-soluble PEG. Mice received nano-BDNF, native-BDNF, or saline treatment after being subjected to 60 minutes of reversible middle cerebral artery occlusion (MCAO). The first dose was given at 3 (early treatment), 6 (intermediate treatment), or 12 hours (delayed treatment) following stroke onset, while a second dose was given in all cohorts at 24 hours after stroke onset. Results showed that early and intermediate nano-BDNF treatment led to a significant reduction in cerebral tissue loss. Delayed treatment led to improved memory/cognition, reduced post-stroke depressive phenotypes, and maintained myelin basic protein and brain BDNF levels, but had no effect on tissue atrophy [106]. A similar approach was used in a different study, where chemically cross-linked polyion complexes of Cu/Zn superoxide dismutase (SOD1) or catalase with a cationic block copolymer were fabricated and tested. Such complexes, termed cross-linked nanozymes (*cl*-nanozymes) can retain catalytic activity and have a decreased cytotoxicity compared to non-cross-linked complexes due to suppression of release of the free block copolymer [107]. SOD1 *cl*-nanozymes exhibit prolonged ability to scavenge experimentally induced ROS in cultured brain microvessel endothelial cells and central neurons. *In vivo* they decrease ischemia/reperfusion-induced tissue injury and improve sensorimotor functions in a rat MCAO model after a single i.v. injection [107]. Other researchers encapsulated curcumin in polymeric NPs in order to protect the brain from oxidative stress and reduce inflammation-related M1 like microglial activation during cerebral ischemia reperfusion injury [108]. After brain ischemic injury, microglia polarized into proinflammatory M1 microglia which are rapidly recruited into the ischemic penumbra. At the same time, M1 microglia can also produce and release proinflammatory factors, such as ROS, tumor necrotic factor α (TNF-α), and interleucin 1β (IL-1β), which amplify the inflammatory response and oxidative stress [108]. In a different study which aimed to protect the brain from ROS, a nanocarrier composed of a dextran polymer core modified with ROS-responsive boronic ester and a red blood cell (RBC) membrane shell with stroke homing peptide (SHp, CLEVSRKNC) inserted was designed [109]. The RBC membrane plays a significant role in camouflaging exogenous NPs and enables NPs with advanced natural characteristics, including biocompatibility, better immune-evading capabilities, and prolonged blood retention as compared to their PEGylated counterparts, while the nanomaterial targeted the ischemic brain site via SHp-mediated transcytosis [109]. In a similar approach, a platelet membrane was used in a nanosystem designed to deliver thrombolytics and neuroprotectants to the ischemic stroke site [110]. While these systems aimed at delivering the active molecules using the nanometric carrier, another study used the active molecule to construct the nanosystem, where the NP itself is being used as the therapeutic agent. This nanosystem was also designed to treat the disease through scavenging excessive ROS and reactive nitrogen species (RONS). This was done using bioinspired melanin NPs (MeNPs) for potent and safer antioxidative therapy, as melanin is known to function as a potential radical scavenger and antioxidant [111]. In another study, the encapsulation of both MRI-enabling IONPs and NGF inside human serum albumin (HSA) NPs, which was modified on the surface with apolipoprotein E to facilitate BBB crossing, created a theranostic particle, enabling both treatment using NGF and imaging using MRI [32].

Other than the delivery of biomolecules such as proteins and the delivery of antioxidants, nucleic acid delivery using nanomaterials was also attempted for stroke. Nucleic acids for gene manipulation are extremely hard to deliver due to low membranal permeability and lack of stability in the blood [112]. Since they are negatively charged, complexation using positively charged nanosystems is being studied extensively [113, 114]. However, not all complexed materials are stable when given systemically. In an attempt to address some of these issues, inducible nitric oxide synthase (iNOS) siRNA was encapsulated in gelatin NPs (GNPs) and was administered intranasally to rats induced after 60 minutes of MCAO [115]. Intranasally delivered GNPs were found in extracellular and intracellular compartments of several brain regions, including the olfactory bulb, cerebral cortex, and striatum at one hour after infusion and continued to be detected for days. To determine the extent of target gene suppression by bare iNOS siRNA or by iNOS siRNA/GNPs, iNOS levels in cortical penumbras and cores were assessed by immuno-blotting at one or two days post-MCAO after IN delivery at six hours post-MCAO. Per the authors, iNOS induction was significantly suppressed in the cortical penumbras of iNOS siRNA/GNP-administered animals. Suppression of iNOS induction by iNOS siRNA/GNPs was also detected in cortical cores at slightly lower levels than in penumbras [115]. Cerebral inflammation also plays a crucial role in the pathology of ischemic stroke. Fullerene derivatives, such as fullerenol (OH-F) are radical scavengers acting as neuroprotective agents while glucosamine (GlcN) attenuates cerebral inflammation after stroke. Therefore, glucosamine-fullerene conjugates (GlcN-F) were created [116]. Fullerene derivatives or vehicle was administered intravenously in normotensive Wistar-Kyoto (WKY) rats and spontaneously hypertensive rats (SHR) immediately after transient middle cerebral artery occlusion (tMCAO). MRI revealed a significant reduction of infarct volume in both WKY and SHR rats that were treated with fullerene derivatives. Treated rats showed an amelioration of neurological symptoms as neuronal loss in the perilesional area was also prevented. Cerebral immunoreactivity was reduced in treated WKY and SHR rats and, the expression of IL-1β and TLR-4 was attenuated in OH-F-treated WKY rats [116].

In a different approach, researchers induced tissue repair following ischemic stroke in rats by delivering BDNF directly to the brain [102]. Using an encapsulation-free methodology, BDNF was dispersed in a hydrogel composed of hyaluronan and methylcellulose with poly(lactic-co-glycolic acid) NPs, and this composite was deposited epicortically, directly above the stroke lesion. An enhanced acute recovery of forepaw dexterity and enhanced hindlimb function at seven weeks postinjury by delivering BDNF locally, with sustained release *in vivo* for up to 21 days, were demonstrated. BDNF delivery augmented plasticity and the presence of the vehicle reduced the lesion volume and prevented neuron loss in perilesional tissue [102]. It remains to be seen, however, how such treatments may be translated to the clinic, since such a procedure is invasive, and the efficiency of local delivery needs to be validated as the distance therapeutic needs to travel in the human brain may be larger. However, if a local delivery approach using hydrogels is considered, then the variety of substances that may be delivered is extended, and biomaterials such as stem cells, which were shown to have the potential to regenerate or repair ischemic tissue, may also be delivered using these platforms [104, 117–119].

Labeling stem cells with superparamagnetic iron oxide (SPIO) particles has been used to provide a noninvasive method for tracking the fate of the administered stem cells by MRI [120, 121]. In one study, a multifunctional nano-seaurchin probe in which multi-gold nanorods (multi-GNRs), crystal-seeded magnetic (SPIO) and mesoporous silica nanobeads (GRMNBs) were fabricated and labeled with umbilical cord mesenchymal stem cells through endocytosis. The nanoplatform enables efficient magnetic remote-controlled guiding for stem cell homing and provides dual modalities of photoacoustic imaging (PAI) and MRI for *in situ* tracking and long-term monitoring of the stem cells [120]. With the goal of also providing reliable magnetic targeting of stem cells and highly stable dual-modal MRI and PAI for stroke therapy, researchers applied intravenous transplantation of nanoplatform-labeled umbilical cord derived mesenchymal stem cells (UMSCs) navigated by a skull magnet to investigate the therapeutic efficiency in stroke mice. The study of *in vivo* stroke mice demonstrated that an external magnet can be successfully used to guide the stem cells labeled with the nanobeads to the targeted stroke areas for prolonged retention. Additionally, one of the nanobeads-UMSCs treatment groups of mice were further transfected with lentivirus-luciferase protein (gMLuc-GRMNBs-UMSCs) and showed a reduced infarct volume in the stroke brain and better functional recovery in the stroke mice after cerebral ischemia compared to other groups [120].

2.5 Nanomaterial Applications for Spinal Cord Injuries

The central nervous system consists of the brain and spinal cord. The spinal cord, made of soft tissue and surrounded by bones (vertebrae), extends downward from the base of the brain and is made up of nerve cells and groups of nerves called tracts, which go to different parts of the body. An SCI is damage to any part of the spinal cord or nerves at the end of the spinal canal (cauda equina) which causes temporary or permanent changes in its function, including strength, sensation, and other body functions below the site of the injury [122]. Anatomically "complete" lesions span the entire breadth of the spinal cord, thereby eliminating all neural communication across the site. Their repair will require restoring neural connectivity across large and hostile non-neural lesion sights, which usually requires invasive procedure. On the other hand, anatomically "incomplete" SCIs are formed by smaller or discontinuous lesions that spare enough neural tissue to support communication across the injury. Anatomically incomplete SCIs are associated with partial recovery and therefore can benefit from strategies to augment spontaneous circuit reorganization and improve the conductivity of residual neural connections [123]. Mature lesions usually exhibit three tissue compartments: (i) non-neural lesion cores that consist of perivascular-fibroblast-derived stromal cells, (ii) meningeal fibroblasts and pericytes that proliferate in areas undergoing CNS cell death, reactive astrocyte scar borders formed by newly proliferated astrocytes surround non-neural lesion cores which protect adjacent functional neural tissue and (iii) spared but reactive neural tissue that surrounds astrocyte scar borders contains all cellular elements found in functioning neural tissue [123]. It is theorized that the scar-forming astrocytes prevent neuronal regrowth at sites of SCI.

One study describes the IN delivery of mesenchymal stem cell derived exosomes loaded with phosphatase and tensin homolog (PTEN) siRNA, which per the authors repairs complete SCI [124]. One of the major intrinsic impediments to axonal growth, PTEN is expressed in neurons, regenerating axons and it also plays a vital role in controlling the regeneration of corticospinal neurons via down-regulating cytoplasmic mammalian target of rapamycin (mTOR) activity. In addition to encapsulating siRNAs inside exosomes (ExoPTEN), some imaging assays were done by encapsulating insulin-covered AuNPs. To assess the neuroprotective potential of ExoPTEN treatment, rats with complete SCI were divided into five groups: (i) no treatment, (ii) intralesional PTEN-siRNA treatment, (iii) intralesional ExoPTEN given once, (iv) IN exosomes for five consecutive days, and (v) IN ExoPTEN for five consecutive days starting from the day of injury, two to three hours postinjury. After treatments, rats were subjected to BBB locomotor scoring every week, for eight weeks. Five days of IN ExoPTEN led to significant locomotor recovery compared to all other treatment groups. More specifically, at week 8, the average BBB locomotor score in the ExoPTEN (IN) group reached 7.75 ± 2.14, a sharp contrast to the mean 0.39 ± 0.14 score in untreated transection rats. Per the authors, the multiple-dose IN administration regimen for five consecutive days was more efficacious than the direct single-dose, intralesional administration (Figure 2.4) [124].

A different study describes exosome-mimetic delivery to SCI using an i.v. injection combined with magnetic targeting [34]. This was done by fabricating IONP-incorporated exosome-mimetic nanovesicles (NV-IONPs) from IONP-treated human mesenchymal stem cells (hMSCs). *In vivo* systemic injection of NV-IONPs with magnetic guidance significantly increased the amount of NV-IONPs accumulating in the injured spinal cord. Accumulated NV-IONPs enhanced blood vessel formation, attenuated inflammation and apoptosis in the injured spinal cord, and, consequently, improved spinal cord function [34].

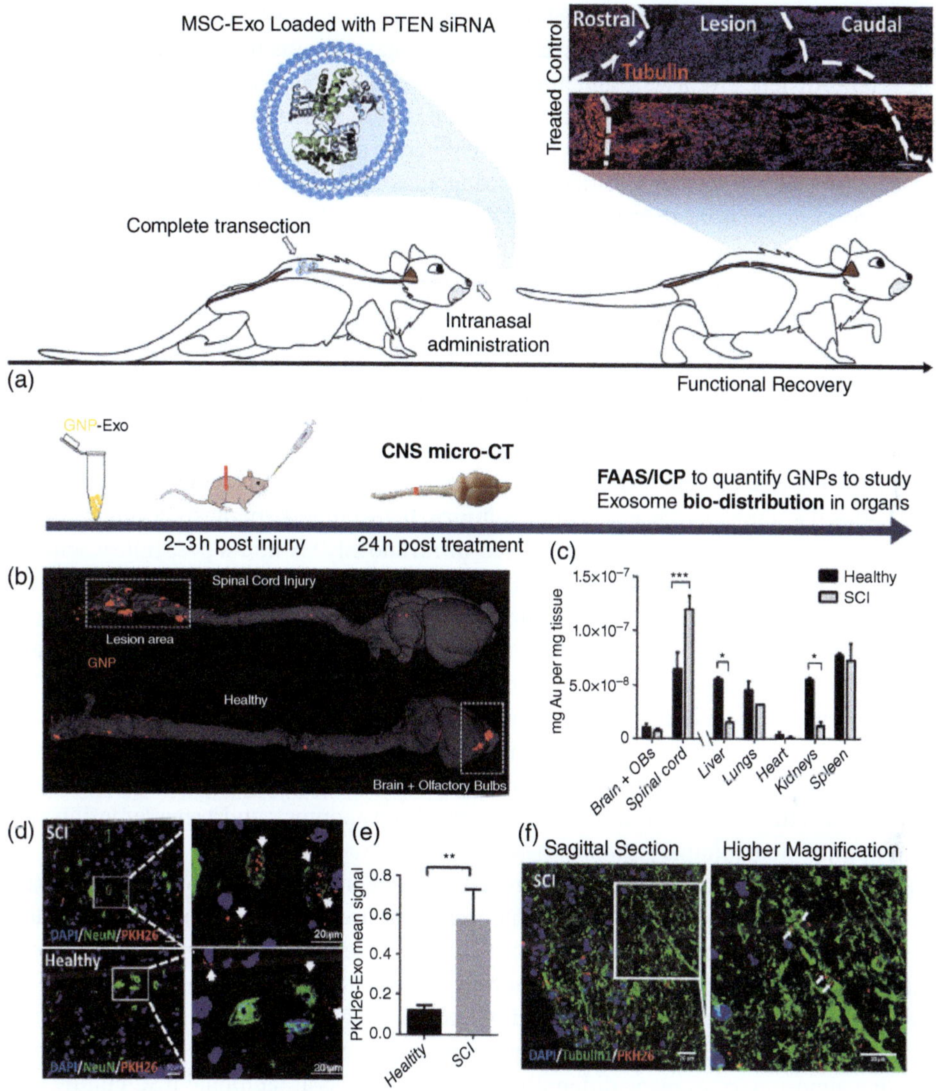

Figure 2.4 Top: Schematic representation of the nanoparticles and methodology. Bottom: Intranasal MSC-Exo cross the BBB and home to spinal cord lesion. (a) Schematic presentation of the steps to examine the migration and biodistribution of IN-administered MSC-Exo. (b) Micro-CT scanning of all the CNS tissues in the injured (upper panel) and healthy (lower panel) rats. GNPs are indicated as red signals. (c) Quantification of GNPs in the CNS, using ICP (left panel), and major organs (right panel), using FAAS, in healthy and injured rats. (d) Immunofluorescent staining of the T10 spinal segment area in the injured (upper panel) and intact (lower panel) rats. NeuN (green), PKH26-Exo (red), nuclei (blue). (e) Quantitative PKH26-Exo mean signal intensity in the T10 spinal segment in intact and injured rats. (f) PKH26-Exo colocalization with Tuj1+ cells in sagittal sections in injured rats. Tuj1 (green), PKH26-Exo (red), nuclei (blue). *Source:* Reprinted (adapted) with permission from Ref. [124]: ACS Nano. 2019;13:10015–28. Copyright (2017) American Chemical Society.

In a study, a functionalized nanogel (NG)-based nanovector was selectively internalized in activated mouse or human astrocytes [125]. The nanogel was based on two main polymers, polyethylene glycol 8000 and linear PEI 2500. Rolipram, an anti-inflammatory drug, when administered by these nanovectors limited the inflammatory response in A1 astrocytes, reducing iNOS and Lcn2, which in turn reverses the toxic effect of proinflammatory astrocytes on motor neurons *in vitro*, showing advantages over conventionally administered anti-inflammatory therapy. *In vivo*, after spinal cord compression, dorsal muscles were juxtaposed using absorbable sutures, and the skin was sutured. Two hours or one day after surgery, the spinal cord of SCI mice was exposed and injected intraparenchymally with NG, NG-Cy5 loaded with Rolipram, or free Rolipram. The results showed that the Rolipram-loaded NG improved motor performance, but only in the early stage after injury, reducing the astrocytosis and preserving neuronal cells [125].

In a study, biodegradable NPs (bovine serum albumin [BSA] based) were used to encapsulate antioxidant enzymes—SOD and catalase (CAT) [126]. Per the authors, delivery of the enzymes at the lesion site would protect mitochondria from oxidative stress and hence the spinal cord from secondary injury. To test their hypothesis, a single dose of nano-SOD/CAT in saline was administered intravenously six hours post-injury, and the spinal cords were analyzed one-week post-treatment. Mitochondria isolated from the affected region of the spinal cord of nano-SOD/CAT-treated animals demonstrated significantly reduced mitochondrial ROS activities, increased mitochondrial membrane potential, reduced calcium levels, and higher adenosine triphosphate (ATP) production capacity than those isolated from the spinal cords of untreated control or SOD/CAT solution-treated animals. Further, immunohistochemical analyses of the spinal cords of treated animals showed significantly lower ROS, cleaved caspase-3, and cytochrome c activities, leading to reduced spinal cord neuronal cell apoptosis and a smaller lesion area than in untreated animals [126].

Finally, one study tried to manage both neurodegenerative diseases and traumatic brain injuries by transplantation of differentiated neurons from stem cells to help restore lost functions [127]. Plasmonic gold nanorods (AuNRs) were integrated in growth surfaces to stimulate and modulate neural cells in order to tune cell physiology. An AuNR nanocomposite system was utilized to study the differentiation of embryonic rat neural stem cells (NSCs). Results demonstrated that this plasmonic surface accelerated cell differentiation yielding almost twice as many differentiated neural cells as a traditional NSC culture surface coated with poly-D-lysine and laminin for the same time period, as well as promoted differentiation of NSCs into neurons and astrocytes in a 2:1 ratio [127].

2.6 Nanomaterial Applications for Brain Tumors

A brain tumor consists of a mass of abnormal cells in the brain. The cells can often cause a change in the surrounding microenvironment by influencing the immune system, the extracellular matrix, and blood vessels. For example, a fast angiogenesis process may occur, in which new blood vessels are formed allowing the delivery of oxygen and nutrients to the tumor. Additionally, some immune cells may change phenotype to a tumor favoring one while other immune cells are disabled [128–131]. Brain tumors are categorized as primary or secondary. While a primary brain tumor originates in the brain, a secondary brain tumor, also known as a metastatic brain tumor, occurs when cancer cells spread from another organ, such as the lung or breast, to the brain. Malignant gliomas remain as aggressive and lethal as primary brain tumors in adults. Glioblastoma (GBM) is the most common and highest-grade malignant glioma. It is characterized by necrosis, intensive vascular proliferation (angiogenesis), and migration of cancer cells (invasion). The BBB of intracranial tumors can be disrupted during the fast stage of angiogenesis. The BBB is then compromised, resulting in a BTB [132]. To some extent, leaky blood vessels of the growing tumor allow some larger-sized therapeutic agents including NPs and antibodies to access the interstitial region of the tumor. The phenomenon, which is generally known in tumors as the enhanced permeability and retention (EPR) effect, has also been studied as a delivery route for nanotherapeutics to brain tumors and is generally referred to as "passive" targeting. However, most of these formulations did not pass clinical trials due to lack of efficacy, indicating that EPR alone was not effective enough to provide therapeutic levels in brain tumors while keeping toxicity and side-effects at a minimum [9, 13, 133]. Many investigators, therefore, are decreasing the sizes of the nanosystems in order to improve BBB penetration [27].

There are multiple options to increase BBB crossing and tumor uptake when nanomaterials are involved [134]. For example, IONPs are magnetic, and magnetic force could be used for delivery [22, 135]. Other nanosystems may use "active" targeting, adding a peptide or antibody ligand that targets overexpressed BBB receptors such as LRP-1 and TfR, facilitating transcytosis [19, 136–138]. In one recent study, a polymeric nanoconjugate (β-poly(L-malic acid)) was used to test different shuttle peptides such as (AP2, via LRP-1 receptor), miniap-4 (M4, based on bee venom structure), and B6 (TfR) ability to cross the normal BBB, where there is no change in receptor density and therefore the different peptides can be compared. Surprisingly, the system showed a successful BBB-crossing ability mainly with AP2 with the presence of trileucine (LLL), which was originally attached to the conjugate as an endosomal escape unit [19] (Figure 2.5).

One unique recent method used nanofibers for the fiber-controlled prolonged delivery of the anti-cancer drug Temozolomide in an orthotopic brain tumor for one month using a flexible polymeric nano-implant [139]. A library of drug-loaded polyester nanofibers of poly(lactic-*co*-glycolic acid)- polylactic acid-polycaprolactone (PLGA-PLA-PCL) electrospun blends (3D) composite nanofiber implant, capable of releasing the anti-glioma drug Temozolomide (TMZ), was administered continuously for one month at a constant rate into the brain tumor. Survival experiments were done using orthotopic C6 glioma models in rats. While the untreated and placebo controls registered median survival of 24 and 27 days, respectively, TMZ-FR (one week release) showed 74 days (P value = 0.0001) and in the case of TMZ-SR (one month release) implanted animals (P value = 0.0001), 85.7% remained long-term (>90 days) survivors [139].

In a different study, multiplexed RNAi therapy against brain tumor-initiating cells via lipopolymeric NP infusion was employed [140]. Brain tumor-initiating cells (BTICs) have been identified as key contributors to therapy resistance, recurrence, and progression of diffuse gliomas and are an elusive therapeutic target that resides across the BBB. The study demonstrates BTIC growth arrest *in vitro* and *in vivo*, and it is accomplished via concurrent siRNA knockdown of four transcription factors (SOX2, OLIG2, SALL2, and POU3F2) that

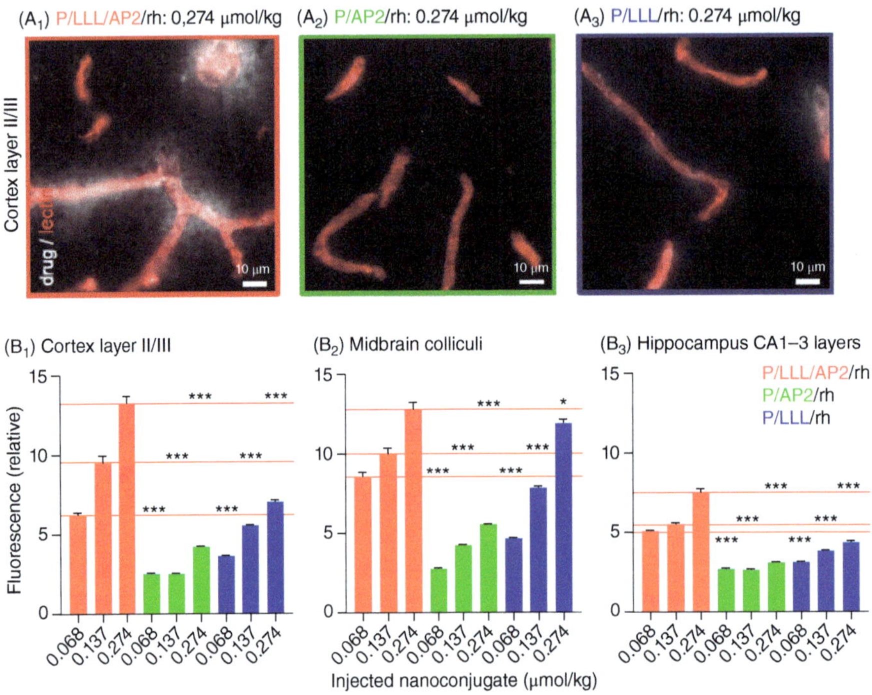

Figure 2.5 Nanoconjugate composition determines degree and locus of BBB penetration. (A1–3) Optical imaging data showing nanoconjugate permeation of the cerebral cortex: Nanoconjugate fluorescence is gray, and the vasculature is red. Different nanoconjugates are indicated with different colors. (B) Average nanoconjugate fluorescence in layers II/III of the somatosensory cortex (B1), the midbrain colliculi (B2), and the hippocampal CA1–3 cell layers (B3) as a function of nanoconjugate composition and concentration: P/LLL/AP2/rh is shown in red, P/AP2/rh in green, and P/LLL/rh in blue. Average nanoconjugate fluorescence measurements were obtained from 20 randomly sampled ROIs explicitly outside of the cerebral vasculature (4 mice with 4 images each, for each measurement). Statistical tests were conducted between nanoconjugate types (e.g., red vs green) within different concentrations. The results are indicated with asterisks where * = $p < 0.01$, ** = $p < 0.001$, and *** = $p < 0.0001$; the red lines show the concentration of P/LLL/AP2/rh against which each comparison was made. *Source:* Reprinted (adapted) with permission from Ref. [19]. Copyright (2019) American Chemical Society.

drive the proneural BTIC phenotype delivered by multiplexed siRNA encapsulation in the lipopolymeric NP ("7C1"). The NPs were fabricated using polyethyleneimine with a molecular weight of 600 (PEI600) which was combined with 200-proof anhydrous ethanol and an epoxide-terminated C15 lipid at a lipid:PEI molar ratio equal to 14:1. Later, a microfluidic device was used to complex the siRNA in the NPs. To test the *in vivo* benefits of the formulation, and because repeated surgery introduces stress and pain that may impact the survival of the experimental animals, researchers opted for a convection-enhanced delivery (CED) strategy using an osmotic pump to deliver a continuous supply of the nano RNAi combination for 14 days. MGG8 cells were used; then the pump was implanted delivering NPs via a brain infusion kit 5 days after the tumor xenograft for the high dose experiment (1.5 $mg\,kg^{-1}$ nano combination RNAi). Per the authors, a clear statistical survival benefit was seen when the combination nano RNAi was compared with nontargeting nano siRNA controls, with a median survival extension of 8 days (median survival of 28.5 days vs. 20.5 days in the control group, ** $p = 0.03$, log-rank test) [140].

Another study, focused on the systemic delivery of "checkpoint inhibitors" antibodies to cytotoxic T-lymphocyte associated antigen 4 (a-CTLA-4) and programmed cell death-1 (a-PD-1) across the BBB and into glioblastoma. These antibodies were chosen in order to induce the local immune system to act against the tumor cells [141]. A versatile drug carrier, poly(β-L-malic acid) (PMLA), a natural polymer obtained from the slime mold *Physarum polycephalum*, is used to deliver covalently conjugated CTLA-4 and PD-1 antibodies (aCTLA-4 and a-PD-1) to brain tumor cells. PMLA-based nanotherapeutics target brain tumors by crossing the BBB using transferrin receptor (TfR)-mediated transcytosis, achieved by the conjugation of a TfR antibody to the polymeric chain. An alternative mechanism of delivery through the BBB with PMLA-conjugated AP-2 peptide was also used. Nanoscale immunoconjugate (NIC) treatment of mice bearing intracranial GL261 glioblastoma (GBM) results in an increase of CD8+ T cells, NK cells, and macrophages with a decrease of regulatory T cells (Tregs) in the brain tumor area. Survival of GBM-bearing mice treated with the NIC combination is significantly longer compared to animals treated with single checkpoint inhibitor-bearing NICs or free a-CTLA-4 and a-PD-1 [141]. The same polymeric backbone was also used to deliver antisense oligo nucleotides (AONs) against ECM laminins [142]. The ECM, including laminins, in the tumor microenvironment is important for tumor invasion and progression. In a panel of 226 patient brain glioma samples, authors have found a clinical correlation between the expression of tumor vascular laminin-411 (α4β1γ1) with higher tumor grade and with expression of CSC markers. The PMLA-based nanobioconjugate was designed to block laminin-411 expression using AONs, attached using a disulfide bond that is released in the cell cytoplasm due to the presence of glutathione. Nanobioconjugate treatment of mice carrying intracranial GBM significantly increased animal survival and inhibited multiple CSC markers, including the Notch axis, in two humanized (LN 229 and U87) tumor mice models. Kaplan-Meier analyses showed that mice with the laminin-411 a4 and b1 DKO (double knockout) tumors exhibited a significantly longer survival than mice carrying the WT GBM cells in both U87MG (median survival WT 67 days [$n = 8$] vs. DKO 93 days [$n = 10$]; *, $P = 0.02$ by log-rank test) and LN229 cohorts (median survival WT 38 days [$n = 8$] vs. DKO 70 days [$n = 10$]; *, $P = 0.015$ by log-rank test) [142].

In another study [143], investigators developed a cyclic peptide iRGD (CCRGDKGPDC)-conjugated solid lipid NP (SLN) to deliver small interfering RNAs (siRNAs) against both epidermal growth factor receptor (EGFR) and programmed cell death ligand-1 (PD-L1) for combined targeted and immunotherapy against glioblastoma. The SLN was fabricated using the microemulsion dilution method and characterized by pseudoternary phase diagram construction studies. In addition, low-dose radiation primes the targeted SLN uptake into the brain tumor region, leading to enhanced down-regulation of PD-L1 and EGFR. Bioluminescence imaging revealed that radiation therapy followed by systemic administration of targeted SLN leads to a significant decrease in glioblastoma growth and prolonged mouse survival in a GL261 animal model. f(SLN)-iRGD:siEGFR/PDL1 (f-(SLN) defined as DSPE-PEG(2000)-DBCO containing functionalized SLN) without radiation had a moderate effect on tumor growth and mouse survival (median survival 24 days versus 21 days for control, **$P = 0.0072$). Similarly, radiation plus nontargeted f(SLN)-scriRGD:siEGFR/PDL1 also has a moderate effect on tumor growth and mouse median survival (24.5 days, **P = 0.0070 versus control). Importantly, mice that received radiation therapy and injected with f(SLN)-iRGD:siEGFR/PDL1 had the most therapeutic benefit with decreased tumor growth and increased mouse survival (median survival 38 days, ***P = 0.0001 versus control) (Figure 2.6) [143].

A different group of investigators fabricated a nanocomposite which in the cancer cell produced ROS, which kills the cancer cell [144]. They termed it "ferroptosis therapy" (FT), a form of cancer therapy, and, per their hypothesis, the FT efficacy can be significantly improved via accelerating the Fenton reaction by simultaneously increasing the local concentrations of all reactants (Fe^{2+}, Fe^{3+}, and H_2O_2) in cancer cells. Thus, Fenton-reaction-accelerable magnetic NPs (MNPs), (cisplatin [CDDP]-loaded Fe_3O_4/Gd_2O_3 hybrid NPs) with conjugation of lactoferrin (LF) and RGD dimer (RGD_2) (FeGd-HN@Pt@LF/RGD2), were exploited in this study for FT of orthotopic brain tumors. The small sized (6.6 nm) FeGd-HN@Pt@LF/RGD2 NPs were able to cross the BBB using LF-receptor-mediated transcytosis. FeGd-HN@Pt@LF/RGD2 can be internalized into cancer cells by integrin $\alpha_v\beta_3$-mediated endocytosis and can then release Fe^{2+}, Fe^{3+}, and CDDP upon endosomal uptake and degradation. Fe^{2+} and Fe^{3+} can directly participate in the Fenton reaction, whereas the CDDP can indirectly produce H_2O_2 to further accelerate the Fenton reaction. The acceleration of the Fenton reaction generates reactive oxygen species to induce cancer cell death. The Fe_3O_4/Gd_2O_3 hybrid NPs (FeGd-HN) were synthesized by a two-step coprecipitation technique under stabilization of poly(acrylic acid) (PAA). Per the authors, to evaluate the FT efficacy of FeGd-HN@Pt2@ LF/RGD2 NPs on an orthotopic brain tumor model, luciferase imaging and diffusion-weighted MRI (DW-MRI) were used. Based on bioluminescence images, FeGd-HN@Pt2@LF/RGD2 was able to inhibit tumor growth compared to control without LF and RGD2, as well as saline, while the survival of the mice treated with FeGd-HN@Pt2@LF/RGD2 was also shown to improve [144].

Since MNPs such as iron oxide have many benefits in tumor treatment and imaging such as magnetic targeting, imaging using MRI and MPI, as well as therapy, many research efforts in the field of nanotechnology for brain tumors include iron oxide in their design and strategy [22]. An additional example is the development of a magnetic nanographene oxide carrier for improved glioma-targeted drug delivery and imaging [145]. Graphene is a two-dimensional sp2 carbon networking nanomaterial that offers excellent potential in the drug delivery field because of its hexagonal carbon ring structure that allows hydrophobic interactions, electrostatic interaction, and π-π stacking with anticancer drugs with aromatic ring structures. In this study, a magnetic multifunctional NP (which contain super paramagnetic IONPs [SPIONs]) based on graphene oxide (NGO/SPION/PLGA) and contains a therapeutic agent (radio-sensitizing 5-iodo-2-deoxyuridin, IUdR) was synthesized and evaluated. To evaluate the magnetic field targeting effect *in vivo* and confirm whether NGO/SPION/PLGA NPs could enhance the contrast of tumor imaging in rats, MRIT2-weighted images of rat brains were obtained before and one hour post intravenous injection of MNPs in the axial and coronal directions. The negative contrast improvement in the applied electromagnetic field (EMF) was more significant than without it, indicating that EMFs improved accumulation and retention of the NGO/SPION/PLGA in glioma. To evaluate the antitumor effects of IUdR/MNPs on glioma-bearing rats, tumor

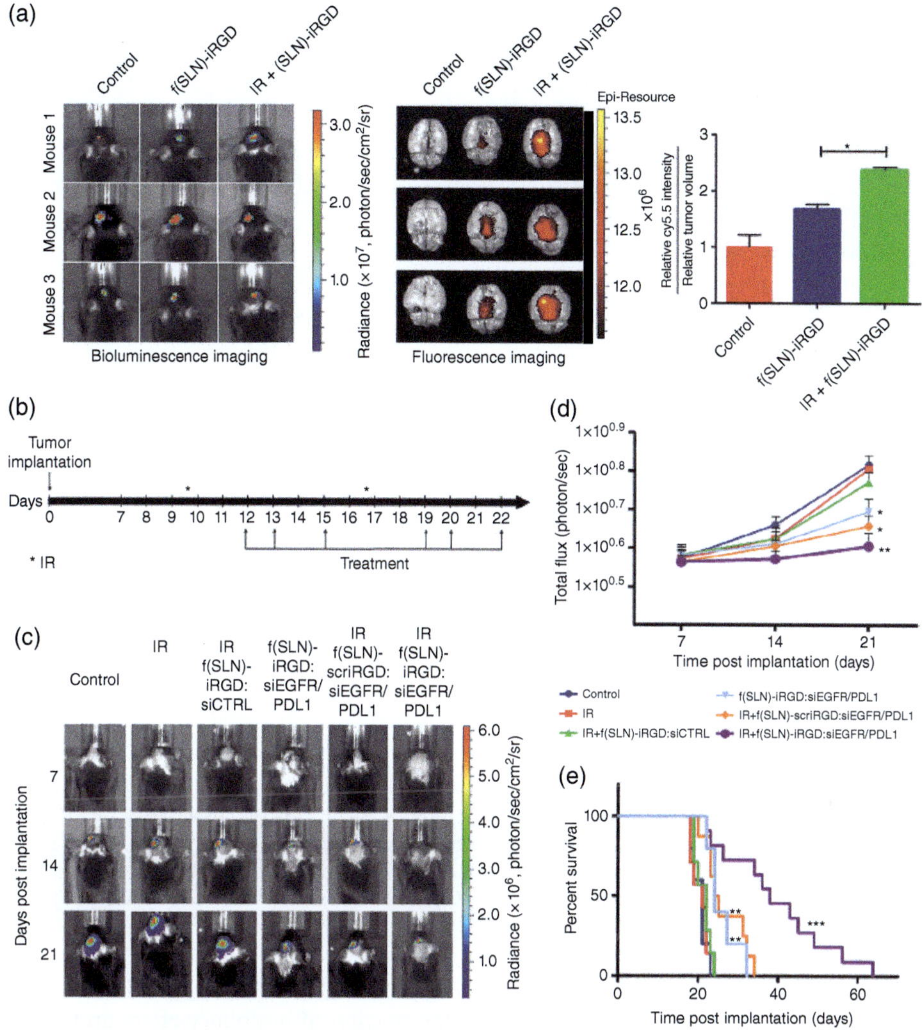

Figure 2.6 Radiation primes glioblastoma for SLN targeted delivery. (a) Mice bearing GL261-Fluc tumors were irradiated (or not as control) and three days later received retro-orbital administration of f(SLN)-iRGD:Cy5.5 or PBS control. Twenty-four hours postinjection, tumor volume was first evaluated by Fluc imaging (left), and brains were removed and imaged ex vivo for Cy5.5 (middle). The mean fluorescence intensity was calculated and normalized to tumor volume (right; $n = 3$, $^{*}P < 0.05$). (b–f) Mice bearing GL261-Fluc tumors were irradiated (or not as a control) and retro-orbitally injected with either f(SLN)-iRGD:siCTRL, f(SLN)-iRGD:siEGFR/PDL1, or f(SLN)-scriRGD:siEGFR/PDL1 according to the scheme in (b). Tumor growth was monitored weekly by Fluc imaging, and survival was recorded. Images from a representative mouse from each group are shown over time (c). Quantification of tumor-associated Fluc radiance intensity with data presented as mean ± SD; $^{*}P < 0.05$ control vs f(SLN)-iRGD:siRNA and control vs IR + f(SLN)-scriRGD:siRNA; $^{**}P < 0.01$ control vs IR + f(SLN)-iRGD:siRNA by ANOVA (d). Kaplan–Meier survival curves are shown ($n = 5–12$); $^{**}P < 0.01$ control vs f(SLN)-iRGD:siRNA; $^{**}P < 0.01$ control vs IR + f(SLN)-scriRGD:siRNA; $^{***}P < 0.001$ control vs IR + f(SLN)-iRGD:siRNA; $^{**}P < 0.01$ IR + f(SLN)-scriRGD:siRNA vs IR + f(SLN)-iRGD:siRNA; by Mantel–Cox (log-rank) test (e). *Source:* Reprinted (adapted) with permission from Ref. [143]: ACS Nano. 2019;13:4028–40. Copyright (2019) American Chemical Society.

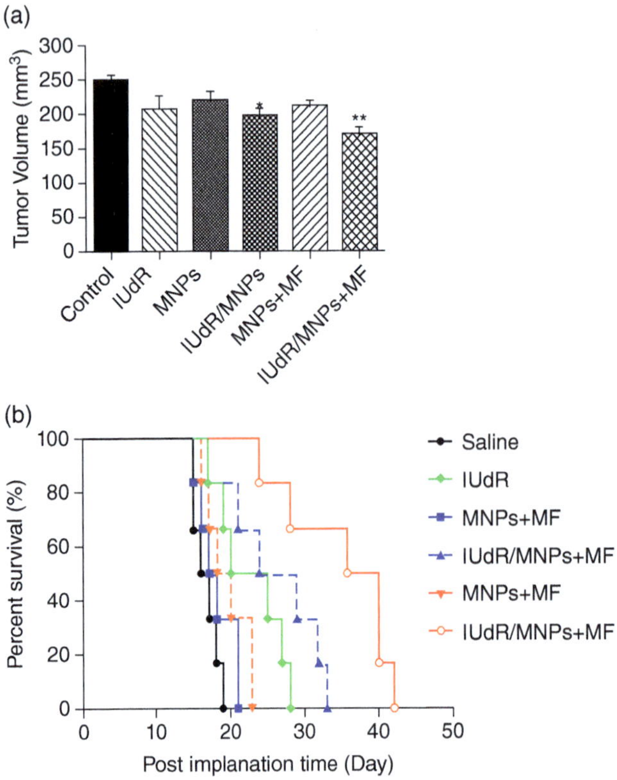

Figure 2.7 Anti-glioma effects of IUdR/NGO/SPION/PLGA + MF vs IUdR/NGO/SPION/PLGA, NGO/SPION/PLGA + MF, NGO/SPION/PLGA, free IUdR, or control group after i. v. of 5 mg kg^{-1} IUdR on 14th day of post-intracranial implantation. (a) glioma tumor volume in different treatment groups on the 17th day, $p < 0.01$ ($n = 3$), (b) percentage of survival (Kaplan–Meier plot) (mean ± SD, $n = 6$). *Source:* Reproduced from Ref. [145]: Chemico-Biological Interactions. 2018;295:97–108.

volume was determined using MRIs. Post hoc analysis confirmed that IUdR/NGO/SPION/PLGA at the present magnetic field indicated the maximum inhibitory effect on tumor growth compared to other treatment groups [145] (Figure 2.7).

2.7 Adverse Effects of Nanomaterials

One of the advantages of using nanomaterials for CNS delivery is the ability to target the nanomaterials to the diseases site and therefore reduce adverse effects elsewhere. However, although nanomaterials improved targeted CNS delivery and better pharmacokinetics, some will still be found in high quantities in other organs. Additionally, adverse effects in the CNS should be considered. Oddly, the main effects that may cause nanomaterials toxicity may be similar to effects found in neurodegeneration; however, the same effects may be an advantage for brain tumor pathology [146]. The toxicity of nanomaterials, however, often depends on the type of nanomaterial, surface charge, and surface functionalities as well as the load. For some materials, if the effect can be limited to the treated area, it would also decrease toxicity by far. Since biocompatibility is an essential property in the design of drug delivery systems, one very general and brief definition of a biocompatible surface is that it cannot trigger an undesired response from the organism. Keck and Müller proposed a nanotoxicological classification system (NCS) (Figure 2.8) that ranks the nanomaterials into four classes according to their respective size and biodegradability [147].

One example are IONPs, which may be used for heat therapy in a brain tumor system [148]. The increased temperature in the tumor can cause cell death, but can also recruit the local immune system to fight the tumor or influence gene expression [149, 150]. However, not all iron oxide particles will reach the target; some will be cleared mainly to the liver by macrophages via the reticuloendothelial system (RES). In that case, "activating" the NPs that are found in the tumor only, using a focused magnetic field, will cause a temperature increment in the tumor only and not in other organs, potentially reducing "off-target" effects and damage to normal tissue. Additionally, when SPIONs are used for biomedical applications, the oxidized crystalline form or maghemite (γ-Fe_2O_3) will not pose the same toxic effect of another common form—magnetite (Fe_3O_4)—since maghemite is less susceptible to oxidation [22, 151, 152]. The oxidation process involving Fe_3O_4 (or other transition metals such as Cu^+ or Mn^{2+}) can lead to the generation of ROS, in a manner similar to neurodegenerative diseases ("oxidative stress"); however, it can be used to abolish cancer cells as mentioned earlier [144, 146]. Several studies were done to evaluate the safety of IONPs in the brain. In one study, Fe_3O_4 was investigated, demonstrating cytotoxicity and proliferative impairment induced in human brain cell cultures after short- and long-term exposure to Fe_3O_4 NPs [153]. For two representative cell types of human CNS in culture (D384 astrocytes and SH-SY5Y neurons), it was shown that Fe_3O_4 NPs can accumulate in astrocytes in a concentration- and time-dependent manner, and functions of mitochondria could be affected, while SH-SY5Y neurons were less susceptible to Fe_3O_4 NPs after either short- or long-term exposure. In another study, however, it was demonstrated that partial oxidation of nano-sized zerovalent iron in magnetite/maghemite as well as

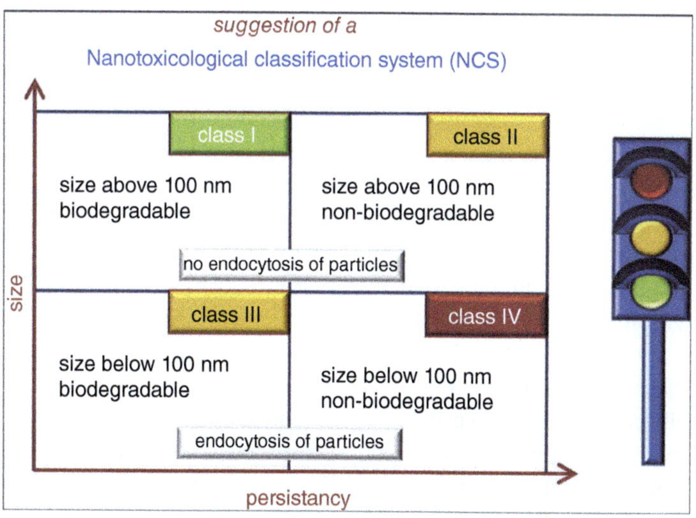

Figure 2.8 Suggestion of nanotoxicological classification system. *Source:* Reproduced from Ref. [147].

surface modifications, for example by reaction with sodium polyaspartate, decreased neurotoxicity, highlighting the importance of the Fe oxidative state when it comes to the biosafety of IONPs [154].

Other crystalline NPs may have a different type of problem. While in some cases iron is familiar to the body and cells may digest and use it, gold by its nature is a very inert metal, and so it is not biodegradable [146], while QDs are often made of toxic metals such as cadmium [155, 156].

Noncrystalline systems may also have issues. For example, micelles and liposomes, which are designed to carry therapeutics, should be able to cross biological barriers and release the loaded drugs back into active forms in order to exert the therapeutic effect in the target [157]. In addition, when crossing the BBB, the size of the nanocarrier is critical, and the smaller the nanosystem is, the better [27].

Polymeric nanoconjugates, or polymeric NPs which are cross-linked, have a better chance to reach the target intact. Biodegradable polymers, such as PMLA which degrades to CO_2 and H_2O, are preferable to reduce adverse effects caused by the nanocarrier itself. Researchers tested the toxicity of the previously mentioned poly(β-L-malic acid) (PMLA) platform, which was shown to target GBM [158]. PMLA and nanodrug biocompatibility and toxicity at low and high dosages were evaluated *in vitro* and *in vivo*. The dual action nanodrug and single action precursor nanoconjugates were assessed under *in vitro* conditions and *in vivo* with multiple treatment regimens (6 and 12 treatments). Per the authors, the 12-treatment regimen (chronic exposure) of nude mice by nanoconjugates does not induce toxic changes in blood cell count or in metabolic and chemistry panels. In the quantitative enzyme immunoassay, complement activation (forming iC3b) by AON and antibody-containing nanoconjugates was weakly elevated at the highest concentration of $1\,mg\,ml^{-1}$, corresponding to only 20% of the level measured for FDA-approved Doxil®. PMLA and mimic nanoconjugates did not induce proliferation of human leukocytes. Additionally, PMLA and mimic nanoconjugates tested at $0.008–1.0\,mg\,ml^{-1}$ did not induce oxidative burst indicated by the absence of nitric oxide secretion (no inflammation-related response). Only insignificant induction of macrophage chemotaxis was observed for PMLA and mimic nanoconjugates ($0.008–1.0\,mg\,ml^{-1}$). PMLA and mimic nanoconjugates (up to $1.0\,mg\,ml^{-1}$) were not internalized by phagocytosis and had no significant effect on phagocytosis of zymosan. Finally, PMLA did not induce maturation markers CD80, CD83, and CD86 and did not affect CD14 expression. [158]

When considering a polymer of a particle, biodegradability and chemical structure are two of several important factors. If the nanocarrier is given systemically (that is, by intravenous injection), stability in the blood stream is important. Positively charged NPs or polymers may react with membranes as well as biomolecules found in the blood and may cause coagulation and toxicity [114]. Therefore, negatively charged carriers are generally preferred [159]. One example of a popular cationic polymer for intracellular delivery is polyethyleneimine (PEI) because it facilitates endosomal escape. For that reason, it is being used in a verity of nanosystems [43, 113]. In one study, it was used to coat maghemite NPs. Even though the maghemite NPs did not show toxicity, acute toxicity was observed with the PEI coating [113, 114, 151, 152]. In order to overcome this issue, the researcher chose a controlled oxidation method to slightly oxidize the PEI while maintaining the biological activity. This approach indeed decreased the acute toxicity of the nanosystem [114].

2.8 Regulatory Issues

The translation of nanotechnology from the bench to the market possesses several steps and challenges: physicochemical characterization, biocompatibility and nanotoxicology evaluation, pharmacokinetics and pharmacodynamics assessment, process control, as well as scaleup and reproducibility [160]. While many nanoformulations are FDA approved and used for different pathologies, nanomaterials for CNS use are rare. One example is Copaxone (by Teva Pharmaceutical Industries), which is a Glatimer acetate-based polymeric nanodrug for multiple sclerosis (MS). Another example of a non-CNS-approved nanomedicine is Feraheme (by AMAG Pharmaceuticals) which is an IONP (Ferumoxytol) and treats iron deficiency [161].

Some of the physicochemical characteristics of nanomaterials can change during the manufacturing process, which can compromise the quality and safety of the final nanomedicine. According to the FDA, the application of nanotechnology may result in product attributes that differ from those of conventionally manufactured products and thus may merit particular examination. Therefore, the FDA does not categorically judge all products that involve the application of nanotechnology as intrinsically benign or harmful. The FDA will regulate nanotechnology products under existing statutory authorities, in accordance with the specific legal standards applicable to each type of product under its jurisdiction [162]. By enhancing its scientific expertise and tools necessary to assess the safety and, as applicable, effectiveness of products, the FDA can enable the responsible development of products with new and beneficial properties [163]. Quality by design (QbD), supported by process analytical technologies (PAT), is one of the pharmaceutical development approaches that was recognized for the systematic evaluation and control of nanomedicines. This new approach allows better focus on the relevant relationships between the characteristics of the formulation, the parameters of the formulation, and the fabrication process in order to develop effective manufacturing processes to ensure the quality ofthe nanomedicines [160].

Although nanomaterials may have an identical composition, slight differences in the surface charge, size, or shape could have an impact on their respective activity and, as a result, on their biodistribution and cellular uptake, leading to different biological responses. Another challenge when it comes to nanomaterials in pharmaceutical development is the scale-up and reproducibility of the nanomedicines when scaling up [160].

The traditional manufacturing processes do not create three-dimensional medicines in the nanometer scale. Nanomedicine manufacturing processes may be comprised of top-down and bottom-down approaches, which include multiple steps such as homogenization, sonication, milling, emulsification, and occasionally the use of organic solvents and further evaporation. At a small scale, it is easier to control and optimize the formulation. However, at a large scale it becomes challenging due to slight variations during the manufacturing process that can originate critical changes in the physicochemical characteristics. That may compromise the quality and safety of the nanomedicines or even affect the therapeutic outcomes [160].

2.9 Conclusions

The delivery of therapeutic and imaging agents into the brain is still a great challenge due to biological barriers aimed at protecting and keeping the brain homeostasis. While small molecule-based drugs are somewhat limited since they need to have both biological activity as well as BBB crossing ability, nanomedicinal methods and systems show promise in the field of brain delivery. The ability to modify a carrier to cross the BBB and target sites in the brain parenchyma while carrying therapeutic and imaging moieties is a huge advantage of nanomedicines. Numerous types of nanosystems were described, starting from biodegradable polymers; to metallic NPs, such as MRI- and MPI-enabling iron oxides that can also be used for hyperthermia and magnetic targeting; to CT-enabling AuNPs. Unique "biological" origin nanosystems such as exosomes were also studied in the field, as well as "nanoghosts," which possess the origin cell's natural targeting ability. The therapeutic method that can be used with nanosystems also includes nucleic acid delivery such as siRNA, otherwise impossible to deliver to the brain cells without a carrier because of low stability. Therefore, new therapeutic options may potentially be available thanks to nanomedicines. Another unique brain delivery method is IN delivery. Even though research in the field is relatively new, it does show promise. The nanomedicinal approach may potentially reduce toxicity, since the targeting and crossing abilities may increase the quantity of therapeutic agent in the brain target, which in turn may decrease the dose needed to reach therapeutic levels and therefore decrease unwanted effects.

Despite nanomaterials' advantage in the field, the more "complex" nanosystems are also often more difficult to characterize, and synthesis robustness is not always easy to achieve. These issues can cause regulatory problems. Moreover, the larger the nanosystem becomes, the more difficult it becomes for it to reach the brain and cross the biological barriers on the way to its target. Additionally, the material used for delivery needs to be chosen carefully as clearance from the brain may also pose a difficulty, as well as potential toxicity to the brain's innate cells. Therefore, toxicity and biocompatibility need to be assessed carefully when it comes to nanosystems used for brain delivery.Despite numerous challenges, nanomaterials have numerous advantages when it comes to brain delivery. They possess great potential in imaging and treating brain-related pathologies.

References

1 Engelhardt, B. and Sorokin, L. (ed.) (2009). The blood–brain and the blood–cerebrospinal fluid barriers: function and dysfunction. In: *Seminars in Immunopathology*. Springer.

2 Lockman, P.R., Mittapalli, R.K., Taskar, K.S. et al. (2010). Heterogeneous blood-tumor barrier permeability determines drug efficacy in mouse brain metastases of breast cancer. *Clinical Cancer Research* 1564.

3 Furtado, D., Björnmalm, M., Ayton, S. et al. (2018). Overcoming the blood-brain barrier: the role of nanomaterials in treating neurological diseases. *Advanced Materials* 30 (46):1801362.

4 Blanchette, M. and Daneman, R. (2015). Formation and maintenance of the BBB. *Mechanisms of Development* 138: 8–16.

5 Zhao, Z., Nelson, A.R., Betsholtz, C., and Zlokovic, B.V. (2015). Establishment and dysfunction of the blood-brain barrier. *Cell* 163 (5): 1064–1078.

6 Pardridge, W.M. (2015). Blood-brain barrier endogenous transporters as therapeutic targets: a new model for small molecule CNS drug discovery. *Expert Opinion on Therapeutic Targets* 19 (8): 1059–1072.

7 Carmeliet, P. and Jain, R.K. (2011). Principles and mechanisms of vessel normalization for cancer and other angiogenic diseases. *Nature Reviews Drug Discovery. [Perspective]* 10: 417.

8 Anchordoquy, T.J., Barenholz, Y., Boraschi, D. et al. (2017). Mechanisms and barriers in cancer nanomedicine: addressing challenges, looking for solutions. *ACS Nano* 11 (1): 12–18.

9 Danhier, F. (2016). To exploit the tumor microenvironment: since the EPR effect fails in the clinic, what is the future of nanomedicine? *Journal of Controlled Release* 244: 108–121.

10 Maeda, H. and Khatami, M. (2018). Analyses of repeated failures in cancer therapy for solid tumors: poor tumor-selective drug delivery, low therapeutic efficacy and unsustainable costs. *Clinical and Translational Medicine* 7 (1): 11.

11 Youn, Y.S. and Bae, Y.H. (2018). Perspectives on the past, present, and future of cancer nanomedicine. *Advanced Drug Delivery Reviews* 130: 3–11.

12 Huynh, E. and Zheng, G. (2015). Cancer nanomedicine: addressing the dark side of the enhanced permeability and retention effect. *Nanomedicine.* 10 (13): 1993–1995.

13 Nichols, J.W. and Bae, Y.H. (2014). EPR: evidence and fallacy. *Journal of Controlled Release* 190: 451–464.

14 Cho, C.-F., Wolfe, J.M., Fadzen, C.M. et al. (2017). Blood-brain-barrier spheroids as an in vitro screening platform for brain-penetrating agents. *Nature Communications* 8: 15623. https://doi.org/10.1038/ncomms15623.

15 Sakamoto, K., Shinohara, T., Adachi, Y. et al. (2017). A novel LRP1-binding peptide L57 that crosses the blood brain barrier. *Biochemistry and Biophysics Reports* 12: 135–139.

16 Boado, R.J., Lu, J.Z., Hui, E.K.-W. et al. (2016). Insulin receptor antibody–α-N-acetylglucosaminidase fusion protein penetrates the primate blood-brain barrier and reduces glycosoaminoglycans in Sanfilippo type B fibroblasts. *Molecular Pharmaceutics* 13 (4): 1385–1392.

17 Demeule, M., Currie, J.C., Bertrand, Y. et al. (2008). Involvement of the low-density lipoprotein receptor-related protein in the transcytosis of the brain delivery vector Angiopep-2. *Journal of Neurochemistry* 106 (4): 1534–1544.

18 Johnsen, K.B., Burkhart, A., Melander, F. et al. (2017). Targeting transferrin receptors at the blood-brain barrier improves the uptake of immunoliposomes and subsequent cargo transport into the brain parenchyma. *Scientific Reports* 7 (1): 10396.

19 Israel, L.L., Braubach, O., Galstyan, A. et al. (2019). A combination of tri-leucine and angiopep-2 drives a poly-anionic polymalic acid Nanodrug platform across the blood-brain barrier. *ACS Nano* 13 (2): 1253–1271.

20 Agrawal, M., Saraf, S., Saraf, S. et al. (2018). Nose-to-brain drug delivery: an update on clinical challenges and progress towards approval of anti-Alzheimer drugs. *Journal of Controlled Release* 281: 139–177.

21 Jahangiri, A., Chin, A.T., Flanigan, P.M. et al. (2017). Convection-enhanced delivery in glioblastoma: a review of preclinical and clinical studies. *Journal of Neurosurgery* 126 (1): 191–200.

22 Israel, L.L., Galstyan, A., Holler, E., and Ljubimova, J.Y. (2020). Magnetic iron oxide nanoparticles for imaging, targeting and treatment of primary and metastatic tumors of the brain. *Journal of Controlled Release* 320: 45–62.

23 Sitburana, O. and Ondo, W.G. (2009). Brain magnetic resonance imaging (MRI) in parkinsonian disorders. *Parkinsonism & Related Disorders* 15 (3): 165–174.

24 O'Sullivan, C., Lindenberg, M., Bryla, C. et al. (2016). ANG1005 for breast cancer brain metastases: correlation between 18F-FLT–PET after first cycle and MRI in response assessment. *Breast Cancer Research and Treatment* 160 (1): 51–59.

25 Yan, Q., Nho, K., Del-Aguila, J.L. et al. (2018). Genome-wide association study of brain amyloid deposition as measured by Pittsburgh compound-B (PiB)-PET imaging. *Molecular Psychiatry* 1–13.

26 Zanganeh, S., Hutter, G., Spitler, R. et al. (2016). Iron oxide nanoparticles inhibit tumour growth by inducing pro-inflammatory macrophage polarization in tumour tissues. *Nature Nanotechnology* 11 (11): 986–994.

27 Ljubimova, J.Y., Sun, T., Mashouf, L. et al. (2017). Covalent nano delivery systems for selective imaging and treatment of brain tumors. *Advanced Drug Delivery Reviews* 113: 177–200.

28 Kilinc, D., Dennis, C.L., and Lee, G.U. (2016). Bio-nano-magnetic materials for localized Mechanochemical stimulation of cell growth and death. *Advanced Materials* 28 (27): 5672–5680.

29 Betzer, O., Shilo, M., Motiei, M., and Popovtzer, R. (ed.) (2019). Insulin-coated gold nanoparticles as an effective approach for bypassing the blood-brain barrier. In: *Nanoscale Imaging, Sensing, and Actuation for Biomedical Applications XVI*. International Society for Optics and Photonics.

30 Roy, U., Drozd, V., Durygin, A. et al. (2018). Characterization of nanodiamond-based anti-HIV drug delivery to the brain. *Scientific Reports* 8 (1): 1–12.

31 Moscariello, P., Raabe, M., Liu, W. et al. (2019). Unraveling in vivo brain transport of protein-coated fluorescent nanodiamonds. *Small* 15 (42): 1902992. https://doi.org/10.1002/smll.201902992.

32 Feczkó, T., Piiper, A., Ansar, S. et al. (2019). Stimulating brain recovery after stroke using theranostic albumin nanocarriers loaded with nerve growth factor in combination therapy. *Journal of Controlled Release* 293: 63–72.

33 Betzer, O., Perets, N., Angel, A. et al. (2017). In vivo neuroimaging of exosomes using gold nanoparticles. *ACS Nano.* 11 (11): 10883–10893.

34 Kim, H.Y., Kumar, H., Jo, M.-J. et al. (2018). Therapeutic efficacy-potentiated and diseased organ-targeting Nanovesicles derived from mesenchymal stem cells for spinal cord injury treatment. *Nano Letters* 18 (8): 4965–4975.

35 Michalet, X., Pinaud, F., Bentolila, L. et al. (2005). Quantum dots for live cells, in vivo imaging, and diagnostics. *Science* 307 (5709): 538–544.

36 Wu, T., Zhang, T., Chen, Y., and Tang, M. (2016). Research advances on potential neurotoxicity of quantum dots. *Journal of Applied Toxicology.* 36 (3): 345–351.

37 Huang, N., Cheng, S., Zhang, X. et al. (2017). Efficacy of NGR peptide-modified PEGylated quantum dots for crossing the blood-brain barrier and targeted fluorescence imaging of

glioma and tumor vasculature. *Nanomedicine: Nanotechnology, Biology and Medicine* 13 (1): 83–93.

38 Wang, X.-J., Peng, C.-H., Zhang, S. et al. (2019). Polysialic-acid-based micelles promote neural regeneration in spinal cord injury therapy. *Nano Letters* 19 (2): 829–838.

39 Kafa, H., Wang, J.T.-W., Rubio, N. et al. (2016). Translocation of LRP1 targeted carbon nanotubes of different diameters across the blood-brain barrier in vitro and in vivo. *Journal of Controlled Release: Official Journal of the Controlled Release Society.* 225: 217–229.

40 Zhou, L., Fan, L., Yi, X. et al. (2018). Soft conducting polymer hydrogels cross-linked and doped by tannic acid for spinal cord injury repair. *ACS Nano* 12 (11): 10957–10967.

41 Neamtu, I., Rusu, A.G., Diaconu, A. et al. (2017). Basic concepts and recent advances in nanogels as carriers for medical applications. *Drug Delivery* 24 (1): 539–557.

42 Toledano Furman, N.E., Lupu-Haber, Y., Bronshtein, T. et al. (2013). Reconstructed stem cell nanoghosts: a natural tumor targeting platform. *Nano Letters.* 13 (7): 3248–3255.

43 Kaneti, L., Bronshtein, T., Malkah Dayan, N. et al. (2016). Nanoghosts as a novel natural nonviral gene delivery platform safely targeting multiple cancers. *Nano Letters* 16 (3): 1574–1582.

44 Kumar, P. and Bose, P.P. (2019). Macrophage ghost entrapped amphotericin B: a novel delivery strategy towards experimental visceral leishmaniasis. *Drug Delivery and Translational Research* 9 (1): 249–259.

45 Krishnamurthy, S., Gnanasammandhan, M., Xie, C. et al. (2016). Monocyte cell membrane-derived nanoghosts for targeted cancer therapy. *Nanoscale* 8 (13): 6981–6985.

46 Schoen, B. and Machluf, M. (2016). Cell ghosts: cellular membranes for drug delivery. *Perspectives in Micro-and Nanotechnology for Biomedical Applications: World Scientific.* 225–273.

47 Khait, N.L., Malkah, N., Kaneti, G. et al. (2019). Radiolabeling of cell membrane-based nano-vesicles with 14C-linoleic acid for robust and sensitive quantification of their biodistribution. *Journal of Controlled Release* 293: 215–223.

48 Tan, A., Rajadas, J., and Seifalian, A.M. (2013). Exosomes as nano-theranostic delivery platforms for gene therapy. *Advanced Drug Delivery Reviews* 65 (3): 357–367.

49 Batrakova, E.V., Haney, M., Klyachko, N. et al. (2018). Exosomes as drug delivery vehicles for therapeutic proteins to the brain. *Journal of Extracellular Vesicles* 7: 266.

50 Haney, M.J., Klyachko, N.L., Zhao, Y. et al. (2015). Exosomes as drug delivery vehicles for Parkinson's disease therapy. *Journal of Controlled Release* 207: 18–30.

51 Yuan, D., Zhao, Y., Banks, W.A. et al. (2017). Macrophage exosomes as natural nanocarriers for protein delivery to inflamed brain. *Biomaterials* 142: 1–12.

52 Bunggulawa, E.J., Wang, W., Yin, T. et al. (2018). Recent advancements in the use of exosomes as drug delivery systems. *Journal of Nanobiotechnology* 16 (1): 81.

53 Familtseva, A., Jeremic, N., and Tyagi, S.C. (2019). Exosomes: cell-created drug delivery systems. *Molecular and Cellular Biochemistry* 1–6.

54 Luan, X., Sansanaphongpricha, K., Myers, I. et al. (2017). Engineering exosomes as refined biological nanoplatforms for drug delivery. *Acta Pharmacologica Sinica* 38 (6): 754.

55 Gaugler, J., James, B., Johnson, T. et al. (2019). 2019 Alzheimer's disease facts and figures. *Alzheimer's & Dementia* 15 (3): 321–387.

56 Takeda, S., Sato, N., and Morishita, R. (2014). Systemic inflammation, blood-brain barrier vulnerability and cognitive/non-cognitive symptoms in Alzheimer disease: relevance to pathogenesis and therapy. *Frontiers in Aging Neuroscience* 6: 171.

57 Zipser, B., Johanson, C., Gonzalez, L. et al. (2007). Microvascular injury and blood–brain barrier leakage in Alzheimer's disease. *Neurobiology of Aging* 28 (7): 977–986.

58 Montagne, A., Barnes, S.R., Sweeney, M.D. et al. (2015). Blood-brain barrier breakdown in the aging human hippocampus. *Neuron* 85 (2): 296–302.

59 Sweeney, M.D., Sagare, A.P., and Zlokovic, B.V. (2018). Blood-brain barrier breakdown in Alzheimer disease and other neurodegenerative disorders. *Nature Reviews Neurology* 14 (3): 133–150.

60 Halliday, M.R., Rege, S.V., Ma, Q. et al. (2016). Accelerated pericyte degeneration and blood–brain barrier breakdown in apolipoprotein E4 carriers with Alzheimer's disease. *Journal of Cerebral Blood Flow and Metabolism.* 36 (1): 216–227.

61 Zenaro, E., Piacentino, G., and Constantin, G. (2017). The blood-brain barrier in Alzheimer's disease. *Neurobiology of Disease* 107: 41–56.

62 Montagne, A., Nation, D.A., Pa, J. et al. (2016). Brain imaging of neurovascular dysfunction in Alzheimer's disease. *Acta Neuropathologica* 131 (5): 687–707.

63 Pan, Y. and Nicolazzo, J.A. (2018). Impact of aging, Alzheimer's disease and Parkinson's disease on the blood-brain barrier transport of therapeutics. *Advanced Drug Delivery Reviews* 135: 62–74.

64 Umlauf, B.J. and Shusta, E.V. (2019). Exploiting BBB disruption for the delivery of nanocarriers to the diseased CNS. *Current Opinion in Biotechnology* 60: 146–152.

65 Ramanathan, A., Nelson, A.R., Sagare, A.P., and Zlokovic, B.V. (2015). Impaired vascular-mediated clearance of brain amyloid beta in Alzheimer's disease: the role, regulation and restoration of LRP1. *Frontiers in Aging Neuroscience* 7: 136.

66 Solbach, C., Uebele, M., Reischl, G., and Machulla, H.-J. (2005). Efficient radiosynthesis of carbon-11 labelled uncharged Thioflavin T derivatives using [11C] methyl triflate for β-amyloid imaging in Alzheimer's disease with PET. *Applied Radiation and Isotopes* 62 (4): 591–595.

67 Kaushik, A., Jayant, R.D., Tiwari, S. et al. (2016). Nano-biosensors to detect beta-amyloid for Alzheimer's disease management. *Biosensors and Bioelectronics* 80: 273–287.

68 Miao, Q., Lyu, Y., Ding, D., and Pu, K. (2016). Semiconducting oligomer nanoparticles as an activatable photoacoustic probe with amplified brightness for in vivo imaging of pH. *Advanced Materials* 28 (19): 3662–3668.

69 Patil, R., Gangalum, P.R., Wagner, S. et al. (2015). Curcumin targeted, polymalic acid-based MRI contrast agent for the detection of Aβ plaques in Alzheimer's disease. *Macromolecular Bioscience* 15 (9): 1212–1217.

70 Mpambani, F., Åslund, A.K., Lerouge, F. et al. (2018). Two-photon fluorescence and magnetic resonance specific imaging of Aβ amyloid using hybrid nano-GdF3 contrast media. *ACS Applied Bio Materials* 1 (2): 462–472.

71 Cui, Z., Bu, W., Fan, W. et al. (2016). Sensitive imaging and effective capture of Cu2+: towards highly efficient theranostics of Alzheimer's disease. *Biomaterials* 104: 158–167.

72 Zhang, K., Yang, Q., Fan, Z. et al. (2019). Platelet-driven formation of interface peptide nano-network biosensor enabling a non-invasive means for early detection of Alzheimer's disease. *Biosensors and Bioelectronics* 145: 111701.

73 Di Santo, S.G., Prinelli, F., Adorni, F. et al. (2013). A meta-analysis of the efficacy of donepezil, rivastigmine, galantamine, and memantine in relation to severity of Alzheimer's disease. *Journal of Alzheimer's Disease.* 35 (2): 349–361.

74 Chintamaneni, P.K., Krishnamurthy, P.T., Rao, P.V., and Pindiprolu, S.S. (2017). Surface modified nano-lipid drug conjugates of positive allosteric modulators of M1 muscarinic acetylcholine receptor for the treatment of Alzheimer's disease. *Medical Hypotheses* 101: 17–22.

75 Kheradmand, E., Hajizadeh Moghaddam, A., and Zare, M. (2018). Neuroprotective effect of hesperetin and nano-hesperetin on recognition memory impairment and the elevated oxygen stress in rat model of Alzheimer's disease. *Biomedicine & Pharmacotherapy* 97: 1096–1101.

76 Kwon, H.J., Cha, M.-Y., Kim, D. et al. (2016). Mitochondria-targeting ceria nanoparticles as antioxidants for Alzheimer's disease. *ACS Nano* 10 (2): 2860–2870.

77 Hettiarachchi, S.D., Zhou, Y., Seven, E. et al. (2019). Nanoparticle-mediated approaches for Alzheimer's disease pathogenesis, diagnosis, and therapeutics. *Journal of Controlled Release* 314: 125–140.

78 Zhao, Y., Cai, J., Liu, Z. et al. (2019). Nanocomposites inhibit the formation, mitigate the neurotoxicity, and facilitate the removal of β-amyloid aggregates in Alzheimer's disease mice. *Nano Letters* 19 (2): 674–683.

79 Carradori, D., Balducci, C., Re, F. et al. (2018). Antibody-functionalized polymer nanoparticle leading to memory recovery in Alzheimer's disease-like transgenic mouse model. *Nanomedicine: Nanotechnology, Biology and Medicine* 14 (2): 609–618.

80 Elnaggar, Y.S., Etman, S.M., Abdelmonsif, D.A., and Abdallah, O.Y. (2015). Intranasal piperine-loaded chitosan nanoparticles as brain-targeted therapy in Alzheimer's disease: optimization, biological efficacy, and potential toxicity. *Journal of Pharmaceutical Sciences* 104 (10): 3544–3556.

81 Sun, J., Wei, C., Liu, Y. et al. (2019). Progressive release of mesoporous nano-selenium delivery system for the multi-channel synergistic treatment of Alzheimer's disease. *Biomaterials* 197: 417–431.

82 Chen, Q., Du, Y., Zhang, K. et al. (2018). Tau-targeted multifunctional nanocomposite for combinational therapy of Alzheimer's disease. *ACS Nano* 12 (2): 1321–1338.

83 Ghaemi, M., Hilker, R., Rudolf, J. et al. (2002). Differentiating multiple system atrophy from Parkinson's disease: contribution of striatal and midbrain MRI volumetry and multi-tracer PET imaging. *Journal of Neurology, Neurosurgery, and Psychiatry* 73 (5): 517.

84 Péran, P., Cherubini, A., Assogna, F. et al. (2010). Magnetic resonance imaging markers of Parkinson's disease nigrostriatal signature. *Brain* 133 (11): 3423–3433.

85 Haller, S., Badoud, S., Nguyen, D. et al. (2013). Differentiation between Parkinson disease and other forms of parkinsonism using support vector machine analysis of susceptibility-weighted imaging (SWI): initial results. *European Radiology* 23 (1): 12–19.

86 Bartzokis, G., Cummings, J.L., Markham, C.H. et al. (1999). MRI evaluation of brain iron in earlier-and later-onset Parkinson's disease and normal subjects. *Magnetic Resonance Imaging* 17 (2): 213–222.

87 Beitollahi, H. and Garkani, N.F. (2016). Graphene oxide/ZnO Nano composite for sensitive and selective electrochemical sensing of levodopa and tyrosine using modified graphite screen printed electrode. *Electroanalysis* 28 (9): 2237–2244.

88 Lill, C.M. (2016). Genetics of Parkinson's disease. *Molecular and Cellular Probes* 30 (6): 386–396.

89 Chu, Y., Dodiya, H., Aebischer, P. et al. (2009). Alterations in lysosomal and proteasomal markers in Parkinson's disease: relationship to alpha-synuclein inclusions. *Neurobiology of Disease* 35 (3): 385–398.

90 Bose, A. and Beal, M.F. (2016). Mitochondrial dysfunction in Parkinson's disease. *Journal of Neurochemistry* 139 (S1): 216–231.

91 Kwon, H.J., Kim, D., Seo, K. et al. (2018). Ceria nanoparticle systems for selective scavenging of mitochondrial, intracellular, and extracellular reactive oxygen species in Parkinson's disease. *Angewandte Chemie International Edition* 57 (30): 9408–9412.

92 Tang, S., Wang, A., Yan, X. et al. (2019). Brain-targeted intranasal delivery of dopamine with borneol and lactoferrin co-modified nanoparticles for treating Parkinson's disease. *Drug Delivery* 26 (1): 700–707.

93 LeWitt, P.A., Lipsman, N., and Kordower, J.H. (2019). Focused ultrasound opening of the blood-brain barrier for treatment of Parkinson's disease. *Movement Disorders* 34 (9): 1274–1278.

94 Mead, B.P., Kim, N., Miller, G.W. et al. (2017). Novel focused ultrasound gene therapy approach noninvasively restores dopaminergic neuron function in a rat Parkinson's disease model. *Nano Letters* 17 (6): 3533–3542.

95 Zhao, Y., Haney, M.J., Mahajan, V. et al. (2011). Active targeted macrophage-mediated delivery of catalase to affected brain regions in models of Parkinson's disease. *Journal of Nanomedicine & Nanotechnology* S4: 003.

96 Niu, S., Zhang, L.-K., Zhang, L. et al. (2017). Inhibition by multifunctional magnetic nanoparticles loaded with alpha-synuclein RNAi plasmid in a Parkinson's disease model. *Theranostics* 7 (2): 344–356.

97 Kim, D., Yoo, J.M., Hwang, H. et al. (2018). Graphene quantum dots prevent α-synucleinopathy in Parkinson's disease. *Nature Nanotechnology* 13 (9): 812–818.

98 about-stroke wso. About Stroke | American Stroke Association.

99 Moskowitz, M.A., Lo, E.H., and Iadecola, C. (2010). The science of stroke: mechanisms in search of treatments. *Neuron* 67 (2): 181–198.

100 Muir, K.W., Buchan, A., von Kummer, R. et al. (2006). Imaging of acute stroke. *The Lancet Neurology* 5 (9): 755–768.

101 Leigh, R., Knutsson, L., Zhou, J., and van Zijl, P.C.M. (2018). Imaging the physiological evolution of the ischemic penumbra in acute ischemic stroke. *Journal of Cerebral Blood Flow & Metabolism* 38 (9): 1500–1516.

102 Obermeyer, J.M., Tuladhar, A., Payne, S.L. et al. (2019). Local delivery of brain-derived neurotrophic factor enables behavioral recovery and tissue repair in stroke-injured rats. *Tissue Engineering Part A*. 1175–1187. http://doi.org/10.1089/ten.tea.2018.0215.

103 Obermeyer, J.M., Ho, E., Gracias, A., and Shoichet, M.S. (2019). Influencing neuroplasticity in stroke treatment with advanced biomaterials-based approaches. *Advanced Drug Delivery Reviews* 148: 204–218.

104 Tapeinos, C., Larrañaga, A., Tomatis, F. et al. (2020). Advanced functional materials and cell-based therapies for the treatment of ischemic stroke and Postischemic stroke effects. *Advanced Functional Materials* 30 (1): 1906283.

105 Zhao, Y., Xin, Z., Li, N. et al. (2018). Nano-liposomes of lycopene reduces ischemic brain damage in rodents by regulating iron metabolism. *Free Radical Biology and Medicine* 124: 1–11.

106 Harris, N.M., Ritzel, R., Mancini, N.S. et al. (2016). Nano-particle delivery of brain derived neurotrophic factor after focal cerebral ischemia reduces tissue injury and enhances behavioral recovery. *Pharmacology Biochemistry and Behavior* 150–151: 48–56.

107 Manickam, D.S., Brynskikh, A.M., Kopanic, J.L. et al. (2012). Well-defined cross-linked antioxidant nanozymes for treatment of ischemic brain injury. *Journal of Controlled Release* 162 (3): 636–645.

108 Wang, Y., Luo, J., and Li, S.-Y. (2019). Nano-curcumin simultaneously protects the blood–brain barrier and reduces M1 microglial activation during cerebral ischemia–reperfusion injury. *ACS Applied Materials & Interfaces* 11 (4): 3763–3770.

109 Lv, W., Xu, J., Wang, X. et al. (2018). Bioengineered boronic ester modified dextran polymer nanoparticles as reactive oxygen species responsive nanocarrier for ischemic stroke treatment. *ACS Nano* 12 (6): 5417–5426.

110 Xu, J., Wang, X., Yin, H. et al. (2019). Sequentially site-specific delivery of thrombolytics and neuroprotectant for enhanced treatment of ischemic stroke. *ACS Nano* 13 (8): 8577–8588.

111 Liu, Y., Ai, K., Ji, X. et al. (2017). Comprehensive insights into the multi-antioxidative mechanisms of melanin nanoparticles and their application to protect brain from injury in ischemic stroke. *Journal of the American Chemical Society* 139 (2): 856–862.

112 Jafari, M., Soltani, M., Naahidi, S. et al. (2012). Nonviral approach for targeted nucleic acid delivery. *Current Medicinal Chemistry* 19 (2): 197–208.

113 Israel, L.L., Lellouche, E., Kenett, R. et al. (2014). $Ce^{3/4+}$ cation-functionalized maghemite nanoparticles towards siRNA-mediated gene silencing. *Journal of Materials Chemistry B* 2 (37): 6215–6225.

114 Israel, L.L., Lellouche, E., Ostrovsky, S. et al. (2015). Acute in vivo toxicity mitigation of PEI-coated Maghemite nanoparticles using controlled oxidation and surface modifications towards siRNA delivery. *ACS Applied Materials & Interfaces* https://doi.org/10.1021/acsami.5b02743.

115 Kim, I.-D., Sawicki, E., Lee, H.-K. et al. (2016). Robust neuroprotective effects of intranasally delivered iNOS siRNA encapsulated in gelatin nanoparticles in the postischemic brain. *Nanomedicine: Nanotechnology, Biology and Medicine* 12 (5): 1219–1229.

116 Fluri, F., Grünstein, D., Cam, E. et al. (2015). Fullerenols and glucosamine fullerenes reduce infarct volume and cerebral inflammation after ischemic stroke in normotensive and hypertensive rats. *Experimental Neurology* 265: 142–151.

117 Wang, H., Agarwal, P., Xiao, Y. et al. (2017). A nano-in-micro system for enhanced stem cell therapy of ischemic diseases. *ACS Central Science* 3 (8): 875–885.

118 Nih, L.R., Sideris, E., Carmichael, S.T., and Segura, T. (2017). Brain transplantation of microporous particle hydrogel after stroke promotes tissue repair and endogenous NPC recruitment to the lesion. *Advanced Materials (Deerfield Beach, Fla)* 29 (32). https://doi.org/10.1002/adma.201606471.

119 Moshayedi, P., Nih, L.R., Llorente, I.L. et al. (2016). Systematic optimization of an engineered hydrogel allows for selective control of human neural stem cell survival and differentiation after transplantation in the stroke brain. *Biomaterials* 105: 145–155.

120 Chen, P.-J., Kang, Y.-D., Lin, C.-H. et al. (2015). Multitheragnostic multi-GNRs crystal-seeded magnetic nanoseaurchin for enhanced in vivo mesenchymal-stem-cell homing, multimodal imaging, and stroke therapy. *Advanced Materials* 27 (41): 6488–6495.

121 Adams, C., Israel, L.L., Ostrovsky, S. et al. (2016). Development of multifunctional magnetic nanoparticles for genetic engineering and tracking of neural stem cells. *Advanced Healthcare Materials* .

122 Mayo-Clinic. (2021). Spinal cord injury. Available from: https://www.mayoclinic.org/diseases-conditions/spinal-cord-injury/symptoms-causes/syc-20377890.

123 Courtine, G. and Sofroniew, M.V. (2019). Spinal cord repair: advances in biology and technology. *Nature Medicine.* 25 (6): 898–908.

124 Guo, S., Perets, N., Betzer, O. et al. (2019). Intranasal delivery of mesenchymal stem cell derived exosomes loaded with phosphatase and tensin homolog siRNA repairs complete spinal cord injury. *ACS Nano* 13 (9): 10015–10028.

125 Vismara, I., Papa, S., Veneruso, V. et al. (2020). Selective modulation of A1 astrocytes by drug-loaded nano-structured gel in spinal cord injury. *ACS Nano* 14 (1): 360–371.

126 Andrabi, S.S., Yang, J., Gao, Y. et al. (2020). Nanoparticles with antioxidant enzymes protect injured spinal cord from neuronal cell apoptosis by attenuating mitochondrial dysfunction. *Journal of Controlled Release* 317: 300–311.

127 Pandanaboina, S.C., Alghazali, K.M., Nima, Z.A. et al. (2019). Plasmonic nano surface for neuronal differentiation and manipulation. *Nanomedicine: Nanotechnology, Biology and Medicine* 21: 102048.

128 Charles, N.A., Holland, E.C., Gilbertson, R. et al. (2011). The brain tumor microenvironment. *Glia* 59 (8): 1169–1180.

129 Hide, T., Shibahara, I., and Kumabe, T. (2019). Novel concept of the border niche: glioblastoma cells use oligodendrocytes progenitor cells (GAOs) and microglia to acquire stem cell-like features. *Brain Tumor Pathology.* 36 (2): 63–73.

130 Xia, S., Lal, B., Tung, B. et al. (2015). Tumor microenvironment tenascin-C promotes glioblastoma invasion and negatively regulates tumor proliferation. *Neuro-Oncology* 18 (4): 507–517.

131 Markovic, D.S., Vinnakota, K., Chirasani, S. et al. (2009). Gliomas induce and exploit microglial MT1-MMP expression for tumor expansion. *Proceedings of the National Academy of Sciences.* 106 (30): 12530.

132 Khaitan, D., Reddy, P.L., Narayana, D.S., and Ningaraj, N.S. (2018). Recent advances in understanding of blood–brain tumor barrier (BTB) permeability mechanisms that enable better detection and treatment of brain tumors. In: *Drug Targeting and Stimuli Sensitive Drug Delivery Systems* (ed. A.M. Grumezescu), 673–688. William Andrew Publishing.

133 Baetke, S.C., Lammers, T., and Kiessling, F. (2015). Applications of nanoparticles for diagnosis and therapy of cancer. *British Journal of Radiology.* 88 (1054): 20150207.

134 Meng, J., Agrahari, V., and Youm, I. (2017). Advances in targeted drug delivery approaches for the central nervous system tumors: the inspiration of Nanobiotechnology. *Journal of Neuroimmune Pharmacology* 12 (1): 84–98.

135 Marie, H., Lemaire, L., Franconi, F. et al. (2015). Superparamagnetic liposomes for MRI monitoring and external magnetic field-induced selective targeting of malignant brain tumors. *Advanced Functional Materials* 25 (8): 1258–1269.

136 Chou, S.T., Patil, R., Galstyan, A. et al. (2016). Simultaneous blockade of interacting CK2 and EGFR pathways by tumor-targeting nanobioconjugates increases therapeutic efficacy against glioblastoma multiforme. *Journal of Controlled Release* 244 (Pt A): 14–23.

137 Lu, F., Pang, Z., Zhao, J. et al. (2017). Angiopep-2-conjugated poly (ethylene glycol)-co-poly (ε-caprolactone) polymersomes for dual-targeting drug delivery to glioma in rats. *International Journal of Nanomedicine* 12: 2117.

138 Chen, G.-J., Su, Y.-Z., Hsu, C. et al. (2014). Angiopep-pluronic F127-conjugated superparamagnetic iron oxide nanoparticles as nanotheranostic agents for BBB targeting. *Journal of Materials Chemistry B* 2 (34): 5666–5675.

139 Ramachandran, R., Junnuthula, V.R., Gowd, G.S. et al. (2017). Theranostic 3-dimensional nano brain-implant for prolonged and localized treatment of recurrent glioma. *Scientific Reports* 7: 43271.

140 Yu, D., Khan, O.F., Suvà, M.L. et al. (2017). Multiplexed RNAi therapy against brain tumor-initiating cells via lipopolymeric nanoparticle infusion delays glioblastoma progression. *Proceedings of the National Academy of Sciences.* 114 (30): E6147–E6156.

141 Galstyan, A., Markman, J.L., Shatalova, E.S. et al. (2019). Blood-brain barrier permeable nano immunoconjugates induce local immune responses for glioma therapy. *Nature Communications* 10 (1): 3850.

142 Sun, T., Patil, R., Galstyan, A. et al. (2019). Blockade of a Laminin-411–notch Axis with CRISPR/Cas9 or a nanobioconjugate inhibits glioblastoma growth through tumor-microenvironment cross-talk. *Cancer Research* 79 (6): 1239–1251.

143 Erel-Akbaba, G., Carvalho, L.A., Tian, T. et al. (2019). Radiation-induced targeted nanoparticle-based gene delivery for brain tumor therapy. *ACS Nano* 13 (4): 4028–4040.

144 Shen, Z., Liu, T., Li, Y. et al. (2018). Fenton-reaction-Acceleratable magnetic nanoparticles for Ferroptosis therapy of orthotopic brain tumors. *ACS Nano* 12 (11): 11355–11365.

145 Shirvalilou, S., Khoei, S., Khoee, S. et al. (2018). Development of a magnetic nano-graphene oxide carrier for improved glioma-targeted drug delivery and imaging: in vitro and in vivo evaluations. *Chemico-Biological Interactions* 295: 97–108.

146 Catalan-Figueroa, J., Palma-Florez, S., Alvarez, G. et al. (2016). Nanomedicine and nanotoxicology: the pros and cons for neurodegeneration and brain cancer. *Nanomedicine* 11 (2): 171–187.

147 Keck, C.M. and Müller, R.H. (2013). Nanotoxicological classification system (NCS)–a guide for the risk-benefit assessment of nanoparticulate drug delivery systems. *European Journal of Pharmaceutics and Biopharmaceutics.* 84 (3): 445–448.

148 Silva, A.C., Oliveira, T.R., Mamani, J.B. et al. (2011). Application of hyperthermia induced by superparamagnetic iron oxide nanoparticles in glioma treatment. *International Journal of Nanomedicine.* 6: 591.

149 Frey, B., Weiss, E.-M., Rubner, Y. et al. (2012). Old and new facts about hyperthermia-induced modulations of the immune system. *International Journal of Hyperthermia.* 28 (6): 528–542.

150 Lohr, F., Hu, K., Huang, Q. et al. (2000). Enhancement of radiotherapy by hyperthermia-regulated gene therapy. *International Journal of Radiation Oncology, Biology, Physics* 48 (5): 1513–1518.

151 Israel, L.L., Kovalenko, E.I., Boyko, A.A. et al. (2015). Towards hybrid biocompatible magnetic rHuman serum albumin-based nanoparticles: use of ultra-small $(CeL_n)^{3/4+}$ cation-doped maghemite nanoparticles as functional shell. *Nanotechnology.* 26 (4): 045601.

152 Israel, L.L., Karimi, F., Bianchessi, S. et al. (2015). Surface metal cation doping of maghemite nanoparticles: modulation of MRI relaxivity features and chelator-free 68Ga-radiolabelling for dual MRI-PET imaging. *Materials Research Express* 2 (9): 095009.

153 Coccini, T., Caloni, F., Ramírez Cando, L.J., and De Simone, U. (2017). Cytotoxicity and proliferative capacity impairment induced on human brain cell cultures after short- and long-term exposure to magnetite nanoparticles. *Journal of Applied Toxicology.* 37 (3): 361–373.

154 Phenrat, T., Long, T.C., Lowry, G.V., and Veronesi, B. (2008). Partial oxidation ("aging") and surface modification decrease the toxicity of nanosized zerovalent iron. *Environmental Science & Technology* 43 (1): 195–200.

155 Pons, T., Pic, E., Lequeux, N. et al. (2010). Cadmium-free CuInS2/ZnS quantum dots for sentinel lymph node imaging with reduced toxicity. *ACS Nano* 4 (5): 2531–2538.

156 Valizadeh, A., Mikaeili, H., Samiei, M. et al. (2012). Quantum dots: synthesis, bioapplications, and toxicity. *Nanoscale Research Letters* 7 (1): 480.

157 Bae, Y., Nishiyama, N., Fukushima, S. et al. (2005). Preparation and biological characterization of polymeric micelle drug carriers with intracellular pH-triggered drug release property: tumor permeability, controlled subcellular drug distribution, and enhanced in vivo antitumor efficacy. *Bioconjugate Chemistry* 16 (1): 122–130.

158 Ljubimova, J.Y., Portilla-Arias, J., Patil, R. et al. (2013). Toxicity and efficacy evaluation of multiple targeted polymalic acid conjugates for triple-negative breast cancer treatment. *Journal of Drug Targeting. [Article]* 21 (10): 956–967.

159 Asati, A., Santra, S., Kaittanis, C., and Perez, J.M. (2010). Surface-charge-dependent cell localization and cytotoxicity of cerium oxide nanoparticles. *ACS Nano* 4 (9): 5321–5331.

160 Soares, S., Sousa, J., Pais, A., and Vitorino, C. (2018). Nanomedicine: principles, properties, and regulatory issues. *Frontiers in Chemistry* 6: 360.

161 Ventola, C.L. (2017). Progress in nanomedicine: approved and investigational Nanodrugs. *Pharmacy and Therapeutics* 42 (12): 742–755.

162 FDA. (2014). Considering Whether an FDA-Regulated Product Involves the Application of Nanotechnology-Guidance for Industry. FDA-2010-D-0530. https://www.fda.gov/regulatory-information/search-fda-guidance-documents/considering-whether-fda-regulated-product-involves-application-nanotechnology

163 FDA. (2018). FDAs Approach to Regulation of Nanotechnology Products' https://www.fda.gov/science-research/nanotechnology-programs-fda/fdas-approach-regulation-nanotechnology-products.

3

Nanomolecular Diagnostics

Anila Fariq[1,2], Ayesha Selhaba[1], Anum Zulfiqar[2], and Azra Yasmin[2]

[1] Department of Biotechnology, University of Kotli Azad Jammu and Kashmir, Pakistan
[2] Department of Biotechnology, Fatima Jinnah Women University, Rawalpindi, Pakistan

3.1 Introduction

Although the history and examination of patients play a key role for the further clinical assessment of patients, numerous symptoms are inadequate to either phase in or out particular diagnoses or to track chronic diseases in patients. The quick and accurate diagnosis of diseases is vital for the suitable treatment of patients. Technologies have already been established in primary care practices as they are quick and small, and they provide accurate measurements of different microbiological, hematological, and biochemical parameters. Glucometers, urine dipsticks, blood pressure devices, common blood tests, lipid profiles, and urine albumin-creatinine ratios are easily available in general health care centers [1]. Recently, molecular-level interaction and interrogation technologies are progressing, thereby paving new ways of diagnosis, with higher speed and accuracy.

3.1.1 Molecular Diagnostics

Molecular diagnostics is defined as the detection of genomic variations for the identification of disease, prediction and monitoring of treatment. The integration of genomics, molecular biology techniques, and laboratory medicine has led to the foundation of molecular diagnostics. Together these elements provide an accurate characterization of inherited diseases essential for an accurate diagnosis. High-throughput next-generation sequencing methods and genome-wide association studies provide instrumental discernments into disease mechanisms and allow physicians to assess disease predilection and devise precise diagnostic methods to personalise a cure [2]. In recent years, the advancement of molecular technologies has remarkably accelerated the growth of molecular diagnostics. Diagnostic molecular biology has been extensively used in immunology, hematology, microbiology, and clinical biochemistry [3].

The basic principle of molecular diagnostics starts from localization to identification and then characterization of genes accountable for human disease. It also takes into

Nanomaterials and Nanotechnology in Medicine, First Edition. Edited by P.M. Visakh.

account how genes and proteins interact in a cell. It unravels the set of changes in gene and protein activity patterns and captures them as expression patterns called "molecular signatures," thereby improving disease diagnoses. Molecular diagnostics involves novel technologies in clinical tests to provide a molecular impression of tumors that reflects the response to a specific treatment. It also offers potential insights to move further from diagnostics to prognostics [4].

3.1.2 Molecular Diagnostic Techniques

The molecular analysis of diseases has transformed disease diagnoses. Genomic technologies incorporate extensive laboratory techniques that provide nucleic acid sequences, their functions, and other genome information at relatively high speed and low cost. Some major molecular diagnostic techniques include fluorescence *in situ* hybridization (FISH), mass spectrometry for analysis of DNA mutation, polymerase chain reaction (PCR), microarray-based assays, and next generation sequencing methods. Advanced technologies along with clinical knowledge enabled significant progress in disease diagnosis and management [5].

FISH is a rapid, highly sensitive, cytogenetic technique used for the detection of chromosomal anomalies, hematological cancers, and solid tumors. It can be applied to the detection of a genetic disorder or to prognostic conclusions. FISH can also be used for gene mapping, unique oncogene identification, or cancer-causing genetic abnormalities. The principle of FISH is based on DNA probes forging to specific target DNA sequence. Fluorescent reporter molecules are attached to probes which spot the particular genetic aberration under fluorescence microscopy. It can also screen the whole genome through multicolor whole-chromosome probing methods like array-based comparative genomic hybridization, multiplex FISH, or spectral karyotyping [6].

PCR is a biochemical tool to amplify DNA fragments and generates millions of copies of a specific DNA sequence. PCR is commonly used in laboratories for the diagnoses of various diseases including autoimmune and genetic nervous disorders, malignant cancers, and pathogenic infections. Various forms of PCR techniques are used for detection of different diseases. Allele-specific PCR is used to diagnose specific disease on the basis of single-nucleotide variations. Recent developments in PCR techniques have revolutionized the medical diagnostic fields like mycology, virology, microbiology, parasitology, and dentistry. It is an efficient technique for the quick detection of pathogens difficult to culture otherwise. Nowadays, Real-Time PCR is playing a substantial role in clinical diagnostics. It is considered a fast and accurate platform to provide qualitative as well as quantitative results [7, 8].

DNA microarrays are oligonucleotide probes immobilized on nylon or silicon chips. The probes are made from a complementary DNA library, PCR products, or synthetic oligonucleotides. There are several ways of immobilizing the oligonucleotides on the solid support and creating DNA chips. A sample DNA to be analyzed is labeled, placed on the chip, and then hybridized with the complementary sequence to an oligonucleotide probe. Spots appear on the chip after hybridization are observed by autoradiography, and then analyzed with software to interpret the patterns of signals [9]. DNA microarrays are widely used in clinical diagnostics right from the discovery of disease-causing genes to diagnosis using particular biomarkers, screening of single nucleotide polymorphisms (SNPs), mutation analysis, identifying chromosome aberrations, and posttranslational modifications [10].

Mass spectrometry coupled with PCR and Matrix Assisted Laser Desorption/Ionisation–Time-of-Flight mass spectrometry (MALDI-TOF MS) is a rapid method to determine the genetic target material and derived variants. This technology uses multiplex assays and high-throughput analyses to identify SNPs and diagnose mutations in human diseases, genotyping, methylation, and quantitative gene expression. MALDI-TOF is also a highly advanced diagnostic technique analogous to other classic sequencing technologies for direct genetic characterization of viruses, bacteria, and fungi. It is an accurate, quick, and sensitive method; requires minimal sample preparation; and is cost-efficient [11].

Next-generation sequencing (NGS) includes a collection of technologies capable of executing and capturing data from millions of sequencing reactions instantaneously and acquiring the individual sequences of hundreds of millions of base pairs. NGS is an efficient and cost-effective method commonly used to detect the genomic aberrations causing a variety of cancers like lung, breast, melanoma, and brain cancers. In clinical settings, NGS biomarkers are used for the diagnosis of specific diseases. Predisposition biomarkers are used to detect the germline variants associated with disease risk. Pharmacogenetic biomarkers predict human drug interaction. Profile biomarkers interrogate variations within a diseased tissue like tumors. Prognostic biomarkers give information on the effect of the disease while predictive biomarkers detect the response of patients to specific therapies. Pharmacotherapeutic biomarkers identify gene mutations or expression patterns in disease tissue pliable to therapeutic targets [12].

3.1.3 Molecular Diagnostics of Infectious Diseases

Over the past few decades, development and application of molecular diagnostic techniques has revolutionized the diagnosis of infectious diseases. Common lab practices for identification and differentiation of pathogens include their phenotypic characterization, i.e., protein, bacteriophage, and chromatographic profiles, biotyping and susceptibility testing, as well as nucleic acid–based tenchniques, such as plasmid profiling, direct detection of drug resistant genes, restriction fragment length polymorphisms, and the PCR to rapidly detect the etiologic agents of disease directly from clinical samples without culturing. Extensive use of automation and user-friendly bioinformatics software makes these technologies more readily available. Advancements in molecular-based techniques have provided a crux to clinical microbiology techniques [13].

3.2 Nanodiagnostics

Nanodiagnostics is defined as the use of nanotechnology for state of the art clinical diagnostic purposes to meet the demands of enhanced sensitivity and prognosis of disease in clinical diagnostics. The increased need for accuracy can be met only through a diagnostically significant interaction between analyte molecules and signal-generating particles to detect a single analyte molecule. Nanotechnology has empowered one-to-one interaction between analytes and signal-generating particles in the size range of proteins (1–20 nm) and other biomolecules [14].

3.3 Nanoparticles for Molecular Diagnostics

Nanotechnology features the matter with dimensions between 1 and 100 nm. This field is playing a substantial role in molecular diagnostics by incorporating new methods based on the inimitable properties of nanometer-scale materials. They include biobarcodes, dendrimers, nanowires, nanocantilevers, nanopores, and nanofluidic devices [13].

3.3.1 Nanowires

A nanostructure with a diameter of the order of a nanometer is known as a nanowire. They are also known as nanowhiskers. Nanowires are enabling a new generation of computing, integrated photonics, energy, and biomedical technology. In early nineties they are of great observance including oxides and semiconductors. One of the growth methods that was highly appreciated was Wagner's vapor–liquid–solid crystal method with the introduction of metal nanoparticles (NPs) for the determination of the diameter of nanowires [15, 16]. Nanowires are categorized on the basis of the metals used—that is, silver nanowires, gold nanowires, and magnetic nanowires. One of the nanowires of great interest is gold nanoparticles-decorated silicon nanowires (AuNPs@SiNW). These nanowires are highly efficient near infrared hyperthermia agents for the destruction of cancerous cells. Three different types of cancerous cells are destructed in three minutes when these cells are treated with AuNPs@SiNWs by producing efficient heat [17].

Recently, a low cost and reliable method for the synthesis of nanowires is electrospinning. It is a promising method to produce quality nanowires. This process not only is cost-friendly but also gives a promising control over many processes like viscosity, flow rate, electric potential and solution, which has a direct influence on the quality of nanofibers. The process involves the fabrication of very thin fibers through an electrically charged jet of a liquid forerunner [18]. The main components of an electrospinning apparatus consist of injection, pump, syringe needle, high voltage power supply, and collector plate. The solution that contains the polymer is subjected to the tip of the needle. The electric field is generated through a high voltage power supply between the tip of the needle and the collector plate in the system. By the introduction of an electric field, the surface tension of the droplet is weakened, and it causes the distortion of the droplet shaping the so-called Taylor's cone of the droplet is formed. This distortion conducted to an electrically charged jet ejection that is introduced to collector for the formation of nanofibers. For the production of aligned polymer fiber, a rotating collector is used [19]. The main applications of nanowires are in nanocatalysts, tissue scaffolds, protective clothing, filtration, and optical electronics. These nanowires are preferred over regular wires because of their smaller pores and higher surface area.

3.3.2 Nanotubes

Carbon is a versatile element on Earth because of its four-atom bonding property. Carbon is the most important element in the history of mankind from the beginning, in its different way. Carbon nanotubes (CNT) are one of the allotropic forms of carbon. CNTs were first isolated by Iijima, Ebbese, and Ajayan in large quantities [20]. CNTs are defined as tubular structures with their diameter typically measured in nanometers. The most promising feature of nanotubes is the target delivery of drug toward the tumor [21].

CNTs are of great interest because they are super materials which can be stronger than steel and have a greater tendency to conduct electricity than copper. There are two types of CNTs:

- Single-walled carbon nanotubes (SWCNTs)
- Multi-walled carbon nanotubes (MWCNTs)

SWCNTs consist of a single cylindrical carbon layer, with a diameter of about 0.4–2 nm, depending on the temperature. The temperature has a direct relation with the diameter of CNTs. Their structure may be armchair, zigzag, chiral, or helical [22].

SWCNTs have an ultra-high surface area of about 1300 m^2g^{-1} and are more preferable than MWCNTs. SWCNTs render enough space for drug-loading and bioconjugation. The SWCNT drug complex has more potential to stay in blood circulation than the anticancerous drug on its own. Therefore, it takes to more prolonged and sustained uptake of drug by cancerous cells. One more promising quality of SWCNTs is drug delivery at specific site, it excretes out of body through biliary pathways and then in the feces [23].

MWCNTs consist of several coaxial cylinders that are composed of a single graphene sheet surrounding a hollow core. They have two diameters, with outer ranges from 2 to 100 nm and inner ranges from 1 to 3 nm. Their length is in micrometers. A cloud of delocalized electrons along the wall is generated as a result of sp2 hybridization that causes interactions between adjacent cylindrical layers, which is a major reason for loss of flexibility and structural defects [24]. On the arrangements of graphite layers, their structure can be split into two categories: the Russian doll model, in which the layers of graphite are arranged in a concentric structure; and the Parchment-like structure, which involves the rolling up of a graphene sheet around itself [25]. These MWCNTs are decorated with NPs, either on the walls or on the ends. MWCNTs have vast applications in catalysis, electronic devices, biomedical devices, biosensors, and magnetic storage devices [26]. There are several methods for the synthesis of CNTs depending upon the range of temperature, nanotube length, dimeter, purity, density, alignments, and orientation of CNTs [27]. The most commonly methods used for the synthesis of CNTs are discussed below.

3.3.2.1 Arc Discharge Evaporation Method

CNTs are synthesized by the arc discharge evaporation method on a large scale. The arc discharge method is done in two ways, either by using carbon electrodes or by using metal electrodes. In the carbon electrode method, graphite electrodes are used as a source of carbon atoms to produce nanotubes. The anode part is vaporized and rotated at high speed to produce CNTs, and this rotation has a direct relation to the production of CNTs.

This rotation of anode uniformly dispersed the micro-discharge, and a stable plasma is generated. The turbulence is caused by this rotation, which speeds the carbon vapor perpendicular to the anode. The carbon vapor does not get condensed on the carbon surface, but it is collected from the graphite collector that was set at the periphery of plasma. This method fulfills two optimized conditions of production: the density of the carbon vapor and the sufficient temperature for the production of CNTs in the collector [28]. The second method is achieved by using metal electrodes. The electrodes are made of metals like Ni, and hydrocarbon solvents that may be toluene or cyclohexane are used. Thus, solvents act as the source of carbon for the production of nanotubes. Rod-shaped nanotubes are made at the anode through oxidation.

3.3.2.2 Laser Ablation Method

The laser ablation method is the same as the arc discharge method in the aspect of high temperature generation and optimum background gas and catalyst mix. The laser ablation method uses a continuous and pulsed laser for vaporization o graphite target with the help of oven and for pressure noble gases like He or Ar in the oven [29]. By using continuous wave carbon dioxide laser ablation, SWCNTs are prepared without the introduction of additional heat to the target. It was found that the average diameter of SWCNTs is directly related to the laser power used [30].

3.3.2.3 Thermal Synthesis Method

The thermal synthesis method is not based on a plasma-based synthesis of CNTs like laser ablation and the arc discharge method. The thermal synthesis method depends on thermal energy, and the hot zone does not go beyond 1200 °C. Active catalyst species such as Fe, Ni, and Co are used for the synthesis of CNTs. Sometime promoters like Mo and Ru are also added for the greater production of CNTs [31]. CNTs have wide applications in the medical field because of their versatile structure, consisting of an inner and an outer core, both of which can be modified by bonding different functional groups according to need. In the field of biomedicine, CNTs are used in four main fields: drug delivery, biomedical imaging, scaffolds in tissue engineering, and biosensors [32]. CNTs have the potential to carry NPs for the delivery of drug, genes, and proteins. It has been demonstrated that cancerous cells over-express folic acid (FA) receptors, and some researchers have developed NPs with engineered FA derivative surfaces [33].

Acute lymphoblastic leukemia is one of the four major types of leukemia. It is treated through a concentrated targeted delivery of daunorubicin (Dau). The drug is released in a slightly acidic pH of 5.5 from complex in 72 hours at 37 °C [34]. Breast cancer is a worldwide malignancy among females. Chemotherapy is used for the treatment of breast cancer, but researchers also focus on targeted drug delivery through NPs. The NPs can efficiently deliver the drug at the target site as well as enhance the hydrophilicity of anti-cancer drugs [35].

3.3.3 Quantum Dots

Quantum dots (QDs) are zero-dimensional semiconductor nanocrystals with very unique properties [36]. Carbon-based QDs (CQDs) consist of graphite QDs and carbon QDs. They were first observed during the purification of SWCNTs in 2004 and in 2006 from the laser ablation of graphite and cement [37]. They are also known as carbon nanolights because of their good solubility and strong luminescence [38]. In the field of biomedicines, CQDs have potential application in biosensing, drug delivery, and bioimaging because of less toxicity and good biocompatibility [39].

Many methods have been used for the synthesis of CQDs, but the methods face three main problems:

1) During carbonization, carbonaceous aggregation
2) Uniformity and size control
3) Surface properties

These problems are sorted through a few techniques. There are some systematic methods for the synthesis of CQDs [40] discussed below.

3.3.3.1 Chemical Ablation

Small organic molecules can be carbonized by strong oxidizing acids and these carbonaceous materials are cut into small sheets by controlled oxidation. This method has harsh and drastic conditions. A simple route reported in which an aqueous solution prepared by dehydrating carbohydrates with concentrated sulfuric acid. It is further proceeded by breaking the carbonaceous materials into QDs by HNO_3 and the final step is passivation with amine-terminated compounds [41].

3.3.3.2 Electrochemical Carbonization

Electrochemical carbonization is a very powerful method for the synthesis of CQDs by using carbon bulk materials as precursors; it is also known as chemical soaking. This method produces eco-friendly as well as cost-friendly QDs. In this method, QDs with uniform size of 3, 5, and 8.2 nm diameter were synthesized from MWCNTs soaked in the solvent propylene carbonate by using LiClO at 90 °C [42, 43].

3.3.3.3 Laser Ablation

A simple approach for the laser synthesis of QDs is using nanocarbon materials as precursors and a liquid media such as a simple solvent. In a typical method for the synthesis of nanocarbon materials, about 0.02 g was dispersed in solvents such as ethanol, water, and acetone. After ultrasonication, the suspension about 4 ml poured into a glass cell for irradiation. Yttrium Aluminum Garnet (YAG) pulsed laser having second harmonic wavelength of 532 nm used to irradiate the suspension. In the end, the solution as centrifuged to obtain the supernatant that contains the CQDs [44].

3.3.3.4 Microwave Irradiation

Microwave irradiation is a bottom-up approach for the synthesis of QDs. In microwave irradiation, 3 g of succinic acid and 5.572 g of tris(2-aminoethyl) amine are completely dissolved in 10 ml of deionized water in a 250 ml flat-bottom flask. The flask is loosely wrapped with wrapping film and placed in an oven for five minutes. After the microwave treatment, a brown-colored solution is obtained that signifies the presence of CQDs. After this, the CQDs are extracted out by using 100 ml deionized water to the solid and it is ultrasonicates in a bath sonicator for 30 minutes to remove insoluble precipitates; it is centrifuged at 4000 rpm for 30 minutes. By freezing the solution, 1.48 g of brown CQD powder is obtained [45].

Carbon QDs have low toxicity, and good biocompatibility provides promising features for fluorescent bioimaging ad multimodal bioimaging of cells and tissues. Carbon nanodots are also present in our daily-life foods such as coffee and breads, proving that they are more biocompatible than the inorganic nanodots. Due to high solubility in water, nontoxicity, excellent biocompatibility, good cell absorbency, multicolor emission, flexibility in surface modification, and high photostability, nanodots are also used as biosensors. They can be effective for fluorescent sensing of nucleic acid detection with selectively single-base mismatch [46]. Drug delivery using biocompatible nanodots is very effective. One example of using this approach is theragnostic agent (CD-Oxa) was prepared by the combination of an anticancer agent named oxidized oxaliplatin,oxa(iv)-COOH on the surface of CQDs for efficient drug delivery to target site [45].

3.3.4 Nanopores

The nanometer holes or pores that are formed naturally by proteins or cells that allow the passage of ions between nerve cells are called nanopores. These pores can be synthetically produced. Nanopores are high-throughput ion microscope and open the mysterious doors of DNA hidden information [47].

3.3.4.1 Template Synthesis

One of the methods for the synthesis of nanopores is the template synthesis method. In various fields of nanotechnology, template synthesis has developed independently and has positive feedbacks. In this method, electrodeposition or magnetophoretic deposition is introduced for the synthesis of granular nanowires. When the precursors for the template synthesis are of atomic size (such as ions for electrodeposition or molecular vapors), then the filling of pores is relatively easy [48].

3.3.4.2 Track-Etch Method

The track-etch method is a commercially based method to produce nanopore filters with monodispersed pores. The track-etch method involves the bombardment of a polymer membrane with high-energy particles. These high-energy particles are obtained either from a nuclear reactor or from a cyclotron. It produces damage paths that run through the complete thickness (about 5–10 μm) of the membrane. These damaged paths are then chemically etched to produce monodisperse pores. Nanopore membranes with a range of pore size down to ~10 nm composed of polycarbonate, polyethylene terephthalate (PET), and polyimide are commercially available [49]. Because of the ion-transport property, nanopores are used for the mimicry of biological ion transport systems such as voltage-gating [50]. Nanopores are widely used to study the protein translocation and protein-probe interactions, protein-antibody interactions, and probe-single catalytic interactions [51].

3.3.5 Nanoneedles

Needles made of stainless steel with nanoscale diameter are termed nanoneedles; in the near future, they can be used in cell surgery. Different kinds of nanoneedles are available depending upon their use. The four main kinds of nanoneedles are solid nanoneedles, coated nanoneedles, polymer nanoneedles, and hollow nanoneedles [52].

3.3.5.1 Solid Nanoneedles

Solid nanoneedles are synthesized for the pre-treatment of skin which leads to increased permeability through the skin. These needles contain the polymer solution that is homogenized with drug. Examples are amantadine and pramipexole.

3.3.5.2 Coated Nanoneedles

Coated nanoneedles are coated over the normal polymer surface. The drug is used to mix with skin after insertion. Examples are vitamin-B, calcein, bovine, and serum albumin.

3.3.5.3 Polymer Nanoneedles

Polymer nanoneedles contain the drug that eventually dissolves in the skin. The most used polymer is PVC that is non-toxic in nature and strength. Examples are calcein and human serum albumin.

3.3.5.4 Hollow Nanoneedles

Hollow nanoneedles contain the drug placed in a hollow cavity inside the nanoneedle and used for drug infusion. An example is phenylephrine.

Nanoneedles can be synthesized by following ways

3.3.5.5 Electrochemical Deposition

Electrochemical deposition has been widely used for the production of nanoneedles in porous structure.

3.3.5.6 Sol-Gel Technique

The sol-gel technique has been widely used for the synthesis of NPs that are associated with a gel that contains sol. In the first step, the colloidal suspension is prepared by desired precursor materials. Then an anodic aluminum oxide (AAO) template is introduced in the sol for attachment of sol particles on the AAO template. With the passage of deposition time, the channels will be filled with sol particles. By applying thermal annealing, gel will be removed to obtain products.

3.3.5.7 Surfactant-Assisted Growth

Surfactant-assisted growth is a trial-based method to produce NPs. It mainly depends on the precursor species; surfactant, pH, and temperature must be selected very carefully. It is a micro-emulsion system that consists of three phases: oil, surfactant, and aqueous [53]. Nanoneedles have vast application in target drug delivery with chemicals, proteins, or nucleic acids [54]. Nanoneedles may be used in cell surgery in the near future.

3.3.6 Nanorobots

Nanorobots are also known as computerized surgical nanorobots and can function like a surgeon inside the body under the control of experts [54]. These "nanobots" are extensions of injectables that go into the body and gather information about the body through images. Properties like chemical sensing; receiving information from, to, and outside the body; locomotion; and computation play very crucial roles in giving output. The picowatts are important for propelling a nanobot at a speed of 1 mm s^{-1} [55]. The main interesting property of nanobots is drug delivery at a specific site to minimize the effect on other healthy parts of the body. Nanobots are designed to detect the defective parts of the body and deliver the drug. The most important parts of nanobots are sensors and propulsion equipment [56].

3.3.6.1 Sensors

The most important part of nanobots are sensors that may be thermal, mechanical, magnetic, biological, and chemical sensors which have been detected in the applications of nanobots. These sensors are termed nanosensors because their diameter lies in nanometers. The organic part of sensor served as biosensors which uses biological reactions for detection of defect.

3.3.6.2 Propulsion Equipment

The in charge of movements of nanobots is known as propulsion and it varies with the variation in nanobots. Propulsion equipment consists of nanomotor as name indicate that the size of motor lies in the range of nm. Nanomotors gain their energy from chemical reactions that occur in medium, electrical, acoustic, or magnetic fields. The movements of nanobots depends upon viscosity and Brownian movement. Brownian movements cause hindrance to the movement of nanobots, so here we require a nanobot with high energy. Two approaches are used for the movement of nanobots: external and internal. In the external approach for nanobot propulsion, MRI devices are used to obtain real feedback of the nanobot's behavior. In the internal approach for nanobot propulsion, naturally present biomotors keep in check the activities of nanobots.

Nanorobots play a significant role in the science of medicine like targeted drug delivery and may become a nanosurgeon soon. In the present, DNA nanobots specified 12 different kinds of cancer in humans. It is assumed that within a very short period of time, these nanobots will be capable of curing cancer and repairing damage to the spinal cord. Nanobots known as respirocytes, which are artificial red blood cells, can deliver oxygen 236 times more than natural red blood cells. Nanobots called microbivores can destroy the pathogenic bacteria present in the body. Like respirocytes, clottocytes are capable of healing wounds by making fresh fiber mesh. Chromallocytes are hypothetical mobile cell repair nanobots that are capable of reversing the effect of genetic diseases by replacing the defective chromosomes [57].

3.3.7 Nanoscale Cantilevers

Nanocantilevers are nanomechanical-based mass sensors that play an important role in micro/nanoelectrochemical systems [58]. The nanocantilevers resemble tiny diving boards composed of silicon and vibrate at different frequencies [59]. Nanocantilevers are being introduced in nanorobots. The main significant property of nanocantilevers is real-time detection. In simple terms, they can be used to control objects at the single-molecule level. Nanocantilevers were fabricated in array and selected according to biomolecules.

The nanocantilevers vary with the variation in biomolecules. In the fabrication of DNA nanocantilevers, in solution, different sequences of a DNA probe were adjusted on the surface of cantilevers and target DNA was injected. Only the nanocantilevers that have a matched sequence will have surface stress [59, 88]. Nanocantilevers are working in two different modes: static and dynamic.

3.3.7.1 Static Mode

In the static mode of action, nanocantilevers depend on the absorption of measured analytes from a static cantilever which causes bending and results in deflection regarding reference point. These cantilevers can be used to detect the presence of target biomolecules in a small amount of sample, which later on, in drug can implicit an early detection of diseases like cancer.

3.3.7.2 Dynamic Mode

In the dynamic mode of action, the resonance frequency of the target molecule plays a crucial role in the action of nanocantilevers. Nanocantilevers act when there is a change in rigidity or mass of the target molecule due to resonance frequency. In this way, the frequency is used to detect the different molecular bonds and to analyze the area of the surface.

Nanocantilevers have many applications in the diagnosis of various diseases. For example, nanocantilevers coated with antibodies are used to detect certain viruses depending upon the size of the nanocantilevers. Nanocantilevers are used as biosensors as well. As in the case of the diagnosis of prostate cancer, prostate marker proteins and prostate specific antigens (PSAs) are coated on the nanocantilevers for the detection of cancer. Such PSA nanocantilevers are also used for the detection of breast cancer in females. Nanocantilevers are also used for the detection of proteins as well. The protein-detecting nanocantilevers can detect the proteins in ultralow amounts in the body. This detection will help in the diagnosis of diseases such as cancer [60].

3.4 Applications of Nanoparticles for Molecular Diagnostics

Nanotechnology in integration with biology, and photonics has opened the revenues for detection of atoms and molecules using nanodevices with potential applications (Figure 3.1) in a wide variety of medical uses at the cellular level [61].

3.4.1 Nanoparticle-Based Molecular Diagnostics for Cardiovascular Diseases

For the diagnosis of cardiovascular diseases, cardiovascular imaging is the most reliable diagnostic tool. Due to the unique features of NPs, they could be efficiently utilized in the field of medical imaging. The biocompatible size distribution of NPs helps them to house different biocomponents within the subcellular organelle. High penetration

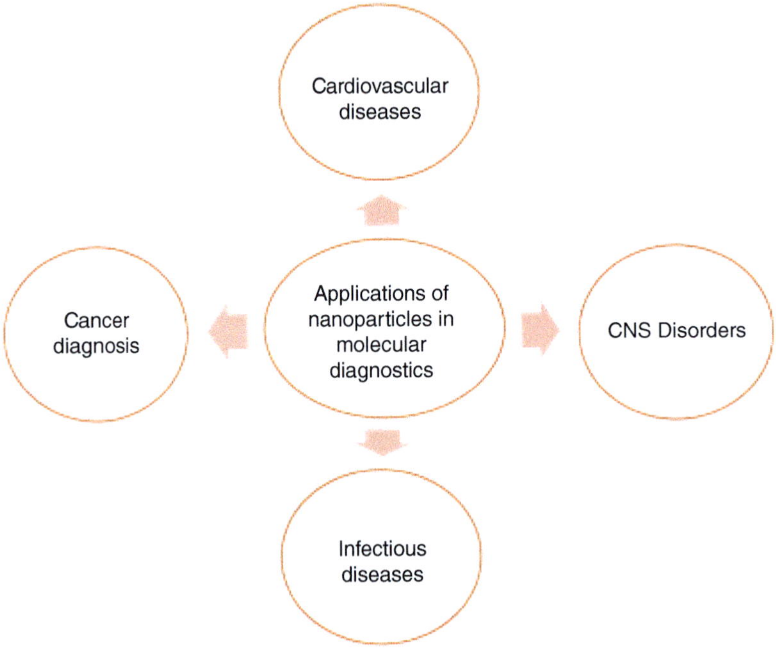

Figure 3.1 Applications of nanoparticles in molecular diagnostics.

power is another characteristic making them more favorable for biomedical imaging. They also possess high contrasting capabilities. Paramagnetic NPs are efficiently utilized for magnetic resonance imaging, iodinated NPs are used in computed tomography contrast agents, and QDs are fluorescent enhancers. Moreover, nanoforms are much more stable than a chemical image probe and have increased half-life. Nano-based cardiovascular imaging is used for monitoring the live physiological system with high sensitivity and no pain. This live imaging is used for the understanding of pathological conditions and proper diagnosis. The majority of the nano-based cardiovascular imaging modules includes thrombus imaging, theranostic imaging, stem cell imaging, and Graft imaging [62].

NPs with therapeutic properties and diagnostic ligands, also referred to as theranostic NPs, are used for improved diagnosis, targeted therapy, and the monitoring of thrombosis and atherosclerosis. Particularly NPs coated with specific targeted molecules alongside disease biomarkers, the delivery, accumulation, and elution of therapeutic agent within pathological tissue is done by stimuli-responsive release [63]. Stem cells labeled with NPs are used for the monitoring of myocardium stem cells. They are used to label the transplanted cells and carry out cardiac repair by improving cell retention [64].

Ferromagnetic iron oxide particles with dispersion properties are used as contrasting agents for MRI and for drug delivery as well at low magnetic field gradients. They also reduce the concentration of NPs required for cell separation. Recognition molecules such as radiotracers are conjugated with NPs and therefore increase the signal-to-noise ratio for imaging. Photoacoustic imaging involves AuNPs as contrast agents because of their strong light-scattering properties. These AuNPs are used in optical imaging of coronary blood vessels. NP-based blood pool contrasting agents are used for the visualization of vasculature *in vivo*. Membrane protein labeling and cell tracking studies have been done using QDs. Fluorescence resonance energy transfer (FRET) is used to track enzymatic activity in the pathogenesis of atherosclerosis [65].

3.4.2 Nanoparticle-Based Molecular Diagnostics for Cancer Diagnosis

A biomarker is a quantifiable biological molecule found in blood and tissues or other body fluids like saliva and urine to detect cancers in the body. Cancer biomarkers include cell surface proteins and secreted proteins, carbohydrates, miRNA, and circulating tumor DNA (ctDNA), secreted by the body if cancer is present. The quantification of cancer biomarkers helps in the early detection of cancer or tumor recurrence and aids in monitoring the efficacy of the therapeutic treatment. However, there are several limitations in the use of biomarkers, including the low concentrations of biomarkers in body fluids and variation in the timing and abundance of biomarkers within patients. Nanotechnology provides highly selective, more sensitive, and simultaneous measurements of multiple targets. Biosensors are improved with nanomaterials for specific targeting. QDs, AuNPs, and polymer dots are the common NP probes used in cancer diagnostics [66].

A few proteins have been used for cancer detection, including colorectal cancer, liver cancer, prostate cancer, and ovarian cancer. Specific interactions with antibodies, antibody fragments, or aptamers can help in the detection of these properties. The interaction event will then be converted into a quantifiable signal that can be measured. Carcinoembryonic antigens are the most popular cancer biomarker used for the monitoring of anticancer treatment and for the prediction of tumor recurrence following surgical resection in late-stage cancer patients. QD-based biosensors have been used for detecting cancer biomarkers. QD-conjugated antibodies against neuron-specific enolase and carcinoembryonic antigens were used to detect two biomarkers for lung cancers [67]. With zinc oxide QDs as electrochemical and fluorescent labels, a sandwich immunoassay was established to detect carbohydrate antigen 19–9, which is a biomarker for pancreatic cancer [68].

Tumor-derived DNA fragments of approximately 100–200 base-pairs present in the bloodstream are termed as ctDNA. A study described DNA-silver nanocluster (NC) fluorescence-based detection strategy for mutation in breast cancer related gene BRCA1. The hybridization of DNA-silver NCs fluorescent probe to target DNAs act as efficient templates for enhancing AgNCs fluorescence to differentiate the deletion of BRCA1 [69]. A microRNA, i.e. miR-141, is a potential biomarker for prostate cancer. A study developed two-step sensing platform for the sensitive detection of miR-141. It includes the use of semiconductor CdSe/ZnS QDs with recognition sequence for miR-141 in the first step and a telomerase primer sequence in the second step of the analysis [70]. Different nanomaterials like magnetic NPs, AuNPs, QDs, nanowires, silicon nanopillars, CNTs, dendrimers, graphene oxide, and polymers have been used for circulating tumor cell detection. These nanomaterials enhance the sensitivity and specificity of circulating tumor cell capture devices and can be used to expedite cancer prognosis and diagnosis [66].

3.4.3 Nanoparticle-Based Molecular Diagnostics for CNS Diseases

CNS diseases refers to neurological disorders affecting the structure and function of the brain or spinal cord. Common CNS disorders including Parkinson's disease (PD), Alzheimer's disease (AD), and stroke are characterized by progressive cellular dysfunction and death. There are still difficulties in the therapeutic development for these diseases because of hitches in early diagnosis, long disease progressions, and drug delivery challenges. Nanotechnology holds considerable potential for the diagnosis, treatment, and monitoring of CNS diseases. Tailored nanomaterials target damaged cells and tissues of the CNS through molecular-scale interactions. Modified nanomaterials concurrently maintain drug release, enhance bioavailability, deliver multiple therapeutic agents, and protect compounds from degradation. In addition, they have potential applications for AD diagnosis & treatment. Nanomaterials applied to target amyloid-β (Aβ) peptide and its aggregated forms aided in modulation of Aβ production, inhibited Aβ aggregation, enhanced Aβ clearance, and even reversed Aβ-induced learning and memory defects in AD model mice. Controlling AD pathogenesis using nanotechnology is another promising approach to treat it. Moreover, nanomaterials can be applied to detect AD biomarkers. In the case of PD, nanotechnology can be applied to detect α-synuclein, dopamine, and mitochondrial dysfunction. In the case of stroke, NPs are used to enhance the permeability of the blood–brain barrier and controlled drug delivery. Hence, nanomaterials can be efficiently utilized not only for the detection of neuronal diseases associated biomarkers but for targeted drug delivery as well [71].

3.4.4 Nanoparticle-Based Molecular Diagnostics for Infectious Diseases

Infectious diseases are mainly caused by pathogenic microorganisms including viruses, fungi, bacteria, and parasites that have a profound impact on humankind because of their unique characteristics, such as the ability to rapidly multiply in number, unpredictability, and evolutionary advantage. Although tremendous efforts have been put into the treatment of infectious diseases to prevent epidemics and mortality, it is still one of the major health care issues that have a profound impact on humankind. Therefore, the development of specific, sensitive, accurate, rapid, low-cost, and easy-to-use diagnostic tools is still in urgent demand [72].

Various NPs, particularly magnetic, metallic, and fluorescent NPs have been effectively employed for the diagnosis of infectious diseases. Fluorescent NPs act as sensitive and photostable fluorescent probes for disease diagnosis in real time through bioimaging. QD nanobeads have been used for the diagnosis of hepatitis B virus (HBV) through highly sensitive dot immunoassays. The HBV surface antigens are immobilized on the porous polyvinylidene difluoride, the antibody-conjugated QDNBs as molecular probes were incubated with the antigens together, illumination of the fluorescent dots is observed under UV lamp and recorded [73]. AuNPs have been successively utilized for the detection of DNA. Change in color and other physical and chemical properties of AuNPs are unique properties making them ideal for nanodiagnostics. Several biomolecules like antibodies, antigens, and enzymes function as electrochemical labels, optical probes, and signal transfer amplifiers upon conjugation with AuNPs to detect various diseases. The gold nanorods have been reported to diagnose HIV virus DNA with high sensitivity and selectivity through Rayleigh scattering spectroscopy [74].

AuNPs have also been used to diagnose HCV. HCV ssRNA tagged with rhodamine 6G conjugated with AuNPs has been used for the detection of HCV ssRNA [75]. The magnetic NPs containing iron oxide core and silver shell have been used for the early diagnosis of malaria. These magnetic NPs are capable of detecting β-hematin crystals—that is, hemozoin biocrystals (a biomarker for malaria infection)—through the magnetic field-enriched surface-enhanced resonance Raman spectroscopy [76]. In another study, AuNPs have been used to diagnose tuberculosis using DNA probes coupled with AuNPs through colorimetric detection of *Mycobacterium tuberculosis* [77].

3.5 Comparison Between Nanomaterials and Other Materials in Molecular Diagnostics

Nanomaterials possess distinct magnetic, optical, thermal, chemical, and electrical properties as compared to other materials of similar composition. Moreover, they interact differently with molecules in the same size range, such as proteins and nucleic acids. Hence these nanomaterials offer improved multiplexed assay sensitivity, specificity, portability, speed, affordability, and ease of use. They are used as specific molecular diagnostic probes and become functional when attached with bio-recognition molecules. Multifunctional nanostructures in combination with molecular diagnostics, imaging, and therapeutics have become vital in early disease detection, *in vivo* diagnosis, and personalized, predictive medicine [13].

3.6 Prospects of Nanodiagnostics

Nanodiagnostics have great potential to reform the field of personalized medicine and needs further exploitation. This technique can lead to the development of various new treatment strategies resulting in therapeutic decision making. The choice of proper strategy for handling a disease is crucial in making patient treatment successful. These high-cost complex systems can prove to be economic when precision in treatment is developed. Unnecessary expenses of false treatments provided to the wrong patients and the additional costs for treatment of their adverse effects would be eliminated by proper disease diagnosis and its treatment. Nanodiagnostics would have a huge impact on precision nanomedicine by playing a role in image-guided surgery, in the early detection of different diseases, and in the preselection of diseased patients and their treatment as a result of the development of clinical drugs [78].

Nanodiagnostics will become more effective in the future by contracting biochip technology to a nanoscale level. An important diagnostic application of this technology would be in the analysis of proteins in blood. The assessment of health would be more rapid as circulating blood components would be detected in the form of molecular fingerprints. Nanodiagnostics would be helpful in reducing the time period for receiving results as they would be available in no time so that treatment could be done immediately. The biomedical application of nanotechnology would be vast in the coming years with significant roles in the prediagnosis and treatment of various diseases as well as in developing personalized medicine [79]. Genetic profiling would be commonly used for the detection of cancer in the next decade to overcome current delays due to the lack of proper diagnostic tools in cancer detection. Some nanodevices such as nanorobotics would have the potential for combined use both in diagnosis and as therapeutics. These devices could have multiple functions including implantation in cancer patients as a protective measure and involvement in cancer investigation by external monitoring. These nanodevices would be helpful in detecting cancer during its early stages, leading to proper and timely therapeutic intervention. Continuous *in vivo* analysis by these nanodevices would provide accurate results related to any level of disease. Further developments in nanodiagnostics tools including firm formulations, better localization, biocompatibility, bio-distribution, and efficiency would result in precision of disease diagnosis and therapy [79, 80].

Nanodiagnostics would not be effective for cancer detection, but it will have a marked role in the detection, control, and cure of many other diseases, including infectious diseases [72]. In the neuroimaging field, nanodiagnostics would be applied for single-neuron detection, and other techniques such as phase-contrast X-ray imaging would be devised. The chances of visualizing previously undetectable and obscure brain areas would be enhanced. Nanodiagnostic tools might overcome flaws in conventional CT and MRI like poor soft tissue contrast and poor spatial resolution [81]. Current healthcare challenges related to disease diagnosis and treatment, including cancer and cardiovascular diseases, would be addressed with nanotechnological development. Bacterial and viral diseases would also be detected and treated using this technique. Further reduction in the size of diagnostic tools and implants is one of the significant aims of this field. Tuberculosis management would be made possible with improved technology, leading to rapid and easy testing at cheap rates. This would be especially convenient for countries with limited resources to combat different diseases [82].

3.7 Regulatory Issues

Breakthroughs in nanotechnology research need a sound translation into health care services. This translation is not simple as it requires the conversion of basic research into clinical trials for its broad application. To make this transition effective, rigorous screening is required. The field of nanodiagnostics is based on nanosized particles whose application pertains to a series of clinical, social, and ethical issues. Generally, no proper regulations have been implemented for incorporating nanodiagnostics in clinics. A lack of standardized protocols for the characterization of NPs and sensing product limits the application of nanodiagnostics. Properties of NPs and their interactions with biological systems are complex and ambiguous. Many attributes including size, shape, alignment, and rate of accumulation need consideration at cellular or molecular levels. These complex characteristics and mechanisms of biodetection tools are not properly understood. Strict regulations regarding safety and reliability should be implemented specifically for nanodiagnostic tools, systems, and their applications. Ambiguity in using standards is also a constraint as this technique requires a specific standard for detection. Rapid generation of more sensitivity in results affects the use of previous standards disturbing relevance and precision efficiency [83].

The use of nanomedicine is problematic due to various constraints related to their biodistribution, physiochemical properties, stability and distribution in cells, effects on the immune system, and genotoxicity. The safety of NPs to be used in diagnostics need knowledge of their interaction with biological systems. The effect of NPs differs by varying chemical and physical characteristics. They may adsorb with proteins, penetrate in the cells, or get dissolved. Such factors are critical in the regulation of NPs for bio-diagnostic purposes [84]. Apart from these regulatory issues, various ethical issues also evolve in nanodiagnostics. In order to generate effective diagnostic techniques for the protection and well-being of patients, risks associated with individuals involved in research and the public are often neglected. Concerns are focused more on patients than on the impact of this development on society, the environment, and future generations. This leads to human rights violation and ignore medical ethical principles of dignity, justice, truth, and trust, etc. Factors involved in this moral blindness include the power of the researcher, the lack of proper ethical training, population dependency, government pressure for conducting research that is supposed to be a national priority, necessity, and the lack of sympathy for research subjects. In addition, safety issues are generated by the toxicity of nanomaterials used which can harm patients, medical staff, and the environment. Translocation of NPs to untargeted body parts and people may have adverse effects [85].

In the case of nanoimaging and diagnostics, privacy issues occur during the exchange of information. The use and storage of such information are major concerns. The immoral use of private information raises ethical concerns. Wireless technology that makes use of such diagnostic information widely available that could be misused. Furthermore, the large volume of data obtained by such devices makes the handling of data and privacy arrangements difficult. It also confuses patients with a lot of irrelevant data [86]. Other moral and ethical concerns are also involved in the development of nanodiagnostics. Equity, or the even distribution of benefits, associated with this technique is an important concern globally. The benefits provided by these diagnostics are available exclusively to rich nations as poor ones cannot afford them. Issues of such inequity also arise within a nation due to differences in financial status. Diagnostics at the molecular and cellular level is done by nanotechnology and that information contradicts with patient's personal experience of illness. Nanomedicine is aimed at the early detection of diseases to combat their future outcomes, but this early diagnosis is problematic because it deals with the detection of molecular changes at an early stage. The treatment of such diseases may be unnecessary as molecular changes at an early stage may not lead to harm later. This results in anxiety, stress, and discomfort in patients, violating their basic rights. Other ethical concerns related to patients' and doctors' responsibilities and basic rights of autonomy are also critical in nanodiagnostics [87].

3.8 Conclusion

The application of NPs in molecular diagnostics is a rapidly developing area of nanotechnology that fosters new prospects in the diagnosis and treatment of various diseases. However, the application of nanotechnology in disease diagnosis must be in compliance with the requirements of public health. In this chapter, we provided a comprehensive overview of recent trends in the nanotechnology-based molecular diagnosis of various diseases. Although NPs have revolutionized medical imaging, diagnostics, and therapeutics, these advancements may also underline their toxicity, which is a point of major concern. A comprehensive evaluation of the biocompatibility and toxicity of NPs is imperative for their safe and sustainable use in clinical diagnostics.

References

1 Thompson, M., Plüddemann, A., Price, C.P., Heneghan, C. (2013). Emerging diagnostic technologies in primary care: what's on the horizon? *British Journal of General Practice* 63 (609): 177–178.

2 Goebel, E.A., Vidal, A., Matias-Guiu, X., and Gilks, C.B. (2017). The evolution of endometrial c.arcinoma classification through application of immunohistochemistry and molecular diagnostics: past, present and future. *Virchows Archiv* 1–12.

3 Beilby, J. (2006). Diagnostic molecular biology. *Clinical Biochemist Reviews.* 27 (1): 3.

4 Debnath, M., Prasad, G.B., and Bisen, P.S. (2010). Introduction to Molecular Diagnostics. *Molecular Diagnostics: Promises and Possibilities*, (ed. Mousumi Debnath Godavarthi B.K.S. Prasad Prakash S. Bisen) 1–10. Springer Science & Business Media.

5 Choe, J.Y., Kim, J.K., and Han, H.S. (2017). Simple Molecular Diagnostic Tools in Clinical Medicine. *Journal of Clinical Diagnosis and Research* 6: 41. https://doi.org/10.4172/2376-0311.1000141.

6 Bishop, R. (2010). Applications of fluorescence in situ hybridization (FISH) in detecting genetic aberrations of medical significance. *Bioscience Horizons* 3 (1): 85–95.

7 DeBiasi, R.L. and Tyler, K.L. (1999). Polymerase chain reaction in the diagnosis and management of central nervous system infections. *Archives of Neurology.* 56 (10): 1215–1219.

8 Valones, M.A.A., Guimarães, R.L., Brandão, L.A.C. et al. (2009). Principles and applications of polymerase chain reaction in medical diagnostic fields: a review. *Brazilian Journal of Microbiology* 40 (1): 1–11.

9 Fosbrink, M.D. (2000). Professional perspectives: clinical diagnosis with gene expression microarrays. *Laboratory Medicine* 31 (12): 658–659.

10 Lee, S., Yoo, N.C., Choi, J.H. et al. (2009). Applications of DNA microarray in disease diagnostics. *Journal of Microbiology and Biotechnology.* 19 (7): 635–646.

11 Ong, B., Chong, B., Phelan, D. et al. (2010). Mass-spectrometry in DNA diagnostics. *Pathology.* 42: S31.

12 Ilyas, M. (2017). Next-generation sequencing in diagnostic pathology. *Pathobiology* 84 (6): 292–305.

13 Tang, Y.W., Procop, G.W., and Persing, D.H. (1997). Molecular diagnostics of infectious iseases. *Clinical Chemistry.* 43 (11): 2021–2038.

14 Azzazy, H.M., Mansour, M.M., and Kazmierczak, S.C. (2006). Nanodiagnostics: a new frontier for clinical laboratory medicine. *Clinical Chemistry.* 52 (7): 1238–1246.

15 Wu, Y. and Yang, P. (2001). Direct observation of vapor–liquid–solid nanowire growth. *Journal of the American Chemical Society.* 123: 3165–3166.

16 Garnett, E., Mai, L., and Yang, P. (2019). Introduction: 1D nanomaterials/nanowires. *Chemical Reviews.* 119: 8955–8957.

17 Su, Y., Wei, X., Peng, F. et al. (2012). Gold nanoparticles-decorated silicon nanowires as highly efficient near-infrared hyperthermia agents for cancer cells destruction. *Nano Letters.* 12: 1845–1850.

18 Khalil, A., Singh Lalia, B., Hashaikeh, R., and Khraisheh, M. (2013). Electrospun metallic nanowires: synthesis, characterization, and applications. *Journal of Applied Physics.* 114. https://doi.org/10.1063/1.4822482.

19 Subbiah, T., Bhat, G.S., Tock, R.W. et al. (2005). Electrospinning of nanofibers. *Journal of Applied Polymer Science.* 96: 557–569.

20 Pang, L.S., Saxby, J.D., and Chatfield, S.P. (1993). Thermogravimetric analysis of carbon nanotubes and nanoparticles. *Journal of Physical Chemistry.* 97: 6941–6942.

21 Fabbro, C., Ali-Boucetta, H., Ros, T.D. et al. (2012). Targeting carbon nanotubes against cancer. *Chemical Communications.* 48: 3911–3926.

22 Danailov, D., Keblinski, P., Nayak, S., and Ajayan, P. (2002). Bending properties of carbon nanotubes encapsulating solid nanowires. *Journal of Nanoscience and Nanotechnology* 2: 503–507.

23 Saito, N., Haniu, H., Usui, Y. et al. (2014). Safe clinical use of carbon nanotubes as innovative biomaterials. *Chemical Reviews.* 114: 6040–6079.

24 Rastogi, V., Yadav, P., Bhattacharya, S.S. et al. (2014). Carbon nanotubes: an emerging drug carrier for targeting cancer cells. *Journal of Drug Delivery.* 2014. http://dx.doi.org/10.1155/2014/670815

25 Xia, Z., Guduru, P., and Curtin, W. (2007). Enhancing mechanical properties of multiwall carbon nanotubes via s p 3 interwall bridging. *Physical Review Letters.* 98: 245501.

26 Madani, S.Y., Naderi, N., Dissanayake, O. et al. (2011). A new era of cancer treatment: carbon nanotubes as drug delivery tools. *International Journal of Nanomedicine.* 6: 2963.

27 Pandey, P. and Dahiya, M. (2016). Carbon nanotubes: types, methods of preparation and applications. *International Journal of Pharmaceutical Sciences and Research* 1: 15–21.

28 Lee, S.J., Baik, H.K., J-e, Y., and Han, J.H. (2002). Large scale synthesis of carbon nanotubes by plasma rotating arc discharge technique. *Diamond and Related Materials.* 11: 914–917.

29 Sarangdevot, K. and Sonigara, B. (2015). The wondrous world of carbon nanotubes: structure, synthesis, properties and applications. *Journal of Chemical and Pharmaceutical Research.* 7: 916–933.

30 Golnabi, H. (2012). Carbon nanotube research developments in terms of published papers nd patents, synthesis and production. *Scientia Iranica.* 19. 2012–2022.

31 Varshney, K. (2014). Carbon nanotubes: a review on synthesis, properties and applications. *International Journal of Engineering Research and General Science.* 2: 660–677.

32 Beg, S., Rizwan, M., Sheikh, A.M. et al. (2011). Advancement in carbon nanotubes: basics, biomedical applications and toxicity. *Journal of Pharmacy and Pharmacology.* 63: 141–163.

33 Elhissi, A., Ahmed, W., Hassan, I.U. et al. (2012). Carbon nanotubes in cancer therapy and drug delivery. *Journal of Drug Delivery* 2012. https://doi:10.1155/2012/837327.

34 Taghdisi, S.M., Lavaee, P., Ramezani, M., and Abnous, K. (2011). Reversible targeting and controlled release delivery of daunorubicin to cancer cells by aptamer-wrapped carbon nanotubes. *European Journal of Pharmaceutics and Biopharmaceutics* 77: 200–206.

35 Lee, J.J., Yazan, L.S., and Abdullah, C.A.C. (2017). A review on current nanomaterials and their drug conjugate for targeted breast cancer treatment. *International Journal of Nanomedicine* 12: 2373.

36 Bera, D., Qian, L., Tseng, T.-K., and Holloway, P.H. (2010). Quantum dots and their multimodal applications: a review. *Materials* 3: 2260–2345.

37 Xu, X., Ray, R., Gu, Y. et al. (2004). Electrophoretic analysis and purification of fluorescent single-walled carbon nanotube fragments. *Journal of the American Chemical Society* 126: 12736–12737.

38 Baker, S.N. and Baker, G.A. (2010). Luminescent carbon nanodots: emergent nanolights. *Angewandte Chemie International Edition* 49: 6726–6744.

39 Shen, J., Zhu, Y., Yang, X., and Li, C. (2012). Graphene quantum dots: emergent nanolights for bioimaging, sensors, catalysis and photovoltaic devices. *Chemical Communications* 48: 3686–3699.

40 Wang, Y. and Hu, A. (2014). Carbon quantum dots: synthesis, properties and applications. *Journal of Materials Chemistry C.* 2: 6921–6939.

41 Peng, H. and Travas-Sejdic, J. (2009). Simple aqueous solution route to luminescent carbogenic dots from carbohydrates. *Chemistry of Materials* 21: 5563–5565.

42 Ming, H., Ma, Z., Liu, Y. et al. (2012). Large scale electrochemical synthesis of high quality carbon nanodots and their photocatalytic property. *Dalton Transactions* 41: 9526–9531.

43 Shinde, D.B. and Pillai, V.K. (2012). Electrochemical preparation of luminescent graphene quantum dots from multiwalled carbon nanotubes. *Chemistry–A European Journal* 18: 12522–12528.

44 Li, X., Wang, H., Shimizu, Y. et al. (2010). Preparation of carbon quantum dots with tunable photoluminescence by rapid laser passivation in ordinary organic solvents. *Chemical Communications* 47: 932–934.

45 Cheng, Z., Al Zaki, A., Hui, J.Z. et al. (2012). Multifunctional nanoparticles: cost versus benefit of adding targeting and imaging capabilities. *Science* 338: 903–910.

46 Chae, A., Choi, Y., Jo, S. et al. (2017). Microwave-assisted synthesis of fluorescent carbon quantum dots from an A 2/B 3 monomer set. *RSC Advances* 7: 12663–12669.

47 Wanunu, M. (2012). Nanopores: a journey towards DNA sequencing. *Physics of Life Reviews* 9: 125–158.

48 Wade, T. and Wegrowe, J.E. (2005). Template synthesis of nanomaterials. *European Physical Journal of Applied Physics* 29: 3–22.

49 Baker, L.A., Choi, Y., and Martin, C.R. (2006). Nanopore membranes for biomaterials synthesis, biosensing and bioseparations. *Current Nanoscience* 2: 243–255.

50 Sexton, L.T., Horne, L.P., and Martin, C.R. (2007). Developing synthetic conical nanopores for biosensing applications. *Molecular BioSystems* 3: 667–685.

51 Oukhaled, A., Bacri, L., Pastoriza-Gallego, M. et al. (2012). Sensing proteins through nanopores: fundamentals to applications. *ACS Chemical Biology* 7: 1935–1949.

52 Shende, P., Sardesai, M., and Gaud, R. (2018). Micro to nanoneedles: a trend of modernized transepidermal drug delivery system. *Artificial Cells, Nanomedicine, and Biotechnology* 46: 19–25.

53 Zhuiykov, S. (2014). Structural and chemical modification of semiconductor nanocrystals. In: *Nanostructured Semiconductor Oxides for the Next Generation of Electronics and Functional Devices* (ed. Serge Zhuiykov), 50–94. Woodhead Publishing.

54 Krishnamurthy, S. and Vijayasarathy, S. (2016). Role of nanomaterials in clinical dentistry. In: *Nanobiomaterials in Dentistry*, (ed. Alexandru Mihai Grumezescu) 211–240. Elsevier.

55 Ramsden, J.J. The nano/bio interface. In: *Nanotechnology*, 2e (ed. J.J. Ramsden), 61–90. Oxford: William Andrew Publishing.

56 Gutierrez, B., Villalobos Bermúdez, C., Corrales Ureña, Y.R. et al. (2017). Nanobots: development and future. *International Journal of Biosensors & Bioelectronics* 2 (5): 1–6

57 Ahmad, U. and Md, F. (2016). Smart nanobots: the future in nanomedicine and biotherapeutics. *Journal of Nanomedicine and Biotherapeutic Discovery* 6: e140.

58 Stachiv, I. and Sittner, P. (2018). Nanocantilevers with adjustable static deflection and significantly tunable spectrum resonant frequencies for applications in nanomechanical mass sensors. *Nanomaterials* 8: 116.

59 Jain, A. and Alam, M.A. (2013). Emerging ideas in nanocantilever based biological sensors. arXiv preprint. arXiv:13055729.

60 Hwang, K.S., Lee, S.-M., Kim, S.K. et al. (2009). Micro- and nanocantilever devices and systems for biomolecule detection. *Annual Review of Analytical Chemistry* 2: 77–98.

61 Mohanty, C., Arya, G., Verma, R.S., and Sahoo, S.K. (2009). Nanobiotechnology: application of nanotechnology in therapeutics and diagnosis. *International Journal of Green Nanotechnology: Biomedicine* 1 (1): B24–B38.

62 Deb, S., Ghosh, K., and Shetty, S.D. (2015). Nanoimaging in cardiovascular diseases: current state of the art. *Indian Journal of Medical Research* 141 (3): 285.

63 Karagkiozaki, V., Pappa, F., Arvaniti, D. et al. (2016). The melding of nanomedicine in thrombosis imaging and treatment: a review. *Future Science OA* 2 (2).

64 Yi, D.K., Nanda, S.S., Kim, K., and Selvan, S.T. (2017). Recent progress in nanotechnology for stem cell differentiation, labeling, tracking and therapy. *Journal of Materials Chemistry B* 5 (48): 9429–9451.

65 Ambesh, P., Campia, U., Obiagwu, C. et al. (2017). Nanomedicine in coronary artery disease. *Indian Heart Journal* 69 (2): 244–251.

66 Zhang, Y., Li, M., Gao, X. et al. (2019). Nanotechnology in cancer diagnosis: progress, challenges and opportunities. *Journal of Hematology & Oncology* 12 (1): 137.

67 Li, H., Cao, Z., Zhang, Y. et al. (2011). Simultaneous detection of two lung cancer biomarkers using dual-color fluorescence quantum dots. *Analyst* 136 (7): 1399–1405.

68 Gu, B., Xu, C., Yang, C. et al. (2011). ZnO quantum dot labeled immunosensor for carbohydrate antigen 19-9. *Biosensors and Bioelectronics* 26 (5): 2720–2723.

69 Borghei, Y.S., Hosseini, M., and Ganjali, M.R. (2017). Detection of large deletion in human BRCA1 gene in human breast carcinoma MCF-7 cells by using DNA-silver nanoclusters. *Methods and Applications in Fluorescence* 6 (1): 015001.

70 Jou, A.F.J., Lu, C.H., Ou, Y.C. et al. (2015). Diagnosing the miR-141 prostate cancer biomarker using nucleic acid-functionalized CdSe/ZnS QDs and telomerase. *Chemical Science* 6 (1): 659–665.

71 Zhang, W., Wang, W., Yu, D.X. et al. (2018). Application of nanodiagnostics and nanotherapy to CNS diseases. *Nanomedicine* 13 (18): 2341–2371.

72 Wang, Y., Yu, L., Kong, X., and Sun, L. (2017). Application of nanodiagnostics in point-of-care tests for infectious diseases. *International Journal of Nanomedicine* 12: 4789.

73 Zhang, P., Lu, H., Chen, J. et al. (2014). Simple and sensitive detection of HBsAg by using a quantum dots nanobeads based dot-blot immunoassay. *Theranostics.* 4 (3): 307.

74 Darbha, G.K., Rai, U.S., Singh, A.K., and Ray, P.C. (2008). Gold-nanorod-based sensing of sequence specific HIV-1 virus DNA by using hyper-Rayleigh scattering spectroscopy. *Chemistry–A European Journal.* 14 (13): 3896–3903.

75 Griffin, J., Singh, A.K., Senapati, D. et al. (2009). Sequence-specific HCV RNA quantification using the size-dependent nonlinear optical properties of gold nanoparticles. *Small* 5 (7): 839–845.

76 Yuen, C. and Liu, Q. (2012). Magnetic field enriched surface enhanced resonance Raman spectroscopy for early malaria diagnosis. *Journal of Biomedical Optics.* 17 (1): 1–8. 017005.

77 Baptista, P., Pereira, E., Eaton, P. et al. (2008). Gold nanoparticles for the development of clinical diagnosis methods. *Analytical and Bioanalytical Chemistry.* 391 (3): 943–950.

78 Blau, R., Krivitsky, A., Epshtein, Y., and Satchi-Fainaro, R. (2016). Are nanotheranostics and nanodiagnostics-guided drug delivery stepping stones towards precision medicine? *Drug Resistance Updates.* 27: 39–58.

79 Moffatt, S. (2016). Nanodiagnostics: a revolution in biomedical nanotechnology. *MOJ Proteomics & Bioinformatics* 3 (2): 00080.

80 Ali, E.S., Sharker, S.M., Islam, M.T. et al. (2020). Targeting cancer cells with nanotherapeutics and nanodiagnostics: current status and future perspectives. *Seminars in Cancer Biology.* 69: 52–68.

81 Teleanu, D.M., Chircov, C., Grumezescu, A.M. et al. (2019). Contrast agents delivery: an up-to-date review of nanodiagnostics in neuroimaging. *Nanomaterials.* 9 (4): 542.

82 Gupta, A.K., Singh, A., and Singh, S. (2020). Diagnosis of tuberculosis: nanodiagnostics approaches. In: *NanoBioMedicine*, 261–283. Singapore: Springer.

83 Baptista, P.V. (2014). Nanodiagnostics: leaving the research lab to enter the clinics? *Diagnosis.* 1 (4): 305–309.

84 Kelly, B. (2010). Nanomedicines: regulatory challenges and risks ahead. *Regulatory Affairs Pharma* 10: 14–17.

85 Allon, I., Ben-Yehudah, A., Dekel, R. et al. (2017). Ethical issues in nanomedicine: tempest in a teapot? *Medicine, Health Care and Philosophy.* 20 (1): 3–11.

86 Marchant, G.E. (2009). Small is beautiful: what can nanotechnology do for personalized medicine? *Current Pharmacogenomics and Personalized Medicine (Formerly Current Pharmacogenomics).* 7 (4): 231–237.

87 Fisher, E., Boenink, M., Van Der Burg, S., and Woodbury, N. (2012). Responsible healthcare innovation: anticipatory governance of nanodiagnostics for theranostics medicine. *Expert Review of Molecular Diagnostics.* 12 (8): 857–870.

88 Wang, J., Liu, G., and Lin, Y. (2007). Nanotubes, nanowires, and nanocantilevers in biosensor development. In: *Nanomaterials in Biosensors* (ed. C.S.S.R. Kumar), 56–100. Weinheim.

4

Nanopharmaceuticals

David Quintanar-Guerrero[1], *Gerardo Leyva-Gómez*[2], *Nancy Evelyn Magaña Vergara*[3], *and Néstor Mendoza Muñoz*[3]

[1] *Facultad de Estudios Superiores Cuautitlán, Universidad Nacional Autónoma de México, Mexico, Mexico*
[2] *Facultad de Química, Universidad Nacional Autónoma de México, Ciudad de México, Mexico*
[3] *Facultad de Ciencias Químicas, Universidad de Colima, Colima, Mexico*

4.1 Introduction

The first conceptualization of nanopharmaceuticals could be fairly attributed to Paul Ehrlich when, at the beginning of the twentieth century, he introduced the concept of "magic bullets," a term used to refer to "ideal therapeutic agents" that act specifically against a particular pathogen without causing damage to host cells. This idea has been put into practice through the development of nanotechnology, a novel science that involves the ability to observe and control individual atoms and molecules at dimensions between approximately 1 and 100 nm.

The application of nanotechnology to medicine has given rise to the development of nanomedicine, an emerging field of science that seeks to improve diagnostic procedures and the prevention and treatment of diseases through the design and implementation of systems and devices on the nanometer scale. One of the specific applications of nanotechnology has been in the creation of advanced drug-delivery systems, which have given rise to the term "nanopharmaceuticals." In order to be classified as a "nanopharmaceutical," Weissing and colleagues [1] suggest that the drug product has to meet two major criteria: (i) nanoengineering has to be employed during the manufacturing process, and (ii) the nanomaterial used has to be either essential for the therapeutic activity or has to confer additional and unique properties to the active drug entity. These systems are mainly oriented toward increasing the bioavailability of drugs, reducing or protecting against degradation, modifying the release rate and the target, and targeting the drug to the required site.

There are two approaches used in the construction of nanopharmaceuticals: the top-down approach consists of the step-wise size reduction of large pieces by mechanical or physicochemical methods to create smaller structures, and the bottom-up approach involves the construction of nanometric structures through the self-assembly of atoms or molecules induced by physical or chemical forces. Several nanoscale drug-delivery systems have been developed and constructed from different types of materials

Nanomaterials and Nanotechnology in Medicine, First Edition. Edited by P.M. Visakh.

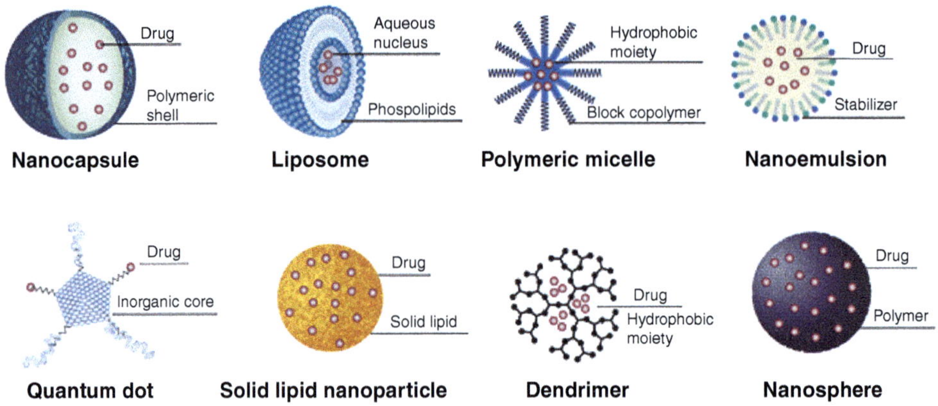

Figure 4.1 Nanopharmaceuticals in drug delivery.

(Figure 4.1). Remarkable structures include inorganic particles such as the following: quantum dots (QDs), magnetic nanoparticles (NPs), and silica particles; polymeric NPs such as nanospheres, nanocapsules, and polymeric mixed micelles; lipid-based carriers—for example, liposomes, solid lipid nanoparticles (SLNPs), nanostructured lipid carriers, self-microemulsifying systems, and microemulsions; and others such as dendrimers, a class of branched macromolecules or carbon nanotubes.

Each of these previously mentioned systems has its own advantages and disadvantages with respect to others; however, each of the templates represents a completely viable option as a drug carrier and has shown to improve the efficacy of drugs compared to those formulated in conventional dosage forms. From 1995 to 2017, there have been 50 nanopharmaceuticals approved by the U.S. Federal Drug Administration (FDA) that are available for clinical use [2], indicating that nanopharmaceuticals are now a reality. At the beginning, liposomal technology was the pioneer in the field; however, currently, other platforms have been developed and approved for human use: in addition to liposomes, polymeric nanoparticles and nanocrystals are among the most frequently approved nanopharmaceuticals. During the last decade, the diversification of nanoplatforms has allowed other types of nanopharmaceuticals that are already in clinical trials or approved; we refer here to protein-based, micelle-based, and inorganic NPs including pure metals (Au, Ag, etc.) and metal oxides.

The majority of nanopharmaceuticals have been employed directly for therapeutic use. Most of the drugs included among nanopharmaceuticals are pre-approved drugs; the discovery of novel drugs is not mandatory. Nanoencapsulation solely improves the pharmacokinetic and/or pharmacodynamic characteristics of the drug, increasing the efficacy of the treatments. Based on the approved list of nanopharmaceuticals, the number and type of ailments or clinical indications are very broad, and chemotherapeutic agents appear to be dominant among them; however, applications for immune diseases have been acquiring relevance in recent years. One of the most important technological advances in nanopharmaceuticals represents the superficial modification of NPs with a hydrophilic polymer to increase half-life in blood circulation in the body. The best-known polymer is Poly(Ethylene Glycol) (PEG) and, without a doubt, PEGylation has allowed nanopharmaceuticals to become a viable therapeutic option in clinical practice. However, nanotechnology also has offered the opportunity to integrate diagnostic capabilities with pharmaceuticals in a single agent. This class of nanosystems are denominated theranostic agents, due to the ability to provide diagnostic and therapeutic skills. For example, QDs comprise a class of luminescent nanocrystals results that become a multifunctional platform with imaging capabilities; due to their intensive fluorescence, they also possess drug-delivery abilities, even for the intracellular controlled release of drugs [3].

Nanopharmaceuticals face a series of challenges, including ensuring low toxicity in patients as well as in the environment, designing manufacturing methods that represent a good cost-benefit ratio, and, finally, generating a regulatory framework that provides the rapid arrival on the market of products proven to be safe and effective. Regulatory agencies (e.g., the FDA and the European Medicines Agency [EMA]) have made significant efforts to advance from general approval standards (common for conventional pharmaceutical products, as in the case with the first generation of products) to more specific standards, as in the case of novel nanopharmaceutical products. Some special considerations that must be considered include the following: changes in toxicity, the interaction of cells or organs, a profile of bioavailability, critical physical and chemical properties, residual solvents, and impurities and excipients. The latter are some of the modifications that nanotechnology introduces into pre-existing drugs during the reformulation of nanopharmaceuticals or during the development of new products.

In summary, nanopharmaceuticals must prove that they can be a revolutionary technological advance to improve the performance of already available or novel therapeutic agents. Novel applications are now in the pipeline and will soon appear on the market; gene therapy, stimuli-response nanopharmaceuticals, and theranostic agents are examples of the next generation of nanodrugs. This chapter aims to offer an overview of some of the more promising nanopharmaceuticals, and some of them are now in the market, including liposomes, polymeric NPs, SLNPs, dendrimers, and QDs. The synthesis, application, and future of nanopharmaceuticals are also covered. Finally, regulatory issues are discussed, including the concern of establishing a harmonized definition of terms such as nanomaterials and nanopharmaceuticals. This section also presents the characteristics of the reflections issued by the EMA with respect to liposomal systems, iron-based formulations, and nanocrystal-based formulations in an effort to establish guidelines for the demonstration of the nanosimilarity of products whose patents have expired.

4.2 Liposomes in Nanopharmaceuticals

Liposomes are a submicron and vesicular system for the transport and release of drugs; they are composed of a bilayer or multiple layers of natural or synthetic lipids with a central aqueous compartment. The development of formulations with liposomes dates from a considerable time back: 1965 [4]. This period has made it possible for them to be one of the drug- delivery systems possessing a broad impact in the clinic.

One of the best examples is the development of Doxil® [5], the first FDA-approved nanodrug (1995). Research on liposomes even established the route for the research and industrial development of other drug-delivery systems. Moreover, it also opened the way for regulatory governmental systems, health systems, and finally, a new frontier in knowledge for medical personnel with new horizons and hopes for emerging diseases.

For liposomes, as with other modified drug release systems, there are some specific requirements that the FDA requires for possible registration. These comprise morphology, surface loading, final system viscosity, drug loading, particle size, release profiles, and stability studies. The FDA places special emphasis on the quality control of the parameters involved in the production process, even those that may appear to be "less significant," such as a change in the volume of the production batch and, in addition, to the stability phenomena, particularly the effect of temperature on the structure of the liposome. Moreover, biological interactions are fundamental to explaining the mode of action of the proposed new formulation [6]. Some representative examples of commercial products with liposomes include the following: Doxil®; liposomes with Doxorubicin, approved in 1995 for the treatment of ovarian cancer, breast cancer, Kaposi's sarcoma from Sequus Pharmaceuticals; Abelcet®; liposomes with Amphotericin B, approved in 1995 for the treatment of invasive severe fungal infections from Sigma-Tau Pharmaceuticals; DepoDur™, liposomes with morphine sulfate, approved in 2004 for the pain management, a treatment of SkyPharma, Inc. Mepact®; liposomes with Mifamurtide, approved in 2004 for the treatment of high-grade resectable non-metastatic osteosarcoma from Takeda Pharmaceutical, Ltd.; and Onivyde™, liposomes with Irinotecan, approved in 2015 for the treatment of metastatic adenocarcinoma of the pancreas from Merrimack Pharmaceuticals, Inc. [1].

One of the strategies that has been used in marketed liposome systems is PEGylation. With this modification, it has been possible to extend time in circulation and increase contact time at the target site. Moreover, this type of transporter has offered a greater recording of information on the concept of PEGylation, which was later transferred to other submicron systems for the transport and modified release of drugs. However, this has not been a condition for all liposome systems. There is even a tendency in some drug-delivery systems not to include the modification of the surface by PEGylation; with these characteristics, the systems are known as non-PEGylated liposome technology [6, 7]. The great trajectory in liposome research allows us to have an enriching panorama concerning its impact in clinical phases (Table 4.1) [8]. The number of clinical trials involving the administration of liposomes for any health condition is 2252 to date. The category of completed studies with the largest number of studies, according to the registries, is cancers and other neoplasms, followed by symptoms and general pathology, and blood and lymph conditions.

Clinical trials include the following recruitment status: not yet recruiting, recruiting, enrollment by invitation, active (not recruiting), suspended, terminated, completed, withdrawn, and unknown status. The total number of studies with status not yet recruiting is 74, recruiting 372, enrollment by invitation 1, active not recruiting 202, suspended 16, terminated 225, and completed 1047. Each condition has a different clinical trial progress.

Table 4.1 Number of studies completed by condition involving the administration of liposomes.

Conditions by category	Number of studies
Cancers and other neoplasms	286
Symptoms and general pathology	102
Blood and lymph conditions	95
Immune system diseases	68
Urinary tract, sexual organs, and pregnancy conditions	65
Digestive system diseases	60
Nervous system diseases	56
Skin and connective tissue diseases	53
Bacterial and fungal diseases	48
Respiratory tract (lung and bronchial) diseases	41
Wounds and injuries	37
Heart and blood diseases	5
Muscle, bone, and cartilage diseases	33
Viral diseases	30
Eye diseases	29
Nutritional and metabolic diseases	28
Diseases and abnormalities at or before birth	26
Gland and hormone-related diseases	24
Mouth and tooth diseases	18
Ear, nose, and throat diseases	14
Behaviors and mental disorders	12
Parasitic diseases	9
Substance-related disorders	8
Occupational diseases	1

Clinical trials are usually classified into four phases. Phase 1 is characterized by focusing on the safety of a drug. It is performed in healthy persons and the objective is to determine the most frequent adverse effects, metabolism, and elimination pathways. Phase 2 is characterized by evaluating the effectiveness of the drug under real conditions of the disease, in addition to the safety of the substance and the adverse effects in the short term. Phase 3 is characterized by continuing the study of effectiveness and safety in different populations, and at different doses, administering the drug in combination with other drugs. The group of participants is larger. Phase 4 is characterized by commercialization.

The category of studies completed for liposomes in phase 1 has a total number of 246, of which 51% were conducted by industries and 19.5% by other organizations. In phase 2, there is a registry of 488 completed studies, of which 38.5% were conducted by industries and 29.7% by other organizations. In phase 3, there is a registry of 231 completed studies, of which 35% were conducted by industries and 32% by other organizations, while in phase 4, there is a record of 60 completed studies, of which 35% were conducted by industries and 61% by other organizations. Currently, there is a total of 386 studies that are recruiting participants, of which 88 are phase 1, 190 are phase 2, 68 are phase 3, and 40 are phase 4. The majority of the studies are concentrated in North America, with a total number of 1578, and in the United States, where more studies are carried out (Table 4.2) [8].

The drug that predominates in registered clinical studies is Doxorubicin, with a total of 1097 studies, of which 535 are listed as completed. Among these, 144 correspond to phase 1, 276 to phase 2, 105 to phase 3, and 10 to phase 4. Phase 4 studies are directed at breast neoplasms, ovarian neoplasms, Kaposi's sarcoma, non-Hodgkin's lymphoma, breast cancer, and hepatocellular carcinoma. Cancer is a kind of disease that is hardest to cure and requires the use of innovative tools such as those based on nanotechnology. In this regard, studies involving liposomes for the treatment of some type of cancer account for 1390 studies, of which 661 have been completed, 179 are in phase 1, 341 are in phase 2, 140 are in phase 3, and 4 are in phase 4. The conception of liposome-based drugs offers several advantages, including well-established methods of large-scale production, various modifications at the surface level to allow adequate degrees of vectorization, and, mainly, documentary record routes before governmental entities in several countries. Applications are limited not only to known drugs, but also to the new needs of gene therapy.

Table 4.2 Number of studies completed by region involving the administration of liposomes.

World region name	Number of studies
Africa	57
Central America	77
East Asia	219
Japan	38
Europe	634
Middle East	88
North America	1578
Canada	219
Mexico	26
United States	1333
North Asia	62
Pacifica	129
South America	77
South Asia	33
Southeast Asia	45

4.3 Polymeric Nanoparticles in Nanopharmaceuticals

Nanopharmaceuticals comprise a new generation of therapeutic-containing nanomaterials with unique properties and high potential that are capable of solving the ongoing challenge of generating novel therapeutic treatments to improve patients' quality of life [9]. Several of the approximately 100 FDA-approved nanomedicine applications and products involve the use of polymeric nanosystems, and many more are in development or in clinical phases. Polymer nanomedicines are the simplest form of nanomedicines and contain soft materials to increase solubility, biocompatibility, half-life, and bioavailability, as well as to control the release of active pharmaceutical gradients from nanomedicines in the body [10]. The greatest number of polymer-based nanoproducts in the pharmaceutical market is linked to PEGylated drug conjugates, particularly linear or slightly branched molecules. PEGylation is a lyophilic process in which a biomolecule (e.g., an antibody, enzyme, cytokine, or nucleotide) is conjugated by a linker to one or more chains of PEG [11]. PEG is a polymer with the formula HO-(CH 2- CH 2- O) n-H. PEG polymers possess many properties, such as biocompatibility, low toxicity, and high solubility in aqueous and organic media, for which they are widely employed in many biomedical fields [12].

The commercial introduction of PEGylated nanomedicaments is directly impacted by the development of new biotechnological molecules that are more effective and stable than their non-PEGylated counterparts. PEGylation has been adopted to improve the pharmacokinetic properties of several biomolecular therapeutics. Recent studies on highly branched PEG suggest superior properties in several areas including cellular uptake, gene regulation efficacy, the reduction of side-effects, and biodistribution [13]. Table 4.3 summarizes some polymer-based nanopharmaceuticals that are currently available on the world pharmaceutical market that are based on PEGylation or other polymer functions (e.g., micelles and nanosuspensors).

Although PEGylation possesses several advantages, particularly biopharmaceuticals, for formulating drugs using polymer molecules, it is not strictly an encapsulation technique and it does not completely satisfy the goal of serving as vehicles to protect drug molecules and to deliver them, in a controlled manner, into targeted diseased sites. In this regard, polymeric NPs are, to a greater degree, more functional drug-delivery systems. In 2008, CALAA-01 was the first NP to be approved for cancer clinical trials to treat patients with solid tumors that were refractory to standard-of-care therapies [9]. A significant number of pre-clinical and clinical trials can be expected to result in the development of new polymeric NPs intended for non-parenteral routes of delivery, such as pulmonary, ocular, nasal, vaginal, and dermal [14]. Polymeric NPs have been proposed in practically all medical applications as follows: (i) protection of active substances from light, oxygen, stomach acid condition, premature degradation, and other degradation pathways; (ii) diagnostic imaging of diseases (including theranostics); (iii) tissue engineering; (iv) drug delivery (controlled drug delivery within a predictable and limited time at therapeutically effective concentrations); (v) for site-specific systems (targeted to a particular tissue or cell type or targeted to a specific diseased tissue but not to healthy tissue), which are usually far from the absorption site; and (vi) in smart (stimulus-responsive) drug-delivery systems [9, 15, 16].

Table 4.3 Examples of polymer-based nanopharmaceuticals available in the world pharmaceutical market.

Medicament name	Pharmaceutical laboratory	Drug	FDA approval (year)	Disease	Polymer function
Cimzia®	UCB (Belgium)	Fab fragment (IgG without Fc region)	2008	Rheumatoid arthritis, Crohn's disease, psoriatic arthritis and ankylosing spondylitis	PEGylation
Adagen®	Enzon (USA)	Adenosine deaminase (ADA)	1990	Immunodeficiency disorders	PEGylation
Neulasta	Amgen (USA)	Filgrastim	2002	Febrile neutropenia	PEGylation
Oncaspar®	Enzon Pharmaceuticals (USA)	Pegaspargase	1994	Acute lymphoblastic leukemia, chronic myelogenous leukemia	PEGylation
Pegasys®	Pfizer Pharmaceuticals (USA)	Recombinant human alfa-2a interferon	2002	Hepatitis C, HBeAg positive chronic hepatitis	PEGylation
Somavert®	Pfizer Pharmaceuticals (USA)	Analog of human growth hormone	2003	Acromegaly	PEGylation
Macugen®	EyeTech Pharmaceuticals	Pegatinib sodium	2004	Macular degeneration	PEGylation
Mircera®	Roche (Swiss)	Eritropoyetine beta	2007	Anemia	PEGylation
PEG-Intron®	Schering-Plough (USA)	Interferon alfa-2b	2001	Hepatitis C	PEGylation

Krystexxa®	Savient Pharmaceuticals (USA)	PEGloticase	2010	Refractory chronic gout	PEGylation
Plegridy®	Biogen (USA)	Interferon-β-1a	2014	Relapsing remitting multiple sclerosis (RRMS)	PEGylation
Adynovate®	Shire (USA)	Anti-hemophilic factor	2015	Hemophilia A	PEGylation
Copaxone®	Teva Pharmaceuticals	Glatiramer acetate	1996	Multiple sclerosis	Copolymerization
Eligard®	Atrix Laboratories (USA)	Leuprolide acetate	2002	Prostate cancer	Nanosuspensor agent
Renagel®	Genzyme (USA)	Sevelamer	2000	End-stage renal disease, chronic kidney disease	Polymeric network structuring
Estrasorb®	Novavax (USA)	17β-estradiol	2003	Menopause	Micellar encapsulant
Zilretta®	Flexion Therapeutics (USA)	Triamcinolone acetonide	2017	Knee osteoarthritis	Nanosuspensor agent (poly (lactic-co-glycolic acid)-hydrogel)

Sources: Choi YH, Han H-K [10]; Farjadian F, et al. [11].

One of the main advantages of using polymeric NPs with respect to other nanosystems is the possibility of utilizing biocompatible and biodegradable polymers such as poly(alpha-hydroxy acids), poly(anhydrides), poly(ortho esters), and poly(amino acids), avoiding the requirement to remove the residual polymer within the patient's body. Polymeric NPs show the following advantages in relation to other colloidal systems (e.g., liposomes): (i) better stability in biological fluids and during storage; (ii) easy preparation and diversity in preparation techniques; (iii) reproducibility and easy large-scale manufacturing; and (iv) batch-to-batch reproducibility. Furthermore, nanoformulation with polymeric NPs is one of the better strategies to deliver drugs (in a controlled way) more precisely to the target tissue and to reduce the overall dose and the potential toxic adverse effects of the drug. These aspects can have important therapeutic implications in the treatment of several diseases, such as cancer, chronic inflammations, infections, neurodegenerative diseases, and atherosclerotic cardiovascular disease with maximal clinical benefit. For example, it has been reported that polymeric NPs have the potential to accumulate in specific tissues, such as tumors, through an enhanced permeation and retention (EPR) effect, which is largely attributed to leaky vasculature and the lack of efficient lymphatic drainage [17, 18]. It is noteworthy that the term NP has been employed in a very general manner and often covers materials or structures (e.g., QDs, nanocrystals, quantum rods, liposomes, dendrimers, etc.) that are better described using other names. According to the FDA, the dimension of NP is 1–100 nm [19].

For polymeric NPs, this size is controversial because these structures are generally reported as ranging from 10 to 1000 nm. They consist of macromolecular materials (preferable synthetic and natural polymers) in which the active principal (drug or biologically active substance) is dissolved, entrapped, encapsulated, adsorbed, or attached to the surface. Two types of polymeric NPs can be described in terms of morphology and architecture: nanospheres and nanocapsules. Nanospheres are made up of a dense polymeric matrix, while nanocapsules are composed of an oil core cover by means of a unique polymeric membrane, forming a reservoir or membrane system that exhibits a typical core-shell structure in which the drug is confined to a reservoir (oily or aqueous nucleus) or within a cavity surrounded by a thin polymeric wall. Despite the fact that polymeric NPs could be prepared by polymerization reactions from monomers (e.g., interfacial polymerization), these techniques are usually avoided due to the presence of residual toxic compounds, including remaining monomers, oligomers, and catalysts that can be toxic and the unlikelihood of cross-reactions with the drug. Therefore, methods that utilize synthetic pre-formed polymers, particularly biodegradables, are preferred [20]; thus, they are free of immunogenicity, and their physical-chemical properties are predictable and reproducible.

In contrast, due to variations in purity and batch-to-batch consistency, natural polymers result in poor reproducibility and controlled release behavior for the entrapped drug(s) [14]. Several techniques for polymeric NP preparation have been described in patents and research papers, beginning in the 1980s. Five techniques for the preparation of polymeric NPs from pre-formed polymers have been described to date: (i) emulsification-evaporation; (ii) solvent displacement or nanoprecipitation; (iii) salting-out; (iv) emulsification-solvent diffusion; and (v) double emulsion-solvent evaporation [20]. Figure 4.2 presents a schematic description of these polymeric NP-preparation techniques.

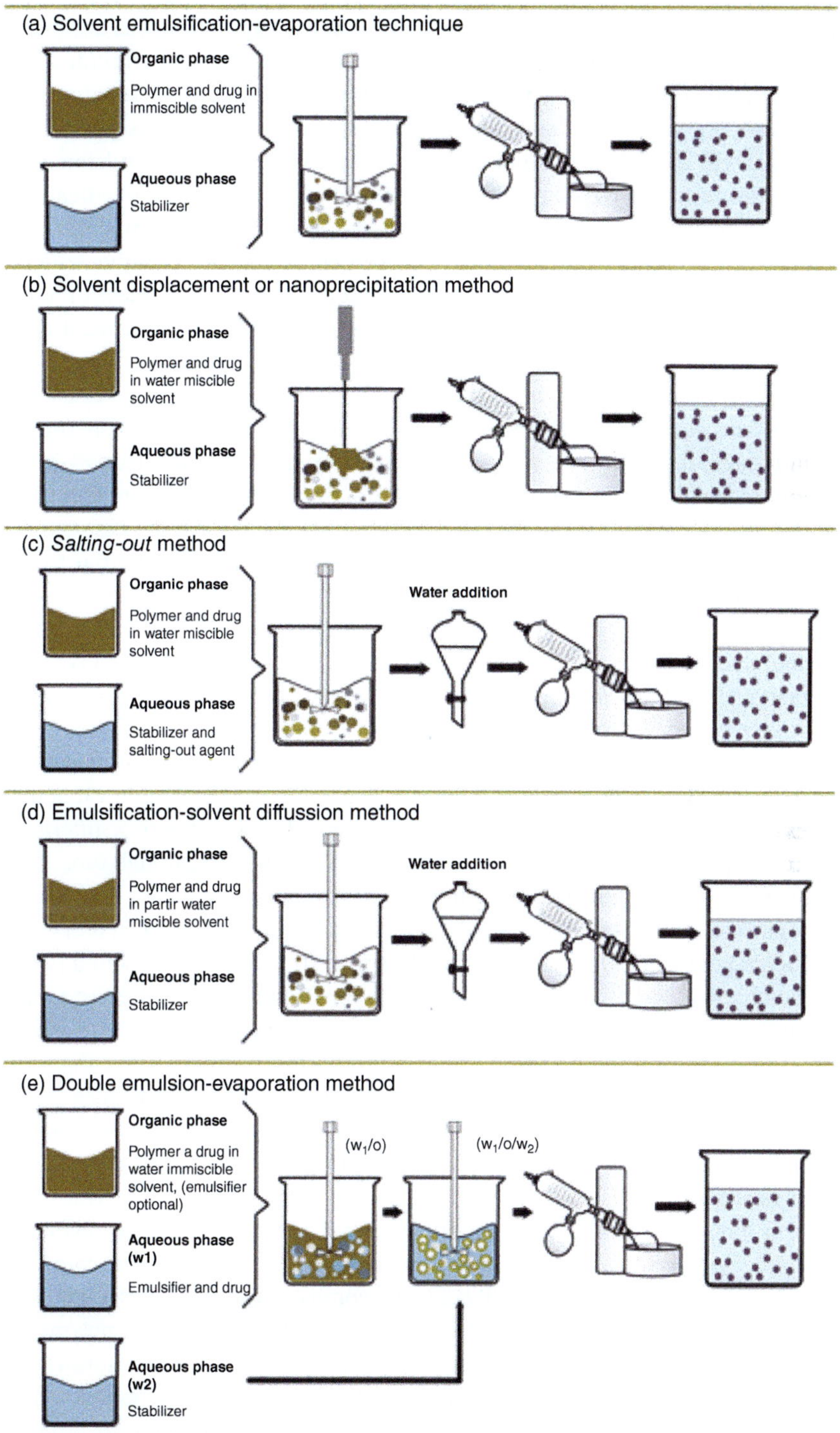

Figure 4.2 Methods for the preparation of polymeric nanoparticles.

These techniques include lyophobic bottom-up processes to encapsulate drugs of a lipophilic nature. The technique of double emulsion-solvent evaporation was designed to encapsulate hydrophilic drugs including biomolecules. All of these techniques are similar in that they involve an organic solution (typically the solvent phase) containing the NP components and an aqueous solution (typically the non-solvent phase) containing stabilizers that will constitute the dispersion medium. They form an emulsion as a step prior to the formation of NP dispersion. Also, these methods require purification and drying operations in order to obtain a powder amenable to administration. Knowledge of the NP preparation methods is a key issue for the pharmaceutical formulator involved with drug-delivery research and development. The selection of a preparation technique depends on the type of application, the physicochemical and biopharmaceutical characteristics of the drug to be encapsulated, regulatory and economic restrictions, the desired physicochemical and morphological parameters of the polymeric NPs (size, surface charge, hydrophilicity, etc.), the administration route, therapeutic purposes, and, more recently, the environmental outlook (e.g., Green Chemistry). Furthermore, the preparation of polymeric NPs requires a more complex engineering process to engineer in comparison with bulk materials. All of these issues need to be considered prior to adopting a robust process that has the potential to be scaled up to an industrial level [16, 21, 22].

PEG can be employed as a coating of polymeric NPs; it has become the most widely used "stealth" polymer in drug delivery. Due to their flexible and hydrophilic nature, PEG polymers on polymeric NP surfaces can form a thick and dynamic hydration shell, which prevents the absorption of serum proteins on surfaces. As a result, PEG polymers grafted onto NP surfaces are able to reduce opsonization and clearance by the mononuclear phagocyte system, extending blood circulation time. This behavior can increase the possibility that the drug molecules carried into polymeric NPs can be delivered to diseased sites (e.g., internalized by tumor cells) [12]. Additionally, active targeting can be achieved through functionalization of the nanosystems with specific ligands for receptors on target cells [23]. More recently, a new generation of more effective polymeric NPs for targeting has been proposed based on stimuli-responsive (SR) polymers. Among various stimuli (pH, ultrasound [US], temperature, sound, oxidizing/reducing condition, etc.), pH and temperature are those most widely studied for enhanced drug release at the site-of-action. Researchers are focusing on dual (pH and temperature) responsive polymeric NPs for controlled and enhanced drug release at the site-of-action. Two main applications of SR-polymeric NPs have been reported for cancer therapy, considering that certain malignancies can cause a slight increase in temperature and a decrease in extracellular pH around the tumor site and for imaging purposes [18, 24].

Another recent innovative proposal in nanodrug-delivery systems comprises polymeric lipid hybrid NPs, or lipomers, which are nanocontainers composed of a polymeric core and lipid shell that impart physicochemical stability and biocompatibility to the NPs. Lipomers have been reported as a blend of the positive attributes of both liposomes and polymeric NPs wherein their individual innate flaws are negated. These lipomers have demonstrated good results as drug-delivery systems to treat cancer and infectious diseases, as well as for theranostic purposes [25, 26].

The development of these novel nanotherapeutics with multi-functionalities possesses considerable potential to fill the lacunae existing in the present therapeutic domain (e.g., effective targeting in affected tissues, specific cancer treatments, improving specificity, lowering toxicity, increasing the imaging elements to diagnose, designing nanomachines and nanorobots, and optimizing physiotherapy applications) [14]. One main challenge to the introduction of polymeric NPs on the market is satisfying safety valuation and regulation, starting with the initial research, the manufacturing process, quality control, and the pre-clinical and clinical trials. The FDA, the EMA, and other regulatory agencies then inspect each novel NP- formulated drug on a product-by-product basis. The files need to include in-depth efficacy, physical, chemical, physicochemical, pharmacological, biopharmaceutical, technological, and toxic studies. Despite these regulatory demands, it is possible to state that polymeric NPs continue to possess a prominent dominance over other NP types with regard to safety, stability, and targeting objectives [26]. Thus an important increase in the number of FDA approvals and clinical trials utilizing polymeric NPs has been manifested in the last lustrum [27].

4.4 Solid Lipid Nanoparticles in Nanopharmaceuticals

One of the first registries of SLNPs dates from 1991 and has emerged as an alternative to polymeric nanoparticles, mainly in terms of biocompatibility and biodegradation [28]. One of the major attractions of SLNPs has been the issue of toxicity, attributed to lipid composition [29, 30]. It is understandable that the administration of lipids, in their majority derived from triglycerides, entertains high biocompatibility. This same representation is even indistinguishable for our human bodies if it derives from a medicine or from a food. While our bodies have had extensive experience with food, with respect to SLNPs, the difference is that the composition is not only lipid. There is also a percentage of stabilizer (surfactant) on the surface. The surface will be the first line of contact with biological systems, as described for the corona phenomenon in any type of nanoparticle. The corona of an NP can be described as the formation of a layer of proteins, ions, and water on the surface [31]. The interaction between an NP and the corona is determined by the type of interaction with the stabilizer. Although different cores have the same stabilizer, the corona may be different, because the stabilizer arrangement is also determined by interaction with the core. For this reason, each system will be different and nearly unique— hence the need for the monitoring of nanotoxicology for systems administered to complex biological systems.

Some typical components of SLNP formulations include glyceryl behenate, glyceryl palmitate, glyceryl trimyristate, stearic acid, and beeswax. An additional feature is that in the majority of SLNP production methods, a temperature of at least 5 °C higher than the lipid melting point is required. In practice, this involves greater consumption of energy and resources. In small-scale production in the laboratory, this also includes special care to avoid lipid precipitation, in addition to the limitation for the incorporation of thermolabile drugs. In the scale-up of production of submicron systems, it is possible to have production equipment for systems such as liposomes and lipid NPs; however, for the conditions that the SLNPs require, there are only prototypes under development to date. The difference is the temperature control for lipid fusion. One of these prototypes in development belongs to that studied by Gasco and collaborators [32] for the production of SLNPs with a size of 25 nm and a polydispersity index (PDI) of 0.1, with a stirring rate of 300 rpm. In scaling processes, a limiting factor may be the speed of agitation. Achieving a stirring rate of 25,000 rpm for a hypothetical 10-kg system may be difficult for controlling particle sizes, for example, those close to 25 nm. In this regard, Dingler and Gohla [33] also optimized a production process for a capacity of up to 50 kg, reaching, with 10 kg, a size close to 200 nm and a processing capacity of $60\,\text{L}\,\text{h}^{-1}$. With respect to the recent advances of SLNPs, findings in chemical modifications to improve vectorization in neuronal treatments and in cancers of other types are prominent. These modifications include combinations of fragments of surfactants with ligands, which would facilitate transit to the site-of-action and traveling in only one direction. In smaller numbers, some examples also include changes in lipid chains to offer greater interaction with the drug. Or even more so, some lipid modification has served to form complexes only with drugs or, to an even greater degree, with biological systems. The modifications are oriented to form a unique chain that allows hydrophobic/hydrophilic interactions, interactions by electrical charge with the drug, and in turn a certain degree of vectorization toward a target tissue, all in nanometric groups (Figure 4.3).

Among administration routes for SLNPs, in addition to those shared with the other submicron systems, oral administration can be highlighted. This administration route has been preferred for the majority of conditions, as long as the patient's condition allows it. Oral medications permit a controlled dosage form and a high amount of the active agent. In the case of SLNPs, the route can be long; therefore, the probability of reaching the blood circulation decreases. This is often described as one of the easiest routes for NPs, and indeed it is, if one thinks about ease of administration. However, the long transit time can cause loss of the formulation and degradation of the lipid matrix. Thus, one can also observe a reduced number

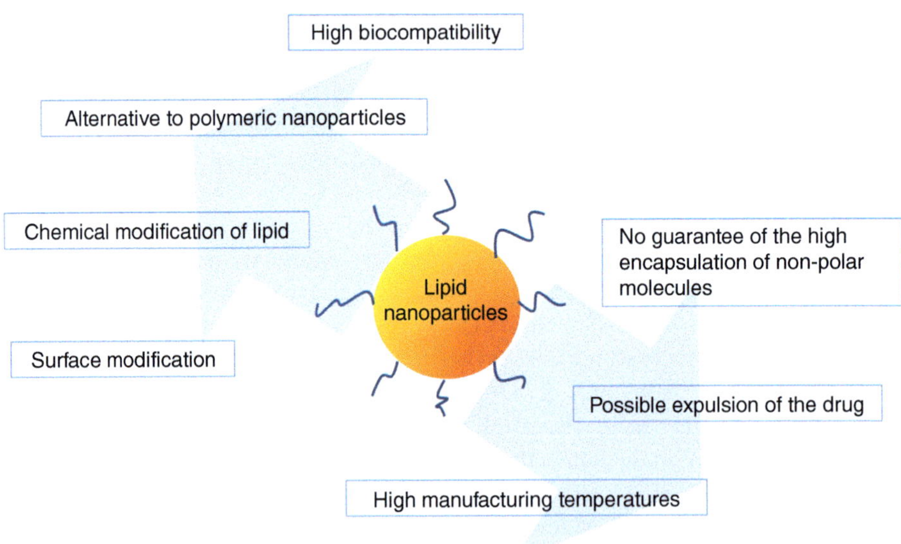

Figure 4.3 Attractive features of SLNPs represented on the left side and real-life characteristics of SLNPs on the right side.

of NPs that are generally administered orally with respect to the others. If the same logic were to be employed as that of conventional drugs, the numbers would be the other way around. Lipid degradation, however, can also help improve the absorption of SLNPs in combination with bile salts by the formation of mixed micelles. SLNPs have been paid special attention, among other emerging needs, for cosmetic applications. Lipid composition offers high biocompatibility and allows incorporation in all dermatological forms. The composition of lipids has permitted different applications on the skin, such as occlusive properties, an increase in skin hydration, a modified release profile, and an increase in skin penetration. The combination of these properties has also facilitated their incorporation into various commercial products. Another useful application in cosmetology is for active agents that undergo degradation, such as coenzyme Q10, ascorbyl palmitate, tocopherol, and retinol. The incorporation into the interior of the SLNP diminishes these effects.

At http://ClinicalTrials.gov, a database of privately and publicly funded clinical studies conducted worldwide [8], there is no precision about tracking SLNPs. This is because the closest approach is attained with the term "lipid nanoparticle," with 12 study records, of which three studies are recruiting, two are not yet recruiting, three are finished, and four are completed. To date, clinical studies with lipid NPs have been restricted. Shaimaa El-Housiny and colleagues [34] reported in 2017 one of the first studies of SLNPs with Fluconazole in patients with the administration of a topical gel for the treatment of pityriasis versicolor. One of the formulations evaluated Precirol® ATO5 (Glyceryl distearate/palmitostearate) with Poloxamer 407. NPs had a size of 292 nm, 0.22 of PDI, a zeta potential of −22.9 mV, wand an entrapment efficiency of 79%. To facilitate the administration of NPs and increase the retention time of the formulation on the skin, the authors chose Carpobol® 934 at 1% [35], in addition to methyl paraben for conservation reasons, with a pH of 5.5–6.7 and a drug content of 8–9 $mg\,g^{-1}$ of gel.

The resulting images of the study with pityriasis versicolor revealed a significant improvement in one of the patients enrolled in the clinical trial. The clinical trial was characterized by 30 patients for four weeks, with two applications per day. The gel with NPs showed an even greater beneficial effect than a commercial formulation containing Fluconazole. The authors mentioned that the patients did not report any discomfort due to the formulation. Apparently, at least in the short term, there were no considerable adverse reactions. Other formulations of lipid composition correspond to nanostructured lipid carriers and lipid mixtures. A similar study developed previously was that of Agrawal and colleagues [36] in 2010 with Acitretin-loaded nanostructured lipid carriers for the topical treatment of psoriasis. The formulation consisted of Precirol® ATO5, oleic acid, and Tween 80. The authors also utilized the same base formulation of Carbopol® 934 P at 1%. The report noted an improvement in the clinical response of psoriasis with the new formulation.

4.5 Dendrimers in Nanopharmaceuticals

The application of nanotechnology in nanopharmaceutical products is an emerging field that provides novel approaches to revolutionizing conventional medical treatments. These nanosystems are designed to modify the properties of drug molecules such as solubility, half-life, hall biocompatibility, and release characteristics. They can also increase the possibility of crossing the cell membranes and decreasing unwanted adverse effects [37]. Taking this into consideration, several nanomaterial systems have been developed as powerful strategies for delivering drugs, recombinant proteins, vaccines, small interfering RNA (siRNA), nucleotides, and genes. Dendrimers are a type of versatile three-dimensional, hyperbranched nanostructure whose size, shape, and topology can be controlled precisely during the synthesis process and prepared by means of iterative methods [38].

These molecules exhibit three main architectural components: (i) the core, which determines the shape and steric hindrance present in the molecule; (ii) the layers, termed generations, formed by the repeating units with branched structures, forming empty cavities [39] that are able to encapsulate small molecules; and (iii) the multiple functional groups at the periphery of the dendrimer, which offer multivalent surfaces, improving binding affinities through covalent bonds or electrostatic interactions between ligands and receptors. These three architectural components (Figure 4.4) determine their physical and chemical properties, as well as the overall size, shape, and flexibility of a dendrimer.

Dendrimers have been synthesized employing divergent and convergent methods or combinations thereof. These synthetic strategies have been developed to vary the terminal groups, the internal blocks, and the core of the dendrimeric structure. The divergent method (cascade-type reaction) involves the growth of the dendrimer from a core or focal point to the periphery, layer by layer, and "inside-out." In the first step, the core molecule reacts with monomers, forming the first-generation (G1) dendrimer. The periphery of the G1 dendrimer may be activated for reactions with more monomers, and the process can be repeated for several generations (G2, G3, G*n*). An advantage of this method is the possibility of stopping the reaction at any step and, in this way, obtaining dendrimers of different generations as well as the possibility of automating the repetitive steps. Therefore, this strategy of synthesis is utilized to obtain all commercially available dendrimers, such as Poly(AMidoAMine) (PAMAM) or Poly(Propylene Imine) (PPI). On the other hand, in the convergent method [40], the individual branches of the polymeric arms (dendrons) are synthesized first and then attached to a multi-functional core molecule. This approach offers advantages, such as minimizing the occurrence of structural

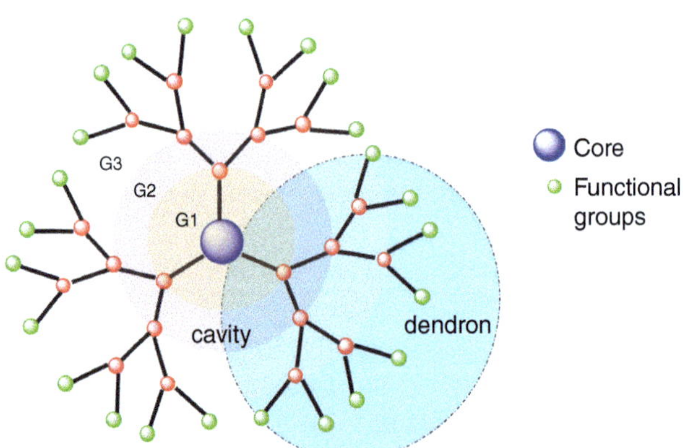

Figure 4.4 Schematic representation of dendrimer components.

defects and easier purification of the final product. However, the convergent approach rarely allows the obtention of high-generation dendrimers due to steric issues in the reactions of dendrons with the multi-functional core.

Dendrimers offer several mechanisms of interactions with different drugs or biomolecules. The interior is well suited for host-guest interaction, for the incorporation of hydrophobic/hydrophilic molecules inside their empty cavities (nanoscale containers), which present around the core through hydrophobic interactions and hydrogen bonding; thus, these can be used as dendritic boxes and unimolecular micelles [41]. In addition, terminal functionalities provide a platform for conjugation by covalent bonding (the prodrug approach) of the drug and targeting moieties. These peripheral functional groups can be employed to tailor-make the properties of dendrimers, enhancing their versatility.

Attractive features of dendrimers, such as nanoscopic size, low polydispersity, high control over molecular structure, the functional groups at the periphery, and the cavities formed during the synthesis process, render them molecules with high potential applications in the biomedical and pharmaceutical field. Dendrimers present novel reformulation opportunities for active agents (drugs, proteins, nucleic acids) that were previously insoluble or that could not be targeted to the specific site of the body where they were needed. This approach entertains the goal of reducing toxicity and enhancing bioavailability, thereby improving therapeutic efficacy [42].

However, even though drug conjugation by covalent bonding provides prolonged retention of therapeutic agents compared with physical encapsulation, it is not always practical because the reduction in drug release can diminish the efficacy. To overcome the issue of low efficacy, have been used cleavable linkers for controlling drug release, these can be activated by the use of SR such as pH and temperature [43] changes, and ultraviolet (UV) light. Common linkers used to covalently attach drugs to dendrimers include acid labile groups and ester and amide groups. In the last decade, many studies have demonstrated the ability of dendrimers as antibacterial agents [44] and as topical antimicrobial agents against herpes simplex virus (HSV), due to its efficacy as a vaginal microbicide for the prevention of HIV and HSV infections [45] VivaGel® (Gel SPL7013) developed by Starpharma Pty Ltd. [46, 47].

Also, with dendrimers as anticancer drug carriers, many researchers have compared the anticancer activity of conventional drugs (Methotrexate, Paclitaxel) [48], steroidal molecules [49]. When conjugated to a dendrimer, conjugates resulted with slower release and higher apoptosis and cytotoxic activity against cancer cells, in comparison with the free drug. Magnetic resonance imaging (MRI) is a powerful technique for visualizing tissues with high resolution. Currently, MRI examinations use contrast agents such as gadolinium (III) chelates to enhance image contrast between diseased and normal tissues. However, these agents rapidly extravasate from the blood circulation; an alternative to this was the use of dendrimer-gadolinium (III) complexes, which prolonged the time of blood circulation and enhanced contrast imaging in cancer pre-clinical studies [50].

Dendrimers have also been employed to track the drug in specific sites by MRI; for instance, Gonawala and colleagues developed water-insoluble Combretastatin conjugated to a water-soluble G3-succinamic acid PAMAM dendrimer, obtaining nanomolecules ranging in size from 3 to 5 nm to monitor cerebral blood flow across the blood brain tumor barrier, observing that the intratumor blood vessels collapsed, leading to necrosis at the core of the tumor [51]. The intracellular delivery of nanocarriers involves both extracellular drug release at the interstitial delivery (tissue site) and intracellular delivery upon internalization. The dendrimer PAMAM is a commercially available dendrimer with amine terminal groups appropriate for interacting with biomolecules through charge-based interactions. Several recent attempts have combined dendrimers with protein-RNA complex, siRNA for gene transfer [52] or oligonucleotide delivery [53]. The PAMAM dendrimer was employed as a carrier to co-deliver oligonucleotide (as-miR-21) and 5-Fluorouracil (5-FU; a cytotoxic agent), forming a complex smaller than 100 nm to achieve delivery to human glioblastoma cells. Both the chemotherapeutant and as-miR-21 could be efficiently introduced into tumor cells, improving the cytotoxicity of 5-FU, while the migration ability of the tumor cells was decreased.

4.6 Quantum Dots in Nanopharmaceuticals

Multifunctional NPs, composed of organic and inorganic materials, have been developed as a promising tool with applications in nanomedicine. Many fundamental biological processes (the dynamics of intracellular networks, signal transduction, drug and cell interactions) consist of the interplay of biomolecules with each another or with various ionic or molecular species. To visualize these biological processes in real time would aid in attaining a clear understanding of interaction and drug release at the target site, characterizing the molecular status of cancer tumors. Nanostructured materials with optoelectronic properties, such as dendrimers, liposomes, QDs, carbon nanotubes, and gold and iron oxide NPs [54] have been developed as novel diagnostic imaging methods and targeted therapies as promising tools in nanomedicine. Fluorescence methods encompass several experimental parameters—for instance, excitation and emission wavelength, intensity, and fluorescence lifetime, obtaining nanometer-scale resolution and sensitivity. Some of the optical imaging methods for analysis include MRI, US, computed tomography (CT), positron emission tomography (PET), single-photon emission computed tomography (SPECT), and confocal microscopy. The potential of a detection or imaging method is determined by the physicochemical properties of the chromophore, its chemical nature and size, its biocompatibility, and its interplay with the biological unit. QDs are nanocrystals, initially prepared by Reed [55], composed of a core made up of elements from groups II, III, V, and VI, as a semiconductor material, enclosed within a shell of another semiconductor that has a larger spectral band gap such as ZnS. The size and shape of QDs can be controlled during the reaction process, varying the temperature and the ligand molecules used in the synthesis. At present, commercial products made of CdSe, CdTe, and InP or InGaP are available. QD design comprises three steps: (i) synthesizing the quantum dot, (ii) modifying its surface so that it is biologically compatible, and (iii) adding to its surface some target agent such as antibodies, peptides, or small molecules that can be directed at a specific site (Figure 4.5).

There are several strategies for synthesizing QDs with various properties depending on the required applications, categorized either as a top-down approach (e.g., electron beam lithography, focused ion beam) or a bottom-up approach (e.g., wet-chemical methods, sol-gel process, microemulsion, hot-solution decomposition) [56]. Prior to 1993, QDs were mainly prepared in aqueous solution with added stabilizing agents; this procedure yielded

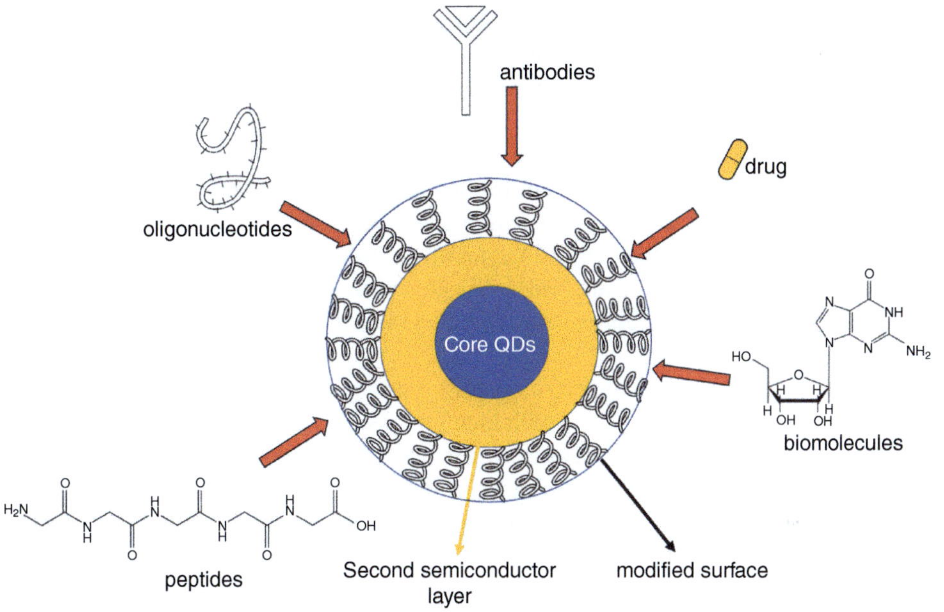

Figure 4.5 Schematic representation of a quantum dot.

low-quality QDs with poor fluorescence efficiencies and large size variations. To get around this problem, the core of one semiconductor was wrapped in the shell of another. Also, some monodisperse QDs are usually synthesized by the injection of a liquid precursor into hot (300 °C) nonpolar coordinating organic solvent, and the size and shape of the QDs are controlled by the temperature and crystal-growth duration, as well as by the ligand molecules employed in their synthesis [57]. The size and composition of QDs determine their emission wavelength and color: a typical QD has a diameter of about −10 nm and its emission ranges from about 450 to 850 nm. QDs are relatively efficient with regard to conversion of excitation light into emission, where the quantum yield is generally over 50% [58].

In comparison to organic and organometallic dyes, such as fluorescein derivatives and rhodamine derivatives, QDs have strong light absorbance and a fluorescence lifetime of 10–40 ns [59], which is significantly longer than that of typical organic dyes, bright fluorescence, photostability, broad excitation wavelength, continuous and broad absorption spectra, and high susceptibility to surface functionalization [60]. In order to utilize QDs in a biological environment, they need to be conjugated to biological molecules to increase biocompatibility. This conjugation can be through electrostatic attraction, covalent linkage, adsorption, and mercapto (–SH) exchange. The main strategies to render QDs biocompatible include silanization [61], coating with amphiphilic polymers and mercaptohydrocarbonic acids, and the encapsulation of QDs within phospholipids, polymer shells, and amphiphilic polysaccharides.

The potential of QDs for biological applications includes immunolabeling, cell motility assays, and *in situ* hybridization, as well as live cell markers and as a fluorescent stain. QD tracking and detection methods are numerous and include confocal microscopy, wide-field epifluorescence microscopy, and fluorometry. QDs have garnered attention owing to their remarkable imaging property; however, their use is limited due to their cellular toxicity, mainly when selenium and cadmium are used. Currently, several types of QDs have been developed, for instance, near infrared QDs, metal nanoclusters, carbon dots, and graphene and silicon dots, in order to reduce their toxicity, maintaining high-resolution imaging [62]. Near infrared QDs (QDNIRs) have been used as stains in tumor imaging, blood vessel imaging, and lymph-node mapping [63]. The first generation of QDNIRs were composed of cadmium, mercury, or lead, with high toxicity, while the second generation of QDs was based on indium and silver, reducing cellular toxicity in animal imaging studies. Fluorescent metal nanoclusters (FNCs) of silver, gold, and copper derivatives possess strong photoluminescence, photostability, and less toxicity, rendering them promising lumiphores for bioimaging applications [64].

Gold nanoclusters (AuNCs) protected by bovine serum albumin (BSA) as sensor to the determination of the hydroxyl radical with high sensitivity in live cells have been reported. The AuNCs exhibited long-term stability against light illumination and pH, good cell permeability, and low cytotoxicity. In the absence of the OH radical, the AuNCs demonstrated only one emission peak at 637 nm; however, after the reaction with OH dianionic fluorescein, the emission changed to 515 nm [65]. Luminescent carbon dots are promising probes for bioimaging applications due to their low toxicity, chemical inertness, and versatile surface chemistry, exhibiting different emission colors when excited with different emission wavelengths. QDs have been applied to different bio-applications including cancer cell targeting [66], intracellular pH sensing, and Cu 2+ sensing [67]. Li and colleagues reported that three types of carbon dots with varying functional polymer chains at the periphery showed no apparent cytotoxicity on their own and revealed successfully cellular internalization by conjugation with transferrin. Graphene Quantum Dots (GQDs) have lower toxicity and cell compatibility than bulk graphene, rendering them constructive and effective materials for targeted drug-delivery agents. GQDs encapsulated with β-CycloDextrin (β-CD) as a drug-delivery system with a diameter of less than 5 nm have been reported. These systems are promising drug-delivery platforms for delivery profiles for the management of arthritic pain with high compatibility to the cells, with more than 90% cell viability [68]. Oxidized silicon nanospheres (O-SiNSs), which possess excellent aqueous dispersibility, high photoluminescent quantum yield (PLQY), ultraphotostability, wide pH stability, and favorable biocompatibility are conjugated with the antibody, and the resultant O-SiNS/antibody bioconjugates are successfully applied in real-time cellular labeling and in immunofluorescent cell imaging [62].

4.7 Regulatory Issues

The development of new therapeutic agents based on nanotechnology has been accompanied by a series of parallel challenges. These changes are due to technological advances in the techniques of synthesis, control, and characterization of nanosystems. However, these advances have not been the only ones, they have been accompanied by the implementation of a regulatory framework that has allowed the development of nanomedicine focused on the quality, safety, efficacy, and risk management of nanoproducts. Regulatory issues always appear to have retreated to the rear guard of the implementation of new technologies; progress apparently always exceeds regulatory frameworks. This has not been the exception in the case of nanomedicine. The first generation of nanomedicines was introduced into the market following the traditional approach, where approval was based solely on a risk/benefit analysis [69]. At present, in a broader approach, nanomedicines are required to prove their superiority in terms of bioavailability, speed, delivery site, and the mechanism of action regarding their non-nanometric equivalents. Physical and chemical properties also must be established, as well as critiques; characterization methods; detection methods in tissues, the environment, and equipment; and other aspects such as toxicology and ecotoxicology. To date, there remains the challenge of establishing a harmonized definition of what a nanomaterial, nanomedicine, nanopharmaceutical, or nanodevice represents. As mentioned in this chapter, the fact that the material in question, referred to as a nanomaterial, has novel physicochemical characteristics that are different from those of bulk material is accepted. However, the precise size is still subject to debate. One of the most widespread definitions of what a nanomaterial is corresponds to that declared by the European Commission in October 2011, where the size of 1–100 nm is the accepted standard.

However, there are systems declared as nanopharmaceutical that are receiving great attention and that do not necessarily meet this criterion. For example, polymeric NPs generally range in size from 10 to 1000 nm and are more usually between 200 and 400 nm [70]. Therefore, and even though the European Commission itself recognizes that there are "special circumstances" in the pharmaceutical sector, there is still a general definition that allows the inclusion of pharmaceutical systems and medical devices smaller than 1 μm, as long as their special characteristics are due to miniaturization.

Another issue of special attention is the expiration of the patents of the first generation of nanomedicines and the development of nanosimilar systems. These nanosimilar systems claim to be like reference nanomedicines, and studies are needed that show that they possess similarity in terms of quality, safety, and efficacy. The problem comes at the moment of defining the proofs that sustain the similarity between the innovative system and the similar one: due to the complexity of the systems, the establishment of the tests must be conducted case-by-case, which complicates the regulatory agencies seeking the development of general guidelines that reduce time and workload. Talking about the establishment of a biowaiver in the case of nanosimilars is one of the subjects that is not considered to occur in the immediate future due to the complexity of the mechanisms of action of the nanopharmaceuticals.

The EMA has engaged in very important efforts in the establishment of general principles for the evaluation of nanosimilars through the Marketing Authorization Application, which specifies that nanomedicines be considered as similar. For example, in the case of liposomal formulations, which have been on the market since 1990 and whose main use is in the oncological area, the EMA in 2013 published a reflectional document on liposomes that indicates the requirements for intravenous liposomal products with reference to an innovative liposomal product. The document seeks to facilitate decisions concerning the pharmaceutical data necessary to demonstrate the comparability between a tested product and its reference in terms of safety and efficacy. The document also seeks to clarify the need for pre-clinical and clinical studies and the circumstances that can allow the exemption of certain studies. Finally, it also determines the design of non-clinical *in vivo* studies and the potential role of *in vitro* models. To date, in addition to the reflection regarding liposomal systems, the EMA has published reflections on other systems, such as iron-based formulations and nanocrystal-based formulations whose patents are about to expire. The documents consist of three main sections. The first section details the characteristics of the materials, such as the purity of the materials entailed in nanomedicines and other critical excipients.

Also included are some physicochemical parameters that nanopharmaceuticals must possess, such as particle size and distribution, morphology, encapsulated and non-encapsulated active fraction, *in vitro* release profile, surface charge, and pH of the internal compartment, among others. The second part of the document includes the characteristics of the non-clinical tests required to demonstrate similarity, emphasizing the need to validate the methods of analysis and the importance of conducting distribution studies using the most relevant animal model to determine the distribution, metabolism, and excretion of NPs. Finally, the last section describes the clinical studies that indicate the pharmacokinetics, as well as the safety and efficacy studies that should be carried out. For example, in the case of liposomal formulations, the pharmacokinetic characteristics of the encapsulated and non- encapsulated drug must be determined.

It is accepted that there is a lack of information on the toxicology of nanosystems. The current task of nanotoxicology should be contributing to clarify if safety studies are sufficiently robust and must show if the nanosystems do not put humans or the environment at risk. It is considered that the effect of nanomaterials on inflammatory and pro-oxidant processes has not yet been widely evaluated, as well as considering that the cytotoxicity of nanomaterials has not been consciously evaluated: the concern has been directed to a greater extent toward determining whether experimental procedures should provide sufficient and clear information in order to obtain valid data [71].

In addition to the latter, another concern is the long-term effects of nanomaterials, without neglecting the acute and sub-acute effects; however, the latter are by their nature faster to obtain, while the former are obtained years after the nanosystem is in the environment. In the same way, damage to the environment by nanosystem should not stop being part of the surveillance by the regulatory agencies. Undoubtedly, the identification of potential risks and their mitigation should be a joint task among government entities, academia, and commercial organizations. Efforts should be directed toward establishing adequate protocols that allow the following: (i) the identification of nanomaterials with the potential to be considered unsafe, based on knowledge of their physical, chemical, and physicochemical properties; (ii) the application of toxicological tests of exposure by all routes in which it is considered that the nanomaterial can be absorbed, based on the characteristics of the size, that is, differentiated from the bulk material; (iii) manufacturing processes based on good manufacturing practices that ensure low risk of exposure; and (iv) the development of tools for detecting and monitoring levels of exposure to nanomaterials in the workplace, air/water, humans, other organisms, and the environment [72].

In 2013, the EMA and the Japanese Ministry of Health, Labor, and Welfare (MHLW) launched a public consultation reflecting on the development of block-copolymer-micelle medicinal products, one of the first examples of a joint collaboration between regulatory agencies. The path has been drawn, and groups of experts should be created consisting of academics, corporate representatives, and government representatives, working in turn with agencies in other countries to achieve the harmonization and generation of guidelines for the development and evaluation of nanopharmaceuticals.

4.8 Conclusion

Since the establishment of liposomal therapy, we have entered into the era of nanopharmaceutics. At present, the use of nanomaterials in the pharmaceutical field is a reality, and we can observe the use of increasingly complex and multifunctional platforms that will be administered by a wide range of routes. In this chapter, we have highlighted some of the most promising nanotechnological platforms such as liposomes, polymeric NPs, SLNPs, dendrimers, and QDs), there no doubt that liposomes deserve a special mention they are an example of continuous improvement due to substantial advances that have been introduced since they entered into the market. The superficial modification of nanopharmaceutics has brought important benefits, such as the minimization of aggregation, improvement of stability, greater biocompatibility, and, in some cases, the reduction of clearance due to the reticuloendothelial system. Currently, the most sophisticated coatings of nanomedicines seek to facilitate specific targeting of the systems at the cellular level and at the more advanced intracellular level. Undoubtedly, there are important challenges regarding manufacturing methods; these should be attractive from the industrial point of view, without neglecting the implementation of development processes based on Quality by Desing (QbD) to ensure quality, safety, and efficiency during each stage of the process. This implies the development of real-time characterization and process-control tests. The field of the application of nanomedicines has become extensive, and not only for oncological applications. At present, it is clear that nanomedicines can be applied in practically any medical area, and in the future, we will see nanomedicines with multiple applications administered by different routes. Therefore, the development of methodologies for the evaluation of interaction with targeted biological structures, the determination of their biocompatibility, their pharmacokinetics, and their pharmacodynamics, as well as their toxicology, will be relevant.

References

1 Weissig, V., Pettinger, T., and Murdock, N. (2014). Nanopharmaceuticals (part 1): products on the market. *International Journal of Nanomedicine* 9: 4357. https://doi.org/10.2147/IJN.S46900.

2 Ventola, C.L. (2017). Progress in nanomedicine: approved and investigational nanodrugs. *P & T: A Peer-Reviewed Journal for Formulary Management* 42 (12): 742–755. https://doi.org/10.1016/j.psychres.2007.07.030.

3 Matea, C., Mocan, T., Tabaran, F. et al. (2017). Quantum dots in imaging, drug delivery and sensor applications. *International Journal of Nanomedicine* 12: 5421–5431. https://doi.org/10.2147/IJN. S138624.

4 Allen, T.M. and Cullis, P.R. (2013). Liposomal drug delivery systems: from concept to clinical applications. *Advanced Drug Delivery Reviews* 65 (1): 36–48. https://doi.org/10.1016/j.addr.2012.09.037.
5 Barenholz, Y. (2012). Doxil®–the first FDA-approved nano-drug: lessons learned. *Journal of Controlled Release* 160 (2): 117–134. https://doi.org/10.1016/j.jconrel.2012.03.020.
6 U.S. Department of Health and Human Services Food and Drug Administration Center for Drug Evaluation and Research (2018). Guidance for industry: liposome drug products: chemistry, manufacturing, and controls; human pharmacokinetics and bioavailability; and labeling documentation.
7 Bulbake, U., Doppalapudi, S., Kommineni, N. et al. (2017). Liposomal formulations in clinical use: an updated review. *Pharmaceutics* 9 (2): 1–33. https://doi.org/10.3390/pharmaceutics9020012.
8 ClinicalTrials.gov (2019). https://clinicaltrials.gov/
9 Squillaro, T., Cimini, A., Peluso, G. et al. (2018). Nano-delivery systems for encapsulation of dietary polyphenols: an experimental approach for neurodegenerative diseases and brain tumors. *Biochemical Pharmacology*. Elsevier; 154: 303–317. https://doi.org/10.1016/j.bcp.2018.05.016.
10 Choi, Y.H. and Han, H.-K. (2018). Nanomedicines: current status and future perspectives inaspect of drug delivery and pharmacokinetics. *Journal of Pharmaceutical Investigation* 48 (1): 43–60. https://doi.org/10.1007/s40005-017-0370-4.
11 Farjadian, F., Ghasemi, A., Gohari, O. et al. (2019). Nanopharmaceuticals and nanomedicines currently on the market: challenges and opportunities. *Nanomedicine* 14 (1): 93–126. https://doi.org/10.2217/nnm-2018-0120.
12 Shen, Z., Fisher, A., Liu, W.K. et al. (2018). PEGylated "stealth" nanoparticles and liposomes. *Engineering of Biomaterials for Drug Delivery Systems*. Elsevier 1–26. https://doi.org/10.1016/B978-0-08-101750-0.00001-5.
13 Lu, X. and Zhang, K. (2018). PEGylation of therapeutic oligonucletides: from linear to highly branched PEG architectures. *Nano Research* 11 (10): 5519–5534. https://doi.org/10.1007/s12274-018-2131-8.
14 Prasad, M., Lambe, U.P., Brar, B. et al. (2018). Nanotherapeutics: an insight into healthcare and multi-dimensional applications in medical sector of the modern world. *Biomedicine & Pharmacotherapy* 97: 1521–1537. https://doi.org/10.1016/j.biopha.2017.11.026.
15 Mendoza-Muñoz, N., Quintanar-Guerrero, D., and Allémann, E. (2012). The impact of the salting-out technique on the preparation of colloidal particulate systems for pharmaceutical applications. *Recent Patents on Drug Delivery & Formulation* 6 (3): 236–229. https://doi.org/10.2174/187221112802652688.
16 Piñón-Segundo, E., Llera-Rojas, V.G., Leyva-Gómez, G. et al. (2018). The emulsification-diffusion method to obtain polymeric nanoparticles. In: *Nanoscale Fabrication, Optimization, Scale-Up and Biological Aspects of Pharmaceutical Nanotechnology* (ed. A.M. Grumezescu), 51–83. Elsevier https://doi.org/10.1016/B978-0-12-813629-4.00002-4.
17 Matoba, T., Koga, J., Nakano, K. et al. (2017). Nanoparticle-mediated drug delivery system for atherosclerotic cardiovascular disease. *Journal of Cardiology* 70 (3): 206–211. https://doi.org/10.1016/j.jjcc.2017.03.005.
18 Tayo, L.L. (2017). Stimuli-responsive nanocarriers for intracellular delivery. *Biophysical Reviews* 931–940. https://doi.org/10.1007/s12551-017-0341-z.

19 Tyagi, P. and Subramony, J.A. (2018). Nanotherapeutics in oral and parenteral drug delivery: key learnings and future outlooks as we think small. *Journal of Controlled Release*. Elsevier 272: 159–168. https://doi.org/10.1016/j.jconrel.2018.01.009.

20 Quintanar-Guerrero, D., Allémann, E., Doelker, E. et al. (1998). Preparation and characterization of nanocapsules from preformed polymers by a new process based on emulsification-diffusion technique. *Pharmaceutical Research* 15 (7): 1056–1062.

21 Quintanar-Guerrero, D, Zambrano-Zaragoza, M. L., Gutiérrez-Cortez, E. et al. (2012). Impact of the emulsification-diffusion method on the development of pharmaceutical nanoparticles. *Recent Patents on Drug Delivery & Formulation* 6 (3): 184–194. https://doi.org/10.2174/187221112802652642.

22 Mendoza-Muñoz, N., Alcalá-Alcalá, S., and Quintanar-Guerrero, D. (2016). Preparation of polymer nanoparticles by the emulsification-solvent mvaporation method: from Vanderhoff's pioneer approach to recent adaptations. In: *Polymer Nanoparticles for Nanomedicines* (ed. C. Vauthier and G. Ponchel), 87–121. Switzerland: Springer International Publisher.

23 Fernandez-Piñeiro, I., Badiola, I., and Sanchez, A. (2017). Nanocarriers for microRNA delivery in cancer medicine. *Biotechnology Advances* 35 (3): 350–360. https://doi.org/10.1016/j.biotechadv.2017.03.002.

24 Kalhapure, R.S. and Renukuntla, J. (2018). Thermo- and pH dual responsive polymeric micelles and nanoparticles. *Chemico-Biological Interactions* 295: 20–37. https://doi.org/10.1016/j.cbi.2018.07.016.

25 Jain, A., Sharma, G., Kushwah, V. et al. (2017). Methotrexate and beta-carotene loaded-lipid polymer hybrid nanoparticles: a preclinical study for breast cancer. *Nanomedicine* 12 (15): 1851–1872. https://doi.org/10.2217/nnm-2017-0011.

26 Jose, C., Amra, K., Bhavsar, C. et al. (2018). Polymeric lipid hybrid nanoparticles: properties and therapeutic applications. *Critical Reviews in Therapeutic Drug Carrier Systems* 35 (6): 555–588. https://doi.org/10.1615/CritRevTherDrugCarrierSyst.2018024751.

27 El-Say, K.M. and El-Sawy, H.S. (2018). Polymeric nanoparticles: promising platform for drug delivery. *International Journal of Pharmaceutics* 528 (1–2): 675–691. https://doi.org/10.1016/j.ijpharm.2017.06.052.

28 Mishra, V., Bansal, K., Verma, A. et al. (2018). Solid lipid nanoparticles: emerging colloidal nano drug delivery systems. *Pharmaceutics* 10 (4): 191. https://doi.org/10.3390/pharmaceutics10040191.

29 Cortés, H., Alcala-Alcala, S., Avalos-Fuentes, A. et al. (2017). Nanotechnology as potential tool for siRNA delivery in Parkinson's disease. *Current Drug Targets* 18 (16): https://doi.org/10.2174/1389450118666170321130003.

30 Tapeinos, C., Battaglini, M., and Ciofani, G. (2017). Advances in the design of solid lipid nanoparticles and nanostructured lipid carriers for targeting brain diseases. *Journal of Controlled Release*. Elsevier 264: 306–332. https://doi.org/10.1016/j.jconrel.2017.08.033.

31 Leyva-Gómez, G., Cortés, H., Magaña, J.J. et al. (2015). Nanoparticle technology for treatment of Parkinson's disease: the role of surface phenomena in reaching the brain. *Drug Discovery Today* 20 (7): 824–837. https://doi.org/10.1016/j.drudis.2015.02.009.

32 Gasco, M.R., Priano, L., and Zara, G.P. (2009). Solid lipid nanoparticles and microemulsions for drug delivery: the CNS. *Progress in Brain Research* 180: 181–192. https://doi.org/10.1016/S0079-6123(08)80010-6.

33 Dingler, A. and Gohla, S. (2002). Production of solid lipid nanoparticles (SLN): scaling up feasibilities. *Journal of Microencapsulation* 19 (1): 11–16. https://doi.org/10.1080/02652040010018056.

34 El-Housiny, S., Shams Eldeen, M.A., El-Attar, Y.A. et al. (2018). Fluconazole-loaded solid lipid nanoparticles topical gel for treatment of pityriasis versicolor: formulation and clinical study. *Drug Delivery* 25 (1): 78–90. https://doi.org/10.1080/10717544.2017.1413444.

35 Desfrançois, C., Auzély, R., and Texier, I. (2018). Lipid nanoparticles and their hydrogel composites for drug delivery: a review. *Pharmaceuticals* 11 (4): 118. https://doi.org/10.3390/ph11040118.

36 Agrawal, Y., Petkar, K.C., and Sawant, K.K. (2010). Development, evaluation and clinical studies of acitretin loaded nanostructured lipid carriers for topical treatment of psoriasis. *International Journal of Pharmaceutics* 401 (1–2, 102): 93. https://doi.org/10.1016/j.ijpharm.2010.09.007.

37 Bawa, R. (2010). Nanopharmaceuticals: nanopharmaceuticals. *European Journal of Nanomedicine* 3 (1): https://doi.org/10.1515/EJNM.2010.3.1.34.

38 Tomalia, D.A. and Fréchet, J.M.J. (2002). Discovery of dendrimers and dendritic polymers: a brief historical perspective. *Journal of Polymer Science Part A: Polymer Chemistry* 40 (16): 2719–2728. https://doi.org/10.1002/pola.10301.

39 Natarajan, B., Gupta, S., Ramamurthy, V. et al. (2011). Interfacial regions governing internal cavities of dendrimers. Studies of poly(alkyl aryl ether) dendrimers constituted with linkers of varying alkyl chain length. *Journal of Organic Chemistry* 76 (10): 4018–4026. https://doi.org/10.1021/jo200464g.

40 Hawker, C.J. and Frechet, J.M.J. (1990). Preparation of polymers with controlled molecular architecture. A new convergent approach to dendritic macromolecules. *Journal of the American Chemical Society* 112 (21): 7638–7647. https://doi.org/10.1021/ja00177a027.

41 Tomalia, D.A. (2005). Birth of a new macromolecular architecture: dendrimers as quantized building blocks for nanoscale synthetic polymer chemistry. *Progress in Polymer Science* 30 (3–4, 324): 294. https://doi.org/10.1016/j.progpolymsci.2005.01.007.

42 Hsu, H., Bugno, J., Lee, S. et al. (2017). Dendrimer-based nanocarriers: a versatile platform for drug delivery. *Wiley Interdisciplinary Reviews: Nanomedicine and Nanobiotechnology* 9 (1): e1409. https://doi.org/10.1002/wnan.1409.

43 Sideratou, Z., Agathokleous, M., Theodossiou, T.A. et al. (2018). Functionalized hyperbranched polyethylenimines as thermosensitive drug delivery nanocarriers with controlled transition temperatures. *Biomacromolecules* 19 (2): 315–328. https://doi.org/10.1021/acs.biomac.7b01325.

44 Bosnjakovic, A., Mishra, M.K., Ren, W. et al. (2011). Poly(amidoamine) dendrimer-erythromycin conjugates for drug delivery to macrophages involved in periprosthetic inflammation. *Nanomedicine: Nanotechnology, Biology and Medicine* 7 (3): 284–294. https://doi.org/10.1016/j.nano.2010.10.008.

45 McCarthy, T.D., Karellas, P., Henderson, S.A. et al. (2005). Dendrimers as drugs: discovery and preclinical and clinical development of dendrimer-based microbicides for HIV and STI prevention. *Molecular Pharmaceutics* 2 (4): 312–318. https://doi.org/10.1021/mp050023q.

46 Rupp, R., Rosenthal, S.L., and Stanberry, L.R. (2007). VivaGel (SPL7013 Gel): a candidate dendrimer – microbicide for the prevention of HIV and HSV infection. *International Journal of Nanomedicine* 2 (4): 561–566.

47 Limited, S.H. (2018). Starpharma. https://starpharma.com/vivagel/vivagel_availability (accessed 20 January 2018).

48 Medina, S.H. and El-Sayed, M.E.H. (2009). Dendrimers as carriers for delivery of chemotherapeutic agents. *Chemical Reviews* 109 (7): 3141–3157. https://doi.org/10.1021/cr900174j.

49 Magaña-Vergara, N.E., Rárová, L., Soto-Castro, D. et al. (2013). Synthesis and antiproliferative activity of novel steroidal dendrimer conjugates. *Steroids* 78 (12–13): 1254–1262. https://doi.org/10.1016/j.steroids.2013.09.001.

50 Xu, Aryal M., Ewing J. et al. (2007). in vivo evaluation of a PAMAM-Cystamine-(Gd-DO3A) conjugate as a biodegradable macromolecular MRI contrast agent. *Experimental Biology and Medicine* 232 (8): 1081–1089. https://doi.org/10.3181/0702-RM-33.

51 Gonawala, S., Aryal, M., Ewing, J. et al. (2018). MRI monitoring of cerebral blood flow after the delivery of nanocombretastatin across the blood brain tumor barrier. *Journal of Nanomedicine & Nanotechnology* 9 (5): 1–15. https://doi.org/10.4172/2157-7439.1000516.

52 Krasheninina, O., Apartsin, E., Fuentes, E. et al. (2019). Complexes of pro-apoptotic siRNAs and carbosilane dendrimers: formation and effect on cancer cells. *Pharmaceutics* 11 (1): 25. https://doi. org/10.3390/pharmaceutics11010025.

53 Ren, Y., Kang, C., Yuan, X. et al. (2010). Co-delivery of as-miR-21 and 5-FU by poly(amidoamine) dendrimer attenuates human glioma cell growth in vitro. *Journal of Biomaterials Science*. Polymer Edition 21 (3): 303–314. https://doi.org/10.1163/156856209X415828.

54 Alivisatos, A.P. (1996). Semiconductor clusters, nanocrystals, and quantum dots. *Science* 271 (5251): 933–937. https://doi.org/10.1126/science.271.5251.933.

55 Reed, M., Randall, J., Aggarwal, R. et al. (1988). Observation of discrete electronic states in a zero-dimensional semiconductor nanostructure. *Physical Review Letters* 60 (6): 535–537. https://doi.org/10.1103/PhysRevLett.60.535.

56 Bera, D., Qian, L., Tseng, T. et al. (2010). Quantum dots and their multimodal applications: a review. *Materials* 3 (4): 2260–2345. https://doi.org/10.3390/ma3042260.

57 Peng, Z.A., Peng, X. (2001). Formation of high-1uality CdTe, CdSe, and CdS nanocrystals nsing CdO as precursor. *Journal of the American Chemical Society* 123:183–184. https://doi-org/10.1021/ja003633m.

58 Dabbousi, B.O., Rodriguez-Viejo, J., Mikulec F.V. et al. (1997). (CdSe)ZnS core–shell quantum dots: synthesis and characterization of a size series of highly luminescent nanocrystallites. *Journal of Physical Chemistry B* 101 (46): 9463–9475. https://doi.org/10.1021/jp971091y.

59 Azzazy, H.M.E., Mansour, M.M.H., and Kazmierczak, S.C. (2007). From diagnostics to therapy: prospects of quantum dots. *Clinical Biochemistry* 40 (13–14): 917–927. https://doi.org/10.1016/j.clinbiochem.2007.05.018.

60 Resch-Genger, U., Grabolle, M., Cavaliere-Jaricot S. et al. (2008). Quantum dots versus organic dyes as fluorescent labels. *Nature Methods* 5 (9): 763–775. https://doi.org/10.1038/nmeth.1248.

61 Alivisatos, A.P., Gu, W., and Larabell, C. (2005). Quantum dots as cellular probes. *Annual Review of Biomedical Engineering* 7 (1): 55–76. https://doi.org/10.1146/annurev.bioeng.7.060804.100432.

62 He, X. and Ma, N. (2014). An overview of recent advances in quantum dots for biomedical applications. *Colloids and Surfaces B: Biointerfaces* 124: 118–131. https://doi.org/10.1016/j.colsurfb.2014.06.002.

63 Shibu, E.S., Sugino, S., Ono K. et al. (2013). Singlet-oxygen-sensitizing near-infrared-fluorescent multimodal nanoparticles. *Angewandte Chemie International Edition* 52 (40): 10559–10563. https://doi.org/10.1002/anie.201304264.

64 Hu, L., Zhang, C., Zeng, G. et al. (2016). Metal-based quantum dots: synthesis, surface modification, transport and fate in aquatic environments and toxicity to microorganisms. *RSC Advances. Royal Society of Chemistry* 6 (82): 78595–78610. https://doi.org/10.1039/C6RA13016J.

65 Zhuang, M., Ding, C., Zhu, A. et al. (2014). Ratiometric fluorescence probe for monitoring hydroxyl radical in live cells based on gold nanoclusters. *Analytical Chemistry* 86 (3): 1829–1836. https://doi. org/10.1021/ac403810g.

66 Li, Q., Ohulchanskyy, T.Y., Liu, R. et al. (2010). Photoluminescent carbon dots as biocompatible nanoprobes for targeting cancer cells in vitro. *Journal of Physical Chemistry C* 114 (28): 12062–12068. https://doi.org/10.1021/jp911539r.

67 Zhu, A., Qu, Q., Shao, X. et al. (2012). Carbon-dot-based dual-emission nanohybrid produces a ratiometric fluorescent sensor for in vivo imaging of cellular copper ions. *Angewandte Chemie International Edition* 51 (29): 7185–7189. https://doi.org/10.1002/anie.201109089.

68 Qian, C., Yan, P., Wan, G. et al. (2018). Facile synthetic photoluminescent graphene quantum dots encapsulated β-cyclodextrin drug carrier system for the management of macular degeneration: detailed analytical and biological investigation. *Journal of Photochemistry and Photobiology B: Biology* 189: 244–249.

69 Soares, S., Sousa, J., Pais, A. et al. (2018). Nanomedicine: principles, properties, and regulatory issues. *Frontiers in Chemistry* 6(August):: 360. https://doi.org/10.3389/fchem.2018.00360.

70 Swarnavalli, G.C.J., Kannappan, V., Roopsingh, D. et al. (2017). Effect of solvent on the size and properties of polymeric nanoparticles of poly(ε-caprolactam). *Journal of Molecular Liquids* 236: 61–67. https://doi.org/10.1016/J.MOLLIQ.2017.04.006.

71 Kong, B., Seog, J.H., Graham, L.M. et al. (2011). Experimental considerations on the cytotoxicity of nanoparticles. *Nanomedicine* 6 (5): 929–941. https://doi.org/10.2217/nnm.11.77.

72 Chan, V.S.W. (2006). Nanomedicine: an unresolved regulatory issue. *Regulatory Toxicology and Pharmacology.* 46 (3): 218–224. https://doi.org/10.1016/j.yrtph.2006.04.009.

5

Role of Nanotechnology in Biological Therapies

Blanca Ocampo-García[1], Liliana Aranda Lara[2], Guillermina Ferro-Flores[1], Enrique Morales-Avila[3], and Keila Isaac-Olivé[2]

[1] *Departamento de Materiales Radiactivos, Instituto Nacional de Investigaciones Nucleares, Ocoyoacac, Estado de México, Mexico*
[2] *Facultad de Medicina, Universidad Autónoma del Estado de México, Toluca, Estado de México, Mexico*
[3] *Facultad de Química, Universidad Autónoma del Estado de México, Toluca, Estado de México, Mexico*

5.1 Introduction

Biological therapy is defined as a type of treatment that employs our own organic capabilities (biomolecules) as a source of drugs to treat various diseases—for example, the therapies based on biomolecules for the treatment of infections, cancer, immune system suppression or stimulation, and biosystems to directly or indirectly inhibit the proliferation of cancer cells.

The developments in the design of receptor-specific and biocompatible nanoparticles (NPs) have demonstrated the potential of nanotechnology applications in the field of biological therapies. The understanding of interaction kinetics of different nanosystems at a cellular level indicates an achievable improvement in the treatment of cancer.

Several organic NPs are approved for human use by the U.S. Food and Drug Administration (FDA). Not only have nanosystems for drug delivery, based on polymeric NPs (PNPs) and liposomes, shown successful results, but metal NPs, dendrimers, and fullerenes designed as functionalized biological nanoplatforms for targeted therapy also have shown success. The applicability of these inorganic core NPs, directed to specific molecular targets, is based on the multifunctionality and ability of the nanosystems to produce multimodal therapies in cancer tissues.

In this context of progress, we can mention gold nanoparticles (AuNPs) attached to target-specific biomolecules (peptides, antibodies, siRNA, aptamers, etc.), which can be luminescent and plasmonic [1]. Their luminescent properties are suitable for acquiring biomedical images, while the photothermal effect in plasmonic AuNPs produces localized heating under irradiation or exposition to laser light or to radiofrequency fields, which results in thermal destruction of malignant tumors [2–4]. When therapeutic radionuclides and several antiangiogenic biomolecules are linked to AuNPs, it is possible to produce

Nanomaterials and Nanotechnology in Medicine, First Edition. Edited by P.M. Visakh.

multivalent nanosystems that work concomitantly as photothermal therapy (PTT) systems, radiotherapy systems, and antiangiogenic drugs [5, 6]. If AuNPs and drugs are trapped within radiolabeled dendrimer cavities, then a plasmonic-photothermal nanomedicine with radiotherapeutic and fluorescent properties is produced [7].

Biomolecule-coated carbon nanotubes (fullerenes) also release heat under radiofrequency fields [8–10]. The thermal properties of nanotubes conjugated to different biomolecules could be useful in the therapy of deep-tissue tumors, where heat can induce necrosis or apoptosis, since radiofrequency easily penetrates the tissues [10].

The combination of polymeric and magnetic NPs functionalized with biomolecules produces smart nanosystems with a response to pH changes, redox fluctuations, or thermal stimuli for drug release, as well as thermal therapy for metastatic cancer, if they are subjected to alternating magnetic fields (AMFs) [11–13]. Therefore, they can be used against tumors by merging hyperthermia with stimuli-dependent controlled drug delivery [14]

The developments in drug delivery nanosystems based on extracellular vesicles and high-density lipoproteins (HDLs), pre-loaded with therapeutic radionuclides and drugs, have the advantage of functioning as natural vectors that may give an added value for translation of nanotechnology to the clinic [15–18]. As will be described in the following review, there is no doubt that nanotechnology based on multifunctional target-specific nanosystems for multimodal therapy of cancer and other illnesses is a promising path to future biological therapies.

5.2 Biological Therapies

5.2.1 Immunotherapy

5.2.1.1 Antibodies and Fusion Proteins

Biological therapies with humanized, chimeric, or human antibodies, and in general fusion proteins, modulate the immunosystem response of the immune system to treat diseases [19]. Antibodies and protein fusion have been developed for B or T cell-directed therapies. B cell-directed treatments are based on the depletion and inhibition of costimulation and the survival or differentiation of B cells, while T cell-directed therapies are based on T cell activation [20]. In particular, antibodies can induce target cell death through different mechanisms: direct targeting (antibody-dependent cellular cytotoxicity by immune effector cells, antibody-mediated direct signaling, and complement-mediated cell lysis), altering the host response (angiogenesis inhibition or T cell checkpoint blockade), delivering cytotoxic agents (radioimmunoconjugates or antibody-drug conjugates), and redirecting T cells (bispecific/trispecific antibodies or chimeric antigen receptor T cells) [19, 21–24]. Depending on the antibody action mecnhanism, the antibody is modified to improve the interaction with target cell antigens and/or immune cells associated with a pathological microenvironment (T cells, B cells, and NK cells) [21, 23, 25]. Approaches for enhancing this interaction include the exchange of amino acids in a constant region, the glycosylation of IgG, the use of different human antibody isotypes (IgG 1, IgG 2, and IgG 4), the linking of β or α particle emitters to the antibody with a stable linker, and the linking of cytotoxic drugs (300–1000 Da with nanomolar IC50 values) to the antibody via optimal linkers [23, 26]. The selection of target antigens, cytotoxic agents, linkers, and conjugation chemistries determine antibody characteristics such as stability, homogeneity, low aggregation propensity, pharmacokinetic efficiency, and therapeutic efficiency [21, 25].

5.2.1.2 Cytokines

Cytokines are secreted proteins that regulate the growth, differentiation, and activation of immune cells [27–29]. Biological therapy with cytokines is mainly used for modulating the response of immune cells [27, 29, 30]. Several cytokines (IFN-α) can directly kill cells through their anti-proliferative, pro-apoptotic or antiangiogenic activity, by NK cells and T cells (granulocyte-macrophage colony-stimulating factor (GM-CSF), IL-15, IL-12, IL-21, IL-2), or through the production of antiangiogenic cytokines (IFN-γ); however, their low half-life value in blood circulation and high systemic toxicity limit their use as therapeutic agents [27–30]. To overcome such characteristics and enhance the pharmacokinetic and pharmacodynamic properties, the cytokines have been modified using recombinant DNA technology and PEGylation techniques [27, 31–33]. Such cytokines can be used in combination with other therapies or as single agents (cytotoxic drugs, radiation, checkpoint inhibitor antibodies, vaccines, and other cytokines) to increase the activity of cytokines targeted at the pathological microenvironment or of immune cells [27, 32, 34]. Interferon α-2b, interferon α-2a, and PEGylated granulocyte colony-stimulating factor have the approval of the FDA. Other alternative polymers of the poly(2-oxazoline) types were used in cytokine-polymer conjugates [35, 36].

5.2.1.3 Vaccines

The fundamental principle of vaccination is to administer a dead or attenuated form of an infectious microorganism or a component that does not cause disease, but rather promotes the formation of antibodies to prevent a disease caused by a living pathogenic microbe. Vaccines have a very wide range of application (a great example is the vaccine against human papilloma, positive for cervical cancer). In addition, dendritic cells (DCs) have been used as antitumor vaccines for different types of cancer. Several attempts are currently being made to develop a vaccine against breast cancer in a murine model, since DCs are the main antigen-presenting cells and are key molecules in the control of immune tolerance and activation. DCs can regulate tolerogenic immune responses, which, in a tumor, facilitate their uncontrolled growth. The efficacy of the CD103 + cDC1 vaccine (which causes inhibition of STAT3 and IL-10) against breast cancer was assessed through tumor growth and survival of the test subject [37]. Efforts are currently being made to develop vaccines based on the use of DCs against breast, lung, melanoma, prostate, and oral cancer [38].

5.2.2 Gene Therapy

Gene therapy consists of modulating gene expression by introducing non-coding RNA or DNA into target cells. It was recently employed to treat diseases by modifying their genetic sources, either by silencing the defective gene (knockout), by replacing defective genes, or by completing missing genes [39, 40]. Currently, small interfering ribonucleic acid (siRNA), antisense RNA, and microRNA (miRNA) are the most efficient ways to inhibit specific gene expression by interfering with mRNA translation within the cytoplasm [40–43]. Recombinant DNA plasmids (pDNA) are the most efficient strategy for encoding specific proteins once their nuclear internalization occurs [39, 44, 45]. Other DNA-based therapeutics such as DNA-aptamers (which intervene in protein biosynthesis), antisense DNA (which directly targets mRNA within the nucleus or cytosol), and Unmethylated cytosine-phosphate-guanosine dinucleotide (CpG) oligodeoxynucleotide (which induces interferon type I production and NK cell activation) have been evaluated as potential gene therapy approaches [45, 46]. However, the highly negative charge of non-coding RNA or DNA reduces its transport through the cell membrane and facilitates its degradation within the cytoplasm and excretion from the blood stream and kidneys (short *in vivo* half-life), limiting its therapeutic activity [36]. To overcome such problems, passively targeted non-coding RNA or DNA delivery systems, such as lipids and PNPs, are the most used strategy for improving their delivery [40–42, 46–48].

5.2.3 Targeted Therapy

For increasing the uptake/retention at the target site and to decrease the non-specific distribution of biological therapies, active targeting ligands with high affinity are conjugated [36, 47, 48]. Particularly, targeting ligands such as pro-apoptotic peptides, antimicrobial peptides, apolipoproteins, immunoregulatory cytokines, toll Like receptor (TLR) ligands, folate, mannose, and epidermal growth factor have been used to bind receptors that are commonly overexpressed on the target cells (cancer or immune cells). These specific recognition systems have shown high affinity with high therapeutic efficiency [22, 47, 49–51].

5.3 Nanoparticles in Biological Therapies

5.3.1 Biocompatible Nanoparticles

5.3.1.1 Organic Nanoparticles

Due to the large variety of NPs, a simple classification is difficult. It is well known that the literature lacks a standardized nomenclature [52]. For example, although organic NPs are made from organic matter, the classification excludes carbon-coated NPs. When designing NPs, size is the most important factor to be considered. For drug delivery, organic NPs are in the size range from 5 to 200 nm and include liposomes, PNPs, dendrimers, micelles, and nanocapsules. Liposomes, lipid-based NPs, and PNPs are the most studied nanosystems [53–55]. NPs are used as vehicles and are expected to have longer circulation times, which would allow them to accumulate at the target site. A great challenge is to selectively bind or reach a target tissue for drug delivery, making it necessary to functionalize the NP surface with specific biomolecules. In general, biocompatible NPs for drug delivery can be divided into several groups.

Polymeric Nanoparticles Polymeric nanoparticles (PNPs) are submicron-sized colloidal particles formed by biodegradable or non-degradable polymers [56, 57]. This type of NP has promising properties that make it useful as a delivery vehicle. The encapsulation is relatively easy, with high capacity and low toxicity [58]. It is well known that encapsulation of hydrophobic drugs improves its bioavailability and tolerability. Also, PNPs are functionalized with specific ligands to make them able to reach a specific receptor site, thus improving the tumor-targeting properties with practically no toxic effects [56]. The main properties of polymeric nanosystems are biocompatibility, biodegradability, and the ability to encapsulate a great number of therapeutic drugs (alone or combined) [57].

PNPs can be nanocapsules (the shape is non-spherical; the drug is encapsulated) or nanospheres (a spherical form with drugs embedded within the matrix) [56]. PNPs are chemically synthetized or obtained from natural polymers: poly(lactic-co-glycolic acid (PLGA), cyclodextrins, poly(β-amino ester) (PBAE), poly(ε-caprolactone) (PCL), poly(styrene-maleic anhydride) copolymer, poly(lactic acid) (PLA), and N-(2-hydroxypropyl)-methacrylamide copolymer (HPMA). Natural polymers include gelatin, alginate, poly(L-lysine), dextran, guar gum, chitosan, and collagen [26]. The most common materials for the synthesis of PNPs are PLGA and poly(ethylene glycol) (PEG).

To synthesize PNPs, there are several methods including: solvent evaporation or solvent extraction and emulsification, nanoprecipitation, and the supercritical anti-solvent method. Many types of molecules such as aptamers, proteins, peptides, antibody fragments, and small molecules are attached to the PNP surface in order to accomplish specific targeting delivery [57].

PLGA has been functionalized to target tumors in murine models. Several biomolecules with specific recognition have been bound to the surface for preclinical evaluation. Also, PLGA NPs have been functionalized with aptamers to evaluate the *in vivo* therapeutic efficacy. To specifically reach nucleolin proteins overexpressed in the membrane of malignant cells, PLGA NPs were used to encapsulate superparamagnetic iron oxide nanoparticles (SPIONs) and doxorubicin. Additionally, they have suitable properties for Magnetic Resonance Imaging (MRI) diagnosis [59].

Liposomes Liposomes are amphipathic vesicles based on phospholipids, which can self-assemble into a bilayer vesicle if it is in aqueous media. Thus, the non-polar end forms an internal bilayer and the polar end faces the water. Vesicles can be unilamellar or multilamellar [60, 61]. Liposome size is found to be between 20 and 1000 nm [53].

To synthetize liposomes, some common methods have been reported: the Bangham or thin film hydration method, the solvent injection technique, and the reverse-phase evaporation method. Recently, with all the emerging technologies, the supercritical anti-solvent method and the reverse-phase evaporation methods have gained importance [60, 62].

Liposomes were first described in 1961 [63]. Doxil® was the first approved liposomal system for cancer treatment. Onivyde® was approved in 2015. In 2017, the FDA approved a new pharmaceutical formulation containing a mixture of daunorubicin and cytarabine to treat patients with acute myeloid leukemia [64]. Between these approvals, a large number of liposomal constructs have been reported in well-organized published reviews [63].

Dendrimers PNPs can also be prepared from PAMAM poly(amidoamines), which are dendritic polymers termed dendrimers. A dendrimer has three main components: a core, branching components, and surface groups [60]. The dendrimer surface has numerous water-soluble functional moieties, while the inner core and branches allow them to encapsulate hydrophobic molecules [65]. When compared to linear polymers, dendrimers are able to prevent renal excretion [66].

The frequently reported methods to produce dendrimers are divergent, convergent, and combined convergent-divergent synthesis. In divergent synthesis, the initiator core grows toward the outside; this process is initiated with a multifunctional core, which is coupled to monomer units. This in turn is coupled to other monomer units, making the structure grow outward by adding building blocks to the surface of the parental dendrimer. PAMAM dendrimers have been synthetized by using N-(2-minoethyl)acrylamide monomers [67].

The convergent method initiates from the outside, in the direction of the central focal point. In this manner, the surface units are attached to additional blocks to form the branch structure. Through this method, more symmetric dendrimers can be synthesized, but with the disadvantage of lesser purity [67]. There are different subfamilies of dendrimers. The most-known dendrimer used as a drug delivery system is the poly(amidoamine) (PAMAM) dendrimer and polypropyleneimine (PPI) dendrimers. Other families are arborols and polyether dendrimers [60, 67, 68]. For targeted drug delivery, dendrimers have been bound to specific biomolecules as nanovectors [68].

Recently, a PPI dendrimer was functionalized with the mAbK1 antibody, which targets the mesothelin protein in ovarian cancer. Paclitaxel was encapsulated into PPI dendrimers (4 : 1 drug : dendrimer ratio). The mAbK1-PPI-PTX nanosystem was then evaluated in mice with induced ovarian cancer. After treatment, tumor volume was reduced and the survival time of the treated mice improved, attributed to the specific drug uptake by the tumor [69, 70].

A nanosystem with epirubicine trapped within the dendrimer cavities and target-specific aptamers attached to the dendrimer was recently designed for epirubicine delivery in response to physiological pH changes [71].

Endogenous High-Density Lipoproteins High-density lipoproteins (HDLs) are cholesterol transporters. The HDL structure contains a non-polar core of esterified cholesterol enveloped by phospholipids, cholesterol (in free form), and apolipoprotein (apo A-I) formed by 11 amphiphilic alpha-helical segments. Synthetic high-density lipoprotein (rHDL) NPs based on ApoA proteins and the ApoA mimetic are capable of actively targeting tumor cells via SR-B1 (B scavenger receptors) specific receptors; therefore, it is possible to employ them as anti-cancer drug delivery agents to malignant cells and tumors at the preclinical stage [72, 73]. The release mechanism involves the recognition of SR-B1, which is overexpressed in cancer cells. The specific interaction of SR-B1/Apo-A has been exploited to transport and release several drugs [73].

Exosomes Exosomes are made of a bilayer membrane containing several types of phospholipids and proteins that are involved in several biological and pathological processes. They provide specialized cell-cell communications useful for drug delivery systems in targeting cells. Exosomes originate from endosomes and have a size between 40 and 100 nm. An exhaustive investigation has been carried out to use exosomes as drug delivery approaches [74, 75].

5.3.1.2 Inorganic Core Nanoparticles

Gold Nanoparticles Therapy applications of functionalized gold nanoparticles (AuNPs) are widely used because of their multifunctional properties, as well as their simple synthesis, controllable size, strength dispersity, easy ligand exchange, biocompatibility, and absence of toxicity. Due to the "gentle" chemical nature of gold, functionalization with target-specific ligands is relatively easy, mainly via gold-thiolate bonding. Figure 5.1 displays the main strategies in the design of AuNP conjugates for biological therapy.

Biological therapies using AuNPs are based on multimeric and multivalence principles. AuNPs are suitable platforms for radiobioconjugate targeting, using monoclonal antibodies, known as radioimmunotargeting (RIT), and are also suitable tools for molecular targeting coupled with a cytotoxic mechanism (chemical or physical stimuli). For example, AuNPs (~39 nm) conjugated to cetuximab and radiolabeled with iodine-131 [76] or Indium-111 [77] have shown specific localization in tumor cells within the intracellular compartment.

A useful strategy is the use of immuno-nanoconjugates (mAB-AuNPs) as sensitizers, followed by an alternative therapy. For example, the trastuzumab-PEG-AuNP conjugate was designed and evaluated as a nanotechnology-based radiosensitizer directed to the human epidermal growth factor receptor-2 (HER2). HER2 contributes to nanosystem internalization and allows the enhancement of external X-radiation [79]. Radiosensitizers based on AuNPs have been widely studied *in vitro* and *in vivo* (e.g., BSA-AuNPs for glioblastoma tumors [87] and goserelin-AuNPs for prostate tumor radiation therapy [88]). Due to the electron collective oscillation of AuNPs with light, another alternative strategy is the use of biomolecule-AuNPs (e.g., immuno-nanoconjugates) against cancer tissues to be treated by PTT procedures [80].

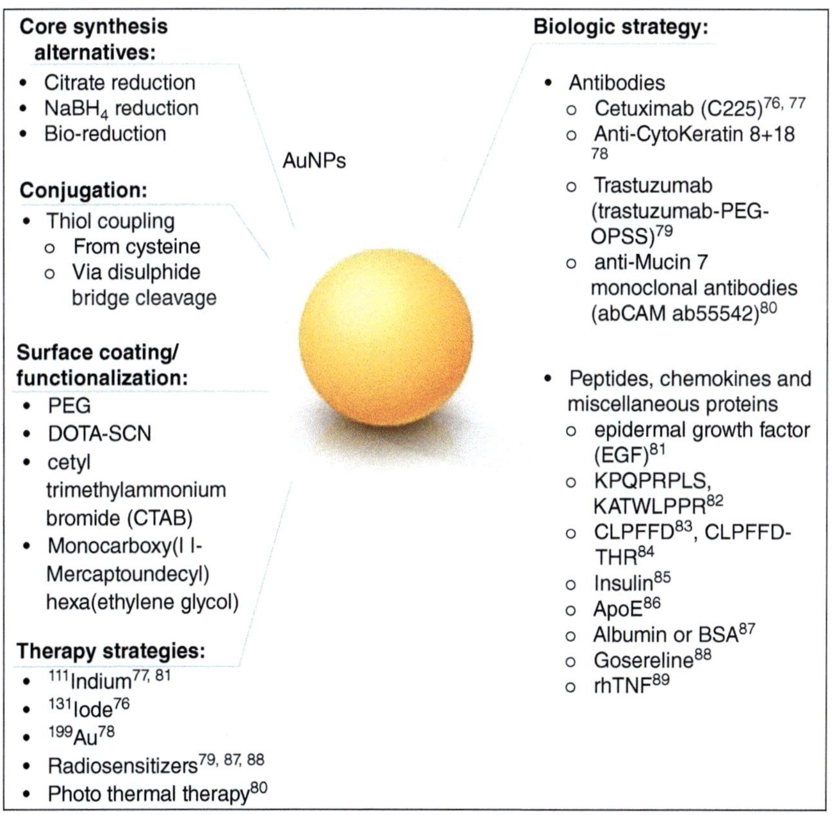

Figure 5.1 Main strategies in AuNP-conjugates for biological therapy [78, 81–86, 89].

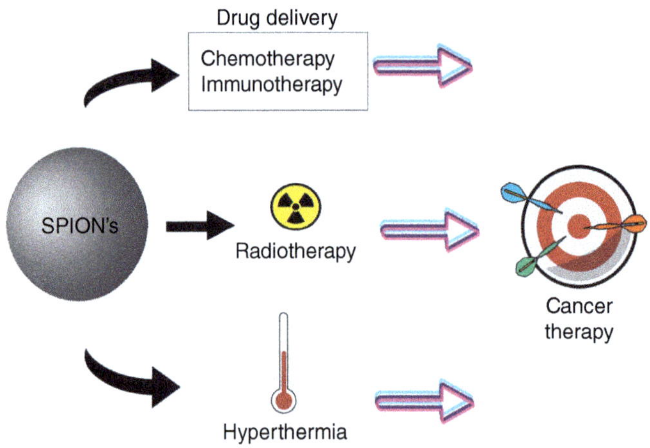

Figure 5.2 Superparamagnetic NPs as promising tools in cancer therapy.

Iron Oxide Nanoparticles Nanoparticles with magnetic or superparamagnetic properties are interesting tools in cancer therapy as drug delivery systems (Figure 5.2) and radiotherapy and primarily for hyperthermia. Metal NPs absorb the energy of AMFs and convert them into heat, due to hysteresis loss, in selected biological tissue (i.e., cancer), causing an increase in temperature, which induces changes in the physiology of the tumor, increasing susceptibility or leading to cell death. In order to enhance the therapeutic effect, functionalization strategies and multifunctional approaches have been reported. Despite the large amount of research in the biomedical field, the applications that involve combination with biological therapies are limited.

Multiple synthesis procedures or stabilization and functionalization methods with biomolecules such as albumin, PEG, dextran, have been reported to minimize systemic toxicity and to maximize the therapeutic effect.

Although different antibodies and peptides conjugated to NPs have been used extensively, many of them have the function of disease site transport, without any activity depending on the biomolecules. The therapeutic activity is related to hyperthermia or an alternative mechanism such as radioactivity.

Multifunctional radiolabeled superparamagnetic iron oxide nanoparticles (SPIONs) were reported by Zolata and colleagues [90, 91] for targeted hyperthermia, multimodal-guided imaging, drug delivery (controlled chemotherapy), and radio-immunotherapy (e.g., Trastuzumab and Doxorubicin [DOX]-conjugated APTES-PEG-coated SPIONs [92]).

A radioconjugate of particular interest was described by Azadbakht and colleagues [93], in which the ^{188}Re-anti-CD20 monoclonal antibody was attached to the modified surface of (amino-silane-PEG)-SPIONs. The *in vivo* and *in vitro* assay demonstrated therapeutic ability against CD20+ cells, with a significant enhancement in tumor uptake with regard to non-conjugate SPIONs.

The production of cetuximab-SPIONs for targeted MRI was described by Tseng and colleagues [94]. The system showed suitable properties as a theranostic nanomedicine for the therapeutic targeting of epidermal growth factor receptor (EGFR)-expressing tumor cells, with active tumor targeting and rapid internalization.

In neurodegenerative therapy, the design nanosystems based on SPIONS (~21 nm) conjugated to fibrinogen-derived γ377–395 peptide, gives stability and maintains the bioavailability of peptide without undesirable systemic effects and toxicity. Considering that γ377–395 peptide is a specific inhibitor of microglial activation via macrophage antigen (MAC)-1 receptor, the nanosystem was proposed as a therapeutic approach in neurodegenerative tauopathies.

Quantum Dots Quantum dots (QDs) are inorganic nanomaterials (e.g., CdSe, CdS, CdTe/CdS, ZnS, PbS, and InP) with optical properties for different bioapplications due to their emissions in the near-infrared (NIR) region.

Although QDs have been widely used for diagnosis, therapeutic and theranostic alternatives are possible. The conjugation of QDs to biomolecules such as antibodies has also been explored [96] for biocompatible systems such as anti-VEGFR2-QD655 (QD with emission at 655 nm) [97], bevacizumab-QD [98], and Anti-HER2-QD [99]. In addition to conventional targeting and delivery, QDs have also been used to deliver peptides and proteins (cell-penetrating peptide and His6-tagged beta-D-galactosidase: JB577-QDs-GALC) [100], nucleic acids (siRNA) [101], drugs, polymers [102], polysaccharides [103], and so on. Although it is already possible to control the physicochemical properties and inorganic synthesis of QDs, the development of strategies focused on biological applications, specifically stabilization, conjugation, and biocompatibility strategies, remain a considerable challenge.

Lanthanide Nanoparticles Lanthanides have paramagnetic and optical properties (sharp-band emissions, large Stokes/anti-Stokes shifts, long luminescent lifetimes, photostability), suitable for biological applications.

Just as in the case of SPIONs, biomedical applications are mostly oriented to imaging techniques (MRI, biosensing through up-conversion emissions and luminescence) [104–106]. The success of the diagnostic, therapeutic, or theranostic applications is based on suitable conjugation, stability, and target ability. Additionally, the increase in cellular response could be obtained by physical means such as photo-dynamic therapy, photo-thermal therapy, radiolabeling drug delivery, and chemotherapy [107] (Figure 5.3).

5.3.2 Nanoparticle-Based Drug Delivery Systems

Nanoparticle-based drug delivery systems (NDDSs) are natural or synthetic nanomaterials designed for the controlled delivery of drugs in order to deposit the optimal amount at the desired site [108]. They consist of three components: the carrier, the route, and the target site. The carrier is composed of a matrix phase (the NP) and a dispersed phase (the drug). The selection of route is crucial for the drug to arrive intact at the desired site. The most used is parenteral, although the oral route is also employed.

Conventional NPs are used for the transport of biomolecules such as proteins and drugs based on nucleic acids [44, 109–116], interleukins [116–119], monoclonal antibodies [120–122], enzymes [123, 124], peptides [125, 126], and substances of natural origin [127, 128]. The *in vivo* stability of these biological substances—in particular, proteins (of easy renal clearance and enzymatic degradation) [111]—issignificantly improved by

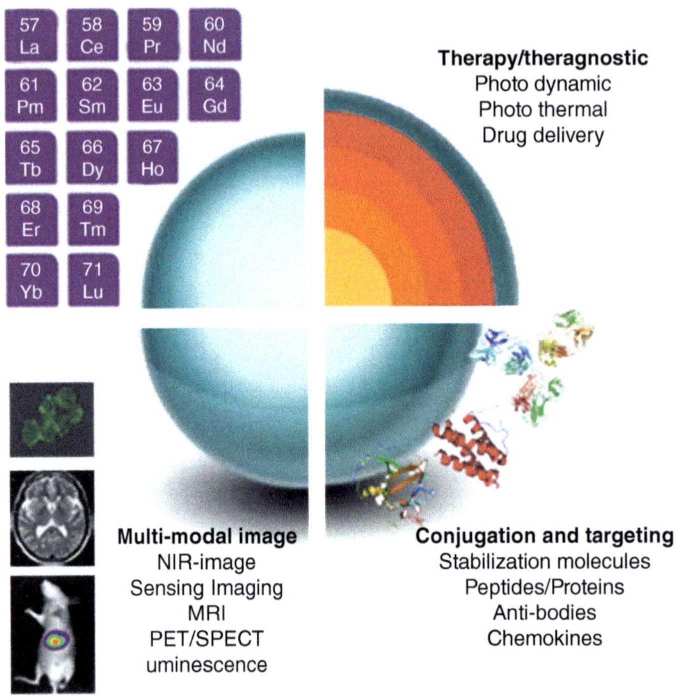

Figure 5.3 Lanthanide NPs; preparation strategies for medical applications.

being protected by the NP [109, 129]. The best NDDSs for biological molecules are silica NPs of diverse structure [112, 126, 129, 130], although good results have been obtained with polymer NPs/graphene oxide [111, 124], polymeric nanogels [113], AuNPs [129], lipid NPs [131, 132], micelles [124, 133], and supramolecular architectures such as NP/protein [109, 134].

To reach the target site, the NP must go through a series of stages [135]: blood circulation and accumulation/penetration in the tumor.

The blood circulation time of the nanosystem should be prolonged. Plasma proteins tend to eliminate NPs through the reticuloendothelial system (RES), by leading them to the liver, spleen, or bone marrow [135]. The blood circulation time depends on the intrinsic characteristics of the particle (composition, surface charge, size, among others). To prolong their stay in circulation, various strategies are used, such as covering the surface of the NP with inert polymer molecules [122, 136] or proteins [137], changing the surface charge of the NP [135], and using multifunctional NPs [124], among others.

The accumulation in the tumor is carried out through two mechanisms: passive targeting and active targeting. In conventional NDDSs, the NP accumulates in the tumor by passive targeting (convection or diffusion) through tumor capillary fenestrations, due to enhanced permeability and retention effect (EPR) of the tumor tissue [135, 138]. In this manner, it is retained by the tumor. Passive targeting depends on tumor vascularization, angiogenesis, interstitial fluid pressure, temperature and pH of the biological environment, and of course physicochemical properties of the NP, such as lipophilic characteristics, molecular weight, size, and surface charge [139, 140].

Penetration into tumor tissue depends on the microenvironment of the tumor, whose structure is very complex. Perfusion is low, blood flow is irregularly distributed, the vessels' permeability is heterogeneous, and the microenvironment is dense [53]; all of these factors cause steric, hydrodynamic, and electrostatic impediments that the NP must overcome to enter the tumor and reach the target cells [141]. At this stage, the shape, charge, and size of the NP are fundamental. Smaller particles (30 nm) penetrate tumors much more easily than larger ones [135, 142], but they are cleared more quickly by the kidneys and liver when they enter the bloodstream, so only a small part reaches the tumor [122]. Non-spherical particles are suitable for tumors with small pore sizes, but not for tumors with large pores. Positively charged ones penetrate tumor tissue better [140]. Those that have a good accumulation of enhanced permeability and retention (EPR) do not have a favorable tumor penetration [141]. On the other hand, the structure of the tumor tissue widely differs from one tumor to another, so the barriers also differ. Some of these disadvantages can be overcome by modifying the microenvironment of the tumor by physical or chemical methods [142, 143], by structural changes in the NP [135, 144], or by the use of stimuli-response NPs [143, 145].

Cellular internalization is vital for the therapy to be effective. The primary role is played by the cell membrane [146, 147], although the size, shape, load, and stiffness of the NP also play a significant role [147]. The main mechanism of internalization of NDDSs is endocytosis (phagocytosis and pinocytosis); particles larger than 500 nm are internalized by phagocytosis, while smaller ones are internalized by pinocytosis (caltrin-independent or not) [147] that may or not be mediated by receptors [148]. The type of endocytosis determines the degree of internalization. In caltrin mediated endocytosis, the particle reaches the lysosomal compartment, but if mediated by caveolin, the particle does not reach the compartment [144, 147]. When the surface of the NPs is functionalized with ligands for targeted therapy, the fundamental mechanism of internalization is receptor-mediated endocytosis [140].

The controlled release of the drug is produced by different mechanisms, but the main mechanisms are diffusion, erosion, and swelling [108, 149]. Depending on the nature of the particle and the drug, some have more influence than others. In general, the release level depends on (i) the solubility of the encapsulated drug, (ii) the desorption of the drug that is bound or adsorbed onto the surface of the NP, (iii) the diffusion of the molecule through the particle matrix, (iv) degradation/erosion of the particle matrix, and (v) the combination of diffusion and erosion. Of these, the most important are solubility, diffusion, and degradation/erosion [149, 150]. In polymeric particles, for example, the swelling of the polymer affects the diffusion of the drug through the matrix and produces changes in its structure; the pH varies due to hydrolysis, which further accelerates the degradation. The polymer is thus destroyed, and the drug is released. These processes depend on the bonds of the polymeric structure and the composition of the monomers, copolymers, and conjugated molecules [108, 149]. In PNPs, enzymatic degradation and the composition of the lipid layer play a fundamental role [151]. Another influential factor is the method of incorporating the drug into the NP. In the nanospheres (where the drug is homogeneously distributed), the release is due to the diffusion and erosion of the matrix. While in the case of nanocapsules, the solubility of the drug plays a more fundamental role [149, 150].

The substances incorporated onto the surface of the NP are released faster than those encapsulated in the inner core. If the NP is covered by a polymer, the release is controlled by the diffusion of the drug through the polymeric membrane [149]. Smaller NPs release the drug faster than those that are larger [150].

The changes that occur as the NP penetrates the tumor tissue influence the release. Tumor tissue is characterized by low oxygen levels and low perfusion, which cause hypoxia. The pH is more acidic as a result of the production of carbon dioxide and lactic acid [152]; this difference in pH causes variation in the concentrations of several enzymes that contribute to particle degradation and drug release [138, 140].

The five stages described depend on the physicochemical properties of the NDDSs, which must overcome biological barriers before exerting their therapeutic effect [122]. In particular, the internalization and release of the drug are facilitated and accelerated by internal and external stimuli using the so-called stimuli-responsive drug delivery systems, which undergo changes in their physicochemical properties (size reduction, surface load, exposure or surface ligand concealment, and particle permeability adjustment) when exposed to such stimuli [135, 145, 153]. The use of these intelligent systems improves tumor penetration, cell internalization, and controlled drug release. Among the external stimuli, there is electricity [154], magnetism [128, 155], ultrasound [156, 157], visible/ultraviolet light [124, 135, 145, 158], and temperature [159, 160]. Internal stimuli consist of enhancing/decreasing factors such as pH [144, 145, 160], internal mechanical forces [153], ionic strength redox conditions [145, 160, 161], concentration of adenosin triphosphate (ATP) [135], and enzymes [9, 160, 161]. Such stimuli-response systems can be used alone or in combination [162].

5.3.3 Nanoparticle-Based DNA and RNA Delivery Systems

Naked genetic material has little internalization in the target cell due to the action of nucleases, repulsion between negative charges of DNA and the cell membrane, rapid renal elimination, and phagocyte uptake. To eliminate these difficulties, the genetic material is carried inside a vector that "protects" it. The first vectors used were viral, but many of them produced an immune response [114]. The development of nanotechnology has allowed the design of non-viral vectors (NPs), which has promoted research in this field [163].

In gene therapy, the biological compounds (genetic material) that are carried to the target cell are mainly DNA nucleotides, DNA plasmids, siRNA, miRNA, ribozymes, aptamers, and chemically synthesized molecules [114, 164]. Organic or inorganic NPs have been used as non-viral vectors. Among the organic NPs are the lipid NPs and the polymeric NPs (PNPs). Lipids are subclassified into cationic lipids, liposomes, asymmetric liposomes, stable nucleic acid lipid particles, solid lipid particles, bubble liposomes, lipoproteins, extracellular vesicles, self-nanoemulsifying drug delivery systems (SNEDDSs), and carbon. Regarding carbon, nanotubes and carbon dots are usually employed. The PNPs, in turn, are subclassified into polymeric micelles, dendrimers, and nanogels. The inorganic vectors are classified into QDs, ceramics, metals, and graphene oxides. Metals, in turn, are subclassified into magnetic and non-magnetic.

To allow the genetic material to cross the cell membrane, its charge must be modified. To do this, derivatives, compounds, adducts, or ion pair complexes are prepared, adding a compound bearing cationic sites. This complex is formed by the electrostatic attraction between the positive charges of the added compound and the negative charges of the genetic material. Depending on the final balance of the charges, the complex formed is hydrophobic or cationic, which guarantees the absence of repulsion by the cell membrane. This complex has the function of condensing and protecting the genetic material and is present in all nanoparticulate systems used for gene therapy.

5.3.3.1 Organic Nanoparticles

If a cationic lipid is used for complex formation, the complex formed is called a lipoplex. A lipoplex is the common element in the preparation of lipid release systems: cationic lipids [165], liposomes [114, 166], asymmetric liposomes [167], stable nucleic acid lipid particles [114], solid lipid particles [114], bubble liposome-assisted transfection [114], and SNEDDSs [168, 169]. In low-density lipoproteins (LDLs) and HDLs, the genetic material can be incorporated in the form of a lipoplex, but also by using cationic amino acid derivatives [170–172]. In all these systems, the lipoplex or ion pair complex is incorporated into the NP structure during assembly.

Extracellular vesicles are also lipid nanosystems and use a lipoplex to carry genetic material. The difference is in the incorporation of the lipoplex to the NP. Vesicles are not synthesized but are isolated from cell cultures. The genetic material must be introduced into them, and this can be done in two ways: the first route is by direct transfection of the vesicle using a lipoplex complex, primarily with lipofectamine; the other route is by incubating the genetic material with the vesicle donor cell in a cell culture for a certain time, followed by an incubation in the normal culture medium for the secretion of the vesicles with the genetic material already incorporated. These secreted vesicles are subsequently isolated [173, 174]. The lipoplex transfection efficiency is high because of its easy "dissociation" once it reaches the cellular environment.

If the cationic compound used for complex formation is polymeric in nature, the complex formed is called a polyplex. In the polymeric micelles, the polyplex is housed in the hydrophobic nucleus of the NP [166, 175]. The polymer that is used to form the polyplex is a molecule with a greater number of cationic sites than a lipid. The polyplex is therefore not easy to "dissociate" once the cellular environment has been reached, as it is a complex with more genetic material binding sites than a lipid. For this reason, the process of transfection from a polyplex is more difficult and generally requires the assistance of external or internal stimuli such as the use of light, ultrasound, magnetism, electricity, temperature, pH, redox properties, and reactive oxygen species (ROS), among others [175, 176]. Nanogels also form polyplex structures with the genetic material [177], since, in their composition, there are polymeric molecules. Once the cellular environment is reached, they have the inherent property of changing their structure and architecture (swelling and deflation) in response to changes in pH, temperature, redox conditions, ionic strength, light, and so on, between the extra- and intracellular environment [177]. Dendrimers are other NPs classified as polymeric; these possess cationic sites within their structure (generally, amino groups), so they naturally form a complex with the genetic material. Although the nature of the complex formed in dendrimers is also polyplex, it is called dendriplex. As the synthesis of dendrimers is controlled, the formation of dendrimex is also controlled, achieving high transfection efficiencies *in vitro* [178] without the need to use stimuli. The genetic material in the polymeric micelles can be loaded during the assembly of the NPs or after the preparation of them. In the case of nanogels and dendrimers, the loading of the genetic material is done once the NPs are prepared.

Carbon nanotubes, as well as carbon dots, lack cationic functional groups in their structure; therefore, these must be functionalized [179, 180]. Covalent functionalization increases the biocompatibility of the NP but limits the number of cationic centers in the structure and weakens the physical properties of the nanotube. The non-covalent functionalization coats the nanotube with an amphiphilic molecule to form the complex with the positively charged genetic material. In this variant, the NP does not lose the physical properties (fundamentally, optical). Although the transfection efficiency of carbon nanotubes and QDs is high, it can be increased by their optical properties, such as photothermal properties. These same properties make these systems attractive as theranostics and dual therapy based on genetics and phototherapy [180, 181].

5.3.3.2 Inorganic Nanoparticles

Although inorganic particles that have positively charged centers like magnets (such as iron oxide) and ceramics (such as hydroxyapatite) allow the formation of a direct electrostatic complex with the genetic material, these are usually coated with an amphiphilic material [182]. This coating allows the control of crystal size (difficult to control with the direct complex) and increases the transfection efficiency [163, 182, 183]. The rest of the inorganic NPs which can carry genetic material (QDs, ceramics such as silica and calcium carbonates, metal such as gold, and carbon such as graphene oxide) are all coated with amphiphilic compounds that provide cationic sites to form the lipoplex or polyplex complexes. The coating is performed with the surfactants, lipids, or polymers that are used in the preparation of the lipid and PNPs [164, 173, 182, 184]. The function of the coating is not only to form the electrostatic complex (which increases transfection efficiency), but also to decrease the toxicity of inorganic NPs such as QDs. An advantage of inorganic NPs is the possibility of preparing theragnostic systems for gene therapy taking advantage of the optic features of these NPs.

From all of these, the lipid and polymer systems seem to be the most effective and have products approved by the FDA and the European Commission. Therefore, the preparation of hybrid systems, such as inorganic NPs loaded with genetic material and internalized or adsorbed into/onto an organic NP, could be an option to boost the arrival of inorganic NPs to the clinical trial phase.

5.3.4 Nanoparticle-Based Targeted Therapy

Difficulties of conventional NDDSs, such as low tumor penetration, low specificity, and uncontrolled drug release can be solved through active targeting therapy. This mechanism is based on the different features of tumor cells with regard to healthy cells and is important for the internalization of drugs; particularly, those based on biological molecules that cannot otherwise be internalized into the cell [185, 186]. For this purpose, tumor-specific ligands or biomarkers capable of binding to homogeneously overexpressed receptors or antigens of the malignant cells or tumor vasculature cells [135, 140], which are conjugated to the surface of the carrier NP expressed or inaccessible in normal cells, have been designed [139, 144]. These ligands, prone to internalization, contribute to increasing therapeutic efficacy and minimizing the side effects of the drugs carried [128]. The ligand is coupled to the NP in two ways: (i) by physical adsorption or chemical conjugation after the NP is formed and (ii) chemical conjugations to molecules prior to "bottom up" nanoparticle synthesis [138].

Because cancer biomarkers include cellular molecules and components (mutant genes, RNA, proteins, lipids, and small metabolites), the variety of ligands for targeted therapy is considerable. It includes monoclonal antibodies (MAbs) or fragments of these (Afbs), peptides, aptamers, vitamins, carbohydrates, and other molecules that promote cellular uptake of NDDSs by means of receptor-mediated endocytosis [124, 135, 138]. Some ligands prevent the action of enzymes, proteins, and other molecules that participate in the growth and spread of cells [140]; others help the immune system in destroying cancer cells or carrying toxic substances directly to them with the purpose of destroying them [185], while others, such as the Tat peptide (derived from HIV-1) [144], the AS-1411 aptamer [138], the VP22 protein of the type 1 herpes simplex virus, and the transcription factor of homeodomain Antennapedia [140], reach the cell nucleus.

For the targeted therapy to be successful, a strong interaction between the NP surface groups and the surface receptors/antigens is necessary. To achieve this, there must be (i) a large number of receptors/antigens accessible on the cell surface; (ii) a high ligand-receptor/antigen affinity, and (iii) the ability of the ligand to induce receptor-mediated endocytosis [187].

In targeted therapy, two types of targets must be recognized: (i) the tumor cell and (ii) the tumor microenvironment cells [139, 144]. Ligands that target the tumor cell interfere with cell function and signaling and inhibit pro-cancer cell pathways through different mechanisms [144, 188]. Ligands that target tumor endothelial cells interfere with angiogenesis and induce cell death due to oxygen deficiency and nutrients [135, 144, 152, 189]. To transfer the NP to the tumor cell, transferrin, folic acid (FA), non-immunogenic glycoproteins (which bind to carbohydrate residues on the surface of the membrane), antibodies/fragments, and specific peptides are used [138, 144]. Vascular endothelial growth factor and its receptors (VEGFR), prostate-specific membrane antigen (PSMA), and peptides with the sequence Arginine-Glycine-Aspartic Acid (RGD) that bind to integrin asp3 [144] are used to direct the nanosystem toward endothelial cells [190]. Table 5.1 shows some of the most-used ligands in targeted therapy to functionalize the surface of NPs.

Monoclonal antibodies are biocompatible and very specific for cell receptors but are immunogenic. The surface density of the NP is low due to its size, which also influences the low permeability to the cell membrane [138], although methods such as membrane fusion have been developed, which allow direct access to the cytosol (and potentially, to the nucleus) without passing through the cell membrane [134]. Afbs and peptides are more stable than MAbs, they are produced at a lower cost, their spatial structure allows them to be more easily incorporated into the surface of the NP, they are less immunogenic and with specificity similar to that of MAbs [137, 138], they penetrate the cell membrane better, they are more resistant to enzymatic hydrolysis, and some are able to directly inhibit tumors [191]. Peptides that recognize the interleukin 4 receptor (IL-4R) can be used to direct the NP to

Table 5.1 Some ligands used in targeted therapy with NPs and their corresponding cell antigens/receptors.

Ligand	Receptor/antigen to which it is directed	Ligand	Receptor/antigen to which it is directed
Monoclonal antibodies and antibody fragments			
Chimeric L6	L6 glycoprotein	anti-VEGF	VEGF
anti-CD-20	CD-20 antigen	anti-Fas	Fas receptor
di-scFv-c-antiMUC-1	MUC-1 glycoprotein	anti-EGFR	EGFR
anti-CD-47	CD-47 receptor	PR81	MUC-1 receptor
Peptides			
RGD	$\alpha_v\beta_3$ integrin receptor	Tat	Nuclear internalization
IL4RPep-1	IL-4 receptor	NR7	EGFR
AG73	AG73 receptor	LHRH	LHRH receptor
H2009.1	$\alpha_v\beta_6$ integrin	Angiopep-2	LRP
CXCR4-L	CXCR4 receptor	TbFGF	FGFR
iPSMA	PSMA	IL-13	IL-13 receptor
Bombesin	GRP receptor		
Aptamers			
AS-1411	Receptor de nucleolin	GBI-10	Tenascin C
Pegaptanib	VEGF receptor		
Vitamins			
Folic acid	Folate receptor (FR)	Vitamin D	Vitamin D receptor (VDR, NR111)
Proteins			
Transferrin	Transferrin receptor	Lectin	
APO-AI	SRB1 receptor	APO-AII	LDL receptor
Carbohydrates and polysaccharides			
Mannose	Mannose receptor	Galactosamine	ASGPR
Lactose	Receptor asialoglycoprotein (ASGPR)	Hyaluronic acid	CD44 receptor
Galactose	ASGPR	Lactobionic acid	ASGPR
Other molecules			
EGF	EGFR	CSA	CD44 receptor
EGal	EGFR	BSA-folic acid	Folate receptor and secreted protein, acidic and rich in cysteine

both tumor and endothelial cells [138]. Aptamers can target more cell surface receptors/antigens than antibodies and peptides [192]. They are small and very specific, with high binding capacity; they are biodegradable and can be designed at will and modified by chemical and/or enzymatic means [138, 192]. Some are immune system modulators, so they can be used in multi-purpose particles [138, 193]. Vitamins—in particular, folic acid (FA)—constitute favorable ligands. FA receptors are overexpressed on the surface of solid and macrophage tumors [194], so they are widely used as a ligand in many NDDSs, including stimuli-response systems [195]. Transferrin (Tf) receptors are highly expressed in many cells, so this glycoprotein, responsible for iron transport in the body and with great capacity for cellular internalization [138], is widely used as a ligand for targeted therapy while lectin is used to bind to carbohydrates present on the cell surface [140]. Among the polysaccharides, hyaluronic acid (HA) stands out. This molecule, together with collagen, is one of the main components of the extracellular matrix [191]; it binds to CD44, which is overexpressed on the surface of cancer cells [138]. It has the peculiarity that it can be simultaneously used as a superficial ligand and as part of the NP structure [191]. It is degraded by hyaluronidase 1 (Hyal-1) and accelerates the release of the hauled drug [138, 191]. Sugars such as lactose, galactose, and galactosamine are used to direct the NPs to the asialoglycoprotein receptor (ASGPR), while mannose 6-phosphate targets specific receptors expressed in various types of tumors [196]. Most of these ligands have been widely used for targeted therapy with cytotoxic drugs, in gene therapy, and to guide stimuli-responsive drug delivery systems that employ external stimuli.

5.4 Application of Nanotechnology in Biological Therapies

5.4.1 Drug Delivery Nanosystems in Oncology

Drug delivery nanosystems based on polymeric and lipidic NPs have been applied in preclinical and clinical oncology studies. In particular, these NPs, in combination with biological therapies (antibodies, siRNA, and DNA plasmids), are used to release cytotoxic drugs onto the target cells and stimulate the immune system for the enhancement of drug therapy; however, few systems have been approved by the FDA. Table 5.2 shows polymeric and lipidic NPs under clinical trials.

The only clinical trial (Phase I) that involves dendrimers is the AZ0466 system (AstraZeneca), to treat solid tumors, lymphoma, and multiple myeloma, as well as intermediate risk or high TLS risk with hematologic malignancies. AZ0466 is an inhibitor of Bcl-X protein [197].

For NP-based regenerative therapy, the OP-101dendrimer (useful for immunomodulation in cerebral adrenoleukodystrophy) is still being clinically evaluated in healthy volunteers to evaluate its safety (Protocol NCT03500627, 2018).

Immunoengineering technology has also taken advantage of PNPs to generate cancer vaccine formulations to treat human papillomavirus, among a large number of designed, prepared, and evaluated biological PNPs [60].

Table 5.2 Polymeric, lipidic, and micellar NPs in clinical trials.

Nanoparticle	Description	Phase	Reference
PLGA	Cancer vaccine	Preclinical Human papillomavirus	[198]
pLHMGA	BIND-014 Docetaxel onto NPs functionalized with PSMA (prostate-specific membrane antigen)	First-in-human phase I Prostate cancer	[199, 200]
Paclitaxel (PTX) albumin-stabilized NP (Nab-paclitaxel) with bevacizumab	Albumin-micelle NP with monoclonal antibody bevacizumab (anti-VEGF monoclonal antibody)	Phase II completed Breast cancer	NCT00733408 [201]
PTX Albumin NP with Gemcitabine, and Bevacizumab	Paclitaxel-albumin-polymeric NP with monoclonal antibody bevacizumab and Gemcitabine	Phase II completed Breast cancer	NCT00662129 [202]
PTX Albumin NP with Cetuximab/Cisplatin	Paclitaxel-albumin-polymeric NP with monoclonal antibody cetuximab and Cisplatin	Phase I and II completed Head and neck cancer	NCT00851877 [203]
NC604 micellar NP PEG-poly (amino acid) block co-polymer	Nontargeted system NP, containing docetaxel	Phase Ib/II Phase I Lung, breast, colon, and pancreatic adenocarcinomas	[204–206]
NK105 polymeric micelles	Nontargeted system paclitaxel (PTX) enhanced permeability and retention (EPR) effect	Phase I Breast/gastric cancer	[207, 208]
AZ0466 Dendrimer	Advanced hematologic or solid tumors	Phase I Diseases with advanced hematologic or solid tumors	NCT04214093
ImDendrim 5th generation [188Re]Rhenium complex coupled to an Imidazolic ligand and associated with a dendrimer	Inoperable liver cancer	Treatment of non-responding to conventional therapy inoperable liver cancers by *in situ* introduction of ImDendrim	NCT03255343 [209, 210]
EGFR-immunoliposomes loaded with doxorubicin	Solid tumors	Phase I completed EGFR-positive solid tumors	NCT01702129 [211]

Table 5.2 (Continued)

Nanoparticle	Description	Phase	Reference
Pegylated liposomal doxorubicin hydrochloride	PEG-PEI-cholesterol Liposomal loaded with doxorubicin hydrochloride in combination PEG-PEI-cholesterol with IL-12 DNA plasmid vector GEN-1	Phase I completed Treatment of persistent ovarian epithelial, peritoneal, or fallopian tube cancer	NCT01489371 [212]
Nano-liposomal CPT-11	NP/liposome containing CPT-11 irinotecan	Phase I completed Recurrent malignant glioma	NCT00734682 [213]
Liposomal irinotecan, fluorouracil, and leucovorin	Nanoliposome irinotecan with fluorouracil and leucovorin	Phase II recruiting refractory advanced high-grade neuroendocrine cancer of gastrointestinal, unknown, or pancreatic origin	NCT03736720 [214]
ALN-VSP02 Liposome loaded with KSP-siRNA and VEGF-siRNA	siRNAs into Lipidic NP against the kinesin spindle protein and VEGF mRNAs	Phase I completed Solid tumors	NCT 00882180 NCT01158079 [215]
siRNA-EphA2-DOPC	A liposome loaded with a siRNA directed against EPHA2 mRNA	Phase I recruiting Advanced cancers	NCT01591356 [216]
AtuPLEX Liposome	Cationic liposome loaded with siRNA against CD31 and the tyrosine kinase receptor for angiopoietins (TIE-2) mRNAs	Phase I and Phase II completed Advanced solid tumors	NCT 00938574 I NCT 01808638 I/II [217, 218]
TKM-080301 Liposome	Liposome loaded with siRN against Polo-like kinase 1 (PLK1)	Phase II completed patients with advanced hepatocellular carcinoma	NCT 01262235 [219]
DCR-MYC MYC Lipid NP	Lipid NP with siRNA against c-Myc mRNA	Phase I terminated Phase I/II terminated Patients with hepatocellular carcinoma, lymphoma, solid tumors, multiple myeloma, non-Hodgkin lymphoma	NCT 02110563 I NCT02314052 I/II [220]

Source: U.S. Food and Drug Administration. https://clinicaltrials.gov website.

5.4.2 Nanoparticle-Based Targeted Radiotherapy

Cancer remains one of the leading health problems worldwide. Nanomedicine has new routes for the advancement of biological and target-specific therapies based on smart radionuclide-bearing NPs, which selectively deliver high doses of ionizing radiation in malignant cells for tumor ablation without significantly damaging healthy tissues. As mentioned earlier, therapeutic nanosystems are multivalent and multifunctional. Table 5.3 presents a number of radiotherapeutic nanosystems with successful results already demonstrated in preclinical studies. As can be observed, all of them are prepared as multimeric and/or drug-loaded NPs capable of generating local and selective multimodal therapies, such as immunotherapy, chemotherapy, gene therapy, antiangiogenic therapy, radiotherapy, and plasmonic PTT. For example, the multimodal ^{177}Lu-nanoplatform composed by AuNPs, nuclear localization peptides, antiangiogenic aptamers (anti-VEGF, 28 RNA bases), and RGD target peptides shown a synergistic effect induced by antiangiogenic therapy, plasmonic PTT and radiotherapy [6, 221]. Nevertheless, the most significant impact is that produced by the ionizing radiation generated by Lu, which at the cancer-cell level, imparts a dense ionizing energy pattern related to tumor ablation [222] Another highlight is plasmonic PTT, which can feasibly be produced by all nanosystems containing AuNPs (AuNP-nanorods and AuNP-nanospheres, with sizes ranging from 2 to 100 nm), when tissues containing AuNPs are exposed to laser irradiation [1, 2].

Table 5.3 Radiotherapeutic nanosystems evaluated in preclinical studies.

Radionanosystem for therapeutic applications	Description of the potential multifunctional properties	References
^{99m}Tc/^{177}Lu-AuNP-Tat-BN	Radiotherapy and Plasmonic Photothermal Therapy (PPT). Dual radiolabeling of AuNPs (AuNPs labeled with ^{99m}Tc and ^{177}Lu) conjugated to the Tat(49–57)-Lys^3-bombesin peptide with internalization to the cell nucleus (reaching t DNA) and molecular recognition by gastrin-releasing peptide receptors (GRPRs) overexpressed in prostate cancer cells.	[4, 223]
^{166}Ho-MSN (mesoporous silica NPs)	Radiotherapy, Cerenkov Luminescence. MSN with ^{165}Ho(AcAc)$_3$ and activated by neutron irradiation.	[224]
^{177}Lu-AuNP-RGD	Radiotherapy, antiangiogenic therapy and Plasmonic Photothermal Therapy. ^{177}Lu-labeled AuNPs conjugated to the RGD peptide with molecular recognition by alpha(v)beta(3) integrins overexpressed in the tumor-associated neovasculature.	[3, 5]
^{177}Lu-DOTA-DN(PTX)-BN	Radiotherapy and Chemotherapy. ^{177}Lu-labeled dendrimer loaded with PTX and conjugated to bombesin with molecular recognition by GRPRs overexpressed in breast cancer.	[225]
^{177}Lu-Bombesin-PLGA (paclitaxel)	Radiotherapy and Chemotherapy. ^{177}Lu-labeled PGLA polymeric NPs loaded with paclitaxel and conjugated to bombesin with molecular recognition by GRPRs overexpressed in breast cancer.	[226]

Table 5.3 (Continued)

Radionanosystem for therapeutic applications	Description of the potential multifunctional properties	References
^{177}Lu-HA-PLGA(MTX)	Radiotherapy and Chemotherapy. ^{177}Lu-labeled PGLA-hyaluronic acid (HA) NPs loaded with methotrexate (MTX) showing molecular recognition by hyaluronan receptors and CD44 overexpressed in the inflamed tissue of joints. Designed for the local treatment of rheumatoid arthritis.	[227]
^{165}Ho-IG-Pt	Radiotherapy, Chemotherapy and Thermotherapy. Magnetic NPs doped with ^{165}Ho activated by neutron irradiation and loaded with platinum radiosensitizers. An external magnet is used for selective delivery to tumors.	[228]
^{177}Lu-DN-AuNP-Folate-Bombesin	Radiotherapy and Plasmonic Photothermal Therapy produced by AuNPs in the dendritic cavity of ^{177}Lu-dendrimer conjugated to bombesin and folate, showing plasmonic-photothermal and fluorescent properties, with molecular recognition by two overexpressed receptors in triple-negative breast cancer cells.	[7, 229]
^{188}Re-BMEDA-loaded immunoliposomes and ^{188}Re-DOX-liposome	Radiotherapy, Immunotherapy and Chemotherapy. Antibodies conjugated to liposomes (immunoliposomes) bearing high specific activities of ^{188}Re. ^{188}Re-BMEDA-labeled PEGylated doxorubicin-loaded liposomes.	[230, 231]
^{177}Lu-T-AuNPs	Radiotherapy and Plasmonic Photothermal Therapy. ^{177}Lu-labeled AuNPs with molecular recognition by the EGFR overexpressed in several cancer types.	[232, 233]
^{211}At-AuNP-S-PEG-SP	Radiotherapy and Plasmonic Photothermal Therapy. ^{211}At-labeled AuNPs with molecular recognition by the NK1 receptors overexpressed on glioma cells.	[234]
^{177}Lu-trastuzumab-AuNP	Radiotherapy, Immunotherapy, and Plasmonic Photothermal Therapy. ^{177}Lu-labeled AuNPs with molecular recognition by breast cancer cells expressing HER2 (epidermal growth factor receptors-2).	[235]
^{198}AuNP-RGD and ^{198}AuNP-Bombesin	Radiotherapy, Cerenkov Luminescence. AuNPs activated by neutron irradiation and molecular recognition by integrins and gastrin-releasing peptide receptor (GRPR) overexpressed in several cancer types. From ^{198}Au decay in tissue, Cerenkov radiation is produced.	[236, 237]
^{131}I-HSA-PTX	Radiotherapy and Chemotherapy. ^{131}I-labeled albumin NPs containing paclitaxel (PTX).	[238]
^{131}I-AuNR-PEG-cRGD	Radiotherapy, Antiangiogenic Therapy and Plasmonic Photothermal Therapy. ^{131}I-labeled gold nanorods conjugated to RGD for molecular recognition by integrins overexpressed in the tumor neovasculature.	[239]

(*Continued*)

Table 5.3 (Continued)

Radionanosystem for therapeutic applications	Description of the potential multifunctional properties	References
^{225}Ac-SWCNT-E4G10	Radiotherapy, Immunotherapy, and Thermotherapy. ^{225}Ac-labeled single-wall carbon nanotubes (SWCNT) conjugated to the E4G10 antivascular antibody.	[240]
^{90}Y-DTPA-SiO_2@Fe_3O_4	Radiotherapy and Thermotherapy. ^{90}Y-labeled magnetic NPs of SiO_2@Fe_3O_4.	[241]
^{225}Ac-immunoliposomes	Radiotherapy, Immunotherapy. Antibody conjugated to liposomes (immunoliposomes) bearing ^{225}Ac.	[242]
^{177}Lu-AuNP-RGD-NLS-aptamer	Radiotherapy, dual Antiangiogenic Therapy and Plasmonic Photothermal Therapy. ^{177}Lu-AuNP-nanoradiopharmaceutical: an antiangiogenic agent for targeting both α(v)β(3) integrin and the VEGF pathway.	[6, 221]

Although traditional therapeutic radionuclides are based on β-particle emitters such as yttrium-90, iodine-131, lutetium-177, and copper-67, in recent years, α-particle-emitting radionuclides have shown impressive results for the ablation of malignant tumors in preclinical and clinical research protocols [243–245]. The short range and high energies of alpha particles produce cytotoxicity directed to the target cells, minimizing damage to surrounding healthy cells. Compared to the conventional monomeric alpha-emitting radiotracers used in targeted radiotherapy so far, the application of nanotechnology can allow specific multivalent binding of the biomolecules attached to the NPs with the specific protein receptors associated with the cancer cells or the tumor microenvironment. In this context, alpha-emitting radionuclides such as Ac [224] and 211At may be the most suitable in the design of future nanosystems, in combination with specific biological therapies based on peptides, antibodies, RNAs, or aptamers.

5.5 Advantages and Disadvantages of Nanoparticles in Biological Therapies

One of the main advantages of NPs, both organic and inorganic, is their ability to functionalize (linking to the coating or directly) with different targeting ligands. The coating and functionalization decrease the innate toxicity of the NPs and enzymatic degradation and oxidation of the biological material, guaranteeing their release and accumulation at the site of interest, with the consequent protection of healthy tissue. The ability to functionalize several target ligands on the NP surface allows the design of multimodal systems (theranostics and combined therapies). These systems can be enhanced by taking advantage of the physical properties of the NPs themselves, which can be enhanced by different stimuli.

Control of size and solubility are among the disadvantages of nanoparticulate systems. These parameters depend mostly on the method of synthesis and preparation. Therefore, the development of new preparation methods is required. Optimization of the routes of administration is another need, since the intravenous route produces the capture of the NP by the endothelial reticulum system. It is desirable to increase the efficiency of the loading and releasing of the material carried in the nanostructure, so new and improved techniques are required.

Although there is data on the cytotoxicity and genotoxicity of each element that makes up the nanosystem, there is still much work to be done to establish the cytotoxic and genotoxic effects of complete nanosystems. These data are essential to establish the rules and regulations concerning nanotechnology for clinical use.

5.6 Conclusion

Advances in the knowledge of the processes by which cancer cells, viruses, and other entities involved in diseases evade or suppress the immune and biological responses of the human body have undoubtedly improved the understanding (at a molecular level) of the causes of cancer progression and other diseases, which, in turn, has allowed progress in the development of biological therapies. This chapter has shown that nanotechnology, combined with biological therapies, makes innovation possible in the field of multimodal treatments by applying a single nanosystem. While these advances are well-established at the preclinical level, there is still much research to be done in the clinical field. Nevertheless, nanotechnology based on multifunctional target-specific nanosystems for multimodal treatment of cancer and other diseases will surely be the biological therapy of the future.

References

1 Zhou, C., Yang, S., Liu, J. et al. (2013). Luminescent gold nanoparticles: a new class of nanoprobes for biomedical imaging. *Experimental Biology and Medicine* 238 (11): 1199–1209.

2 Letfullin, R.R., Iversen, C.B., and George, T.F. (2011). Modeling nanophotothermal therapy: kinetics of thermal ablation of healthy and cancerous cell organelles and gold nanoparticles. *Nanomedicine: Nanotechnology, Biology and Medicine* 7 (2): 137–145.

3 Luna-Gutiérrez, M., Ferro-Flores, G., Ocampo-García, B.E. et al. (2013). A therapeutic system of 177Lu-labeled gold nanoparticles-RGD internalized in breast cancer cells. *Journal of the Mexican Chemical Society* 57 (3): 212–219.

4 Jiménez-Mancilla, N., Ferro-Flores, G., Santos-Cuevas, C. et al. (2013). Multifunctional targeted therapy system based on 99mTc/177Lu-labeled gold nanoparticles-tat (49–57)-Lys3-bombesin internalized in nuclei of prostate cancer cells. *Journal of Labelled Compounds and Radiopharmaceuticals* 56 (13): 663–671.

5 Vilchis-Juárez, A., Ferro-Flores, G., Santos-Cuevas, C. et al. (2014). Molecular targeting radiotherapy with cyclo-RGDFK (C) peptides conjugated to 177Lu-labeled gold nanoparticles in tumor-bearing mice. *Journal of Biomedical Nanotechnology* 10 (3): 393–404.

6 González-Ruíz, A., Ferro-Flores, G., Azorín-Vega, E. et al. (2017). Synthesis and in vitro evaluation of an antiangiogenic cancer-specific dual-targeting 177 Lu-au-nanoradiopharmaceutical. *Journal of Radioanalytical and Nuclear Chemistry* 314 (2): 1337–1345.

7 Mendoza-Nava, H., Ferro-Flores, G., Ramírez, F.M. et al. (2017). Fluorescent, plasmonic, and radiotherapeutic properties of the 177Lu–dendrimer-AuNP–folate–bombesin nanoprobe located inside cancer cells. *Molecular Imaging* 16: 1–10. doi: 1536012117704768.

8 Dong, X., Sun, Z., Wang, X. et al. (2017). An innovative MWCNTs/DOX/TC nanosystem for chemo-photothermal combination therapy of cancer. *Nanomedicine: Nanotechnology, Biology and Medicine* 13 (7): 2271–2280.

9 Wang, D., Ren, Y., Shao, Y. et al. (2017). Facile preparation of doxorubicin-loaded and folic acid-conjugated carbon nanotubes@ poly (N-vinyl pyrrole) for targeted synergistic chemo–Photothermal cancer treatment. *Bioconjugate Chemistry* 28 (11): 2815–2822.

10 Gannon, C.J., Cherukuri, P., Yakobson, B.I. et al. (2007). Carbon nanotube-enhanced thermal destruction of cancer cells in a noninvasive radiofrequency field. *Cancer: Interdisciplinary International Journal of the American Cancer Society* 110 (12): 2654–2665.

11 Wolski, P., Nieszporek, K., and Panczyk, T. (2018). Multimodal, pH sensitive, and magnetically assisted carrier of doxorubicin designed and analyzed by means of computer simulations. *Langmuir* 34 (7): 2543–2550.

12 DeNardo, S.J., DeNardo, G.L., Miers, L.A. et al. (2005). Development of tumor targeting bioprobes (111In-chimeric L6 monoclonal antibody nanoparticles) for alternating magnetic field cancer therapy. *Clinical Cancer Research* 11 (19): 7087s–7092s.

13 Qiu, L., Hu, Q., Cheng, L. et al. (2016). cRGDyK modified pH responsive nanoparticles for specific intracellular delivery of doxorubicin. *Acta Biomaterialia* 30: 285–298.

14 Mai, B.T., Fernandes, S., Balakrishnan, P.B. et al. (2018). Nanosystems based on magnetic nanoparticles and thermo- or pH-responsive polymers: an update and future perspectives. *Accounts of Chemical Research* 51 (5): 999–1013.

15 Kittel, Á, Falus, A, Buzás, E. (2013). Microencapsulation technology by nature: cell derived extracellular vesicles with therapeutic potential. *European Journal of Microbiology & Immunology* 3 (2): 91–96.

16 van der Meel, R., Fens, M.H., Vader, P. et al. (2014). Extracellular vesicles as drug delivery systems: lessons from the liposome field. *Journal of Controlled Release* 195: 72–85.

17 Ng, K.K., Lovell, J.F., and Zheng, G. (2011). Lipoprotein-inspired nanoparticles for cancer theranostics. *Accounts of Chemical Research* 44 (10): 1105–1113.

18 Zhang, X. and Huang, G. (2017). Synthetic lipoprotein as nano-material vehicle in the targeted drug delivery. *Drug Delivery* 24 (2): 16–21.

19 Schirrmacher, V. (2019). From chemotherapy to biological therapy: a review of novel concepts to reduce the side effects of systemic cancer treatment. *International Journal of Oncology* 54 (2): 407–419.

20 Smith, R.M., Clatworthy, M.R., and Jayne, D.R. (2010). Biological therapy for lupus nephritis—tribulations and trials. *Nature Reviews Rheumatology* 6 (9): 547.

21 Li, J. and Zhu, Z. (2010). Research and development of next generation of antibody-based therapeutics. *Acta Pharmacologica Sinica* 31 (9): 1198–1207.

22 Paramsothy, S., Rosenstein, A.K., Mehandru, S., Colombel, J. F. (2018). The current state of the art for biological therapies and new small molecules in inflammatory bowel disease. *Mucosal Immunology* 11 (6): 1558–1570.

23 Weiner GJ. Building better monoclonal antibody-based therapeutics. 2015.
24 Wu, L., Seung, E., Xu, L. et al. (2019). Trispecific antibodies enhance the therapeutic efficacy of tumor-directed T cells through T cell receptor co-stimulation. *Nature Cancer* 1: 86–98.
25 Beck, A., Goetsch, L., Dumontet, C., and Corvaïa, N. (2017). Strategies and challenges for the next generation of antibody–drug conjugates. *Nature Reviews Drug Discovery* 16 (5): 315.
26 Senapati, S., Mahanta, A.K., Kumar, S. et al. (2018). Controlled drug delivery vehicles for cancer treatment and their performance. *Signal Transduction and Targeted Therapy* 3 (1): 1–19.
27 Berraondo, P., Sanmamed, M.F., Ochoa, M.C. et al. (2019). Cytokines in clinical cancer immunotherapy. *British Journal of Cancer* 120 (1): 6–15.
28 Showalter A, Limaye A, Oyer JL, et al. Cytokines in immunogenic cell death: applications for cancer immunotherapy. 2017.
29 Waldmann, T.A. (2018). Cytokines in cancer immunotherapy. *Cold Spring Harbor Perspectives in Biology* 10 (12): a028472.
30 Lee, S. and Margolin, K. (2011). Cytokines in cancer immunotherapy. *Cancers (Basel)* 3 (4): 3856–3893.
31 Bootz, F. and Neri, D. (2016). Immunocytokines: a novel class of products for the treatment of chronic inflammation and autoimmune conditions. *Drug Discovery Today* 21 (1): 180–189.
32 Conlon, K.C., Miljkovic, M.D., and Waldmann, T.A. (2019). Cytokines in the treatment of cancer. *Journal of Interferon & Cytokine Research* 39 (1): 6–21.
33 Murer, P. and Neri, D. (2019). Antibody-cytokine fusion proteins: a novel class of biopharmaceuticals for the therapy of cancer and of chronic inflammation. *New Biotechnology* 52: 42–53.
34 Müller, D. (2015). Antibody fusions with immunomodulatory proteins for cancer therapy. *Pharmacology & Therapeutics* 154: 57–66.
35 Lühmann, T., Schmidt, M., Leiske, M.N. et al. (2017). Site-specific POxylation of interleukin-4. *ACS Biomaterials Science & Engineering* 3 (3): 304–312.
36 Xie, Y.Q., Wei, L., and Tang, L. (2018). Immunoengineering with biomaterials for enhanced cancer immunotherapy. *Wiley Interdisciplinary Reviews: Nanomedicine and Nanobiotechnology* 10 (4): e1506.
37 Chrisikos, T.T., Zhou, Y., Li, H.S. et al. (2020). STAT3 inhibits CD103+ cDC1 vaccine efficacy in murine breast cancer. *Cancers (Basel)* 12 (1): 128.
38 Gowda, V.M.A. and Smitha, T. (2019). The dendritic cell tool for oral cancer treatment. *Journal of Oral and Maxillofacial Pathology* 23 (3): 326.
39 Ibraheem, D., Elaissari, A., and Fessi, H. (2014). Gene therapy and DNA delivery systems. *International Journal of Pharmaceutics* 459 (1–2): 70–83.
40 Setten, R.L., Rossi, J.J., and Han, S.-P. (2019). The current state and future directions of RNAi-based therapeutics. *Nature Reviews Drug Discovery* 18 (6): 421–446.
41 Bardin, P., Sonneville, F., Corvol, H., and Tabary, O. (2018). Emerging microRNA Therapeutic Approaches for Cystic Fibrosis. *Frontiers in Pharmacology* 9: 1113.
42 Godfrey, C., Desviat, L.R., Smedsrød, B. et al. (2017). Delivery is key: lessons learnt from developing splice-switching antisense therapies. *EMBO Molecular Medicine*.
43 Meng, Z. and Lu, M. (2017). RNA interference-induced innate immunity, off-target effect, or immune adjuvant? *Frontiers in Immunology* 8: 331.

44 Buck, J., Grossen, P., Cullis, P.R. et al. (2019). Lipid-based DNA therapeutics: hallmarks of non-viral gene delivery. *ACS Nano* 13 (4): 3754–3782.
45 Saraswat, P., Soni, R., Bhandari, A. et al. (2009). DNA as therapeutics; an update. *Indian Journal of Pharmaceutical Sciences* 71 (5): 488.
46 Xiang, D., Shigdar, S., Qiao, G. et al. (2015). Aptamer-mediated cancer gene therapy. *Current Gene Therapy* 15 (2): 109–119.
47 Zhu, Y., Choi, S.H. et al. (2015). Multifunctional receptor-targeting antibodies for cancer therapy. *Lancet Oncology* 16 (15): e543–e554.
48 Minko, T., Khandare, J., Vetcher, A. et al. (2008). In: Multifunctional nanotherapeutics for cancer. *Multifunctional Pharmaceutical Nanocarriers*. 309–336. Springer.
49 Cieslewicz, M., Tang, J., Yu, J.L. et al. (2013). Targeted delivery of proapoptotic peptides to tumor-associated macrophages improves survival. *Proceedings of the National Academy of Sciences* 110 (40): 15919LP–15924LP.
50 Tang X., Mo C., Wang Y., Wei D., Xiao H. (2013). Anti-tumour strategies aiming to target tumour-associated macrophages. *Immunology* 138 (2): 93–10.
51 Lim, S., Park, J., Shim, M.K. et al. (2019). Recent advances and challenges of repurposing nanoparticle-based drug delivery systems to enhance cancer immunotherapy. *Theranostics* 9 (25): 7906.
52 Pelaz, B., Alexiou, C., Alvarez-Puebla, R.A. et al. (2017). Diverse applications of nanomedicine. *ACS Nano* 11 (3): 2313–2381.
53 Kumari, P., Ghosh, B., and Biswas, S. (2016). Nanocarriers for cancer-targeted drug delivery. *Journal of Drug Targeting* 24 (3): 179–191.
54 Xin, Y., Yin, M., Zhao, L. et al. (2017). Recent progress on nanoparticle-based drug delivery systems for cancer therapy. *Cancer Biology & Medicine* 14 (3): 228.
55 Jeevanandam, J., Barhoum, A., Chan, Y.S. et al. (2018). Review on nanoparticles and nanostructured materials: history, sources, toxicity and regulations. *Beilstein Journal of Nanotechnology* 9 (1): 1050–1074.
56 Sur, S., Rathore, A., Dave, V. et al. (2019). Recent developments in functionalized polymer nanoparticles for efficient drug delivery system. *Nano-Structures & Nano-Objects* 20: 100397.
57 Masood, F. (2016). Polymeric nanoparticles for targeted drug delivery system for cancer therapy. *Materials Science and Engineering: C* 60: 569–578.
58 Duan, X., He, C., Kron, S.J. et al. (2016). Nanoparticle formulations of cisplatin for cancer therapy. *Wiley Interdisciplinary Reviews: Nanomedicine and Nanobiotechnology* 8 (5): 776–791.
59 Mosafer, J., Abnous, K., Tafaghodi, M. et al. (2017). in vitro and in vivo evaluation of anti-nucleolin-targeted magnetic PLGA nanoparticles loaded with doxorubicin as a theranostic agent for enhanced targeted cancer imaging and therapy. *European Journal of Pharmaceutics and Biopharmaceutics* 113: 60–74.
60 Hossen, S., Hossain, M.K., Basher, M. et al. (2019). Smart nanocarrier-based drug delivery systems for cancer therapy and toxicity studies: a review. *Journal of Advanced Research* 15: 1–18.
61 Belfiore, L., Saunders, D.N., Ranson, M. et al. (2018). Towards clinical translation of ligand-functionalized liposomes in targeted cancer therapy: challenges and opportunities. *Journal of Controlled Release* 277: 1–13.

62 Has, C. and Sunthar, P. (2019). A comprehensive review on recent preparation techniques of liposomes. *Journal of Liposome Research*: 1–30.
63 Bulbake, U., Doppalapudi, S., Kommineni, N. et al. (2017). Liposomal formulations in clinical use: an updated review. *Pharmaceutics* 9 (2): 12.
64 Alfayez, M., Kantarjian, H., Kadia, T. et al. (2019). Emerging drug profile: CPX-351 (vyxeos) in AML. *Leukemia & Lymphoma*: 1–10.
65 Sharma, A.K., Gothwal, A., Kesharwani, P. et al. (2017). Dendrimer nanoarchitectures for cancer diagnosis and anticancer drug delivery. *Drug Discovery Today* 22 (2): 314–326.
66 Zhang, C., Pan, D., Li, J. et al. (2017). Enzyme-responsive peptide dendrimer-gemcitabine conjugate as a controlled-release drug delivery vehicle with enhanced antitumor efficacy. *Acta Biomaterialia* 55: 153–162.
67 Gupta, V. and Nayak, S.K. (2015). Dendrimers: a review on synthetic approaches. *Journal of Applied Pharmaceutical Science* 5 (3): 117–122.
68 Kesharwani, P. and Iyer, A.K. (2015). Recent advances in dendrimer-based nanovectors for tumor-targeted drug and gene delivery. *Drug Discovery Today* 20 (5): 536–547.
69 Jain, N.K., Tare, M.S., Mishra, V. et al. (2015). The development, characterization and in vivo anti-ovarian cancer activity of poly (propylene imine)(PPI)-antibody conjugates containing encapsulated paclitaxel. *Nanomedicine: Nanotechnology, Biology and Medicine* 11 (1): 207–218.
70 Yang, H. (2016). Targeted nanosystems: advances in targeted dendrimers for cancer therapy. *Nanomedicine: Nanotechnology, Biology and Medicine* 12 (2): 309–316.
71 Taghdisi, S.M., Danesh, N.M., Ramezani, M. et al. (2016). Double targeting and aptamer-assisted controlled release delivery of epirubicin to cancer cells by aptamers-based dendrimer in vitro and in vivo. *European Journal of Pharmaceutics and Biopharmaceutics* 102: 152–158.
72 Henrich, S.E. and Thaxton, C.S. (2019). An update on synthetic high-density lipoprotein-like nanoparticles for cancer therapy. *Expert Review of Anticancer Therapy* 19 (6): 515–528.
73 Isaac-Olivé, K., Ocampo-García, B.E., Aranda-Lara, L. et al. (2019). [99m Tc-HYNIC-N-dodecylamide]: a new hydrophobic tracer for labelling reconstituted high-density lipoproteins (rHDL) for radioimaging. *Nanoscale* 11 (2): 541–551.
74 Ha, D., Yang, N., and Nadithe, V. (2016). Exosomes as therapeutic drug carriers and delivery vehicles across biological membranes: current perspectives and future challenges. *Acta Pharmaceutica Sinica B* 6 (4): 287–296.
75 Batrakova, E.V. and Kim, M.S. (2015). Using exosomes, naturally-equipped nanocarriers, for drug delivery. *Journal of Controlled Release* 219: 396–405.
76 Kao, H.-W., Lin, Y.-Y., Chen, C.-C. et al. (2013). Evaluation of EGFR-targeted radioimmuno-gold-nanoparticles as a theranostic agent in a tumor animal model. *Bioorganic & Medicinal Chemistry Letters* 23 (11): 3180–3185.
77 Kao, H.-W., Lin, Y.-Y., Chen, C.-C. et al. (2014). Biological characterization of cetuximab-conjugated gold nanoparticles in a tumor animal model. *Nanotechnology* 25 (29): 295102.
78 Garg, P. and Hazra, D.K. (2017). Conjugation of antibodies with radiogold nanoparticles, as an effector targeting agents in radiobioconjugate cancer therapy: optimized labeling and biodistribution results. *Indian Journal of Nuclear Medicine* 32 (4): 296.
79 Chattopadhyay, N., Cai, Z., Pignol, J.-P. et al. (2010). Design and characterization of HER-2-targeted gold nanoparticles for enhanced X-radiation treatment of locally advanced breast cancer. *Molecular Pharmaceutics* 7 (6): 2194–2206.

80 Chen, C.H., Wu, Y.J., and Chen, J.-J. (2015). Gold nanotheranostics: photothermal therapy and imaging of Mucin 7 conjugated antibody nanoparticles for urothelial cancer. *BioMed Research International* 2015: 813632.

81 Song, L., Falzone, N., and Vallis, K.A. (2016). EGF-coated gold nanoparticles provide an efficient nano-scale delivery system for the molecular radiotherapy of EGFR-positive cancer. *International Journal of Radiation Biology* 92 (11): 716–723.

82 Roma-Rodrigues, C., Heuer-Jungemann, A., Fernandes, A.R. et al. (2016). Peptide-coated gold nanoparticles for modulation of angiogenesis in vivo. *International Journal of Nanomedicine* 11: 2633.

83 Guerrero, S., Herance, J.R., Rojas, S. et al. (2012). Synthesis and in vivo evaluation of the biodistribution of a 18F-labeled conjugate gold-nanoparticle-peptide with potential biomedical application. *Bioconjugate Chemistry* 23 (3): 399–408.

84 Prades, R., Guerrero, S., Araya, E. et al. (2012). Delivery of gold nanoparticles to the brain by conjugation with a peptide that recognizes the transferrin receptor. *Biomaterials* 33 (29): 7194–7205.

85 Schäffler, M., Sousa, F., Wenk, A. et al. (2014). Blood protein coating of gold nanoparticles as potential tool for organ targeting. *Biomaterials* 35 (10): 3455–3466.

86 Shilo, M., Motiei, M., Hana, P. et al. (2014). Transport of nanoparticles through the blood–brain barrier for imaging and therapeutic applications. *Nanoscale* 6 (4): 2146–2152.

87 Chen, N., Yang, W., Bao, Y. et al. (2015). BSA capped Au nanoparticle as an efficient sensitizer for glioblastoma tumor radiation therapy. *RSC Advances* 5 (51): 40514–40520.

88 Wolfe, T., Chatterjee, D., Lee, J. et al. (2015). Targeted gold nanoparticles enhance sensitization of prostate tumors to megavoltage radiation therapy in vivo. *Nanomedicine: Nanotechnology, Biology and Medicine* 11 (5): 1277–1283.

89 Libutti, S.K., Paciotti, G.F., Byrnes, A.A. et al. (2010). Phase I and pharmacokinetic studies of CYT-6091, a novel PEGylated colloidal gold-rhTNF nanomedicine. *Clinical Cancer Research* 16 (24): 6139–6149.

90 Zolata, H., Davani, F.A., and Afarideh, H. (2015). Synthesis, characterization and theranostic evaluation of Indium-111 labeled multifunctional superparamagnetic iron oxide nanoparticles. *Nuclear Medicine and Biology* 42 (2): 164–170.

91 Zolata, H., Afarideh, H., and Davani, F.A. (2016). Triple therapy of HER2+ cancer using radiolabeled multifunctional iron oxide nanoparticles and alternating magnetic field. *Cancer Biotherapy and Radiopharmaceuticals* 31 (9): 324–329.

92 Truffi, M., Colombo, M., Sorrentino, L. et al. (2018). Multivalent exposure of trastuzumab on iron oxide nanoparticles improves antitumor potential and reduces resistance in HER2-positive breast cancer cells. *Scientific Reports* 8 (1): 1–11.

93 Azadbakht, B., Afarideh, H., Ghannadi-Maragheh, M. et al. (2017). Preparation and evaluation of APTES-PEG coated iron oxide nanoparticles conjugated to rhenium-188 labeled rituximab. *Nuclear Medicine and Biology* 48: 26–30.

94 Tseng, S.-H., Chou, M.-Y., and Chu, I.-M. (2015). Cetuximab-conjugated iron oxide nanoparticles for cancer imaging and therapy. *International Journal of Nanomedicine* 10: 3663.

95 Glat, M., Skaat, H., Menkes-Caspi, N. et al. (2013). Age-dependent effects of microglial inhibition in vivo on Alzheimer's disease neuropathology using bioactive-conjugated iron oxide nanoparticles. *Journal of Nanobiotechnology* 11 (1): 32.

96 Wu, Y.-Z., Sun, J., Zhang, Y. et al. (2017). Effective integration of targeted tumor imaging and therapy using functionalized InP QDs with VEGFR2 monoclonal antibody and miR-92a inhibitor. *ACS Applied Materials & Interfaces* 9 (15): 13068–13078.

97 Carbary-Ganz, J.L., Barton, J.K., and Utzinger, U. (2014). Quantum dots targeted to vascular endothelial growth factor receptor 2 as a contrast agent for the detection of colorectal cancer. *Journal of Biomedical Optics* 19 (8): 086003.

98 Gazouli, M., Bouziotis, P., Lyberopoulou, A. et al. (2014). Quantum dots-bevacizumab complexes for in vivo imaging of tumors. *in vivo* 28 (6): 1091–1095.

99 Nifontova, G., Ramos-Gomes, F., Baryshnikova, M. et al. (2019). Cancer cell targeting with functionalized quantum dot-encoded polyelectrolyte microcapsules. *Frontiers in Chemistry* 7: 34.

100 Dawson, G. (2016). Quantum dots and potential therapy for K rabbe's disease. *Journal of Neuroscience Research* 94 (11): 1293–1303.

101 Lin, G., Chen, T., Zou, J. et al. (2017). Quantum dots-siRNA nanoplexes for gene silencing in central nervous system tumor cells. *Frontiers in Pharmacology* 8: 182.

102 Omidi, M., Hashemi, M., and Tayebi, L. (2019). Microfluidic synthesis of PLGA/carbon quantum dot microspheres for vascular endothelial growth factor delivery. *RSC Advances* 9 (57): 33246–33256.

103 Joshi, P.N., Agawane, S., Athalye, M.C. et al. (2017). Multifunctional inulin tethered silver-graphene quantum dots nanotheranostic module for pancreatic cancer therapy. *Materials Science and Engineering: C* 78: 1203–1211.

104 Zheng, X.-Y., Li, L.-D., Sun, L.-D. et al. (2016). Lanthanide nanoparticles: promising candidates for magnetic resonance imaging contrast enhancement. In: *Handbook on the Physics and Chemistry of Rare Earths*, vol. 50, 301–335. Elsevier.

105 Lee, S.Y., Lin, M., Lee, A. et al. (2017). Lanthanide-doped nanoparticles for diagnostic sensing. *Nanomaterials* 7 (12): 411.

106 Wen, S., Zhou, J., Zheng, K. et al. (2018). Advances in highly doped upconversion nanoparticles. *Nature Communications* 9 (1): 1–12.

107 Dong, H., Du, S.-R., Zheng, X.-Y. et al. (2015). Lanthanide nanoparticles: from design toward bioimaging and therapy. *Chemical Reviews* 115 (19): 10725–10815.

108 Tangboriboon, N. (2019). Carbon and carbon nanotube drug delivery and its characterization, properties, and applications. In: *Nanocarriers for Drug Delivery*, 451–467. Elsevier.

109 Chou, M.-J., Yu, H.-Y., Hsia, J.-C. et al. (2018). Highly efficient intracellular protein delivery by cationic polyethyleneimine-modified gelatin nanoparticles. *Materials* 11 (2): 301.

110 Das, P. and Jana, N.R. (2016). Length-controlled synthesis of calcium phosphate nanorod and nanowire and application in intracellular protein delivery. *ACS Applied Materials & Interfaces* 8 (13): 8710–8720.

111 Emadi, F., Amini, A., Gholami, A. et al. (2017). Functionalized graphene oxide with chitosan for protein nanocarriers to protect against enzymatic cleavage and retain collagenase activity. *Scientific Reports* 7: 42258.

112 Liu, Y. and Cao, A. (2017). Encapsulating proteins in nanoparticles: batch by batch or one by one. *Methods in Enzymology* 590: 1–31.

113 Suner, S.S., Sahiner, M., Sengel, S.B. et al. (2018). Responsive biopolymer-based microgels/nanogels for drug delivery applications. In: *Stimuli Responsive Polymeric Nanocarriers for Drug Delivery Applications*, vol. 1, 453–500. Elsevier.

114 Barba, A.A., Bochicchio, S., Dalmoro, A. et al. (2019). Lipid delivery systems for nucleic-acid-based-drugs: from production to clinical applications. *Pharmaceutics* 11 (8): 360.
115 Elsana, H., Olusanya, T.O., Carr-Wilkinson, J. et al. (2019). Evaluation of novel cationic gene based liposomes with cyclodextrin prepared by thin film hydration and microfluidic systems. *Scientific Reports* 9 (1): 1–17.
116 Chen, E.Y., Liu, W.F., Megido, L. et al. (2018). Understanding and utilizing the biomolecule/nanosystems interface. In: *Nanotechnologies in Preventive and Regenerative Medicine*, 207–297. Elsevier.
117 Rangel-Corona, R., Soto-Vázquez, R., Weiss-Steider, B. et al. (2014). Propuesta Biotecnológica para el Tratamiento del Cáncer de Cérvix Uterino, Empleando Liposomas. *Anales de la Real Academia Nacional de Farmacia* 80 (1): 179–191.
118 Christian, D.A. and Hunter, C.A. (2012). Particle-mediated delivery of cytokines for immunotherapy. *Immunotherapy* 4 (4): 425–441.
119 Duncan, S.A., Dixit, S., Sahu, R. et al. (2019). Prolonged release and functionality of interleukin-10 encapsulated within PLA-PEG nanoparticles. *Nanomaterials* 9 (8): 1074.
120 Gdowski, A., Ranjan, A., Mukerjee, A. et al. (2015). Development of biodegradable nanocarriers loaded with a monoclonal antibody. *International Journal of Molecular Sciences* 16 (2): 3990–3995.
121 Sousa, F., Fonte, P., Cruz, A. et al. (2018). Polyester-based nanoparticles for the encapsulation of monoclonal antibodies. In: *Recombinant Glycoprotein Production*, 239–253. Springer.
122 Çalış, S., Atar, K.Ö., Arslan, F.B. et al. (2019). Nanopharmaceuticals as drug-delivery systems: for, against, and current applications. In: *Nanocarriers for Drug Delivery*, 133–154. Elsevier.
123 Dean, S.N., Turner, K.B., Medintz, I.L. et al. (2017). Targeting and delivery of therapeutic enzymes. *Therapeutic Delivery* 8 (7): 577–595.
124 Bamburowicz-Klimkowska, M., Poplawska, M., and Grudzinski, I.P. (2019). Nanocomposites as biomolecules delivery agents in nanomedicine. *Journal of Nanobiotechnology* 17 (1): 48.
125 Choonara, B.F., Choonara, Y.E., Kumar, P. et al. (2014). A review of advanced oral drug delivery technologies facilitating the protection and absorption of protein and peptide molecules. *Biotechnology Advances* 32 (7): 1269–1282.
126 Duo, Y., Yang, M., Du, Z. et al. (2018). CX-5461-loaded nucleolus-targeting nanoplatform for cancer therapy through induction of pro-death autophagy. *Acta Biomaterialia* 79: 317–330.
127 Dhule, S.S., Penfornis, P., He, J. et al. (2014). The combined effect of encapsulating curcumin and C6 ceramide in liposomal nanoparticles against osteosarcoma. *Molecular Pharmaceutics* 11 (2): 417–427.
128 Muhamad, N., Plengsuriyakarn, T., and Na-Bangchang, K. (2018). Application of active targeting nanoparticle delivery system for chemotherapeutic drugs and traditional/herbal medicines in cancer therapy: a systematic review. *International Journal of Nanomedicine* 13: 3921.
129 Tu, J., Boyle, A.L., Friedrich, H. et al. (2016). Mesoporous silica nanoparticles with large pores for the encapsulation and release of proteins. *ACS Applied Materials & Interfaces* 8 (47): 32211–32219.

130 Prasetyanto, E.A., Bertucci, A., Septiadi, D. et al. (2016). Breakable hybrid organosilica nanocapsules for protein delivery. *Angewandte Chemie International Edition* 55 (10): 3323–3327.

131 Gu, Z., Biswas, A., Zhao, M. et al. (2011). Tailoring nanocarriers for intracellular protein delivery. *Chemical Society Reviews* 40 (7): 3638–3655.

132 Xu, X., Costa, A., and Burgess, D.J. (2012). Protein encapsulation in unilamellar liposomes: high encapsulation efficiency and a novel technique to assess lipid-protein interaction. *Pharmaceutical Research* 29 (7): 1919–1931.

133 Suthiwangcharoen, N. and Nagarajan, R. (2017). Nanoarmoring of proteins by conjugation to block copolymer micelles. In: *Methods in Enzymology*, vol. 590, 277–304. Elsevier.

134 Scaletti, F., Hardie, J., Lee, Y.-W. et al. (2018). Protein delivery into cells using inorganic nanoparticle–protein supramolecular assemblies. *Chemical Society Reviews* 47 (10): 3421–3432.

135 He, Q., Chen, J., Yan, J. et al. (2019). Tumor microenvironment responsive drug delivery systems. *Asian Journal of Pharmaceutical Sciences*.

136 Wu, T.-T. and Zhou, S.-H. (2015). Nanoparticle-based targeted therapeutics in head-and-neck cancer. *International Journal of Medical Sciences* 12 (2): 187.

137 Oh, J.Y., Kim, H.S., Palanikumar, L. et al. (2018). Cloaking nanoparticles with protein corona shield for targeted drug delivery. *Nature Communications* 9 (1): 1–9.

138 Yoo, J., Park, C., Yi, G. et al. (2019). Active targeting strategies using biological ligands for nanoparticle drug delivery systems. *Cancers (Basel)* 11 (5): 640.

139 Danhier, F., Feron, O., and Préat, V. (2010). To exploit the tumor microenvironment: passive and active tumor targeting of nanocarriers for anti-cancer drug delivery. *Journal of Controlled Release* 148 (2): 135–146.

140 Bahrami, B., Hojjat-Farsangi, M., Mohammadi, H. et al. (2017). Nanoparticles and targeted drug delivery in cancer therapy. *Immunology Letters* 190: 64–83.

141 Chauhan, V.P. and Jain, R.K. (2013). Strategies for advancing cancer nanomedicine. *Nature Materials* 12 (11): 958.

142 Chauhan, V.P., Stylianopoulos, T., Martin, J.D. et al. (2012). Normalization of tumour blood vessels improves the delivery of nanomedicines in a size-dependent manner. *Nature Nanotechnology* 7 (6): 383.

143 Park, J., Wrzesinski, S.H., Stern, E. et al. (2012). Combination delivery of TGF-β inhibitor and IL-2 by nanoscale liposomal polymeric gels enhances tumour immunotherapy. *Nature Materials* 11 (10): 895–905.

144 Li, Y. and Zhang, H. (2019). Nanoparticle-based drug delivery systems for enhanced tumor-targeting treatment. *Journal of Biomedical Nanotechnology* 15 (1): 1–27.

145 Deirram, N., Zhang, C., Kermaniyan, S.S. et al. (2019). pH-responsive polymer nanoparticles for drug delivery. *Macromolecular Rapid Communications* 40 (10): 1800917.

146 Bareford, L.M. and Swaan, P.W. (2007). Endocytic mechanisms for targeted drug delivery. *Advanced Drug Delivery Reviews* 59 (8): 748–758.

147 Zhao, J. and Stenzel, M.H. (2018). Entry of nanoparticles into cells: the importance of nanoparticle properties. *Polymer Chemistry* 9 (3): 259–272.

148 Juárez, B.H. (2011). *Nanoparticulas semiconductoras coloidales y aplicaciones.* Anales de Química.

149 Lee, J.H. and Yeo, Y. (2015). Controlled drug release from pharmaceutical nanocarriers. *Chemical Engineering Science* 125: 75–84.

150 Singh, R. and Lillard, J.W. Jr. (2009). Nanoparticle-based targeted drug delivery. *Experimental and Molecular Pathology* 86 (3): 215–223.

151 Shukla, R., Singh, A., Pardhi, V. et al. (2019). Dendrimer-based nanoparticulate delivery system for cancer therapy. In: *Polymeric Nanoparticles as a Promising Tool for Anti-Cancer Therapeutics* (eds. P. Kesharwani, K.M. Paknikar and V. Gajbhiye), 233–255. Academic Press.

152 Uthaman, S., Huh, K.M., and Park, I.-K. (2018). Tumor microenvironment-responsive nanoparticles for cancer theragnostic applications. *Biomaterials Research* 22 (1): 22.

153 Gao, S., Tang, G., Hua, D. et al. (2019). Stimuli-responsive bio-based polymeric systems and their applications. *Journal of Materials Chemistry B* 7 (5): 709–729.

154 Hosseini-Nassab, N., Samanta, D., Abdolazimi, Y. et al. (2017). Electrically controlled release of insulin using polypyrrole nanoparticles. *Nanoscale* 9 (1): 143–149.

155 Qiu, Y., Tong, S., Zhang, L. et al. (2017). Magnetic forces enable controlled drug delivery by disrupting endothelial cell-cell junctions. *Nature Communications* 8 (1): 1–10.

156 Huebsch, N., Kearney, C.J., Zhao, X. et al. (2014). Ultrasound-triggered disruption and self-healing of reversibly cross-linked hydrogels for drug delivery and enhanced chemotherapy. *Proceedings of the National Academy of Sciences* 111 (27): 9762–9767.

157 Chowdhury, S.M., Lee, T., and Willmann, J.K. (2017). Ultrasound-guided drug delivery in cancer. *Ultrasonography* 36 (3): 171.

158 Linsley, C.S. and Wu, B.M. (2017). Recent advances in light-responsive on-demand drug-delivery systems. *Therapeutic Delivery* 8 (2): 89–107.

159 Matsumoto, K., Kimura, S.-I., Itai, S. et al. (2019). in vivo temperature-sensitive drug release system trigged by cooling using low-melting-point microcrystalline wax. *Journal of Controlled Release* 303: 281–288.

160 Rideau, E., Dimova, R., Schwille, P. et al. (2018). Liposomes and polymersomes: a comparative review towards cell mimicking. *Chemical Society Reviews* 47 (23): 8572–8610.

161 Zangabad, P.S., Mirkiani, S., Shahsavari, S. et al. (2018). Stimulus-responsive liposomes as smart nanoplatforms for drug delivery applications. *Nanotechnology Reviews* 7 (1): 95–122.

162 Raza, A., Rasheed, T., Nabeel, F. et al. (2019). Endogenous and exogenous stimuli-responsive drug delivery systems for programmed site-specific release. *Molecules* 24 (6): 1117.

163 Chen, J., Guo, Z., Tian, H. et al. (2016). Production and clinical development of nanoparticles for gene delivery. *Molecular Therapy-Methods & Clinical Development* 3: 16023.

164 Marzbali, M.Y. and Khosroushahi, A.Y. (2017). Polymeric micelles as mighty nanocarriers for cancer gene therapy: a review. *Cancer Chemotherapy and Pharmacology* 79 (4): 637–649.

165 Koynova, R. and Tenchov, B. (2010). Cationic lipids: molecular structure/transfection activity relationships and interactions with biomembranes. In: *Nucleic Acid Transfection*, 51–93. Springer.

166 Rai, R., Alwani, S., and Badea, I. (2019). Polymeric nanoparticles in gene therapy: new avenues of design and optimization for delivery applications. *Polymers* 11 (4): 745.

167 de Matos, M.B., Miranda, B.S., Rizky Nuari, Y. et al. (2019). Liposomes with asymmetric bilayers produced from inverse emulsions for nucleic acid delivery. *Journal of Drug Targeting* 27 (5–6): 681–689.

168 Hauptstein, S., Prüfert, F., and Bernkop-Schnürch, A. (2015). Self-nanoemulsifying drug delivery systems as novel approach for pDNA drug delivery. *International Journal of Pharmaceutics* 487 (1–2): 25–31.

169 Mahmood, A., Prüfert, F., Efiana, N.A. et al. (2016). Cell-penetrating self-nanoemulsifying drug delivery systems (SNEDDS) for oral gene delivery. *Expert Opinion on Drug Delivery* 13 (11): 1503–1512.

170 Jin, H., Lovell, J.F., Chen, J. et al. (2012). Mechanistic insights into LDL nanoparticle-mediated siRNA delivery. *Bioconjugate Chemistry* 23 (1): 33–41.

171 Ding, Y., Wang, Y., Zhou, J. et al. (2014). Direct cytosolic siRNA delivery by reconstituted high density lipoprotein for target-specific therapy of tumor angiogenesis. *Biomaterials* 35 (25): 7214–7227.

172 Lacko, A.G., Sabnis, N.A., Nagarajan, B. et al. (2015). HDL as a drug and nucleic acid delivery vehicle. *Frontiers in Pharmacology* 6: 247.

173 Estanqueiro, M., Vasconcelos, H., Lobo, J.M.S. et al. (2018). Delivering miRNA modulators for cancer treatment. In: *Drug Targeting and Stimuli Sensitive Drug Delivery Systems*, 517–565. Elsevier.

174 Tang, T.-T. and Liu, B.-C. (2019). Extracellular vesicles: opportunities and challenges for the treatment of renal fibrosis. In: *Renal Fibrosis: Mechanisms and Therapies*, 693–709. Springer.

175 Dan, N. (2018). Structure and kinetics of synthetic, lipid-based nucleic acid carriers: lipoplexes. In: *Lipid Nanocarriers for Drug Targeting*, 529–562. Elsevier.

176 Mathew, A.P., Cho, K.-H., Uthaman, S. et al. (2017). Stimuli-regulated smart polymeric systems for gene therapy. *Polymers* 9 (4): 152.

177 Kandil, R. and Merkel, O.M. (2019). Recent progress of polymeric nanogels for gene delivery. *Current Opinion in Colloid & Interface Science* 39: 11–23.

178 Santander-Ortega, M., Lozano, M., Uchegbu, I. et al. (2016). Dendrimers for gene therapy. In: *Polymers and Nanomaterials for Gene Therapy*, 113–146. Elsevier.

179 Cao, X., Wang, J., Deng, W. et al. (2018). Photoluminescent cationic carbon dots as efficient non-viral delivery of plasmid SOX9 and chondrogenesis of fibroblasts. *Scientific Reports* 8 (1): 1–11.

180 Riley, M.K. and Vermerris, W. (2017). Recent advances in nanomaterials for gene delivery—a review. *Nanomaterials* 7 (5): 94.

181 Kammari, R., Das, N.G., and Das, S.K. (2017). Nanoparticulate systems for therapeutic and diagnostic applications. In: *Emerging Nanotechnologies for Diagnostics, Drug Delivery and Medical Devices*, 105–144. Elsevier.

182 Thomas, S.C., Kumar Mishra, P., and Talegaonkar, S. (2015). Ceramic nanoparticles: fabrication methods and applications in drug delivery. *Current Pharmaceutical Design* 21 (42): 6165–6188.

183 Lee, M.S., Lee, J.E., Byun, E. et al. (2014). Target-specific delivery of siRNA by stabilized calcium phosphate nanoparticles using dopa–hyaluronic acid conjugate. *Journal of Controlled Release* 192: 122–130.

184 Mohammadinejad, R., Dadashzadeh, A., Moghassemi, S. et al. (2019). Shedding light on gene therapy: carbon dots for the minimally invasive image-guided delivery of plasmids and noncoding RNAs—a review. *Journal of Advanced Research* 18: 81–93.

185 Trapani, G., Denora, N., Trapani, A., and Laquintana, V. (2012). Recent advances in ligand targeted therapy. *Journal of Drug Targeting* 20 (1): 1–22.

186 Ragelle, H., Danhier, F., Préat, V. et al. (2017). Nanoparticle-based drug delivery systems: a commercial and regulatory outlook as the field matures. *Expert Opinion on Drug Delivery* 14 (7): 851–864.

187 Karra, N. and Benita, S. (2012). The ligand nanoparticle conjugation approach for targeted cancer therapy. *Current Drug Metabolism* 13 (1): 22–41.

188 Wujcik, D. (2014). Science and mechanism of action of targeted therapies in cancer treatment. In: *Seminars in Oncology Nursing*, 139–146. Elsevier.

189 Linton, S.S., Sherwood, S.G., Drews, K.C., and Kester, M. (2016). Targeting cancer cells in the tumor microenvironment: opportunities and challenges in combinatorial nanomedicine. *Wiley Interdisciplinary Reviews: Nanomedicine and Nanobiotechnology* 8 (2): 208–222.

190 Nieberler, M., Reuning, U., Reichart, F. et al. (2017). Exploring the role of RGD-recognizing integrins in cancer. *Cancers (Basel)* 9 (9): 116.

191 Zhao, Q., Liu, J., Zhu, W. et al. (2015). Dual-stimuli responsive hyaluronic acid-conjugated mesoporous silica for targeted delivery to CD44-overexpressing cancer cells. *Acta Biomaterialia* 23: 147–156.

192 Zhu, G. and Chen, X. (2018). Aptamer-based targeted therapy. *Advanced Drug Delivery Reviews* 134: 65–78.

193 Jo, H. and Ban, C. (2016). Aptamer–nanoparticle complexes as powerful diagnostic and therapeutic tools. *Experimental & Molecular Medicine* 48 (5): e230.

194 Zwicke, G.L., Ali Mansoori, G., and Jeffery, C.J. (2012). Utilizing the folate receptor for active targeting of cancer nanotherapeutics. *Nanotechnology Reviews* 3 (1): 18496.

195 Pang, J., Gao, Z., Tan, H. et al. (2019). Fabrication, investigation and application of light-responsive self-assembled nanoparticles. *Frontiers in Chemistry* 7: 620.

196 Prakash, J., Beljaars, L., Harapanahalli, A.K. et al. (2010). Tumor-targeted intracellular delivery of anticancer drugs through the mannose-6-phosphate/insulin-like growth factor II receptor. *International Journal of Cancer* 126 (8): 1966–1981.

197 Abergel, R. and Arnedo-Sanchez, L. (2019). Challenges of actinium coordination chemistry for nuclear medicine. *Journal of Medical Imaging and Radiation Sciences* 50 (1): S39.

198 Rahimian, S., Fransen, M.F., Kleinovink, J.W. et al. (2015). Polymeric nanoparticles for co-delivery of synthetic long peptide antigen and poly IC as therapeutic cancer vaccine formulation. *Journal of Controlled Release* 203: 16–22.

199 Von Hoff, D.D., Mita, M.M., Ramanathan, R.K. et al. (2016). Phase I study of PSMA-targeted docetaxel-containing nanoparticle BIND-014 in patients with advanced solid tumors. *Clinical Cancer Research* 22 (13): 3157–3163.

200 Autio, K.A., Dreicer, R., Anderson, J. et al. (2018). Safety and efficacy of BIND-014, a docetaxel nanoparticle targeting prostate-specific membrane antigen for patients with metastatic castration-resistant prostate cancer: a phase 2 clinical trial. *JAMA Oncology* 4 (10): 1344–1351.

201 Symonds, L., Linden, H., Gadi, V. et al. (2019). Combined targeted therapies for first-line treatment of metastatic triple negative breast cancer—a phase II trial of weekly nab-paclitaxel and bevacizumab followed by maintenance targeted therapy with bevacizumab and erlotinib. *Clinical Breast Cancer* 19 (2): e283–e296.

202 Northfelt, D., Dueck, A., Flynn, T. et al. (2011). Phase II trial combining nab-paclitaxel (NP), gemcitabine (G), and bevacizumab (B) in patients (pts) with metastatic breast cancer (MBC): NCCTG N0735. *Journal of Clinical Oncology* 29 (15 suppl): 1126–1126.

203 Chun, S.G., Hughes, R., Sumer, B.D. et al. (2017). A phase I/II study of nab-paclitaxel, cisplatin, and Cetuximab with concurrent radiation therapy for locally advanced squamous cell cancer of the head and neck. *Cancer Investigation* 35 (1): 23–31.

204 Subbiah, V., Grilley-Olson, J.E., Combest, A.J. et al. (2018). Phase Ib/II trial of NC-6004 (nanoparticle cisplatin) plus gemcitabine in patients with advanced solid tumors. *Clinical Cancer Research* 24 (1): 43–51.

205 Hamaguchi, T., Doi, T., Shitara, K. et al. (2016). A phase I study of NC-6004, a micellar formulation of cisplatin, in combination with gemcitabine in Japanese patients with advanced solid tumors. *American Society of Clinical Oncology*.

206 Doi, T., Hamaguchi, T., Shitara, K. et al. (2017). NC-6004 phase I study in combination with gemcitabine for advanced solid tumors and population PK/PD analysis. *Cancer Chemotherapy and Pharmacology* 79 (3): 569–578.

207 Mukai, H., Kato, K., Esaki, T. et al. (2016). Phase I study of NK105, a nanomicellar paclitaxel formulation, administered on a weekly schedule in patients with solid tumors. *Investigational New Drugs* 34 (6): 750–759.

208 Fujiwara, Y., Mukai, H., Saeki, T. et al. (2019). A multi-national, randomised, open-label, parallel, phase III non-inferiority study comparing NK105 and paclitaxel in metastatic or recurrent breast cancer patients. *British Journal of Cancer* 120 (5): 475.

209 Yang, G., Sadeg, N., Belhadj-Tahar, H. (2017). New potential in situ anticancer agent derived from [188Re] rhenium nitro-imidazole ligand loaded 5th generation poly-L-lysine dendrimer for treatment of transplanted human liver carcinoma in nude mice. *Drug Designing*: Open Access 6: 1–7.

210 Belhadj-Tahar, H., Chen, A., Jia, Y. et al. (2018). In situ anti-cancer agent derived from [188re] rhenium nitro-imidazole ligand loaded poly-l-lysine dendrimer administrated by direct CT guided stereotactic intrahepatic injection. *Journal of Clinical Oncology* 36 (15_suppl): e15569–e15569.

211 Mamot, C., Ritschard, R., Wicki, A. et al. (2012). Tolerability, safety, pharmacokinetics, and efficacy of doxorubicin-loaded anti-EGFR immunoliposomes in advanced solid tumours: a phase 1 dose-escalation study. *Lancet Oncology* 13 (12): 1234–1241.

212 Thaker, P.H., Brady, W.E., Lankes, H.A. et al. (2017). A phase I trial of intraperitoneal GEN-1, an IL-12 plasmid formulated with PEG-PEI-cholesterol lipopolymer, administered with pegylated liposomal doxorubicin in patients with recurrent or persistent epithelial ovarian, fallopian tube or primary peritoneal cancers: an NRG oncology/gynecologic oncology group study. *Gynecologic Oncology* 147 (2): 283–290.

213 Clarke, J.L., Molinaro, A.M., Cabrera, J.R. et al. (2017). A phase 1 trial of intravenous liposomal irinotecan in patients with recurrent high-grade glioma. *Cancer Chemotherapy and Pharmacology* 79 (3): 603–610.

214 Macarulla, T., Blanc, J.-F., Wang-Gillam, A. et al. (2019). Liposomal irinotecan and 5-fluorouracil/leucovorin in older patients with metastatic pancreatic cancer–a subgroup analysis of the pivotal NAPOLI-1 trial. *Journal of Geriatric Oncology* 10 (3): 427–435.

215 Tabernero, J., Shapiro, G.I., LoRusso, P.M. et al. (2013). First-in-humans trial of an RNA interference therapeutic targeting VEGF and KSP in cancer patients with liver involvement. *Cancer Discovery* 3 (4): 406–417.

216 Wagner, M.J., Mitra, R., McArthur, M.J. et al. (2017). Preclinical mammalian safety studies of EPHARNA (DOPC nanoliposomal EphA2-targeted siRNA). *Molecular Cancer Therapeutics* 16 (6): 1114–1123.

217 Schultheis, B., Strumberg, D., Kuhlmann, J. et al. (2016). A phase Ib/IIa study of combination therapy with gemcitabine and Atu027 in patients with locally advanced or metastatic pancreatic adenocarcinoma. *Journal of Clinical Oncology* 34 (4_suppl): 385.

218 Schultheis, B., Strumberg, D., Santel, A. et al. (2014). First-in-human phase I study of the liposomal RNA interference therapeutic Atu027 in patients with advanced solid tumors. *Journal of Clinical Oncology* 32 (36): 4141–4148.

219 El Dika, I., Lim, H.Y., Yong, W.P. et al. (2019). An open-label, multicenter, phase I, dose escalation study with phase II expansion cohort to determine the safety, pharmacokinetics, and preliminary antitumor activity of intravenous TKM-080301 in subjects with advanced hepatocellular carcinoma. *The Oncologist* 24 (6): 747.

220 Tolcher, A.W., Papadopoulos, K.P., Patnaik, A. et al. (2015). Safety and activity of DCR-MYC, a first-in-class dicer-substrate small interfering RNA (DsiRNA) targeting MYC, in a phase I study in patients with advanced solid tumors. *American Society of Clinical Oncology*.

221 González-Ruíz, A., Ferro-Flores, G., Jiménez-Mancilla, N. et al. (2018). in vitro and in vivo synergistic effect of radiotherapy and plasmonic photothermal therapy on the viability of cancer cells using 177 Lu–au-NLS-RGD-aptamer nanoparticles under laser irradiation. *Journal of Radioanalytical and Nuclear Chemistry* 318 (3): 1913–1921.

222 Ferro-Flores, G., Ocampo-García, B.E., Santos-Cuevas, C.L. et al. (2015). Theranostic radiopharmaceuticals based on gold nanoparticles labeled with 177Lu and conjugated to peptides. *Current Radiopharmaceuticals* 8 (2): 150–159.

223 Jimenez-Mancilla, N., Ferro-Flores, G., Ocampo-Garcia, B. et al. (2012). Multifunctional targeted radiotherapy system for induced tumours expressing gastrin-releasing peptide receptors. *Current Nanoscience* 8 (2): 193–201.

224 Di Pasqua, A.J., Yuan, H., Chung, Y. et al. (2013). Neutron-activatable holmium-containing mesoporous silica nanoparticles as a potential radionuclide therapeutic agent for ovarian cancer. *Journal of Nuclear Medicine* 54 (1): 111–116.

225 Gibbens-Bandala, B., Morales-Avila, E., Ferro-Flores, G. et al. (2019). Synthesis and evaluation of 177Lu-DOTA-DN (PTX)-BN for selective and concomitant radio and drug-therapeutic effect on breast cancer cells. *Polymers* 11 (10): 1572.

226 Gibbens-Bandala, B., Morales-Avila, E., Ferro-Flores, G. et al. (2019). 177Lu-Bombesin-PLGA (paclitaxel): a targeted controlled-release nanomedicine for bimodal therapy of breast cancer. *Materials Science and Engineering: C* 105: 110043.

227 Trujillo-Nolasco, R.M., Morales-Avila, E., Ocampo-García, B.E. et al. (2019). Preparation and in vitro evaluation of radiolabeled HA-PLGA nanoparticles as novel MTX delivery system for local treatment of rheumatoid arthritis. *Materials Science and Engineering: C* 103: 109766.

228 Munaweera, I., Shi, Y., Koneru, B. et al. (2015). Chemoradiotherapeutic magnetic nanoparticles for targeted treatment of nonsmall cell lung cancer. *Molecular Pharmaceutics* 12 (10): 3588–3596.

229 Mendoza-Nava, H., Ferro-Flores, G., Ramírez, F.M. et al. (2016). ^{177}Lu-dendrimer conjugated to folate and bombesin with gold nanoparticles in the dendritic cavity: a potential theranostic radiopharmaceutical. *Journal of Nanomaterials* 2016.

230 Chen, L.C., Chang, C.-H., Yu, C.-Y. et al. (2007). Biodistribution, pharmacokinetics and imaging of 188Re-BMEDA-labeled pegylated liposomes after intraperitoneal injection in a C26 colon carcinoma ascites mouse model. *Nuclear Medicine and Biology* 34 (4): 415–423.

231 Chen, L.-C., Chang, C.-H., Yu, C.-Y. et al. (2008). Pharmacokinetics, micro-SPECT/CT imaging and therapeutic efficacy of 188Re-DXR-liposome in C26 colon carcinoma ascites mice model. *Nuclear Medicine and Biology* 35 (8): 883–893.

232 Yook, S., Cai, Z., Lu, Y. et al. (2015). Radiation nanomedicine for EGFR-positive breast cancer: panitumumab-modified gold nanoparticles complexed to the **β**-particle-emitter, 177Lu. *Molecular Pharmaceutics* 12 (11): 3963–3972.

233 Yook, S., Cai, Z., Lu, Y. et al. (2016). Intratumorally injected 177Lu-labeled gold nanoparticles: gold nanoseed brachytherapy with application for neoadjuvant treatment of locally advanced breast cancer. *Journal of Nuclear Medicine* 57 (6): 936–942.

234 Dziawer, L., Koźmiński, P., Męczyńska-Wielgosz, S. et al. (2017). Gold nanoparticle bioconjugates labelled with 211 at for targeted alpha therapy. *RSC Advances* 7 (65): 41024–41032.

235 Cai, Z., Yook, S., Lu, Y. et al. (2017). Local radiation treatment of HER2-positive breast cancer using trastuzumab-modified gold nanoparticles labeled with 177 Lu. *Pharmaceutical Research* 34 (3): 579–590.

236 Chanda, N., Kattumuri, V., Shukla, R. et al. (2010). Bombesin functionalized gold nanoparticles show in vitro and in vivo cancer receptor specificity. *Proceedings of the National Academy of Sciences* 107 (19): 8760–8765.

237 Chakravarty, R., Chakraborty, S., Guleria, A. et al. (2019). Clinical scale synthesis of intrinsically radiolabeled and cyclic RGD peptide functionalized 198Au nanoparticles for targeted cancer therapy. *Nuclear Medicine and Biology* 72: 1–10.

238 Tian, L., Chen, Q., Yi, X. et al. (2017). Radionuclide I-131 labeled albumin-paclitaxel nanoparticles for synergistic combined chemo-radioisotope therapy of cancer. *Theranostics* 7 (3): 614.

239 Eskandari, N., Yavari, K., Outokesh, M. et al. (2013). Iodine-131 radiolabeling of poly ethylene glycol-coated gold nanorods for in vivo imaging. *Journal of Labelled Compounds and Radiopharmaceuticals* 56 (1): 12–16.

240 Ruggiero, A., Villa, C.H., Holland, J.P. et al. (2010). Imaging and treating tumor vasculature with targeted radiolabeled carbon nanotubes. *International Journal of Nanomedicine* 5: 783.

241 Khameneh, E.S., Amini, M.M., Kakaei, S., and Khanchi, A. (2018). Preparation of dual-modality yttrium-90 radiolabeled nanoparticles for therapeutic investigation. *Radiochimica Acta* 106 (11): 897–907.

242 Chang, M.-Y., Seideman, J., and Sofou, S. (2008). Enhanced loading efficiency and retention of 225Ac in rigid liposomes for potential targeted therapy of micrometastases. *Bioconjugate Chemistry* 19 (6): 1274–1282.

243 Kratochwil, C., Bruchertseifer, F., Rathke, H. et al. (2018). Targeted **α**-therapy of metastatic castration-resistant prostate cancer with 225Ac-PSMA-617: swimmer-plot analysis suggests efficacy regarding duration of tumor control. *Journal of Nuclear Medicine* 59 (5): 795–802.

244 Sathekge, M.M., Bruchertseifer, F., Lawal, I.O. et al. (2019). Treatment of brain metastases of castration-resistant prostate cancer with 225 ac-PSMA-617. *European Journal of Nuclear Medicine and Molecular Imaging* 46 (8): 1756–1757.

245 Sathekge, M., Bruchertseifer, F., Vorster, M. et al. (2020). Predictors of overall and disease-free survival in metastatic castration-resistant prostate cancer patients receiving 225Ac-PSMA-617 Radioligand therapy. *Journal of Nuclear Medicine* 61 (1): 62–69.

6

Nanomaterials for Gene Therapy

V. Karthik[1], A. Vigneshwaran[2], D. Dharun Daniel Raj[3], T.G.N. Nagarjun[4], S. Poornima[1], R. Subbaiya[5], and M. Saravanan[6]

[1] *Department of Biotechnology, K.S. Rangasamy College of Technology, Tiruchengode, Tamil Nadu, India*
[2] *School of Biosciences and Technology, Vellore Institute of Technology, Vellore, Tamil Nadu, India*
[3] *Department of Biotechnology, Maulana Abul Kalam Azad University of Technology, Haringhata, West Bengal, India*
[4] *Department of Food Technology, K.S. Rangasamy College of Technology, Tiruchengode, Tamil Nadu, India*
[5] *Department of Biological Sciences, School of Mathematics and Natural Sciences, The Copperbelt University, Riverside, Kitwe, Zambia*
[6] *AMR and Nanotherapeutics Laboratory, Department of Pharmacology, Saveetha Dental College, Saveetha Institute of Medical and Technical Sciences, Chennai, India*

6.1 Introduction and Recent Advances

Nanotechnology has found increased applications in biomedical therapies, due to their nano size and the ability to mimic native molecules of the biological system of human body. The current understanding of potential toxicity of nanomaterials is very limited due to their extremely small size. The have the potential to be retained longer within the biological system; not only in therapies but also in day-to-day applications. Nanotechnology is used in many applications, which could lead to the exposure to nanoparticles, which can accumulate in biological system by ingestion, inhalation, and penetration. Nanomaterials are garnering high attention for their applications in gene therapies. However, there is not, yet, a standard recognized protocol for the assessment of toxicity in the use of these nanomaterials. Gene delivery is a process of introducing a new gene into the host environment that can be expressed to bring the desired effect on the host's genome in order to treat diseases of genetic origin, cancer, heart diseases, and neurodegenerative diseases.

6.2 Nanomaterials and their Physicochemical Properties

Physicochemical characteristics of nanoparticles can be exploited for various biomedical applications such as gene delivery (transfection), drug delivery (therapy), drug screening (labelling), tissue engineering by virtue of their controlled size, high surface-to-volume ratio, differential shape, well-ordered composition, surface charge, solubility, specific hydrophilic and hydrophobic nature, and aggregation [1].

Nanomaterials and Nanotechnology in Medicine, First Edition. Edited by P.M. Visakh.

6.2.1 Size and Shape

With reference to the therapeutic applications of nanoparticles, the size of the nanoparticles plays a vital role in achieving the desired type of selected biological actions. Correspondingly, the size of the nanoparticles regulates the kinetics, absorption, distribution, and metabolism of the biological obstacles. The shape of the nanomaterial also has consequential potential to influence nano–bio interfaces [1]. Nanomaterials exhibit varied shapes including fibers, rings, tubes, spheres, and planes. It has been recognized that endocytosis of spherical nanoparticles is easier when compared to rod-shaped and fiber nanoparticles and more importantly, spherical nanoparticles are relatively less toxic irrespective of whether they are homogenous or heterogeneous, while nonspherical nanomaterials are more disposed to flow through capillaries causing other biological consequences [2].

6.2.2 Surface Charge

Surface charge also plays an important role in interactions with biological systems. Various aspects of nanomaterials such as selective adsorption of nanoparticles, colloidal behavior, plasma protein binding, and blood–brain barrier integrity are primarily regulated by the surface charge of nanoparticles. By the aggregate and agglomerate formation of nanomaterials, the surface charge influences the organism's response by changing the shape and size of the nanomaterials [2].

6.2.3 Surface-to-Volume Ratio

At the nanoscale level, the surface-to-volume ratio is also one of the most significant physical properties. Since the nanomaterial's size and the number of surface-expressed molecules show an inverse relationship, it is mandatory to recognize the number of surface molecules [1, 2].

6.3 Methods of Characterizing the Physicochemical Properties of Nanomaterials

The various physicochemical properties of nanomaterials such as, particle size, size distribution, shape, agglomeration, aggregation, and crystal structure can be characterized by different methods, namely, Transmission Electron Microscopy, Atomic Force Microscopy, Field Flow Fractionation, Differential Electrical Mobility Analysis, Dynamic Light Scattering, Particle Tracking Analysis, and Scanning Electron Microscopy. Though all these characterization methods are more efficient in characterization of nanomaterials, each method has its own limitations in terms of state of the sample and duration of the characterization process. The limitations and the possible physicochemical properties of nanomaterials for different methods are shown in Table 6.1 [3].

Table 6.1 Physicochemical properties of nanomaterials.

Method	Physicochemical properties of nanomaterials that can be characterized	Limitations of the method
Transmission electron microscopy	Particle size and size distribution, shape, agglomeration.	Operation in high-vacuum, time consuming and expensive
Atomic force microscopy	Agglomeration, aggregation, surface properties.	Samples must adhere to a substrate
Field flow fractionation	Particle size distribution.	Particles in agglomerates or aggregates are not determined.
Differential electrical mobility analysis	Particle size distribution.	Only aerosol samples
Dynamic light scattering	Particle size distribution.	Bad resolution for polydisperse samples.
Particle tracking analysis	Particle size and size distribution, agglomeration, aggregation.	The sample must be a suspension.
Scanning electron microscopy	Agglomeration, aggregation, crystal structure.	Complex sample preparation.

6.4 Target Organ Biocompatibility/Toxicity

In vivo studies on rodents and humans by [4], focused on the clearance of nanomaterials from organs of the human body after exposure by inhalation. From this study it was evident that the lungs (as the port for entry of the nanomaterial), the liver, and the kidneys are the major organs affected by nanomaterials. If the nanomaterial enters the human body via ingestion or inhalation, they could cause tissue injury by accumulating in the tissue. Mostly, when they enter the bloodstream, they can be translocated to different parts of the body via blood circulation [5]. In case of the lungs, the airway and alveolar macrophages are the first line of defense against nanomaterials. It was observed in case studies that any nanomaterial aggregates less than 100 nm were not cleared by phagocytosis [6, 7]. If the nanomaterial gets opsonized, then clearance can be assured by opsonin-dependent phagocytosis.

However, opsonization can cause the agglomerates to become loose and escape phagocytosis [8]. It has been observed that prolonged deposition of nanomaterials in the lungs can cause inflammation, fibrosis, and cancer [9]. The liver, being the main organ for the clearance of drugs and xenobiotics, was also found to help clear nanomaterials from the blood. The liver's structural unit – hepatic lobules – consist of parenchymal cells (hepatocytes),

Kupffer cells (KCs), sinusoidal endothelial cells, and radially arranged lymphocytes. Any nanomaterial agglomerates larger than 200 nm can be easily removed by KCs through phagocytosis and pinocytosis, as the blood flow into liver sinusoids are sufficiently slow for the removal process to occur. Anything smaller than 200 nm can easily enter the endothelium and can be taken up by the hepatocytes [10]. The hydrolase enzyme in hepatocytes and lysosomes can also help in degrading nanomaterials when they enter the hepatocytes. However, inert nanomaterial such as gold nanoparticles (AuNPs), can lead to deposition. Prolonged deposition of nanomaterials can cause significant hepatotoxicity [11].

The kidney is the organ least involved with nanomaterial deposition as the liver takes care of most of the nanomaterial. Renal clearance consists of three steps: glomerular filtration, tubular secretion, and tubular reabsorption. The specialized glomerular filtration can be categorized into three compartments: (i) fenestrated endothelium (80–100 nm diameter), (ii) the glomerular basement membrane (200–400 nm depending on species), and (iii) slit diaphragm (30–40 nm wide with a pore size of 4–5 nm). Thus in glomerular filtration, the molecule will be segregated based on size due to the size gradient in the compartments of the glomerular filter [12].

Thus, the molecular weight (MW) cut-off point, in order to pass through the glomerular filter, is 70 kDa, or any molecule with a diameter under 5 nm [13]. Therefore, most nanomaterial will not be retained by the kidney and due to their larger size will be processed by the liver. There are, however, a few exceptional cases where the threshold cut-off point (to pass through glomerular filter) seems inconsistent. Wang et al. [14] synthesized SWCNTs (single-walled carbon nanotubes) and labeled it with iodine-125, by hydroxylation. They injected the nanomaterial via the peritoneum. The average MW of the SWCNT was 600 kDa with an average length of 340 nm. A 75% clearance of the injected nanomaterial was observed in 11 days. Thus, the actual mechanism for expulsion and passing through the filters in the kidney is still not clear. In the case where nanomaterial deposition was observed, the kidney might become injured severely and prolonged accumulation or exposure can lead to a degeneration of the kidney and necrosis [15].

6.4.1 Mechanism of Potential Cytotoxic Effects of Nanomaterials

The nanomaterial because of its reactivity, can create reactive oxygen species (ROS), or indirectly initiate an oxidative enzymatic pathway, which may cause oxidative stress to the cell. As a result of the oxidative stress, the nuclei material inside the cell can be damaged, which might in turn trigger the intrinsic pathway for cellular apoptosis [16]. Nanomaterials upon internalization were frequently found in the lysosome of the cell, which might trigger mitochondrial cellular apoptosis [17]. There have been reported cases where the nanomaterials once internalized into a cell stuck to the tubulins in the cytoskeleton of the cell. Upon mitotic division of the cell, the nanomaterials stuck to the microtubule can interrupt mitosis, leading to apoptosis or necrosis [18]. In another cases perturbation caused by the nanomaterial to the centromere during division can cause, diploidy, triploidy, or aneuploidy, resulting in carcinogenesis. Thus, the accumulation of nanomaterial can cause a storm in the cell during the mitotic phase, eventually leading to cancer or cell death by apoptosis or necrosis [19].

Pyroptosis is programmed cell death caused by inflammation, which may occur as a response to bacterial infection. In pyroptosis, caspase 1 gets activated, resulting in the generation of pyrogenic factors such as interleukin-1β [20]. It has been reported that carbon black nanomaterials activate caspase 1 resulting in the formation of interleukin-1β, which inflames the cell and leads to its death. In some cases, it is reported that interleukin-1α causes the pyroptosis. The actual mechanism is still being studied, but in general, cell death by nanomaterial accumulation is highly possible [21]. Thus, in short, nanomaterial in a biological system can cause oxidative stress, or interfere with the signaling within the cell leading to cell apoptosis, pyroptosis, or mitotic catastrophe, which can lead to health-related complications.

6.4.2 Biological Defense System against Nanomaterial

The biological defense system is the first barrier to any foreign particles that try to enter the biological system. Physiological barriers, which include the blood–brain barrier and the air–lung blood barrier, are specially designed barriers that protects the specific organ and allow only suitable candidates to pass through it. Physical barriers such as mucus and hair have their own role to play barrier creation. However, nanomaterials administered for therapies can bypass most of the barriers, but face difficulties in blood/serum and liver enzymes.

a) **Blood/serum as a barrier to nanomaterials**

Our blood carries multiple cells such as red blood cells, platelets, macrophages, and complementary proteins. When nanomaterials bind to platelets they cause its activation resulting in the formation of thrombosis or blood clots, which can restrict the flow of cells to the desired location. This defense mechanism can cause severe health issue based on the size of the blood clots [22]. The peripheral blood mononuclear cells (PBMC), are responsible for defense in a biological system. The PBMC includes B, T, and NK cells. When nanomaterials enter the bloodstream, they can get opsonized and be attacked by macrophages, in which the nanomaterial can be agglomerated and can be consumed by phagocytosis [23]. In case the nanomaterial gets bound to B cell or T cells, they start the cascade signaling that results in alerting NK cells, which the track bound B and T cells and eliminate them. Thus, the nanomaterial gets deposited in a tissue of an organ, reducing the circulation time, resulting in ineffective therapy.

b) **Liver enzymes as a barrier for nanomaterials**

As discussed in Section 6.4, the liver is the major organ in which nanomaterial gets deposited. When nanomaterials get deposited in the lysozyme of KCs in the liver, the liver can activate lysozyme-associated degradation of the nanomaterial. When the nanomaterial enters the acidic pH of the KCs a lot of hydrolase enzymes are generated, thus degrading the nanomaterial, making nanotherapy unsuccessful [24]. It was observed that haemoxygenase-1, peroxidase, and α-glucosidase facilitated in the degradation of nanomaterials is internalized in liver cells. More enzymes such as monooxygenase, transferases, esterases, and epoxide hydrolase, that aid in degradation of nanomaterials, internalize in hepatic cells. However, the proof of evidence *in vivo* is still lacking. It is quite clear that the degradation by liver enzymes is impossible on inert nanomaterials such as gold and is only possible with noninert metallic nanomaterial [25].

6.5 Gene Delivery

In order to carry out the successful gene delivery, the delivery vector must be chosen carefully. The vectors are the agents that carry the genetic material that is intended to cure the disease and transport it to the target cell or tissue. The success of gene delivery lies in decreasing the protein expressions that can cause harm to the host and down-regulation of genes that cause tumors and other side effects [26]. Other factors that influence gene delivery are pharmacokinetic properties and gene transfer conditions. The ability of the chosen vector to deliver the gene to the desired location by crossing all the barriers is still a challenge [27].

6.5.1 Gene-delivery Vectors

The gene-delivery vector can be classified into two types: viral vectors and nonviral vectors. In the case of viral vectors, there are a lot of safety concerns that need to be taken into account. The use of viral vectors may impart an excessive immune response or insertion mutagenesis in the host. In order to overcome such difficulties, nonviral vectors are used for successful gene delivery. Particularly, the nanoparticle-based gene-delivery system could be exploited for successful gene insertion. The nanoparticle-based gene-delivery systems possess properties that no other nonviral vector systems possess. The size of the nanoparticle-based vector system is a great advantage because it can penetrate cellular membranes. The surface area of such vectors is also large so it can carry larger payloads. The physical properties of the vectors can also be modified for the successful delivery of genes [27].

The first of its kind is the lipid nanoparticle-based gene-delivery systems. The lipid nanoparticles are very often used in drug-delivery applications as most of them have been approved by the US Food and Drug Administration (FDA). The physical property of the lipid nanoparticle is well suited to carrying genetic materials [28]. It has a solid lipid core that is surrounded by the surfactants, which are positively charged, in order to bind electrostatically to the negatively charged nucleus. It can also be inversed, i.e. an anionic surfactant can be used if the nucleic acids are previously coated with cationic ingredients [29]. N-[1-(2,3-dioleyloxy)propyl]-N,N,N-trimethylammoniumchloride(DOTMA),[1,2-bis(oleoyloxy)-3-(trimethylammonio)propane] (DOTAP), 3β[N-(N′, N′ -dimethylaminoethane)-carbamoyl] cholesterol (DCChol), and dioctadecylamidoglycylspermine (DOGS) are the commonest cationic liposomes that are used for gene-delivery applications.

The neutral lipids for transfection include dioleoylphosphatidylcholine (DOPC) and dioleoylphosphatidylethanolamine (DOPE). DOPE can be conjugated with cationic lipids for endolysosomal escape at a lower pH. Phosphatidic acid, phosphatidylglycerol, and phosphatidylserine are categorized as anionic lipids because of the presence of a negatively charged head and they are not very efficient in gene-delivery applications [30].

The lipid nanoparticles used in the ocular gene delivery for treating retinal disease are liposome nanoparticles, solid lipid nanoparticles, lipoplexes, and CK30-PEG (polyethylene glycol). They deliver the genes locally on the retina of the infected person [31]. The p53 genes were successfully transferred to lung cancer cells with the help of solid lipid nanoparticles. It produced better results compared to Lipofectin®, a commercial transfection agent. It induced apoptosis and cell growth inhibition better than the Lipofectin-mediated transfection. *O*, *O′*-dimyristyl-N-lysyl glutamate (DMKE), a cationic lipid component, was made to form a complex with a small interfering anionic RNA (siRNA) that can deliver the genes into the cancer cells for treatment [32]. The lipid nanoparticles were also used to deliver siRNA into the leukocytes, the hardest to transfect cell types for treating multiple myeloma cells [33]. Second are inorganic nanoparticles. Inorganic nanoparticles possess highly advanced chemistry. They exhibit good storage stability, are nontoxic, and have the potential to resist microbial attack. They can carry payloads to the target without undergoing any changes [34].

Commonly used inorganic nanoparticles for gene delivery are AuNPs, carbon nanotubes (CNT), fullerenes, magnetic nanoparticles (MNP), quantum dots, silica nanoparticles, calcium phosphate nanoparticles, etc. AuNPs were used for the delivery of genes into mesenchymal stem cells. AuNPs are cationic so that they can carry the genes to the target location without undergoing any changes. Also, they have antibacterial peptides, which enhance the transfection efficiency and provide an antibacterial activity [35]. Peptide-modified dendrimer-entrapped AuNP was also used to transfect the genes into human mesenchymal stem cells (hMSC). Antimicrobial peptide (LL37) was conjugated with ultra-small AuNPs for accelerating wound healing in diabetic patients. This imparts more antibacterial ability than a topical treatment. The PEG-modified silver nanoparticles (AgNPs) are also used in the gene-delivery applications for treating cancer [36].

CNTs are used as a gene-delivery vehicle because it they better biocompatibility, high length-to-diameter ratio and more surface area. The CNT surface can be modified so that it can deliver siRNA, Plasmid DNA (pDNA), and micro-RNA (miRNA) into the target mammalian cells. The CNTs can be single-walled or multiwalled [37]. They are employed to treat cancer, HIV, and diabetes [38]. The process of gene transfer using MNP is called magnetofection. The plasmids form complexes with the polymer-coated MNP. The complex formation is called magnetoplex. The common MNPs used for gene-delivery applications are iron oxide nanoparticles and magnetite [39]. The complex is injected into the host and the magnetic field is set up near the target region using a small rare earth magnet. This method has an advantage over other gene-delivery techniques because the gene delivery to the target can be accomplished with better efficiency due to the application of an external magnetic field [40, 41]. The p53 tumor-suppressor gene is transfected using magnetic iron oxide nanoparticles. This method has led to higher levels of expression of the tumor-suppressor gene, which is traceable through magnetic resonance imaging [42].

Calcium nanoparticle-mediated gene delivery is the other mode of gene delivery that can used alone for transfection of the gene or it can be associated with other viral and nonviral vectors for better transfection efficiency. The ionic complexes can be formed between DNA and the other divalent metal cation (Mg^{2+}, Ca^{2+}, and Mn^{2+}) to form stable structures and to cross cell membranes for transfection [43]. The calcium nanoparticle complex can also be coated with lipid to give protection to the payload for treating breast cancer [44]. The calcium phosphate–pDNA complex exhibited very low toxicity when targeted toward the mesenchymal stem cells [45]. The quantum dots are semiconductor particles with core-shell structures. They can be utilized for transporting gene-associated drugs. The gene-delivery system that employs quantum dots is called Super Paramagnetic iron oxide nanoparticles (SPION). The SPION-loaded polymersome that carries the genetic material is made to get accumulated and expressed to treat colon cancer [46]. The polymer-based nanoparticles are the most significant type of vectors for gene delivery. They are cationic substances that form complexes with anionic genetic material, which are called as polyplexes. They possess an advantage over lipoplexes as they are much smaller in size. The commonly used polymeric nanoparticles include poly(lactic-co-glycolic-acid), polyethylenimine (PEI), polysaccharide (dextran, chitosan, etc.), dendrimers, PIC, poly(2-N (dimethylaminoethyl) methacrylate) (pDMAEMA), poly(amidoamine) (PAMAM), Poly-L-lysine (PLL) [27, 47]. Poly(β-amino ester)s (PBAEs) are used to deliver genes for treating brain tumor cells and Glioblastoma cells. The noncondensing polymeric nanoparticles made up of the aforementioned polymer can be used for gene delivery into cancer cells and oral gene deliver [48]. Hybrid nanoparticles are combinations of polymer-lipid-DNA, which is the most efficient type of nanoparticles that help in better gene delivery to the target. The combination is called lipid-polyplexes or lipopolyplexes. They form a highly condensed DNA structure that is formed through layer-by-layer assembly [27].

6.5.2 Intracellular Uptake and Trafficking

The cellular membrane is a permeable one that allows only selected small molecules to pass through it. The small molecules include ions and amino acids, which are carried inside the cellular environment through active transport. Whereas the payload carriers like nanoparticles are carried into the cell by means of endocytosis. The intracellular uptake profile is an important component in measuring the delivery efficiency of the carrier and the bioavailability of the carrier. The surface charge on the nanoparticle also plays an important role in reaching the target site. The receptor targeting for cellular uptake is also the method used by the nanoparticles. The receptors targeted are folate receptor (FR) (overexpressed by ovarian cancer, lung cancer, brain cancer, etc.). The isoforms include FR-α and FR-β, FR-γ. The epidermal growth factor receptor (EGFR) also called HER1 is one of the receptors that is commonly targeted. Other targeted receptors include the transferrin receptor (TfR), the prostate-specific membrane antigen (PSMA), integrin, and the neonatal Fc-receptor (FcRn). The uptake pathways/mechanism followed by the nanoparticle endocytosis are clathrin-mediated endocytosis where clathrin-coated vesicles containing plasma membrane proteins specific to nanoparticles are internalized. It is an energy-dependent mechanism where very small nanoparticles are carried. The other mode is caveolae-dependent endocytosis where they are carried through flask-shaped pits after binding with cholesterol-binding protein caveolin. The nanoparticles in the size range of 200 to 500 nm are carried through this method. The nonspecific mechanism called pinocytosis may also be followed for the internalization of larger nanoparticles [49].

When the nanoparticles enter the cells, after the internalization, the early encounter membrane bound intracellular vesicles are called endosomes. The endosomes in the plasma membrane are categorized into early endosomes, late endosomes, and recycling endosomes. The early endosomes carry the cargo to the target cellular destination. This early endosome is transformed into late endosomes due to maturation and differentiation processes [50]. A small section of the cargo is recycled to the plasma membrane through recycling endosomes. Through the lysosomal degradation process, the nanoparticles are able to escape from the late endosomes pathway and are released into cytoplasm [51]. This intracellular degradation pathway by nanoparticles is referred as autophagy. In addition, the autophagy degrades the compacted proteins and dysfunctional organelles to maintain homeostasis. The recent study shows the intracellular trafficking of Tat peptide-conjugated quantum dots leads to internalization by macropinocytosis. These quantum dots containing the vesicles are actively transported to the perinuclear region known as the microtubule-organizing center [52]. Quantum dots bind to cellular membrane structures, like filopodia. Vesicles containing quantum dots are released from the filopodia tips through vesicle shedding [52].

6.5.3 Gene Regulation

Gene regulation is a kind of vast mechanism used by all prokaryotic and eukaryotic cells to control the production of specific gene products like proteins and RNAs. It increases the adaptability of the cells by expressing protein when it is needed. In multicellular organisms, the regulation of gene expression in the same genome sequence leads to cell differentiation (creation of different cell types) and embryo morphogenesis [53]. This increases understanding of how the biological evolution occurs at a molecular level.

During the process of protein synthesis from DNA, all the gene expression stages from transcription to post-translational modification are regulated. For example, the modification of histone in the chromatin structure by DNA methylation, ncRNA, or DNA binding proteins regulates the gene expression as inheritable and is commonly known as epigenetic regulation [54]. In DNA transcription, a DNA segment is wound around by tightly packed histone protein leads to DNA supercoiling, this process is temporarily modified by phosphorylation or permanently modified by methylation. Such modifications create a direct impact on the regulation of gene expressions [55]. As a chemical modification, DNA methylation and histone deacetylation (process of association of DNA from the histone complex) combine to send a signal for DNA to densely pack, this controls gene expression. These regulations happen naturally but some of the gene expression regulations deal with synthetic materials like nanoparticles for gene therapy. There are various mechanisms within the cell that can regulate gene expression during cellular growth, cell differentiation, and metabolism. If it does not function properly, the cells will undergo apoptosis or develop into tumor cells [56]. For these genetic disorders, several exogenous and endogenous approaches were developed to influence transcription and translation for regulating gene expression. Nowadays, most of the anticancer strategies targeted on abnormally overexpressed genes called oncogenes are treated with advanced nanotechnology methods. In most cases, synthetic nanomaterials are directly used as gene regulatory systems but, in some cases, they depend on the physiochemical properties and pliable shape and size, these nanomaterials are used as nanocarriers for silencing RNAs or antisense oligonucleotides. Especially for gene regulation, new varieties of nanomaterials have been developed. Due to their unique properties most of these are carbon- and polymer-based materials [57]. These polymer-based nanomaterials are mimicked naturally due to their biocompatibility, intrinsic lower toxicity, and easy decomposition inside the biological system compared to other available materials. Due to these specific features, in the biomedical field they are used for the regulation of gene expression within the cell, valid metabolic regulatory delivery, and to improve the performance of high-resolution cellular-tracking imaging [58].

The carbon-based nanomaterials with high efficiency have many concerns, mainly to avoid immune response while performing cellular metabolic regulations. With the help of nanocarriers we can avoid immune responses by wrapping the functional component, which regulates gene expression, over the surface of nanoparticles, so it can easily recognize target cells and aid intracellular entry [59]. AuNPs are a perfect example of a gene regulatory nanocarrier, these AuNPs are widely using materials in medical applications because of its remarkable properties such as high biocompatibility, low cytotoxicity, easily bind with thiol derivatives by covalent bond [60]. From several studies in photothermal therapy or chemotherapy, AuNPs were used as a secondary material to increase efficacy through gene regulation in cancer cells. During the gene regulation process in cancer cells, antisense sequences coated over the AuNPs, which complement the target mRNAs, are incorporated in the cancer cell that degrades mRNAs [61]. The fluorescence emitted by the nanocarrier helps to monitor and confirm the mRNAs' reduction.

MNPs are another material used for regulation of gene expression. They include nickel, iron, or cobalt and have a distinctive superparamagnetic property while at nano size. According to the recent studies, these MNPs, or carriers, can efficiently deliver drug molecules into the cells for gene regulation. siRNAs have been developed and enveloped by iron oxide nanoparticles with a mesoporous silica layer [62]. This iron oxide nanoparticles were supervised by an external magnetic field and efficiently silenced the target oncogene PLK1. It leads the tumor cells into apoptosis. In the same way as some other nanoparticles, like quantum dots and silica nanoparticles, they are also used in incorporation of siRNAs or antisense RNA interference for gene regulation [63]. Nano-sized hydroxyapatites are a new method of inducing DNA methylation for the control of gene expression. Hydroxyapatites are mainly used to reduce or inhibit cell differentiation, especially in bone cells. When hydroxyapatite particles are injected into bone marrow stromal cells, the genes that are responsible for cell differentiation, like alkaline phosphatase, osteopontin, and bone sialoprotein, are arrested, which results in a sudden reduction of gene expression by the promotion of DNA methylation [64].

The most convenient and cost-effective nanomaterial is carbon based. Graphene, single, and multiwalled CNT, and carbon dots are widely used carbon-based nanomaterials. Due to its low toxicity and higher biocompatibility compared to other metals, it is most suitable for biomedical application after AuNPs. The bonding between the gene and the carbon molecules (a ring structure) is based on noncovalent bond interaction. Several studies show the modified graphene oxide with PEG and folic acid incorporated with siRNA molecules. This modified graphene oxide with siRNAs was injected into pancreatic cancer cell leading to simultaneous silencing of the HDAC1 (histone deacetylase1) and K-Ras. Instead of siRNA, an antisense specific for miR21 delivered with peptide nucleic acid incorporated with graphene can be used. These complexes have targeted breast cancer cells to knockdown the miR-21, this leads to controlling the proliferation of cancer cells and leads to apoptosis. The fluorescence from nanomaterials confirms the reduction in gene expression. The advantages of these carbon-based materials are the transfection rate (over 90%) and photothermal properties (over 80%), which have the ability to control gene expressions. Out of all nanomaterials used for gene regulation, CNT has the highest transfection efficiency for gene delivery. The modified PEI SWCNT induced with siRNA targets the vascular endothelial growth factor (VEGF), which is responsible for angiogenesis. This

CNT attached on the cell surface with the help of PEI and internalization of the siRNA complex inhibits the VEGF expression and will reduce tumor growth and control angiogenesis. The polymer-based nanomaterials are mostly composed of natural compounds, for example the composition of hyaluronic acid and calcium phosphate incorporated with siRNAs targets Bcl2 genes in melanoma cancer cells leading to silencing of the targeted genes, resulting in apoptosis. Another polymeric structure made of N-acetyl-D-glucosamine and deacetylated glucosamine targeting the PCSK9 gene may completely inhibit gene expression.

6.6 Regulatory Issues

Though nanomedicine has had more promising results in gene therapy, it also entails a great challenge to public health in terms of regulatory issues. Many research studies state that not all nanomaterial is toxic for a human system so is available for use in therapeutic applications, but it is important to screen those that may contain undesirable properties in order to produce bionanomaterial with the minimum risk to human health. In the development of novel drug applications nanotechnology contributes especially in the areas of cancer therapy and treatment of neurological disorders. Biological vectors like viruses are one of the ways that nanotechnology may facilitate gene therapy and immunologically, since the nanomaterials are much less reactive than the vectors. Nanotherapy allows the delivery of a maximum quantity of the genetic material. Recent studies on MNP and copper nanoparticles have more *in vivo* therapeutical information on tissue distribution, morphological changes, and pathological examinations and are beginning to inform risk assessments that will be used to regulate the use of bionanomaterial in consumer products [65]. The European Medicines Agency have made available some reflection papers regarding nanomedicines with surface coating, intravenous liposomal, block copolymer micelle, and iron-based nanocolloidal nanomedicines. The principles outlined in these documents contribute general issues regarding the complexity of nanosystems in pharmaceutical development.

References

1 Navya, P.N. and Daima, H.K. (2016). Rational engineering of physicochemical properties of nanomaterials for biomedical applications with nanotoxicological perspectives. *Nano Convergence* https://doi.org/10.1186/s40580-016-0064-z.

2 Gatoo, M.A., Naseem, S., Arfat, M.Y. et al. (2014). Physicochemical properties of nanomaterials: implication in associated toxic manifestations. *BioMed Research International* https://doi.org/10.1155/2014/498420.

3 Soares, S., Sousa, J., Pais, A., and Vitorino, C. (2018). Nanomedicine: principles, properties, and regulatory issues. *Frontiers in Chemistry* https://doi.org/10.3389/fchem.2018.00360.

4 Wang, B., He, X., Zhang, Z. et al. (2013). Metabolism of nanomaterials in vivo: blood circulation and organ clearance. *Accounts of Chemical Research* 46 (3): 761–769. https://doi.org/10.1021/ar2003336.

5 Kermanizadeh, A., Gaiser, B.K., Johnston, H. et al. (2014). Toxicological effect of engineered nanomaterials on the liver. *British Journal of Pharmacology* 171 (17): 3980–3987. https://doi.org/10.1111/bph.12421.

6 He, X., Zhang, H., Ma, Y. et al. (2010). Lung deposition and extrapulmonary translocation of nano-ceria after intratracheal instillation. *Nanotechnology* 21 (28): 285103. https://doi.org/10.1088/0957-4484/21/28/285103.

7 Geiser, M. (2010). Update on macrophage clearance of inhaled micro-and nanoparticles. *Journal of Aerosol Medicine and Pulmonary Drug Delivery* 23: 207–217.

8 Ruge, C.A., Kirch, J., Cañadas, O. et al. (2011). Uptake of nanoparticles by alveolar macrophages is triggered by surfactant protein A. *Nanomedicine: Nanotechnology, Biology, and Medicine* 7 (6): 690–693. https://doi.org/10.1016/j.nano.2011.07.009.

9 Zhu, M.T., Feng, W.Y., Wang, Y. et al. (2009). Particokinetics and extrapulmonary translocation of intratracheally instilled ferric oxide nanoparticles in rats and the potential health risk assessment. *Toxicological Sciences: An Official Journal of the Society of Toxicology* 107 (2): 342–351. https://doi.org/10.1093/toxsci/kfn245.

10 Braet, F. and Wisse, E. (2002). Structural and functional aspects of liver sinusoidal endothelial cell fenestrae: a review. *Comparative Hepatology* 1 (1): 1. https://doi.org/10.1186/1476-5926-1-1.

11 Abdelhalim, M.A. and Jarrar, B.M. (2011). Gold nanoparticles administration induced prominent inflammatory, central vein intima disruption, fatty change and Kupffer cells hyperplasia. *Lipids in Health and Disease* 10: 133. https://doi.org/10.1186/1476-511X-10-133.

12 Ota, Z., Makino, H., Miyoshi, A. et al. (1979). Molecular sieve in glomerular basement membrane as revealed by electron microscopy. *Journal of Electron Microscopy* 28 (1): 20–28.

13 Choi, H.S., Liu, W., Misra, P. et al. (2007). Renal clearance of quantum dots. *Nature Biotechnology* 25 (10): 1165–1170. https://doi.org/10.1038/nbt1340.

14 Wang, H., Wang, J., Deng, X. et al. (2004). Biodistribution of carbon single-wall carbon nanotubes in mice. *Journal of Nanoscience and Nanotechnology* 4 (8): 1019–1024. https://doi.org/10.1166/jnn.2004.146.

15 Chen, Z., Meng, H., Xing, G. et al. (2006). Acute toxicological effects of copper nanoparticles in vivo. *Toxicology Letters* 163 (2): 109–120. https://doi.org/10.1016/j.toxlet.2005.10.003.

16 Nel, A., Xia, T., Mädler, L., and Li, N. (2006). Toxic potential of materials at the nanolevel. *Science (New York, NY)* 311 (5761): 622–627. https://doi.org/10.1126/science.1114397.

17 Repnik, U., Stoka, V., Turk, V., and Turk, B. (2012). Lysosomes and lysosomal cathepsins in cell death. *Biochimica et Biophysica Acta* 1824 (1): 22–33. https://doi.org/10.1016/j.bbapap.2011.08.016.

18 Vitale, I., Galluzzi, L., Castedo, M., and Kroemer, G. (2011). Mitotic catastrophe: a mechanism for avoiding genomic instability. *Nature Reviews: Molecular Cell Biology* 12 (6): 385–392. https://doi.org/10.1038/nrm3115.

19 Sargent, L.M., Hubbs, A.F., Young, S.H. et al. (2012). Single-walled carbon nanotube-induced mitotic disruption. *Mutation Research* 745 (1–2): 28–37. https://doi.org/10.1016/j.mrgentox.2011.11.017.

20 Galluzzi, L., Vitale, I., Abrams, J.M. et al. (2012). Molecular definitions of cell death subroutines: recommendations of the nomenclature committee on cell death. *Cell Death*

and Differentiation 19 (1): 107–120. https://doi.org/10.1038/cdd.2011.96. Epub 2011 Jul 15. PMID: 21760595; PMCID: PMC3252826.

21 Yazdi, A.S., Guarda, G., Riteau, N. et al. (2010). Nanoparticles activate the NLR pyrin domain containing 3 (Nlrp3) inflammasome and cause pulmonary inflammation through release of IL-1α and IL-1β. *Proceedings of the National Academy of Sciences of the United States of America* 107 (45): 19449–19454. https://doi.org/10.1073/pnas.1008155107.

22 Laloy, J., Minet, V., Alpan, L. et al. (2014). Impact of silver nanoparticles on Haemolysis, platelet function and coagulation. *Nanobiomedicine* 1: 4. https://doi.org/10.5772/59346.

23 Andersson-Willman, B., Gehrmann, U., Cansu, Z. et al. (2012). Effects of subtoxic concentrations of TiO2 and ZnO nanoparticles on human lymphocytes, dendritic cells and exosome production. *Toxicology and Applied Pharmacology* 264 (1): 94–103. https://doi.org/10.1016/j.taap.2012.07.021.

24 Zhu, M., Feng, W., Wang, B. et al. (2008). Comparative study of pulmonary responses to nano-and submicron-sized ferric oxide in rats. *Toxicology* 247: 102–111.

25 Kagan, V.E., Konduru, N.V., Feng, W. et al. (2010). Carbon nanotubes degraded by neutrophil myeloperoxidase induce less pulmonary inflammation. *Nature Nanotechnology* 5 (5): 354–359. https://doi.org/10.1038/nnano.2010.44.

26 Katragadda, C.S., Choudhury, P.K., and Murthy, P.N. (2010). Nanoparticles as non-viral gene delivery vectors. *Indian Journal of Pharmaceutical Education and Research* 44 (2): 109–120.

27 Shim, M.S. and Kwon, Y.J. (2012). Stimuli-responsive polymers and nanomaterials for gene delivery and imaging applications. *Advanced Drug Delivery Reviews* 64 (11): 1046–1059. https://doi.org/10.1016/j.addr.2012.01.018.

28 Pozo-Rodríguez, A., Solinís, M.Á., and Rodríguez-Gascón, A. (2016). Applications of lipid nanoparticles in gene therapy. *European Journal of Pharmaceutics and Biopharmaceutics* 109: 184–193. https://doi.org/10.1016/j.ejpb.2016.10.016.

29 Pozo-Rodríguez, A., Delgado, D., Solinís, M.A., and Pedraz, J.L. (2007). Solid lipid nanoparticles: formulation factors affecting cell transfection capacity. *International Journal of Pharmaceutics* 339: 261–268. https://doi.org/10.1016/j.ijpharm.2007.03.015.

30 Balazs, D.A. and Godbey, W.T. (2011). Liposomes for use in gene delivery. *Journal of Drug Delivery* 2011: 326497. https://doi.org/10.1155/2011/326497.

31 Wang, Y., Rajala, A., and Rajala, R.V.S. (2015). Lipid nanoparticles for ocular gene delivery. *Journal of Functional Biomaterials* 6 (2): 379–394. https://doi.org/10.3390/jfb6020379.

32 Kim, M.W., Jeong, H.Y., Kang, S.J. et al. (2017). Cancer-targeted nucleic acid delivery and quantum dot imaging using EGF receptor aptamer-conjugated lipid nanoparticles. *Nature: Scientific Reports* 7: 1–11. https://doi.org/10.1038/s41598-017-09555-w.

33 Ramishetti, S., Hazan-Halevy, I., Palakuri, R. et al. (2020). A combinatorial library of lipid nanoparticles for RNA delivery to leukocytes. *Advanced Materials* 32 (12): 1–8. https://doi.org/10.1002/adma.201906128.

34 Sokolova, V. and Epple, M. (2008). Inorganic nanoparticles as carriers of nucleic acids into cells. *Angewandte Chemie International Edition* 1382–1395. https://doi.org/10.1002/anie.200703039.

35 Peng, L.-H., Huang, Y.-F., Zhang, C.-Z. et al. (2019). Integration of antimicrobial peptides with gold nanoparticles as unique non-viral vectors for gene delivery to mesenchymal stem cells with antibacterial activity. *Biomaterials* 103 (May): 137–149. https://doi.org/10.1016/j.biomaterials.2016.06.057.

36 Sarkar, K., Banerjee, S.L., Kundu, P.P. et al. (2015). Biofunctionalized surface-modified silver nanoparticles for gene delivery. *Journal of Materials Chemistry B* 3 (26): 5266–5276. https://doi.org/10.1039/C5TB00614G.

37 Behnam, B., Shier, W.T., Hashem, A. et al. (2013). Non-covalent functionalization of single-walled carbon nanotubes with modified Polyethyleneimines for efficient gene delivery. *International Journal of Pharmaceutics* 454 (1): 204–215. https://doi.org/10.1016/j.ijpharm.2013.06.057.

38 Bates, K. and Kostarelos, K. (2013). Carbon nanotubes as vectors for gene therapy : past achievements, present challenges and future goals. *Advanced Drug Delivery Reviews* 65 (15): 2023–2033. https://doi.org/10.1016/j.addr.2013.10.003.

39 Kami, D., Takeda, S., Itakura, Y. et al. (2011). Application of magnetic nanoparticles to gene delivery. *International Journal of Molecular Sciences* 12 (6): 3705–3722. https://doi.org/10.3390/ijms12063705.

40 Li, C., Li, L., and Keates, A.C. (2012). Targeting cancer gene therapy with magnetic nanoparticles. *Oncotarget* 3 (4): 365–370.

41 Mykhaylyk, O., Sobisch, T., Almstätter, I. et al. (2012). Silica-iron oxide magnetic nanoparticles modified for gene delivery: a search for optimum and quantitative criteria. *Pharmaceutical Research* 29: 1344–1365. https://doi.org/10.1007/s11095-011-0661-9.

42 Lee, H.J., Nguyen, Y.T.C., Muthiah, M., and Namgung, R. (2012). MR traceable delivery of P53 tumor suppressor gene by PEI-functionalized superparamagnetic iron oxide nanoparticles. *Journal of Biomedical Nanotechnology* 8 (3): 361–371. https://doi.org/10.1166/jbn.2012.1407.

43 Giger, E.V., Puigmartí-luis, J., Schlatter, R. et al. (2011). Gene delivery with bisphosphonate-stabilized calcium phosphate nanoparticles. *Journal of Controlled Release* 150 (1): 87–93. https://doi.org/10.1016/j.jconrel.2010.11.012.

44 Tang, J., Li, L., Howard, C.B. et al. (2015). Preparation of optimized lipid-coated calcium phosphate nanoparticles for enhanced in vitro gene delivery to breast cancer cells. *Journal of Materials Chemistry B* 3 (33): https://doi.org/10.1039/C5TB00912J.

45 Cao, X., Deng, W., Wei, Y. et al. (2011). Encapsulation of plasmid DNA in calcium phosphate nanoparticles : stem cell uptake and gene transfer efficiency. *International Journal of Nanomedicine* 6: 3335–3349.

46 Lee, S.J., Lee, H.J., Moon, M.-J. et al. (2011). Superparamagnetic iron oxide nanoparticles-loaded polymersome-mediated gene delivery guided by enhanced magnetic resonance signal. *Journal of Nanoscience and Nanotechnology* 11 (8): 7057–7060. https://doi.org/10.1166/jnn.2011.4861.

47 Lin, G., Zhang, H., and Huang, L. (2015). Smart polymeric nanoparticles for cancer gene delivery. *Molecular Pharmaceuticals* 12 (1): 314–321. https://doi.org/10.1021/mp500656v.

48 Xu, J., Ganesh, S., and Amiji, M. (2012). Non-condensing polymeric nanoparticles for targeted gene and SiRNA delivery. *International Journal of Pharmaceutics* 427 (1): 21–34. https://doi.org/10.1016/j.ijpharm.2011.05.036.

49 Iversen, T.-G., Skotland, T., and Sandvig, K. (2011). Endocytosis and intracellular transport of nanoparticles: present knowledge and need for future studies. *Nano Today* 6 (2): 176–185. https://doi.org/10.1016/j.nantod.2011.02.003.

50 Stern, S.T. and Johnson, D.N. (2008). Role for nanomaterial-autophagy interaction in neurodegenerative disease. *Autophagy* 4 (8): 1097–1100. https://doi.org/10.4161/auto.7142.

51 Wang, J., Yu, Y., Lu, K. et al. (2017). Silica nanoparticles induce autophagy dysfunction via lysosomal impairment and inhibition of autophagosome degradation in hepatocytes. *International Journal of Nanomedicine* 12: 809–825. https://doi.org/10.2147/IJN.S123596.

52 Ruan, G., Agrawal, A., Marcus, A.I., and Nie, S. (2007). Imaging and tracking of Tat peptide-conjugated quantum dots in living cells: new insights into nanoparticle uptake, intracellular transport, and vesicle shedding. *Journal of the American Chemical Society* 129 (47): 14759–14766. https://doi.org/10.1021/ja074936k.

53 Monod, J. and Jacob, F. (1961). Teleonomic mechanisms in cellular metabolism, growth, and differentiation. *Cold Spring Harbor Symposia on Quantitative Biology* 26: 389–401. https://doi.org/10.1101/sqb.1961.026.01.048.

54 Sierra, M.I., Valdés, A., Fernández, A.F. et al. (2016). The effect of exposure to nanoparticles and nanomaterials on the mammalian epigenome. *International Journal of Nanomedicine* 11: 6297–6306. https://doi.org/10.2147/IJN.S120104.

55 Moustakas, A., Pardali, K., Gaal, A., and Heldin, C.H. (2002). Mechanisms of TGF-β signaling in regulation of cell growth and differentiation. *Immunology Letters* 82 (1–2): 85–91. https://doi.org/10.1016/S0165-2478(02)00023-8.

56 Miller, D.M., Thomas, S.D., Islam, A. et al. (2012). c-Myc and cancer metabolism. *Clinical Cancer Research* 18 (20): 5546–5553. https://doi.org/10.1158/1078-0432.CCR-12-0977.

57 Palui, G., Aldeek, F., Wang, W., and Mattoussi, H. (2015). Strategies for interfacing inorganic nanocrystals with biological systems based on polymer-coating. *Chemical Society Reviews* 44 (1): 193–227. https://doi.org/10.1039/c4cs00124a.

58 Rai, R., Alwani, S., and Badea, I. (2019). Polymeric nanoparticles in gene therapy: new avenues of design and optimization for delivery applications. *Polymers*, 11 (4): 745. https://doi.org/10.3390/polym11040745.

59 Bhat, M., Robichaud, N., Hulea, L. et al. (2015). Targeting the translation machinery in cancer. *Nature Reviews Drug Discovery* 14 (4): 261–278. https://doi.org/10.1038/nrd4505.

60 Ma, X., Wu, Y., Jin, S. et al. (2011). Gold nanoparticles induce autophagosome accumulation through size-dependent nanoparticle uptake and lysosome impairment. *ACS Nano* 5 (11): 8629–8639. https://doi.org/10.1021/nn202155y.

61 Rudnitzki, F., Feineis, S., Rahmanzadeh, R. et al. (2018). siRNA release from gold nanoparticles by nanosecond pulsed laser irradiation and analysis of the involved temperature increase. *Journal of Biophotonics* 11 (9): https://doi.org/10.1002/jbio.201700329.

62 Möller, K., Müller, K., Engelke, H. et al. (2016). Highly efficient siRNA delivery from core-shell mesoporous silica nanoparticles with multifunctional polymer caps. *Nanoscale* 8 (7): 4007–4019. https://doi.org/10.1039/c5nr06246b.

63 Lin, S., Zhang, Q., Zhang, T. et al. (2018). Tetrahedral DNA nanomaterial regulates the biological behaviors of adipose-derived stem cells via DNA methylation on Dlg3. *ACS Applied Materials and Interfaces* 10 (38): 32017–32025. https://doi.org/10.1021/acsami.8b12408.

64 Zhang, H., Zhang, H., Aldalbahi, A. et al. (2017). Fluorescent biosensors enabled by graphene and graphene oxide. *Biosensors and Bioelectronics* 89: 96–106. https://doi.org/10.1016/j.bios.2016.07.030.

65 Dizaj, S.M., Jafari, S., and Khosroushahi, A.Y. (2014). A sight on the current nanoparticle-based gene delivery vectors. *Nanoscale Research Letters* 9 (1): 1–9. https://doi.org/10.1186/1556-276X-9-252.

7

Nanotools for the Treatment of Ocular Diseases

Elisa J. Campos[1], *António Campos*[1,2], *João Martins*[1,3], *and António Francisco Ambrósio*[1,4]

[1] *University of Coimbra, Coimbra Institute for Clinical and Biomedical Research (iCBR), Faculty of Medicine, Coimbra, Portugal; University of Coimbra, Center for Innovative Biomedicine and Biotechnology (CIBB), Coimbra, Portugal; Clinical Academic Center of Coimbra (CACC), Coimbra, Portugal*
[2] *Department of Ophthalmology, Centro Hospitalar Leiria E.P.E., Leiria, Portugal; ciTechCare, Center for Innovative Care and Health Technology, Polytechnic Institute of Leiria, Leiria, Portugal*
[3] *University of Coimbra, Coimbra Institute for Biomedical Imaging and Translational Research (CIBIT), Institute for Nuclear Sciences Applied to Health (ICNAS), Coimbra, Portugal*
[4] *Association for Innovation and Biomedical Research on Light and Image (AIBILI), Coimbra, Portugal*

7.1 Introduction

Vision plays a crucial role in every stage of our lives. In July 2021, the United National General Assembly has adopted the resolution "Vision for Everyone; accelerating action to achieve the Sustainable Development Goals". The resolution sets targets for "vision for everyone" by 2030 and aims to tackle avoidable sight loss, committing 193 countries to ensure full access to eye care services and making eye health "integral" to their nation's commitments to achieve the development goals.

While some ocular diseases cause vision impairment, many lead to personal and financial hardships. More than 2.2 billion people worldwide have a vision impairment, and of these, at least 1 billion have a vision impairment that could have been prevented. Ageing, lifestyle exposure and behaviors, infections, and a several health conditions, will dramatically increase these figures in the next decades. Ocular drug delivery is still a major challenge to researchers and pharmaceutical technology. Conventional formulations are unable to efficiently deliver drugs into the eye because of inherent anatomical and physiological ocular barriers which result in low ocular drug bioavailability. Overcoming these barriers thus enhancing the permeation of ocular drugs will constitute an invaluable increment on the quality of life of those patients.

This chapter provides an insight into nanoparticles-based solutions for ocular therapy, the present scenario, and strategies for improvement. In order to give the reader a better understanding on the advances in ocular therapy, the anatomy of the eye, its main functional barriers, and the conventional methods used in clinical practice for drug administration in ocular diseases are briefly described.

Nanomaterials and Nanotechnology in Medicine, First Edition. Edited by P.M. Visakh.

7.2 Ocular Anatomy

Vision is by far the most important of the five senses. Up to 80% of the information we receive from the outside world is provided by the sense of sight. In the human eye, two segments can be distinguished (Figure 7.1a): (i) the anterior segment of the eye, consisting in the cornea, iris, ciliary body, and lens, and (ii) the posterior segment of the eye, including the retina, choroid, and optic nerve head. Furthermore, the eye has three different fluid chambers: (i) the anterior chamber, limited by the cornea and iris, (ii) the posterior chamber, limited by the iris, zonule fibers, and lens, and (iii) the vitreous chamber limited by the lens and the retina. The first two chambers are filled with aqueous humor, that supplies nutrients and oxygen to the cornea and lens, whereas the vitreous chamber is filled with the vitreous humor.

The neurons of the retina are divided into three major layers (Figure 7.1b): (i) the outer nuclear layer (ONL) is constituted by the photoreceptor cell bodies, (ii) the inner nuclear layer (INL), contains the cell bodies of the bipolar, horizontal, and amacrine cells, (iii) the ganglion cell layer (GCL) is composed by the nuclei of retinal ganglion cells and displaced amacrine cells. The synapses are confined to the two synaptic, or plexiform layers: the outer plexiform layer (OPL) and the inner plexiform layer (IPL). In the retina, besides neurons, there are also glial cells (astrocytes, microglia, and Müller cells) and the cells that constitute the retinal vessels (endothelial cells and pericytes) (Figure 7.1b) [3].

Light passes through the lens and vitreous to reach the retina. In the retina, light crosses through several cell layers, before it finally reaches the rods and cones located on the outer retina. The photoreceptors are able to convert light energy into neural activity and, thus,

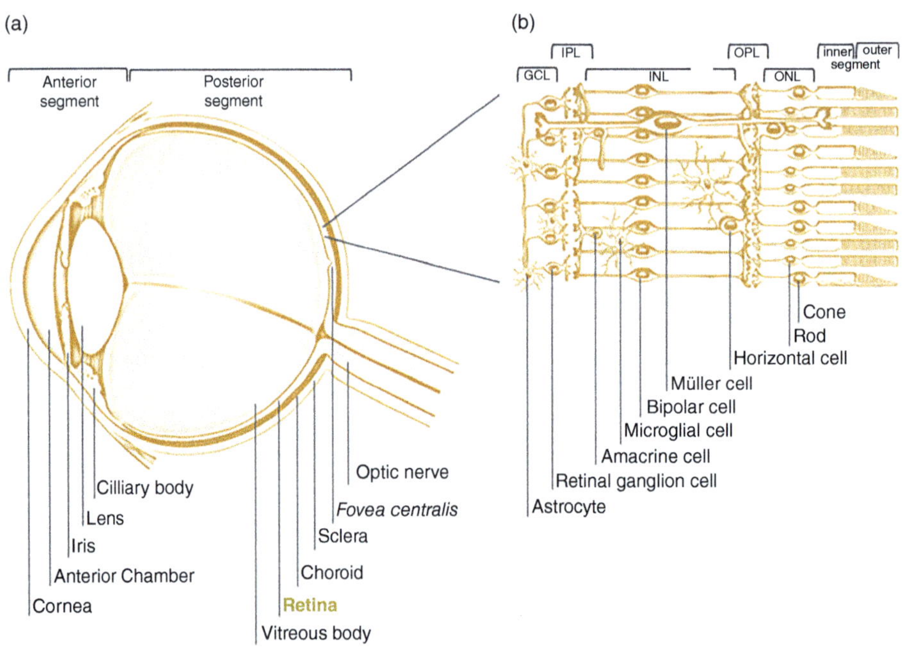

Figure 7.1 Schematic representation of (a) the sagittal section of the human eye, and (b) the major retinal cell types and their organization in the retina (see text for abbreviations). *Source:* © Sara Boia.

are known to be the first neurons in the visual pathway. Then, neural activity is transmitted to the second and third order neurons in the visual pathway: bipolar and ganglion cells, respectively. The lateral connections formed by horizontal cells in the OPL and by amacrine cells in the IPL are crucial for visual signal processing optimization. Ganglion cell axons then exit the retina, bundling together to form the optic nerve, which targets main thalamic and midbrain visual areas, from where visual information is conveyed to visual cortex [4].

7.3 Physiological Barriers in the Eye

The physiological barriers in the eye (Figure 7.2) have a protective function by defining a microenvironment, but also regulate the integrity of cells and tissues, in order to maintain the ocular homeostasis. However, these barriers strongly limit efficient drug delivery into the eye as they limit the diffusion and penetration of any therapeutic agent. How to overcome or minimize this barriers effect, in order to improve ocular drug delivery, has become a challenging task under investigation in recent decades.

7.3.1 Tear Fluid (Lacrimal Fluid)

The precorneal tear film is continuously produced and drained by the lacrimal system; the eyelids play an important role in the distribution of the tear fluid as they spread it all over the anterior surface of the eye, that is, the cornea and the conjunctiva. The tear film

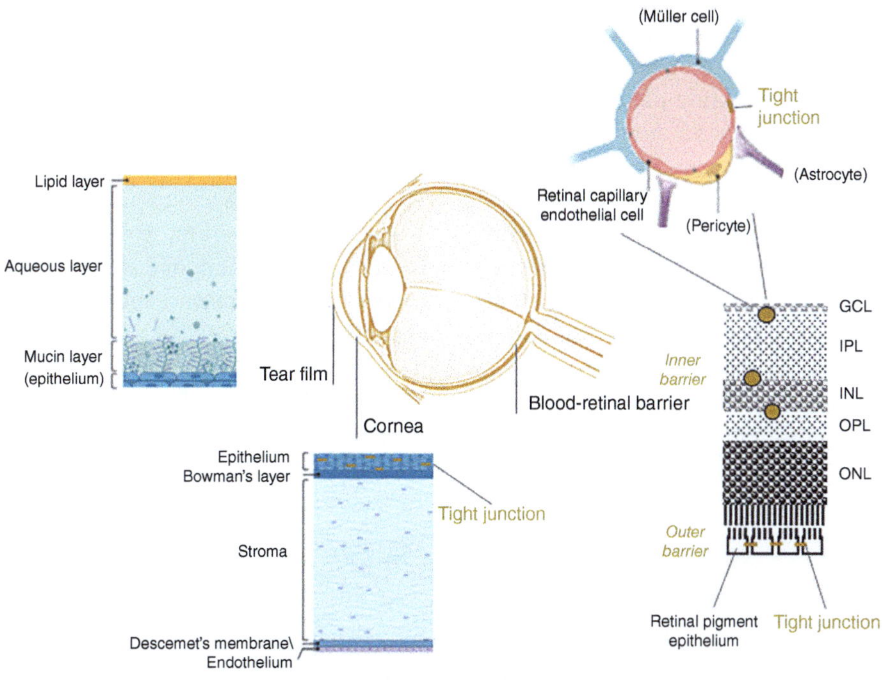

Figure 7.2 The main physiologic barriers in ocular drug delivery.

is made of three layers (lipid, aqueous, and mucus), but actually there is no clear-cut barrier between these layers. The inner mucus layer is in direct contact with the epithelial cells of the cornea and conjunctiva, and is mainly composed of water (95%) and high molecular weight (0.5–20 MDa) glycoproteins (mucins). Mucins reduce the surface tension between the aqueous layer and the cornea, thus allow the aqueous layer to adhere to the epithelial cells of the cornea [5]. This central aqueous layer represents the bulk of the tear film. It is composed of 98% water, but also water-soluble gases, electrolytes, and proteins, such as lysozyme and immunoglobulins. It lubricates the eye, hydrates and nourishes the cornea, and carries waste products away. The outermost layer is a lipid layer that reduces water evaporation from the aqueous layer, preventing drying of the eye.

In humans, the normal resident tear volume is about 7 μl and the normal tear turnover is about 1.2 μl min^{-1}. After the instillation of an eyedrop (ca. 35 μl), the precorneal half-life of the drug in the surface of the eye is between one and three minutes. The majority of the drug (50–100%) topically administered is likely to be absorbed into the systemic circulation [6].

7.3.2 Cornea

Physiologically, the cornea has five layers: three cellular layers (epithelium, stroma, and endothelium), and two acellular layers (Bowman's layer and the Descemet's membrane) [7]. The epithelium has a single layer of basal cells and four to five cell layers of epithelial cells, which are held together by tight junctions, desmosomes, adherens junctions, and gap junctions [8], to form the most critical barrier to the permeability of drugs. It contributes up to 90% of the barrier to hydrophilic drugs, and 10% to lipophilic drugs. The Bowman's layer, a thin homogeneous condensation of the anterior stroma, is not considered to be a barrier to drugs diffusion. The stroma represents about 90% of the total corneal thickness. It has a hydrophilic gel structure of collagens, glycosaminoglycans, and is populated by keratocytes. The stroma may act as the rate-limiting barrier for lipophilic drugs that permeate the epithelium relatively rapidly. The Descemet's membrane is a 7 μm tough homogeneous band continuously secreted by the endothelium. The endothelium is a 5 μm monolayer of cells, which appears as a honeycomb-like mosaic when viewed from the posterior side, important in regulating the water content of the cornea. Furthermore, the endothelium can be seen as a leaky lipophilic barrier that offers no permeation resistance toward hydrophilic drugs but may offer some resistance toward lipophilic drugs [6, 9, 10].

This structure makes the cornea a unique barrier to most lipophilic and hydrophilic drugs. Overall, the molecular weight, the lipophilicity, and the charge of the drug heavily affect its permeation trough the cornea. This is why small lipophilic drugs (log D: 2–3), and cationic drugs are able to more efficiently permeate the cornea. However, for most topically applied drugs, only up to 5% but often much less is absorbed through the cornea [10].

7.3.3 Vitreous Humor

The vitreous humor is a clear aqueous gel that consists mainly of water (99%), glycosaminoglycans, mainly hyaluronan, an anionic hydrophilic polymer with a high molecular weight (2–4 MDa), and collagen fibers that provide strength and resistance to tractional forces. A relatively low content of protein in the human vitreous is also known, including proteases, protease inhibitors, growth factors, cytokines, and proteins related to signaling activity and visual perception. Although the vitreous has almost no effect on the diffusion of anionic drugs, the diffusion of cationic drugs is dramatically restricted [11].

7.3.4 Blood–Retinal Barrier

The blood–retinal barrier (BRB) is located at two levels: at the endothelial cells of the retinal capillary vessels (inner BRB), and at the epithelial cells of the retinal pigment epithelium (outer BRB) [12, 13]. The inner BRB mediates highly selective diffusion of molecules from and to the blood to the retina and is essential in maintaining retinal homeostasis, whereas the outer BRB regulates the transport of solutes and nutrients from the choroid to the sub-retinal space, and also the removal of metabolic wastes and water from the retina to the choroid [14–16]. In the inner BRB, pericytes communicate with adjacent endothelial cells, microglia, astrocytes, and neurons in a functional unit, called the neurovascular unit [17]. It has been considered that diabetic retinopathy is initiated by an alteration of the inner BRB, particularly, in the retinal neurovascular unit [18–20].

7.4 Methods of Ocular Disease Treatment

Ocular drug delivery is achieved by several ways, mainly, topical, intravitreal, and systemic routes (Figure 7.3) [21, 22].

Topical administration is the most common and patient friendly route for ocular disease treatment, particularly for eye diseases that affect the anterior segment. Nevertheless, it is known that less than 5% of administrated drug is able to reach the aqueous humor. The limited contact time with the ocular surface due to elimination mechanisms such, as tear turnover, nasolacrimal drainage, and systemic absorption in the precorneal area, and the simultaneous restricted entry into the eye due to the cornea barrier account for this ineffectiveness. Together with the long diffusional path from the front to the back of the eye, this approach is even less efficient in the treatment of posterior segment diseases, even when frequent dosage regimens are followed. The main ocular penetration routes for topically applied drugs are: (i) cornea → aqueous chamber → lens/iris → vitreous → retina,

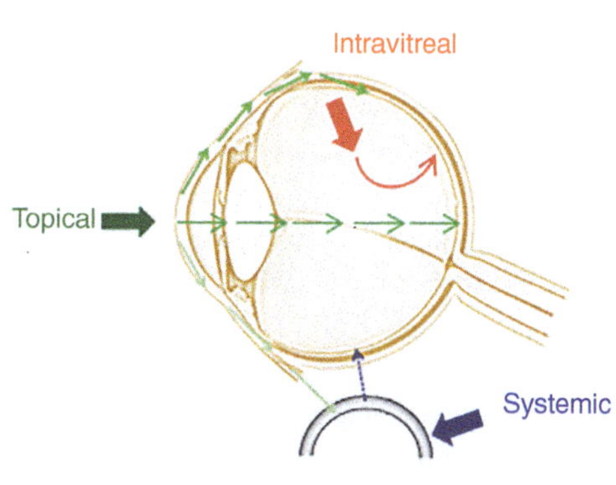

Figure 7.3 Delivery routes for the main methods of drug administration in the treatment of ocular diseases.

(ii) cornea → conjunctiva → sclera → choroid → retina, and (iii) nasolacrimal drainage → systemic circulation → BRB → retina [23, 24].

Intravitreal administration is the most used route for the treatment of posterior segment diseases, since the aforementioned elimination mechanisms and barriers are overcome. However, this is an invasive approach, that requires an operating room setting and an experienced clinician. Furthermore, frequent injections are generally needed, which may cause pain, endophthalmitis, sterile inflammation, glaucoma, retinal detachment, and cataract. Pharmacokinetic insights concerning intravitreal drug administration were deeply discussed elsewhere [11, 25].

Systemic administration of drugs for ocular therapy is delivered by intravenous injection and oral dosing. The later is the least invasive route for ocular drug administration, but both routes of administration cause important ocular and systemic side effects. The drug delivery to the posterior segment of the eye, particularly, to the retina is hampered by both the inner and the outer BRB via the tight junctions, adherens junctions, and the desmosomes [26] in the retinal pigment epithelium cells. Actually, only 1–2% of administrated drugs is able to reach the retina. Therefore, relatively high doses are required to achieve therapeutic intraocular concentrations, which may account for considerable undesired side effects.

7.5 Nanomedicine in Ocular Therapy

Currently, the use of nanotechnology has been investigated for several ophthalmic applications in the treatment of eye diseases [27]. Nanoparticles have been receiving attention as one of the novel drug-delivery systems that overcome the barriers of the eye including the cornea, conjunctiva, and BRB [28]. Also, the number of intravitreal injections is expected to decrease by using nanoparticles [29], when comparing to the use of intravitreal implants and intravitreal injections.

7.5.1 Nanoparticle-based Systems

The main nanosystems that are under investigation for eye diseases treatment are: (i) nanoliposomes, (ii) nanoparticles, (iii) cyclodextrins, and (iv) dendrimers, as revised before [30]. Here, special focus will be given to nanoparticles.

Nanoparticles are defined as nanometer-sized particles, in the range of 1–100 nm in the three dimensions [31]. Nanoparticles can be produced from various materials (e.g. natural and synthetic polymers, metal oxides, silica, and noble metals). These nanoparticles have various sizes, shapes, and stabilities for drug delivery, allowing the encapsulation of either hydrophilic or hydrophobic therapeutic agents [32]. Several properties ascribed to nanoparticles, such as composition, size, surface charge, and shape, are adjustable, resulting in a diverse spectrum of nanoparticles with specific characteristics and performance, which allow researchers to find the appropriate design of nanoparticles according to their purposes of application (Figure 7.4).

Nanoparticles may have potential for application for several medical purposes, such as gene therapy, drug delivery (treatment and/or prevention of diseases), and imaging [31, 34].

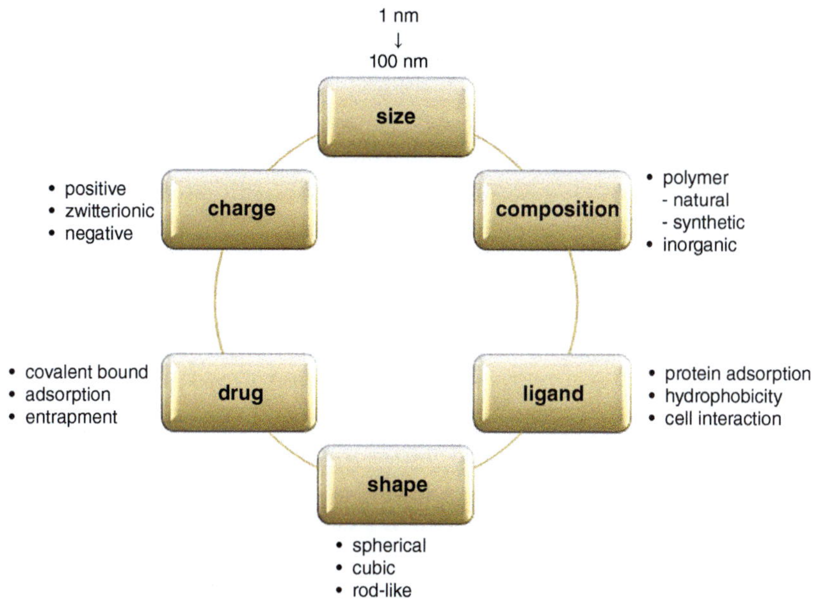

Figure 7.4 Main nanoparticle properties influencing their application and performance. *Source:* Adapted with permission from Elsevier [33].

Furthermore, these systems are able to combine both therapy and diagnosis, delivering therapeutic drugs and diagnostic imaging agents simultaneously, and are remarkable useful tools for theranostics [35]. Particularly, as drug-delivery systems, nanoparticles can provide: (i) sustained delivery, (ii) targeted delivery to specific cells or tissues, (iii) improved delivery of water-insoluble and large molecules, and (iv) reduced side effects, minimizing toxicological reactions [34].

Concerning the size of nanoparticles, it is assumed that nanoparticles in the range of 2 and 200 nm are the most suitable for therapeutic purposes, via systemic administration. If too small, nanoparticles are vulnerable to renal excretion and clearance from target tissues [36]. However, only smaller nanoparticles (e.g. 20 nm) were reported to pass through the BRB, whereas larger nanoparticles (e.g. 100 nm) were not observed in any of the retinal layers, when administered intravenously [28]. The surface charge defines cellular uptake, biodistribution, and interaction with environment. Generally, positively charged nanoparticles are known to be more easily internalized than neutral and negatively charged nanoparticles. Finally, shape also affects biodistribution, as well as the adhesion pattern [36].

Nevertheless, toxicity of nanoparticles on neuronal cells is still questionable. Size, dose, interaction time, chemical composition (core and coating), surface charge (zeta potential), and water solubility are factors that contribute to neuronal toxicity of nanoparticles. Therefore, nanoparticles must be thoroughly characterized before their application [28]. Neuronal toxicity due to nanoparticles is likely to occur via the formation of reactive oxygen species [37], and the alteration of gene expression pattern [38]. Microglial cells are able to phagocyte nanoparticles thus increasing production of intracellular reactive oxygen species and reactive nitrogen species [39], as well secretion levels of proinflammatory cytokines such as TNF (tumor necrosis factor), IL-1β, and IL-6 [40].

A brief final note on the limitations of these nanoparticle-based drug-delivery systems. In reality, these systems are not devoid of pitfalls and there are still many limitations requiring resolution. Improvements need to be made in drug loading, release, biodistribution, toxicity, ocular irritation, and patient compliance [31, 32].

7.5.2 Improving Strategies in Nanomedicine for Ocular Therapy

7.5.2.1 Light as an External Stimulus

The eye is the sole organ system that allows the penetration of light deep into the human body (Figure 7.5). Several mechanisms, such as the anatomy of the head, the eyebrows and the eyelashes, the constriction of the pupil and the squinting reflex, protect this very sensitive organ against the harmful effects of light. However, almost every electromagnetic radiation can cause damage to the eye. In natural environments, the most deleterious radiations are the UV-A, UV-B, and blue light radiation of the visible spectrum (380–500 nm). Whereas UV radiation is thought to cause damage to the anterior eye (e.g. cataracts), since the cornea and crystalline lens absorbs almost all UV radiation, blue light can damage the retina [42]. The light is thought to cause damage to the retina via photothermal, photomechanical, and photochemical mechanisms [43]. Concerning UV and blue light radiation, oxidative stress to the retinal pigment epithelium, damage to the photoreceptors, and cell death are likely to occur [44, 45]. Besides UV and visible light, IR radiation also interacts with the eye, thus referred to as optical radiation. It is known that IR-A radiation induces molecular vibrations and rotations that generate heat, thus causing retinal burns; since the regenerative capabilities of the retina are very limited, these damages are expected to cause severe loss of visual acuity. Intense IR-A exposure may also cause cataractogenesis, although the mechanism is not clearly understood [46]. On the other hand, it has already been shown that IR-A (670 nm) improves membrane potential and adenosine triphosphate production in aged mitochondria, and also reduces age-related inflammation [47]. It was also shown that 670 nm light can significantly improve aged retinal function [48]. Authors found that 670 nm exposure significantly improves both the photoreceptor-generated *a*-wave and the postreceptoral *b*-wave; furthermore 670 nm exposure effects were clearer when animals were older.

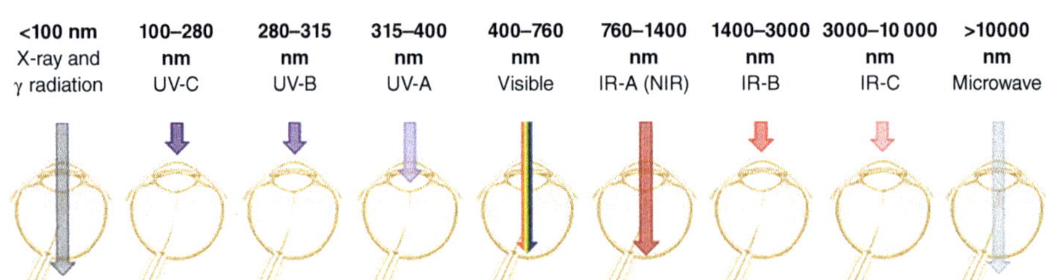

Figure 7.5 Interaction of the different radiation ranges (according to the International Commission on Illumination, CIE) with the eye. *Source:* Adapted from [41].

7.5.2.2 Light-responsive Nanoparticles

Polymeric materials that can be degraded in a controlled fashion upon an external stimulus such as pH, temperature, and light, are being explored. Light stimulus is particularly attractive since it can be remotely applied for a short period of time with high spatial and temporal precision. As an external stimulus for drug targeting and release, light has gained much interest because of its clinical relevance and excellent spatiotemporal control [49]. The low penetration depth resulting from the strong scattering properties of soft tissues in the UV– visible region restricts the use of light-triggered drug delivery to regions of the body that can be directly irradiated, such as the eye and the skin. In order to overcome this limitation, photosensitive groups that respond to higher wavelengths, such as near-infrared (NIR), are under development [50]. There are various mechanisms responsible for the light-triggered release of therapeutic agents, including photocleavage, photoisomerization, and photo-cross-linking [51, 52]. Light-responsive nanoparticles have also been used for the intracellular delivery of biomolecules, namely miRNAs [53, 54], retinoic acid [55–57], and proteins [58].

Light-responsive nanoparticles for ophthalmic application are still in their early stages of development. Almutairi and collaborators described a light-sensitive degradable polymer containing a quinone-methide self-immolative moiety, which was triggered to degrade by UV irradiation, through a light-sensitive group (4,5-dimethoxy-2-nitrobenzyl) along the backbone [59]. Later, for the first time, they reported the *in vivo* light-triggered drug release in the eye [60]. Nanoparticles were prepared based on the same light-sensitive degradable polymer and loaded with nintedanib, a small angiogenesis inhibitor molecule. Loaded nanoparticles were injected intravitreally into Brown Norway rats before laser induction of choroidal neovascularization (CNV). In eyes irradiated 10 weeks after the nanoparticles injection and two weeks after CNV induction, the CNV spot area was significantly smaller than in nonirradiated eyes, as well as in rats receiving poly(lactic-*co*-glycolic acid)-encapsulated nintedanib or free nintedanib. Authors also found that these light-sensitive nanoparticles were biocompatible and caused no adverse effects on the eye. Moreover, despite polymer hydrolysis, nanoparticles were present in the vitreous chamber 30 weeks after injection. Since most current antiangiogenics are administered monthly or bimonthly, this platform has promising clinical relevance for intravitreal drug delivery. More recently, Wang and co-authors developed nanoparticles to be administered intravenously and allow local doxorubicin delivery in the choroid via light-triggered targeting [61]. Photo-targeted nanoparticles were formed by poly(ethylene oxide)-poly(D,L-lactic acid) (PEG-PLA) block copolymer and surface-modified with cell penetrating peptide (CPP). The cellular uptake was inhibited by covalently bind a photocleavable group, 7-(diethylamino) coumarin-4-yl]methyl carboxyl (DEACM). Upon irradiation at 400 nm, DEACM was removed so that the CPP could readily bind to cells thus enhancing the accumulation of doxorubicin in the choroid. Doxorubicin-encapsulated nanoparticles were injected via tail vein in mouse CNV model. The reduction in neovessel area was observed to be 46% compared to the group treated with PBS, and approximately 25% compared to the group treated with free doxorubicin. This approach may minimize off target drug delivery, and most importantly, obviate the need for intraocular injections, improving patient compliance.

As for light-responsive nanoparticles for ophthalmic application, much remains to be done, as advances are made in related fields progress in this topic will accelerate.

The innovative design and properties of light-responsive nanoparticles will serve as a platform for effective long-term treatment of ocular diseases. Light, as an external stimulus, will hereby enhance the overall performance of nanoparticles, thus resulting in creative and promising ocular drug-delivery platforms in the near future.

7.5.3 Contact Lenses

Ocular drug-delivery improvements rely on overcoming the physiological barriers in the eye. Although ocular barriers naturally have a protective function, they also limit drug entry into the eye. Accordingly, designing strategies to weaken or even suppress those barriers, in order to delivery therapeutic agents effectively to both the anterior and the posterior segments of the eye, is currently a challenging task. The enhancement of drug diffusion through the eye structures, mainly the cornea, and the drug residence time in the precorneal region, is of most significance.

Contact lenses are mainly used to correct visual acuity. However, several studies have proposed the use of contact lenses for ocular drug delivery aiming to enhance ocular bioavailability and reduce side effects [62–65]. Contact lenses as drug-delivery systems have been considered mostly for diseases of the anterior segment the eye. Concerning diseases of the posterior segment the eye, further studies need to be carried out. However, several drugs are already successfully delivered to the back of the eye following topical application, including brimonidine, dexamethasone, flunarizine, and voclosporin [23]. Contact lenses are known to extend the residence time of drugs on the precorneal region to more than 30 minutes, thus improving drug bioavailability on the cornea. Also, drug loss via systemic circulation is reduced, and so are the potential side effects. Contact lenses can be used for relatively longer period of time, thus reducing the frequency of administration of the drug. However, extended wear may lead to reduced oxygen permeability, thus causing corneal edema. The incorporation of drugs into contact lenses can be obtained using different methods, such as (i) soaking, (ii) ion interaction, (iii) cyclodextrins complexing, (iv) molecular imprinting, and (v) nanoparticles loading. Different methods can be used to incorporate drug-loaded nanoparticles into contact lens: (i) to disperse the drug-loaded nanoparticles into the contact lens pre-monomer mixture prior to polymerization; (ii) to soak the contact lens in nanoparticle suspensions; and (iii) to immobilize drug-loaded nanoparticles on the surface of the contact lens through chemical bonds. The drug delivery from a nanoparticle-laden contact lens is more complex than from the nanoparticles or the contact lens per se. It is assumed that the drug diffuses through the nanoparticles, reaching the contact lens matrix, and then diffuses through the contact lens matrix to reach the precorneal region [64].

Several studies investigating the use of nanoparticle-laden contact lens have been published. Nasr and co-authors prepared contact lenses embedded with poly ε-caprolactone-based nanoparticles for potential ocular drug delivery of loteprednol, a corticosteroid with anti-inflammatory activity [66]. Hybrid nanoparticles of poly(ε-caprolactone) core, poly(2-hydroxy ethyl methacrylate) (PHEMA) inner shell, and poly(ethylene glycol) diacrylate (PEG-DA) outer shell were prepared via single-step, surfactant-free mini-emulsion polymerization process using Irgacure 2959 as the UV photoinitiator. Nanoparticles were then incorporated in hydrogels, prepared with HEMA (2-hydroxyethyl methacrylate), *N*-vinylpyrrolidone, and PEG-DA, by a free radical photopolymerization using Irgacure 2959, by direct entrapment during the polymerization process. *In vitro* drug release studies showed that the release rate of loteprednol from the hydrogel contact lens loaded with 7.5% hybrid nanoparticles is lower than that for the drug-loaded nanoparticles and hydrogel matrix-containing free drug. As mentioned, the presence of nanoparticles in the contact lens hydrogel slows drug release. The results further suggested that loteprednol release from 7.5% nanoparticle-loaded contact lens was about $80\,\mu g\,d^{-1}$ for 12 days, which is roughly in the therapeutic range. The authors concluded that the release time of loteprednol from the nanoparticle-laden contact lens increased, meaning improvement in the control of drug delivery. This can occur due to (i) a better distribution of the drug in the contact lens matrix, and (ii) the control of the amount of drug included in the system. *in vivo* release studies are needed to support the advantage of loteprednol release from the nanoparticle-laden contact lens system described here.

Behl and co-authors synthesized PHEMA-based contact lenses containing chitosan nanoparticles for ocular delivery of dexamethasone, a corticosteroid used for the treatment of inflammation and after cataract surgery [67]. The dexamethasone-loaded chitosan nanoparticles were mixed with a HEMA monomer followed by the addition of an ethylene glycol dimethacrylate cross-linker and trimethylbenzoyl-diphenyl-phosphineoxide, the photoinitiator. The addition of the drug-loaded nanoparticles to PHEMA was found to maintain the optical transparency of the contact lens, keeping an average transmittance of 95–98% relative to the neat contact lens. A maximum of 48 μg (55.73%) of drug was released over a period of 22 days. The drug release was found to increase continuously for 10 days; by that time, about 43 μg of the drug (ca. 50%) was released. After that time, the drug release slowed down and, by the end of the study, the plateau was reached. Again, the drug release rate from the nanoparticle-laden contact lens was found to be lower when compared to chitosan nanoparticles. Although dexamethasone release from this nanoparticle-laden contact lens was sufficient to meet the anti-inflammatory levels, thus being a viable alternative to eye drops, it should also be taken into account that the PHEMA-based lenses are not recommended for continuous use because of their low oxygen permeability. Cytotoxicity and *in vivo* release studies are further needed to confirm that the system here described might be a promising candidate for the delivery of dexamethasone and other ophthalmic therapeutics agents.

Besides the drug-loaded nanoparticles, pure nanoparticles (without drugs), can also be dispersed into contact lenses. Silver nanoparticles, whose antimicrobial [68] and toxic [69] activity is well known, have been incorporated into contact lens. Rad and co- authors evaluated the antibacterial efficacy of silver-nanoparticles-laden contact lenses [70]. Keratitis is described as an infectious ophthalmic inflammation, which can develop from bacteria, such as *Pseudomonas aeruginosa*. The biofilm formation on contact lens surface is assumed to be an example of a hydrogel contact lens induced infection. Silver can be loaded on the contact lens surface and released as an antimicrobial agent, thus reducing bacterial colonization. Contact lenses were prepared by co- polymerization of HEMA and methacrylic acid, using, ethylene glycol dimethacrylate as the cross-linker. Silver nanoparticles were incorporated by soaking. The authors observed that these silver-nanoparticle-laden contact lenses demonstrated excellent antimicrobial effect against *P. aeruginosa*, for up to 72 hours. The concentration of silver nanoparticles in the soaking suspension plays an important role in silver nanoparticle loading and, thus, in the antibacterial effect of the system. Authors obtained maximum bacterial activity after soaking hydrogel in 10 and 20 ppm silver nanoparticles suspensions. *in vivo* and toxicity studies are necessary to evaluate and optimize antibacterial effect of silver-nanoparticle-laden contact lenses, aiming clinical application.

Despite the promising studies previously cited, further research needs to be carried out on the application of nanoparticle-laden contact lenses in the clinic. The physical properties of the contact lens per se, such as transparency, surface roughness, oxygen permeability, and water content, as well as the physical properties of the incorporated nanoparticles, such as the size (that limit the transparency of the nanoparticle-laden contact lens) must be taken into account in future studies. For drug-delivery systems, an important issue that needs to be considered is the drug release profile, and also the leaching of the drug loaded during storage and distribution [65]. In this topic, to design external stimuli-responsive systems might be an exciting alternative.

7.6 Closing Remarks

People affected by ocular diseases will significantly increase in the next decades, and, consequently, a substantial increase in health costs is expected. The anatomy, physiology, and biochemistry of the eye make ocular drug delivery a challenging topic, both to research and to industry. Current therapies for ocular diseases require repeated and regular administrations, over a period of several months to years. Accordingly, new ocular drug-delivery approaches enabling less-invasive routes and less-frequent administrations, precise drug targeting within the eye, higher therapeutic efficacy, less side effects, and good patient compliance, are of paramount importance.

Nowadays, several nanotechnology-based solutions can be found in different fields, from chemistry to engineering. Concerning nanomedicine, it is expected to ameliorate human health, through superior diagnosis and therapy quality. The use of nanometer-scaled particles seems to be a promising alternative for the treatment of ocular diseases, since they are able to bypass the BRB or other barriers in the eye, and remain in the eye for longer periods, reducing treatment administrations. Furthermore, new drug-delivery nanosystems need to be designed aiming to improve the bioavailability of drugs, to allow better cellular targeting, to provide a better release profile control, to lower side effects, and to improve patient compliance. Despite the promising results already reported with nanoparticles in preclinical studies, most of them still need further research. The cost of development and manufacture, the ability for scale up, the lack of knowledge about the impact of nanoparticles on living systems, and the need for approval by the regulatory authorities has been also hampering their entrance in the pharmaceutical market. However, the pharmaceutical products available on the market in the United States and other countries whose efficacy and safety has been enhanced by nanotechnology, and the development of other products that would not even exist without nanotechnology, do encourage and push research in nanomedicine.

Conflict of Interest

The authors declare no competing financial interests.

References

1 Flaxman, S.R. et al. (2017). Global causes of blindness and distance vision impairment 1990–2020: a systematic review and meta-analysis. *The Lancet Global Health* 5 (12): e1221–e1234.

2 Sabanayagam, C. and Cheng, C.-Y. (2017). Global causes of vision loss in 2015: are we on track to achieve the vision 2020 target? *The Lancet Global Health* 5 (12): e1164–e1165.

3 Madeira, M.H., Boia, R., Santos, P.F. et al. ((2015). Contribution of microglia-mediated neuroinflammation to retinal degenerative diseases. *Mediators of Inflammation* 2015: 15.

4 Erskine, L. and Herrera, E. (2014). Connecting the retina to the brain. *ASN Neuro* 6 (6): 1759091414562107. https://doi.org/10.1177/759091414562107.

5 Bansil, R. and Turner, B.S. (2006). Mucin structure, aggregation, physiological functions and biomedical applications. *Current Opinion in Colloid & Interface Science* 11 (2): 164–170.

6 Jóhannesson, G., Stefánsson, E., and Loftsson, T. (2016). Microspheres and nanotechnology for drug delivery. *Developments in Ophthalmology* 55: 93–103.

7 Sridhar, M.S. (2018). Anatomy of cornea and ocular surface. *Indian Journal of Ophthalmology* 66 (2): 190–194.

8 Mantelli, F., Mauris, J., and Argüeso, P. (2013). The ocular surface epithelial barrier and other mechanisms of mucosal protection: from allergy to infectious diseases. *Current Opinion in Allergy and Clinical Immunology* 13 (5): 563–568.

9 Morrison, P.W.J. and Khutoryanskiy, V.V. (2014). Enhancement in corneal permeability of riboflavin using calcium sequestering compounds. *International Journal of Pharmaceutics* 472 (1): 56–64.

10 Huang, D., Chen, Y.-S., and Rupenthal, I.D. (2018). Overcoming ocular drug delivery barriers through the use of physical forces. *Advanced Drug Delivery Reviews* 126: 96–112.

11 del Amo, E.M., Rimpelä, A.-K., Heikkinen, E. et al. (2017). Pharmacokinetic aspects of retinal drug delivery. *Progress in Retinal and Eye Research* 57: 134–185.

12 Cunha-Vaz, J., Faria de Abreu, J.R., and Campos, A.J. (1975). Early breakdown of the blood-retinal barrier in diabetes. *The British Journal of Ophthalmology* 59 (11): 649–656.

13 Do Carmo, A., Ramos, P., Reis, A. et al. (1998). Breakdown of the inner and outer blood retinal barrier in streptozotocin-induced diabetes. *Experimental Eye Research* 67 (5): 569–575.

14 Campbell, M. and Humphries, P. (2012). The blood–retina barrier: tight junctions and barrier modulation. *Advances in Experimental Medicine and Biology* 763: 70–84.

15 Xu, H.-Z. and Le, Y.-Z. (2011). Significance of outer blood–retina barrier breakdown in diabetes and ischemia. *Investigative Ophthalmology & Visual Science* 52 (5): 2160–2164.

16 Díaz-Coránguez, M., Ramos, C., and Antonetti, D.A. (2017). The inner blood-retinal barrier: cellular basis and development. *Vision Research* 139: 123–137.

17 Klaassen, I., Schlingemann, R.-O., and Van Noorden, C.J.F. (2013). Molecular basis of the inner blood-retinal barrier and its breakdown in diabetic macular edema and other pathological conditions. *Progress in Retinal and Eye Research* 34: 19–48.

18 Cunha-Vaz, J. (2009). The blood-retinal barrier in retinal disease. *European Ophthalmic Review* 3 (2): 105–108.

19 Cunha-Vaz, J., Bernardes, R., and Lobo, C. (2011). Blood-retinal barrier. *European Journal of Ophthalmology* 21 (Suppl. 6): S3–S9.

20 Gardner, T.W. and Davila, J.R. (2017). The neurovascular unit and the pathophysiologic basis of diabetic retinopathy. *Graefe's Archive for Clinical and Experimental Ophthalmology* 255 (1): 1–6.

21 Weng, Y., Liu, J., Jin, S. et al. (2017). Nanotechnology-based strategies for treatment of ocular disease. *Acta Pharmaceutica Sinica B* 7 (3): 281–291.

22 Bisht, R., Mandal, A., and Jaiswal, J.K. (2018). Nanocarrier mediated retinal drug delivery: overcoming ocular barriers to treat posterior eye diseases. *Wiley Interdisciplinary Reviews: Nanomedicine and Nanobiotechnology* 10 (2): e1473.

23 Boddu, S.H., Gupta, H., and Patel, S. (2014). Drug delivery to the back of the eye following topical administration: an update on research and patenting activity. *Recent Patents on Drug Delivery & Formulation* 8 (1): 27–36.

24 Tsai, C.-H., Wang, P.-Y., Lin, I.-C. et al. (2018). Ocular drug delivery: role of degradable polymeric nanocarriers for ophthalmic application. *International Journal of Molecular Sciences* 19 (9): 2830.

25 del Amo, E.M. and Urtti, A. (2015). Rabbit as an animal model for intravitreal pharmacokinetics: clinical predictability and quality of the published data. *Experimental Eye Research* 137: 111–124.

26 Bonilha, V.L. (2014). Retinal pigment epithelium (RPE) cytoskeleton in vivo and in vitro. *Experimental Eye Research* 126: 38–45.

27 Kompella, U.B., Amrite, A.C., Ravi, P.R. et al. (2013). Nanomedicines for back of the eye drug delivery, gene delivery, and imaging. *Progress in Retinal and Eye Research* 36: 172–198.

28 Jo, D.H., Lee, T.G., and Kim, J.H. (2011). Nanotechnology and nanotoxicology in retinopathy. *International Journal of Molecular Sciences* 12 (11): 8288–8301.

29 Moshfeghi, A.A. and Peyman, G.A. (2005). Micro- and nanoparticulates. *Advanced Drug Delivery Reviews* 57 (14): 2047–2052.

30 Campos, E.J., Campos, A., Martins, J. et al. (2017). Opening eyes to nanomedicine: where we are, challenges and expectations on nanotherapy for diabetic retinopathy. *Nanomedicine: Nanotechnology, Biology and Medicine* 13 (6): 2101–2113.

31 Sahoo, S.K., Dilnawaz, F., and Krishnakumar, S. (2008). Nanotechnology in ocular drug delivery. *Drug Discovery Today* 13 (3–4): 144–151.

32 Park, W. and Na, K. (2015). Advances in the synthesis and application of nanoparticles for drug delivery. *Wiley Interdisciplinary Reviews: Nanomedicine and Nanobiotechnology* 7 (4): 494–508.

33 Saraiva, C., Praça, C., Ferreira, R. et al. (2016). Nanoparticle-mediated brain drug delivery: overcoming blood–brain barrier to treat neurodegenerative diseases. *Journal of Controlled Release* 235: 34–47.

34 Fangueiro, J.F., Silva, A.M., Garcia, M.L. et al. (2015). Current nanotechnology approaches for the treatment and management of diabetic retinopathy. *European Journal of Pharmaceutics and Biopharmaceutics* 95 (Part B): 307–322.

35 Kelkar, S.S. and Reineke, T.M. (2011). Theranostics: combining imaging and therapy. *Bioconjugate Chemistry* 22 (10): 1879–1903.

36 Jo, D.H., Kim, J.H., Lee, T.G. et al. (2015). Size, surface charge, and shape determine therapeutic effects of nanoparticles on brain and retinal diseases. *Nanomedicine: Nanotechnology, Biology and Medicine* 11 (7): 1603–1611.

37 Iavicoli, I., Bergamaschi, A., Leso, V. et al. (2012). Toxicological effects of titanium dioxide nanoparticles: a review of in vivo studies. *Journal of Nanomaterials* 2012: 1–36.

38 Wang, J., Rahman, M.F., Duhart, H.M. et al. (2009). Expression changes of dopaminergic system-related genes in PC12 cells induced by manganese, silver, or copper nanoparticles. *NeuroToxicology* 30 (6): 926–933.

39 Wang, Y., Wang, B., Zhu, M.-T. et al. (2011). Microglial activation, recruitment and phagocytosis as linked phenomena in ferric oxide nanoparticle exposure. *Toxicology Letters* 205 (1): 26–37.

40 Xue, Y., Wu, J., Sun, J. et al. (2012). Four types of inorganic nanoparticles stimulate the inflammatory reaction in brain microglia and damage neurons in vitro. *Toxicology Letters* 214 (2): 91–98.

41 Sukhamburg Safety at Work: Laser Safety Goggles, from https://www.sukhamburg.com/products/Laser_Safety_Goggles.html (accessed 11 June 2019).

42 Roberts, J.E. (2011). Ultraviolet radiation as a risk factor for cataract and macular degeneration. *Eye & Contact Lens* 37 (4): 246–249.

43 Youssef, P.N., Sheibani, N., Albert, D.M. et al. (2011). Retinal light toxicity. *Eye (London, England)* 25 (1): 1–14.

44 Chalam, K.V., Khetpal, V., Rusovici, R. et al. (2011). A review: role of ultraviolet radiation in age-related macular degeneration. *Eye & Contact Lens* 37 (4): 225–232.

45 Jurja, S., Hîncu, M., Dobrescu, M.A. et al. (2014). Ocular cells and light: harmony or conflict? *Romanian Journal of Morphology and Embryology* 55 (2): 257–261.

46 Kourkoumelis, N. and Tzaphlidou, M. (2011). Eye safety related to near infrared radiation exposure to biometric devices. *ScientificWorldJournal* 11: 520–528.

47 Fitzgerald, M., Hodgetts, S., van den Heuvel, C. et al. (2013). Red/near-infrared irradiation therapy for treatment of central nervous system injuries and disorders. *Reviews in the Neurosciences* 24 (2): 205–226.

48 Sivapathasuntharam, C. et al. (2017). Aging retinal function is improved by near infrared light (670 nm) that is associated with corrected mitochondrial decline. *Neurobiology of Aging* 52: 66–70.

49 Rwei, A.Y., Wang, W., Kohane, D.S. et al. (2015). Photoresponsive nanoparticles for drug delivery. *Nano Today* 10 (4): 451–467.

50 Mura, S., Nicolas, J., and Couvreur, P. (2013). Stimuli-responsive nanocarriers for drug delivery. *Nature Materials* 12: 991.

51 Fomina, N., Sankaranarayanan, J., and Almutairi, A. (2012). Photochemical mechanisms of light-triggered release from nanocarriers. *Advanced Drug Delivery Reviews* 64 (11): 1005–1020.

52 Karimi, M., Zangabad, P.S., Baghaee-Ravari, S. et al. (2017). Smart nanostructures for cargo delivery: uncaging and activating by light. *Journal of the American Chemical Society* 139 (13): 4584–4610.

53 Lino, M.M., Susana Simões, S., Vilaça, A. et al. (2018). Modulation of angiogenic activity by light-activatable miRNA-loaded nanocarriers. *ACS Nano* 2 (6): 5207–5220.

54 Blersch, J., Francisco, V., Rebelo, C. et al. (2020). A light-triggerable nanoparticle library for the controlled release of non-coding RNAs. *Angewandte Chemie International Edition* 59 (5): 1985–1991.

55 Boto, C., Quartin, E., Cai, Y. et al. (2017). Prolonged intracellular accumulation of light-inducible nanoparticles in leukemia cells allows their remote activation. *Nature Communications* 8: 15204–15204.

56 Santos, T., Ferreira, R., Quartin, E. et al. (2017). Blue light potentiates neurogenesis induced by retinoic acid-loaded responsive nanoparticles. *Acta Biomaterialia* 59: 293–302.

57 Francisco, V., Lino, M., and Ferreira, L. (2019). A near infrared light-triggerable modular formulation for the delivery of small biomolecules. *Journal of Nanobiotechnology* 17 (1): 97.

58 Lino, M.M., Simões, S., Pinho, S. et al. (2017). Intracellular delivery of more than one protein with spatio-temporal control. *Nanoscale* 9 (47): 18668–18680.

59 Fomina, N., McFearin, C., Sermsakdi, M. et al. (2010). UV and near-IR triggered release from polymeric nanoparticles. *Journal of the American Chemical Society* 132 (28): 9540–9542.

60 Huu, V.A.N., Luo, J., Zhu, J. et al. (2015). Light-responsive nanoparticle depot to control release of a small molecule angiogenesis inhibitor in the posterior segment of the eye. *Journal of Controlled Release* 200: 71–77.

61 Wang, Y., Liu, C.-H., Ji, T. et al. (2019). Intravenous treatment of choroidal neovascularization by photo-targeted nanoparticles. *Nature Communications* 10 (1): 804.

62 Maulvi, F.A., Soni, T.G., and Shah, D.O. (2016). A review on therapeutic contact lenses for ocular drug delivery. *Drug Delivery* 23 (8): 3017–3026.

63 Hui, A. (2017). Contact lenses for ophthalmic drug delivery. *Clinical and Experimental Optometry* 100 (5): 494–512.

64 Xu, J., Xue, Y., Hu, G. et al. (2018). A comprehensive review on contact lens for ophthalmic drug delivery. *Journal of Controlled Release* 281: 97–118.

65 Choi, S.W. and Kim, J. (2018). Therapeutic contact lenses with polymeric vehicles for ocular drug delivery: a review. *Materials* 11 (7): 1125.

66 Nasr, F.H., Khoee, S., Dehghan, M.M. et al. (2016). Preparation and evaluation of contact lenses embedded with polycaprolactone-based nanoparticles for ocular drug delivery. *Biomacromolecules* 17 (2): 485–495.

67 Behl, G., Iqbal, J., O'Reilly, N.J. et al. (2016). Synthesis and characterization of poly(2-hydroxyethylmethacrylate) contact lenses containing chitosan nanoparticles as an ocular delivery system for dexamethasone sodium phosphate. *Pharmaceutical Research* 33 (7): 1638–1648.

68 Lemire, J.A., Harrison, J.J., Turner, R.J. et al. (2013). Antimicrobial activity of metals: mechanisms, molecular targets and applications. *Nature Reviews Microbiology* 11: 371.

69 Vazquez-Muñoz, R., Borrego, B., Juárez-Moreno, K. et al. (2017). Toxicity of silver nanoparticles in biological systems: does the complexity of biological systems matter? *Toxicology Letters* 276: 11–20.

70 Rad, M.S., Khameneh, B., Sabeti, Z. et al. (2016). Antibacterial activity of silver nanoparticle-loaded soft contact lens materials: the effect of monomer composition. *Current Eye Research* 41 (10): 1286–1293.

8

Nanotechnology Applications in Food and Nutrition Science

Kobra S. Rizi[1], Majid Rezayi[2,3], Ehsan Aryan[1], Zahra Meshkat[1], and Majid G. Mobarhan[4]

[1] *Antimicrobial Resistance Research Center, Department of Medical Bacteriology and Virology, Qaem University Hospital, School of Medicine, Mashhad University of Medical Sciences, Mashhad, Iran*
[2] *Medical Toxicology Research Center, Mashhad University of Medical Sciences, Mashhad, Iran*
[3] *Department of Medical Biotechnology and Nanotechnology, School of Medicine, Mashhad University of Medical Sciences, Mashhad, Iran*
[4] *Metabolic Syndrome Research Center, School of Medicine, Mashhad University of Medical Sciences, Mashhad, Iran*

8.1 Introduction

Nanotechnology maintains the covenant of improving human life quality and advancing knowledge in a magnitude of methods. This science that will determine the investigation of materials that are one-billionth of a meter in size (smaller than 100 nm) has seen remarkable growth in recent decades [1]. The development, usage, and manipulation of materials, devices, or systems at the nanometer scale have been carried out in the field of nanotechnology [2]. For human societies, with a growing population, the need for minimum food and drink in order to obtain sufficient micro- and macro-nutrition and an improvement of the nutrition value of humans' daily food is important, particularly so for poor and low-income countries. Nanotechnology through the increase of surface-to-volume ratio nanomaterials can: improve the nutritional value; modify human nutritional defects; cure hard diseases of humans; help to provide more effective delivery of drugs to their target in the human body so that a minimum dose provides maximum benefit. According to the opinion of one of the scientists from Johns Hopkins University [3], one of the greatest challenges is identifying novel strategies to specifically deliver high doses of chemotherapeutics to a tumor or cancer cell/tissue while minimizing systemic toxicity [3]. Examples of the potential applications of nanotechnology in human nutrition including the production of nano-size powders for increasing absorption of nutrients, cellulose nanocrystal composites, nanocochleates (coiled nanoparticles [NPs]) for more effective nutrient delivery to cells without affecting the color or taste of food; and vitamin sprays that disperse nanodroplets with better absorption [4]. Nanocochleates provides a new drug-delivery method that is based on trapping an optimal quantity of drug molecules into the multilayered structure, including the solid–lipid bilayer in the form of a

Nanomaterials and Nanotechnology in Medicine, First Edition. Edited by P.M. Visakh.

sheet rolled up in a spiral form [5]. Nanomaterials due to the high surface-to-volume ratio and new physicochemical effects, can influence the color, solubility, and thermodynamics of food substances and so improve and modify the texture, taste, and appearance of food [6, 7]. These new effects supply opportunities to rectify the sensory qualities of food such as taste, texture, and color. Also, nanomaterials can be employed to improve preservation mechanisms for food. Recruiting nanosensors and nano-packaging materials provide fast, sensitive, and reliable detection of microbial contamination, harmful chemicals, and pesticides. On the other hand, nano-encapsulation systems can increase the quality of food processing by enabling the delivery of bioactive compounds for increasing bioavailability in foods [8]. Larger particles often discharge their exemplified mixes gradually and over longer periods, while decreased molecule size presents a few biocement change variables, including expanded glue power and delayed gastrointestinal travel time, prompting higher bioavailability of the embedded compound. So, enrichment of human foods with vitamin D, calcium, omega-3 unsaturated fats, natural antioxidants, and so on can produce controllable conveyance rate outputs and with less side effects [9, 10]. The term "nano food" describes food that has been cultivated, produced, processed, or packaged using nanotechnological methods or tools, or food to which manufactured nanomaterials have been added [11]. In this chapter, the classification, methods of preparation, and safety issues of nanomaterials are described. The main focus of this chapter is on nanotechnological applications for foods and includes the controlled release of flavors, targeted delivery of bioactive compounds for enhancing the bioavailability, and nanosensors for poisoning and pathogenic microorganisms, as well as for chemical detection in foods.

8.2 Nanostructured Delivery Systems

8.2.1 Nanocarriers

The maintenance of the bioactive ingredients in the food industry that protect against non-standard processes and environmental conditions that are correctly delivered to the target organs and cells should be formulated under nano-encapsulation techniques [12]. Nanocarriers are nanomaterials that are used as a transport module for another substance, such as a drug (Figure 8.1) [13]. Conventional classification of nanocarriers includes organic-based, inorganic-based, or a combination of both [14].

Organic nanocarriers contain polymeric NPs and lipid-based NPs such as liposomes, nanoemulsions (e.g. micelles and reversed micelles), dendrimers, and carbon-based nanocarriers (e.g. fullerenes and carbon nanotubes [CNTs]) [15]. Inorganic nanocarrier structures compromise of metallic nanostructures such as quantum dots (QDs) (Figure 8.2) [13]. Different types nanocarriers are used for different nutraceuticals: those containing phenolic substrates and antioxidants, natural food colorants, antimicrobial agents and essential oils, vitamins, minerals, flavors, fish oils and essential fatty acids [12]. A systematic classification for nanocarriers involved in nano-encapsulation includes lipid-based nanocarriers, nature-inspired nanocarriers, special-equipment-based nanocarriers, biopolymer nanocarriers, and other miscellaneous nanocarriers [12] (Table 8.1).

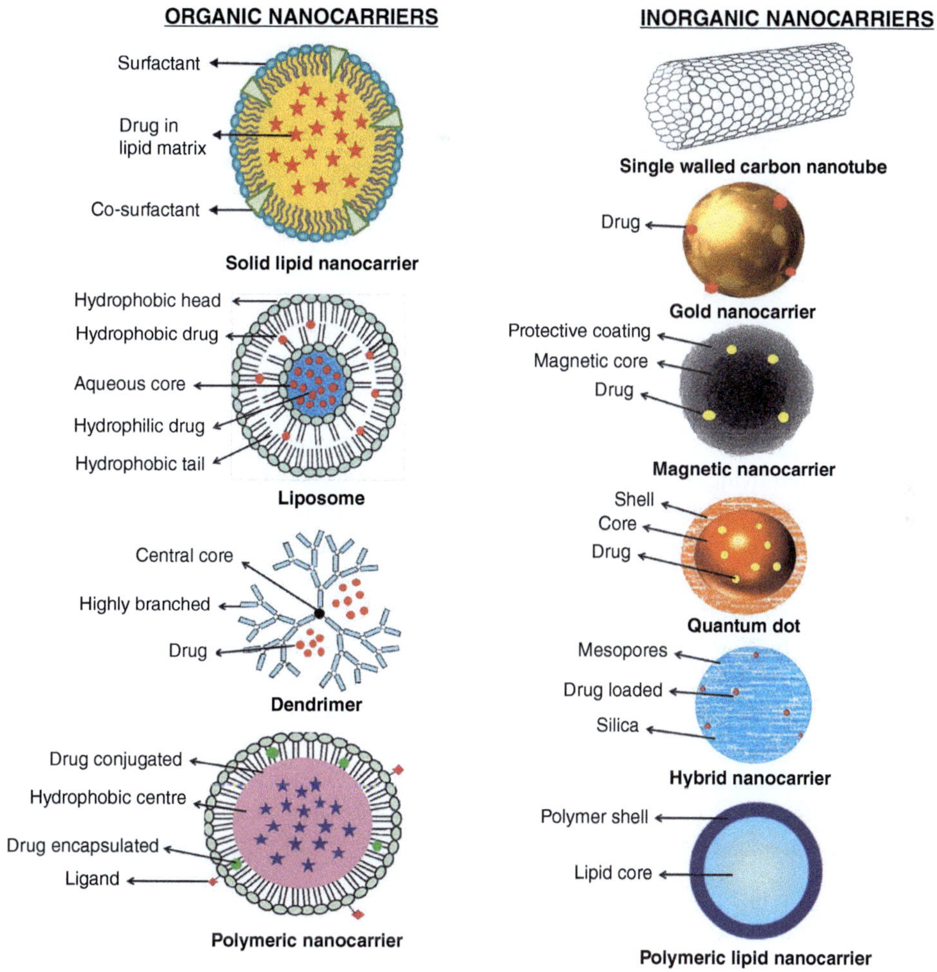

Figure 8.1 Examples of the different types of nanocarriers. *Source:* Adapted and modified from [13].

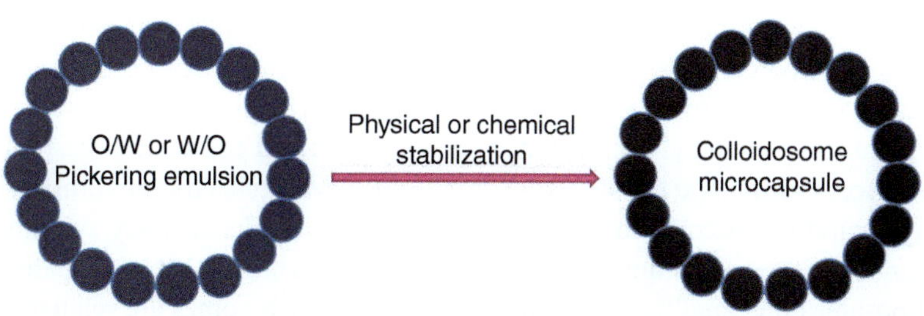

Figure 8.2 The structure of a colloidsome. *Source:* Adapted and modified from [16].

Table 8.1 Types of nano-functional food delivery systems [17, 18].

Association colloids	Biopolymeric nanoparticles	Lipid-based nano-encapsulators	Nanolaminates	Nanofibers	Nanoemulsions
• Micelles • Vesicles • Colloidosomes • Cubosomes	• Protein nanoparticless • Polysaccharide nanoparticless	• Liposomes • Archaesomes • Nanocochleates	• Sublayers and/or • Bilayers of binary compounds	• Carbon nanofibers • Polymer nanofibers	• Single-layer emulsion • Multilayer emulsion

8.2.2 Colloidosomes

Effective encapsulation of active components such as drugs, proteins, vitamins, flavors, gas bubbles, or even living cells is becoming increasingly important for a wide variety of uses and technologies, ranging from functional foods to drug delivery to biomedical uses. Colloidosomes are selectively permeable capsules included in colloidal particles [19]. These aggregates with self-assembled shells containing colloidal particles, provide remarkable flexibility for microcapsule functionality [16].

In the study by Niaz et al. [20], the colloidosomes loaded with nisin as a bacteriocin for evaluating their antimicrobial activity against milk-borne resistant pathogens e.g. *Listeria monocytogenes*, *Staphylococcus aureus*, and *Enterococcus faecalis* were presented and have shown good results [20]. So, due to enhanced controlled release, stability, and biocompatibility these multicomponent colloidosomes can be beneficial for incorporating antimicrobial agents into functional foods, beverages, and pharmaceutical products to combat pathogenic and bacteria responsible for spoiling.

8.2.3 Cubosomes

Cubosomes are NPs but in place of the solid particles, cubosomes are self-assembled liquid crystalline particles of specified surfactants with the proper ratio of water with a microstructure [21]. They are widely used in drug-delivery systems such as anticancer drugs [22–24].

8.2.4 Archaesomes

Archaea (Archaeobacteria) are considered to be distinct from eubacteria and eukaryotes, and they include aerobic, anaerobic, thermophilic, extremely thermophilic, thermoacidophilic, and halophilic microorganisms. All lipid extracts from individual species of *Archaea* consist of polar ether lipids and from 5% to 20% neutral lipids. The polar ether lipids of *Archaea* consist of branched phytanyl chains that are usually saturated and are attached via ether bonds to the glycerol carbons at the sn-2,3 positions [25]. Novel archaeosome compositions and their application in vaccine formulations as adjuvants and/or delivery systems increase the immune response to immunogens in an animal such as a human [26]. Compared to conventional liposomes prepared from ester phospholipids, archaeosomes are relatively more stable to extremes of pH, high temperature, and action of lipases and bile salts [27, 28].

8.2.5 Nanocochleates

Nanocochleates consist of purified soy phospholipids that contain about 75% by weight of lipids. The phospholipids may be phosphatidylserine, dioleoylphosphatidyl serine, phosphatidic acid, phosphatidylinositol, phosphatidylglycerol, or a mixture of one or more of these lipids [29].

8.2.6 Biopolymeric Nanoparticles

Biopolymer NPs were first produced using albumin and nonbiodegradable synthetic polymers such as polyacrylamide and poly(methyl acrylate). The risks of chronic toxicity due to intracellular and/or tissue overloading of nondegradable polymers were soon considered as a major limitation for the systemic management of polyacrylamide and poly(methyl acrylate) NPs in humans. As a consequence, the type of NPs that received much attention was designed with synthetic biodegradable polymers including polyalkylcyanoacrylate, poly (lactic-co-glycolic acid) (PLGA), and polyanhydride [30–35].

Solid particles with diameters <100 nm form the basis of modern anticancer drug-delivery systems. These polymers have effects such as extremely high bioactivity and they are able to act as carriers of antimicrobial components.

8.2.7 Liposomes

The chemical structure of liposomes includes phospholipids or synthetic amphiphiles merged with sterols, such as cholesterol, to improve membrane permeability. A thin-film hydration method (followed by extrusion) is the main way used to prepare liposomes, in which lipid components, with or without a drug, are dissolved in an organic solvent [36]. Liposomes can be categorized according to their structure: multilamellar vesicles (MLVs), oligolamellar vesicles (OLVs), and unilamellar vesicles (ULVs). ULVs are further divided into small unilamellar vesicles (SUVs) of 20 to 100 nm diameter, medium unilamellar vesicles (MUVs), large unilamellar vesicles (LUVs) that are larger than 100 nm diameter, and giant unilamellar vesicles (GUVs) that are larger than 1000 nm diameter [37]. The immunoliposomes, virosomes, stealth liposomes, and archaeosomes as several examples of carrier systems, which compromise of lipid bilayers that are biocompatible and may improve the solubility and stability of core materials [38, 39]. Several recent studies on liposomes show their high potential in the encapsulation of vitamins, antioxidants, proteins, peptides, antimicrobials, essential oils, flavors, enzymes, minerals, minerals, and fatty acids [40–44]. Liposomes have also been used for improving the flavor of ripened cheese, and for liposome-encapsulated micronutrients (such as iron) in milk [45].

8.2.8 Nanolaminate

A nanolaminate consists of two or more layers of material with nanometer dimensions that are physically or chemically bonded to each other (Figure 8.3).

Nanolaminates provide food scientists with some advantages, over conventional technologies, in the production of edible coatings and films and may thus have several important applications

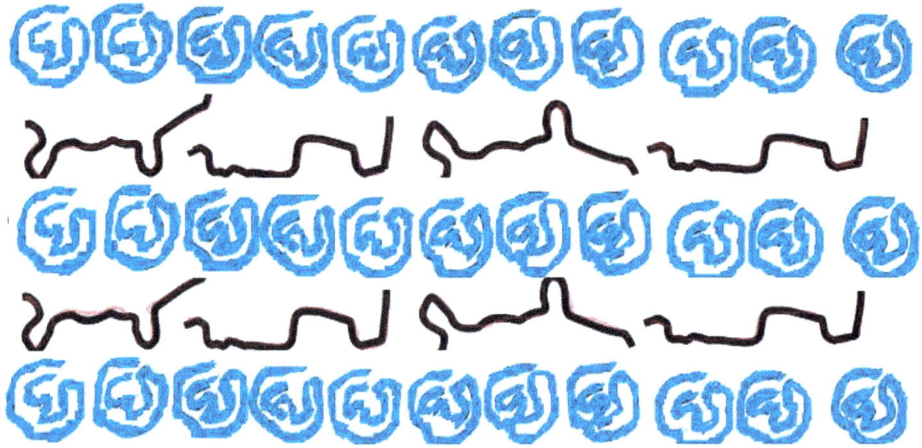

Figure 8.3 Example of a possible nanolaminate material formed from a globular protein and a polysaccharide. Each layer is approximately 1 to 100 nm. *Source:* Adapted and modified from [46].

within the food industry. These coatings or films could serve as moisture, lipid, and gas barriers. Alternatively, they could improve the textural properties of foods or serve as carriers of functional agents such as colors, flavors, antioxidants, nutrients, and antimicrobials [47, 48].

8.2.9 Nanofibers

Nanofibers are characterized as nanomaterials that have at least one dimension, that is, 100 nm or less, based on the definition of the American National Science Foundation. Nanofibers are grouped as one-dimensional nanomaterials (having a degree of freedom in one direction). Apart from the specific definition used in textiles, nanofibers are defined as fibers that have a minimum ratio of length to thickness of 1000 : 1 [49, 50]. Typical layer thicknesses range from several Å to up to 100 nm, which depends on the adsorption conditions and material properties [51]. The effects of nanofibers can be specified as having a low unit length (diameter), high surface-area-to-volume ratio, high strength value, low basis weight, high porosity, and small pore size. These make them indispensable in numerous applications [52–55]. Examples of the most important applications of cellulosic nanofibers (a type of nanofiber) in food science are: (i) immobilization of bioactive substances such as enzymes, vitamins, and antimicrobials; (ii) nutraceutical delivery systems and controlled release of materials; (iii) as biosensors; (iv) filtration; and (v) for reinforcing composites and in films [56]. Various polymers such as keratin, collagen, silk fibroin, cellulose, gelatin, poly(lactic acid) (PLA), PLGA, poly(ethylene-co-vinyl acetate) (PEVA), and polysaccharides such as alginate and chitosan can be utilized in the production of nanofibers [57].

8.2.10 Dendrimers

Dendrimers are artificial macromolecules that are determined by a combination of a high number of surface functional groups and a compact molecular structure [58, 59].

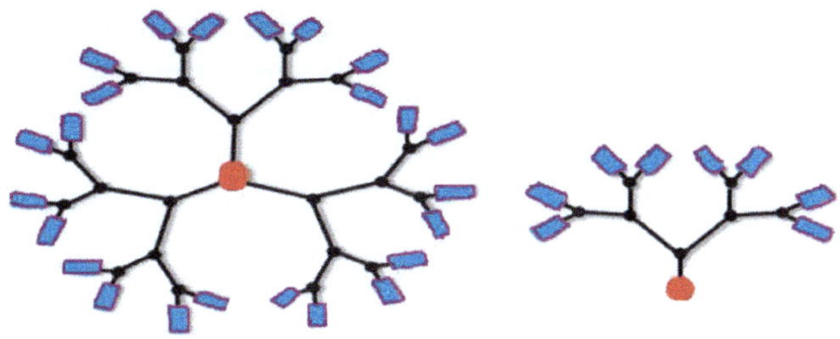

Figure 8.4 Chemical structure of dendrimers. *Source:* Adapted and modified from [60].

Dendrimers, however, as single molecules can contain high numbers of functional groups on their surfaces (Figure 8.4).

Due to the antibacterial effects of dendrimers, they can be used for water and wastewater disinfection with minimum adverse side effects. For example, the PPI-G2 dendrimer with end amine groups shows a positive effect for the removal of main bacterial isolated strains [61].

8.2.11 Hydrogel Nanoparticles[1]

Hydrogel polymers are typically soft and elastic due to their thermodynamic compatibility with water and can form a three-dimensional polymeric network able to carry cells and drugs. Indeed, hydrogel polymers exhibit many additive properties that make them very suitable as delivery vehicles [62]. A combination of a hydrogel system (e.g. hydrophilicity and extremely high water content) with a NP (e.g. very small size) produces a favorable delivery system [62]. In addition, they have beneficial properties such as being mucoadhesive, gastro-adhesive, and stimuli-responsive polymers. The natural polymers chitosan and alginate have been extensively investigated in the preparation of hydrogel NPs and from a synthetic group, hydrogel NPs based on poly(vinyl alcohol), poly(ethylene oxide), poly(ethyleneimine), poly(vinyl pyrrolidone), and poly-*N*-isopropyl acrylamide [63].

The use of alginate hydrogels in food include:

- in edible films and food coatings [64, 65];
- as dual-function emulsifiers and liposome stability enhancers [66, 67];
- as antioxidants [68];
- as antibacterial agents [69];
- as dietary fibers [70];
- as heavy metals sorbents [71]; and
- as carriers for macro-, micro-, and nano-encapsulation of active ingredients [72].

Routine methods for the formation of alginate hydrogel microparticles for food use involves ionotropic gelation via the manipulation of laminar flow, spray-drying, micro-sized liposome entrapment, emulsification, lyophilization, and homogenization. Alginate and chitosan layer-by-layer coating of nanoliposomes and electrospinning provide superior stability of vitamin B9 and vitamin C in an acidic environment, relating to the gastric environment and natural fruit juices [73–75]. Current challenges for nano-encapsulation of

1 Polymeric nanogels or macromolecular micelles.

bioactive food ingredients that can be addressed by complexation with chitosan or modification of initial alginate characteristics are: (i) interdiction of NP aggregation, (ii) delivery of imperfectly adsorbed or hydrophobic ingredients, and (iii) elimination of burst release of encapsulated materials [72].

8.2.12 Polymeric Nanoparticles

Polymeric NPs include colloidal particles of size range 20 to 1000 nm including solid particles. These NPs contain a polymeric core with the active component entrapped in the polymeric matrix, mostly surrounded by a surfactant layer that stabilizes the system [76, 77]. A limitation of the usage of polymeric NPs in the food industry is that the is host exposed repeatedly to nanoengineered particles over long periods [77]. It is important that only food-grade particles are used in polymeric particle synthesis processes planned for food utilization in the form of natural polymers, alongside the use of biodegradable and biocompatible surfactants [78]. Furthermore, chitosan and many other natural polymers such as gelatin and sodium alginate have many advantages over conventional synthetic pathways [79]. Chitosan, alginate, pectin, cellulose, carrageenan, hyaluronate, gelatin, dextran, zein (a corn protein rich in prolamines), xanthan, albumin, milk proteins (casein), and whey proteins (β-lactoglobulin and α-lactalbumin) are natural polymers that are the most widely used biopolymers in this field [80]. Among synthetic polymers, the polyesters family (i.e. PLA, poly(caprolactone) [PCL]), poly(glycolic acid) [PGA] have high biocompatibility and biodegradability properties [78]. An important issue in the stabilization of polymer NPs during the synthesis process is the requirement of the addition of surfactants during synthesis to stabilize the polymeric NPs. Examples of surfactants (emulsifiers) available for food applications are diacetyl tartaric esters of monoglycerides (DATEM), lecithin, mono- and diglycerides, monosodium phosphate derivatives of mono- and diglycerides, calcium stearoyl lactylate, citric acid esters of mono- and diglycerides, sorbitan monolaurate, stearyl tartrate, and sucrose acetate isobutyrate [81].

8.2.13 Carbon-based Nanocarriers

Carbon-based nanomaterials have been used in the development of high-performance sensing devices for food safety inspection to produce, identify, and increase the sensing signals [82]. According to their spatial dimensions, carbon-based nanomaterials can be roughly divided into fullerenes (zero-dimensional), CNTs (one-dimensional), graphene (two-dimensional), graphene coil (multidimensional), etc. [82]. Activated carbon has been widely used as a sorbent for conventional wastewater treatment due to its large surface area and high potential to adsorb an extended spectrum of organic and inorganic contaminants [83–86].

8.2.14 Quantum Dots

QDs of NPs are defined as inorganic semiconductor fluorescent NPs with a size in the range of 1 to 10 nm [87]. Utilization of them as a marker to either detect pathogenic microorganisms or track proteins in the food matrix provides a better understanding of their role in food science [88]. Top-down and bottom-up synthetic routes are synthesis methods for carbon quantum dots (CQDs) [89]. Electro-oxidation, acid-assisted chemical oxidation, or laser ablation are processes that involved in CQD synthesis and these conditions are complex and extreme. So, researchers are trying to use bottom-up methods [89] because they use simple chemicals such as glucose [90], amino acids [91], or natural materials such as fresh tomato [92], garlic [93], and green tea [94], and without the presence of any synthetic chemical substance. Most synthesis methods use biomass as a precursor substrate. Nowadays, food waste is used as a carbon source in CQDs technology production [89]. For example, Xu et al. [95] synthesized QDs from apple juice and involved them to measure and detect Hg^{2+}. Also, QDs have been used in food science as a marker for detecting pathogenic bacteria, for example, QDs include a Cd-Se core functionalized in different ways and coated with polymers for suitable adherence to bacteria and proteins [87, 96].

8.2.15 Nanoemulsions

Nanoemulsions are kinetically uniform liquid-in-liquid dispersions with droplet sizes of about 100 nm [97]. Due to the small particle size and the high surface-to-volume ratio of nanoemulsions, delivery systems based on nanoemulsions can also improve the bioavailability of the encapsulated components [98]. Lutein, β-carotene, lycopene, vitamins A, D, and E3, the co-enzyme Q10, and omega-3-fatty acids are some of examples of nano-sized functional molecules that are encapsulated in self-assembled nanoemulsions [99]. Lipid-based nanoemulsions are better compared with traditional nanoemulsions for the delivery of constituents within biological systems [100]. But several recent investigations have shown that the high lipid content of these nanoemulsions results in adverse effects on the body, such as obesity and cardiovascular diseases [100]. So, the production of nanoemulsions with a low-fat content is an important issue.

In 2015 Nestlé and Unilever produced nanoemulsion-based ice cream with a decreased fat content [101]. In Table 8.2 we provide details of several examples of organic, inorganic, metal NPs, and conjugation formulations of these in nutrition and food industries.

8.2.15.1 Curcumin: The Golden Nutraceutical

Curcumin or feruloylmethane (1,7-bis[4-hydroxy-3-methoxyphenyl]-1,6-heptadiene-3,5-dione) (Figure 8.5) is a hydrophobic polyphenol compound naturally concentrated in the rhizome of the herb *Curcuma longa*, commonly known as turmeric, with the chemical name diferuloylmethane, was first isolated two centuries ago in 1815 by two German Scientists, Vogel and Pelletier [140, 141]. The first signs related to curcumin's potential beneficial applications came from India, where Alzheimer's rates are low and people use a lot of turmeric [139].

Monotargeted therapies or "smart drugs" have been designed over the past few years for the treatment of some chronic diseases (e.g. cardiovascular diseases, metabolic diseases, cancer, and neurological diseases). The main reason for these chronic disorders is a defect in signaling pathways and so in the treatment of them the targeting of a single pathway among many of the pathways involved is not likely to be effectual for the prevention and treatment of these diseases [142]. Several studies over the past five decades indicate that curcumin has antioxidant, anti-inflammatory, antiviral, antimicrobial, antihypertensive,

Table 8.2 The overview of different types of nanoparticles and their applications in medicine.

Organic NPs	Form	Disease type/effect application/comment	References
Curcumin	Liposomal, micelle, noisome, cyclodextrin, dendrimer, nano gel, chitosan nanocapsules curcumin with chitosan	Alzheimer's disease, diabetes, asthma, hyperlipidemia (reduction of cholesterol, triglyceride, and low-density lipoprotein cholesterol), antidepressant, gastric cancer	[102–105]
Saffron	• Nanoparticles based on polymeric carriers (natural/ synthetic) • Nanoparticles based on lipid carriers	Different types of cancers, antidepressant, anxiolytic and hypnotic, against infectious diseases (measles, smallpox, and scarlet fever) and antibacterial, antiseptic, anti-Parkinson, anti-Alzheimer, improves memory and learning skills,	[106–109]
Inorganic NPs			
Silica NPs	Colloidal silica, silica gels, precipitated silica, silicon quantum dots (SiQDs)	Bone health, flavor carriers (vitamins, fish oils, lycopene and coenzyme Q10), drug delivery,	[110]
TiO_2 & Ti-N nanoparticles	TiO_2 nanoparticles	Positive and negative effect on the quality of intestinal villi, intensification of an existing intestinal cancer (colon cancer)	[111, 112]
Carbon nanotubes (CNTs)	Nanotubes	Treatment of osteoporosis and broken bones, antibacterial effect, vaccine, gene and drug delivery, breast cancer detection, tumor therapy	[113–115]
Metal NPs			
Nano ZnO (as a metal oxide nanoparticles)	Nanoparticles, quantum dots (ZnO QDs)	Antimicrobial, anticorrosive, bedsore therapy, more than 30 times more selective cytotoxicity toward cancer cells compared to healthy cells, tumor suppressor, antioxidant activity, antibacterial effect, toxic effect on lung human cells, Anti-MDR bacteria	[116–120]
Selenium nanoparticles	Selenium nanocapsules, a liposomal formulation of SeNPs	Antimicrobial, antioxidant, anticancer, fatty liver disease, anti-diabetic	[121–125]

Table 8.2 (Continued)

Organic NPs	Form	Disease type/effect application/comment	References
Iron-oxide nanoparticles (IONPs)	Nanoparticles	Cancer therapy, targeting amyloid-beta (Aβ) in the arteries, inhibiting the microglial cells, delivering drugs, and increasing contrast in magnetic resonance imaging	[126, 127]
Silver NPs (AgNPs)	Nanoparticles	Anti-inflammatory, antibacterial, antiviral antifungal, anticancer agents (e.g., endothelial cells, IMR-90 lung fibroblasts, U251 glioblastoma cells, and MDA-MB-231 breast cancer cells), metal-coated implants	[128–134]
Conjugation			
Crocin-SeNPs		Drug delivery, anticancer synergistic effect especially human lung cancer cell lines	[135]
Saffron-SiNPs		Cancer and cardiovascular diseases	[136]
IONPs with a phospholipid–polyethylene glycol (PEG) coating anddoxorubicin (DXR)		Drug release and increased apoptosis of cancer cells, which was further enhanced by hyperthermia	[137]
PLGA nanoparticles containing crocetin		Anti-cancer therapy	[138]

Note: nano-ZnO; nano zinc-oxide, SeNPs; selenium nanoparticles, SiNPs; silicon nanoparticles, MDR; multi-drug resistance

Figure 8.5 The chemical structure of curcumin. It is a powder found in the golden yellow spice turmeric (top). *Source:* Adapted and modified from [139]. Popova Olga / Adobe Stock.

and anticancer effects [143–146]. According to the opinion of the toxicology scientists, “only 15% or 20% of uptake curcumin from diet reaches the brain – the residual amount of it is metabolized in different organs” [15, 139]. If curcumin is to be effective medically, a person needs to eat 12–20 g of curcumin daily [147].

So, curcumin must be given in large doses to animals to have any effect. Regrettably, curcumin may be toxic in high quantities. NP-based drug-delivery approaches [148], in which curcumin is encapsulated in liposomes [149], solid lipid microparticles [150], such as bovine serum albumin [151], and chitosan [152], or complexed with phospholipids and cyclodextrin were developed due to the poor water solubility and low bioavailability of curcumin [153]. NPs such as liposomes, micelles, nanogels, and polymeric NPs can be used to deliver a therapeutic concentration of curcumin that increases the therapeutic and medical efficacy of curcumin. Curcumin is soluble in ethanol (10 mg ml^{-1}), DMSO (25 mg ml^{-1}), acetic acid, 0.1 M NaOH, 10 mM Na_2CO_3, acetone, water, and chloroform [154, 155] (Figure 8.6).

To remove the delivery problem, researchers are aiming to encapsulate the curcumin in NPs made from the biodegradable, drug-releasing PLGA [15, 156]. Today, PLGA (Figure 8.7) is used widely in the nanotechnology field. It is a biodegradable and biocompatible polymer, however, its biodegradability (due to the hydrolysis of ester bonds to produce lactic and glycolytic acids) is perhaps problematic for the shelf life of the formulation [157].

By conjugating curcumin to metal oxide NPs or encapsulation in lipid NPs, dendrimers, nanogels, and polymeric NPs, the water solubility and bioavailability of curcumin can be improved and thus increase its pharmacological effectiveness [158] (Table 8.3).

We continue with several examples of the medical application of nanocurcumin particles. The blood–brain barrier (BBB) creates a nonpenetrating matrix for drugs in Alzheimer's disease (AD) therapy and so most conventional strategies failed to treat AD in clinical trials.

Figure 8.6 The solubility of the nanocurcumin particles (CURNs) (left) and curcumin (right) in water. *Source:* Adapted and modified from [147], Basniwal, R.K., et al. (2011) Figure 02, p. 02 / with permission of American Chemical Society.

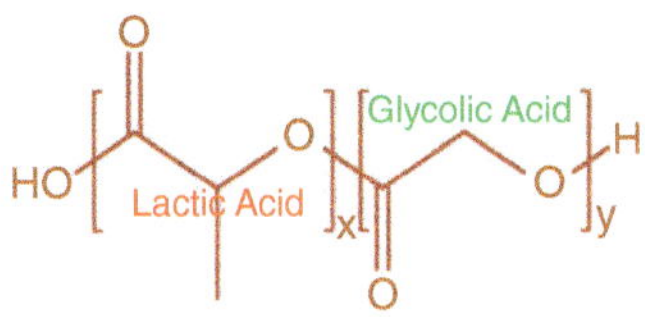

Figure 8.7 The chemical structure of PLGA.

The formation of the extracellular senile plaques and neuronal loss induced by amyloid β (Aβ) collections was found in the brains of AD patients [195]. Huang et al. demonstrated a combined therapeutic formulation for AD that included a new type of PLGA NPs by coating with Aβ generation inhibitor S1 (PQVGHL peptide) and curcumin to target the detrimental factors in AD development and by conjugating with a brain-targeting peptide CRT (cyclic CRTIGPSVC peptide) and an iron-mimicking peptide that targets transferrin receptor (TfR) in order to improve BBB penetration [156]. The antidiabetic property mechanism of nanocurcumin particles is due to: decreased hepatic glucose level, increasing glucose catch by upregulating GLUT2, GLUT3, and GLUT4 gene expressions, increasing the secretion of insulin from pancreatic cells, and decreasing insulin resistance [196]. So, enrichment of the human material food (especially diabetic food products) that has a high glycemic index such as bread, pasta, all kinds of sweets, honey, jams and so on with these curcumin NPs could be an important issue in diabetic patients' nutrition. One method of synthesis used for nanocurcumin is the sonication method. The synthesis of nanocurcumin was evaluated for solubility in water and by UV-visible spectrophotometry. Further, the NPs were characterized by Fourier-transform infrared spectroscopy, transmission electron microscopy, NP tracking and analysis, and zeta potential analysis [197].

Curcumin shows poor solubility and low absorption rates. Therefore, its silver (Ag) NPs were formed to increase solubility, absorption as well as its effectiveness against tumor [198]. Another aspect of curcumin NPs is their protective property against heavy-metals-induced liver damage in humans [199]. Toxic results of the heavy-metal ions for example arsenic (As), cadmium (Cd), chromium (Cr), copper (Cu), lead (Pb), and mercury (Hg) are showed primarily in oxidative stress [200]. So, with more investigations about this subject, we can see this NP has potential to be used in the nutrition of high-risk persons that are exposed to heavy-metal ions. In a study by Pourreza et al. a highly selective and portable chemosensor was designed for mercury detection in water samples using curcumin NPs in a paper-based analytical device (Figure 8.8) [201], showing that curcumin can be employed as a diagnostic tool due to its high affinity for heavy metal ions.

In a study by Yadav et al. [171] curcumin encapsulated in chitosan NPs was employed as a novel strategy for the treatment of arsenic toxicity. We can enrich food products with this combination for arsenic detoxification in clinical cases [171]. Chen et al. [202] studied the effects of several liposomal formulations on curcumin NP stability. Their finding was that liposomal curcumin is more stable than free curcumin in phosphate-buffered saline (PBS). Liposomal curcumin and free curcumin showed similar stability in human blood plasma and in a culture medium [202].

8.2.15.2 Saffron: Its Nanoparticles and Human Nutrition

Saffron (*Crocus Sativus* L.) is a dietary spice derived from the flowers of *C. sativus*-Iridaceae [109]. It is known as "red gold" [203]. There are more than 150 substances in saffron that include carbohydrates, polypeptides, lipids, H_2O, minerals, and vitamins. Besides these constituents, saffron has four main bioactive materials including crocin ($C_{44}H_{64}O_{24}$), crocetin ($C_{20}H_{24}O_4$), picrocrocin ($C_{16}H_{26}O_7$), and safranal ($C_{10}H_{14}O$) [109, 204] (Figure 8.9).

Table 8.3 Different nanocomposite formulations of curcumin/nano curcumin and their therapeutic targets.

Composite NP used	Size (nm)	Dosage	Target disease	Therapeutic mechanism	References
Nano-curcumin ligated to ZnO NPs	53	0.1–0.25 $\mu g ml^{-1}$	AGS gastric cancer	Enhanced apoptosis of cancerous cells	[159]
Curcumin encapsulated in PLGA-NPs	200 ± 20	0.5–5 μM	Alzheimer disease	Activates the Wnt/β-catenin pathway	[139]
Curcumin conjugated to iron oxide nanoparticles	10.5	100 μg	Pancreatic cancer treatment	Curcumin nanocomposite increases β-catenin expression leading to reduced cell proliferation and metastasis	[160]
Curcumin encapsulated in cellular exosomes NPs	30–100	100 $mg\,kg^{-1}$ body weight	LPS-induced septic shock	Enhanced apoptosis of CD11b + Gr1 + cells that mediate the inflammatory response in mice	[161]
Curcumin-loaded transferrin-conjugated superparamagnetic iron oxide NPs	~200	10^4–10^5 $ng\,ml^{-1}$	Anti-leukemic response	Downregulation of Bcr-Abl protein activates intrinsic apoptotic pathways	[162]
Curcumin-conjugated solid lipid nanoparticles	190	330 $\mu g\,kg^{-1}$	Asthma therapy	Suppressed airway hyper-responsiveness and inflammatory cell infiltration	[163]
β-cyclodextrincurcumin inclusion complexes	—	—	Lung and colon cancer	All curcumin-containing formulations were effective in inhibiting cell proliferation *in vitro* cell culture	[164]
Water-soluble liposomal curcumin	—	—	Lung fibrosis	Diminish radiation pneumonitis and fibrosis of the lung and sensitize LL/2 cells to irradiation	[165]
Gold-conjugated curcumin	200–250	—	Cancerous cells	Improved solubilityEnhanced antioxidant and anticancer effects	[104, 166]
Silver-conjugated curcumin	~15	—	Infectionsskin wounds		[167]
Chitosan-conjugated curcumin	100–250	—	Wounds Melanoma tumors	Improved chemical stabilityShowed wound healing effectsIncreased antitumor effectsImproved antioxidant effectsProlonged blood circulation	[168–173]

Dendrimer-conjugated curcumin	15–150	—	Breast cancer Colon cancer	Improved stabilityIncreased antitumor and anti-proliferative effects	[174–179]
Solid lipid-conjugated curcumin	5–1000	—	Cerebral ischemia[a] Colitis[a] Allergy[b] Breast cancer[b]	Prolonged circulation of bloodIncreased anti-inflammatory effects Improved brain delivery	[180–182] [163, 183, 184]
Lamellar form of noisome-conjugated curcumin	190–1140	—	Albino rat skin[c] Cancerous cells[d]	Increased skin penetration Prolonged delivery systemAnti-infection and anticancer effectsEnhanced fluorescence intensity	[185–187]
Spherical micelle included in curcumin NPs	10–100	—	Lung tumor[e] Breast cancer[f]	Increased solubility and bioavailabilityImproved anti-oxidative and antitumor effectsProlonged circulation timeEnhanced fluorescence effect	[188–194]

[a] *In-vitro.*
[b] *In-vivo* (rat and mice).
[c] *In-vitro.*
[d] *In-vivo* (mice and snake), NPs.
[e] *In-vitro.*
[f] *In-vivo* (mice).
Wnt/β-catenin; Wnt cellular signaling pathway, AGS; adenocarcinoma gastric *cell line.*
Source: [158]

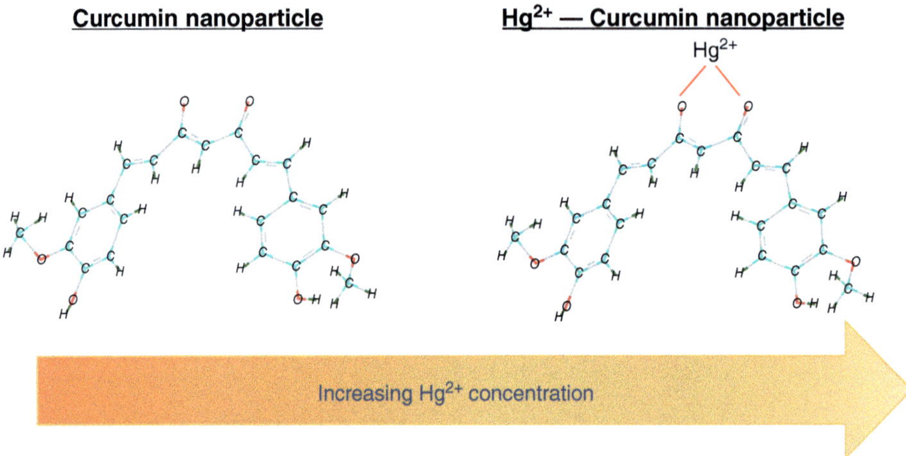

Figure 8.8 The chemosensor for mercury determination in water samples using curcumin nanoparticles. *Source:* Adapted and modified from [201].

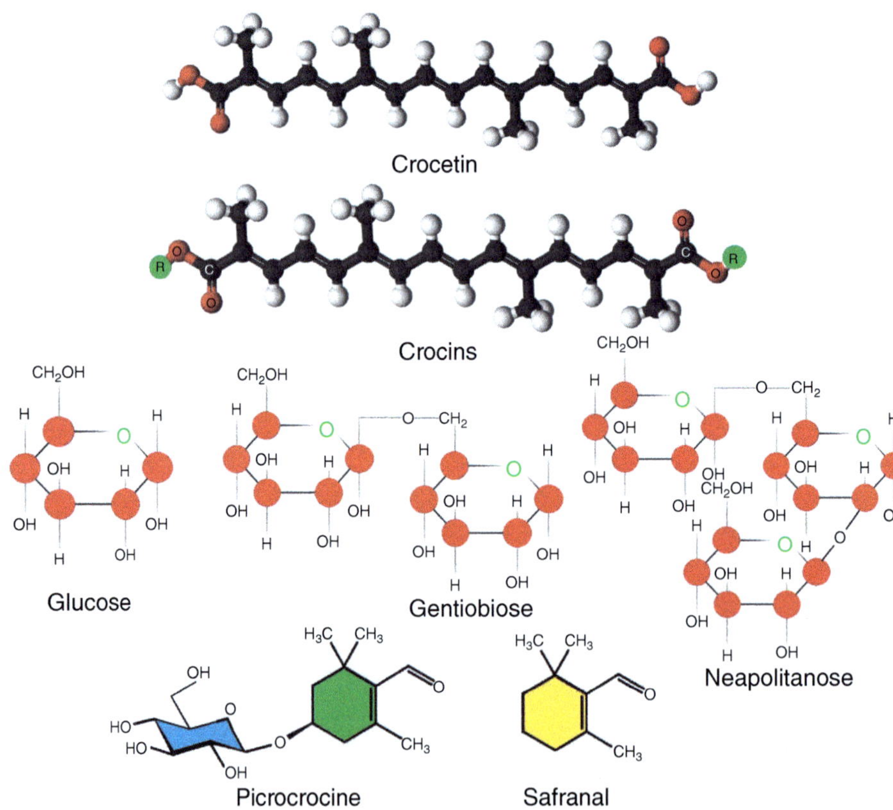

Figure 8.9 The chemical structure of bioactive compounds present in *Crocus sativus* L. dried stigmas. *Source:* Adapted and modified from [204].

Saffron is likely safe for most individuals when used at levels commonly found in the diet. In all the preclinical and clinical studies so far, saffron has been administered orally or intraperitoneally. The common effective doses in clinical trials (30–50 $mg\,d^{-1}$) were much lower than the $LD5_0$ (median lethal dose) of saffron. The latter depends on the form of intervention and can range from around 1.5 $g\,kg^{-1}$ (after injection in mice) to >5 $g\,kg^{-1}$ (after human consumption) indicating a high safety margin for the exploitation of saffron in pharmaceutical preparations [205, 206]. The most significant use for saffron is in food, where it is valued for its coloring, flavoring, and aroma in the production of some traditional dishes [207]. The neuroprotective effects of saffron have been revealed in animal and human studies. Administering crocins at 30 $mg\,kg^{-1}$ to rats with AD induced by intra-cerebroventricular streptozocin showed a significant attenuation in learning and memory performance and spatial cognition evaluation [208]. Crocin acts as an antioxidant by quenching free radicals and protecting cells and tissues against oxidation [209, 210]. Administration of 50 mg of saffron dissolved in 100 ml milk for sixweeks in 10 healthy volunteers and 10 patients showed a significant decrease in lipoprotein oxidation susceptibility [211]. Safranal is a monoterpene aldehyde that has a molecular weight of 150.21 $g\,mol^{-1}$ and is responsible for saffron's aroma. Safranal is the main component of saffron's essential oil [212]. The anticancer activity of the aqueous extract of saffron shows effectiveness in cancers such as lung cancer, pancreatic cancer, colorectal cancer, breast cancer, leukemia, liver cancer, cervical cancer, skin cancer, ovarian cancer, prostate cancer, reducing opioid withdrawal syndrome, neurodegenerative diseases such as epilepsy, brain ischemia, Parkinson's disease, and memory impairment (Table 8.4) [204, 213, 214].

Table 8.4 The molecular mechanisms involved in the main carotenoids of saffron in tumor models for cancer chemotherapy.

	Disease type	Mechanism
Crocin Crocetin	Esophageal cancer Cervical cancer	Amplification of DNA fragmentation
	Ovarian cancer Prostate cancer	Inhibition of invasiveness and metastasis
	Leukemia cancer Liver cancer	Cell cycle arrest
	Pancreatic cancer Bladder cancer	Apoptosis
	Breast cancer	Inhibition of angiogenesis
	Lung cancer Gastric cancer	Antioxidant activity
	Osteosarcoma cancer Colorectal cancer	Anti-inflammatory effect
	Skin cancer Tongue carcinoma	Cytotoxic activity and inhibition of proliferation

Source: [286].

Trans-Crocetin is the only active metabolite that has been shown to be capable of crossing the BBB and reach the central nervous system (CNS), regardless of whether pure crocetin or saffron extract is administered. Yoshino et al. identified that this occurs 90 minutes after the pure crocetin has been orally administered to rats [215]. Linardaki et al. found crocetin in the mice's brains after intra-peritoneal administration of saffron extract [216]. Saffron's anti-adiposity effects were also identified. Sheng et al. proposed a lipid-lowering property induced by the crocetin esters [217]. In a study by Rikabad et al. [218], the use of TiO2 (titanium dioxide) NPs increased the total phenolics and total flavonoids in saffron stigmas. TiO_2 NPs enhanced the antioxidant activity of saffron stigmas [218]. Encapsulation of crocetin in appropriate NP systems could be considered as an ideal solution to overcome the aforementioned limitations, increase water solubility, and improve its pharmacokinetics and bioavailability [138].

8.3 Nanoparticles Based on Inorganic Materials

8.3.1 TiO_2 and Ti-N Nanoparticles

Nano titanium particles used in white plastic packaging and white ink materials are manufactured with TiO_2 for protection from UV light or for aesthetic reasons [219]. These transparent nano-form particles are UV-resistant and UV-absorbent [220] and are used as UV-protected plastic (e.g. co polyether-ester) packaging for food [221–223] as filler materials and plastic containers, bottles, and boxes [224]. Peiro et al. [221]studied the behavior of photocatalytic oxygen consumption in the form of nanocrystalline TiO_2 composite films. They exhibited high scavenging activity and fast oxygen reduction in the presence of ethanol due to the photocatalytic redox reaction [225]. Mills et al. also prepared ethylcellulose film composited with TiO_2 NPs and reported a remarkable reduction in the amount of oxygen in the photocatalytic reaction system under UV irradiation [226]. TiO_2 NPs are typically produced with a crystalline structure such as anatase, rutile, or brookite, each one having a particular effect [227]. According to the toxicological properties of TiO_2, a practical method for determining how much titanium was present in the food samples was introduced by Weir and colleagues [228]. The organic substance from the food was broken down by adding 10 ml of hydrogen peroxide and 0.5 ml of HNO_3 and then heating on a hot plate at 110 °C. Peroxide and HNO_3 did not change the size of P25 (TiO_2 without pigment characteristics) or E171 (referred as food-grade TiO_2, a European Union designation for a white food color additive that is known elsewhere by other designations (CI77891, Pigment White 6) parameters), but they mainly digested the organic structures into which the TiO_2 was embedded. The exceptions were paint samples, which could not be entirely digested by this practical technique [228–230]. The main routine manner for the synthesis of TiO_2 NPs consumes the hydrolysis of titanium (Ti) salts in an acidic solution [231]. TiO_2 particles are used as functional ingredients in certain foods to provide characteristic optical properties such as increased lightness and brightness [228]. Typically, the TiO_2 ingredients utilized in the food industry as lightening agents are optimized to have particle sizes of 100 to 300 nm to increase their light-scattering properties [232]. Nano-titanium nitride (Ti-N NPs) are used as mechanical strength, which can establish better protection and safety of foods and also function as a reheating additive. This nano compound is widely used in the production of PET (polyethylene terephthalate) beverage bottles [233]. In the United States, TiO_2 NPs can be used in food if its content does not exceed 1% of the total weight of the product [234–236]. In Europe, the quantum states principle is in force, which means it can be used in amounts not exceeding the intended target level [237, 238]. Ti-N NPs have a high melting point (2950 °C), high hardness, high-temperature chemical stability, high-performance infra-red absorption, and UV shielding of more than 80% [219, 239, 240].

8.3.2 Silica Nanoparticles

Nano silica particles are sorted into P-types and S-types according to their structure and morphology. P-type particles are determined by numerous nanopores, that have a pore rate of 0.61 ml g^{-1} and show a higher UV reflectivity in contrast to the S-type; the S-type also has a comparatively smaller surface area.

The main application of the nano compound silicon dioxide (SiO2) in food packaging is as a coating on plastic film (e.g. PET, PP [polypropylene], nylon) to make transparent high-gas barrier films [219, 241]. To improve the elastic properties of PP [242], starch [243], starch/polyvinyl alcohol [244], reducing water absorption by starch [243, 244], and perfecting oxygen barrier of polypropylene [242], these SiO_2 NPs were widely used. The beneficial effects of nano silicon dioxide (size about 12 nm) on the seed germination of tomato (*Lycopersicum esculentum* Mill. cv Super Strain B) were documented [245]. SiO_2 NPs remarkably raised the effects of seed germination. Also, SiO_2 NPs are added to some powdered foods and act as antiaging agents to increase flow effects, e.g. salts, icing sugar, spices, dried milk, and dry mixes [232]. Of course, silica NPs react with immunocompetent cells and induce immunotoxicity. So, the over-permission dose usage and absorption of SiNPs in food and food packaging materials is a dangerous issue. The physicochemical characteristics, such as particle size, shape, composition and crystallinity, affect toxicity for immune cells and organs [246, 247]. Autophagy dysfunction is believed to be a potential toxic mechanism of SiNPs [247].

8.3.3 Carbon Nanotubes

These conductive nanomaterials have been widely composited into polymers (e.g. PVOH [poly(vinyl alcohol)], PP, nylon, and PLA) and studied for packaging, especially for antimicrobial and smart-detection sensor applications. Two types of CNTs have been presented including the single-atom-thick nanotubes and several concentric nanotubes [248]. The small size (to passing through holes in tumors or to transport DNA), the high surface-area-to-volume ratio, and the ability to contain chemicals makes CNTs suitable for the removal of different pollutants, in agricultural, medical, and nutritional fields [249–251]. The antimicrobial effect of the highly pure single-walled carbon nanotubes (SWCNTs) with low metal content are toxic toward bacteria e.g. *Escherichia coli* [252–254]. Several studies showed the potential usage of CNTs in interactive smart packaging [255, 256]. These packages are sprayed with CNT-based gas sensors to form a thin, clear film inserted with a disposable wireless transmissions chip. This illustrated that this packaging would be able to communicate with the customer or market manager when an element or compound (e.g. O_2) has attained a specific threshold and the meat or fresh food is about to spoil [219, 257].

8.4 Metal Nanoparticles

8.4.1 Nano Zinc Oxide

Zinc oxide (ZnO) NPs (a semiconducting material) with the size of 20–30 nm show a wide range of antimicrobial activity (gram-positive and negative bacteria, algae, fungi, and possibly some viruses) for food and beverage storage and have been used for many years [258]. A permissive maximum dose for Ag nitrate as an agent with antimicrobial activity in bottled water was 17 g kg^{-1} [219, 259, 260]. Jeong et al. [261] demonstrated that Ag ions can kill more than 650 disease-causing pathogens. Ag NPs (metallic Ag atom clusters) with a larger surface area per mass than microscale Ag particles or bulk Ag material have a greater potential to release Ag ions [261]. Nano ZnO particles are more easily taken up by the body so, nano ZnO is usually consumed as a food additive. Also, ZnO is categorized as a GRAS (generally recognized as safe) material by the US Food and Drug Administration (FDA) [262]. The production of ZnO NPs by the chemical precipitation method is simple, easily controlled, and also easily industrialized [263].

8.4.2 Iron Oxide Nanoparticles

Iron oxide (IO) NPs consist of maghemite (γ-Fe_2O_3) and/or magnetite (Fe_3O_4) particles with diameters ranging from 1 and 100 nm [264]. Enzyme immobilization, protein purification, and food analysis are examples of IO NP applications in the food industry [265]. The enzyme immobilization has been used in the process of immobilization in food packaging materials [266]. The use of enzymes has many benefits such as stabilizing temperature and resistance to protease and other denaturing formulations. Also, catalysis of a reaction by enzymes could induce the reduction of an unfavorable material and raise the nutritional value of food [267]. Soares et al. [268] improved the nutritional value of grapefruit juice through immobilizing and incorporating the enzyme naringinase in plastic packaging. Naringinase hydrolyzed naringin, which is accountable for the bitter taste of the citrus juices [268]. The co-precipitation method is an easy method to construct magnetic NPs of iron with a large surface and small particle size, which increases the potential of these particles to act as an appropriate carrier for enzyme stabilization [269].

8.4.3 Selenium Nanoparticles

Selenium is an essential trace element in an animal's diet. It is required for the maintenance of health and growth; however, its toxicity could cause serious damage depending on the dose and chemical form. Selenium nanoparticles (SeNPs) represent what we believe to be a novel prospect for nutritional supplementation due to their lower toxicity and ability to gradually release selenium after ingestion [124] (Figure 8.10). Selenium compounds are believed to be detoxifying agents, playing an antagonistic role toward mercury [270–272], methylmercury [273, 274], cadmium [275], Ag [276], lead [277], and many other elements [278]. Interactions of selenium with other toxic elements in living organisms are continuing to be reported. As a major advantage over other selenium forms, elemental (zero-valent) NPs are a high Se-density formulation with the potential for local delivery of high doses into cancer cells [279]. Further researches on intracellular mechanisms indicated that SeNPs induced

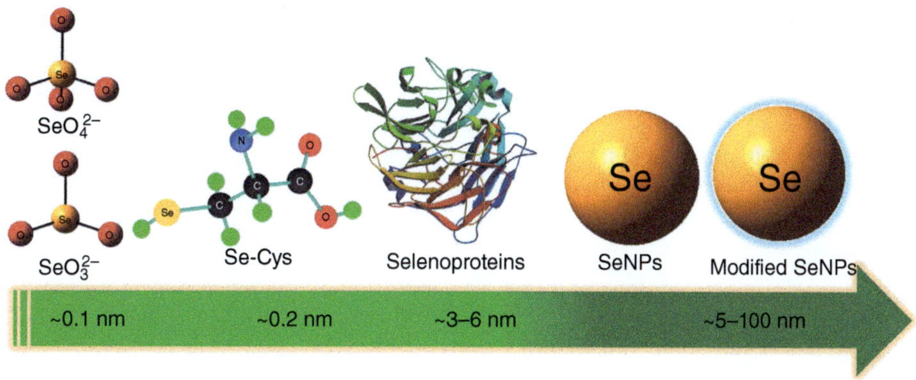

Figure 8.10 Comparison of the size of various selenium forms and selenium nanoparticles. *Source:* Adapted and modified from [124].

apoptosis of A375 cells with the participation of oxidative stress and mitochondrial function failure. SeNPs demonstrate that they are potential chemopreventive and chemotherapeutic agents for human cancers [280]. Also, the conjugation of SeNPs and chemotherapeutic agent Adriamycin could be a promising approach to cancer treatment. It was found that Adriamycin and SeNPs could inhibit the proliferation of cancer Bel7402 cells and the combination resulted in higher inhibition efficiencies than each part alone [281]. The SeNPs modified with folate-chitosan to form smart-shell nanocapsules could target cancer cells overexpressing the folate receptor and enter the cells via folate receptor-mediated endocytosis. Internalized folate chitosan smart-shell selenium nanocapsules (FAC@CurP-SeNPs) targeted intracellular reactive oxygen species overproduction and promoted apoptosis. These results suggested that FAC@CurP-SeNPs may be a candidate for further evaluation as a therapeutic agent for human cancers. The strategy to use transformable nanocapsules could be a highly efficient way to increase controlled drug release and anticancer efficacy [282].

The biological properties of SeNPs depend upon their size; smaller particles have more significant activity [124, 283]. NPs synthesis using bacteria is more effective than chemical synthesis, thanks to: (i) the high purity of selenium spheres (that are relatively regular and uniform and their size depends on the bacterium), (ii) a cheaper and high speed and rate production process, and (iii) a better possibility of controlling the parameters [284]. It is noteworthy that various analyses were conducted on the obtained NPs consisting of UV–vis, FT-IR, NMR, and XRD. UV–vis (absorption maximum [λmax] at 390 nm) and color of solution (changing from colorless to red) established the synthesis of PEG-SeNPs. Additional peaks at 470 and 475 nm proved conjugation of crocin onto SeNPs. The size of these spherical NPs was in the range of 40 to 50 nm [285].

8.5 Conclusion

Nanotechnology has developed in human nutrition and the food industry in the areas of improvement of the quality of nutrition, control, and prophylaxis of a wide range of diseases. By delivering of high concentrations of nano-formula drugs to disease sites the effective materials can: produce a quicker dissolution in the human blood; persist at the target sites for the longer periods; and as a result, decrease the side effects. Of course, a decrease in size could be responsible for several material interactions that could lead to toxicological effects. So, excessive use of NPs in human food and related fields can become out of control and cause problems. Try to obtain nontoxic, good quality, effective nanomaterials in a good quality nano formulation to help humans achieve these goals.

References

1 Nickols-Richardson, S. and Piehowski, K. (2008). Nanotechnology in nutritional sciences. *Minerva Biotecnologica* 20 (3): 117.

2 Yu, H., Park, J.-Y., Kwon, C.W. et al. (2018). An overview of nanotechnology in food science: preparative methods, practical applications, and safety. *Journal of Chemistry* 2018: 20180101.

3 Levy, O., Brennen, W.N., Han, E. et al. (2016). A prodrug-doped cellular Trojan horse for the potential treatment of prostate cancer. *Biomaterials* 91: 140–150.

4 Yaktine, A. and Pray, L. (2009). *Nanotechnology in Food Products: Workshop Summary*. National Academies Press.

5 Bhosale, R.R., Ghodake, P.P., Mane, A. et al. (2013). Nanocochleates: a novel carrier for drug transfer. *Journal of Scientific and Industrial Research* 2: 964–969.

6 Zhu, M., Nie, G., Meng, H. et al. (2012). Physicochemical properties determine nanomaterial cellular uptake, transport, and fate. *Accounts of Chemical Research* 46 (3): 622–631.

7 Singh, T., Shukla, S., Kumar, P. et al. (2017). Application of nanotechnology in food science: perception and overview. *Frontiers in Microbiology* 8: 1501.

8 Chaudhry, Q., Scotter, M., Blackburn, J. et al. (2008). Applications and implications of nanotechnologies for the food sector. *Food Additives and Contaminants* 25 (3): 241–258.

9 Diarrassouba, F., Garrait, G., Remondetto, G. et al. (2015). Food protein-based microspheres for increased uptake of vitamin D3. *Food Chemistry* 173: 1066–1072.

10 Yao, M., McClements, D.J., and Xiao, H. (2015). Improving oral bioavailability of nutraceuticals by engineered nanoparticle-based delivery systems. *Current Opinion in Food Science* 2: 14–19.

11 Momin, J.K., Jayakumar, C., and Prajapati, J.B. (2013). Potential of nanotechnology in functional foods. *Emirates Journal of Food and Agriculture (EJFA)* 25 (1): 10–19.

12 Assadpour, E. and Jafari, S.M. (2019). A systematic review on nanoencapsulation of food bioactive ingredients and nutraceuticals by various nanocarriers. *Critical Reviews in Food Science and Nutrition* 59 (19): 3129–3151.

13 Chamundeeswari, M., Jeslin, J., and Verma, M.L. (2019). Nanocarriers for drug delivery applications. *Environmental Chemistry Letters* 17 (2): 849–865.

14 Mishra, D.K., Shandilya, R., and Mishra, P.K. (2018). Lipid based nanocarriers: a translational perspective. *Nanomedicine: Nanotechnology, Biology and Medicine* 14 (7): 2023–2050.

15 Fan, S., Zheng, Y., Liu, X. et al. (2018). Curcumin-loaded PLGA-PEG nanoparticles conjugated with B6 peptide for potential use in Alzheimer's disease. *Drug Delivery* 25 (1): 1091–1102.

16 Thompson, K.L., Williams, M., and Armes, S.P. (2015). Colloidosomes: synthesis, properties and applications. *Journal of Colloid and Interface Science* 447: 217–228.

17 Thiruvengadam, M., Rajakumar, G., and Chung, I.-M. (2018). Nanotechnology: current uses and future applications in the food industry. *3 Biotech* 8 (1): 74.

18 Azadmanjiri, J., Berndt, C.C., Wang, J. et al. (2016). Nanolaminated composite materials: structure, interface role and applications. *RSC Advances* 6 (111): 109361–109385.

19 Dinsmore, A., Hsu, M.F., Nikolaides, M. et al. (2002). Colloidosomes: selectively permeable capsules composed of colloidal particles. *Science* 298 (5595): 1006–1009.

20 Niaz, T., Shabbir, S., Noor, T. et al. (2018). Polyelectrolyte multicomponent colloidosomes loaded with nisin Z for enhanced antimicrobial activity against foodborne resistant pathogens. *Frontiers in Microbiology* 8: 2700.

21 Garg, G., Saraf, S., and Saraf, S. (2007). Cubosomes: an overview. *Biological and Pharmaceutical Bulletin* 30 (2): 350–353.

22 Caltagirone, C., Falchi, A.M., Lampis, S. et al. (2014). Cancer-cell-targeted theranostic cubosomes. *Langmuir* 30 (21): 6228–6236.

23 Karami, Z. and Hamidi, M. (2016). Cubosomes: remarkable drug delivery potential. *Drug Discovery Today* 21 (5): 789–801.

24 Saber, M.M., Al-Mahallawi, A.M., Nassar, N.N. et al. (2018). Targeting colorectal cancer cell metabolism through development of cisplatin and metformin nano-cubosomes. *BMC Cancer* 18 (1): 822.

25 Kates, M. (ed.) (1992). Archaebacterial lipids: structure, biosynthesis and function. *Biochemical Society Symposium* 58: 51–72.

26 Sprott, G.D., Patel, G.B., and Makabi-Panzu, B (2002). Archaesomes, archaeosomes containing coenzyme Q10 and other types of liposomes containing coenzyme Q10 adjuvants and as delivery vehicles. Google Patents: US Patent no. 6403117B1.

27 Patel, G.B. and Sprott, G.D. (1999). Archaeobacterial ether lipid liposomes (archaeosomes) as novel vaccine and drug delivery systems. *Critical Reviews in Biotechnology* 19 (4): 317–357.

28 Patel, G.B., Agnew, B.J., Deschatelets, L. et al. (2000). in vitro assessment of archaeosome stability for developing oral delivery systems. *International Journal of Pharmaceutics* 194 (1): 39–49.

29 Ahmad, M.U. (2015). *Lipids in Nanotechnology*. Elsevier.

30 Scheffel, U., Rhodes, B.A., Natarajan, T. et al. (1972). Albumin microspheres for study of the reticuloendothelial system. *Journal of Nuclear Medicine* 13 (7): 498–503.

31 Birrenbach, G. and Speiser, P. (1976). Polymerized micelles and their use as adjuvants in immunology. *Journal of Pharmaceutical Sciences* 65 (12): 1763–1766.

32 Kreuter, J. and Speiser, P.P. (1976). New adjuvants on a polymethylmethacrylate base. *Infection and Immunity* 13 (1): 204–210.

33 Couvreur, P., Kante, B., Roland, M. et al. (1979). Polycyanoacrylate nanocapsules as potential lysosomotropic carriers: preparation, morphological and sorptive properties. *Journal of Pharmacy and Pharmacology* 31 (1): 331–332.

34 Gurny, R., Peppas, N., Harrington, D. et al. (1981). Development of biodegradable and injectable latices for controlled release of potent drugs. *Drug Development and Industrial Pharmacy* 7 (1): 1–25.

35 Vauthier-Holtzscherer, C., Benabbou, S., Spenlehauer, G. et al. (1991). Methodology for the preparation of ultra-dispersed polymer systems. *STP Pharma Sciences* 1 (2): 109–116.

36 Zhang, H. (2017). Thin-film hydration followed by extrusion method for liposome preparation. In: *Liposomes*, 17–22. Springer.

37 Amaranth, S. and Uma, S. (1997). Liposome in drug delivery: progress and limitation. *International Journal of Pharmaceutics* 154: 123–140.

38 Merino, M., Zalba, S., and Garrido, M.J. (2018). Immunoliposomes in clinical oncology: state of the art and future perspectives. *Journal of Controlled Release* 275: 162–176.

39 Fleddermann, J., Diamanti, E., Azinas, S. et al. (2016). Virosome engineering of colloidal particles and surfaces: bioinspired fusion to supported lipid layers. *Nanoscale* 8 (15): 7933–7941.

40 Zhou, W., Liu, W., Zou, L. et al. (2014). Storage stability and skin permeation of vitamin C liposomes improved by pectin coating. *Colloids and Surfaces. B, Biointerfaces* 117: 330–337.

41 Tan, C., Xue, J., Abbas, S. et al. (2014). Liposome as a delivery system for carotenoids: comparative antioxidant activity of carotenoids as measured by ferric reducing antioxidant power, DPPH assay and lipid peroxidation. *Journal of Agricultural and Food Chemistry* 62 (28): 6726–6735.

42 Sherry, M., Charcosset, C., Fessi, H. et al. (2013). Essential oils encapsulated in liposomes: a review. *Journal of Liposome Research* 23 (4): 268–275.

43 Liu, W., Liu, J., Liu, W. et al. (2013). Improved physical and in vitro digestion stability of a polyelectrolyte delivery system based on layer-by-layer self-assembly alginate–chitosan-coated nanoliposomes. *Journal of Agricultural and Food Chemistry* 61 (17): 4133–4144.

44 Gao, W., Vecchio, D., Li, J. et al. (2014). Hydrogel containing nanoparticle-stabilized liposomes for topical antimicrobial delivery. *ACS Nano* 8 (3): 2900–2907.

45 Khanniri, E., Bagheripoor-Fallah, N., Sohrabvandi, S. et al. (2016). Application of liposomes in some dairy products. *Critical Reviews in Food Science and Nutrition* 56 (3): 484–493.

46 Weiss, J., Takhistov, P., and McClements, D.J. (2006). Functional materials in food nanotechnology. *Journal of Food Science* 71 (9): R107–R116.

47 Cagri, A., Ustunol, Z., and Ryser, E.T. (2004). Antimicrobial edible films and coatings. *Journal of Food Protection* 67 (4): 833–848.

48 Morillon, V., Debeaufort, F., Blond, G. et al. (2002). Factors affecting the moisture permeability of lipid-based edible films: a review. *Critical Reviews in Food Science and Nutrition* 42 (1): 67–89.

49 Rocco, M., William, R., and Alivisiatos, P. (1999). *Nanotechnology Research Directions: IWGN Workshop Report*. Springer/National Science and Technology Council.

50 Frenot, A. and Chronakis, I. (2003). Polymer nanofibers assembled by electrospinning. *Current Opinion in Colloid & Interface Science* 8: 64–75.

51 Clark, S.L. and Hammond, P.T. (1998). Engineering the microfabrication of layer-by-layer thin films. *Advanced Materials* 10 (18): 1515–1519.

52 Visser, J., Melchels, F., Jeon, J. et al. (2015). Reinforcement of hydrogels using three-dimensionally printed microfibres. *Nature Communications* 6: 6933.

53 Graham, K., Gogins, M., and Schreuder-Gibson, H. (2004). Incorporation of electrospun nanofibers into functional structures. *International Nonwovens Journal* 2: https://doi.org/10.1177/1558925004os-1300209.

54 Sen, K. and Babu, K.M. (2004). Studies on Indian silk. I. Macrocharacterization and analysis of amino acid composition. *Journal of Applied Polymer Science* 92 (2): 1080–1097.

55 Huang, C., Wang, S., Zhang, H. et al. (2006). High strength electrospun polymer nanofibers made from BPDA–PDA polyimide. *European Polymer Journal* 42 (5): 1099–1104.

56 Rezaei, A., Nasirpour, A., and Fathi, M. (2015). Application of cellulosic nanofibers in food science using electrospinning and its potential risk. *Comprehensive Reviews in Food Science and Food Safety* 14 (3): 269–284.

57 Smith, G.B. (2011). Green nanotechnology. In: *Proceedings SPIE 8104, Nanostructured Thin Films IV* (ed. R.J. Martín-Palma, Y.-J. Jen and T.G. Mackay). International Society for Optics and Photonics https://doi.org/10.1117/12.893114.

58 Abbasi, E., Aval, S.F., Akbarzadeh, A. et al. (2014). Dendrimers: synthesis, applications, and properties. *Nanoscale Research Letters* 9 (1): 247.

59 Tomalia, D.A. and Fréchet, J.M. (2002). Discovery of dendrimers and dendritic polymers: a brief historical perspective. *Journal of Polymer Science Part A: Polymer Chemistry* 40 (16): 2719–2728.

60 Le, N.T.T., Nguyen, T.N.Q., Cao, V.D. et al. (2019). Recent progress and advances of multi-stimuli-responsive dendrimers in drug delivery for cancer treatment. *Pharmaceutics* 11 (11): 591.

61 Jebelli, M.A., Kalantar, E., Maleki, A. et al. (2015). Antimicrobial activities of the polypropylene imine dendrimer aginst bacteria isolated from rural water resources. *Jundishapur Journal of Natural Pharmaceutical Products* 10 (3).

62 Mauri, E., Negri, A., Rebellato, E. et al. (2018). Hydrogel-nanoparticles composite system for controlled drug delivery. *Gels* 4 (3): 74.

63 Hamidi, M., Azadi, A., and Rafiei, P. (2008). Hydrogel nanoparticles in drug delivery. *Advanced Drug Delivery Reviews* 60 (15): 1638–1649.

64 Rezaei, F. and Shahbazi, Y. (2018). Shelf-life extension and quality attributes of sauced silver carp fillet: a comparison among direct addition, edible coating and biodegradable film. *LWT* 87: 122–133.

65 Tavassoli-Kafrani, E., Shekarchizadeh, H., and Masoudpour-Behabadi, M. (2016). Development of edible films and coatings from alginates and carrageenans. *Carbohydrate Polymers* 137: 360–374.

66 Song, S., Cheong, L.-Z., Falkeborg, M. et al. (2013). Facile synthesis of phosphatidyl saccharides for preparation of anionic nanoliposomes with enhanced stability. *PLoS One* 8 (9): e73891.

67 Falkeborg, M., Cheong, L.-Z., Gianfico, C. et al. (2014). Alginate oligosaccharides: enzymatic preparation and antioxidant property evaluation. *Food Chemistry* 164: 185–194.

68 Falkeborg, M. and Guo, Z. (2015). Dodecenyl succinylated alginate (DSA) as a novel dual-function emulsifier for improved fish oil-in-water emulsions. *Food Hydrocolloids* 46: 10–18.

69 Kraan, S. (2012). Algal polysaccharides, novel applications and outlook. In: *Carbohydrates-Comprehensive Studies on Glycobiology and Glycotechnology* (ed. C.-F. Chang), 489. IntechOpen.

70 Brownlee, I., Allen, A., Pearson, J. et al. (2005). Alginate as a source of dietary fiber. *Critical Reviews in Food Science and Nutrition* 45 (6): 497–510.

71 Idota, Y., Kogure, Y., Kato, T. et al. (2016). Relationship between physical parameters of various metal ions and binding affinity for alginate. *Biological and Pharmaceutical Bulletin* 39 (11): 1893–1896.

72 Črnivec, I.G.O. and Ulrih, N.P. (2019). Nano-hydrogels of alginate for encapsulation of food ingredients. In: *Biopolymer Nanostructures for Food Encapsulation Purposes* (ed. S.M. Jafari), 335–380. Elsevier.

73 Alborzi, S., Lim, L.T., and Kakuda, Y. (2010). Electrospinning of sodium alginate-pectin ultrafine fibers. *Journal of Food Science* 75 (1): C100–C107.

74 Alborzi, S., Lim, L.-T., and Kakuda, Y. (2013). Encapsulation of folic acid and its stability in sodium alginate-pectin-poly (ethylene oxide) electrospun fibres. *Journal of Microencapsulation* 30 (1): 64–71.

75 Liu, W., Tian, M., Kong, Y. et al. (2017). Multilayered vitamin C nanoliposomes by self-assembly of alginate and chitosan: long-term stability and feasibility application in mandarin juice. *LWT.* 75: 608–615.

76 Sharma, M. (2019). Transdermal and intravenous Nano drug delivery systems: present and future. In: *Applications of Targeted Nano Drugs and Delivery Systems* (ed. S.S. Mohapatra, S. Ranjan, N. Dasgupta, et al.), 499–550. Elsevier.

77 Sabliov, C.M. and Astete, C.E. (2015). Polymeric nanoparticles for food applications. In: *Nanotechnology and Functional Foods: Effective Delivery of Bioactive Ingredients*, vol. 272 (ed. C.M. Sabilov, H. Chen and R.Y. Yada), 296.

78 Cristina, M., Sabilo, H.C., and Rickey, Y. (2015). Polymeric nanoparticles for food applications. In: *Nanotechnology and Functional Foods: Effective Delivery of Bioactive Ingredients*. Chichester: Wiley.

79 Rana, V. and Sharma, R. (2019). Recent advances in development of Nano drug delivery. In: *Applications of Targeted Nano Drugs and Delivery Systems* (ed. S.S. Mohapatra, S. Ranjan, N. Dasgupta, et al.), 93–131. Elsevier.

80 Martău, G.A., Mihai, M., and Vodnar, D.C. (2019). The use of chitosan, alginate, and pectin in the biomedical and food sector—biocompatibility, bioadhesiveness, and biodegradability. *Polymers* 11 (11): 1837.

81 Hasenhuettl, G.L. and Hartel, R.W. (2008). *Food Emulsifiers and their Applications*. Springer.

82 Pan, M., Yin, Z., Liu, K. et al. (2019). Carbon-based nanomaterials in sensors for food safety. *Nanomaterials* 9 (9): 1330.

83 Das, R., Ali, M.E., Hamid, S.B.A. et al. (2014). Carbon nanotube membranes for water purification: a bright future in water desalination. *Desalination* 336: 97–109.

84 Qu, X., Alvarez, P.J., and Li, Q. (2013). Applications of nanotechnology in water and wastewater treatment. *Water Research* 47 (12): 3931–3946.

85 Liu, X., Wang, M., Zhang, S. et al. (2013). Application potential of carbon nanotubes in water treatment: a review. *Journal of Environmental Sciences* 25 (7): 1263–1280.

86 Smith, S.C. and Rodrigues, D.F. (2015). Carbon-based nanomaterials for removal of chemical and biological contaminants from water: a review of mechanisms and applications. *Carbon* 91: 122–143.

87 Bonilla, J.C., Bozkurt, F., Ansari, S. et al. (2016). Applications of quantum dots in food science and biology. *Trends in Food Science and Technology* 53: 75–89.

88 Claussen, J.C., Malanoski, A., Breger, J.C. et al. (2015). Probing the enzymatic activity of alkaline phosphatase within quantum dot bioconjugates. *The Journal of Physical Chemistry C* 119 (4): 2208–2221.

89 Fan, H., Zhang, M., Bhandari, B. et al. (2019). Food waste as a carbon source in carbon quantum dots technology and their applications in food safety detection. *Trends in Food Science & Technology*.

90 Zhu, H., Wang, X., Li, Y. et al. (2009). Microwave synthesis of fluorescent carbon nanoparticles with electrochemiluminescence properties. *Chemical Communications* 34: 5118–5120.

91 Pei, S., Zhang, J., Gao, M. et al. (2015). A facile hydrothermal approach towards photoluminescent carbon dots from amino acids. *Journal of Colloid and Interface Science* 439: 129–133.

92 Liu, W., Li, C., Sun, X. et al. (2017). Highly crystalline carbon dots from fresh tomato: UV emission and quantum confinement. *Nanotechnology* 28 (48): 485705.

93 Zhao, S., Lan, M., Zhu, X. et al. (2015). Green synthesis of bifunctional fluorescent carbon dots from garlic for cellular imaging and free radical scavenging. *ACS Applied Materials and Interfaces* 7 (31): 17054–17060.

94 Xu, Y., Fan, Y., Zhang, L. et al. (2019). A novel enhanced fluorescence method based on multifunctional carbon dots for specific detection of Hg2+ in complex samples. *Spectrochimica Acta Part A: Molecular and Biomolecular Spectroscopy* 220: 117109.

95 Yue, X., Chun-Jing, T., Huang, H. et al. (2014). Green synthesis of fluorescent carbon quantum dots for detection of Hg2+. *Chinese Journal of Analytical Chemistry* 42 (9): 1252–1258.

96 Wang, Z., Yu, J., Zhang, X. et al. (2016). Large-scale and controllable synthesis of graphene quantum dots from rice husk biomass: a comprehensive utilization strategy. *ACS Applied Materials and Interfaces* 8 (2): 1434–1439.

97 Komaiko, J.S. and McClements, D.J. (2016). Formation of food-grade nanoemulsions using low-energy preparation methods: a review of available methods. *Comprehensive Reviews in Food Science and Food Safety* 15 (2): 331–352.

98 Sun, Y., Xia, Z., Zheng, J. et al. (2015). Nanoemulsion based delivery systems for nutraceuticals: influence of carrier oil type on bioavailability of pterostilbene. *Journal of Functional Foods* 13: 61–70.

99 Choi, A.-J., Kim, C.-J., Cho, Y.-J. et al. (2011). Characterization of capsaicin-loaded nanoemulsions stabilized with alginate and chitosan by self-assembly. *Food and Bioprocess Technology* 4 (6): 1119–1126.

100 Pradhan, N., Singh, S., Ojha, N. et al. (2015). Facets of nanotechnology as seen in food processing, packaging, and preservation industry. *BioMed Research International* 15: 365672.

101 Singh, N. (2015). An overview of the prospective application of nanoemulsions in foodstuffs and food packaging. *ASIO Journal of Microbiology, Food Science and Biotechnological Innovations* 1 (1): 20–25.

102 Jakubek, M., Kejík, Z., Kaplánek, R. et al. (2019). Strategy for improved therapeutic efficiency of curcumin in the treatment of gastric cancer. *Biomedicine and Pharmacotherapy* 118: 109278.

103 Gera, M., Sharma, N., Ghosh, M. et al. (2017). Nanoformulations of curcumin: an emerging paradigm for improved remedial application. *Oncotarget* 8 (39): 66680.

104 Ghalandarlaki, N., Alizadeh, A.M., and Ashkani-Esfahani, S. (2014). Nanotechnology-applied curcumin for different diseases therapy. *BioMed Research International* 2014: 1–23.

105 Noorafshan, A. and Ashkani-Esfahani, S. (2013). A review of therapeutic effects of curcumin. *Current Pharmaceutical Design* 19 (11): 2032–2046.

106 Hosseinzadeh, H., Karimi, G., and Niapoor, M. (2003). Antidepressant effect of *Crocus sativus* L. stigma extracts and their constituents, crocin and safranal, in mice. *International Symposium on Saffron Biology and Biotechnology* 650.

107 Hosseinzadeh, H. (2014). Saffron: a herbal medicine of third millennium. *Jundishapur Journal of Natural Pharmaceutical Products* 9 (1): 1.

108 Tashakori-Sabzevar, F., Hosseinzadeh, H., Sadat Motamedshariaty, V. et al. (2013). Crocetin attenuates spatial learning dysfunction and hippocampal injury in a model of vascular dementia. *Current Neurovascular Research* 10 (4): 325–334.

109 Khorasany, A.R. and Hosseinzadeh, H. (2016). Therapeutic effects of saffron (*Crocus sativus* L.) in digestive disorders: a review. *Iranian Journal of Basic Medical Sciences* 19 (5): 455.

110 Jugdaohsingh, R., Tucker, K.L., Qiao, N. et al. (2004). Dietary silicon intake is positively associated with bone mineral density in men and premenopausal women of the Framingham offspring cohort. *Journal of Bone and Mineral Research* 19 (2): 297–307.

111 Chaudhry, Q., Watkins, R., and Castle, L. (2010). Nanotechnologies in the food arena: new opportunities, new questions, new concerns. *Nanotechnologies in Food* 14: 1–17.

112 Urrutia-Ortega, I.M., Garduño-Balderas, L.G., Delgado-Buenrostro, N.L. et al. (2016). Food-grade titanium dioxide exposure exacerbates tumor formation in colitis associated cancer model. *Food and Chemical Toxicology* 93: 20–31.

113 Chen, H., Wang, B., Gao, D. et al. (2013). Broad-spectrum antibacterial activity of carbon nanotubes to human gut bacteria. *Small* 9 (16): 2735–2746.

114 Zhao, B., Hu, H., Mandal, S.K. et al. (2005). A bone mimic based on the self-assembly of hydroxyapatite on chemically functionalized single-walled carbon nanotubes. *Chemistry of Materials* 17 (12): 3235–3241.

115 Shao, N., Lu, S., Wickstrom, E., and Panchapakesan, B. (2007). Integrated molecular targeting of IGF1R and HER2 surface receptors and destruction of breast cancer cells using single wall carbon nanotubes. *Nanotechnology* 18 (31): 315101.

116 Zhao, C.-Y., Tan, S.-X., Xiao, X.-Y. et al. (2014). Effects of dietary zinc oxide nanoparticles on growth performance and antioxidative status in broilers. *Biological Trace Element Research* 160 (3): 361–367.

117 Ng, K.W., Khoo, S.P., Heng, B.C. et al. (2011). The role of the tumor suppressor p53 pathway in the cellular DNA damage response to zinc oxide nanoparticles. *Biomaterials* 32 (32): 8218–8225.

118 Taccola, L., Raffa, V., Riggio, C. et al. (2011). Zinc oxide nanoparticles as selective killers of proliferating cells. *International Journal of Nanomedicine* 6: 1129.

119 Thurber, A., Wingett, D.G., Rasmussen, J.W. et al. (2012). Improving the selective cancer killing ability of ZnO nanoparticles using Fe doping. *Nanotoxicology* 6 (4): 440–452.

120 Tiwari, V., Mishra, N., Gadani, K. et al. (2018). Mechanism of anti-bacterial activity of zinc oxide nanoparticle against carbapenem-resistant Acinetobacter baumannii. *Frontiers in Microbiology* 9: 1218.

121 Hosnedlova, B., Kepinska, M., Skalickova, S. et al. (2018). Nano-selenium and its nanomedicine applications: a critical review. *International Journal of Nanomedicine* 13: 2107.

122 Sarkar, B., Bhattacharjee, S., Daware, A. et al. (2015). Selenium nanoparticles for stress-resilient fish and livestock. *Nanoscale Research Letters* 10 (1): 371.

123 Ahmed, H.H., El-Maksoud, M.D.A., Moneim, A.E.A. et al. (2017). Pre-clinical study for the antidiabetic potential of selenium nanoparticles. *Biological Trace Element Research* 177 (2): 267–280.

124 Skalickova, S., Milosavljevic, V., Cihalova, K. et al. (2017). Selenium nanoparticles as a nutritional supplement. *Nutrition* 33: 83–90.

125 Al-Quraishy, S., Dkhil, M.A., and Moneim, A.E.A. (2015). Anti-hyperglycemic activity of selenium nanoparticles in streptozotocin-induced diabetic rats. *International Journal of Nanomedicine* 10: 6741.

126 Yarjanli, Z., Ghaedi, K., Esmaeili, A. et al. (2017). Iron oxide nanoparticles may damage to the neural tissue through iron accumulation, oxidative stress, and protein aggregation. *BMC Neuroscience* 18 (1): 51.

127 Jhan, H.-J., Liu, J.-J., Chen, Y.-C. et al. (2015). Novel injectable thermosensitive hydrogels for delivering hyaluronic acid–doxorubicin nanocomplexes to locally treat tumors. *Nanomedicine* 10 (8): 1263–1274.

128 Burduşel, A.-C., Gherasim, O., Grumezescu, A.M. et al. (2018). Biomedical applications of silver nanoparticles: an up-to-date overview. *Nanomaterials* 8 (9): 681.

129 Alshareef, A., Laird, K., and Cross, R. (2017). Shape-dependent antibacterial activity of silver nanoparticles on *Escherichia coli* and *Enterococcus faecium* bacterium. *Applied Surface Science* 424: 310–315.

130 Etemadzade, M., Ghamarypour, A., Zabihollahi, R. et al. (2016). Synthesis and evaluation of antiviral activities of novel sonochemical silver nanorods against HIV and HSV viruses. *Asian Pacific Journal of Tropical Disease* 6 (11): 854–858.

131 Kalaivani, R., Maruthupandy, M., Muneeswaran, T. et al. (2018). Synthesis of chitosan mediated silver nanoparticles (ag NPs) for potential antimicrobial applications. *Frontiers in Laboratory Medicine* 2 (1): 30–35.

132 Zhang, X.-F., Liu, Z.-G., Shen, W. et al. (2016). Silver nanoparticles: synthesis, characterization, properties, applications, and therapeutic approaches. *International Journal of Molecular Sciences* 17 (9): 1534.

133 Thapa, R.K., Kim, J.H., Jeong, J.-H. et al. (2017). Silver nanoparticle-embedded graphene oxide-methotrexate for targeted cancer treatment. *Colloids and Surfaces. B, Biointerfaces* 153: 95–103.

134 Rajeshkumar, S., Malarkodi, C., Vanaja, M. et al. (2016). Anticancer and enhanced antimicrobial activity of biosynthesizd silver nanoparticles against clinical pathogens. *Journal of Molecular Structure* 1116: 165–173.

135 Khurana, A., Tekula, S., Saifi, M.A. et al. (2019). Therapeutic applications of selenium nanoparticles. *Biomedicine and Pharmacotherapy* 111: 802–812.

136 O'Farrell, N., Houlton, A., and Horrocks, B.R. (2006). Silicon nanoparticles: applications in cell biology and medicine. *International Journal of Nanomedicine* 1 (4): 451.

137 Quinto, C.A., Mohindra, P., Tong, S. et al. (2015). Multifunctional superparamagnetic iron oxide nanoparticles for combined chemotherapy and hyperthermia cancer treatment. *Nanoscale* 7 (29): 12728–12736.

138 Hafezi Ghahestani, Z., Alebooye Langroodi, F., Mokhtarzadeh, A. et al. (2017). Evaluation of anti-cancer activity of PLGA nanoparticles containing crocetin. *Artificial Cells, Nanomedicine, and Biotechnology* 45 (5): 955–960.

139 Tiwari, S.K., Agarwal, S., Seth, B. et al. (2014). Curcumin-loaded nanoparticles potently induce adult neurogenesis and reverse cognitive deficits in Alzheimer's disease model via canonical Wnt/β-catenin pathway. *ACS Nano* 8 (1): 76–103.

140 Kunnumakkara, A.B., Bordoloi, D., Padmavathi, G. et al. (2017). Curcumin, the golden nutraceutical: multitargeting for multiple chronic diseases. *British Journal of Pharmacology* 174 (11): 1325–1348.

141 Altenburg, J.D., Bieberich, A.A., Terry, C. et al. (2011). A synergistic antiproliferation effect of curcumin and docosahexaenoic acid in SK-BR-3 breast cancer cells: unique signaling not explained by the effects of either compound alone. *BMC Cancer* 11 (1): 149.

142 Bordoloi, D., Roy, N.K., Monisha, J. et al. (2016). Multi-targeted agents in cancer cell chemosensitization: what we learnt from curcumin thus far. *Recent Patents on Anti-Cancer Drug Discovery* 11 (1): 67–97.

143 Tsang, C.K., Kelly, T.L., Sailor, M.J. et al. (2012). Highly stable porous silicon–carbon composites as label-free optical biosensors. *ACS Nano* 6 (12): 10546–10554.

144 Bhullar, K.S., Jha, A., Youssef, D. et al. (2013). Curcumin and its carbocyclic analogs: structure-activity in relation to antioxidant and selected biological properties. *Molecules* 18 (5): 5389–5404.

145 Pacher, P., Beckman, J.S., and Liaudet, L. (2007). Nitric oxide and peroxynitrite in health and disease. *Physiological Reviews* 87 (1): 315–424.

146 Kukongviriyapan, U., Pannangpetch, P., Kukongviriyapan, V. et al. (2014). Curcumin protects against cadmium-induced vascular dysfunction, hypertension and tissue cadmium accumulation in mice. *Nutrients* 6 (3): 1194–1208.

147 Basniwal, R.K., Buttar, H.S., Jain, V. et al. (2011). Curcumin nanoparticles: preparation, characterization, and antimicrobial study. *Journal of Agricultural and Food Chemistry* 59 (5): 2056–2061.

148 Anand, P., Nair, H.B., Sung, B. et al. (2010). *RETRACTED: Design of Curcumin-Loaded PLGA Nanoparticles Formulation with Enhanced Cellular Uptake, and Increased Bioactivity in vitro and Superior Bioavailability in vivo*. Elsevier.

149 Wang, D., Veena, M.S., Stevenson, K. et al. (2008). Liposome-encapsulated curcumin suppresses growth of head and neck squamous cell carcinoma in vitro and in xenografts through the inhibition of nuclear factor κB by an AKT-independent pathway. *Clinical Cancer Research* 14 (19): 6228–6236.

150 Gupta, V., Aseh, A., Ríos, C.N. et al. (2009). Fabrication and characterization of silk fibroin-derived curcumin nanoparticles for cancer therapy. *International Journal of Nanomedicine* 4: 115.

151 Das, R.K., Kasoju, N., and Bora, U. (2010). Encapsulation of curcumin in alginate-chitosan-pluronic composite nanoparticles for delivery to cancer cells. *Nanomedicine: Nanotechnology, Biology and Medicine* 6 (1): 153–160.

152 Maiti, K., Mukherjee, K., Gantait, A. et al. (2007). Curcumin–phospholipid complex: preparation, therapeutic evaluation and pharmacokinetic study in rats. *International Journal of Pharmaceutics* 330 (1–2): 155–163.

153 Yallapu, M.M., Jaggi, M., and Chauhan, S.C. (2010). Poly (β-cyclodextrin)/curcumin self-assembly: a novel approach to improve curcumin delivery and its therapeutic efficacy in prostate cancer cells. *Macromolecular Bioscience* 10 (10): 1141–1151.

154 Shin, G.H., Li, J., Cho, J.H. et al. (2016). Enhancement of curcumin solubility by phase change from crystalline to amorphous in cur-TPGS nanosuspension. *Journal of Food Science* 81 (2): N494–N501.

155 Modasiya, M. and Patel, V. (2012). Studies on solubility of curcumin. *International Journal of Pharmaceutical and Life Sciences* 3: 1490–1497.

156 Huang, N., Lu, S., Liu, X.-G. et al. (2017). PLGA nanoparticles modified with a BBB-penetrating peptide co-delivering Aβ generation inhibitor and curcumin attenuate memory deficits and neuropathology in Alzheimer's disease mice. *Oncotarget* 8 (46): 81001.

157 Betbeder, D., Lipka, E., Howsam, M. et al. (2015). Evolution of availability of curcumin inside poly-lactic-co-glycolic acid nanoparticles: impact on antioxidant and antinitrosant properties. *International Journal of Nanomedicine* 10: 5355.

158 Shome, S., Talukdar, A.D., Choudhury, M.D. et al. (2016). Curcumin as potential therapeutic natural product: a nanobiotechnological perspective. *The Journal of Pharmacy and Pharmacology* 68 (12): 1481–1500.

159 Dhivya, R., Ranjani, J., Rajendhran, J. et al. (2015). pH responsive curcumin/ZnO nanocomposite for drug delivery. *Advanced Materials Letters* 6: 505–512.

160 Yallapu, M.M., Ebeling, M.C., Khan, S. et al. (2013). Novel curcumin-loaded magnetic nanoparticles for pancreatic cancer treatment. *Molecular Cancer Therapeutics* 12 (8): 1471–1480.

161 Sun, D., Zhuang, X., Xiang, X. et al. (2010). A novel nanoparticle drug delivery system: the anti-inflammatory activity of curcumin is enhanced when encapsulated in exosomes. *Molecular Therapy* 18 (9): 1606–1614.

162 Dilnawaz, F., Singh, A., and Sahoo, S.K. (2011). Retraction notice to "transferrin-conjugated curcumin-loaded superparamagnetic iron oxide nanoparticles induce augmented cellular uptake and apoptosis in K562 cells". *Acta Biomaterialia* 8: 704–719. Elsevier; 2012.

163 Wang, W., Zhu, R., Xie, Q. et al. (2012). Enhanced bioavailability and efficiency of curcumin for the treatment of asthma by its formulation in solid lipid nanoparticles. *International Journal of Nanomedicine* 7: 3667.

164 Rahman, S., Cao, S., Steadman, K.J. et al. (2012). Native and β-cyclodextrin-enclosed curcumin: entrapment within liposomes and their in vitro cytotoxicity in lung and colon cancer. *Drug Delivery* 19 (7): 346–353.

165 S.-H., S., Gao, X., Li, D. et al. (2012). A systemic administration of liposomal curcumin inhibits radiation pneumonitis and sensitizes lung carcinoma to radiation. *International Journal of Nanomedicine* 7: 2601.

166 Singh, D.K., Jagannathan, R., Khandelwal, P. et al. (2013). In situ synthesis and surface functionalization of gold nanoparticles with curcumin and their antioxidant properties: an experimental and density functional theory investigation. *Nanoscale* 5 (5): 1882–1893.

167 Varaprasad, K., Mohan, Y.M., Vimala, K. et al. (2011). Synthesis and characterization of hydrogel-silver nanoparticle-curcumin composites for wound dressing and antibacterial application. *Journal of Applied Polymer Science* 121 (2): 784–796.

168 Zhang, J., Tang, Q., Xu, X. et al. (2013). Development and evaluation of a novel phytosome-loaded chitosan microsphere system for curcumin delivery. *International Journal of Pharmaceutics* 448 (1): 168–174.

169 Akhtar, F., Rizvi, M.M.A., and Kar, S.K. (2012). Oral delivery of curcumin bound to chitosan nanoparticles cured Plasmodium yoelii infected mice. *Biotechnology Advances* 30 (1): 310–320.

170 Li, X., Chen, S., Zhang, B. et al. (2012). In situ injectable nano-composite hydrogel composed of curcumin, N, O-carboxymethyl chitosan and oxidized alginate for wound healing application. *International Journal of Pharmaceutics* 437 (1–2): 110–119.

171 Yadav, A., Lomash, V., Samim, M. et al. (2012). Curcumin encapsulated in chitosan nanoparticles: a novel strategy for the treatment of arsenic toxicity. *Chemico-Biological Interactions* 199 (1): 49–61.

172 Shukla, S.K., Mishra, A.K., Arotiba, O.A. et al. (2013). Chitosan-based nanomaterials: a state-of-the-art review. *International Journal of Biological Macromolecules* 59: 46–58.

173 Karewicz, A., Bielska, D., Loboda, A. et al. (2013). Curcumin-containing liposomes stabilized by thin layers of chitosan derivatives. *Colloids and Surfaces. B, Biointerfaces* 109: 307–316.

174 Debnath, S., Saloum, D., Dolai, S. et al. (2013). Dendrimer-curcumin conjugate: a water soluble and effective cytotoxic agent against breast cancer cell lines. *Anti-Cancer Agents in Medicinal Chemistry (Formerly Current Medicinal Chemistry-Anti-Cancer Agents)* 13 (10): 1531–1539.

175 Yallapu, M.M., Ebeling, M.C., Chauhan, N. et al. (2011). Interaction of curcumin nanoformulations with human plasma proteins and erythrocytes. *International Journal of Nanomedicine* 6: 2779.

176 Cao, J., Zhang, H., Wang, Y. et al. (2013). Investigation on the interaction behavior between curcumin and PAMAM dendrimer by spectral and docking studies. *Spectrochimica Acta Part A: Molecular and Biomolecular Spectroscopy* 108: 251–255.

177 Mollazade, M., Zarghami, N., Nasiri, M. et al. (2011). Polyamidoamine (PAMAM) encapsulated curcumin inhibits telomerase activity in breast cancer cell line. *Clinical Biochemistry* 13 (44): S217.

178 Sarbolouki, M.N., Alizadeh, A.M., Khaniki, M. et al. (2012). Protective effect of dendrosomal curcumin combination on colon cancer in rat. *Tehran University Medical Journal* 69 (11): 678–685.

179 Alizadeh, A.M., Khaniki, M., Azizian, S. et al. (2012). Chemoprevention of azoxymethane-initiated colon cancer in rat by using a novel polymeric nanocarrier – curcumin. *European Journal of Pharmacology* 689 (1–3): 226–232.

180 Ekambaram, P. and Sathali, A.A.H. (2011). Formulation and evaluation of solid lipid nanoparticles of ramipril. *Journal of Young Pharmacists* 3 (3): 216–220.

181 Kakkar, V., Singh, S., Singla, D. et al. (2011). Exploring solid lipid nanoparticles to enhance the oral bioavailability of curcumin. *Molecular Nutrition & Food Research* 55 (3): 495–503.

182 Kakkar, V., Muppu, S.K., Chopra, K. et al. (2013). Curcumin loaded solid lipid nanoparticles: an efficient formulation approach for cerebral ischemic reperfusion injury in rats. *European Journal of Pharmaceutics and Biopharmaceutics* 85 (3): 339–345.

183 Yadav, V.R., Suresh, S., Devi, K. et al. (2009). Novel formulation of solid lipid microparticles of curcumin for anti-angiogenic and anti-inflammatory activity for optimization of therapy of inflammatory bowel disease. *The Journal of Pharmacy and Pharmacology* 61 (3): 311–321.

184 Mulik, R.S., Mönkkönen, J., Juvonen, R.O. et al. (2010). Transferrin mediated solid lipid nanoparticles containing curcumin: enhanced in vitro anticancer activity by induction of apoptosis. *International Journal of Pharmaceutics* 398 (1–2): 190–203.

185 Jain, S., Singh, P., Mishra, V. et al. (2005). Mannosylated niosomes as adjuvant–carrier system for oral genetic immunization against Hepatitis B. *Immunology Letters* 101 (1): 41–49.

186 Mandal, S., Banerjee, C., Ghosh, S. et al. (2013). Modulation of the photophysical properties of curcumin in nonionic surfactant (Tween-20) forming micelles and niosomes: a comparative study of different microenvironments. *The Journal of Physical Chemistry B* 117 (23): 6957–6968.

187 Rungphanichkul, N., Nimmannit, U., Muangsiri, W. et al. (2011). Preparation of curcuminoid niosomes for enhancement of skin permeation. *Die Pharmazie-An International Journal of Pharmaceutical Sciences* 66 (8): 570–575.

188 Raveendran, R., Bhuvaneshwar, G., and Sharma, C.P. (2013). in vitro cytotoxicity and cellular uptake of curcumin-loaded pluronic/polycaprolactone micelles in colorectal adenocarcinoma cells. *Journal of Biomaterials Applications* 27 (7): 811–827.

189 Yu, H., Li, J., Shi, K. et al. (2011). Structure of modified ε-polylysine micelles and their application in improving cellular antioxidant activity of curcuminoids. *Food & Function* 2 (7): 373–380.

190 Podaralla, S., Averineni, R., Alqahtani, M. et al. (2012). Synthesis of novel biodegradable methoxy poly (ethylene glycol)–zein micelles for effective delivery of curcumin. *Molecular Pharmaceutics* 9 (9): 2778–2786.

191 Song, Z., Feng, R., Sun, M. et al. (2011). Curcumin-loaded PLGA-PEG-PLGA triblock copolymeric micelles: preparation, pharmacokinetics and distribution in vivo. *Journal of Colloid and Interface Science* 354 (1): 116–123.

192 Leung, M.H., Colangelo, H., and Kee, T.W. (2008). Encapsulation of curcumin in cationic micelles suppresses alkaline hydrolysis. *Langmuir* 24 (11): 5672–5675.

193 Adhikary, R., Carlson, P.J., Kee, T.W. et al. (2010). Excited-state intramolecular hydrogen atom transfer of curcumin in surfactant micelles. *The Journal of Physical Chemistry B* 114 (8): 2997–3004.

194 Began, G., Sudharshan, E., and Rao, A.A. (1998). Inhibition of lipoxygenase 1 by phosphatidylcholine micelles-bound curcumin. *Lipids* 33 (12): 1223.

195 Sala Frigerio, C. and De Strooper, B. (2016). Alzheimer's disease mechanisms and emerging roads to novel therapeutics. *Annual Review of Neuroscience* 39: 57–79.

196 Ghorbani, Z., Hekmatdoost, A., and Mirmiran, P. (2014). Anti-hyperglycemic and insulin sensitizer effects of turmeric and its principle constituent curcumin. *International Journal of Endocrinology and Metabolism* 12 (4): e18081.

197 Pandit, R.S., Gaikwad, S.C., Agarkar, G.A. et al. (2015). Curcumin nanoparticles: physico-chemical fabrication and its in vitro efficacy against human pathogens. *3 Biotech* 5 (6): 991–997.

198 Garg, S. and Garg, A. (2018). Encapsulation of curcumin in silver nanoparticle for enhancement of anticancer drug delivery. *International Journal of Pharmaceutical Sciences and Research* 9: 1160–1166.

199 García-Niño, W.R. and Pedraza-Chaverrí, J. (2014). Protective effect of curcumin against heavy metals-induced liver damage. *Food and Chemical Toxicology* 69: 182–201.

200 Flora, S., Mittal, M., and Mehta, A. (2008). Heavy metal induced oxidative stress & its possible reversal by chelation therapy. *Indian Journal of Medical Research* 128 (4): 501.

201 Pourreza, N., Golmohammadi, H., and Rastegarzadeh, S. (2016). Highly selective and portable chemosensor for mercury determination in water samples using curcumin nanoparticles in a paper based analytical device. *RSC Advances* 6 (73): 69060–69066.

202 Chen, C., Johnston, T.D., Jeon, H. et al. (2009). An in vitro study of liposomal curcumin: stability, toxicity and biological activity in human lymphocytes and Epstein-Barr virus-transformed human B-cells. *International Journal of Pharmaceutics* 366 (1–2): 133–139.

203 ISO (2003). Saffron (*Crocus sativus* L)—Part 1: Specification and Part 2: Test Methods. Geneva, Switzerland: International Organization for Standardization.

204 Kyriakoudi, A., Ordoudi, S., Roldán-Medina, M. et al. (2015). Saffron, a functional spice. *Austin Journal of Nutrition and Food sciences* 3 (1): 1059.

205 Ulbricht, C., Conquer, J., Costa, D. et al. (2011). An evidence-based systematic review of saffron (*Crocus sativus*) by the natural standard research collaboration. *Journal of Dietary Supplements* 8 (1): 58–114.

206 Schmidt, M., Betti, G., and Hensel, A. (2007). Saffron in phytotherapy: pharmacology and clinical uses. *Wiener Medizinische Wochenschrift* 157 (13–14): 315.

207 José Bagur, M., Alonso Salinas, G.L., Jiménez-Monreal, A.M. et al. (2018). Saffron: an old medicinal plant and a potential novel functional food. *Molecules* 23 (1): 30.

208 Khalili, M. and Hamzeh, F. (2010). Effects of active constituents of *Crocus sativus* L., crocin on streptozocin-induced model of sporadic Alzheimer's disease in male rats. *Iranian Biomedical Journal* 14 (1–2): 59.

209 Mehri, S., Abnous, K., Mousavi, S.H. et al. (2012). Neuroprotective effect of crocin on acrylamide-induced cytotoxicity in PC12 cells. *Cellular and Molecular Neurobiology* 32 (2): 227–235.

210 Mousavi, S.H., Tayarani, N., and Parsaee, H. (2010). Protective effect of saffron extract and crocin on reactive oxygen species-mediated high glucose-induced toxicity in PC12 cells. *Cellular and Molecular Neurobiology* 30 (2): 185–191.

211 Verma, S. and Bordia, A. (1998). Antioxidant property of saffron in man. *Indian Journal of Medical Sciences* 52 (5): 205–207.

212 Carmona, M., Zalacain, A., Salinas, M. et al. (2007). A new approach to saffron aroma. *Critical Reviews in Food Science and Nutrition* 47 (2): 145–159.

213 Nassiri-Asl, M. and Hosseinzadeh, H. (2015). Neuropharmacology effects of saffron (*Crocus sativus*) and its active constituents. In: *Bioactive Nutraceuticals and Dietary Supplements in Neurological and Brain Disease* (ed. R.S. Watson and V.R. Preedy), 29–39. Elsevier.

214 Bie, X., Chen, Y., Zheng, X. et al. (2011). The role of crocetin in protection following cerebral contusion and in the enhancement of angiogenesis in rats. *Fitoterapia* 82 (7): 997–1002.

215 Yoshino, F., Yoshida, A., Umigai, N. et al. (2011). Crocetin reduces the oxidative stress induced reactive oxygen species in the stroke-prone spontaneously hypertensive rats (SHRSPs) brain. *Journal of Clinical Biochemistry and Nutrition* 49 (3): 182–187.

216 Linardaki, Z.I., Orkoula, M.G., Kokkosis, A.G. et al. (2013). Investigation of the neuroprotective action of saffron (*Crocus sativus* L.) in aluminum-exposed adult mice through behavioral and neurobiochemical assessment. *Food and Chemical Toxicology* 52: 163–170.

217 Sheng, L., Qian, Z., Zheng, S. et al. (2006). Mechanism of hypolipidemic effect of crocin in rats: crocin inhibits pancreatic lipase. *European Journal of Pharmacology* 543 (1–3): 116–122.

218 Rikabad, M.M., Pourakbar, L., and Moghaddam, S.S. (2019). Agrobiological, chemical and antioxidant properties of saffron (*Crocus sativus* L.) exposed to TiO_2 nanoparticles and ultraviolet-B stress. *Industrial Crops and Products* 137: 137–143.

219 Bumbudsanpharoke, N., Choi, J., and Ko, S. (2015). Applications of nanomaterials in food packaging. *Journal of Nanoscience and Nanotechnology* 15 (9): 6357–6372.

220 Blasco, C. and Picó, Y. (2011). Determining nanomaterials in food. *TrAC, Trends in Analytical Chemistry* 30 (1): 84–99.

221 Peiro, A.M., Colombo, C., Doyle, G. et al. (2006). Photochemical reduction of oxygen adsorbed to nanocrystalline TiO_2 films: a transient absorption and oxygen scavenging study of different TiO_2 preparations. *The Journal of Physical Chemistry B* 110 (46): 23255–23263.

222 Ghanbarzadeh, B., Oleyaei, S.A., and Almasi, H. (2015). Nanostructured materials utilized in biopolymer-based plastics for food packaging applications. *Critical Reviews in Food Science and Nutrition* 55 (12): 1699–1723.

223 Lawrence, G., Lyons, K., and Wallington, T. (2013). *Food Security, Nutrition and Sustainability*. Earthscan.

224 Sahu, J.K. (2014). *Introduction to Advanced Food Process Engineering*. CRC Press.

225 Rapsomanikis, A., Papoulis, D., Panagiotaras, D. et al. (2014). Nanocrystalline TiO_2 and halloysite clay mineral composite films prepared by sol-gel method: synergistic effect and the case of silver modification to the photocatalytic degradation of basic blue-41 azo dye in water. *Global NEST Journal* 16 (3): 485–498.

226 Mills, A., Doyle, G., Peiro, A.M. et al. (2006). Demonstration of a novel, flexible, photocatalytic oxygen-scavenging polymer film. *Journal of Photochemistry and Photobiology A: Chemistry* 177 (2–3): 328–331.

227 Macwan, D., Dave, P.N., and Chaturvedi, S. (2011). A review on nano-TiO_2 sol–gel type syntheses and its applications. *Journal of Materials Science* 46 (11): 3669–3686.

228 Weir, A., Westerhoff, P., Fabricius, L. et al. (2012). Titanium dioxide nanoparticles in food and personal care products. *Environmental Science and Technology* 46 (4): 2242–2250.

229 Zhang, Y., Chen, Y., Westerhoff, P. et al. (2009). Impact of natural organic matter and divalent cations on the stability of aqueous nanoparticles. *Water Research* 43 (17): 4249–4257.

230 Zhang, Y., Chen, Y., Westerhoff, P. et al. (2008). Stability of commercial metal oxide nanoparticles in water. *Water Research* 42 (8–9): 2204–2212.

231 Mahshid, S., Askari, M., and Ghamsari, M.S. (2007). Synthesis of TiO_2 nanoparticles by hydrolysis and peptization of titanium isopropoxide solution. *Journal of Materials Processing Technology* 189 (1–3): 296–300.

232 McClements, D.J. and Xiao, H. (2017). Is nano safe in foods? Establishing the factors impacting the gastrointestinal fate and toxicity of organic and inorganic food-grade nanoparticles. *NPJ Science of Food* 1 (1): 1–13.

233 Busquets, R. (2017). *Emerging Nanotechnologies in Food Science*. Elsevier.

234 Chen, Z., Wang, Y., Zhuo, L. et al. (2015). Effect of titanium dioxide nanoparticles on the cardiovascular system after oral administration. *Toxicology Letters* 239 (2): 123–130.

235 Faust, J.J., Doudrick, K., Yang, Y. et al. (2014). Food grade titanium dioxide disrupts intestinal brush border microvilli in vitro independent of sedimentation. *Cell Biology and Toxicology* 30 (3): 169–188.

236 Baranowska-Wójcik, E., Szwajgier, D., Oleszczuk, P. et al. (2020). Effects of titanium dioxide nanoparticles exposure on human health—a review. *Biological Trace Element Research* 193 (1): 118–129.

237 Dudefoi, W., Moniz, K., Allen-Vercoe, E. et al. (2017). Impact of food grade and nano-TiO_2 particles on a human intestinal community. *Food and Chemical Toxicology* 106: 242–249.

238 Efsa, A. (2013). Panel (EFSA panel on food additives and nutrient sources added to food). Scientific opinion on the re-evaluation of aspartame (E 951) as a food additive. *EFSA Journal* 11 (12): 3496.

239 Sherman, L.M. (2012). Sub-micron additives make strides(just Don't Say 'nano'). *Plastics Technology* 58 (7): 26–31.

240 Xia, Z., Stafford, S.L., Sims, S. et al. 2008) Polyester polymer and copolymer compositions containing titanium nitride particles. Google Patents.

241 Schneider, J., Akbar, M.I., Dutroncy, J. et al. (2009). Silicon oxide barrier coatings deposited on polymer materials for applications in food packaging industry. *Plasma Processes and Polymers* 6 (S1): S700–S704.

242 Vladimirov, V., Betchev, C., Vassiliou, A. et al. (2006). Dynamic mechanical and morphological studies of isotactic polypropylene/fumed silica nanocomposites with enhanced gas barrier properties. *Composites Science and Technology* 66 (15): 2935–2944.

243 Xiong, H., Tang, S., Tang, H. et al. (2008). The structure and properties of a starch-based biodegradable film. *Carbohydrate Polymers* 71 (2): 263–268.

244 Tang, C. and Liu, H. (2008). Cellulose nanofiber reinforced poly (vinyl alcohol) composite film with high visible light transmittance. *Composites Part A: Applied Science and Manufacturing* 39 (10): 1638–1643.

245 Siddiqui, M.H. and Al-Whaibi, M.H. (2014). Role of nano-SiO_2 in germination of tomato (*Lycopersicum esculentum* seeds Mill.). *Saudi Journal of Biological Sciences* 21 (1): 13–17.

246 Chen, L., Liu, J., Zhang, Y. et al. (2018). The toxicity of silica nanoparticles to the immune system. *Nanomedicine* 13 (15): 1939–1962.

247 Dobrovolskaia, M.A., Aggarwal, P., Hall, J.B. et al. (2008). Preclinical studies to understand nanoparticle interaction with the immune system and its potential effects on nanoparticle biodistribution. *Molecular Pharmaceutics* 5 (4): 487–495.

248 Silvestre, C., Duraccio, D., and Cimmino, S. (2011). Food packaging based on polymer nanomaterials. *Progress in Polymer Science* 36 (12): 1766–1782.

249 Pramudita, D., Iskandar, I., and Indarto, A. (2018). Nano-enhanced materials for reclamation of mine spoils. *Bio-Geotechnologies for Mine Site Rehabilitation*: Elsevier 201–214.

250 Kushwaha, S.K.S., Ghoshal, S., Rai, A.K. et al. (2013). Carbon nanotubes as a novel drug delivery system for anticancer therapy: a review. *Brazilian Journal of Pharmaceutical Sciences* 49 (4): 629–643.

251 Liu, Q., Chen, B., Wang, Q. et al. (2009). Carbon nanotubes as molecular transporters for walled plant cells. *Nano Letters* 9 (3): 1007–1010.

252 Song, J. and Jang, J. (2014). Antimicrobial polymer nanostructures: synthetic route, mechanism of action and perspective. *Advances in Colloid and Interface Science* 203: 37–50.

253 Liu, S., Wei, L., Hao, L. et al. (2009). Sharper and faster "nano darts" kill more bacteria: a study of antibacterial activity of individually dispersed pristine single-walled carbon nanotube. *ACS Nano* 3 (12): 3891–3902.

254 Liu, S., Ng, A.K., Xu, R. et al. (2010). Antibacterial action of dispersed single-walled carbon nanotubes on Escherichia coli and Bacillus subtilis investigated by atomic force microscopy. *Nanoscale* 2 (12): 2744–2750.

255 Abdelhalim, A., Abdellah, A., Scarpa, G. et al. (2013). Fabrication of carbon nanotube thin films on flexible substrates by spray deposition and transfer printing. *Carbon* 61: 72–79.

256 Abdellah, A., Abdelhalim, A., Loghin, F. et al. (2013). Flexible carbon nanotube based gas sensors fabricated by large-scale spray deposition. *IEEE Sensors Journal* 13 (10): 4014–4021.

257 Abdelhalim, A., Abdellah, A., Scarpa, G. et al. (2014). Metallic nanoparticles functionalizing carbon nanotube networks for gas sensing applications. *Nanotechnology* 25 (5): 055208.

258 Wahab, R., Mishra, A., Yun, S.-I. et al. (2010). Antibacterial activity of ZnO nanoparticles prepared via non-hydrolytic solution route. *Applied Microbiology and Biotechnology* 87 (5): 1917–1925.

259 Crocetti, G., Miller, G., and Lowrey, N. (2011). Nano-silver, Report produced for Friends of the Earth, Australia. Friends of the Earth, Australia.

260 Sondi, I. and Salopek-Sondi, B. (2004). Silver nanoparticles as antimicrobial agent: a case study on E. coli as a model for gram-negative bacteria. *Journal of Colloid and Interface Science* 275 (1): 177–182.

261 Jeong, S.H., Yeo, S.Y., and Yi, S.C. (2005). The effect of filler particle size on the antibacterial properties of compounded polymer/silver fibers. *Journal of Materials Science* 40 (20): 5407–5411.

262 Rasmussen, J.W., Martinez, E., Louka, P. et al. (2010). Zinc oxide nanoparticles for selective destruction of tumor cells and potential for drug delivery applications. *Expert Opinion on Drug Delivery* 7 (9): 1063–1077.

263 Jiang, J., Pi, J., and Cai, J. (2018). The advancing of zinc oxide nanoparticles for biomedical applications. *Bioinorganic Chemistry and Applications* 2018: 1–18.

264 Schwertmann, U. and Cornell, R.M. (2008). *Iron Oxides in the Laboratory: Preparation and Characterization*. Wiley.

265 Cao, M., Li, Z., Wang, J. et al. (2012). Food related applications of magnetic iron oxide nanoparticles: enzyme immobilization, protein purification, and food analysis. *Trends in Food Science and Technology* 27 (1): 47–56.

266 Lopez-Rubio, A., Gavara, R., and Lagaron, J.M. (2006). Bioactive packaging: turning foods into healthier foods through biomaterials. *Trends in Food Science and Technology* 17 (10): 567–575.

267 Chellaram, C., Murugaboopathi, G., John, A. et al. (2014). Significance of nanotechnology in food industry. *APCBEE Procedia* 8: 109–113.

268 Soares, N. and Hotchkiss, J. (1998). Naringinase immobilization in packaging films for reducing naringin concentration in grapefruit juice. *Journal of Food Science* 63 (1): 61–65.

269 Salmani, M.H., Mikaie, M., Torabizadeh, H. et al. (2017). Application of magnetic iron oxide nanoparticles in stabilization process of biological molecules. *SSU_Journals* 25 (4): 287–299.

270 Pařízek, J. and Ošťádalová, I. (1967). The protective effect of small amounts of selenite in sublimate intoxication. *Experientia* 23 (2): 142–143.

271 Ganther, H., Goudie, C., Sunde, M. et al. (1972). Selenium: relation to decreased toxicity of methylmercury added to diets containing tuna. *Science* 175 (4026): 1122–1124.

272 Iwata, H., Okamoto, H., and Ohsawa, Y. (1973). Effect of selenium on methylmercury poisoning. *Research Communications in Chemical Pathology and Pharmacology* 5 (3): 673–680.

273 Iwata, H., Masukawa, T., Kito, H. et al. (1981). Involvement of tissue sulfhydryls in the formation of a complex of methylmercury with selenium. *Biochemical Pharmacology* 30 (23): 3159–3163.

274 Naganuma, A., Kojima, Y., and Imura, N. (1980). Interaction of methylmercury and selenium in mouse: formation and decomposition of bis(methylmercuric) selenide. *Research Communications in Chemical Pathology and Pharmacology* 30 (2): 301–316.

275 Kar, A. (1960). Pervention of cadmium induced changes in the gonads of rat by zinc and selenium – a study in antagonism between metals in the biological system. *Proceedings of the National Academy of Sciences, India* B26: 40–50.

276 Wagner, P.A., Hoekstra, W.G., and Ganther, H.E. (1975). Alleviation of silver toxicity by selenite in the rat in relation to tissue glutathione peroxidase. *Proceedings of the Society for Experimental Biology and Medicine* 148 (4): 1106–1110.

277 Rastogi, S.C., Clausen, J., and Srivastava, K. (1976). Selenium and lead: mutual detoxifying effects. *Toxicology* 6 (3): 377–388.

278 Masukawa, T. (1987). Pharmacological and toxicological aspects of inorganic and organic selenium compounds. *Organic Selenium and Tellurium Compounds* 2: 377–392.

279 Sarin, L., Sanchez, V.C., Yan, A. et al. (2010). Selenium-carbon bifunctional nanoparticles for the treatment of malignant mesothelioma. *Advanced Materials* 22 (45): 5207–5211.

280 Chen, T., Wong, Y.-S., Zheng, W. et al. (2008). Selenium nanoparticles fabricated in Undaria pinnatifida polysaccharide solutions induce mitochondria-mediated apoptosis in A375 human melanoma cells. *Colloids and Surfaces. B, Biointerfaces* 67 (1): 26–31.

281 Tan, L., Jiang, X., Zhang, Y. et al. (2009). in vitro study on the individual and synergistic cytotoxicity of adriamycin and selenium nanoparticles against Bel7402 cells with a quartz crystal microbalance. *Biosensors and Bioelectronics* 24 (7): 2268–2272.

282 Yu, B., Li, X., Zheng, W. et al. (2014). pH-responsive cancer-targeted selenium nanoparticles: a transformable drug carrier with enhanced theranostic effects. *Journal of Materials Chemistry B* 2 (33): 5409–5418.

283 Torres, S., Campos, V., León, C. et al. (2012). Biosynthesis of selenium nanoparticles by Pantoea agglomerans and their antioxidant activity. *Journal of Nanoparticle Research* 14 (11): 1236.

284 Eszenyi, P., Sztrik, A., Babka, B. et al. (2011). Elemental, nano-sized (100-500 nm) selenium production by probiotic lactic acid bacteria. *International Journal of Bioscience, Biochemistry and Bioinformatics* 1 (2): 148.

285 Mirhadi, E., Nassirli, H., and Malaekeh-Nikouei, B. (2019). An updated review on therapeutic effects of nanoparticle-based formulations of saffron components (safranal, crocin, and crocetin). *Journal of Pharmaceutical Investigation* 1–12.

286 Colapietro, A., Mancini, A., D'Alessandro, A.M. et al. (2019). Crocetin and crocin from saffron in cancer chemotherapy and chemoprevention. *Anti-Cancer Agents in Medicinal Chemistry (Formerly Current Medicinal Chemistry-Anti-Cancer Agents)* 19 (1): 38–47.

9

Rubber Nanocomposites for Biomedical Applications

Jayalatha Gopalakrishnan

Department of Polymer Science and Rubber Technology, Cochin University of Science and Technology, Kochi, Kerala, India

9.1 Introduction

During the past few years, rubber nanocomposites have gained major attention in the area of biomedical applications. Since the 1940s, elastomers have been known to be used for a wide range of biomedical products. They are generally categorized into three groups: (i) commodity elastomers used in the biomedical field, (ii) medical grade elastomers ideal for short-term physiological contact, and (iii) a small group of elastomers suitable for implants or long-term physiological contact. The prerequisites of rubbers/elastomers for biomedical applications include high purity, biocompatibility, desirable physical, chemical and mechanical properties, ease of fabrication, sterilizability, and high stability. Additionally, when brought in contact with biological tissues and fluids, they must not: induce thrombosis, immune response, or cancer; produce toxic and allergic reactions; alter plasma protein; destroy cellular elements and enzymes; or deplete electrolytes.

For the manufacture of biomedical elastomers, the catalysts selected for polymerization should be less toxic and the additives such as fillers, plasticizers, accelerators, antioxidants, curatives, pigments, stabilizers for modifying the polymer properties should be nontoxic and nonleaching. It has been observed that the presence of residual monomers, oligomers, catalysts. and polymerization additives can lead to adverse biological reactions. Furthermore, the chemical substances extracted from elastomeric products consist of residues from accelerators (e.g. MBT [mercaptobenzothiaole], TMTD [tetramethylthiuram disulfide]), activators (e.g. zinc oxide (ZnO), stearic acid), antioxidants (wax), plasticizers (oil), and products of vulcanization (zinc stearates and zinc salts of dithiocarbamates). In order to reduce the quantity of extracts, the elastomeric products may be cleaned and/or coated. Another problem associated with the rubber manufacturing industry is the concern about toxic nitrosamines that are released during vulcanization by reacting the species viz. nitrates or nitrites $(NO)_x$ with the secondary amines from accelerators, antioxidants etc. Therefore, elimination of amines is commonly recommended for elastomers that are to be used in medical applications. Considering the potential hazards such as toxicity, chemical

Nanomaterials and Nanotechnology in Medicine, First Edition. Edited by P.M. Visakh.

migrants, and contaminants, it has been observed that silicone elastomers have milder characteristics i.e. they are chemically inactive compared to other competing elastomers proposed for the same application.

Biocompatibility is considered the most important property required for elastomers. In addition, elastomer-based devices must be sterilizable to eliminate infectious bacteria. For the sterilization of elastomers, ethylene oxide gas is widely used. The key advantages compared to other sterilizing methods such as using dry heat or steam are that it can be used at low temperatures and the time of exposure is relatively short. The method involves exposure of the device to ethylene oxide gas along with a carrier gas for about three hours at 55 °C followed by degassing in an air chamber at the same temperature for 10 hours. The level of toxic ethylene oxide residues must be within acceptable limits.

9.2 Rubbers for Biomedical Applications

9.2.1 Silicone Rubber

Silicone rubber is a synthetic polymer with a backbone composed of Si-O bonds as repeating units. The substituent R can be methyl, phenyl, vinyl, or trifluoropropyl resulting in various types of silicones. Polydimethylsiloxanes (PDMS) is the most common type of silicone where R = methyl group (Figure 9.1). PDMS have been widely used for biomedical applications such as tubings, catheters, artificial finger joint hinges, drug-delivery systems, breast implants. They offer major advantages such as the possibility for autoclaving, little deterioration of properties with time, minimal tissue reaction, and resistance to attack by the body and metabolism by other organisms. Moreover, they are not absorbed by living tissues and there is little change in mechanical properties when placed in contact with tissue. Another property of significance is their high permeability to gases and vapors. PDMS are 10 times more permeable to oxygen than natural rubber. Hence, they are used as membranes for blood oxygenators in heart–lung circulation outside the body. They are also widely used for long-term implantation as they are not physically modified by soft tissues.

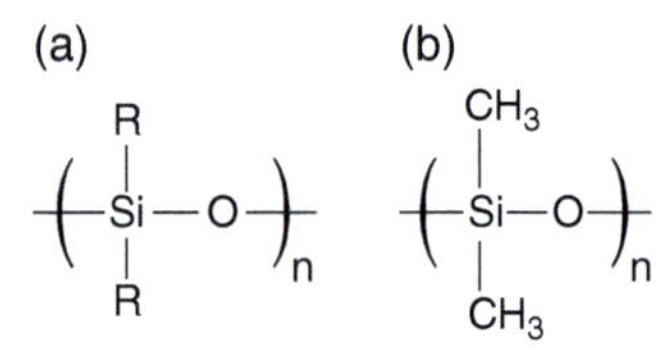

Figure 9.1 (a) General structure of silicone rubber (b) structure of PDMS.

Silicone rubbers can be cured via high temperature vulcanization or room temperature vulcanization. They are generally compounded with pure, finely divided amorphous silica as a filler to increase tensile and tear strength. However, very fine silica can adversely affect the blood compatibility of silicones. Biocompatible fillers have been prepared using two types of modified amorphous silica: (i) silica containing covalently bonded reactive Si-H groups formed by reacting silica with methyldichlorosilane and (ii) silica bonded with a covalently immobilized antithrombogenic heparinoid agent. Incorporation of such fillers in silicones exhibited good mechanical properties and excellent tissue compatibility.

For applications such as catheters with balloon and wound dressings, the manufacturing process proceeds with the dispersion of silicone in solvents such as Freon or trichloroethane so as to be dipped or cast resulting in a uniform smooth, flawless surface. Organosilicone that contains low molecular weight dimethylsiloxane with silanol end groups, tetraethoxysilane curing agent, tin octoate catalyst, and X-ray contrasting substances, are effective for treating embolisms in kidney arteries. Silicone elastomer implants are widely used in plastic surgery that includes reconstruction of breasts, chin, nose, ear armature etc. Mammary prosthesis consists of a bag of silicone rubber with a soft silicone gel. Ear armatures are used to rebuild ears to correct birth defects. They have also gained importance in other specialty areas such as neurosurgery, orthopedic surgery, urology, and cardiovascular surgery. One of the oldest applications of elastomer implants involves a hydrocephalus shunt used to drain cerebrospinal fluid from the ventricle of the brain to either the peritoneal cavity or to the vascular system.

9.2.2 Polyurethanes

Polyurethanes are a class of copolymers having diverse chemical compositions that can be tailored depending on the application. The backbone is composed of aliphatic or aromatic units linked with polar urethane (-NHCOO-) group. Segmented polyurethanes are used in prosthetic devices such as heart-assist systems, cannulas, catheters, heart valves, pacemaker lead-wire insulation, owing to their good biocompatibility, high mechanical strength, and elastomeric properties. These properties are dependent on the semi-crystalline structure of the hard segment and amorphous structure of the soft segment. Poly(ether-urethane) elastomers are produced by the reaction of a diisocyanate such as 4, 4-diphenyl methane diisocyanate (MDI) with a polyol such as polytetramethylene ether (PTMG) glycol to produce a soft segment terminated by isocyanate groups, which is further extended by interaction with a diol e.g. butanediol. The use of a diamine chain extender instead of a diol results in the formation of poly(ether-urethane-urea) segments (Figure 9.2). Polyurethanes have been used for many years for products including blood bags and dialysis tubing. The flexibility of these materials obviates the need for the addition of plasticizers that can leach during use. Also, it resists shrinking, cracking, and

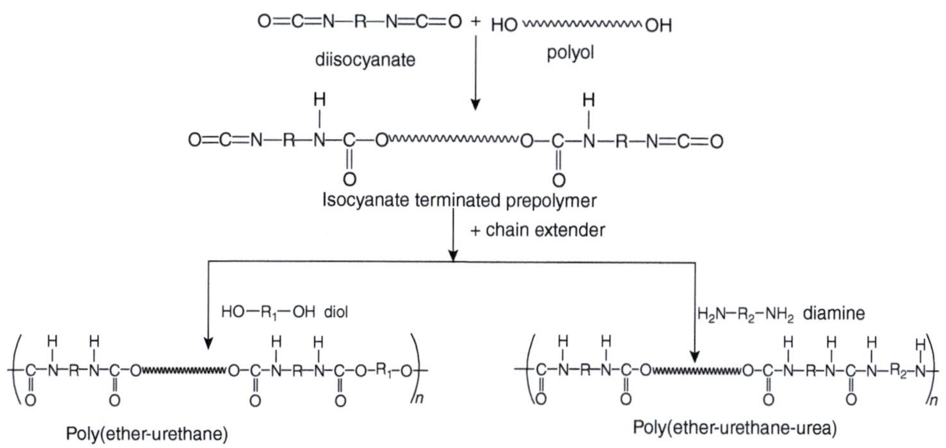

Figure 9.2 Synthesis of segmented poly(ether-urethane) and poly(ether-urethane-urea).

kinking. In addition, they are resistant to hydrolysis and fungus. Moreover, they can be sterilized by conventional steam sterilization processes. Further, polyurethanes can be tailored to modify the rate of moisture transmission, which makes it acceptable for transdermal-drug-delivery patches and surgical drapes, gowns etc.

The poly(ether-urethane-urea) segments, with trade names Biomer and Mitrathane, have been used in cardiovascular applications. Polyurethane-ureas are processed via solution casting since heating of these polymers to softening temperature may lead to degradation. The factors that decide the quality of the product depends on the concentration of polymer in solution, solvent characteristics, mold surface, humidity etc. Biomer exhibits excellent blood compatibility and resilience and hence is notable as coatings for cannulas, catheters, left-ventricular assist blood pumps, intra-aorta balloon pumps etc. Mitrathane is produced as a clear gel-free solution in DMA (Urinary Dimethylamine) containing 25% solids. Poly(ether-urethanes) (Pellethane) can be prepared with varying degree of hardness. The one with lower hardness has been used as an alternative to silicones as insulators for implanted human pacemaker leads, while the harder ones are used as implants that exhibit greater density of adherent leucocytes. Aliphatic polyurethanes consist of poly(oxytetramethylene) glycol, hydrogenated methylene diisocyanate and 1,4-butanediol. As opposed to aromatic polyurethanes, the aliphatic ones on exposure to UV light do not yellow or decompose to carcinogenic products. They find applications in diagnostic and therapeutic catheters, intravenous catheters, interventional radiology, pacemaker insulation leads etc.

TM5 based on PTMG, MDI, and propylene diamine has been supplied as solution in DMF (dimethylformamide). Although its static and dynamic flexibility are less stable than Biomer, it has a higher strength and ductility. They are suitable for blood pump applications on account of reasonable blood compatibility.

Cardiothane is a copolymer composed of 90% polyether-polyurethane and 10% PDMS, which is cross-linked via an interaction of acetoxy-terminated siloxane blocks with the unsubstituted nitrogen in urethane. The copolymer is processed by casting from solution.

It has been reported that interactions of polyurethanes with body fluids can result in failure due to calcification, environmental stress cracking, and chain scission. The deposition of calcium hydroxyapatite (HA) on the implanted device is referred to as calcification. When in contact with body fluid, polyurethane can undergo environmental stress cracking under stress conditions. In addition, polyetherurethanes can develop cracks on surfaces that are in contact with tissues. This may be due to the auto-oxidative degradation process accelerated by the corrosion products of metallic components in implanted devices.

Polyurethanes with varying degrees of hydrophobicity and hydrophilicity are excellent candidates for replacement of articular cartilage, artificial joint prosthesis, and percutaneous implantable devices.

9.2.3 Thermoplastic Elastomers

Styrenic thermoplastic elastomers (TPEs) have been used for medical tubing due to their cleanliness, chemical inertness, dimensional stability, ease of sterilization, and low gas and vapor transmission. It has been reported that tubings for peristaltic pumps made from styrenic TPEs lasts 30 times longer when compared to silicone rubber. Some catheters and tubing made of PVC have been replaced with TPE to prevent plasticizer migration and extraction. TPEs are flexible at low temperatures so are able to replace natural rubber with no chemical irritants/proteins.

SEBS (styrene-ethylene-butylene-styrene) and SBS (styrene-butadiene-styrene) are used in medical applications because of their purity and absence of chemical cross-linking agents that may produce toxic byproducts. They are physically cross-linked, which maintains the elasticity. When heated, the rigid-block domains soften and begin to flow. Upon cooling, the rigid domains restore the network. They can be sterilized by steam, ethylene oxide, and gamma radiation without losing their properties. They are used as elastomeric films in applications such as blood bags and surgical drapes. The styrene content in SBS determines the stability when placed in physiological serum. It has been reported that 30% styrene can undergo considerable degradation in the elastomer part while the product with 40% styrene exhibited no degradation. However, poor tissue compatibility is one of the key issues with styrenic TPEs [1].

9.3 Rubber-based Nanocomposites

During the past decade there has been significant interest among researchers in developing rubber-based nanocomposites suitable for biomedical use. The addition of nanoparticles offers the opportunity to upgrade both the functional and structural properties of rubber, which cannot be attained through the use of rubber alone. By utilizing the innate properties of nanofillers, tuning various parameters, and controlling the interaction between fillers and rubber, unique combinations of properties can be achieved that provide an interesting platform for the design and development of products in the biomedical field.

9.3.1 Silicone Rubber Nanocomposites

The inherent properties of silicone rubbers such as high hydrophobicity and low load-bearing capacity can be problematic for biomedical applications. In addition, the formation of biofilm on implants is a matter of concern. Further, implant-related infections can become a financial burden on healthcare systems. Once infected, the implants are surgically removed and replaced, which is costly and the chance of reinfection cannot be ignored [2]. Although silicone coatings have been used on ophthalmologic biomaterials, artificial finger joints, endotracheal tubes, prosthetic implants, artificial lungs, temporomandibular joints, rhinoplasty, mentoplasty, breast augmentation, ureteral stents, catheters etc. they are prone to bacterial attachment, biofilm formation, and nonadhesion of proteins and biomolecules. Coating implants with bactericide agents or antiadhesives are current approaches for producing an antimicrobial surface. Antimicrobial activity can be assessed by colony forming units (CFUs) upon recovery of bacteria after incubation, bacterial membrane damage, or scanning electron microscopy to detect the formation of biofilm.

9.3.1.1 Effect of Metal and Metal Oxide Nanoparticles

Silver (Ag) nanoparticles are known to be an efficient antibacterial agent as they can potentially obstruct the pathway of respiratory enzymes as well as alter the cell wall and microbial DNA. However, clinically, when it comes in contact with blood, it becomes ineffective and the coating gradually wears [3]. At higher dosages, it may lead to genotoxicity and cytotoxicity in human cells [4]. Moreover, once the surface gets covered with macromolecules and dead microorganisms, their antimicrobial effect is lost [5]. Antiadhesion coatings, being hydrophobic, can reduce the initial bacterial adhesion but are ineffective in preventing biofilm formation in the long term. An effective alternative is to incorporate nanofillers in silicone polymers to generate nanocomposites. Further, nanoparticles can alter mechanical strength, hydrophobicity, and surface roughness.

It has been reported that CuO (copper oxide) and ZnO also possess antibacterial properties [6]. PDMS nanocomposites have been developed using nanofillers (0.1% w/v) including CuO, CTAB (cetyltrimethylammonium bromide) capped CuO and ZnO via mechanical agitation and further spin coating. In the case of CTAB capped CuO, the cationic surfactant CTAB involved in the synthesis of CuO significantly reduces the particle size. The contact angle of PDMS, being hydrophobic is around 93°. The addition of ZnO in PDMS further increases the contact angle to 119°. This may be due to the presence of pits on the surface that can trap air thereby resulting in more hydrophobicity. CuO slightly reduced the contact angle (90°) while CTAB capped CuO is hydrophilic with a contact angle of about 78° due to the CTAB molecule being surface active and charged. The surface hydrophobicity of microorganisms plays a significant role in determining their adhesion to surfaces. *Staphylococcus aureus* (*S. aureus*) being hydrophobic, attaches more on PDMS/ZnO while *Escherichia coli* (*E. coli*) is relatively less hydrophobic and prefers to attach to PDMS/CuO(CTAB) nanocomposites. After 24 hours incubation, PDMS/CuO(CTAB) and PDMS/ ZnO nanocomposites demonstrated the best antibacterial behavior against *S. aureus* and *E. coli*, respectively. Once attached, the bacterial cells are sacrificed due to their contact with the antibacterial nanoparticles present on the PDMS surface. This indicates that fast-growing bacteria are more vulnerable to nanoparticles than slow-growing ones. Through the expression of stress-response genes, slow-growing bacteria can develop resistance to nanoparticles [7]. Further, these nanocomposites are noncytotoxic towards rat L6 myoblast cells as the number of viable cells increased after 48 hours incubation with no significant effect on the shape of the cells [8].

9.3.1.2 Effect of Hydroxyapatite

HA, $Ca_{10}(PO)_4(OH)_2$ belongs to the family of calcium phosphate ceramics. It is chemically similar to the natural bone matrix making it an ideal biomaterial with good biocompatibility and bone integration capability. Roy et al. [9] developed PDMS/HA nanofiller scaffolds using a solution mixing technique followed by curing with a silicone hydride cross-linker. Under creep deformation, the nanocomposite exhibited a higher amount of strain recovery compared with the unfilled one. When subjected to isothermal frequency sweep (0.1–40 Hz), the nanocomposite showed a higher modulus value over the entire frequency range compared to PDMS. The *in vitro* cytotoxic effect analyzed by culturing breast epithelial cells both with and without scaffolds indicated the absence of leached cytotoxic products. The initial cell adhesion was not influenced by the HA nanofiller, however, it supported delayed cell proliferation [9].

Reconstructive surgery is mostly performed when there is deformity or defect in human organs caused by congenital factors or resulting from an injury or degradation process such as osteoarthritis. The prerequisites for a bioimplant material include high strength and flexibility, good biocompatibility, chemical stability, nonmagnetic behavior, and ease of fabrication and disinfection. Although silicone rubber is the most commonly used biomaterial for reconstructive surgery, it has inferior bioactivity. HA has outstanding biological performance in that it promotes new bone formation via degradation in the microenvironment of the organism producing pore-like structures that facilitate adherence and the growth of tissues [10]. In addition, they possess high mechanical strength. Composites of silicone with HA has the potential to exhibit the desired mechanical and bioactive properties. Furthermore, these composites will cause less damage to erythrocytes as the inorganic HA can alter the surface charge in a hydrated environment with the formation of different degrees of electro-ionization. Thein-Han et al. [11] fabricated composites using liquid silicone rubber and 5 wt. % nanohydroxyapatite with octamethyltrisiloxane as the solvent via shear mixing, sonication, and subsequent curing. The tensile strength and modulus of the composite (PDMS/HA) improved compared to pure PDMS while retaining high elongation. Cell culture studies using mouse preosteoblasts are performed and the cell-seeded samples stained with nucleic acid dye are examined using fluorescence microscopy with excitation and emission of 346 and 442 nm. The cell nuclei appear as blue fluorescence. The micrographs indicates that the cell attachment within one hour of culture is higher for the nanocomposite compared to silicone rubber. Cell proliferation measured as cell density within 14 days of cell culture is significantly greater for the composites and the cell nuclei are relatively larger indicating its cytocompatibility (Figure 9.3). Scanning electron microscopic images further confirms that after one day of culture, the osteoblasts spread well on the nanocomposite and have rough surfaces while the cell surface on pure rubber was smooth. After 14 days, cell proliferation is high with extracellular matrices containing small globules of mineral deposits (Figure 9.4). The metabolic activity of preosteoblasts on different surfaces can be measured using a MMT (a yellow tetrazole; 3-(4,5-dimethylthiazol-2-yl)-2,5-diphenyltetrazolium bromide) assay, which is based on the reductase activity in the mitochondria of living cells. Higher cell viability of nanocomposites indicates favorable growth conditions of preosteoblasts on the surface compared to pure rubber (Figure 9.5). The effect of important proteins including fibronectin, vinculin, and actin on cell-surface interaction studies using immunofluorescence microscopy indicates that the expression level of vinculin focal adhesion contacts together with actin cytoskeleton organization and fibronectin in preosteoblasts are higher on the nanocomposites in contrast to pure silicone rubber [11]. Focal contacts are the sites of adhesion between tissue cultured cells and substrate surfaces [12].

9.3.1.3 Catheters

Silicone catheters are widely used in clinical applications for administering drugs, intravenous fluids, blood products, draining urine or fluid from the urethra or gastrointestinal tract, and parenteral nutrition. However, they are associated with the high risk of nosocomial infections such as in the bloodstream [13] and urinary tract, and ventilator-related

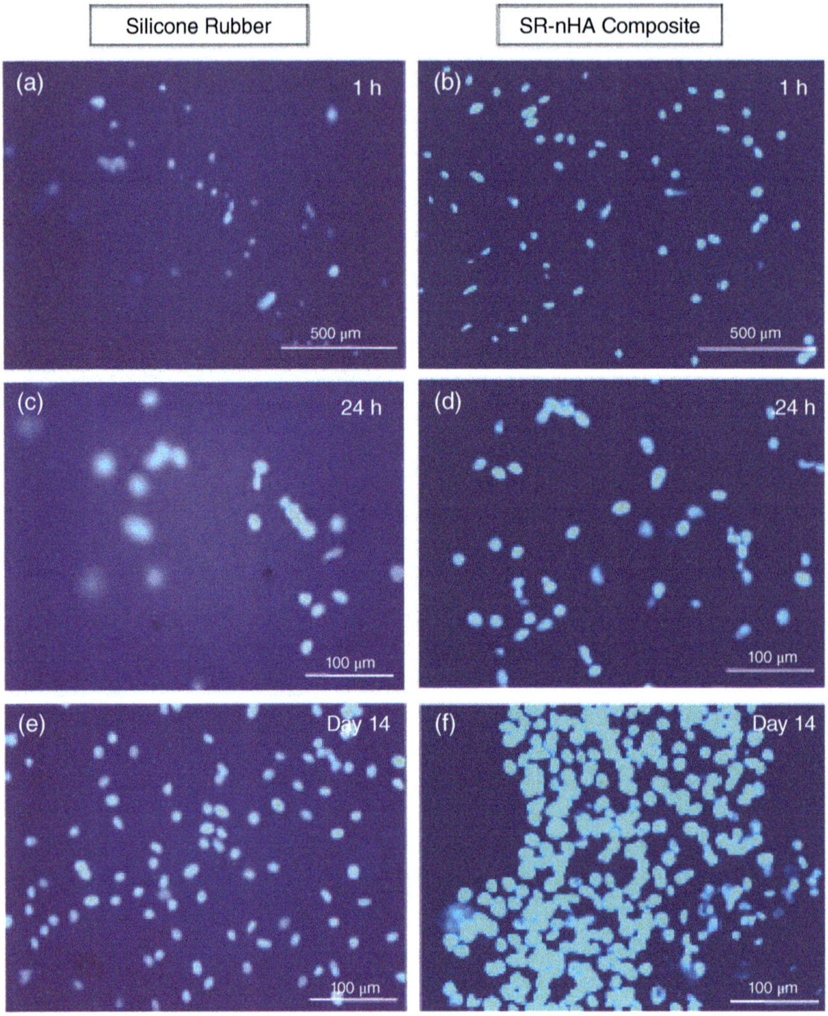

Figure 9.3 Fluorescence micrographs of preosteoblast nuclei stained with Hoechst 33258 illustrating cell attachment and proliferation. Cell proliferation is presented at 1, 24 hours and 14 days of culture on pure SR (a, c, and e) and SR–nHA composite (b, d, and f) surfaces. [11] Thein-Han, W.W., et al., (2009), Figure 03, p 06 / with permission of Elsevier.

pneumonia [14, 15]. It has been reported that titanium dioxide (TiO_2) photocatalyst film coated on silicone catheters when exposed to low intensity UV light (~1 mW cm^{-2}) exhibits a significant antibacterial effect and self-cleaning property [16]. The TiO_2 coating can be achieved by soaking the silicone catheter in sulfuric acid (5 M) for three hours, coating with an adhesive layer, and subsequently dip-coating in an ethanol-water mixture of titanium dioxide sol and silicon oxide compounds. The TiO_2 coating formed by overlapping of TiO_2 particles (20 nm grain size) exhibits a porous structure and strongly adheres to the catheter, both inside and outside. However, the photocatalytic activity is impractical in dark conditions especially inside a human body.

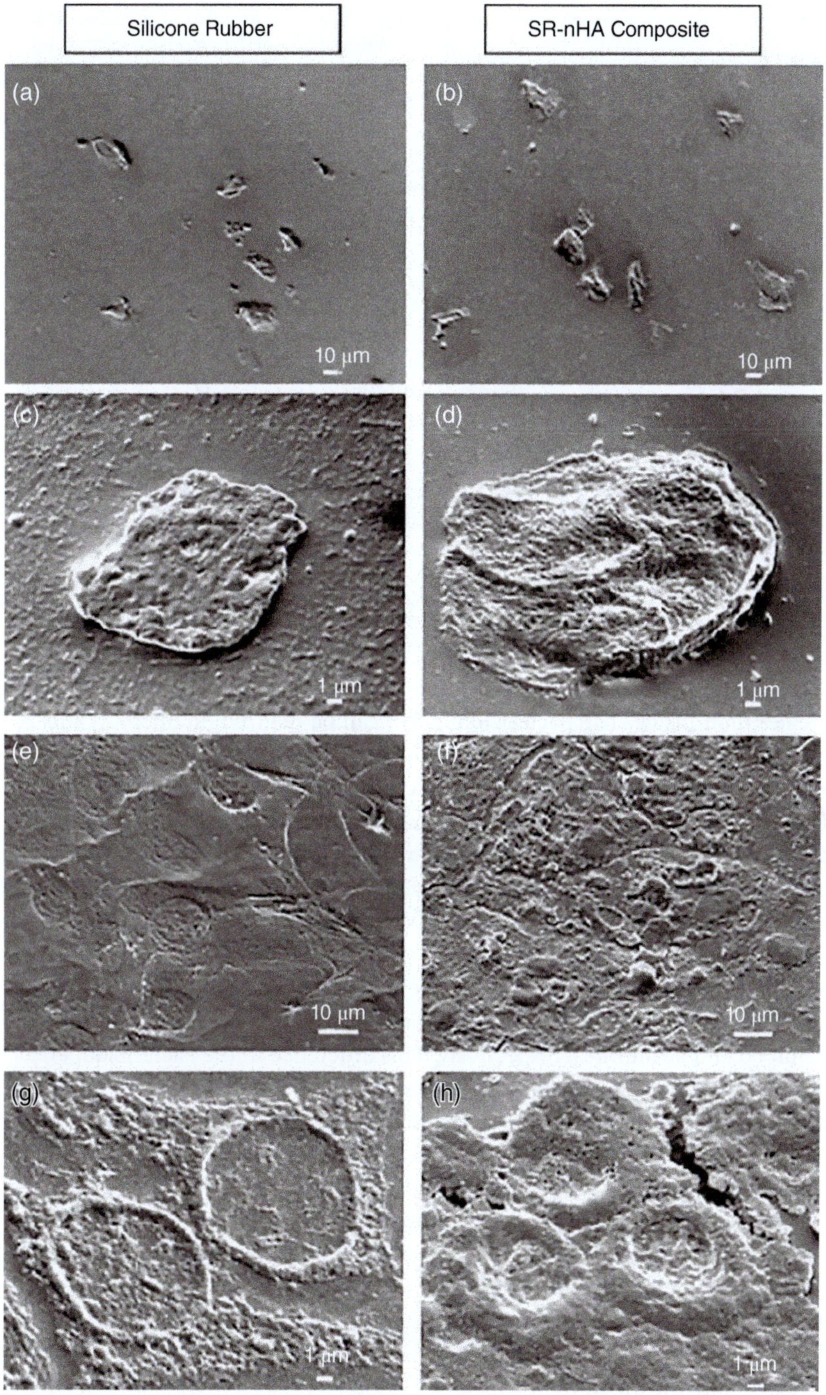

Figure 9.4 Low- and high-magnification scanning electron micrographs of preosteoblasts after 24 hours on pure SR (a and c) and SR–nHA nanocomposite (b and d) surfaces. Cell proliferation and extracellular matrix formation for 14 days of culture are greater on SR–nHA (f and h) in relation to SR (e and g). [11] Thein-Han, W.W., et al., (2009), Figure 04, p 07 / with permission of Elsevier.

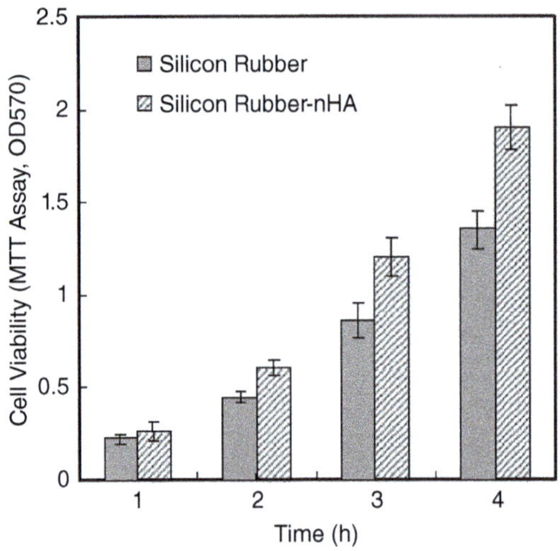

Figure 9.5 Metabolism of preosteoblasts on SR and SR–nHA composite quantified by colorimetric MTT assay. [11] Thein-Han, W.W., et al., (2009), Figure 5b.

Yao et al. [17] suggested that Ag, an antimicrobial metal can be uniformly deposited on a TiO_2-coated catheter via photocatalytic reduction of Ag^+ ions in $AgNO_3$ (2 mM) aqueous solution under UV light irradiation (1 mW cm^{-2}) for 90 seconds. The nanoporous TiO_2 facilitates the infiltration of $AgNO_3$ solution. Ag deposited outside the catheter (8 nmol cm^{-2}) is four times greater than inside. The bactericidal effect can be realized on microorganisms including gram-negative bacteria such as *E. coli* and *Pseudomonas aeruginosa* (*P. aeruginosa*), and the gram-positive strain, *S. aureus*, which are usually associated with nosocomial infections. *E. coli* frequently occurs in urinary tract infections, *P. aeruginosa,* in patients with endotracheal and urinary catheters, and *S. aureus* in venous and peritoneal dialysis catheters. The bactericidal effect of an Ag/TiO_2-coated catheter indicates that the survival fraction of *E. coli* reduces to a negligible level within 20 minutes in dark. This can be ascribed to the diffusion of Ag or Ag^+ through the bacterial cell membrane. A similar efficiency can be observed for TiO_2-coated catheters after 40 minutes of UV irradiation. Compared to *E. coli*, the survival fraction of *P. aeruginosa* and *S. aureus* approaches a nearly zero level within 60 and 90 minutes, respectively. This difference in the bactericidal effect may be related to the lower outer-membrane permeability of *P. aeruginosa* and thickened peptidoglycan cell wall in the case of *S. aureus*, which affects the diffusion process [17].

Among the nosocomial infections, catheter-associated urinary tract infections account for nearly 40% infections worldwide. The urinary catheter allows the entry of bacteria from the outside environment to the urinary tract, weakens the defense mechanism of the bladder thereby making the catheter susceptible to bacterial adhesion and colonization [18]. Though Ag-coated catheters can delay the onset of infection, it is evidenced that bacteriuria increases after one week of catheterization due to the formation of a foundation layer including live and dead bacteria on the coated catheter [19]. Polytetrafluoroethylene (PTFE) coatings, despite the inherent nonstick properties could lead to catheter infections as the rippled surface of these coatings provides sites for bacterial attachment. Zhang et al. developed an Ag-PTFE nanocomposite coating on silicone catheters by incorporating PTFE nanoparticles into an Ag matrix via a wet chemistry method [20]. The process

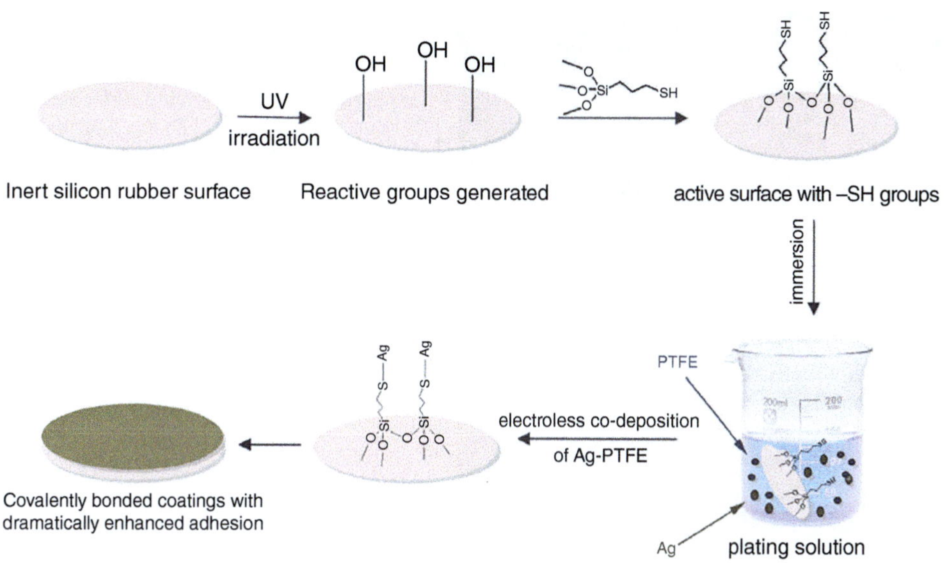

Figure 9.6 Scheme of preparation of Ag/PTFE coatings onto silicone rubber substrate. [21] Guo, R., et al., (2015), Scheme 1.

involves subjecting the silicone rubber to UV irradiation to generate reactive groups on the surface followed by the grafting of silane coupling agent, 3-mercaptopropyl trimethoxysilane (MPTS). The coatings are deposited via electroless plating using mixtures of an $AgNO_3$ solution, reducing agent, and PTFE emulsion (Figure 9.6). The -SH group on the grafted surface reacts with Ag^+ ions to form strong S-Ag interaction while PTFE particles gets uniformly deposited within the coating [21].

Comparing the early bacterial adherence, the Ag-PTFE-coated catheters exhibit the lowest adherence for both *E. coli* and *S. aureus* in contrast to uncoated and Ag-coated catheters. In addition, the antibacterial efficacy of the adhered bacteria is also higher in the case of Ag-PTFE-coated catheters. This is evident from the fluorescence microscopy images (Figure 9.7), which indicates that for both *E. coli* and *S. aureus*, most of the adhered cells are dead (red) while for the uncoated catheter, most of them are alive (green) and only a few dead (red). The antibacterial efficacy can be maintained for up to 14 days of immersion for both *E. coli* and *S. aureus*. The biofilm formation after incubation of the catheters in bacterial suspension for 48 hours is highest for the uncoated catheter while the Ag-PTFE-coated catheters is free from bacterial colonization. The bacterial migration can be understood via an *in vitro* bladder model that mimics the flow of urine through the catheter. Here, sterile artificial urine is dripped via a peristaltic pump at a steady rate into the bladder and drained through the catheter lumen and urethra meatus (Figure 9.8a). With ~10^2 CFU/ml of *E. coli*, the Ag-PTFE-coated catheter requires a significantly longer time (~40 days) to cause bacteriuria in comparison to the uncoated catheter (Figure 9.8b). Interestingly, the silicone catheter surface shows a higher level of biomass accumulation while it increases gradually for the coated catheter demonstrating that Ag-PTFE-coated catheters can maintain a longer period of catheterization.

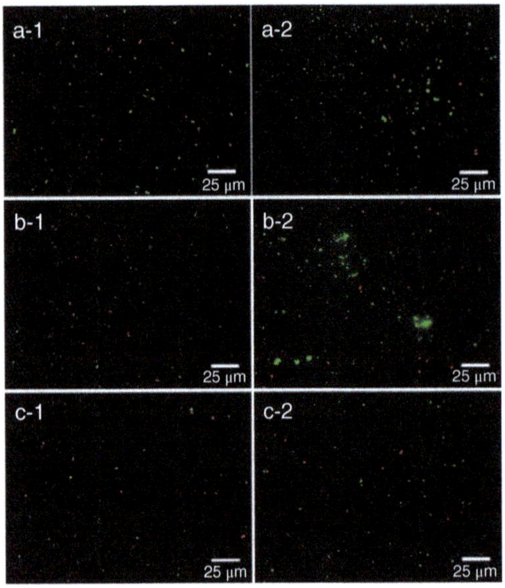

Figure 9.7 Live/dead fluorescent images of adhered (left) *Escherichia coli* and (right) *Staphylococcus aureus* on (a) silicone surface, (b) Ag, and (c) Ag-PTFE-coated surfaces after 24 hours incubation. [21] Zhang, S., et. al., (2019), Figure 03, p 04 / with permission of American Chemical Society.

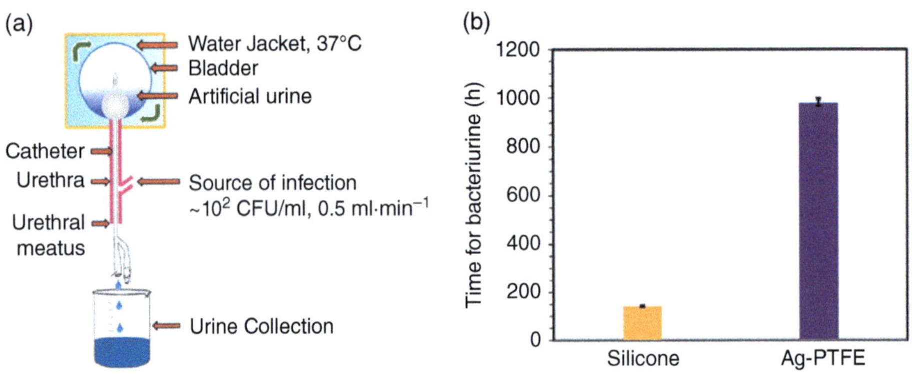

Figure 9.8 (a) Illustrative diagrams of the *in vitro* bladder model; (b) comparisons of time taken for the occurrence of bacteriuria of uncoated versus coated catheters. [21] Zhang, S., et. al., (2019), Figure 9.6.

The cytotoxicity of the catheters evaluated after incubation in mouse fibroblast cells indicated that cell viability is higher for silicone catheters and over 82% for coated catheters. Further, the cell viability persisted throughout the culturing period indicating improved biocompatibility of the Ag-PTFE coating. Moreover, the fibroblast cells cultured on the coated surface spread with a flattened shape due to surface hydrophilicity. The uncoated silicone catheter surface, being hydrophobic, develops cells that have spindle-like morphology with single long arm [21].

Silicone rubber used for peritoneal dialysis catheters are susceptible to infections caused by the main bacteria Coagulase-negative staphylococci (*Staphylococcus epidermis*). Though the use of prophylactic intravenous antibiotics has been attempted to prevent infections, the emergence of antibiotic resistant microbial strains is of major concern [22, 23]. Graphene nanoplatelets (GNPs) also referred to as multilayer graphene (MLG) or few layer graphene (FLG), comprising of stacked (2–10 layers) graphene sheets, have been considered antimicrobial [24]. Although the mechanism is not fully understood, it is agreed that GNPs can act as a membrane piercing structure. In addition, oxidized GNPs can exert oxidative stress resulting in lipid peroxidation, protein inactivation, and mitochondrial disfunction [25, 26]. Gomes et al. [27] have reported the antimicrobial activity of silicone catheters coated with GNP and oxidized GNP (ox-GNP). The particles are immobilized on silicone rubber using dip and spray coating techniques. With dip coating, the individual GNP platelets and agglomerates lie flat on the surface but are not uniformly distributed. In the case of ox-GNP, via spray coating, the wrinkled sheets uniformly cover the silicone surface with some protruding edges. Dip coating do not change the hydrophobicity of GNP-coated silicone while spray coating increases the contact angle with increasing GNP content. The contact angle of ox-GNP-coated samples is almost similar to that of uncoated silicone rubber. To understand the adhesion of the GNPs on the rubber surface, a rubbing test is performed. Dip coating provides good adherence of GNPs and ox-GNPs as the white eraser remains clear after rubbing. Silicone present in the silicone/GNP dispersion effectively binds the GNPs to the silicone film surface. In contrast, spray coating leaves traces of GNPs revealing detachment with increasing concentration while ox-GNPs detaches only at higher concentrations. This may be due to accumulation of multiple layers of GNPs with the topmost layer being less perfectly adhered. The biocompatibility of the materials evaluated *in vitro* by the resazurin assay after 24 hours incubation of NIH3T3 mouse fibroblasts with extracts of silicone/GNP samples show that viability is above 70% in relation to cells incubated in a culture medium. This indicates the noncytotoxic nature of silicone/GNPs though some GNPs detach from the surface. The antibacterial study assessed using *Staphylococcus epidermis* for inoculum reveals that for both dip- and spray-coated samples, the percentage of dead bacteria on silicone/GNPs is almost same as that on uncoated silicones but is significantly higher on silicone/ox-GNPs. Fluorescence microscopy studies reveal that dead bacterial cells are attached to the edges of GNPs suggesting membrane disruption. In contrast, the dead cells are randomly distributed throughout the basal planes in the case of ox-GNPs, indicating the mechanism of membrane damage via oxidative stress [27].

9.3.1.4 *In vitro* Drug Release

Silicone elastomers have been used for fabrication of drug-delivery systems usually for lipophilic drugs where it serves as a carrier. Vasilakos et al. [28] studied the effect of organically modified montmorillonite clay (OMMT) in silicone nanocomposites on the release of hydrophilic drug, metronidazole (2-methyl-5-nitroimidazole-1-ethanol). The drug is an antibiotic for certain bacteria. Large doses of the drug can have adverse effects on the gastrointestinal tract and neurological system. The side effects can be reduced by using a controlled dosage of drugs. It has been reported that 2 phr of OMMT, especially in a low molecular weight PDMS system (~26 000 $g\,mol^{-1}$), can enhance drug release. With the cross-linking of PDMS the propanol produced can locally dissolve the drug in the composite and facilitate its transport in the surrounding buffer solution [28].

9.3.1.5 Surgical Implants

Metallic alloys have been widely accepted for surgical implants. Stainless steel 316L has received great attention owing to its biocompatibility, corrosion resistance, availability, and low cost. For long-term applications, in aggressive biological environments, there are possibilities that stainless steel may corrode and the product accumulates in tissues surrounding the implants. Several *in vitro* and *in vivo* studies have substantiated that the metal ions (corrosion products) released can cause toxic effects on seminiferous epithelium of mouse testis, alter the cellular population of spleen, and, in humans, can alter lymphocyte-surface antigens [29]. In addition, the ions can impair the proliferation and differentiation of osteogenic cells [30]. Besides corrosion, formation of biofilm plays a vital role in implant-associated infections [31]. These issues can be resolved by using a PDMS coating on the metallic substrate. Furthermore, the addition of nanoparticles in PDMS could substantially improve corrosion resistance. Silica, a propitious corrosion inhibitor for stainless steel [32], when added to PDMS can substantially reduce the corrosion rate and improve surface roughness. Nevertheless, these nanocomposites have no antibacterial activity. CuO nanoparticles are well known for their antibacterial property and are cost effective compared to other metal oxides. The antibacterial effect of CuO rises from the intracellular reactive oxygen species (ROS) generation leading to cell damage.

Tavakoli et al. [33] developed PDMS/SiO_2-CuO nanocomposites by incorporating 20 wt.% of silica nanoparticles in PDMS and varying the loading of CuO nanorods (0.5–2 wt.%). They reported that the addition of nano silica promoted the hydrophobicity of PDMS. Further, with the increasing loadings of CuO, the contact angle enhanced significantly. The corrosion resistance of the coatings can be examined using a potentiodynamic polarization test in simulated biological fluid (SBF). Initially, the samples are immersed in SBF for one hour to attain an equilibrium state. The electrochemical cell consists of a platinum sheet as the anode, the sample as the cathode, and Ag/AgCl as the reference electrodes. The test is carried out at a particular potential range versus open circuit potential. Corrosion potential (E_{corr}) and current density (i_{corr}) are the main electrochemical parameters extracted from the data. A more positive value of the corrosion potential of coated samples in contrast to stainless steel reflects the corrosion resistance of PDMS/SiO_2/CuO nanocomposites. The addition of CuO significantly reduces the current density, which may be due to the formation of a denser structure of coating that restricts the diffusion of ions through the surface. This significantly diminishes the electrochemical reactions in the electrolyte and on the surface. Moreover, surface roughness of the coatings could substantially affect the corrosion behavior as air molecules gets trapped in surface pores and protect the surface from a corrosive environment. The nanocomposite coatings exhibit antibacterial activity much better against *E. coli* than *S. aureus*. This is due to the higher thickness of peptidoglycan in the cell wall of gram-positive bacteria than gram-negative bacteria. Further, the coatings containing 0.5 wt.% CuO improve MG63 cell proliferation [33].

Superhydrophobic materials have the potential to be used as blood compatible coatings in biomedical products. They have been shown to exhibit antibacterial behavior. A combination of superhydrophobicity and antibacterial agents are expected to result in highly efficient antibacterial materials [34]. Incorporation of nanoparticles in coatings can induce surface roughness and thereby improve superhydrophobicity. Among various Ag-based compounds, silver phosphate (Ag_3PO_4) possesses strong antibacterial activity. The main advantage of using Ag_3PO_4 is the slow release of Ag^+ ions, which would result in lower cytotoxicity. Seyfi et al. [35] developed superhydrophobic antibacterial coatings using Ag_3PO_4 in a PDMS coating. The wettability of the coating can be evaluated by measuring the water contact angle. With a 1 wt.% addition of Ag_3PO_4, the water contact angle increased from 105° to 123° and the sliding angle reduced from 90° to 65°. The addition of higher contents up to 7 wt.% can lead to aggregation of nanoparticles on the surface layers that in turn creates voids and grooves, thus reducing solid–liquid contact area. It has been noticed that with 1 and 7 wt. % of Ag_3PO_4 in the coatings there is an 80% and 99.9% reduction in the adhesion of *S. aureus* and *E. coli*, respectively. Many mechanisms have been proposed for the antibacterial activity of Ag_3PO_4. One of the proposed mechanisms suggests an electrostatic attraction between the negatively charged bacterial cell wall and positively charged Ag^+ ions released. Moreover, the superhydrophobic surfaces may enhance the distance between the cells and the solid surface so that the bacteria have to overcome the kinetic barrier to adhere to the surface. An alternative mechanism suggests the mechanical rupture of the cell membrane once it encounters the nanostructures on the surface. However, the surface roughness induced by a higher content of nanostructures on coatings can reduce the extent of protein adsorption from fresh human platelet poor plasma (PPP) [35].

9.3.1.6 Bone Regeneration

Bone defects caused by injury, abnormalities in growth, tumor, or infection are one of the key issues in the medical field. Autologous bone grafting is the traditional method of treatment. However, the limitations associated with bone grafting include blood loss, lengthy surgery time, post-operative pain, infection, and changes in bone quality. Bone-tissue engineering is considered to be a superior alternative to bone transplantation, the key advantages being minor damage, vast sources, and no antigenicity. Scaffold is one of the essential elements for bone defect regeneration, which controls the survival, colonization, migration of seed cells, cell proliferation, osteogenesis etc. Scaffold materials should have an osteoinductive property that is, the ability to stimulate differentiation of cells towards osteoblasts. For promoting *in vivo* new bone formation, the conventional approach is to add vascular-endothelial growth factor, platelet-derived growth factor, or cytokine of bone morphogenetic protein 2 to the scaffolds. The addition of zinc and silicon to scaffolds can promote the formation of osteoblast and calcification of bone, respectively.

Recently, graphene oxide foam (GOF) has received much attention as scaffold materials due to their porous structure and large specific surface area that can promote migration, proliferation, and differentiation of bone marrow mesenchymal stem cells. Nonetheless, it has poor mechanical strength to support cell attachment and tissue growth. The addition of PDMS to GOF can improve the mechanical properties without altering the cell compatibility. The hydrophobicity of PDMS renders it nonconducive to cell growth. This can be overcome by providing a dopamine deposition on the PDMS's surface. Further, zinc silicate particles, when added to the scaffold, supply zinc and silicon ions that can induce osteogenic differentiation. In comparison to GOF, the composite exhibits a higher yield strength. Also, the porosity of the composite PDMS/GOF/ZS is significantly higher compared to GOF and PDMS/GOF. The cell viability evaluated by a CCK-8 (cell-counting kit 8) method is higher for the composites indicating that it promotes the proliferation of mouse bone marrow mesenchymal stem cells (mBMSCs). The cells not only spread on the surface of the composite scaffold but also create intercell overlapping with good adhesion indicating cell biocompatibility. Moreover, the composite can guide the mineralization of mesenchymal stem cells thus promoting osteogenesis. *In vivo* analysis of critical bone defects in rabbits revealed new bone-tissue formation after 12 weeks of implantation of GOS/PDMS/ZS scaffold with no significant inflammatory response [36].

9.3.2 Polyurethane Nanocomposites

Polyurethane catheters have been a main choice for implantable devices. The addition of Ag compounds can improve the efficiency as they are antibacterial agents effective against both anaerobic and aerobic bacteria. Ag nanoparticles attach to the surface of the cell membrane and in close proximity to body fluids, secretions, or moisture release Ag ions owing to the ionization of metallic Ag. The extent of ionization of metallic Ag depends on the surface of the nanoparticles. Ag ions being oligodynamic and biologically active bind firmly with electron donor moieties such as amino, phosphate, carboxylate, imidazole, or sulfhydryl groups and bind receptor groups on the surface of cells, tissue debris, and bacterial or fungal cell envelope [37]. An Ag-impregnated PU (polyurethane) catheter of 30 cm length was found to release $0.1\,\mu g\,l^{-1}$ of Ag ion [38].

9.3.2.1 Scaffolds

Filip et al. [39] developed Ag-bionanocomposite membranes based on PU and an extracellular matrix (EM) using two techniques viz. solvent casting and electrospinning onto a PU membrane. The electrospinning technique is an electrohydrodynamic process for acquiring nano or microfiber membranes with a high surface area and a high molecular orientation with tunable fiber diameter, topology, surface chemistry etc. The advantage of using DMF as a solvent for PU is that it can also serve as a reducing agent for Ag nitrate. The addition of an EM such as k-elastin (KEL), hydrolyzed collagen (HC), chondroitin sulphate (CS), and hyaluronic acid improves the biological properties especially antithrombogenicity on long-term contact with blood and tissues. The benefit of using an electrospinning technique is that it results in fiber scaffolds of a submicron level that resembles EM. Moreover, the high surface-area-to-volume ratio of the fiber scaffolds supports cell attachment, growth, and proliferation compared to solvent cast PU. Mouse fibroblasts cells cultivated in the presence of such membranes have revealed that Ag content less than 0.3% assures a high proliferation of cells with rounded and/or spindle-shaped morphology [39]. Considering the antimicrobial activity, the surface-active Ag of electrospun membranes inhibits the bacterial growth in the following order: *Salmonella typhimurium* > *Listeria monocytogenes* > *E. coli*. Ag nanoparticles can strongly interact with the biocomponents or the urethane group of PU thereby restricting the release of metallic Ag in the surrounding biological environment [40].

Bahrami et al. [41] studied the cell-supporting behavior of electrospun mats made from polyurethane/graphene nanocomposites. A PU solution is initially prepared in a mixture of THF (tetrahydrofuran) and DMAc (dimethylacetamide), and GNPs are dispersed via ultrasonication and then electrospun. The high surface area of the fibrous scaffold stimulates cell–cell and cell–matrix interactions indicated by cells having a wider morphology with more filopodia. When incubated with [human umbilical vein] endothelial cells (HUVECs), GNPs at 5 wt. % in PU promoted cell proliferation. At 10 wt.% GNP, an antiangiogenic property is observed, which might be due to the release of GNP in the environment [41].

9.3.2.2 Bone-tissue Engineering

Shape memory polymers (SMPs) are promising candidates for scaffolds for tissue engineering because of their tunable mechanical properties and biodegradability [42]. These are smart materials that can be fixed in a temporary shape by physical or chemical means and when exposed to stimulus, can be recovered to a permanent shape [43]. In tissue engineering, such SMP scaffolds can be seeded with cells and using minimally invasive surgery can be implanted in the injury site. Subsequently, the wound recovers fast and, with time, the scaffolds degrade without the need for a second surgery [44]. Recently, scaffolds for bone-tissue engineering have been fabricated using three-dimensional (3D) printing, which is a layer-by-layer additive manufacturing technique. This technique allows customized production of scaffolds via computer-aided design [45–47]. PU can serve as a 3D-printing material as it exhibits both a shape memory effect as well as biodegradability. The degradation rate and mechanical properties can be tuned by varying the relative fractions of hard and soft segments [48]. Moreover, the use of water-borne polyurethane (WPU) makes the process environment friendly.

The process of bone-tissue engineering includes migration, proliferation, differentiation of osteogenic cells, and the formation of EM. It has been noticed that superparamagnetic iron oxide nanoparticles (SPIO) fosters osteogenic induction of human bone derived mesenchymal stem cells (hBMSCs) [49]. When incorporated in PU, it improves the shape memory effect. Wang et al. [50] prepared 3D-printing ink using PU, SPIO nanoparticles and polyethylene oxide or gelatin for fabricating scaffolds. The soft segment of PU is based on poly(ε-caprolactone) diol and poly(L-lactic acid) diol in the ratio 6 : 4 while the isocyanate and chain extenders are: isophorone diisocyanate (IPDI), 2,2-bis(hydroxymethyl)propionic acid (DMPA), and ethylenediamine (EDA). Polyethylene oxide or gelatin is used to adjust the viscosity suitable for printing ink. The printing of such scaffolds is carried out using a low-temperature fuse-deposition manufacturing platform followed by freeze drying (Figure 9.9). The content of SPIO in the scaffold (Figure 9.10) is 500 ppm.

On analysis of the degradation of the scaffolds by placing in PBS solution, it is found that the weight remaining after first week is 76% for PU/PEO/SPIO and 85% for PU/gelatin/

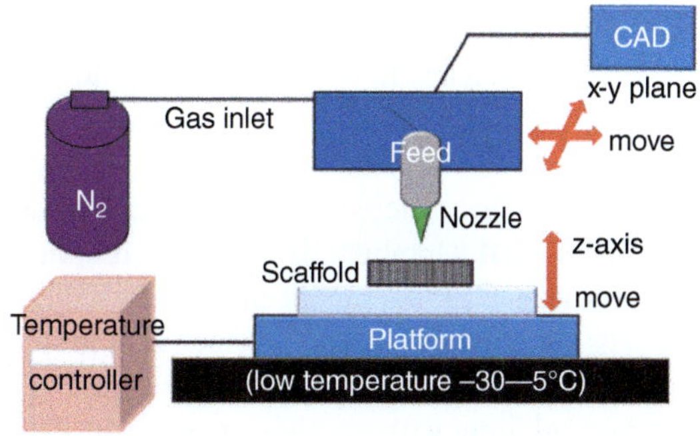

Figure 9.9 Fabrication of shape memory scaffolds by 3D printing using a low-temperature fuse deposition manufacturing (LFDM) platform. [50] Wang, Y. J. et al., (2018), Figure 1 B.

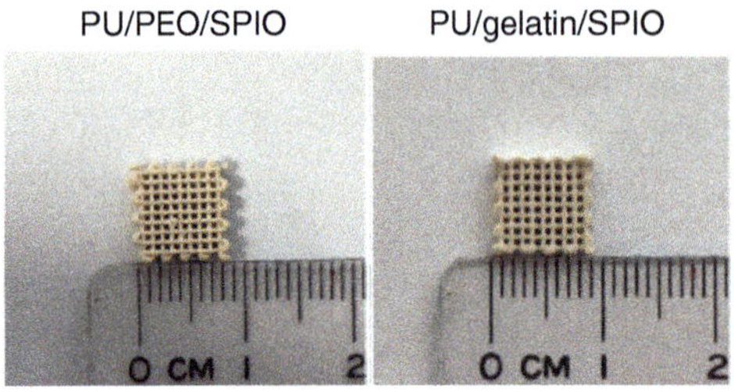

Figure 9.10 Structure of scaffolds. [50] Wang, Y. J., et. al., (2018), Figure 02, p 04 / with permission of American Chemical Society.

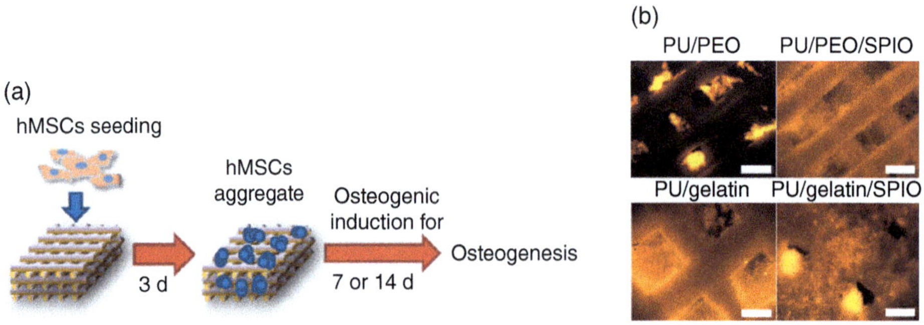

Figure 9.11 (a) Schematics of hMSC seeding procedure (b) Fluorescent images of hMSCs in 3DP scaffolds after 24h of culture. Cells were labelled by PKH26 (Red Fluoresence). Scale bar, 400 μm. [50] Wang, Y. J., et al., (2018), Figure 07B2, p 06 / with permission of American Chemical Society.

SPIO. On seeding the scaffolds with hMSCs (human mesenchymal stem cells), proliferation is better with PU/gelatin/ SPIO than with PU/PEO/SPIO (Figure 9.11). The minerals get deposited not only in the gap areas but also along the stacking fibers. Additionally, collagen secretion is higher for PU/PEO/SPIO indicating significant osteogenesis. Moreover, the gradual release of SPIO from the scaffolds promotes osteogenesis of hMSCs [50].

9.3.2.3 Neural-tissue Engineering

Injury of peripheral nervous system (PNS) arising due to trauma, if untreated, may lead to serious disabilities in later life [51]. The various approaches for nerve reconstruction include suturing, grafting, and tissue engineering. Recent advancements in tissue engineering have led to the development of biodegradable nerve conduits that can promote the regeneration and bridging of defective nerve(s) [52]. For engineering of soft tissues, elastic scaffolds combined with controlled biodegradability is desirable [53]. WPU, based on polycaprolactone diol, is flexible, biocompatible, and biodegradable and is therefore, a potential candidate for the fabrication of nerve conduits. Moreover, the asymmetric microporous structure that favors nerve regeneration can be easily generated in WPU via the freeze drying process [54]. Graphene is known to inhibit the growth of bacteria on its surface. In addition, it enables attachment and differentiation of neural stem cells (NSCs) and endothelial cells (ECs) *in vitro* [55]. The content of graphene in PU beyond 5 wt. % is found to be cytotoxic and can lose its elastomeric nature. The proliferation of ECs after three days decreased slightly with graphene content in PU (PU/G) in contrast to PU. However, the growth and differentiation of NSCs was significantly greater for PU/G5 than PU after three days. When cocultured with NSCs/ECs (1 : 1), the growth and differentiation was faster in the case of PU-G5 without inducing a culture medium. Graphene may promote the differentiation of NSCs to glial-like cells and the crosstalk between NSCs and ECs inclines the differentiation of ECs to vascular-like cells. This may be ascribed to the interaction between the π-π stacking structure of graphene and the cells. After implantation of the nerve conduit in rats, the functional recovery was assessed by walking track analysis. The sciatic functional index (SFI) is calculated using the following formula:

$$SFI = -18.3(\textit{print length function}) + 109.5(\textit{toe spread function}) \\ + 13.3(\textit{intermediate toe spread function}) - 8.8$$

where

$$\textit{print length function} = \frac{(\textit{experimental print length} - \textit{normal print length})}{\textit{normal print length}}$$

$$1st$$

$$\textit{toe spread function} = \frac{(\textit{experimental toe spread} - \textit{normal toe spread})}{\textit{normal toe spread}(5th\ toe)}$$

$$2nd$$

$$\textit{intermediate toe spread function} = \frac{(\textit{experimental intermediate toe} - \textit{normal intermediate toe})}{\textit{normal intermediate toe}(4th\ toe)}$$

The SFI increased from −110 to −55 in six weeks. The footprint analysis indicated that the experimental hind leg supported the body weight of the rat. The axial MRI indicated that axons sprouted from both ends and extended to the middle of the conduit. The regenerated axons in the case of PU/G5 displayed a 15% larger cross-sectional area in comparison to the PU conduit. Moreover, blood vessels and endoneurium were found to be present in PU/G5. During nerve regeneration, first the blood vessels form to transport nutrients and produce growth factors for sprouting of axons [56].

9.3.3 Natural Rubber Nanocomposites

Natural rubber latex (NRL) is composed of approximately 45% cis-polyisoprene, 5% of non-rubber constituents such as proteins, lipids, carbohydrates, and nearly 50% water. NRL is associated with allergic problems when used on the human body, which is a matter of great concern. The Type I allergy is caused by natural proteins while the Type IV allergy is due to the toxic chemicals needed for converting NRL to desirable products [57]. The Type IV allergy can be alleviated by avoiding toxic chemicals, the protein content in NRL can be substantially reduced by centrifugation and proteolytic enzyme treatment [58]. Nevertheless, it finds application as curative bandages in diabetic ulcer patients [59], as an occlusive membrane in guided bone regeneration [60], and as a vascular prosthesis [61] thus offering repair or replacement of defective tissues. NRL membrane induces angiogenesis, which is involved in normal growth and wound healing [62]. The *in vitro* cytocompatibility of NRL has been studied indicating that NRL nanoparticles are toxic to human lung carcinoma (A549) cells. It also promotes cell proliferation of mouse calvaria preosteoblast (MC3T3-E1) cells.

9.3.3.1 Bone Regeneration

A natural rubber/ HA nanocomposite is a suitable candidate for bone-related applications as HA is osteoconductive. For composites with a high content of HA, simple physical mixing in the presence of solvent is not acceptable due to the tendency of agglomeration. A wet chemical synthesis can be adopted by mimicking the biomineralization process where the inorganic phase forms in the presence of organic molecules that work as a template for nucleation and growth. This offers a better control on the morphology of HA particles and interaction between the inorganic HA and organic NR that are preferred in tissue engineering [63, 64].The *in situ* synthesis of HA involves initially the dissolution of NR in THF followed by addition of calcium nitrate tetrahydrate [$Ca(NO_3)_2.4H_2O$] in THF. HA forms with the addition of orthophosphoric acid (H_3PO_4) in the presence of an NH_4OH solution that maintains the $pH > 8$. The presence of proteins and phospholipids in NR acts as nucleation sites for Ca^{2+} leading to precipitation of HA thus facilitating homogenous crystallites and uniform dispersion of HA in NR. The preliminary biocompatibility test using MTT assay with HepG2 cells indicates that 71–91% of cells are viable at 72 hours of culture for composites (20 and 30 wt. % HA) [65]. However, *in vivo* studies are important to evaluate their performance in bone healing.

NR latex nanoparticles covered with HA can be fabricated via biomineralization using simulated body fluid (SBF) that have ion concentrations approximately equal to blood plasma [66]. NRL is mixed with SBF and incubated for 24 hours at 36.5 °C to form NR/HA particles that are centrifuged, dialyzed to remove free ions, and subsequently freeze dried. The electrostatic interaction between the negatively charged layers of NRL and Ca^{2+} ions causes the deposition of calcium ions, which further interacts with phosphate ions $(PO_3)^{4-}$ in SBF to form HA that grows spontaneously. The cell viability assessed by WST-8 assay indicates that MC3T3-E1 cells are not sensitive to NRL/HA particles. However, the HA particles present on the surface of NRL particles enhances the viability of A549 cells even at concentrations of $1.2\,mg\,ml^{-1}$. Moreover, it promotes Alp (alkaline phosphatase) gene expression, which is an early marker for osteoblast phenotype and calcified matrix formation. After incubation with osteoblast supplements for 20 days, there is no significant change in osteogenic differentiation of MC3T3-E1 cells for both nanoparticles compared to control [67].

9.3.3.2 Wound Dressing

There exists a wide range of wound dressings depending on the severity, type, and position of wounds. Besides hydrocolloid and hydrogel that are commonly used, elastic dressings are available for wounds that occur as a result of irregular functioning of venous valves. Such dressings can promote new skin tissue, slowly stretch out the vein walls, improve circulation, and, in addition, restrict swelling of the wound. It has been reported that a moist wound environment is essential for wound healing as it encourages new tissue generation and prevents crusty scabs. Elastic-gel dressings comprised of NR/silk sericin/chitin whiskers have been developed where sericin exhibits skin moisturizing properties. Sericin is the glue protein that surrounds the fibroin fiber in silk. It contains 18 amino acids of which 30% is serine, having the natural moisture factor. The moisturizing effect of sericin has been evidenced by reduction in skin impedance, hydration of epidermal cells, smoothness of upper layer of skin etc. [68]. Furthermore, it is nonimmunogenic [69], and promotes attachment and growth of human skin fibroblasts [70], cell migration [71], and healing of second-degree burn wounds [72]. The major drawbacks of sericin are high water solubility and poor mechanical properties. This can be overcome by binding sericin with highly elastic and flexible NRL. Further, the allergic proteins in NRL could be eliminated by centrifugation followed by enzyme treatment for 48 hours. Besides a moist environment provided by sericin, bioactive molecules such as growth factors and drugs are of prime importance for tissue regeneration. Chitin, a polymer of N-acetyl glucosamine has outstanding tissue regeneration capability for wound healing. Once a wound occurs, healing starts with an immediate release of growth factors and enzymes including lysozyme, which is an antibacterial enzyme. Lysozyme can easily hydrolyze chitin to smaller molecules of glucosamine, which could enhance the rate of wound closure due to the high proliferation of fibroblast and hair follicles [73]. Chitin whiskers with a large surface area can effectively interact with compounds in the living tissues and thus induce granulation tissue formation. Thus, by using a simple solvent casting technique, NR/sericin (SRC)/chitin whisker films (Figure 9.12) can be fabricated as an elastic dressing that are applicable for wounds that require application of pressure to shield the wound. About 30% of sericin can be blended in NR without affecting its film forming ability. The maximum content of chitin whisker that can be incorporated in the NR/SRC is 15% beyond which the strength deteriorates. However, only small amount of chitin whiskers (~2%) eases the removal of sericin protein from the film. A high chitin content contributes to a lower release of SRC, which may be due to the H-bond formation between chitin and SRC. The sericin in the film bestows the water sorption capacity thus functioning as moisturizing agent. The films that are flexible and transparent in the dry state, when wet becomes soft and conforms to the skin (Figure 9.10). They could be stretched up to 300% in the wet state at room temperature. The cell viability of the films is greater than 80% indicating the potential application in elastic dressings [74].

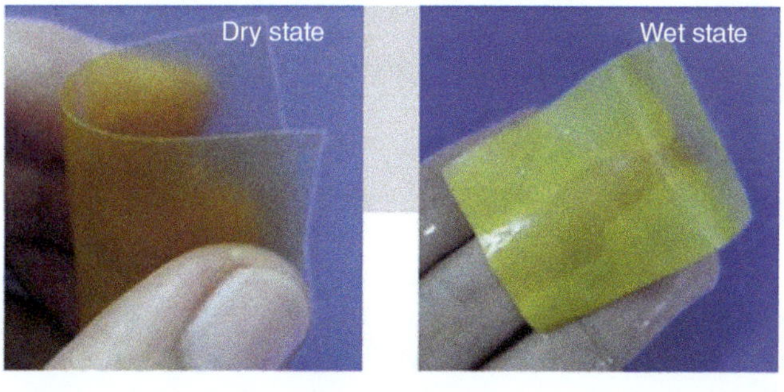

Figure 9.12 Appearance of the ETNR/SRC film containing 30% SRC and the ETNR/SRC/CTW nanocomposite film containing 30% SRC and 15% CTW (%w/w, based on the weight of ETNR) in their dry and wet states. [74] Watthanaphanit, A. & Rujiravanit, R. (2017), Figure 03, p 25 / with permission of Elsevier.

The angiogenic property of NRL and the antibacterial property of Ag nanoparticles can be combined to form a suitable wound dressing material. The dispersion of Ag nanoparticles in an NRL membrane can be ensured as the protein part in NR latex binds on the surface of the nanoparticle and thus prevents their agglomeration. Further, composites containing 0.4 wt.% of nanoparticles can gradually release Ag that are not attached to the membrane thus, augmenting the angiogenic property of the matrix [75].

9.4 Conclusions

There has been enormous advancement in the field of rubber nanocomposites due to the specific properties of nanofillers, which opens new avenues to tailor the structures and explore their use in biomedical applications. The nanofillers can impart specific properties such as antibacterial, hydrophobicity, osteogenesis, angiogenesis to the rubber nanocomposites resulting in a multitude of applications – catheters, scaffolds, bone regeneration, tissue regeneration, and wound dressing. In order to realize the full potential of the materials in these applications, the many challenges associated with the nanocomposites still need to be addressed and relevant preclinical studies carried out. Further, an insight into the interaction of these nanocomposites with the human body needs to be studied in detail.

References

1 Yoda, R. (1998). Elastomers for biomedical applications. *Journal of Biomaterials Science, Polymer Edition* 9 (6): 561–626.

2 Klevens, R.M., Edwards, J.R., Richards, C.L. Jr. et al. (2007). Estimating health care-associated infections and deaths in US hospitals, 2002. *Public Health Reports* 122 (2): 160–166.

3 Rai, M., Yadav, A., and Gade, A. (2009). Silver nanoparticles as a new generation of antimicrobials. *Biotechnology Advances* 27 (1): 76–83.

4 Jena, P., Mohanty, S., Mallick, R. et al. (2012). Toxicity and antibacterial assessment of chitosancoated silver nanoparticles on human pathogens and macrophage cells. *International Journal of Nanomedicine* 7: 1805.

5 Klibanov, A.M. (2007). Permanently microbicidal materials coatings. *Journal of Materials Chemistry* 17 (24): 2479–2482.

6 Sankar, G.G., Sathya, S., Murthy, P.S. et al. (2015). Polydimethyl siloxane nanocomposites: their antifouling efficacy in vitro and in marine conditions. *International Biodeterioration and Biodegradation* 104: 307–314.

7 Mah, T.F.C. and O'Toole, G.A. (2001). Mechanisms of biofilm resistance to antimicrobial agents. *Trends in Microbiology* 9 (1): 34–39.

8 Sankar, G.G., Murthy, P.S., Das, A. et al. (2017). Polydimethyl siloxane based nanocomposites with antibiofilm properties for biomedical applications. *Journal of Biomedical Materials Research Part B: Applied Biomaterials* 105 (5): 1075–1082.

9 Roy, N., Dey, T., Kundu, S.C., and Bhowmick, A.K. (2013). Biocompatible composites of fibrous nanohydroxyapatite embedded in a polydimethylsiloxane. *Journal of Materials Science* 48 (15): 5132–5139.

10 Suzuki, T., Hukkanen, M., Ohashi, R. et al. (2000). Growth and adhesion of osteoblast-like cells derived from neonatal rat calvaria on calcium phosphate ceramics. *Journal of Bioscience and Bioengineering* 89 (1): 18–26.

11 Thein-Han, W.W., Shah, J., and Misra, R.D.K. (2009). Superior in vitro biological response and mechanical properties of an implantable nanostructured biomaterial: nanohydroxyapatite–silicone rubber composite. *Acta Biomaterialia* 5 (7): 2668–2679.

12 Anselme, K. (2000). Osteoblast adhesion on biomaterials. *Biomaterials* 21 (7): 667–681.

13 Stoclin, A., Rotolo, F., Hicheri, Y. et al. (2020). Ventilator-associated pneumonia and bloodstream infections in intensive care unit cancer patients: a retrospective 12-year study on 3388 prospectively monitored patients. *Supportive Care in Cancer* 28 (1): 193–200.

14 Filippo, A.D. and Gaudio, A.D. (2003). Device-related infections in critically ill patients. Part II: prevention of ventilator-associated pneumonia and urinary tract infections. *Journal of Chemotherapy* 15 (6): 536–542.

15 Tenke, P., Mezei, T., Bőde, I., and Köves, B. (2017). Catheter-associated urinary tract infections. *European Urology Supplements* 16 (4): 138–143.

16 Ohko, Y., Utsumi, Y., Niwa, C. et al. (2001). Self-sterilizing and self-cleaning of silicone catheters coated with TiO_2 photocatalyst thin films: a preclinical work. *Journal of Biomedical Materials Research: Applied Biomaterials* 58 (1): 97–101.

17 Yao, Y., Ohko, Y., Sekiguchi, Y. et al. (2008). Self-sterilization using silicone catheters coated with Ag and TiO2 nanocomposite thin film. *Journal of Biomedical Materials Research Part B: Applied Biomaterials* 85 (2): 453–460.

18 Feneley, R.C., Hopley, I.B., and Wells, P.N. (2015). Urinary catheters: history, current status, adverse events and research agenda. *Journal of Medical Engineering and Technology* 39 (8): 459–470.

19 Stickler, D.J. and Morgan, S.D. (2008). Observations on the development of the crystalline bacterial biofilms that encrust and block Foley catheters. *Journal of Hospital Infection* 69 (4): 350–360.

20 Zhang, S., Wang, L., Liang, X. et al. (2019). Enhanced antibacterial and antiadhesive activities of silver-PTFE nanocomposite coating for urinary catheters. *ACS Biomaterials Science and Engineering* 5 (6): 2804–2814.

21 Guo, R., Yin, G., Sha, X. et al. (2015). The significant adhesion enhancement of Ag–polytetrafluoroethylene antibacterial coatings by using of molecular bridge. *Applied Surface Science* 341: 13–18.

22 Nessim, S.J. and Jassal, S.V. (2012). Gentamicin-resistant infections in peritoneal dialysis patients using topical gentamicin exit-site prophylaxis: a report of two cases. *Peritoneal Dialysis International* 32 (3): 339–341.

23 Toleman, M.A. (2017). The future of peritoneal dialysis in a moving landscape of bacterial resistance. *Peritoneal Dialysis International* 37 (2): 134–140.

24 Szunerits, S. and Boukherroub, R. (2016). Antibacterial activity of graphene-based materials. *Journal of Materials Chemistry B* 4 (43): 6892–6912.

25 Henriques, P.C., Borges, I., Pinto, A.M. et al. (2018). Fabrication and antimicrobial performance of surfaces integrating graphene-based materials. *Carbon* 132: 709–732.

26 Zou, X., Zhang, L., Wang, Z., and Luo, Y. (2016). Mechanisms of the antimicrobial activities of graphene materials. *Journal of the American Chemical Society* 138 (7): 2064–2077.

27 Gomes, R.N., Borges, I., Pereira, A.T. et al. (2018). Antimicrobial graphene nanoplatelets coatings for silicone catheters. *Carbon* 139: 635–647.

28 Vasilakos, S.P. and Tarantili, P.A. (2012). In vitro drug release studies from organoclay/poly (dimethyl siloxane) nanocomposite matrices. *Journal of Biomedical Materials Research Part B: Applied Biomaterials* 100 (7): 1899–1910.

29 Tracana, R.B., Sousa, J.P., and Carvalho, G.S. (1994). Mouse inflammatory response to stainless steel corrosion products. *Journal of Materials Science: Materials in Medicine* 5 (9–10): 596–600.

30 Morais, S., Sousa, J.P., Fernandes, M.H. et al. (1998). Decreased consumption of Ca and P during in vitro biomineralization and biologically induced deposition of Ni and Cr in presence of stainless steel corrosion products. *Journal of Biomedical Materials Research* 42 (2): 199–212.

31 Zhao, L., Chu, P.K., Zhang, Y., and Wu, Z. (2009). Antibacterial coatings on titanium implants. *Journal of Biomedical Materials Research Part B: Applied Biomaterials* 91 (1): 470–480.

32 Vasconcelos, D.C.L., Carvalho, J.A.N., Mantel, M., and Vasconcelos, W.L. (2000). Corrosion resistance of stainless steel coated with sol–gel silica. *Journal of Non-crystalline Solids* 273 (1–3): 135–139.

33 Tavakoli, S., Nemati, S., Kharaziha, M., and Akbari-Alavijeh, S. (2019). Embedding CuO nanoparticles in PDMS-SiO_2 coating to improve antibacterial characteristic and corrosion resistance. *Colloid and Interface Science Communications* 28: 20–28.

34 Zhang, X., Wang, L., and Levänen, E. (2013). Superhydrophobic surfaces for the reduction of bacterial adhesion. *RSC Advances* 3 (30): 12003–12020.

35 Seyfi, J., Goodarzi, V., Wurm, F.R. et al. (2020). Developing antibacterial superhydrophobic coatings based on polydimethylsiloxane/silver phosphate nanocomposites: Assessment of surface morphology, roughness and chemistry. *Progress in Organic Coatings* 149: 105944.

36 Li, Y., Zhang, X., Dai, C. et al. (2020). Bioactive three-dimensional graphene oxide foam/polydimethyl siloxane/zinc silicate scaffolds with enhanced osteoinductivity for bone regeneration. *ACS Biomaterials Science and Engineering* 6 (5): 3015–3025.

37 Feng, Q.L., Wu, J., Chen, G.Q. et al. (2000). A mechanistic study of the antibacterial effect of silver ions on Escherichia coli and Staphylococcus aureus. *Journal of Biomedical Materials Research* 52 (4): 662–668.

38 Joyce-Wöhrmann, R.M. and Münstedt, H. (1999). Determination of the silver ion release from polyurethanes enriched with silver. *Infection* 27 (1): S46–S48.

39 Filip, D., Macocinschi, D., Paslaru, E. et al. (2014). Polyurethane biocompatible silver bionanocomposites for biomedical applications. *Journal of Nanoparticle Research* 16 (11): 1–17.

40 Macocinschi, D., Filip, D., Paslaru, E. et al. (2015). Polyurethane–extracellular matrix/silver bionanocomposites for urinary catheters. *Journal of Bioactive and Compatible Polymers* 30 (1): 99–113.

41 Bahrami, S., Solouk, A., Mirzadeh, H., and Seifalian, A.M. (2019). Electroconductive polyurethane/graphene nanocomposite for biomedical applications. *Composites Part B: Engineering* 168: 421–431.

42 Small, W. IV, Singhal, P., Wilson, T.S., and Maitland, D.J. (2010). Biomedical applications of thermally activated shape memory polymers. *Journal of Materials Chemistry* 20 (17): 3356–3366.

43 Berg, G.J., McBride, M.K., Wang, C., and Bowman, C.N. (2014). New directions in the chemistry of shape memory polymers. *Polymer* 55 (23): 5849–5872.

44 Lendlein, A. and Langer, R. (2002). Biodegradable, elastic shape-memory polymers for potential biomedical applications. *Science* 296 (5573): 1673–1676.

45 Wu, G.H. and Hsu, S.H. (2015). Polymeric-based 3D printing for tissue engineering. *Journal of Medical and Biological Engineering* 35 (3): 285–292.

46 Inzana, J.A., Olvera, D., Fuller, S.M. et al. (2014). 3D printing of composite calcium phosphate and collagen scaffolds for bone regeneration. *Biomaterials* 35 (13): 4026–4034.

47 Tarafder, S., Balla, V.K., Davies, N.M. et al. (2013). Microwave-sintered 3D printed tricalcium phosphate scaffolds for bone tissue engineering. *Journal of Tissue Engineering and Regenerative Medicine* 7 (8): 631–641.

48 Chen, Q., Liang, S., and Thouas, G.A. (2013). Elastomeric biomaterials for tissue engineering. *Progress in Polymer Science* 38 (3–4): 584–671.

49 Wang, Q., Chen, B., Cao, M. et al. (2016). Response of MAPK pathway to iron oxide nanoparticles in vitro treatment promotes osteogenic differentiation of hBMSCs. *Biomaterials* 86: 11–20.

50 Wang, Y.J., Jeng, U.S., and Hsu, S.H. (2018). Biodegradable water-based polyurethane shape memory elastomers for bone tissue engineering. *ACS Biomaterials Science and Engineering* 4 (4): 1397–1406.

51 Noble, J., Munro, C.A., Prasad, V.S., and Midha, R. (1998). Analysis of upper and lower extremity peripheral nerve injuries in a population of patients with multiple injuries. *Journal of Trauma and Acute Care Surgery* 45 (1): 116–122.

52 Gaudin, R., Knipfer, C., Henningsen, A. et al. (2016). Approaches to peripheral nerve repair: generations of biomaterial conduits yielding to replacing autologous nerve grafts in craniomaxillofacial surgery. *Bio Med Research International* 2016: 3856262.

53 Guan, J., Fujimoto, K.L., Sacks, M.S., and Wagner, W.R. (2005). Preparation and characterization of highly porous, biodegradable polyurethane scaffolds for soft tissue applications. *Biomaterials* 26 (18): 3961–3971.

54 Hsu, S.H., Chang, W.C., and Yen, C.T. (2017). Novel flexible nerve conduits made of water-based biodegradable polyurethane for peripheral nerve regeneration. *Journal of Biomedical Materials Research Part A* 105 (5): 1383–1392.

55 Park, S.Y., Park, J., Sim, S.H. et al. (2011). Enhanced differentiation of human neural stem cells into neurons on graphene. *Advanced Materials* 23 (36): H263–H267.

56 Lee, T.H., Yen, C.T., and Hsu, S.H. (2020). Preparation of polyurethane-graphene nanocomposite and evaluation of neurovascular regeneration. *ACS Biomaterials Science and Engineering* 6 (1): 597–609.

57 Sussman, G.L., Beezhold, D.H., and Kurup, V.P. (2002). Allergens and natural rubber proteins. *Journal of Allergy and Clinical Immunology* 110 (2): S33–S39.

58 Perrella, F.W. and Gaspari, A.A. (2002). Natural rubber latex protein reduction with an emphasis on enzyme treatment. *Methods* 27 (1): 77–86.

59 Frade, M.A.C., Valverde, R.V., De Assis, R.V.C. et al. (2001). Chronic phlebopathic cutaneous ulcer: a therapeutic proposal. *International Journal of Dermatology* 40 (3): 238–240.

60 Ereno, C., Guimarães, S.A.C., Pasetto, S. et al. (2010). Latex use as an occlusive membrane for guided bone regeneration. *Journal of Biomedical Materials Research Part A* 95 (3): 932–939.

61 Neves-Junior, W.F.P., Ferreira, M., Alves, M.C.O. et al. (2006). Influence of fabrication process on the final properties of natural-rubber latex tubes for vascular prosthesis. *Brazilian Journal of Physics* 36 (2B): 586–591.

62 Ferreira, M., Mendonça, R.J., Coutinho-Netto, J., and Mulato, M. (2009). Angiogenic properties of natural rubber latex biomembranes and the serum fraction of Hevea brasiliensis. *Brazilian Journal of Physics* 39 (3): 564–569.

63 Šupová, M. (2009). Problem of hydroxyapatite dispersion in polymer matrices: a review. *Journal of Materials Science: Materials in Medicine* 20 (6): 1201–1213.

64 Raucci, M.G., D'Antò, V., Guarino, V. et al. (2010). Biomineralized porous composite scaffolds prepared by chemical synthesis for bone tissue regeneration. *Acta Biomaterialia* 6 (10): 4090–4099.

65 Dick, T.A. and Dos Santos, L.A. (2017). In situ synthesis and characterization of hydroxyapatite/natural rubber composites for biomedical applications. *Materials Science and Engineering: C* 77: 874–882.

66 Kokubo, T. and Takadama, H. (2006). How useful is SBF in predicting in vivo bone bioactivity? *Biomaterials* 27 (15): 2907–2915.

67 Furuya, M., Shimono, N., and Okamoto, M. (2017). Fabrication of biocomposites composed of natural rubber latex and bone tissue derived from MC3T3-E1 mouse preosteoblastic cells. *Nanocomposites* 3 (2): 76–83.

68 Padamwar, M.N., Pawar, A.P., Daithankar, A.V., and Mahadik, K.R. (2005). Silk sericin as a moisturizer: an in vivo study. *Journal of Cosmetic Dermatology* 4 (4): 250–257.

69 Zhang, Y.Q., Ma, Y., Xia, Y.Y. et al. (2006). Silk sericin–insulin bioconjugates: synthesis, characterization and biological activity. *Journal of Controlled Release* 115 (3): 307–315.

70 Tsubouchi, K., Igarashi, Y., Takasu, Y., and Yamada, H. (2005). Sericin enhances attachment of cultured human skin fibroblasts. *Bioscience, Biotechnology, and Biochemistry* 69 (2): 403–405.

71 Martínez-Mora, C., Mrowiec, A., García-Vizcaíno, E.M. et al. (2012). Fibroin and sericin from Bombyx mori silk stimulate cell migration through upregulation and phosphorylation of c-Jun. *PLoS One* 7 (7): e42271.

72 Aramwit, P., Palapinyo, S., Srichana, T. et al. (2013). Silk sericin ameliorates wound healing and its clinical efficacy in burn wounds. *Archives of Dermatological Research* 305 (7): 585–594.

73 Amano, K.I. and Ito, E. (1978). The action of lysozyme on partially deacetylated chitin. *European Journal of Biochemistry* 85 (1): 97–104.

74 Watthanaphanit, A. and Rujiravanit, R. (2017). Sericin-binded-deprotenized natural rubber film containing chitin whiskers as elasto-gel dressing. *International Journal of Biological Macromolecules* 101: 417–426.

75 Guidelli, É.J., Kinoshita, A., Ramos, A.P., and Baffa, O. (2013). Silver nanoparticles delivery system based on natural rubber latex membranes. *Journal of Nanoparticle Research* 15 (4): 1–9.

10

Nanomaterials and Nanotechnology in Medicine

Materials Development: Challenges and Opportunities

P. M Visakh

Department of Physical Electronics, TUSUR University, Tomsk, Russia

10.1 Nanomaterials and Scaffolds for Tissue Engineering and Regenerative Medicine

From a tissue-engineering point of view, nanoscale materials could be utilized as *in situ* cell- delivery vehicles, nano-hydrogels, nanofibrous scaffolds, and layer-by-layer nanostructures [1]. Experimental studies proved the biomimetic advantages of nanomaterials; higher protein adsorption and cell attachment regulated by nanostructured scaffolds efficiently stimulate more tissue regeneration. Tissue engineering and regenerative medicine caused a revolution in biomedicine with regard to providing the possibility of fabricating tissue and organ substitutes in the laboratory for maintaining, restoring, or enhancing functions of the injured or diseased counterparts [2].

The history of this field dates back to the early 2000s when the exponential progress in nanotechnology and cell engineering occurred [3–5]. After nearly two decades, substantial advances in this area of science have led to an improved quality of life for huge numbers of patients over the globe. Direct encapsulation, incorporation, covalent, and noncovalent immobilization are among the most important methods and approaches applied to load bioactive molecules to micro- and nano-sized biomaterials as well as 3D (three-dimensional) constructs. The high loading efficiency and a controlled release profile are two constant major targets in drug-delivery strategies for tissue engineering and regenerative medicine. Providing a sustained release of drugs to ensure long-term therapeutic treatment is a worthwhile matter.

In the human body, cells in solid tissues are surrounded by the extracellular matrix (ECM). This intricate 3D network is composed of different molecules including fibrillar proteins, proteoglycans (PGs), and glycosaminoglycans (GAGs); each tissue possesses its own specific ECM composition and topology controlling the activities between cells and the respective microenvironments [6, 7].

Moreover, the ECM provides a supportive environment for a heterogeneous population of cells in tissues by exhibiting appropriate topographical features as well as inducing

Nanomaterials and Nanotechnology in Medicine, First Edition. Edited by P.M. Visakh.

mechanical and biochemical signals [8]. After any severe injuries and damages, the ECM of tissues is disrupted and needs to be replaced with an appropriate replacement (scaffold). ECM-mimicking scaffolds could provide a suitable substrate for attachment, proliferation, and differentiation of cells and subsequent tissue regeneration.

Methods and techniques developed for the fabrication of nanoscale topography on scaffolds might create ordered and nonordered topographies. Soft-lithography techniques such as electron beam lithography and photolithography are mostly used to fabricate ordered patterns on the biomaterials surface, while techniques like polymer de-mixing, phase separation, colloidal lithography, and chemical etching are commonly used for producing unordered features onto biomaterials [9]. Carbon-based nanomaterials are among the most promising drug-delivery systems (DDSs) in tissue- engineering approaches and comprise single or multiwalled carbon nanotubes (MWCNT), nanodiamonds, nanohorns, graphene, and fullerenes [10]. The benefits of these materials as DDSs are high mechanical strength, large specific surface area, ease of preparation, and suitability for functionalization [10–14]. Carbon-based nanomaterials are known as multifunctional drug-delivery platforms; however, they are usually embedded within composites in tissue-engineering approaches [15].

Calcium phosphate nanoparticles (e.g. hydroxyapatite (HAp)) and mesoporous bioactive glasses (MBGs) are other useful inorganic materials in drug delivery and regenerative medicine. HAp nanoparticles with different structures (e.g. hollow mesoporous and solid particles) either in pristine or functionalized forms could serve as DDSs for a variety of biomedical applications, from bone regeneration to anticancer imaging and therapy [16–18].

MBGs are also among the most commonly used porous materials in tissue-engineering applications, especially for the reconstruction of bone and dental tissues. Gold and iron nanoparticles are perhaps the two main metallic candidates for the loading and delivery of therapeutic cargoes (proteins, genes, and drugs). They can be formulated and used in a mixture of polymers (composites) to be responsible for the outside stimulus (e.g. the pH of the surrounding environment) [19–21]. Most studies on these nanomaterials are connected to cancer therapy and diagnostics (theranostics). Nanofibrous scaffolds are mainly fabricated via electrospinning owing to its simplicity, cost-efficacy, and the capability of large-scale production. A major advantage of the electrospun nanofibers is their ECM-mimetic structure, which facilitates cell attachment and migration as well as the ability to carry bioactive molecule cargoes (e.g. growth factors, cytokines) [22–24].

The electrospun nanofibers are commonly fabricated from a wide range of natural polymers (e.g. collagen, gelatin, hyaluronic acid, chitosan, and dextran), synthetic polymers (e.g. polycaprolactone [PCL] polylactic acid [PLA], poly(lactic acid-co-glycolic acid [PLGA]), and their composites [25]. Indeed, natural polymers have not only better biocompatibility and biodegradability in comparison with synthetic polymers, but also a lower immune rejection. Numerous studies have proposed electrospun biodegradable polymers such as PCL [26], PLA [27], and PLGA [28], which benefit from good degradation rates as well as improved mechanical properties [29]. It has well known that surface modification of such nanofibers with gelatin [30], HAp [31], calcium phosphate [32], and type I collagen [33] would have osteoconductive properties and enhance osteoblast differentiation and proliferation as well as cell attachment [34]. The use of nanotopography in tissue-engineering applications is of great importance as regards the nanometric organization of natural ECM (which is hierarchical from the nano- to the macro-scale). The presence of nanoscale topographies on human tissues and organs have been well documented; for example, collagen nanofibrils and HAp nanocrystals encompass osteoblasts and osteoclasts in the bone tissue [35, 36].

10.2 Nanorobotics in Nanomedicine

Nanomedicine is the most promising application of medicinal nanotechnology with the aid of a molar device and microscopic information of human physiology [37]. Nanomedicine covers three kinds of detailed assemblies at a molecular level, nanobiological nanomaterials, biotechnological material, engineered organisms, and nonbiological devices [38]. In the foreseeable future, molecular nanomedicine will comprise biological nanomaterials with a precise nanoscale. In the interim, the information received from genomics and proteomics will help in the design of new personalized therapy, a novel chemical entity that targets the pathogen whose genomes are being deciphered, and stem-cell therapy. In the 2020s and beyond, nanorobotics will provide the most promising combination of three categories of nanomedicine, yet progress continues in how the developments will interpret to actual clinical treatments owing to the complexity of operating inside the human body. This technology is expected to make an impression in the healthcare sector. Animatronics is used to design parts of nanodevices like gears, bearings, and ratchets. The individual nano part might consist of a couple of thousand specifically sited atoms. These nano parts might be assembled to produces micron-scale nanosensors, nano manipulators, nano pumps, nanocomputers, and even nanorobots.

Nanorobotics is a miniature engine intended to achieve definite functions continually with precision at nanoscale dimensions. Nanorobotics, occasionally mentioned as molecular robotics, is an embryonic research area as evinced by the current subject in the literature. These nanodevices will be employed in a human to maintain and defend against pathogens. They will be fabricated of nano-parts and sub-assembly such as bearings and gears, nanomotors and pumps, nano manipulators, nanosensors, and nanocomputers having a diameter of around 0.5–3 μm. Nanorobotics brings different kinds of terms in the literature. Thus, nanorobotics can be divided into two key sectors, the first sector looks at structure, reproduction, control, and management of robots with a nanoscale. Plenty of research performed in this field remains theoretical due to complications with manufacturing such nanodevices. Currently, artificial nanorobots don't exist but biological nanorobotic do exist and offer proof that such a structure is at least conceivable [39]. In consequence, nanorobots possess the utmost components that have been discovered in the biological framework of nanomedicine. The nanorobots break-up the kidney stones with an ultrasound signal and these are then passed out of the body in the urine [40]. A singular terabit (1012 nanorobots) dose of microbivores can completely and safely eradicate blood-borne pathogens. Microbivores are 1000 times faster compared to antibiotic therapy, they also digest 100 times more than a natural white cell in a given period. Microbivores have long durability of phagocytosis when compared to natural white blood cells.

Nanorobots can be a good option for diagnosis and targeted delivery for the treatment of cancer, which may reduce the side effects as they are capable of traveling as blood-borne devices that can then identify cells and deliver an anti-cancer drug at the target site.

Nanorobots with chemical biosensors can detect tumor cells in the early stages of development in the human body. Thus, integrated nanosensors can be used to detect the intensity of E-cadherin signals. In nanorobotics, enzymatic nanolithography plays a crucial role in the targeted delivery of therapeutic agents. The simplest technique for enzymatic nanolithography is where enzymes are immobilized in an atomic force microscope (AFM) tip on the nano parts of nanorobots. These then bind at the targeted site after insertion into the body. An alternative way to deposit mobile enzymes is with dip-pen technology.

Nanorobots can navigate through the human body, compared to a large robot, and reach hard-to-access tissue locations, thus target delivery can be achieved. Current developments have revealed the significant potential to address the challenges using these microdevices for precision surgery. Unfettered nanorobots, ranging from nanodrillers to microgrippers and microbullets, provide a key feature for high-precision minimally invasive surgery. The mobile nanorobots supported with diverse energy with the nanoscale surgical component can pierce or recover cellular tissues for precision surgery. These nanorobots can be directed through the narrowest capillaries of the body and trials down to the cellular level are possible.

Kagan et al. verified ultrasound-driven high-speed propulsion, permitted via rapid evaporation of perfluorocarbon used as biocompatible fuel [41]. In response to an external ultrasound stimulus, conical tube-shaped microbullets comprising energy sources show high-speed mobility with a speed of over $6\,\mathrm{m\,s^{-1}}$. This can deliver nanorobots in deep tissue for penetration, excision, and devastation. Similarly, nanorobots that are ultrasound-driven via rapid evaporation of perfluorocarbon applied in tube-shaped micrometric cannon are proficient in loading and ejecting nanobullets at a notable velocity [42].

Self-driven nanorobots have demonstrated their ability as detoxification devices. Like medical imaging/sensing, detoxification relies on self-driven nanorobots that quickly detect and eradicate the toxin to reduce the environment's toxicity. The collision and binding of toxins, to the motors coated with desired functional materials, are due to effective mobility.

The design of the nanorobot device uses existing techniques. In the past decade, the development of the microfabrication process has been stable. Nanofabrication methods work actively, and a prototype produced via these methods is relatively undiscovered. Most micro- and nanofabrication techniques have roots in standard fabrication procedures established for the semiconductor industry [43–46]. Thus, a sound knowledge of these techniques is essential for the introduction of a research and development path in the micro and nano areas. Physically driven nanorobots are made from rigid and nondegradable material, which may pose a danger. In spite of these difficulties, scientists may discover additional parameters that address the initial toxicity. One of the parameters is the number of motors used for treatment (grams) selected as an escalation of units (inadequacy and toxicity limit). Also, the distribution and toxicity of material used to fabricate nanorobots is an important consideration.

The nanorobotic industry is growing and was expected to grow 6.06% in 2016–2020 with their primary application in medicine [47]. Governments are making efforts to deal with progress in the industry. Regulation may be defined as the intended effort to stimulate the behavior of individuals. This nanodevice industry is moving to increased commercialization regulation, a similar regulatory process exists for nanorobots as other medical devices.

Nanorobots inventors are faced with two significant centralized agencies: (i) the Food and Drug Administration (FDA), which sanctions medical products; and (ii) the Patent and Trademark Office (PTO) which deals with the intellectual property of innovation.

10.3 Nanosensors

A nanosensor is a detecting device that has at least one dimension that is smaller than 100 nm and senses a variable quantity in the nanoscale range [48]. Nanosensors are used for the measurement of physical variables, detection of chemicals, gases, etc. Nanosensors may be in the nanoscale range or may measure a dimension in the nanoscale range. It is not mandatory that a nanosensor must be in the nano range in every dimension. Nanosensors operate in a generalized way while detecting a variable: binding to an analyte, signal generation while interacting with biomolecules, and processing the signal. Nanosensors work like a transducer converting one form of energy to another. Generally, in the first step, they sense a variable and convert it to an electrical signal. Most nanosensors detect an electrical signal and convert it to a readable form. Nanosensors detect changes from external excitations and convey it to other nanoparticles [49]. A recent advancement in nanosensor technology incorporates molecular imprinting technology, which produces molecular matrices from monomers that may act as a receptor to detect foreign molecules [50].

It uses functional monomers with template molecules to form polymer matrices thus enhancing selectivity and affinity of matrices. This specific technology helps nanosensors to detect chemical entity. In order to overcome the limitation of a single nanosensor, in terms of power and effectiveness, a nano network can be used. Nanosensors may be classified as physical nanosensors, chemical nanosensors, or biological nanosensors according to the signal they detect. They may also be classified according to applications, structure, and energy sources.

Biosensors can be considered as a subset of chemical nanosensors. These sensors can be used for cancer detection, specific DNA detection, and various disease detection. Although there are various biosensors available the field-effect transistor is the most efficient due to its mass production capability, low-cost manufacturing, and ultrasensitive detection. Biosensors-on-chip are used in diagnosis as they provide robustness and effectiveness in the human body [51].

Chemical and biological components are merged into a single platform through the integration of biosensor technologies and microfluidic technologies, which offer new directions in biosensing. Graphene-based nanosensors allow positive applications in the field of biomedical and chemical sensing. Chemical vapors change the noise spectra of the graphene transistors allowing them to sense selective gases. In some sectors 3D graphene sheets that have high conductivity are used [52]. Graphene foams are also finding promising applications in various fields. Some nanosensors based on bulk nanostructured materials have superior properties. A nanoparticles' large surface area makes them suitable for biomaterial coated, immobilizing molecules. The catalytic behavior of nanoparticles helps nanosensors be electrochemical sensing devices. In the design of gas diffusion electrodes, platinum nanoparticles with porous carbon have higher effectiveness [53].

Plasma arcing is a popular method of forming nanoparticles and, in turn, nanosensors. It produces positively charged ions to be deposited as nanoparticles. It requires highly ionized gas atoms that possess very high energy. The ample amount of energy leads electrons being released from the valence shell of the material. The end product is positively charged atoms. Due to the high potential difference electron separation from the atoms is possible. It uses inert gas as heat source. Carbon nanotubes (CNTs) have high stability in aqueous and nonaqueous solutions. They have high electronic conductivity. Apart from these properties, CNTs possess many other electronic properties that make these potential perfect nanosensors. CNTs can be manufactured by laser ablation, chemical vapor deposition (CVD), or electrical discharge machining. Gold nanoparticles (AuNPs) have a high surface area, improved optical properties, and a high capacity for surface modification. Gold-based nanosensors are mostly used in environmental applications. Silver-based electrodes have high sensitivity, a higher signal-to-noise ratio, enhanced catalytic action, etc. Silver nanoparticles (AgNPs) present in electrodes facilitates the electron transfer between biomolecules and transducer. It has better bioanalytical property. As a result of the higher stability and complete recovery in chemical redox reaction Au-based nanoparticles are considered effective catalysts [54].

Titanium dioxide (TiO2) nanomaterials are generally used in pharmaceutical and medical applications for electrochemical sensing. A range of proteins can be immobilized into nanoporous TiO_2 film electrodes. This process is used to make optical and electrochemical biosensors. Li et al. studied the performance of TiO_2 to trap biomolecules like hemoglobin, cytochrome, and myoglobin [55]. Polymeric nanomaterials are generally used to determine toxic and chemical contamination in gases and liquids and are used in health applications. Nanocomposites made of CNTs, grapheme, metal nanoparticles, or combinations of these lead to further enhanced electrochemical detection properties. In these composites, the combination of filler material and matrix plays a great role in determining the properties of the nanocomposites or nanosensors. Selectivity, sensitivity, and biocompatibility are also affected by the combinations. In healthcare units the polymeric nanomaterials are extensively used in medical coatings, prosthesis, implantation, medical devices, etc. [56].

In bionanomaterials, biomolecules are formed as nanomaterials. The analytical property of biomolecules with the combination of nanostructures enhances the properties to be used in nanosensors. For electrochemical detection of capsaicin Sabela et al. prepared MWCNTs using a nanobiocomposite of L-phenylalanine ammonia-lyase enzyme [57]. Li et al. also prepared nanosensors for the detection of *Escherichia Coli (E. coli)* [58]. This was done by developing self-assembled monolayers (SAMs). Nanosensors are applied for detecting biomolecules in cellular organelles, monitoring chemical and physical phenomena in regions difficult to reach, and measuring nanoscopic particles in the environment and industry. The ability of nanomaterials to detect nucleic acid, proteins, etc. finds a wide application in the health sector. Carbon-based nanomaterials are used for bioimaging, antibody detection, etc. Noble metals and their metal oxides are extensively used nanomaterials because of their chemical stability and easy fabrication techniques. Nanosensors are greatly used in diagnostic medicines enabling early identification of disease without relying on symptoms. Nanosensors nowadays are equipped with artificial intelligence to perform real-time analysis. Diagnostic devices that use nanosensors have greater speed, and improved sensitivity compared to conventional diagnostic techniques. In the treatment of medical ailments nanosensors are extensively used in various cases. To increase the antimicrobial property and mechanical property of the dental filler material ZnONPs (zinc oxide nanoparticles) are added. Through an ultrasonic process ZnONPs and HAp are mixed to form ZnO–HAP nanocomposite. This can improve antimicrobial activity and can detect *E. coli* DH5a. ZnONPs have the capability of destroying cancer cells. They can also be used against human brain tumor and in anticancer studies [59].

Nanosensors are used for detecting contamination in organic implants by checking whether the implant is healthy and working perfectly. The nanosensor is embedded into the implant for the detection of contamination. Generally, the nanosensors work with piezo-electric effect to self-power. The sensor continually sends electrical signals to the outside world so that the provider can track whether the implant is okay or not. Through interpreting the electrical signal the healthcare provider can identify whether the cells are healthy or contaminated with bacteria [60]. However, in long-term use tissue grows on the implant, which restricts the sending of electrical signals. Thus, the effectiveness of the implant deteriorates. An alternative nanosensor is an optical voltage nanosensor, which uses a noncontact sensing process. It is based on faster resonance energy transfer. A gel-based nanosensor can be used for radiotherapy. This is quite helpful for oncologists and physicists. This also helps in reducing overexposures and also reduces radiation-induced toxicity. This kind of plasmonic nanosensor gel compared to conventional sensors indicates the exposure level of radiation when radiation to the hydrogel increases.

10.4 Inorganic Nanoparticles for Drug-delivery Applications

Metal-oxide particles or particles with a metallic composition are known as inorganic nanomaterials [61]. There is a nanoscale range of inorganic nanomaterials (INPs (inorganic nanoparticles)), INPs enable easy permeation through biological membranes and increased interaction with the cells. Due to unusual mechanical, optical, thermal, physicochemical, and electrical properties, their application in drug-delivery, diagnostic, and theranostic fields is increasing. Compared to organic particle INPs are more stable [62]. AuNPs, AgNPs, mesoporous silica nanoparticles, etc. are the major INPs used nowadays. These nanoparticles can be easily prepared and functionalized with various moieties that helps them to interact easily with biomolecules. The surface modification of INPs with suitable materials enhance their diagnostic, drug-delivery applications [63, 64]. The nanoparticles of Ag were prepared using starch. Interestingly, this nanoparticle showed an increase in absorption from 420 to 440 nm and can be used for the preparation of 2-aryl benzimidazole [65]. The zinc-oxide nanoparticle has been synthesized using an environmentally friendly approach using mini submersible pump-assisted sonochemical reactors [66]. The preparation of rhenium nanoparticles using the sonication method revealed that nanoparticle size, distribution, and morphology are influenced by the intensity and the properties of the solvents used [67].

The main principle involved in the synthesis of INPs is ionizing radiation radiolysis of water and the formation of intermediate products with the compounds dissolved. Varieties of INPs with enhanced properties can be prepared by this method. By changing the composition and dose of the irradiated solution, nanoparticles vary in size from sub-nanometric to tens of nm. This radiation-based method is also responsible for the modification of the nanoparticle structures. This method depends on the irradiation of a solution containing nanoparticle precursors with either accelerated electrons or gamma rays at room temperature.

Cuprous-oxide nanocrystals can be developed by the reduction of copper salts present in the solution induced by radiation. This cuprous-oxide nanocrystals having particle size 14 nm were obtained by the irradiation of copper sulfate present in the deoxygenated aqueous solution by gamma radiation. Apart from copper sulfate, the aqueous solution also contains sodium dodecyl sulfate (SDS), propanol, and a 4–4.5 pH acetate buffer. The surfactant SDS was used to control the formed nanoparticles' size [68]. Uranium-dioxide nanocrystals can be prepared by irradiation of uranyl nitrate by using gamma irradiation or accelerated electron beams in the presence of propanol [69]. The particular characteristics of AuNPs result in easy functionalization with biomolecules, which further helps to expand the applications in drug delivery and diagnosis. Further developments in this area to determine the interaction between the INPs and biological membranes will expand their applications [71]. A polysaccharide-based hybrid nanogel obtained by immobilizing quantum dots (QDs) within a dual controllable polysaccharide copolymer was used for the controlled delivery of temozolomide by a pH-dependent release mechanism [70]. Near-infrared (NIR) emission of micelles loaded with QDs comprising the antitumor drug Herceptin for tumor treatment and monitoring shows high antitumor efficiency [71]. Doxorubicin-loaded (DOX-loaded) immuno-liposomes with QDs and single-chain anti-HER2 Fv fragments bonded to the liposome surface through PEG (polyethylene glycol) chains can be used in a xenograft mouse model for tumor-targeted imaging [72].

AuNPs are mostly used for the treatment of malignant tumors due to their characteristics like unique quenching efficiency, hydrophilicity, biocompatibility, and surface modifiability. The drug efficiency of the anticancer drug is lowered by poor bioavailability and side effects. Hence a combination strategy was developed using dual delivery of anticancer drugs to resolve these limitations. AuNPs loaded with bleomycin and DOX shows high drug loading and increased uptake against HeLa cells [73]. DOX being an intercalating drug blocks the process of replication and results in inhibition of macromolecule synthesis [74] whereas bleomycin produces reactive oxygen species (ROS) and causes DNA damage [75]. AuNPs such as nanorods, nanoshells, and nanocages, having optimum absorption in the NIR, are suitable for photothermal therapy (PTT) since they are capable of absorbing NIR light and releasing the absorbed energy as heat at their localized site of accumulation [76]. The initial study was carried out on Au nanospheres conjugated to immunoglobin G, which selectively targets CD8 receptors on lymphocytic cells causing about a 95% death of cells [77].

Magnetic drug delivery involves the targeted delivery of drugs with aid of a permanent magnet. The magnetic targeting decreases the systemic distribution of drug and delivers drugs specifically to the targeted sites [78]. The various factors that determine the effectiveness of this method involve the strength of the magnetic field, its geometry, depth of the target tissue, rate of blood flow, and vascular supply [79].

10.5 Intelligent Nanomaterials for Medicine

Intelligent nanomaterials have brought a revolution in the area of multiple disciplines, including micro- and nano-engineering, bionanotechnology, biomedical engineering, and nanomedicine, as well as pharmaceutical and biomedical industries [80–82]. During the past decade, nanotechnology has had a steadily increasing impact on preclinical development in medicine, shaping the emerging scientific field of nanomedicine. An important topic is the development of composite nanosystems for diagnosis and therapy within the body. The size of the nanomaterial particles used is around same as that of biological molecules and structures. Various nanomaterials such as dendrimers, liposomes, nanoemulsions, micelles, and metallic nanoparticles can be successfully used as nano DDSs. With innovative nanoscience- and nanotechnology-based approaches, nanomaterials having different functionalities, shapes, and distinct physical, chemical, and biological features were designed and developed for successful applications in the medicine industry [83]. Some intelligent nanomaterials can change their unique features that enable them to be applied in drug-delivery materials. The size of the worldwide nanomedicine market is maintained and grown through cutting-edge technology for drug release, different applications in healthcare, and low-cost therapies. The global nanomedicine-market is categorized into diagnostic imaging, drug release, vaccines, regenerative medicine and implants, etc. Currently, the design of intelligent nanoparticles for cancer therapy is the most desired progress in medicine. For cancer therapy, typical intelligent nanomaterials used are polymers, INPs such as gold, iron oxide, and mesoporous silica, and carbon-based materials such as CNTs or graphene. Polymer nanoparticles can be synthesized by solvent evaporation, nanoprecipitation, emulsion polymerization, and controlled/living radical polymerizations. The most commonly employed polymeric material being PLGA due to relatively non-toxic degradation products. Metal nanoparticles (MeNPs) and metal-oxide nanoparticles (MeONPs) can be fabricated by mechanical attrition, and physical- and chemical-vapor deposition techniques [84].

Liposomes and other nanoparticles have found wide application for the delivery of various payloads. Due to their small size, biocompatibility, stability, and stealth properties, they can extend the half-life of the circulatory system, targeting tumor tissue and uptake by cells. Successful targeting requires sufficient stability to prevent premature drug release into the bloodstream and to avoid degradation of circulating payload before the target is reached. Equally important are adequate stealth properties that allow nanoparticles to avoid interactions with plasma proteins or with the immune system, which could lead to early removal of the material [85]. Nanoparticles extend through the walls of these blood vessels and tend to show prolonged retention in tumor tissue. Nanoparticle extravasation is size dependent. In passive targeting strategies, a long enough nanoparticle perfusion time is important to allow high contrast between tumor accumulation compared to that of the liver, spleen, and kidneys, in which the particles tend to accumulate in absence of a suitable target tissue. CNTs have attracted attention for their use in the controlled release of drugs, as well as the release of nucleic acids, peptides, and antibodies. Its inner core and outer surface allow a certain payload to be introduced into the small inner heart, while the outer surface can be modified to achieve the required biocompatibility in the body or to bind targeting ligands or payloads. The functionalization of the surface of CNTs through the union of suitable molecules has shown a considerable reduction in their toxicity [86].

CNTs are one of the best known one-dimensional carbon nanostructures and were discovered in 1991. CNTs consist of layers of graphene that are combined into a tubular or cylindrical shape with diameters of around 0.5–10 nm and a micrometer length. When it comes to the number of graphene layers in CNTs, they fall into three main categories: single-walled CNT (SWCNT), double-walled CNT (DWCNT), and MWCNT. CNFs are another group of carbon nanostructures. High-energy milling is an innovative approach to producing nanocrystalline NiTi (nickel titanium) materials. During the milling process, nanoscale component interactions take place at atomic levels. Biological properties, including biocompatibility, bioactivity, and biodegradability, appear to be controlled by grinding speed/time, compression pressure, and sintering temperature/times. Shape-memory NiTi alloys can be developed from compact aggregates that are alloyed both elementally and mechanically [87].

Furthermore, liquid laser ablation is known as a technique that generates nanoparticles from colloidal shape-memory alloys. Metallic nanoparticles, such as Au or silicon dioxide-gold (SiO_2-Au) nanoparticles, have been incorporated into temperature-sensitive interpenetrating polymer matrices as novel and intelligent therapeutic nanocomposite systems. Nanoparticles or nanococci absorb light, then convert it to heat and transfer it locally to the surrounding smart therapeutic agent.

10.5.1 Nanocomposite

AuNPs were incorporated into a polyacrylamide/polyacrylic acid coating by *in situ* inverse emulsion polymerization and then their surface was treated with PEG by covalent grafting. Depending on the type and size, the metal nanoparticle core can be adjusted to absorb visible NIR at wavelengths from 530 to 1200 nm. As the temperature increases, the polymer begins to swell, releasing all the beneficial agents encapsulated in the polymer network. Intelligent therapy systems using metals can revolutionize drug delivery and disease treatment based on heat-sensitive metal-polymer nanocomposites. Metallic nanowires have been studied extensively in the past decade due to their better mechanical, electrical, and optical properties owing to their nanoscale size [88].

Mechanisms of drug release include swelling due to the osmotic effects of the aqueous environment followed by an enzyme-induced disorder leading to drug release. However, the end result of targeting nanoparticles depends on certain physical properties. The loading and releasing capacity of the drug depends on the chemical nature of the drug and polymer. Scientists can deduce the binding rate from adsorption isotherms. Furthermore, manipulation of polymeric matrices can change the drug-release effect and there is a direct ratio between loading and release. Research has already correlated the physicochemical properties of polymeric nanoparticles with their biological reactions, in which the morphology and surface charges of biodegradable entities were examined. A recent review highlights the challenges posed by biodegradable entities.

10.5.2 *In vivo* tests

Polymeric nanoparticles are known to offer not only protection against environmental stimuli, but also site-specific deliveries, especially for loaded nanoparticles that are well protected. Parameters such as absorption, bioavailability, and long-term therapeutic efficacy can be optimized by controlling their electrostatic interactions. Investigation of the drug-release properties of diazepam-loaded PLGA-NPs confirmed that parameters such as ultrasound time, polymer content, surfactant, relationship between organic phase and aqueous phase, and amount of drug influenced in its expanded publication [89].

PLA is a biocompatible polymeric nanoparticle that can be produced by evaporating the solvent. Of note is its role in the encapsulation efficiency of antipsychotics, which is increased by approximately 95% and is supported by neuroleptics *in vitro* and *in vivo*. PLA can dramatically improve the blood flow time of oridonin, which targets malignant lymphoid cells for apoptosis [90]. PCL is generally synthesized by nanoprecipitation. Due to its low rate of degradation, this is the best choice in long-lasting implantable instruments. Similar to the aforementioned polymer nanoparticles, PCL is effective in cancer treatments and, in addition to improving tamoxifen encapsulation efficacy, has good drug-release time and shelf life. It has also been used as a carrier vehicle for insulin with a positive glycemic response that preserves the bioactivity of the hormone and ultimately proves to be of great use to diabetics. The encapsulation of PCL nanoparticles with amphotericin-B not only increases drug efficacy through delayed release, but also reduces associated toxic side effects [91]. Chitosan is a natural polymer derived from exoskeletal chitin found in arthropods. Its synthetic methods include ionotropic gelling and microemulsion. Chitosan significantly increases insulin absorption from the gastrointestinal tract after encapsulation with a hormone. AuNPs can be functionalized by thiolated ligands using Au—S bond formation, and many molecules can be covalently attached to Au using this approach. For example, pH-sensitive composite nanostructures with a gold core, an inner layer made of hydrophobic poly (L-aspartate-doxorubicin), and an outer layer made of PEG and folate-conjugated (PEG-OH/FA) were produced successfully. Binding of the acid-cleavable hydrazone binds the cancer drug DOX covalently to the hydrophobic inner layer. The vehicle showed a greater cytotoxic effect on the 4T1 mouse mammary carcinoma cell than with pH-activated drug release and targeting ability.

10.6 Polymer-based Nanocomposites for Biomedical Applications

These nanocomposites can be used for gas barrier performance with flame retardant properties, for solar cell applications, and defense and aerospace use with high mechanical stability such as stiffness, strength, toughness, and for gas separation permeable materials [92]. Interestingly, design of thermo-responsive hydrophobic PNCs (protein nanocontainers) is preferable for the loading of the hydrophilic anticancer drugs and nanomedicines. However, nanocomposite designed for defense applications should have greater mechanical properties like strength, compressive nature, hardness, stress, strain, creep, fatigue, ductile, brittleness, whereas nanocomposite for drug delivery and other biomedical applications should possess biocompatibility, biodegradability, biosafety, with desired size, shape, and morphology with considerable molecular weight and density [93]. The nanocomposite with good engineering properties is not suitable for biomedical applications rather these are a much better fit for other industrial and defense applications. In spite of drug-delivery and biomedical applications, these polymer-based nanocomposites are recommended for many industrial and defense applications in the production of high thermal resistive materials, bullet proof jackets, corrosion free machinery, insulation materials for electrical applications, etc. However, the thermally unstable composites such as ammonium and lactums are not recommended for defense and other applications. Predominantly, PNCs have their own uniqueness suitable for use in biomedical and pharmaceuticals. Understanding of the internal structural pattern of the composite can help to improve the desired properties of material for specific applications [94]. PNCs are unique with their size, shape, structure, morphology, and physicochemical properties as recommended for many biopharma applications. Apart from nanocomposites, nanofibers, nanofillers, hollow capsules, core-shell capsules, nanorods, nanotubes, nanovacuoles, nanofragments, micelles, liposomes, CNTs, and QDs show greater potential and suitability for use in medical applications. However, tissue engineering, molecular imaging, MRI (Magnetic Resonance Image) scanning, CT (Computed Tomography) scanning, drug/gene/cell delivery, patch therapy by using tissue scaffolds are all promising ongoing approaches and methods widely applicable to many biomedical therapeutics [95].

Polymer-based nanocomposites have been reported as providing suitable conditions on which to grow the scaffold grafted tissue on the damaged human organ. These PNCs are suitable for growing the tissue because of its greater affinity to attach with cells and allow them to grow without any hurdles. However, tissue engineering is mainly concerned about tissue implants and replacement of destroyed organs/tissue by means of any unusual activities. For example, damage of cartilage, skin, bone, vessels, and any other parts or organs of body can be replaced with the healthy tissue/organs. Similarly, patch therapy along with tissue engineering also has remarkable impact on organ implantation and replacement without any hazardous issues. The uneven skin has been replaced with even-toned tissue or organs followed by the applicability of patch therapy and tissue engineering [96, 97]. Patch therapy is an advanced nanobiotechnology-based treatment method to treat damaged tissue and organs. The polymer-based nanocomposites are loaded with ointments or nanomedicines or anticancer drugs and the substrate source is fixed on the damaged tissue or skin parts [98].

PNCs with thermo-sensitive properties show morphological and structural disintegration followed by the release of drug or ointment on the wound or scars to heal using the patch therapy approach. Nanocomposites with a thermo-sensitive nature are a boon for many biomedical implications and diagnosis methods. These thermo-sensitive nanocomposites show a significant impact on damaged tissue cells and malignant portion of the body followed by the healing capacity [99]. This is an advanced technique to cure damaged tissue without any diagnosis and surgery requirements. It is a simple approach to curing damaged tissue or organs without any side effects and toxicity issues. Interestingly, core-shell nanoparticles and fibrous nanoparticles with a fluid core and solid shell portion can easily be entrapped with the nanomedicines and anticancer drugs, biological entities like proteins, viruses, bacteria under favorable conditions, which cannot affect their functional activities [100, 101]. Such PNCs are further recommended for biosensor and biomarker detecting tools. For example, green fluorescent protein is immobilized in nanostructured fibers and nanorods by providing a huge surface area and lower distances for the sensor protein analyzer. PNCs along with nanostructured fibrous materials, core-shell nanomaterials, hollow core-shell nanoparticles, nanorods, and nanotubes can provide a basic platform for the usage of them for engineering materials as well life sciences applications. PNCs with different structural complexity lead to various specific functions using their unique characteristics and their usage is not limited to materials science, materials engineering, nanotechnology, biotechnology, pharmaceuticals, and nanobiotech applications [102].

10.7 Toxicity of Nanomaterials

One of the toxicity mechanisms of nanoparticles is the formation of ROS, which results in inflammation, oxidative stress, and damage to DNA, cell membranes, as well as proteins. Thus, in biomedical applications including therapeutic applications, gene delivery, and DDSs, it is necessary to assess the toxicity of nanoparticles [103, 104]. Oxidative stress is induced by prooxidants through inhibiting antioxidants or creating ROS. Through the reaction of cells with nanoparticles, their prooxidant effects are induced via the intracellular generation of ROS [105]. The highest level of cytotoxicity was observed for Au nanostars, which had the greatest anticancer potential, while a lower toxicity level was reported by Au nanospheres, which had low anticancer potential. The impact of size and surface coating of iron oxide nanoparticles and their biological impacts in *in vivo* and *in vitro* have been investigated by Feng et al. [106]. They reported low cytotoxicity and also concluded that the light exposure and the shape of nanoparticles affect the cytotoxicity. Moreover, they observed no substantial lactate dehydrogenase release. Actually, the aggregation tendency of TiO_2NPs and also the cellular uptake influence genotoxicity. Mulenos et al. [107] noticed the low-level toxicity of AgNPs observing the release of just minute quantities of Ag ions. Kassem et al. [108] has reported that the CNT antimicrobial activity depends on different operational factors including temperature, solvent composition, retention time, as well as pH. Based on the study by Mohanta et al. [109], CNTs interaction with heart muscle (myocardium) leads to the toxicity of cardiovascular and induces cell proliferation, muscle damage, bloodstream interruption, as well as atherosclerosis.

Nanomedicine, as a medicine branch, has provided the opportunity of applying tools and knowledge of nanotechnology to diagnose and treat diseases with no side effects. With the application of nanotechnology, the available tools can be improved and the development of entirely new techniques is also possible. The exceptional electronic, magnetic, and optical characteristics of nanomaterials can offer excellent potential for a broad range of biomedical applications such as biosensing, imaging, and drug delivery. Metallic nanoparticles are utilized in several biomedical applications including gene targeting, drug delivery, sensing, as well as imaging [110–112]. Some of these nanoparticles also have been observed to possess considerable potential for therapeutic applications [113–116]. The applications of these nanoparticles are consistently increasing since they have been reported to be comparatively less toxic. Over the past few decades, the utilization of metallic nanoparticles and their alloys in different sectors has been investigated in several research works including cancer treatment [117], drug delivery [118], biological labeling [119], catalysis [120], optical devices [121], and sensor technology [122]. Moreover, because of the outstanding surface plasmon resonance (SPR) characteristics of metallic nanoparticles, they are appropriate as markers for biomolecule optical detection. Furthermore, based on results from previous studies, they provided a promising alternative as antiplatelet [123] and antimicrobial agents [124], and for drug delivery, and the stabilization of proteins. One of the metallic nanoparticles which are widely utilized is nanosilver, which possesses antiparasitic [125] and antimicrobial activity. The biocompatibility of Au particles is comparatively high because they are biologically inert. Primarily, AuNP have been applied for labeling applications. Au particles have a great ability for absorbing and scattering visible light. The free electrons in AuNPs are excited with absorbing light to SPR. In general, SPR dominates the optical characteristics of small metallic nanoparticles [126]. The bulk dielectric constant of Au governs the resonance frequency that lies in the electromagnetic spectrum visible area [127]. Compared to other metallic nanoparticles, AuNPs have been reported to have high surface reactivity, biocompatible catalytic, electronic, and optical properties. Bio-optical imaging is an important application of AuNPs. Drug or active agents are required to be introduced to the considered cell, tissue, and organ to exert their action mode. A significant factor for the efficient employment of nanoparticles is the route of administration. The biodistribution and biokinetics of NPs that are exposed by intravenous injection are dissimilar to that of those exposed via inhalation [128–130]. Several distribution pathways in the body can be used by selecting different routes of administration, which also influence the clearance [131, 132].

Dermal absorption is another administration route that delivers the drug directly to the skin. The largest organ in the body is the skin, which offers an efficient barrier to the environment. The possible nanomaterial interactions with the skin have been reported in previous studies [133, 134]. Additionally, the skin is a possible target for drug delivery by nanocarriers [135]. Retinal drug delivery is another administration route involving the retinal application of various formulations. The low bioavailability of nanoparticles which are utilized as drug carriers can be enhanced partly via optimizing their design [136]. Nanoparticle exocytosis has been discussed in a number of studies in recent years. Two exocytosis pathways are regulated and constitutive pathways. Based on previous studies exocytosis could be considerably influenced by different parameters including the characteristics of nanomaterial (for instance, shape, surface, and size), their exposure time and concentration, and also the cell type [137, 138]. Because of the possible change of the nanoparticles' surface during the processing of lysosomal and uptake, the exocytosed nanoparticles might show dissimilar surfaces, leading to recognizing different receptors or targets on the surface of the cell [139].

10.8 Multifunctional Nanomaterials for Medical Applications

Nowadays, the imaging, therapeutic, and targeted-drug-releasing applications of nanomaterials are receiving wider attention [140–143]. Theranostics is a concept of the combination of diagnostics and therapeutics that aims at personalized monitoring and treatment [144–148]. Recently, nanoparticles have been explored for diagnosis and treatment due to the unique physical–chemical properties [149]. Hence, theranostics now give away to nanotheranostics, which help in providing imaging, diagnosing of disease, and its therapy without effecting healthier cells [150]. The use of nanotheranostics avoids the empirical treatment, improves prognosis, and consequently helps clinicians in making advanced therapeutic decisions. Thus, it ensures early detection and treatment of diseases and makes possible real-time monitoring of drug release with high efficiency [151, 152]. Light-based therapies like photodynamic therapy (PDT) [153], PTT [154], optogenetic therapy [155], and photobiomodulation (PBM) [156] have emerged as photonic healthcare applications. Owing to their biocompatibility and exceptional chemical and physical stability and ease of functionalization, metal-based nanoparticles have potential applications in the medical field [157, 158]. Au nanocages [159], graphene-based materials [160], layered double hydroxides [161], etc. find utility in cancer detection and therapeutics and have advanced application in the field of medicine. Moreover, nanoparticles are adequate in making scaffolds appropriate for the tissue-engineering applications [162]. Nanomaterial-based scaffolds are used to fabricate complex tissue or organs and are being clinically used as a standard treatment. Carbon-based nanomaterials have applications in the field of regenerative medicine and tissue engineering due to their inherent ability to show differentiation and the capacity to attach to stem cells [163, 164]. Nanodiagnostics allows diagnosis to be carried out at a nanoscale, which has led to the development and use of handy devices that are easy to use [165, 166]. The uniqueness of nanomaterials is that their size is similar to most biological molecules and structures and hence they can be useful for both *in vivo* and *in vitro* biomedical research and applications [167]. Magnetic nanoparticles and CNTs have also found utility directly as therapeutic materials for killing cancerous and precancerous cells by thermal ablation, which refers to the destruction of tissue by extreme hyperthermia or hypothermia that involves the temperature change concentrated on a focal zone in and around the tumor via use of an external stimulus [168].

Multifunctional nanoparticles introduced as self-healing materials like hydrogels are recently gaining attention as a cancer treatment due to their drug-loading and releasing characteristics [169–171]. Block copolymer-based nanoparticles are capable of loading the drug by emulsification [172] and it is highly active in making multifunctional nanoparticle-based hybrid gels [173]. Magnetic nanoparticles are capable of producing heat by Neel and Brownian relaxations and the generated heat can be utilized for the destruction of tumor cells [174, 175]. Thus, Wang et al. [176] made Mn—Zn ferrite-based mesoporous silica nanospheres, which are highly efficient in producing hydroxyl radicals against cancer cells by the Fenton reaction. Magnetic-nanoparticle-based core and Au-based shell-structured materials having nanostars are utilized for imaging and NIR-triggered drug delivery. The SPR effect of Au causes a temperature rise under NIR exposure; this light to heat conversion is responsible for the release of the drug [177]. Multifunctional nanoparticles are integrated into cancer therapy due to advanced PTT applications [178, 179]. In order to avoid the lower therapy efficiency due to lower penetration depth Wang et al. [180] made multifunctional Bi-based phthalocyanine manganese composites capable of showing combined PDT/PTT efficiency. Excellent efficiency in converting NIR to heat make them effective against tumor cells. NIR responsive nanodots, bioactive Na-Cd-Si, and inorganic materials are effectively used in PTT applications, especially for tumor treatment [181–183].

10.9 Antimicrobial Applications of Nanoparticles

The antimicrobial activity of metal nanoparticles such as AgNPs, titanium (TiNPs), AuNPs, zinc (ZnNPs), and copper (CuNPs) has been known and applied for centuries [184]. MeONPs such as nanoparticles of titanium dioxide (TiO_2NPs), silver oxide (Ag_2ONPs), copper oxide (CuONPs), ZnONPs, magnesium oxide (MgONPs), and calcium oxide (CaONPs) were recognized to possess antimicrobial activity [185]. Several studies showed that polymeric nanoparticles inhibit several microbial species [186]. Also, polymer nanocomposites have been reported as key materials in many applications, with regard to their antimicrobial activity [187]. MeONPs have attracted considerable interest owing to their unique physical and chemical properties. These MeONPs involve various types of oxides like ZnO, CuO, TiO_2, and Fe_3O_4. MeONPs could be synthesized by a wide variety of methods, which can be categorized into bottom-up or top-down approaches [188]. The bottom-up approach uses atoms and molecules to create the nanoparticles via chemical or biological methods. Conversely, the top-down approach refers to the cutting or slicing of bulk material to get nanoparticles via physical methods (e.g. mechanical milling, physical vapor deposition, thermal evaporation, ultrasonic irradiation, or laser ablation) or chemical methods (e.g. chemical etching) [189]. MeNPs have been widely used as antimicrobial agents for centuries in several countries thanks to their desired antimicrobial properties. Some examples of these nanoparticles are AgNPs, AuNPs, and CuNPs. MeNPs can be prepared through various physical, chemical, and biological methods. MeNPs exhibit several advantages e.g., their small size and larger surface area make them good candidates for many antimicrobial applications [190]. Recently, there has been great interest for the development of polymeric nanoparticles as antimicrobial agents. This is attributed to the fact that the polymers could be formulated into nanoparticles (e.g. vesicles, star polymers, micelles, and inorganic-polymer hybrids) of various sizes and shapes, enabling their use in wide variety of applications [191, 192]. The main advantage of polymeric nanoparticles is their multivalency, where the cluster of (multiple) functional groups from nanoparticle construction allows for higher cell recognition and binding capabilities compared to linear polymers. Additionally, polymer nanoparticles like micelles, vesicles, or star polymers enable the efficient encapsulation of cargo molecules that can be released at target sites. Recently, AgNPs in aqueous solutions have been used as a potential irrigant due to its ability to limit the activities of an array of both gram negative and positive bacteria via binding to the cell walls and releasing Ag ions. Other studies have revealed that the longer contact time of antimicrobial nanoparticles like AgNPs with well-developed biofilms is required to produce significant antimicrobial activity especially at very low concentrations [193]. The combination of $AgNO_3$ and TiO_2 as antimicrobial agents shows 100% reduction in the amount of *E. coli* and *Staphylococcus aureus* (*S. aureus*) after 48 hours of incubation for masks coated with the antimicrobial particles, whereas, the uncoated masks revealed a 25% and 50% increase in the viable count of *E. coli* and *S. aureus*, respectively [194]. Antimicrobial nanoparticles such as ZnONPs embedded within dental composite materials have demonstrated antibacterial activities and lead to a reduction in the growth of bacterial biofilms against *Streptococcus sobrinus* – as much as 80% [195]. Potent and well-known antimicrobial agents such as AgNPs have been shown to be effective when incorporated into acrylic polymer material for denture applications with minimal toxicity toward humans [196].

References

1 Wan, A.C. and Ying, J.Y. (2010). Nanomaterials for in situ cell delivery and tissue regeneration. *Advanced Drug Delivery Reviews* 62 (7–8): 731–740.

2 Khademhosseini, A. and Langer, R. (2016). A decade of progress in tissue engineering. *Nature Protocols* 11 (10): 1775.

3 Smith, L. and Ma, P. (2004). Nano-fibrous scaffolds for tissue engineering. *Colloids and Surfaces B: Biointerfaces* 39 (3): 125–131.
4 Ma, P.X. (2004). Scaffolds for tissue fabrication. *Materials Today* 7 (5): 30–40.
5 Roco, M.C. (2003). Nanotechnology: convergence with modern biology and medicine. *Current Opinion in Biotechnology* 14 (3): 337–346.
6 Bhardwaj, G. and Webster, T.J. (2017). Reduced bacterial growth and increased osteoblast proliferation on titanium with a nanophase TiO_2 surface treatment. *International Journal of Nanomedicine* 12: 363.
7 Cassidy, J.W. (2014). Nanotechnology in the regeneration of complex tissues. *Bone and Tissue Regeneration Insights BTRI* 5: S12331.
8 Xing, H., Lee, H., Luo, L., and Kyriakides, T.R. (2019). Extracellular matrix-derived biomaterials in engineering cell function. *Biotechnology Advances* 107421.
9 Norman, J.J. and Desai, T.A. (2006). Methods for fabrication of nanoscale topography for tissue engineering scaffolds. *Annals of Biomedical Engineering* 34 (1): 89–101.
10 Cha, C., Shin, S.R., Annabi, N. et al. (2013). Carbon-based nanomaterials: multifunctional materials for biomedical engineering. *ACS Nano* 7 (4): 2891–2897.
11 Avani, F., Damoogh, S., Mottaghitalab, F. et al. (2020). Vancomycin loaded halloysite nanotubes embedded in silk fibroin hydrogel applicable for bone tissue engineering. *International Journal of Polymeric Materials and Polymeric Biomaterials* 69 (1): 32–43.
12 Sukhodub, L., Sukhodub, L., Kumeda, M. et al. (2019). C_{60} fullerene loaded hydroxyapatite-chitosan beads as a promising system for prolonged drug release. *Carbohydrate Polymers* 223: 115067.
13 Ekwueme, E.C., Rao, R., Mohiuddin, M. et al. (2019). Single-walled carbon nanohorns modulate tenocyte cellular response and tendon biomechanics. *Journal of Biomedical Materials Research Part B: Applied Biomaterials* 108 (5): 1907–1914.
14 Li, A., Fan, Y., Cao, X. et al. (2019). Morpholino-functionalized phosphorus dendrimers for precision regenerative medicine: osteogenic differentiation of mesenchymal stem cells. *Nanoscale* 11 (37): 17230–17234.
15 Lalwani, G., Henslee, A.M., Farshid, B. et al. (2013). Two-dimensional nanostructure-reinforced biodegradable polymeric nanocomposites for bone tissue engineering. *Biomacromolecules* 14 (3): 900–909.
16 Yang, Y.-H., Liu, C.-H., Liang, Y.-H. et al. (2013). Hollow mesoporous hydroxyapatite nanoparticles (hmHANPs) with enhanced drug loading and pH-responsive release properties for intracellular drug delivery. *Journal of Materials Chemistry B* 1 (19): 2447–2450.
17 Li, D., Huang, X., Wu, Y. et al. (2016). Preparation of pH-responsive mesoporous hydroxyapatite nanoparticles for intracellular controlled release of an anticancer drug. *Biomaterials Science* 4 (2): 272–280.
18 Syamchand, S.S. and Sony, G. (2015). Multifunctional hydroxyapatite nanoparticles for drug delivery and multimodal molecular imaging. *Microchimica Acta* 182 (9–10): 1567–1589.
19 Yavuz, M.S., Cheng, Y., Chen, J. et al. (2009). Gold nanocages covered by smart polymers for controlled release with near-infrared light. *Nature Materials* 8 (12): 935–939.
20 Oka, C., Ushimaru, K., Horiishi, N. et al. (2015). Core–shell composite particles composed of biodegradable polymer particles and magnetic iron oxide nanoparticles for targeted drug delivery. *Journal of Magnetism and Magnetic Materials* 381: 278–284.

21 Liang, R., Wei, M., Evans, D.G., and Duan, X. (2014). Inorganic nanomaterials for bioimaging, targeted drug delivery and therapeutics. *Chemical Communications* 50 (91): 14071–14081.
22 Rim, N.G., Shin, C.S., and Shin, H. (2013). Current approaches to electrospun nanofibers for tissue engineering. *Biomedical Materials* 8 (1): 014102.
23 He, W., Ma, Z., Yong, T. et al. (2005). Fabrication of collagen-coated biodegradable polymer nanofiber mesh and its potential for endothelial cells growth. *Biomaterials* 26 (36): 7606–7615.
24 Cross, M. and Claesson-Welsh, L. (2001). FGF and VEGF function in angiogenesis: Signalling pathways, biological responses and therapeutic inhibition. *Trends in Pharmacological Sciences* 22: 201–207.
25 Bhardwaj, N. and Kundu, S.C. (2010). Electrospinning: a fascinating fiber fabrication technique. *Biotechnology Advances* 28 (3): 325–347.
26 Yoshimoto, H., Shin, Y., Terai, H., and Vacanti, J. (2003). A biodegradable nanofiber scaffold by electrospinning and its potential for bone tissue engineering. *Biomaterials* 24 (12): 2077–2082.
27 Song, X., Ling, F., Ma, L. et al. (2013). Electrospun hydroxyapatite grafted poly (L-lactide)/poly (lactic-co-glycolic acid) nanofibers for guided bone regeneration membrane. *Composites Science and Technology* 79: 8–14.
28 Meng, Z., Wang, Y., Ma, C. et al. (2010). Electrospinning of PLGA/gelatin randomly-oriented and aligned nanofibers as potential scaffold in tissue engineering. *Materials Science and Engineering: C* 30 (8): 1204–1210.
29 Zhang, J., Liu, H., Ding, J.-X. et al. (2015). Annealing regulates the performance of an electrospun poly (ε-caprolactone) membrane to accommodate tissue engineering. *RSC Advances* 5 (41): 32604–32608.
30 Li, X., Xie, J., Yuan, X., and Xia, Y. (2008). Coating electrospun poly (ε-caprolactone) fibers with gelatin and calcium phosphate and their use as biomimetic scaffolds for bone tissue engineering. *Langmuir* 24 (24): 14145–14150.
31 Liao, S., Cui, F., Zhang, W., and Feng, Q. (2004). Hierarchically biomimetic bone scaffold materials: nano-HA/collagen/PLA composite. *Journal of Biomedical Materials Research Part B Applied Biomaterials* 69 (2): 158–165.
32 Yang, F., Wolke, J., and Jansen, J. (2008). Biomimetic calcium phosphate coating on electrospun poly (ε-caprolactone) scaffolds for bone tissue engineering. *Chemical Engineering Journal* 137 (1): 154–161.
33 Chiu, L.-H., Lai, W.-F.T., Chang, S.-F. et al. (2014). The effect of type II collagen on MSC osteogenic differentiation and bone defect repair. *Biomaterials* 35 (9): 2680–2691.
34 Yoo, H.S., Kim, T.G., and Park, T.G. (2009). Surface-functionalized electrospun nanofibers for tissue engineering and drug delivery. *Advanced Drug Delivery Reviews* 61 (12): 1033–1042.
35 Klymov, A., Prodanov, L., Lamers, E. et al. (2013). Understanding the role of nano-topography on the surface of a bone-implant. *Biomaterials Science* 1 (2): 135–151.
36 Chen, Z., Bachhuka, A., Wei, F. et al. (2017). Nanotopography-based strategy for the precise manipulation of osteoimmunomodulation in bone regeneration. *Nanoscale* 9 (46): 18129–18152.

37 Svenson, S. (2014). What nanomedicine in the clinic right now really forms nanoparticles? *Wiley Interdisciplinary Reviews: Nanomedicine and Nanobiotechnology* 6 (2): 125–135.

38 Chen, L., Henein, G., and Luciani, V. (2011). Nanofabrication techniques for controlled drug-release devices. *Nanomedicine* 6 (1): 1–6.

39 Requicha, A. (2003). Nanorobots, NEMS, and nanoassembly. *Proceedings of the IEEE* 9 (11): 1922–1933.

40 Qian, K., Ma, X., Dai, X., and Fang, F. (2010). POMDP navigation of service robots with human motion prediction. *ROBOT* 32 (1): 18–24.

41 Kagan, D., Benchimol, M., Claussen, J. et al. (2012). Acoustic droplet vaporization and propulsion of perfluorocarbon-loaded microbullets for targeted tissue penetration and deformation. *Angewandte Chemie International Edition* 51 (30): 7519–7522.

42 Soto, F., Martin, A., Ibsen, S. et al. (2015). Acoustic microcannons: toward advanced microballistics. *ACS Nano* 10 (1): 1522–1528.

43 Drexler, K. (1992). *Nanosystems: Molecular Machinery, Manufacturing and Computation.* New York: Wiley.

44 Campbell, S.A. (2001). *The Science and Engineering of Microelectronic Fabrication.* New York: Oxford University Press.

45 Jaeger, C.J. (2002). *Introduction to Microelectronic Fabrication.* Upper Saddle River: Prentice Hall.

46 Plummer, J.D., Deal, M.D., and Griffin, P.B. (2000). *Silicon VLSI Technology*. Upper Saddle River: Prentice Hall.

47 Markets, R. (2020). Global Nanorobotics Systems Market Growth at 6% CAGR – Analysis, Technologies & Forecasts to 2020 – Research and Markets. https://www.prnewswire.com/news-releases/global-nanorobotics-systems-market-growth-at-6-cagr---analysis-technologies--forecasts-to-2020---research-and-markets-300375278.html (accessed 27 January 2022).

48 Munawar, A., Ong, Y., Schirhagl, R. et al. (2019). Nanosensors for diagnosis with optical, electric and mechanical transducers. *RSC Advances* 9 (12, 12): 6793–6803.

49 Lu, J. and Bowles, M. (2013). How will nano-sensor technology affect logistics management. *Journal of Convergence Information Technology* 8 (2): 820–828.

50 Crapnell, R.D., Hudson, A., Foster, C.W. et al. (2019). Recent advances in electrosynthesized molecularly imprinted polymer sensing platforms for bioanalyte detection. *Sensors (Basel)* 19 (5): 1204.

51 Bhalla, N., Jolly, P., Formisano, N., and Estrela, P. (2016). Introduction to biosensors. *Essays in Biochemistry* 60 (1): 1–8.

52 Cao, X., Yina, Z., and Zhang, H. (2014). Three-dimensional graphene materials: preparation, structures and application in supercapacitors. *Energy & Environmental Science* 7: 1850–1865.

53 Rassaei, L., Amiri, M., Cirtiu, C.M. et al. (2011). Nanoparticles in electrochemical sensors for environmental monitoring. *TrAC Trends in Analytical Chemistry* 30: 1704–1715.

54 Ansari, A.A., Alhoshan, M.R., Alsalhi, M.S., and Aldwayyan, A.S. (2010). Prospects of nanotechnology in clinical immunodiagnostics. *Sensors* 10 (7): 6535–6581.

55 Topoglidis, E., Campbell, C., Cass, T., and Durrant, J..R. (2006). Nitric oxide biosensors based on the immobilization of hemoglobin on mesoporous titania electrodes. *Electroanalysis* 18: 882–887.

56 Polívková, M., Hubáček, T., Staszek, M. et al. (2017). Antimicrobial treatment of polymeric medical devices by silver nanomaterials and related technology. *International Journal of Molecular Sciences* 18 (2): 419.

57 Sabela, M., Mpanza, T., Kanchi, S. et al. (2016). Electrochemical sensing platform amplified with a nanobiocomposite of L-phenylalanine ammonia-lyase enzyme for the detection of capsaicin. *Biosensors and Bioelectronics* 83.

58 Li, Z., Fu, Y., Fang, W., and Li, Y. (2015). Electrochemical impedance immunosensor based on self-assembled monolayers for rapid detection of Escherichia coli O157:H7 with signal amplification using lectin. *Sensors (Switzerland)* 15: 19212–19224.

59 Siddiqi, K.S., Ur Rahman, A., and Tajuddin, H.A. (2018). Properties of zinc oxide nanoparticles and their activity against microbes. *Nanoscale Research Letters* 13 (1): 141.

60 Majumder, S., Mondal, T., and Deen, M.J. (2017). Wearable sensors for remote health monitoring. *Sensors (Basel)* 17 (1): 130.

61 Fadeel, B. and Garcia-Bennett, A.E. (2010). Better safe than sorry: understanding the toxicological properties of inorganic nanoparticles manufactured for biomedical applications. *Advanced Drug Delivery Reviews [Internet]* 62 (3): 362–374.

62 Neoh, K.G. and Kang, E.T. (2011). Functionalization of inorganic nanoparticles with polymers for stealth biomedical applications. *Polymer Chemistry* 2 (4): 747–759.

63 Salas, G., Costo, R., and Morales, M.d.P. (2012). Synthesis of inorganic nanoparticles. In: *Nanobiotechnology [Internet]* (ed. J.M. de la Fuente and V. Grazu) BT-F of N, 35–79. Elsevier.

64 Pandey, P. and Dahiya, M. (2016). A brief review on inorganic nanoparticles. *Journal of Critical Reviews* 3: 18–26.

65 Kumar, B., Smita, K., Cumbal, L. et al. (2014). Sonochemical synthesis of silver nanoparticles using starch: a comparison. *Bioinorganic Chemistry and Applications* 2014: 784268.

66 Gupta, A. and Srivastava, R. (2019). Mini submersible pump assisted sonochemical reactors: large-scale synthesis of zinc oxide nanoparticles and nanoleaves for antibacterial and anti-counterfeiting applications. *Ultrasonics Sonochemistry* 52: 414–427.

67 Sakkas, P.M., Argirusi, M., Sourkouni, G., and Argirusis, C. (2020). Rhenium oxide nanoparticles – Sonochemical synthesis and integration on anode powders for solid oxide fuel cells. *Ultrasonics Sonochemistry* 69: 105250.

68 Zhu, Y., Qian, Y., Zhang, M. et al. (1994). Preparation and characterization of nanocrystalline powders of cuprous oxide by using γ-radiation. *Materials Research Bulletin* 29 (4): 377–383.

69 Roth, O., Hasselberg, H., and Jonsson, M. (2009). Radiation chemical synthesis and characterization of UO2 nanoparticles. *Journal of Nuclear Materials* 383 (3): 231–236.

70 Wu, W., Aiello, M., Zhou, T. et al. (2010). In-situ immobilization of quantum dots in polysaccharide-based nanogels for integration of optical pH-sensing, tumor cell imaging, and drug delivery. *Biomaterials* 31 (11): 3023–3031.

71 Nurunnabi, M., Cho, K.J., Choi, J.S. et al. (2010). Targeted near-IR QDs-loaded micelles for cancer therapy and imaging. *Biomaterials* 31 (20): 5436–5444.

72 Weng, K.C., Noble, C.O., Papahadjopoulos-Sternberg, B. et al. (2008 Sep). Targeted tumor cell internalization and imaging of multifunctional quantum dot-conjugated immunoliposomes in vitro and in vivo. *Nano Letters* 8 (9): 2851–2857.

73 Farooq, M.U., Novosad, V., Rozhkova, E.A. et al. (2018). Gold nanoparticles-enabled efficient dual delivery of anticancer therapeutics to HeLa cells. *Scientific Reports* 8 (1): 2907.

74 Tacar, O., Sriamornsak, P., and Dass, C.R. (2013 Feb). Doxorubicin: an update on anticancer molecular action, toxicity and novel drug delivery systems. *Journal of Pharmacy and Pharmacology* 65 (2): 157–170.

75 Stubbe, J. and Kozarich, J.W. (1987 Oct). Mechanisms of bleomycin-induced DNA degradation. *Chemical Reviews* 87 (5): 1107–1136.

76 Dreaden, E.C., Austin, L.A., Mackey, M.A., and El-Sayed, M.A. (2012 Apr). Size matters: gold nanoparticles in targeted cancer drug delivery. *Therapeutic Delivery* 3 (4): 457–478.

77 Pitsillides, C.M., Joe, E.K., Wei, X. et al. (2003 Jun). Selective cell targeting with light-absorbing microparticles and nanoparticles. *Biophysical Journal* 84 (6): 4023–4032.

78 Lee, N., Yoo, D., Ling, D. et al. (2015). Iron oxide based nanoparticles for multimodal imaging and magnetoresponsive therapy. *Chemical Reviews* 115 (19): 10637–10689.

79 Neuberger, T., Schöpf, B., Hofmann, H. et al. (2005). Superparamagnetic nanoparticles for biomedical applications: possibilities and limitations of a new drug delivery system. *Journal of Magnetism and Magnetic Materials* 293 (1): 483–496.

80 Patra, J.K., Das, G., Fraceto, L.F. et al. (2018). Nano based drug delivery systems: recent developments and future prospects. *Journal of Nanobiotechnology* 16 (1): 71.

81 Ventola, C.L. (2012). The nanomedicine revolution: part 1: emerging concepts. *Pharmacy and Therapeutics* 37 (9): 512–525.

82 United States National Nanotechnology Initiative (2020). https://www.nano.gov/you/nanotechnology-benefits (accessed 18 July 2020).

83 Sharma, D. and Hussain, C.M. (2020). Smart nanomaterials in pharmaceutical analysis. *Arabian Journal of Chemistry* 13 (1): 3319–3343. ISSN 1878-5352.

84 Navya, P.N., Kaphle, A., Srinivas, S.P. et al. (2019). Current trends and challenges in cancer management and therapy using designer nanomaterials. *Nano Convergence* 6 (1): 23.

85 Aggarwal, P., Hall, J.B., McLeland, C.B. et al. (2009). Nanoparticle interaction with plasma proteins as it relates to particle biodistribution, biocompatibility and therapeutic efficacy. *Advanced Drug Delivery Reviews* 61 (6): 428–437.

86 Vardharajula, S., Ali, S.Z., Tiwari, P.M. et al. (2012). Functionalized carbon nanotubes: biomedical applications. *International Jorunal of Nanomedicine* 7: 5361–5374.

87 Mousavi, T., Karimzadeh, F., and Abbasi, M. (2008). Synthesis and characterization of nanocrystalline NiTi intermetallic by mechanical alloying. *Materials Science and Engineering A-structural Materials Properties Microstructure and Processing* 487: 46–51.

88 Athar, M. and Das, A. (2014). Therapeutic nanoparticles: state-of-the-art of nanomedicine. *Advanced Materials Review* 1: 25–37.

89 Bohrey, S., Chourasiya, V., and Pandey, A. (2016). Polymeric nanoparticles containing diazepam: preparation, optimization, characterization, in-vitro drug release and release kinetic study. *Nano Convergence* 3 (1): 3.

90 Lee, B.K., Yun, Y., and Park, K. (2016). PLA micro- and nano-particles. *Advanced Drug Delivery Reviews* 107: 176–191.

91 Kumari, A., Yadav, S., and Yadav, S. (2009). Biodegradable polymeric nanoparticles based drug delivery systems. *Colloids and Surfaces. B, Biointerfaces* 75: 1–18.

92 Lu, Y., Ye, W., Yang, Q. et al. (2016). Three-dimensional hierarchical porous PtCu dendrites: a highly efficient peroxidase nanozyme for colorimetric detection of H_2O_2. *Sensors and Actuators B: Chemical* 230: 721–730.

93 Luzi, F., Tortorella, I., Di Michele, A. et al. (2020). Novel nanocomposite PLA films with lignin/zinc oxide hybrids: design, characterization, interaction with mesenchymal stem cells. *Nanomaterials (Basel)* 10 (11): 2176.

94 Muthulakshmi, L., Pavithra, U., Sivaranjani, V. et al. (2020). A novel Ag/carrageenan-gelatin hybrid hydrogel nanocomposite and its biological applications: preparation and characterization. *Journal of the Mechanical Behavior of Biomedical Materials* 115: 104257.

95 Negahban, Z., Shojaosadati, S.A., and Hamedi, S. (2021). A novel self-assembled micelles based on stearic acid modified schizophyllan for efficient delivery of paclitaxel. *Colloids and Surfaces B: Biointerfaces* 199.

96 Niu, X., He, Y., Pan, J. et al. (2016). Uncapped nanobranch-based CuS clews used as an efficient peroxidase mimic enable the visual detection of hydrogen peroxide and glucose with fast response. *Analytica Chimica Acta* 947: 42–49.

97 Ounkaew, A., Kasemsiri, P., Jetsrisuparb, K. et al. (2020). Synthesis of nanocomposite hydrogel based carboxymethyl starch/polyvinyl alcohol/nanosilver for biomedical materials. *Carbohydrate Polymers* 248: 116767.

98 Platnieks, O., Gaidukovs, S., Neibolts, N. et al. (2020). Poly(butylene succinate) and graphene nanoplatelet-based sustainable functional nanocomposite materials: structure-properties relationship. *Materials Today Chemistry* 18.

99 Rezvova, M.A., Yuzhalin, A.E., Glushkova, T.V. et al. (2020). Biocompatible nanocomposites based on poly(styrene-block-isobutylene-block-styrene) and carbon nanotubes for biomedical application. *Polymers (Basel)* 12.

100 Xu, Z., Sun, P., Zhang, J. et al. (2020). High-efficiency platinum–carbon nanozyme for photodynamic and catalytic synergistic tumor therapy. *Chemical Engineering Journal* 399.

101 Xu, Z., Xing, W., Hou, Y. et al. (2021). The combustion and pyrolysis process of flame-retardant polystyrene/cobalt-based metal organic frameworks (MOF) nanocomposite. *Combustion and Flame* 226: 108–116.

102 Yang, Q., Li, L., Zhao, F. et al. (2019). Generation of MnO2 nanozyme in spherical polyelectrolyte brush for colorimetric detection of glutathione. *Materials Letters* 248: 89–92.

103 Gupta, R. and Xie, H. (2018). Nanoparticles in daily life: applications, toxicity and regulations. *Journal of Environmental Pathology, Toxicology and Oncology* 37 (3).

104 Wang, F., Yu, L., Monopoli, M.P. et al. (2013). The biomolecular corona is retained during nanoparticle uptake and protects the cells from the damage induced by cationic nanoparticles until degraded in the lysosomes. *Nanomedicine: Nanotechnology, Biology and Medicine* 9 (8): 1159–1168.

105 Regoli, F. and Giuliani, M.E. (2014). Oxidative pathways of chemical toxicity and oxidative stress biomarkers in marine organisms. *Marine Environmental Research* 93: 106–117.

106 Feng, Q., Liu, Y., Huang, J. et al. (2018). Uptake, distribution, clearance, and toxicity of iron oxide nanoparticles with different sizes and coatings. *Scientific Reports* 8 (1): 1–13.

107 Mulenos, M.R., Liu, J., Lujan, H. et al. (2020). Copper, silver, and titania nanoparticles do not release ions under anoxic conditions and release only minute ion levels under oxic conditions in water: evidence for the low toxicity of nanoparticles. *Environmental Chemistry Letters* 1–10.

108 Kassem, A., Ayoub, G.M., and Malaeb, L. (2019). Antibacterial activity of chitosan nano-composites and carbon nanotubes: a review. *Science of the Total Environment* 668: 566–576.

109 Mohanta, D., Patnaik, S., Sood, S., and Das, N. (2019). Carbon nanotubes: evaluation of toxicity at biointerfaces. *Journal of Pharmaceutical Analysis* 9 (5): 293–300.

110 Mandal, A.K. (2017). Copper nanomaterials as drug delivery system against infectious agents and cancerous cells. *Journal of Applied Life Sciences International* 15 (4): 1–8.

111 Zhou, Z., Li, B., Shen, C. et al. (2020). Metallic 1T phase enabling MoS_2 nanodots as an efficient agent for photoacoustic imaging guided photothermal therapy in the near-infrared-II window. *Small* 16 (43): 2004173.

112 Si, Y. and Lee, H.J. (2020). Carbon nanomaterials and metallic nanoparticles-incorporated electrochemical sensors for small metabolites: detection methodologies and applications. *Current Opinion in Electrochemistry* 22: 234–243.

113 Fahimirad, S., Ajalloueian, F., and Ghorbanpour, M. (2019). Synthesis and therapeutic potential of silver nanomaterials derived from plant extracts. *Ecotoxicology and Environmental Safety* 168: 260–278.

114 Shrivastava, S., Bera, T., Singh, S.K. et al. (2009). Characterization of antiplatelet properties of silver nanoparticles. *ACS Nano* 3 (6): 1357–1364.

115 Kim, J.S., Kuk, E., Yu, K.N. et al. (2007). Antimicrobial effects of silver nanoparticles. *Nanomedicine: Nanotechnology, Biology and Medicine* 3 (1): 95–101.

116 Li, Y., Leung, P., Yao, L. et al. (2006). Antimicrobial effect of surgical masks coated with nanoparticles. *Journal of Hospital Infection* 62 (1): 58–63.

117 Cordani, M. and Somoza, Á. (2019). Targeting autophagy using metallic nanoparticles: a promising strategy for cancer treatment. *Cellular and Molecular Life Sciences* 76 (7): 1215–1242.

118 Tran, P.H.L., Duan, W., and Tran, T.T.D. (2020). Fucoidan-based nanostructures: a focus on its combination with chitosan and the surface functionalization of metallic nanoparticles for drug delivery. *International Journal of Pharmaceutics* 575: 118956.

119 Nicewarner-Pena, S.R., Freeman, R.G., Reiss, B.D. et al. (2001). Submicrometer metallic barcodes. *Science* 294 (5540): 137–141.

120 Soldevila-Barreda, J.J. and Metzler-Nolte, N. (2019). Intracellular catalysis with selected metal complexes and metallic nanoparticles: advances toward the development of catalytic metallodrugs. *Chemical Reviews* 119 (2): 829–869.

121 Seemala, B., Therrien, A.J., Lou, M. et al. (2019). Plasmon-mediated catalytic O_2 dissociation on Ag nanostructures: hot electrons or near fields? *ACS Energy Letters* 4 (8): 1803–1809.

122 Su, J., Wang, D., Nörbel, L. et al. (2017). Multicolor gold–silver nano-mushrooms as ready-to-use SERS probes for ultrasensitive and multiplex DNA/miRNA detection. *Analytical Chemistry* 89 (4): 2531–2538.

123 Dakshayani, S.S., Marulasiddeshwara, M.B., Kumar, S. et al. (2019). Antimicrobial, anticoagulant and antiplatelet activities of green synthesized silver nanoparticles using Selaginella (Sanjeevini) plant extract. *International Journal of Biological Macromolecules* 131: 787–797.

124 Jain, S. and Mehata, M.S. (2017). Medicinal plant leaf extract and pure flavonoid mediated green synthesis of silver nanoparticles and their enhanced antibacterial property. *Scientific Reports* 7 (1): 1–13.

125 Marimuthu, S., Rahuman, A.A., Rajakumar, G. et al. (2011). Evaluation of green synthesized silver nanoparticles against parasites. *Parasitology Research* 108 (6): 1541–1549.

126 Daniel, M.-C. and Astruc, D. (2004). Gold nanoparticles: assembly, supramolecular chemistry, quantum-size-related properties, and applications toward biology, catalysis, and nanotechnology. *Chemical Reviews* 104 (1): 293–346.

127 El-Sayed, M.A. (2001). Some interesting properties of metals confined in time and nanometer space of different shapes. *Accounts of Chemical Research* 34 (4): 257–264.

128 Burgess, T.L. and Kelly, R.B. (1987). Constitutive and regulated secretion of proteins. *Annual Review of Cell Biology* 3 (1): 243–293.

129 Wick, P., Manser, P., Limbach, L.K. et al. (2007). The degree and kind of agglomeration affect carbon nanotube cytotoxicity. *Toxicology Letters* 168 (2): 121–131.

130 Sayes, C.M., Fortner, J.D., Guo, W. et al. (2004). The differential cytotoxicity of water-soluble fullerenes. *Nano Letters* 4: 1881–1887.

131 Yu, J., Yin, W., Peng, T. et al. (2017). Biodistribution, excretion, and toxicity of polyethyleneimine modified $NaYF_4$: Yb, Er upconversion nanoparticles in mice via different administration routes. *Nanoscale* 9 (13): 4497–4507.

132 Zhao, Y., Sultan, D., and Liu, Y. (2019). Biodistribution, excretion, and toxicity of nanoparticles. In: *Theranostic Bionanomaterials* (ed. W. Cui and X. Zhao), 27–53.

133 Labouta, H.I. and Schneider, M. (2013). Interaction of inorganic nanoparticles with the skin barrier: current status and critical review. *Nanomedicine: Nanotechnology, Biology and Medicine* 9 (1): 39–54.

134 Alvarez-Román, R., Naik, A., Kalia, Y.N. et al. (2004). Skin penetration and distribution of polymeric nanoparticles. *Journal of Controlled Release* 99 (1): 53–62.

135 Barry, B.W. (2001). Novel mechanisms and devices to enable successful transdermal drug delivery. *European Journal of Pharmaceutical Sciences* 14 (2): 101–114.

136 Bourquin, J., Milosevic, A., Hauser, D. et al. (2018). Biodistribution, clearance, and long-term fate of clinically relevant nanomaterials. *Advanced Materials* 30 (19): 1704307.

137 Chu, Z., Huang, Y., Tao, Q., and Li, Q. (2011). Cellular uptake, evolution, and excretion of silica nanoparticles in human cells. *Nanoscale* 3 (8): 3291–3299.

138 Sakhtianchi, R., Minchin, R.F., Lee, K.-B. et al. (2013). Exocytosis of nanoparticles from cells: role in cellular retention and toxicity. *Advances in Colloid and Interface Science* 201: 18–29.

139 Mahon, E., Salvati, A., Bombelli, F.B. et al. (2012). Designing the nanoparticle–biomolecule interface for "targeting and therapeutic delivery.". *Journal of Controlled Release* 161 (2): 164–174.

140 Chomoucka, J., Drbohlavova, J., Huska, D. et al. (2010). Magnetic nanoparticles and targeted drug delivering. *Pharmacological Research* 62 (2): 144–149.

141 Parveen, S., Misra, R., and Sahoo, S.K. (2012). Nanoparticles: a boon to drug delivery, therapeutics, diagnostics and imaging. *Nanomedicine: Nanotechnology, Biology and Medicine* 8 (2): 147–166.

142 Cheng, L., Wang, X., Gong, F. et al. (2020). 2D nanomaterials for cancer theranostic applications. *Advanced Materials* 32 (13): 1902333.

143 Yew, Y.P., Shameli, K., Miyake, M. et al. (2020). Green biosynthesis of superparamagnetic magnetite $Fe3O_4$ nanoparticles and biomedical applications in targeted anticancer drug delivery system: a review. *Arabian Journal of Chemistry* 13 (1): 2287–2308.

144 Chen, X., Song, J., Chen, X., and Yang, H. (2019). X-ray-activated nanosystems for theranostic applications. *Chemical Society Reviews* 48 (11): 3073–3101.

145 Rizzo, L.Y., Theek, B., Storm, G. et al. (2013). Recent progress in nanomedicine: therapeutic, diagnostic and theranostic applications. *Current Opinion of Biotechnology* 24 (6): 1159–1166.

146 Kojima, R., Aubel, D., and Fussenegger, M. (2015). Novel theranostic agents for next-generation personalized medicine: small molecules, nanoparticles, and engineered mammalian cells. *Current Opinion in Chemical Biology* 28: 29–38.

147 Li, X., Kim, J., Yoon, J., and Chen, X. (2017). Cancer-associated, stimuli-driven, turn on theranostics for multimodality imaging and therapy. *Advanced Materials* 29 (23): 1606857.

148 Vats, S., Singh, M., Siraj, S. et al. (2017). Role of nanotechnology in theranostics and personalized medicines. *Journal of Health Research and Reviews* 4 (1): 1.

149 Yang, K., Feng, L., Shi, X., and Liu, Z. (2013). Nano-graphene in biomedicine: theranostic applications. *Chemical Society Reviews* 42 (2): 530–547.

150 Wong, X.Y., Sena-Torralba, A., Álvarez-Diduk, R. et al. (2020). Nanomaterials for nanotheranostics: tuning their properties according to disease needs. *ACS Nano* 14 (3): 2585–2627.

151 Kim, T.II., Lee, S., and Chen, X. (2013). Nanotheranostics for personalized medicine. *Expert Review of Molecular Diagnostics* 13 (3): 257–269.

152 Mura, S. and Couvreur, P. (2012). Nanotheranostics for personalized medicine. *Advanced Drug Delivery Reviews* 64 (13): 1394–1416.

153 Liu, K., Xing, R., Zou, Q. et al. (2016). Simple peptide-tuned self-assembly of photosensitizers towards anticancer photodynamic therapy. *Angewandte Chemie* 128 (9): 3088–3091.

154 Shao, J., Xie, H., Huang, H. et al. (2016). Biodegradable black phosphorus-based nanospheres for in vivo photothermal cancer therapy. *Nature Communications* 7 (1): 1–13.

155 Deisseroth, K. (2011). Optogenetics. *Nature Methods* 8 (1): 26–29.

156 Hamblin, M.R., Huang, Y., and Heiskanen, V. (2019). Non-mammalian hosts and photobiomodulation: do all life-forms respond to light? *Photochemistry and Photobiology* 95 (1): 126–139.

157 Pissuwan, D., Gazzana, C., Mongkolsuk, S., and Cortie, M.B. (2020). Single and multiple detections of foodborne pathogens by gold nanoparticle assays. *Wiley Interdisciplinary Reviews. Nanomedicine and Nanobiotechnology* 12 (1): e1584.

158 Hu, X., Zhang, Y., Ding, T. et al. (2020). Multifunctional gold nanoparticles: a novel nanomaterial for various medical applications and biological activities. *Frontiers in Bioengineering and Biotechnology* 8: 990.

159 Xia, X. and Xia, Y. (2014). Gold nanocages as multifunctional materials for nanomedicine. *Frontiers of Physics* 9 (3): 378–384.

160 Gong, W., Hu, Z., Liang, Y. et al. (2018). Graphene-based multifunctional nanomaterials in cancer detection and therapeutics. *Journal of Nanoscience and Nanotechnology* 18 (8): 5155–5170.

161 Zümreoglu-Karan, B. and Ay, A. (2012). Layered double hydroxides – multifunctional nanomaterials. *Chemical Papers* 66 (1): 1–10.

162 Abraham, E., Weber, D.E., Sharon, S. et al. (2017). Multifunctional cellulosic scaffolds from modified cellulose nanocrystals. *ACS Applied Materials and Interfaces* 9 (3): 2010–2015.

163 Menaa, F., Abdelghani, A., and Menaa, B. (2015). Graphene nanomaterials as biocompatible and conductive scaffolds for stem cells: impact for tissue engineering and regenerative medicine. *Journal of Tissue Engineering and Regenerative Medicine* 9 (12): 1321–1338.

164 Reddi, A.H., Becerra, J., and Andrades, J.A. (2011). Nanomaterials and hydrogel scaffolds for articular cartilage regeneration. *Tissue Engineering Part B Reviews* 17 (5): 301–305.

165 Baptista, P.V. (2014). Nanodiagnostics: leaving the research lab to enter the clinics? *Diagnosis* 1 (4): 305–309.

166 Jain, K.K. (2003). Nanodiagnostics: application of nanotechnology in molecular diagnostics. *Expert Review of Molecular Diagnostics* 3 (2): 153–161.

167 Curtis, A. and Wilkinson, C. (2001). Nantotechniques and approaches in biotechnology. *Trends in Biotechnology* 19 (3): 97–101.

168 Hao, R., Xing, R., Xu, Z. et al. (2010). Synthesis, functionalization, and biomedical applications of multifunctional magnetic nanoparticles. *Advanced Materials* 22 (25): 2729–2742.

169 Bao, C., Liu, B., Li, B. et al. (2020). Enhanced transport of shape and rigidity-tuned α-lactalbumin nanotubes across intestinal mucus and cellular barriers. *Nano Letters* 20 (2): 1352–1361.

170 Song, J., Yuan, C., Jiao, T. et al. (2020). Multifunctional antimicrobial biometallohydrogels based on amino acid coordinated self-assembly. *Small* 16 (8): 1907309.

171 Mao, C., Xiang, Y., Liu, X. et al. (2017). Photo-inspired antibacterial activity and wound healing acceleration by hydrogel embedded with Ag/Ag@ AgCl/ZnO nanostructures. *ACS Nano* 11 (9): 9010–9021.

172 Vincent, C., Kiran, M.D., Jayan, J.S., and Saritha, A. (2020). Determination of drug loading efficiency of nano capsules formed by the O/W emulsification of block copolymer. *Materials Today Proceedings* 26 (2): 723–723.

173 Jayan, J.S., Saritha, A., Deeraj, B.D.S., and Joseph, K. (2020). Triblock copolymer grafted graphene oxide as nanofiller for toughening of epoxy resin. *Materials Chemistry and Physics* 248: 122930.

174 Zhang, Q., Wu, J., Wang, J. et al. (2020). A neutrophil-inspired supramolecular nanogel for magnetocaloric–enzymatic tandem therapy. *Angewandte Chemie* 132 (9): 3761–3767.

175 Wu, H., Liu, L., Song, L. et al. (2019). Enhanced tumor synergistic therapy by injectable magnetic hydrogel mediated generation of hyperthermia and highly toxic reactive oxygen species. *ACS Nano* 13 (12): 14013–14023.

176 Wang, C., Zhao, N., Huang, Y. et al. (2020). Coordination of injectable self-healing hydrogel with Mn-Zn ferrite@ mesoporous silica nanospheres for tumor MR imaging and

efficient synergistic magnetothermal-chemo-chemodynamic therapy. *Chemical Engineering Journal* 401: 126100.

177 Tomitaka, A., Arami, H., Ahmadivand, A. et al. (2020). Magneto-plasmonic nanostars for image-guided and NIR-triggered drug delivery. *Scientific Reports* 10 (1): 1–10.

178 Yang, J., Dai, D., Lou, X. et al. (2020). Supramolecular nanomaterials based on hollow mesoporous drug carriers and macrocycle-capped CuS nanogates for synergistic chemo-photothermal therapy. *Theranostics* 10 (2): 615.

179 Xu, J., Cheng, X., Chen, F. et al. (2020). Fabrication of multifunctional polydopamine-coated gold nanobones for PA/CT imaging and enhanced synergistic chemo-photothermal therapy. *Journal of Materials Science and Technology*.

180 Wang, Z., Jia, T., Sun, Q. et al. (2020). Construction of bi/phthalocyanine manganese nanocomposite for trimodal imaging directed photodynamic and photothermal therapy mediated by 808 nm light. *Biomaterials* 228: 119569.

181 Li, F., Li, T., Zhi, D. et al. (2020). Novel ultrasmall multifunctional nanodots for dual-modal MR/NIR-II imaging-guided photothermal therapy. *Biomaterials* 256: 120219.

182 Bao, Z., Liu, X., Liu, Y. et al. (2016). Near-infrared light-responsive inorganic nanomaterials for photothermal therapy. *Asian Journal of Pharmaceutical Sciences* 11 (3): 349–364.

183 Ma, L., Zhou, Y., Zhang, Z. et al. (2020). Multifunctional bioactive Nd-ca-Si glasses for fluorescence thermometry, photothermal therapy, and burn tissue repair. *Science Advances* 6 (32): eabb1311.

184 Sánchez-López, E., Gomes, D., Esteruelas, G. et al. (2020). Metal-based nanoparticles as antimicrobial agents: An overview. *Nanomaterials* 10: 1–39.

185 Azam, A., Ahmed, A.S., Oves, M. et al. (2012). Antimicrobial activity of metal oxide nanoparticles against gram-positive and gram-negative bacteria: a comparative study. *International Journal of Nanomedicine* 7: 6003.

186 Lam, S.J., Wong, E.H.H., Boyer, C., and Qiao, G.G. (2018). Antimicrobial polymeric nanoparticles. *Progress in Polymer Science* 76: 40–64.

187 Zahran, M. and Marei, A.H. (2019). Innovative natural polymer metal nanocomposites and their antimicrobial activity. *International Journal of Biological Macromolecules* 136: 586–596.

188 Oskam, G. (2006). Metal oxide nanoparticles: synthesis, characterization and application. *Journal of Sol-Gel Science and Technology* 37: 161–164.

189 Nikam, A.V., Prasad, B.L.V., and Kulkarni, A.A. (2018). Wet chemical synthesis of metal oxide nanoparticles: a review. *CrystEngComm* 20: 5091–5107.

190 Slavin, Y.N., Asnis, J., Häfeli, U.O., and Bach, H. (2017). Metal nanoparticles: understanding the mechanisms behind antibacterial activity. *Journal of Nanobiotechnology* 15: 1–20.

191 Zhu, Y., Yang, B., Chen, S., and Du, J. (2017). Polymer vesicles: mechanism, preparation, application, and responsive behavior. *Progess in Polymer Science* 64: 1–22.

192 Ren, J.M., McKenzie, T.G., Fu, Q. et al. (2016). Star polymers. *Chemical Reviews* 116: 6743–6836.

193 Rodrigues, C.T., De Andrade, F.B., De Vasconcelos, L. et al. (2018). Antibacterial properties of silver nanoparticles as a root canal irrigant against Enterococcus faecalis biofilm and infected dentinal tubules. *International Endodontic Journal* 51: 901–911.

194 Li, Y., Leung, P., Yao, L. et al. (2006). Antimicrobial effect of surgical masks coated with nanoparticles. *Journal of Hospital Infection* 62: 58–63.

195 Aydin Sevinç, B. and Hanley, L. (2010). Antibacterial activity of dental composites containing zinc oxide nanoparticles. *Journal of Biomedical Materials Research Part B Applied Biomaterials* 94: 22–31.

196 Kassaee, M.Z., Akhavan, A., Sheikh, N., and Sodagar, A. (2008). Antibacterial effects of a new dental acrylic resin containing silver nanoparticles. *Journal of Applied Polymer Science* 110: 1699–1703.

11

Nanomaterials and Scaffolds for Tissue Engineering and Regenerative Medicine

Saeid Kargozar[1], Simin Nazarnezhad[1], Farzad Kermani[2], and Francesco Baino[3]

[1] *Tissue Engineering Research Group (TERG), Department of Anatomy and Cell Biology, School of Medicine, Mashhad University of Medical Sciences, Mashhad, Iran*
[2] *Department of Materials Engineering, Faculty of Engineering, Ferdowsi University of Mashhad (FUM), Mashhad, Iran*
[3] *Institute of Materials Physics and Engineering, Department of Applied Science and Technology, Politecnico di Torino, Torino, Italy*

11.1 Introduction and Recent Advances

The need for tissue and organ substitutes has continuously increased due to the high prevalence of chronic diseases, traumas, malignancies, and so on [1]. Traditionally, auto-, allo-, and xenografts are considered as the first options in the matter of tissue replacements. For example, autografts are still considered the gold standard in medicine due to their beneficial properties including excellent biocompatibility and biodegradability as well as lack of rejection. However, there are huge numbers of limitations in the case of autografts in the clinic, summarized as additional surgical procedures, donor shortage, the risk of pathogen transmission and immune-rejection [2]. Therefore, creating new tissues and organs in the laboratory has become an urgent need for the scientific communities over the world. Researchers who have substantial knowledge in materials science, chemistry, biology, and medicine are continually trying to develop tissue replacements to meet the need of patients for the concept of personalized medicine.

The role of nanotechnology in the success of tissue-engineering strategies is undeniable; several studies highlighted the usability of different types of nanostructured materials including organic and inorganic nanoparticles [3–5], self-assembled nanomaterials [6], electrospun nanofibers [7, 8], and nanopatterned surfaces [9]. Based on the definition, nanomaterials have dimensions less than 100 nm, holding great promises in tissue engineering as regards the ability to mimic the surface properties (including topography, energy) of natural tissues [10]. Based on their shapes, nanoscale particles could be classified into zero-dimensional (0D), one-dimensional (1D), two-dimensional (2D), and three-dimensional (3D) structures [11]. There are two main approaches for the synthesis of nanoparticles, i.e. bottom-up and top-down methods [12], which may involve mechanical, chemical, and biological means.

Nanomaterials and Nanotechnology in Medicine, First Edition. Edited by P.M. Visakh.

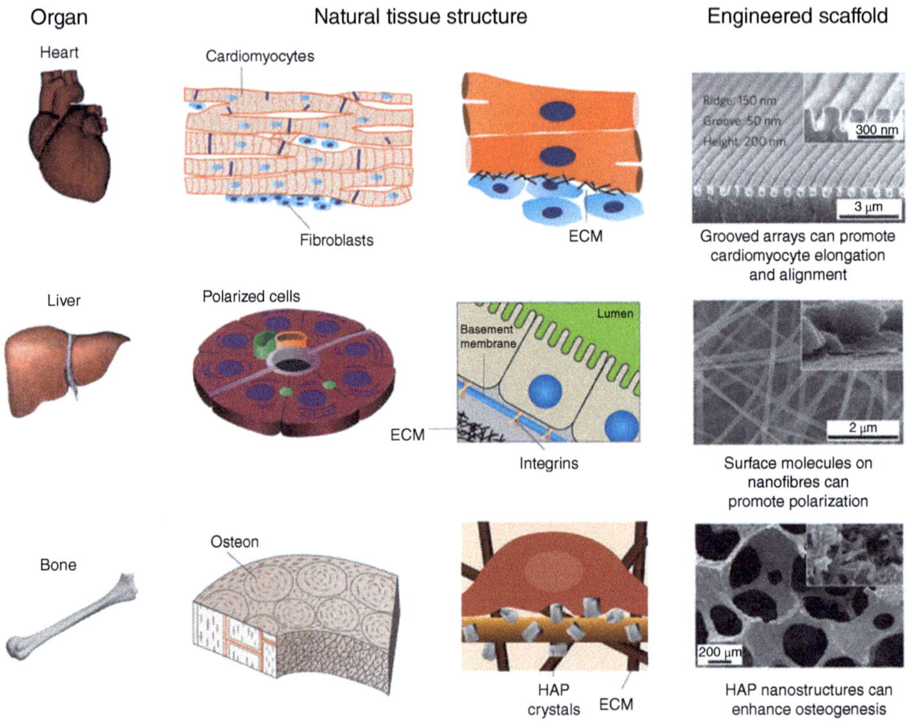

Figure 11.1 Engineered nanostructured scaffolds for human tissues. *Source:* Reproduced with permission, Dvir et al. 2001 [17].

From a tissue-engineering point of view, nanoscale materials could be utilized as *in situ* cell-delivery vehicles, nano-hydrogels, nanofibrous scaffolds, and layer-by-layer nanostructures [13]. Experimental studies have proved the biomimetic advantages of nanomaterials: higher protein adsorption and cell attachment regulated by nano-structured scaffolds efficiently stimulate more tissue regeneration (see Figure 11.1) [14, 15]. It has been well identified that nano-sized structures could improve cell behaviors including adhesion, proliferation, and differentiation, leading to accelerated tissue regeneration *in vivo* [16]. The particle size and alignment, surface roughness, and surface chemistry are some of the critical parameters affecting the final fate of cell activities and subsequent tissue regeneration.

11.2 Tissue Engineering and Regenerative Medicine: General Concepts

Tissue engineering and regenerative medicine caused a revolution in biomedicine as regards providing the possibility of fabricating tissue and organ substitutes in the laboratory for maintaining, restoring, or enhancing functions of the injured or diseased counterparts [18]. The history of this field dates back to the early 2000s when the exponential progress in nanotechnology and cell engineering occurred [19–21]. After nearly two decades, substantial advances in this area of science have led to an improved quality of life of huge numbers of patients over the globe.

Cells, bioactive molecules, and biomaterials are often referred to as the tissue-engineering triad. Up to now, a large number of somatic and stem cells have been evaluated and applied to boost the healing process of damaged tissues *in vivo* [22–24]. Embryonic, fetal, and adult stem cells form the main sources for tissue-reconstruction strategies. Among the various types of stem cells, mesenchymal stem cells (MSCs) derived from the bone marrow, adipose, and umbilical cord are the most commonly used cell types for tissue-engineering applications due to their excellent biological behaviors including easy availability, lack of ethical issues and tumorigenicity, and immuno-privileged properties [22, 25]. However, induced pluripotent stem cells (iPSCs) have gradually attracted much attention in the field thanks to their ability to differentiate into cells of all three primary germ layers without the risk of immunological rejection [26].

Bioactive molecules are soluble molecules, including growth factors, cytokines, chemokines, hormones, angiogenic factors, DNA (deoxyribonucleic acid), and siRNA (small interfering ribonucleic acid) playing critical roles in the healing process of different tissues and organs. It has been well understood that tissue repair and regeneration need a well-orchestrated form of cell migration, infiltration, proliferation, and differentiation [28]. The progress of this complex process strongly depends on effector bioactive molecules regulating the activation or silencing of molecular signaling pathways and thereby modulating cell behaviors. Up to now, several bioactive molecules have been identified and classified in specific families (e.g. transforming growth factor-beta (TGF-β)) regarding their activity and/or impact on cell and tissue types. Loading and delivery of bioactive molecules using biomaterials and 3D scaffolds are often desirable to induce tissue repair and regeneration. Direct encapsulation, incorporation, and covalent and noncovalent immobilization are among the most important methods and approaches applied for loading bioactive molecules to micro- and nano-sized biomaterials as well as 3D constructs. The high loading efficiency and a controlled release profile are two constant major targets in drug-delivery strategies for tissue engineering and regenerative medicine. Besides, providing a sustained release of drugs to ensure long-term therapeutic treatment is a worthwhile activity.

Regarding the literature, providing an appropriate environment for attachment, proliferation, and differentiation of cells is of great importance in tissue-regeneration applications, resulting in better clinical outcomes [29, 30]. In this sense, different types of biomaterials (e.g. polymers and ceramics) in numerous shapes and forms have been designed and developed to mimic the structure and function of the extracellular matrix (ECM) for supporting cells [31–33]. As a general principle, all materials used in the human body must have some essential criteria including biocompatibility and biodegradation to meet the requirements of the biological environment. Polymers of either natural (e.g. collagen) or synthetic origin (e.g. polyvinyl alcohol [PVA]) are mostly used for soft-tissue regeneration, while bioceramics are recognized to be highly suitable materials for hard-tissue engineering [33–37]. Since the ECM structure of the human tissues and organs is actually in the nanoscopic scale (nano-size), researchers are extensively trying to prepare scaffolds by using nanomaterials [38, 39]. In this regard, there are huge numbers of studies clarifying the effectiveness of nanotechnology in tissue repair and regeneration strategies, for example, nanophase surface roughness created by TiO2 (titanium dioxide) treatment could decrease bacterial growth and improve osteoblast proliferation on titanium implants [40].

11.3 Implantable Nanomaterials to Regenerate Living Tissues

In the human body, cells in solid tissues are surrounded by the ECM. This intricate 3D network is composed of different molecules including fibrillar proteins, proteoglycans (PGs), and glycosaminoglycans (GAGs); each tissue possesses its own specific ECM composition and topology controlling the activities between cells and the respective microenvironments [41, 42]. Moreover, the ECM provides a supportive environment for a heterogeneous population of cells in tissues by exhibiting appropriate topographical features as well as inducing mechanical and biochemical signals [43]. After any severe injuries and damages, the ECM of tissues is disrupted and needs to be replaced with an appropriate replacement (scaffold). ECM-mimicking scaffolds could provide a suitable substrate for attachment, proliferation, and differentiation of cells and subsequent tissue regeneration (Figure 11.2).

Over the past years, the use of nanoscale materials for the preparation of the ECM-mimicking scaffolds has led to great achievements in the field of tissue engineering and regenerative medicine [44]. As an illustration, the creation of nano-features on biomaterials surface is a feasible strategy to improve cell–biomaterials interactions and increase the

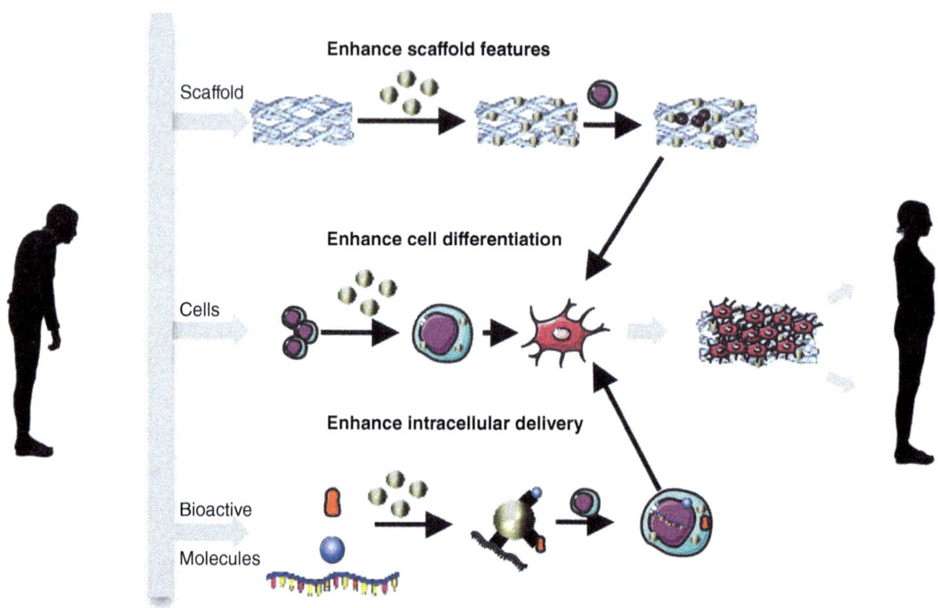

Figure 11.2 Tissue engineering utilizes ECM-mimicking scaffolds, mammalian cells (somatic or stem cells), and bioactive molecules (e.g. proteins and DNA) to fabricate a suitable substitute in the laboratory. *Source:* Reproduced with some modifications from ref. [27].

success of implanted materials in the body [45]. Methods and techniques developed for the fabrication of nanoscale topography on scaffolds might create ordered and nonordered topographies. Soft-lithography techniques such as electron-beam lithography and photolithography are mostly used to fabricate ordered patterns on the biomaterials surface, while techniques like polymer de-mixing, phase separation, colloidal lithography, and chemical etching are commonly used for producing unordered features onto biomaterials [46].

Adding nanoparticles to 3D scaffolds could result in physico-chemical, mechanical, and biological improvements in the final tissue-engineered constructs [47]. Therefore, nano-engineered materials are extensively being used for repair and regeneration of both hard and soft tissues [48–50]. The benefits for nanoparticle usage vary depending on the tissue-engineering fields in which they are used. For example, their advantages for cardiac regeneration could be summarized as follows: (i) increased electrical conductivity of biomaterial scaffolds; (ii) improved cell connections, differentiation, and organization through mimicking the nano-topography of the cardiac ECM; and (iii) enhanced survival of cardiomyocytes by reducing oxidative stress [51, 52]. In the case of skin wound healing, nanoparticle-incorporated scaffolds are mainly intended to enhance antimicrobial, anti-inflammatory, and angiogenic activities [53, 54].

Previous studies showed the ability of the self-assembling peptide systems in the fabrication of different shapes of nanomaterials (nanofibers and nanotubes) [55, 56]. Moreover, self-assembled nanopeptide hydrogel scaffolds, either in pristine and functionalized forms, could be easily applied to improve tissue repair and regeneration [57, 58]. Along with the advent of novel technologies in the tissue-engineering field, the possibility of developing nanomaterials-based hierarchical constructs was also explored to produce more effective tissue substitutes. In this sense, 3D printing of nanomaterials has provided new opportunities in the fabrication of nanocomposite scaffolds [59, 60]. This technique allows the incorporation of functional nanomaterials inside inks that are used to print microscale features in order to create macroscale devices [61].

11.4 Nanomaterials as Carriers for Therapeutic Agents

In order to improve the tissue-healing process, various therapeutic bioactive molecules are delivered to damaged sites by using nano-sized vehicles. To get the best clinical outcomes, it is necessary to provide a controlled release of drugs from tissue-engineering nanoplatforms [62]. Two main strategies have been acknowledged in the tissue-healing process regulated by nanomaterials: (i) nanomaterials themselves could act as therapies and (ii) nanomaterials serve as vehicles for delivering therapeutic agents [63, 64]. The latter is the main topic now discussed.

It has been stated that nano-structured drug-delivery systems (DDSs) ideally have three main components: (i) a core nanoparticulate material, (ii) a therapeutic payload, and (iii) surface modifiers [65]. To date, a large number of organic and inorganic nanoscale materials have been designed and used to accelerate the healing process of tissue defects via the delivery of stimulants for cell and tissue growth. It is worth stressing that the nanoparticulate DDSs employed should have no toxic and immunogenic activity and be easily eliminated from the human body to reduce toxic accumulation and other potential side effects.

Nanoparticulate carriers made of biopolymers (polysaccharides, proteins, and nucleic acids) were among the first DDSs developed for tissue-engineering applications. With reference to previous experiments [66–68], it is possible to use polymeric nanoparticles to load a broad range of either hydrophobic or hydrophilic therapeutics (e.g. proteins, genes, and chemicals); the particle size, charge, and morphology could be key determinants in the loading and delivery efficacy. Collagen, gelatin, albumin, chitosan, and heparin are some examples of natural biopolymeric nanoparticles, while the most used synthetic nanoparticulate polymers comprise poly(ethylene glycol) (PEG), poly(D, L-lactic acid) (PLA), poly(lactic-co-glycolic acid) (PLGA) [69].

The use of nanostructured lipid carriers in drug delivery is full of hopes and promises; they benefit from outstanding features including ease of preparation and scale-up, biocompatibility, site-specific targeted delivery, and the capability of enhanced targeting efficiency [70]. Moreover, they could permeate physiological barriers such as the blood–brain barrier (BBB) due to their lipophilicity without any surface modification [70]. As well-known lipids in DDSs, liposomes have been extensively used in the delivery of therapeutic payloads for tissue-engineering applications [71]. They are identified as the most commonly used DDSs owing to excellent compatibility with the living systems, the capability of sustained drug release, and passive targeting capability [72]. However, liposomes should be used in a combination with other materials in tissue engineering due to its poor mechanical strengths [73, 74].

Carbon-based nanomaterials are among the most promising DDSs in tissue-engineering approaches and comprise single or multiwalled carbon nanotubes (MWCNT), nanodiamonds, nanohorns, graphene, and fullerenes [75]. The benefits of these materials as DDSs are: high mechanical strength, large specific surface area, ease of preparation, and suitability for functionalization [76–79]. Carbon-based nanomaterials are known as multifunctional drug-delivery platforms; however, they are usually embedded within composites in tissue-engineering approaches [80].

Inorganic nanomaterials form a big family of substances having great capacities in drug-delivery and tissue-engineering applications [81–83]. Among various types of nano-sized inorganic materials used for drug delivery, mesoporous materials have attracted much attention for the controlled release of therapeutic molecules to both hard and soft tissues. The first mesoporous material used for drug delivery is MCM-41, which belongs to the M41S family of molecular sieves [84, 85]. Mesoporous materials are extensively being used as both systemic delivery systems and implantable local-delivery devices in tissue-engineering applications [86]. These materials have a porous structure with a highly controllable arrangement and diameters between 2 and 50 nm. Mesoporous silica nanoparticles (MSNs) are among the most widely used mesoporous materials for controlled drug delivery in tissue-engineering applications. MSNs are recognized as flexible platforms for targeted drug delivery in tissue engineering thanks to their relative ease of synthesis in different forms/shapes and functionalization with numerous chemicals as well as high pore volume and surface area [87–90].

Calcium phosphate nanoparticles (e.g. hydroxyapatite (HAp)) and mesoporous bioactive glasses (MBGs) are other useful inorganic materials in drug delivery and regenerative medicine. HAp nanoparticles with different structures (e.g. hollow mesoporous and solid particles) either in pristine or functionalized forms could serve as DDSs for a variety of biomedical applications, from bone regeneration to anticancer imaging and therapy [91–93]. MBGs are also among the most commonly used porous materials in tissue-engineering applications, especially for the reconstruction of bone and dental tissues. Furthermore, this type of material shows great potential in the delivery of various drugs from antibiotics to anticancer agents [94, 95]. As compared to nonmesoporous materials, MBGs show better drug-loading potential thanks to their intrinsic properties including ordered mesoporous channels, large specific surface areas, and high pore volumes [96]. The ability to bond to living tissues (both hard and soft) and deliver a wide range of drugs has made MBGs excellent multifunctional materials in tissue-engineering approaches [97]. In addition to bioactive therapeutic molecules, it is possible to incorporate functional ions (e.g. strontium (Sr) eliciting antiosteoporotic effect and copper (Cu) eliciting antibacterial effect) into MBG to deliver to the desired sites in the body [98–100].

As suitable DDSs, different metal nanoparticles (pristine and functionalized) have shown promises in biomedicine [101–103]. Gold and iron nanoparticles are perhaps the two main metallic candidates for the loading and delivery of therapeutic cargoes (proteins, genes, and drugs). They can be formulated and used in a mixture of polymers (composites) to be responsible for the outside stimulus (e.g. the pH of the surrounding environment) [104–106]. Most studies on these nanomaterials are connected with cancer therapy and diagnostics (theranostics). Silver, copper, and gadolinium nanoparticles are other common nanometals used as DDSs in the biomedical setting to render a broad range of biological functions such as antimicrobial and anticancer activities [107–109].

Drug-loaded nanomaterials could be incorporated into biodegradable 3D scaffolds and implanted into damaged tissues to boost the growth and proliferation of cells and thereby improve the tissue-healing process [110]. By grafting nanoparticles to the surface of scaffolds, it is possible to make multifunctional constructs enabling the delivery of therapeutic cargoes to injured sites. Biodegradable polymeric scaffolds containing drug-loaded nanoparticles are among the most used constructs in tissue engineering; electrospun nanofibers provide great opportunities for efficient hard- and soft-tissue healing promoted by growth-factor-containing nanomaterials [111, 112].

11.5 Nanofibrous Scaffolds

In recent years, nanofibrous mats have been greatly investigated in biomedical applications such as tissue engineering as well as drug delivery due to their excellent features including high surface-area-to-volume ratio, a microscale to nanoscale diameter, suitable porosity and malleability, and ease of physical and chemical modifications to integrate with the desired drugs [113, 114]. Nanofibrous scaffolds are mainly fabricated via electrospinning owing to its simplicity, cost-efficacy, and the capability of large-scale production. A major advantage of the electrospun nanofibers is their ECM-mimetic structure, which facilitates cell attachment and migration as well as the ability to carry bioactive molecule cargoes (e.g. growth factors, cytokines) [115–117].

The main constituents of an electrospinning system include: (i) a high voltage supplier (to charge the electrode); (ii) a syringe with conductive needle (which contains the polymer solution); (iii) a syringe pump (to convey the solution); and (iv) a metallic collector (to deposit the spun nanofibers). In the electrospinning process, the polymer solution is loaded into the syringe and then exposed to a high electric field. The high voltage makes the syringe needle the positive electrode and the collector connected to the earth serves as the negative electrode. By adjusting the desired voltage and flow rate, the charged polymer jet is accumulated onto the collector by evaporation of the solvent [118–121]. As indicated in Table 11.1, a series of determinants can affect the fiber morphology to obtain a fine and seamless nanofibrous mesh. They could be listed as: (i) electrospinning condition like applied voltage, the needle-to-collector distance, flow rate of polymer solution; (ii) solution properties such as the molecular weight of the polymer, viscosity, and conductivity of the solution, and surface tension; and (iii) electrospinning environments involving temperature and humidity [122, 123].

According to the number and structure of nozzles used, the electrospinning technique could be categorized into three groups: single nozzle (monoaxial), coaxial nozzles, and side-by-side nozzle electrospinning [134]. Monoaxial electrospinning is considered the most common and easiest method in which a single polymeric solution becomes electrospun. Although it is possible to incorporate bioactive molecules into the polymer solution and make an electrospinning emulsion [135, 136], the drug could be damaged due to exposure to the harsh environment. Hence, monoaxial electrospinning was further developed to coaxial electrospinning in which the drug can be encapsulated into the core structure thus offering protection from the environment. Additionally, the integration of bioactive

Table 11.1 Parameters affecting the electrospun fiber morphology.

Parameters	Effect on fiber morphology	Reference
Electrospinning parameters		
Applied voltage	Increased voltage leads to decreased fiber diameter	[124]
The needle to collector distance	Too short and too long distances result in bead formation, a minimum optimal distance is suggested	[125]
Flow rate	Reduced fiber diameter is related to decreased flow rate, high flow rate leads to the formation of beads	[126]
Solution parameters		
Polymer concentration	Increased concentration leads to increased fiber diameter	[127]
Molecular weight of the polymer	Increased molecular weight causes less bead formation	[128]
Viscosity	Increased viscosity results in less bead formation and increased fiber diameter	[129]
Conductivity	Higher conductivity leads to a decrease in fiber diameter	[130]
Surface tension	An optimum surface tension leads to proper jet initiation	[131]
Environment parameters		
Temperature	Increased temperature results in decreased viscosity of solution which in turn decreases the fiber diameter	[132]
Humidity	Increased humidity leads to circular pores on the fibers	[133]

molecules into the core structure can lead to the controlled release of the drug, especially when the core solution is hydrophobic and the sheath is hydrophilic, thus creating a diffusive barrier for the bioactive molecules [137, 138]. This structure also results in dual-drug loading in both core and shell layers [139]. In order to enhance the efficiency of drug release as well as provide a multi-DDS, the coaxial electrospinning system might be modified to triaxial electrospinning, which integrates multiple properties into a single multilayer fiber [140]. Side-by-side nozzle electrospinning is a modified form of the single-nozzle system in which separate solutions can be jetted from distinct capillaries [123]. Furthermore, surface immobilization of bioactive molecules onto the electrospun nanofibers can be performed after the electrospinning process by means of chemical or physical treatments including covalent and noncovalent immobilization [141].

The electrospun nanofibers are commonly fabricated from a wide range of natural polymers (e.g. collagen, gelatin, hyaluronic acid, chitosan, and dextran), synthetic polymers (e.g. PCL (polycaprolactone), PLA, PLGA), and their composites [122]. Indeed, natural polymers have not only better biocompatibility and biodegradability in comparison with synthetic polymers, but also a lower immune rejection. However, synthetic polymers have superior mechanical features such as malleability and stiffness. To achieve the advantage of both types of polymers, it has been advised to blend these polymers [142]. For instance, electrospun PCL fibrous mats are biodegradable and biocompatible and exhibit a suitable porous structure for biofluid perfusion; however, the hydrophobic nature of this polymer hinders efficient cell attachment [143]. Combining PCL with other hydrophilic polymers such as gelatin, collagen, or chitosan can overcome this limitation [144].

It has been extensively researched that electrospun nanofibers can be utilized as tissue-engineering constructs owing to the following properties. First, these scaffolds are biocompatible and can generally support cell proliferation. Second, they provide a porous and ECM-mimetic structure allowing the transportation of nutrients and bioactive molecules as well as cellular migration and vascularization. Third, the electrospun scaffolds are typically fabricated from biodegradable polymers that degrade at a rate that is comparable to the kinetics of tissue regeneration. Finally, they possess enough mechanical properties to make them good candidates for tissue regeneration, especially in soft-tissue-reconstruction strategies [145].

In general, electrospun nanofibers have gained much popularity in the field of tissue engineering as they provide a versatile, cost-effective, and biomimetic substitute for regenerating and repairing numerous soft (including skin, cardiac muscle, skeletal muscle, cartilage tissue, blood vessels, and nerve tissue) and hard (e.g. bone and dental tissue) tissues. In this context, various studies have revealed optimistic results to ameliorate the damaged tissue or organ. Up to now, electrospun nanofibers have shown great promise as a biomimetic construct for the management of the skin-tissue injuries such as diabetics wounds and burns [146]. It has also been proved that electrospun scaffolds are excellent candidates for drug delivery to the wound area such as growth factors (e.g. EGF (epidermal growth factor) and VEGF (vascular endothelial growth factor)), antibiotics, and other bioactive molecules [147, 148]. However, the lack of cell infiltration and migration into the construct has limited the utilization of electrospun nanofibrous mats as dermal replacement constructs [149–151]. Repair and regeneration of the nervous system by using nanofibrous scaffolds is a common approach in modern biomedicine; nerve-guide conduits fabricated by electrospinning processes can play a pivotal role in neural-tissue engineering.

Electrospun nerve-guide conduits show appealing features including (i) highly porous and large surface-area-to-volume ratio, (ii) presence of an aligned structure for neurite outgrowth, (iii) the possibility of the encapsulation of neural growth factors for improving survival, proliferation, and differentiation of nerve cells, and (iv) the capability of the integration of conductive polymers (e.g. polyaniline) with electrical stimulation [152]. In addition to soft tissues, electrospun nanofibers are also applied to repair hard tissues such as bone. The natural bone tissue is mainly composed of inorganic nanocrystals like HAp and also an organic matrix including collagen nanofibers [153]. According to this fact, numerous studies have proposed electrospun biodegradable polymers such as PCL [154], PLA [155], and PLGA [156], which benefit from a good degradation rate as well as improved mechanical properties [157]. It has well known that surface modification of such nanofibers with gelatin [158], HAp [159], calcium phosphate [160], and type I collagen [161] have osteoconductive properties and enhance osteoblast differentiation and proliferation as well as cell attachment [162]. However, there are still some drawbacks for conventional electrospun mats including poor cell infiltration to the construct, lack of tensile strength, nonpatterned orientation, and a broad range of fiber diameters [163].

11.6 Nano-topography Techniques for Tissue-engineered Scaffolds

Nano-topography refers to the technology that generates specific features (e.g. pits, grooves, pillars, hemispheres) for reconstructing the sub-micron textural surfaces [164]. Template-assisted and template-free methods are the two main classes of nano-topography developed and used over time (see Table 11.2). As displayed in Table 11.2, five different energy sources – thermal, optical, physical, chemical, and electrical – could be applied to make nanoscale topographies on surfaces.

The use of nano-topography in tissue-engineering applications is of great importance as regards the nanometric organization of natural ECM (which is hierarchical from the nano- to the macroscale). The presence of nanoscale topographies on human tissues and organs has been well documented; for example, collagen nanofibrils and HAp nanocrystals encompass osteoblasts and osteoclasts in the bone tissue [166, 167].

Up to now, a large number of implantable materials and devices have been surface modified to mimic the human environment for cells. Researchers take benefits from nano-topography techniques to improve the biocompatibility of various types of biomaterials. In this regard, adverse reactions between host and implanted materials could be attenuated by surface nano-topography processes [168]. Additionally, surface nano-topographies might be useful for modulating blood cells and platelet responses for particular applications such as cardiovascular regeneration. Given these advantages, a series of biopolymers (e.g. PLA, PLGA, and PCL), bioceramics, and metals have been successfully surface modified and translated to clinical use [169].

Scientifically, topographical modifications may lead to altered cell responses through changes in the geometry of the microenvironment. Experimental research clarified that ECM proteins (e.g. fibronectin) have a central role in mediating cell–nano–topographical interactions [170]. Improved cell adhesion, proliferation, and differentiation have been identified as positive effects of nanometric topographies, resulting in activated

Table 11.2 Different methods applied to generate nonpatterns onto surfaces.

Template	Energy source	Methods	Mechanism
Template-assisted	Thermal	Replica molding (RM)	Thermal cross-linking of the cavity-filled prepolymer
		Nanoimprint lithography (NIL)	Plastic deformation of polymer resin above glass transition temperature (Tg)
	Optical	Photolithography	Solubility change upon selective ultraviolet (UV) exposure
		UV-assisted molding	Photo cross-linking of cavity filled prepolymer
	Physical	Capillary force lithography (CFL)	Capillary rise of thermoplastic polymer resin above Tg
		Micro molding in capillaries (MIMIC)	Capillary-driven microchannel filling
	Chemical	Microcontact printing (μCP)	Transfer of materials by using the relative surface energy difference
	Electrical	Electrochemical deposition	Electrochemical reduction of the materials on the conductive electrode
Template-free	Thermal	Block copolymer (BCP) lithography	Microphase separation of two immiscible polymers
	Optical	E-beam lithography	Solubility change upon selective irradiation of focused electron beam
		Direct laser writing	Selective cross-linking of the polymer by laser irradiation
	Physical	Wrinkle	Mechanical buckling between the elastic substrate and rigid film
		Crack	Mechanical fracture of the stiff film adhered onto the elastic substrate
	Chemical	Dip-pen lithography	Direct writing of molecules with the sharp tip
		Salt leaching/gas foaming	Selective dissolution of salt particulates/bubble formation in the polymer block
	Electrical	Electrospinning	Electric-field-induced fiber drawing from the polymer solution

Source: With some modifications from [165].

tissue-specific functions *in vitro* and subsequently promoted tissue regeneration *in vivo* [171–173]. For instance, it is possible to increase the osseointegration of dental implants by endowing the biomaterial surface with patterned nanoroughness [174]. In the case of stem cells, nanoscale topographies could affect their main features, including orientation, morphology, phenotype, and differentiation [175]. Therefore, researchers and scientists could produce cell culture plates and dishes for the growth and spreading of specific cell types (e.g. embryonic stem cells) through patterned nanorough surfaces [176].

As microbial infections may impair biomaterial–tissue integration and subsequent failure of the implantation, the use of scaffolds with the ability to prevent bacteria growth and proliferation is of the utmost importance in tissue-engineering strategies. Ideally, implants should hinder the adhesion of bacteria without exerting any adverse effect on host cell functions. To date, several attempts have been made to provide antimicrobial implants by applying nanoscale topographies [177]. Most of them mentioned that the optimized surface topography for inhibiting bacterial growth and proliferation is a determinant of microorganism size and shape [178, 179]. As compared to chemical approaches, antibacterial nano-topographies as a physical approach could serve as a permanent structure in the body.

11.7 Regulatory Issues

Tissue engineering and regenerative medicine have recently greatly benefitted from the possibilities provided by nanotechnology [180]. Along with an increasing use of nanostructured materials in tissue-engineering strategies, some concerns on biosafety issues have emerged [181]. Among them, the uptake, distribution and elimination pathways, and metabolism of nanomaterials are considered as critical items in order to determine their short- and long-term effects in the human body [182, 183]. It is worth mentioning that the pharmacokinetics of nanoparticles depend on physicochemical parameters such as size, morphology, crystallinity, and surface charge, which can be controlled at the time of biomaterial synthesis to some extent [184]. Although international regulatory agencies such as the US Food and Drug Administration (FDA) and the European Commission (EC) have determined and presented some criteria in the matter of nanomaterials risk assessments, these substances are still mostly treated as conventional chemicals [185]. In brief, some features require evaluation before using nanoparticles in the body. These include (ii) physicochemical properties (size, shape, surface charge), (ii) microbial load (bacteria, mycoplasma, and virus levels), (iii) immunological features, hematological properties (coagulation time, platelet aggregation, and hemolysis), and (iv) toxicity [186]. Moreover, it should be noted that the manufacturing process might affect the quality of produced nanomaterials and, thereby, alter their final fate in the human body.

Currently, a large number of nanotechnology-based products (e.g. nanopharmaceuticals) are available in the market and used in therapeutic approaches from cancer therapy to tissue engineering and regenerative medicine [187, 188]. However, it is necessary to keep in mind that the financial burden of nanotechnology on health systems could reduce its growth rate.

11.8 Conclusion

The existing literature witnesses that nanomaterials can have a tremendous impact on medicine and healthcare. The use of nanoparticles and nanofibrous scaffolds have paved the way toward new and more effective strategies as regards the repair of injured tissues, controlled drug release, cancer therapy and, in general, applications in tissue engineering and regenerative medicine. However, there still are some barriers to full and safe translation to the clinic. The first problem is related to the partial lack of regulations concerning not only the toxicological aspects but also the production processes of nanomaterials, which is key to ensure reproducibility and safety to all the actors involved (manufacturers and patients). The second point is related to the incomplete knowledge of the relationship between nano-features and biological responses in the human body, and hence the difficulty in tuning the therapeutic effect by wisely manipulating the nano-feature. Therefore, future studies should focus on elucidating the long-term effects and fate of nanomaterials in the body and extending understanding of the underlying signaling cascade activated by nano-stimuli, in the attempt to build a comprehensive library associating nanomaterial characteristics and effects on cells.

References

1 Pina, S., Ribeiro, V.P., Marques, C.F. et al. (2019, 1824). Scaffolding Strategies for tissue engineering and regenerative medicine applications. *Materials* 12 (11).

2 Kargozar, S., Mozafari, M., Hashemian, S.J. et al. (2018). Osteogenic potential of stem cells-seeded bioactive nanocomposite scaffolds: a comparative study between human mesenchymal stem cells derived from bone, umbilical cord Wharton's jelly, and adipose tissue. *Journal of Biomedical Materials Research Part B: Applied Biomaterials* 106 (1): 61–72.

3 Wu, F., Chen, L., Yue, L. et al. (2019). Small-molecule porphyrin-based organic nanoparticles with remarkable photothermal conversion efficiency for in vivo photoacoustic imaging and photothermal therapy. *ACS Applied Materials & Interfaces* 11 (24): 21408–21416.

4 Sahmani, S., Shahali, M., Nejad, M.G. et al. (2019). Effect of copper oxide nanoparticles on electrical conductivity and cell viability of calcium phosphate scaffolds with improved mechanical strength for bone tissue engineering. *The European Physical Journal Plus* 134 (1): 7.

5 Kargozar, S., Ramakrishna, S., and Mozafari, M. (2019). Chemistry of biomaterials: future prospects. *Current Opinion in Biomedical Engineering* 10: 181–190.

6 Dhanwal, V., Katoch, A., Singh, A. et al. (2019). Self-assembled organic nanoparticles of benzimidazole analogue exhibit enhanced uptake in 3D tumor spheroids and oxidative stress induced cytotoxicity in breast cancer. *Materials Science and Engineering: C* 97: 467–478.

7 Naghavi Alhosseini, S., Moztarzadeh, F., Kargozar, S. et al. (2015). Development of polyvinyl alcohol fibrous biodegradable scaffolds for nerve tissue engineering applications: in vitro study. *International Journal of Polymeric Materials and Polymeric Biomaterials* 64 (9): 474–480.

8 Ranjbar-Mohammadi, M., Kargozar, S., Bahrami, S.H., and Joghataei, M. (2017). Fabrication of curcumin-loaded gum tragacanth/poly (vinyl alcohol) nanofibers with optimized electrospinning parameters. *Journal of Industrial Textiles* 46 (5): 1170–1192.

9 Camarero-Espinosa, S. and Cooper-White, J.J. (2019). Combinatorial presentation of cartilage-inspired peptides on nanopatterned surfaces enables directed differentiation of human mesenchymal stem cells towards distinct articular chondrogenic phenotypes. *Biomaterials* 210: 105–115.

10 Zhang, L. and Webster, T.J. (2009). Nanotechnology and nanomaterials: promises for improved tissue regeneration. *Nano Today* 4 (1): 66–80.

11 Khan, I., Saeed, K., and Khan, I. (2019). Nanoparticles: properties, applications and toxicities. *Arabian Journal of Chemistry* 12 (7): 908–931.

12 Lancaster, C.A., Scholl, W.E., Ticknor, M.A., and Shumaker-Parry, J.S. (2020). Uniting top-down and bottom-up strategies using fabricated nanostructures as hosts for synthesis of nanomites. *The Journal of Physical Chemistry C* 124 (12): 6822–6829.

13 Wan, A.C. and Ying, J.Y. (2010). Nanomaterials for in situ cell delivery and tissue regeneration. *Advanced Drug Delivery Reviews* 62 (7–8): 731–740.

14 Xia, L., Lin, K., Jiang, X. et al. (2014). Effect of nano-structured bioceramic surface on osteogenic differentiation of adipose derived stem cells. *Biomaterials* 35 (30): 8514–8527.

15 Lin, K., Xia, L., Gan, J. et al. (2013). Tailoring the nanostructured surfaces of hydroxyapatite bioceramics to promote protein adsorption, osteoblast growth, and osteogenic differentiation. *ACS Applied Materials & Interfaces* 5 (16): 8008–8017.

16 Zhao, C., Tan, A., Pastorin, G., and Ho, H.K. (2013). Nanomaterial scaffolds for stem cell proliferation and differentiation in tissue engineering. *Biotechnology Advances* 31 (5): 654–668.

17 Yang, Y., Chawla, A., Zhang, J. et al. (2019). Applications of nanotechnology for regenerative medicine; healing tissues at the nanoscale. In: *Principles of Regenerative Medicine*, 3e (ed. A. Atala, R. Lanza, A.G. Mikos and R. Nerem), 485–504. Boston: Academic Press.

18 Khademhosseini, A. and Langer, R. (2016). A decade of progress in tissue engineering. *Nature Protocols* 11 (10): 1775.

19 Smith, L. and Ma, P. (2004). Nano-fibrous scaffolds for tissue engineering. *Colloids and Surfaces B: Biointerfaces* 39 (3): 125–131.

20 Ma, P.X. (2004). Scaffolds for tissue fabrication. *Materials Today* 7 (5): 30–40.

21 Roco, M.C. (2003). Nanotechnology: convergence with modern biology and medicine. *Current Opinion in Biotechnology* 14 (3): 337–346.

22 Kargozar, S., Mozafari, M., Hamzehlou, S. et al. (2019). Bone tissue engineering using human cells: a comprehensive review on recent trends, current prospects, and recommendations. *Applied Sciences* 9 (1): 174.

23 Kargozar, S., Lotfibakhshaeish, N., Ebrahimi-Barough, S. et al. (2019). Stimulation of osteogenic differentiation of induced Pluripotent Stem Cells (iPSCs) using bioactive glasses: an in vitro study. *Frontiers in Bioengineering and Biotechnology* 7: 355.

24 Dehkordi, A.N., Babaheydari, F.M., Chehelgerdi, M., and Dehkordi, S.R. (2019). Skin tissue engineering: wound healing based on stem-cell-based therapeutic strategies. *Stem Cell Research & Therapy* 10 (1): 111.

25 Li, H., Shen, S., Fu, H. et al. (2019). Immunomodulatory functions of mesenchymal stem cells in tissue engineering. *Stem Cells International* 2019: 9671206.

26 Hirschi, K.K., Li, S., and Roy, K. (2014). Induced pluripotent stem cells for regenerative medicine. *Annual Review of Biomedical Engineering* 16: 277–294.

27 Vial, S., Reis, R.L., and Oliveira, J.M. (2017). Recent advances using gold nanoparticles as a promising multimodal tool for tissue engineering and regenerative medicine. *Current Opinion in Solid State and Materials Science* 21 (2): 92–112.

28 Aguilar, L.M.C., Silva, S.M., and Moulton, S.E. (2019). Growth factor delivery: defining the next generation platforms for tissue engineering. *Journal of Controlled Release* 306: 40–58.

29 Hyun, J.S., Montoro, D.T., Lo, D.D. et al. (2013). The seed and the soil: optimizing stem cells and their environment for tissue regeneration. *Annals of Plastic Surgery* 70 (2): 235–239.

30 Willerth, S.M. and Sakiyama-Elbert, S.E. (2019). Combining stem cells and biomaterial scaffolds for constructing tissues and cell delivery. *StemJournal* 1 (1): 1–25.

31 Jia, W., Li, M., Weng, H. et al. (2020). Design and comprehensive assessment of a biomimetic tri-layer tubular scaffold via biodegradable polymers for vascular tissue engineering applications. *Materials Science and Engineering: C* 110: 110717.

32 Mahato, A., Sandy, Z., Bysakh, S. et al. (2020). Development of nano-porous hydroxyapatite coated e-glass for potential bone-tissue engineering application: an in vitro approach. *Materials Science and Engineering: C* 110764.

33 Milan, P.B., Lotfibakhshaiesh, N., Joghataie, M.T. et al. (2016). Accelerated wound healing in a diabetic rat model using decellularized dermal matrix and human umbilical cord perivascular cells. *Acta Biomaterialia* 45: 234–246.

34 Kargozar, S., Lotfibakhshaiesh, N., Ai, J. et al. (2016). Synthesis, physico-chemical and biological characterization of strontium and cobalt substituted bioactive glasses for bone tissue engineering. *Journal of Non-Crystalline Solids* 449: 133–140.

35 Kermani, F., Kargozar, S., Tayarani-Najaran, Z. et al. (2019). Synthesis of nano HA/βTCP mesoporous particles using a simple modification in granulation method. *Materials Science and Engineering: C* 96: 859–871.

36 Mozafari, M., Kargozar, S., De Santiago, G. et al. (2018). Synthesis and characterisation of highly interconnected porous poly (ε-caprolactone)-collagen scaffolds: a therapeutic design to facilitate tendon regeneration. *Materials Technology* 33 (1): 29–37.

37 Rahmati, M., Milan, P.B., Samadikuchaksaraei, A. et al. (2017). Ionically crosslinked thermoresponsive chitosan hydrogels formed in situ: a conceptual basis for deeper understanding. *Macromolecular Materials and Engineering* 302 (11): 1700227.

38 Chang, C., Peng, N., He, M. et al. (2013). Fabrication and properties of chitin/hydroxyapatite hybrid hydrogels as scaffold nano-materials. *Carbohydrate Polymers* 91 (1): 7–13.

39 Asadpour, S., Yeganeh, H., Ai, J. et al. (2018). Polyurethane-polycaprolactone blend patches: scaffold characterization and cardiomyoblast adhesion, proliferation, and function. *ACS Biomaterials Science & Engineering* 4 (12): 4299–4310.

40 Bhardwaj, G. and Webster, T.J. (2017). Reduced bacterial growth and increased osteoblast proliferation on titanium with a nanophase TiO_2 surface treatment. *International Journal of Nanomedicine* 12: 363.

41 Cassidy, J.W. (2014). Nanotechnology in the regeneration of complex tissues, bone and tissue regeneration insights. *BTRI* 5: S12331.

42 Badylak, S.F., Freytes, D.O., and Gilbert, T.W. (2009). Extracellular matrix as a biological scaffold material: structure and function. *Acta Biomaterialia* 5 (1): 1–13.

43 Xing, H., Lee, H., Luo, L., and Kyriakides, T.R. (2019). Extracellular matrix-derived biomaterials in engineering cell function. *Biotechnology Advances* 42: 107421.

44 Engel, E., Michiardi, A., Navarro, M. et al. (2008). Nanotechnology in regenerative medicine: the materials side. *Trends in Biotechnology* 26 (1): 39–47.

45 Hasegawa, M., Saruta, J., Hirota, M. et al. (2020). A newly created meso-, micro-, and nano-scale rough titanium surface promotes bone-implant integration. *International Journal of Molecular Sciences* 21 (3): 783.
46 Norman, J.J. and Desai, T.A. (2006). Methods for fabrication of nanoscale topography for tissue engineering scaffolds. *Annals of Biomedical Engineering* 34 (1): 89–101.
47 Mclaughlin, S., Podrebarac, J., Ruel, M. et al. (2016). Nano-engineered biomaterials for tissue regeneration: what has been achieved so far? *Frontiers in Materials* 3 (27).
48 Kargozar, S., Milan, P.B., Baino, F., and Mozafari, M. (2019). Nanoengineered biomaterials for bone/dental regeneration. In: *Nanoengineered Biomaterials for Regenerative Medicine* (ed. M. Mozafari, J. Rajadas and D. Kaplan), 13–38.
49 Urbanska, A.M., Sefat, F., Yousaf, S. et al. (2019). Nanoengineered biomaterials for intestine regeneration. In: *Nanoengineered Biomaterials for Regenerative Medicine* (ed. M. Mozafari, J. Rajadas and D. Kaplan), 363–378. Elsevier.
50 Milan, P.B., Kargozar, S., Joghataie, M.T., and Samadikuchaksaraei, A. (2019). Nanoengineered biomaterials for skin regeneration. In: *Nanoengineered Biomaterials for Regenerative Medicine* (ed. M. Mozafari, J. Rajadas and D. Kaplan), 265–283. Elsevier.
51 Baei, P., Jalili-Firoozinezhad, S., Rajabi-Zeleti, S. et al. (2016). Electrically conductive gold nanoparticle-chitosan thermosensitive hydrogels for cardiac tissue engineering. *Materials Science and Engineering: C* 63: 131–141.
52 Amezcua, R., Shirolkar, A., Fraze, C., and Stout, D.A. (2016). Nanomaterials for cardiac myocyte tissue engineering. *Nanomaterials* 6 (7): 133.
53 Augustine, R., Hasan, A., Patan, N.K. et al. (2019). Cerium oxide nanoparticle incorporated electrospun poly (3-hydroxybutyrate-co-3-hydroxyvalerate) membranes for diabetic wound healing applications. *ACS Biomaterials Science & Engineering* 6 (1): –58, 70.
54 Mohseni, M., Shamloo, A., Aghababaie, Z. et al. (2019). A comparative study of wound dressings loaded with silver sulfadiazine and silver nanoparticles: in vitro and in vivo evaluation. *International Journal of Pharmaceutics* 564: 350–358.
55 Zhao, X. and Zhang, S. (2006). Self-assembling nanopeptides become a new type of biomaterial. In: *Polymers for Regenerative Medicine* (ed. C. Werner), 145–170. Springer.
56 Soler-Botija, C., Bagó, J.R., Llucià-Valldeperas, A. et al. (2014). Engineered 3D bioimplants using elastomeric scaffold, self-assembling peptide hydrogel, and adipose tissue-derived progenitor cells for cardiac regeneration. *American Journal of Translational Research* 6 (3): 291.
57 Wang, T.-W., Chang, K.-C., Chen, L.-H. et al. (2017). Effects of an injectable functionalized self-assembling nanopeptide hydrogel on angiogenesis and neurogenesis for regeneration of the central nervous system. *Nanoscale* 9 (42): 16281–16292.
58 Chiu, Y.-W., Chen, W.-P., Su, C.-C. et al. (2014). The arrhythmogenic effect of self-assembling nanopeptide hydrogel scaffolds on neonatal mouse cardiomyocytes. *Nanomedicine: Nanotechnology, Biology and Medicine* 10 (5): e1065–e1073.
59 Kumar, A., Kargozar, S., Baino, F., and Han, S.S. (2019). Additive manufacturing methods for producing hydroxyapatite and hydroxyapatite-based composite scaffolds: a review. *Frontiers in Materials* 6: 313. https://doi.org/10.3389/fmats.
60 Castro, N.J., O'Brien, J., and Zhang, L.G. (2015). Integrating biologically inspired nanomaterials and table-top stereolithography for 3D printed biomimetic osteochondral scaffolds. *Nanoscale* 7 (33): 14010–14022.

61 Kong, Y.L., Gupta, M.K., Johnson, B.N., and McAlpine, M.C. (2016). 3D printed bionic nanodevices. *Nano Today* 11 (3): 330–350.

62 Rambhia, K.J. and Ma, P.X. (2015). Controlled drug release for tissue engineering. *Journal of Controlled Release* 219: 119–128.

63 Nethi, S.K., Das, S., Patra, C.R., and Mukherjee, S. (2019). Recent advances in inorganic nanomaterials for wound-healing applications. *Biomaterials Sience* 7 (7): 2652–2674.

64 Yadid, M., Feiner, R., and Dvir, T. (2019). Gold nanoparticle-integrated scaffolds for tissue engineering and regenerative medicine. *Nano Letters* 19 (4): 2198–2206.

65 Webster, D.M., Sundaram, P., and Byrne, M.E. (2013). Injectable nanomaterials for drug delivery: carriers, targeting moieties, and therapeutics. *European Journal of Pharmaceutics and Biopharmaceutics* 84 (1): 1–20.

66 Liu, D., Bimbo, L.M., Mäkilä, E. et al. (2013). Co-delivery of a hydrophobic small molecule and a hydrophilic peptide by porous silicon nanoparticles. *Journal of Controlled Release* 170 (2): 268–278.

67 Markwalter, C.E., Pagels, R.F., Wilson, B.K. et al. (2019). Flash nanoprecipitation for the encapsulation of hydrophobic and hydrophilic compounds in polymeric nanoparticles. *Journal of Visualized Experiments* 143: e58757.

68 Xiong, Q., Lim, Y., Li, D. et al. (2020). Photoactive nanocarriers for controlled delivery. *Advanced Functional Materials* 30 (2): 1903896.

69 Bhawani, S.A., Moheman, A., Yakout, A.A., and Ibrahim, M.N.M. (2019). Nanostructured biopolymers for application as drug-delivery vehicles. In: *Nanostructured Polymer Composites for Biomedical Applications* (ed. S.K. Swain and M. Jawaid), 189–210. Elsevier.

70 Khosa, A., Reddi, S., and Saha, R.N. (2018). Nanostructured lipid carriers for site-specific drug delivery. *Biomedicine & Pharmacotherapy* 103: 598–613.

71 Melling, G.E., Colombo, J.S., Avery, S.J. et al. (2018). Liposomal delivery of demineralized dentin matrix for dental tissue regeneration. *Tissue Engineering Part A* 24 (13–14): 1057–1065.

72 Cheng, R., Liu, L., Xiang, Y. et al. (2019). Advanced liposome-loaded scaffolds for therapeutic and tissue engineering applications. *Biomaterials* 232: 119706.

73 Monteiro, N., Martins, A., Reis, R.L., and Neves, N.M. (2014). Liposomes in tissue engineering and regenerative medicine. *Journal of the Royal Society Interface* 11 (101): 20140459.

74 Monteiro, N., Martins, A., Pires, R. et al. (2014). Immobilization of bioactive factor-loaded liposomes on the surface of electrospun nanofibers targeting tissue engineering. *Biomaterials Science* 2 (9): 1195–1209.

75 Cha, C., Shin, S.R., Annabi, N. et al. (2013). Carbon-based nanomaterials: multifunctional materials for biomedical engineering. *ACS Nano* 7 (4): 2891–2897.

76 Avani, F., Damoogh, S., Mottaghitalab, F. et al. (2020). Vancomycin loaded halloysite nanotubes embedded in silk fibroin hydrogel applicable for bone tissue engineering. *International Journal of Polymeric Materials and Polymeric Biomaterials* 69 (1): 32–43.

77 Sukhodub, L., Sukhodub, L., Kumeda, M. et al. (2019). C_{60} fullerene loaded hydroxyapatite-chitosan beads as a promising system for prolonged drug release. *Carbohydrate Polymers* 223: 115067.

78 Ekwueme, E.C., Rao, R., Mohiuddin, M. et al. (2019). Single-walled carbon nanohorns modulate tenocyte cellular response and tendon biomechanics. *Journal of Biomedical Materials Research Part B: Applied Biomaterials* 108 (5): 1907–1914.

79 Li, A., Fan, Y., Cao, X. et al. (2019). Morpholino-functionalized phosphorus dendrimers for precision regenerative medicine: osteogenic differentiation of mesenchymal stem cells. *Nanoscale* 11 (37): 17230–17234.

80 Lalwani, G., Henslee, A.M., Farshid, B. et al. (2013). Two-dimensional nanostructure-reinforced biodegradable polymeric nanocomposites for bone tissue engineering. *Biomacromolecules* 14 (3): 900–909.

81 Khalifehzadeh, R. and Arami, H. (2019). DNA-templated strontium-doped calcium phosphate nanoparticles for gene delivery in bone cells. *ACS Biomaterials Science & Engineering* 5 (7): 3201–3211.

82 Levingstone, T.J., Herbaj, S., and Dunne, N.J. (2019). Calcium phosphate nanoparticles for therapeutic applications in bone regeneration. *Nanomaterials* 9 (11): 1570.

83 Tang, J., Li, B., Howard, C.B. et al. (2019). Multifunctional lipid-coated calcium phosphate nanoplatforms for complete inhibition of large triple negative breast cancer via targeted combined therapy. *Biomaterials* 216: 119232.

84 Vallet-Regi, M., Ramila, A., Del Real, R., and Pérez-Pariente, J. (2001). A new property of MCM-41: drug delivery system. *Chemistry of Materials* 13 (2): 308–311.

85 Kresge, C., Leonowicz, M., Roth, W.J. et al. (1992). Ordered mesoporous molecular sieves synthesized by a liquid-crystal template mechanism. *Nature* 359 (6397): 710–712.

86 Vallet-Regí, M., Balas, F., and Arcos, D. (2007). Mesoporous materials for drug delivery. *Angewandte Chemie International Edition* 46 (40): 7548–7558.

87 Rosenholm, J.M., Zhang, J., Linden, M., and Sahlgren, C. (2016). Mesoporous silica nanoparticles in tissue engineering – a perspective. *Nanomedicine* 11 (4): 391–402.

88 Chen, X., Yuan, P., Liu, Z. et al. (2017). Dual responsive hydrogels based on functionalized mesoporous silica nanoparticles as an injectable platform for tumor therapy and tissue regeneration. *Journal of Materials Chemistry B* 5 (30): 5968–5973.

89 Qiu, K., Chen, B., Nie, W. et al. (2016). Electrophoretic deposition of dexamethasone-loaded mesoporous silica nanoparticles onto poly (L-lactic acid)/poly (ε-caprolactone) composite scaffold for bone tissue engineering. *ACS Applied Materials & Interfaces* 8 (6): 4137–4148.

90 Nik, A.B., Zare, H., Razavi, S.S. et al. (2020). Smart drug delivery: capping strategies for mesoporous silica nanoparticles. *Microporous and Mesoporous Materials* 299: 110115.

91 Yang, Y.-H., Liu, C.-H., Liang, Y.-H. et al. (2013). Hollow mesoporous hydroxyapatite nanoparticles (hmHANPs) with enhanced drug loading and pH-responsive release properties for intracellular drug delivery. *Journal of Materials Chemistry B* 1 (19): 2447–2450.

92 Li, D., Huang, X., Wu, Y. et al. (2016). Preparation of pH-responsive mesoporous hydroxyapatite nanoparticles for intracellular controlled release of an anticancer drug. *Biomaterials Science* 4 (2): 272–280.

93 Syamchand, S.S. and Sony, G. (2015). Multifunctional hydroxyapatite nanoparticles for drug delivery and multimodal molecular imaging. *Microchimica Acta* 182 (9–10): 1567–1589.

94 Kargozar, S., Montazerian, M., Hamzehlou, S. et al. (2018). Mesoporous bioactive glasses: promising platforms for antibacterial strategies. *Acta Biomaterialia* 81: 1–19.

95 Kargozar, S., Mozafari, M., Hamzehlou, S. et al. (2019). Mesoporous bioactive glasses (MBGs) in cancer therapy: full of hope and promise. *Materials Letters* 251: 241–246.

96 Liu, Y.-Z., Li, Y., Yu, X.-B. et al. (2014). Drug delivery property, bactericidal property and cytocompatibility of magnetic mesoporous bioactive glass. *Materials Science and Engineering: C* 41: 196–205.

97 Wu, C. and Chang, J. (2014). Multifunctional mesoporous bioactive glasses for effective delivery of therapeutic ions and drug/growth factors. *Journal of Controlled Release* 193: 282–295.

98 Zhang, J., Zhao, S., Zhu, Y. et al. (2014). Three-dimensional printing of strontium-containing mesoporous bioactive glass scaffolds for bone regeneration. *Acta Biomaterialia* 10 (5): 2269–2281.

99 Kargozar, S., Montazerian, M., Fiume, E., and Baino, F. (2019). Multiple and promising applications of Sr-containing bioactive glasses in bone tissue engineering. *Frontiers in Bioengineering and Biotechnology* 7: 161.

100 Bari, A., Bloise, N., Fiorilli, S. et al. (2017). Copper-containing mesoporous bioactive glass nanoparticles as multifunctional agent for bone regeneration. *Acta Biomaterialia* 55: 493–504.

101 Adeyemi, O.S. and Sulaiman, F.A. (2015). Evaluation of metal nanoparticles for drug delivery systems. *Journal of Biomedical Research* 29 (2): 145.

102 Paramasivam, G., Kayambu, N., Rabel, A.M. et al. (2017). Anisotropic noble metal nanoparticles: synthesis, surface functionalization and applications in biosensing, bioimaging, drug delivery and theranostics. *Acta Biomaterialia* 49: 45–65.

103 Anderson, S.D., Gwenin, V.V., and Gwenin, C.D. (2019). Magnetic functionalized nanoparticles for biomedical, drug delivery and imaging applications. *Nanoscale Research Letters* 14 (1): 1–16.

104 Yavuz, M.S., Cheng, Y., Chen, J. et al. (2009). Gold nanocages covered by smart polymers for controlled release with near-infrared light. *Nature Materials* 8 (12): 935–939.

105 Oka, C., Ushimaru, K., Horiishi, N. et al. (2015). Core–shell composite particles composed of biodegradable polymer particles and magnetic iron oxide nanoparticles for targeted drug delivery. *Journal of Magnetism and Magnetic Materials* 381: 278–284.

106 Liang, R., Wei, M., Evans, D.G., and Duan, X. (2014). Inorganic nanomaterials for bioimaging, targeted drug delivery and therapeutics. *Chemical Communications* 50 (91): 14071–14081.

107 Dos Santos, C.A., Seckler, M.M., Ingle, A.P. et al. (2014). Silver nanoparticles: Therapeutical uses. *Toxicity, and Safety Issues, Journal of Pharmaceutical Sciences* 103 (7): 1931–1944.

108 Chen, W., Yang, W., Chen, P. et al. (2018). Disulfiram copper nanoparticles prepared with a stabilized metal ion ligand complex method for treating drug-resistant prostate cancers. *ACS Applied Materials & Interfaces* 10 (48): 41118–41128.

109 Shen, Z., Liu, T., Yang, Z. et al. (2020). Small-sized gadolinium oxide based nanoparticles for high-efficiency theranostics of orthotopic glioblastoma. *Biomaterials* 235: 119783.

110 Sahai, N., Ahmad, N., and Gogoi, M. (2018). Nanoparticles based drug delivery for tissue regeneration using biodegradable scaffolds: a review. *Current Pathobiology Reports* 6 (4): 219–224.

111 Shao, W., He, J., Sang, F. et al. (2016). Coaxial electrospun aligned tussah silk fibroin nanostructured fiber scaffolds embedded with hydroxyapatite–tussah silk fibroin nanoparticles for bone tissue engineering. *Materials Science and Engineering: C* 58: 342–351.

112 Heidari, M., Bahrami, S.H., Ranjbar-Mohammadi, M., and Milan, P. (2019). Smart electrospun nanofibers containing PCL/gelatin/graphene oxide for application in nerve tissue engineering. *Materials Science and Engineering: C* 103: 109768.

113 Abrigo, M., McArthur, S.L., and Kingshott, P. (2014). Electrospun nanofibers as dressings for chronic wound care: advances, challenges, and future prospects. *Macromolecular Bioscience* 14 (6): 772–792.

114 Stocco, T.D., Bassous, N.J., Zhao, S. et al. (2018). Nanofibrous scaffolds for biomedical applications. *Nanoscale* 10 (26): 12228–12255.

115 Rim, N.G., Shin, C.S., and Shin, H. (2013). Current approaches to electrospun nanofibers for tissue engineering. *Biomedical Materials* 8 (1): 014102.

116 He, W., Ma, Z., Yong, T. et al. (2005). Fabrication of collagen-coated biodegradable polymer nanofiber mesh and its potential for endothelial cells growth. *Biomaterials* 26 (36): 7606–7615.

117 Cross, M. and Claesson-Welsh, L. (2001). FGF and VEGF function in angiogenesis: signalling pathways, biological responses and therapeutic inhibition. *Trends in Pharmacological Sciences* 22: 201–207.

118 Barnes, C.P., Sell, S.A., Boland, E.D. et al. (2007). Nanofiber technology: designing the next generation of tissue engineering scaffolds. *Advanced Drug Delivery Reviews* 59 (14): 1413–1433.

119 L'Heureux, N., Dusserre, N., Konig, G. et al. (2006). Human tissue-engineered blood vessels for adult arterial revascularization. *Nature Medicine* 12 (3): 361–365.

120 Shadwick, R.E. (1999). Mechanical design in arteries. *Journal of Experimental Biology* 202 (23): 3305–3313.

121 Agarwal, S., Wendorff, J.H., and Greiner, A. (2008). Use of electrospinning technique for biomedical applications. *Polymer* 49 (26): 5603–5621.

122 Bhardwaj, N. and Kundu, S.C. (2010). Electrospinning: a fascinating fiber fabrication technique. *Biotechnology Advances* 28 (3): 325–347.

123 Sill, T.J. and von Recum, H.A. (2008). Electrospinning: applications in drug delivery and tissue engineering. *Biomaterials* 29 (13): 1989–2006.

124 Topuz, F. and Uyar, T. (2017). Electrospinning of gelatin with tunable fiber morphology from round to flat/ribbon. *Materials Science and Engineering: C* 80: 371–378.

125 Ki, C.S., Baek, D.H., Gang, K.D. et al. (2005). Characterization of gelatin nanofiber prepared from gelatin–formic acid solution. *Polymer* 46 (14): 5094–5102.

126 Nadri, S., Nasehi, F., and Barati, G. (2017). Effect of parameters on the quality of core-shell fibrous scaffold for retinal differentiation of conjunctiva mesenchymal stem cells. *Journal of Biomedical Materials Research Part A* 105 (1): 189–197.

127 Megelski, S., Stephens, J.S., Chase, D.B., and Rabolt, J.F. (2002). Micro – and nanostructured surface morphology on electrospun polymer fibers. *Macromolecules* 35 (22): 8456–8466.

128 Malik, R., Garg, T., Goyal, A.K., and Rath, G. (2015). Polymeric nanofibers: targeted gastro-retentive drug delivery systems. *Journal of Drug Targeting* 23 (2): 109–124.

129 Ramakrishna, S., Fujihara, K., Teo, W.-E. et al. (2006). Electrospun nanofibers: solving global issues. *Materials Today* 9 (3): 40–50.

130 Lee, J., Kwon, H., Seo, J. et al. (2015). Conductive fiber-based ultrasensitive textile pressure sensor for wearable electronics. *Advanced Materials* 27 (15): 2433–2439.
131 Khalil, H.A., Davoudpour, Y., Bhat, A. et al. (2015). Electrospun cellulose composite nanofibers. In: *Handbook of Polymer Nanocomposites. Processing, Performance and Application* (ed. J.K. Pandey, H. Takagi, A.N. Nakagaito, et al.), 191–227. Springer.
132 Liu, Q., Wang, Y., Dai, L., and Yao, J. (2016). Scalable fabrication of nanoporous carbon fiber films as bifunctional catalytic electrodes for flexible Zn-air batteries. *Advanced Materials* 28 (15): 3000–3006.
133 Bedane, A.H., Eić, M., Farmahini-Farahani, M., and Xiao, H. (2016). Theoretical modeling of water vapor transport in cellulose-based materials. *Cellulose* 23 (3): 1537–1552.
134 Feng, X., Li, J., Zhang, X. et al. (2019). Electrospun polymer micro/nanofibers as pharmaceutical repositories for healthcare. *Journal of Controlled Release* 302: 19–41.
135 Arecchi, A., Mannino, S., and Weiss, J. (2010). Electrospinning of poly (vinyl alcohol) nanofibers loaded with hexadecane nanodroplets. *Journal of Food Science* 75 (6): N80–N88.
136 Zhang, X. and Wang, M. (2012). Effects of emulsion electrospinning parameters on the morphology and structure of core-shell structured PLLA fibers. *Advanced Materials Research*, Trans Tech Publications 410: 386–389.
137 Choe, H., Narayanan, A.S., Gandhi, D.A. et al. (2015). Immunomodulatory peptide IDR-1018 decreases implant infection and preserves osseointegration. *Clinical Orthopaedics and Related Research* 473 (9): 2898–2907.
138 Lu, Y., Huang, J., Yu, G. et al. (2016). Coaxial electrospun fibers: applications in drug delivery and tissue engineering. *Wiley Interdisciplinary Reviews: Nanomedicine and Nanobiotechnology* 8 (5): 654–677.
139 Rychter, M., Baranowska-Korczyc, A., and Lulek, J. (2017). Progress and perspectives in bioactive agent delivery via electrospun vascular grafts. *RSC Advances* 7 (51): 32164–32184.
140 Khalf, A. and Madihally, S.V. (2017). Recent advances in multiaxial electrospinning for drug delivery. *European Journal of Pharmaceutics and Biopharmaceutics* 112: 1–17.
141 Fahimirad, S. and Ajalloueian, F. (2019). Naturally-derived electrospun wound dressings for target delivery of bio-active agents. *International Journal of Pharmaceutics* 566: 307–328.
142 Liu, Y., Zhou, S., Gao, Y., and Zhai, Y. (2019). Electrospun nanofibers as a wound dressing for treating diabetic foot ulcer. *Asian Journal of Pharmaceutical Sciences* 14 (2): 130–143.
143 Bui, H.T., Chung, O.H., Cruz, J.D., and Park, J.S. (2014). Fabrication and characterization of electrospun curcumin-loaded polycaprolactone-polyethylene glycol nanofibers for enhanced wound healing. *Macromolecular Research* 22 (12): 1288–1296.
144 Wang, J. and Windbergs, M. (2017). Functional electrospun fibers for the treatment of human skin wounds. *European Journal of Pharmaceutics and Biopharmaceutics* 119: 283–299.
145 Ding, J., Zhang, J., Li, J. et al. (2019). Electrospun polymer biomaterials. *Progress in Polymer Science* 90: 1–34.
146 Gautam, S., Chou, C.-F., Dinda, A.K. et al. (2014). Surface modification of nanofibrous polycaprolactone/gelatin composite scaffold by collagen type I grafting for skin tissue engineering. *Materials Science and Engineering: C* 34: 402–409.

147 Liu, M., Duan, X.-P., Li, Y.-M. et al. (2017). Electrospun nanofibers for wound healing. *Materials Science and Engineering: C* 76: 1413–1423.

148 Gil, E.S., Panilaitis, B., Bellas, E., and Kaplan, D.L. (2013). Functionalized silk biomaterials for wound healing. *Advanced Healthcare Materials* 2 (1): 206–217.

149 Chouhan, D., Janani, G., Chakraborty, B. et al. (2018). Functionalized PVA–silk blended nanofibrous mats promote diabetic wound healing via regulation of extracellular matrix and tissue remodelling. *Journal of Tissue Engineering and Regenerative Medicine* 12 (3): e1559–e1570.

150 Dong, R.-H., Jia, Y.-X., Qin, C.-C. et al. (2016). In situ deposition of a personalized nanofibrous dressing via a handy electrospinning device for skin wound care. *Nanoscale* 8 (6): 3482–3488.

151 Chouhan, D., Thatikonda, N., Nilebäck, L. et al. (2018). Recombinant spider silk functionalized silkworm silk matrices as potential bioactive wound dressings and skin grafts. *ACS Applied Materials & Interfaces* 10 (28): 23560–23572.

152 Quan, Q., Chang, B., Meng, H.Y. et al. (2016). Use of electrospinning to construct biomaterials for peripheral nerve regeneration. *Reviews in the Neurosciences* 27 (7): 761–768.

153 Zhang, Y., Venugopal, J.R., El-Turki, A. et al. (2008). Electrospun biomimetic nanocomposite nanofibers of hydroxyapatite/chitosan for bone tissue engineering. *Biomaterials* 29 (32): 4314–4322.

154 Yoshimoto, H., Shin, Y., Terai, H., and Vacanti, J. (2003). A biodegradable nanofiber scaffold by electrospinning and its potential for bone tissue engineering. *Biomaterials* 24 (12): 2077–2082.

155 Song, X., Ling, F., Ma, L. et al. (2013). Electrospun hydroxyapatite grafted poly (L-lactide)/ poly (lactic-co-glycolic acid) nanofibers for guided bone regeneration membrane. *Composites Sience and Technology* 79: 8–14.

156 Meng, Z., Wang, Y., Ma, C. et al. (2010). Electrospinning of PLGA/gelatin randomly-oriented and aligned nanofibers as potential scaffold in tissue engineering. *Materials Science and Engineering: C* 30 (8): 1204–1210.

157 Zhang, J., Liu, H., Ding, J.-X. et al. (2015). Annealing regulates the performance of an electrospun poly (ε-caprolactone) membrane to accommodate tissue engineering. *RSC Advances* 5 (41): 32604–32608.

158 Li, X., Xie, J., Yuan, X., and Xia, Y. (2008). Coating electrospun poly (ε-caprolactone) fibers with gelatin and calcium phosphate and their use as biomimetic scaffolds for bone tissue engineering. *Langmuir* 24 (24): 14145–14150.

159 Liao, S., Cui, F., Zhang, W., and Feng, Q. (2004). Hierarchically biomimetic bone scaffold materials: nano-HA/collagen/PLA composite. *Journal of Biomedical Materials Research Part B: Applied Biomaterials* 69 (2): 158–165.

160 Yang, F., Wolke, J., and Jansen, J. (2008). Biomimetic calcium phosphate coating on electrospun poly (ε-caprolactone) scaffolds for bone tissue engineering. *Chemical Engineering Journal* 137 (1): 154–161.

161 Chiu, L.-H., Lai, W.-F.T., Chang, S.-F. et al. (2014). The effect of type II collagen on MSC osteogenic differentiation and bone defect repair. *Biomaterials* 35 (9): 2680–2691.

162 Yoo, H.S., Kim, T.G., and Park, T.G. (2009). Surface-functionalized electrospun nanofibers for tissue engineering and drug delivery. *Advanced Drug Delivery Reviews* 61 (12): 1033–1042.

163 Udomluck, N., Koh, W.-G., Lim, D.-J., and Park, H. (2020). Recent developments in nanofiber fabrication and modification for bone tissue engineering. *International Journal of Molecular Sciences* 21 (1): 99.

164 Skoog, S.A., Kumar, G., Narayan, R.J., and Goering, P.L. (2018). Biological responses to immobilized microscale and nanoscale surface topographies. *Pharmacology & Therapeutics* 182: 33–55.

165 Kim, H.N., Jiao, A., Hwang, N.S. et al. (2013). Nanotopography-guided tissue engineering and regenerative medicine. *Advanced Drug Delivery Reviews* 65 (4): 536–558.

166 Klymov, A., Prodanov, L., Lamers, E. et al. (2013). Understanding the role of nano-topography on the surface of a bone-implant. *Biomaterials Science* 1 (2): 135–151.

167 Chen, Z., Bachhuka, A., Wei, F. et al. (2017). Nanotopography-based strategy for the precise manipulation of osteoimmunomodulation in bone regeneration. *Nanoscale* 9 (46): 18129–18152.

168 Hulander, M., Lundgren, A., Berglin, M. et al. (2011). Immune complement activation is attenuated by surface nanotopography. *International Journal of Nanomedicine* 6: 2653.

169 Richbourg, N.R., Peppas, N.A., and Sikavitsas, V.I. (2019). Tuning the biomimetic behavior of scaffolds for regenerative medicine through surface modifications. *Journal of Tissue Engineering and Regenerative Medicine* 13 (8): 1275–1293.

170 Ngandu Mpoyi, E., Cantini, M., Reynolds, P.M. et al. (2016). Protein adsorption as a key mediator in the nanotopographical control of cell behavior. *ACS Nano* 10 (7): 6638–6647.

171 Yin, H.-M., Mao, C., Liu, W. et al. (2019). Nanotopographical polymeric surface with mussel-inspired decoration to enhance osteoblast differentiation. *Applied Surface Science* 481: 987–993.

172 Kim, H.K., Kim, E., Jang, H. et al. (2017). Neuron–material nanointerfaces: surface nanotopography governs neuronal differentiation and development. *ChemNanoMat* 3 (5): 278–287.

173 Kim, J.I. and Kim, C.S. (2018). Harnessing nanotopography of PCL/collagen nanocomposite membrane and changes in cell morphology coordinated with wound healing activity. *Materials Science and Engineering: C* 91: 824–837.

174 Li, G., Cao, H., Zhang, W. et al. (2016). Enhanced osseointegration of hierarchical micro/nanotopographic titanium fabricated by microarc oxidation and electrochemical treatment. *ACS Applied Materials & Interfaces* 8 (6): 3840–3852.

175 Chen, W., Shao, Y., Li, X. et al. (2014). Nanotopographical surfaces for stem cell fate control: engineering mechanobiology from the bottom. *Nano Today* 9 (6): 759–784.

176 Chen, W., Villa-Diaz, L.G., Sun, Y. et al. (2012). Nanotopography influences adhesion, spreading, and self-renewal of human embryonic stem cells. *ACS Nano* 6 (5): 4094–4103.

177 Mas-Moruno, C., Su, B., and Dalby, M.J. (2019). Multifunctional coatings and nanotopographies: toward cell instructive and antibacterial implants. *Advanced Healthcare Materials* 8 (1): 1801103.

178 Mo, S., Mehrjou, B., Tang, K. et al. (2019). Dimensional-dependent antibacterial behavior on bioactive micro/nano polyetheretherketone (PEEK) arrays. *Chemical Engineering Journal* 392: 123736.

179 Lin, N., Berton, P., Moraes, C. et al. (2018). Nanodarts, nanoblades, and nanospikes: mechano-bactericidal nanostructures and where to find them. *Advances in Colloid and Interface Science* 252: 55–68.

180 Kargozar, S. and Mozafari, M. (2018). Nanotechnology and nanomedicine: start small, think big. *Materials Today: Proceedings* 5 (7): 15492–15500.
181 Soares, S., Sousa, J., Pais, A., and Vitorino, C. (2018). Nanomedicine: principles, properties, and regulatory issues. *Frontiers in Chemistry* 6: 360.
182 Sharifi, S., Behzadi, S., Laurent, S. et al. (2012). Toxicity of nanomaterials. *Chemical Society Reviews* 41 (6): 2323–2343.
183 Ganguly, P., Breen, A., and Pillai, S.C. (2018). Toxicity of nanomaterials: exposure, pathways, assessment, and recent advances. *ACS Biomaterials Science & Engineering* 4 (7): 2237–2275.
184 Adabi, M., Naghibzadeh, M., Adabi, M. et al. (2017). Biocompatibility and nanostructured materials: applications in nanomedicine. *Artificial Cells, Nanomedicine, and Biotechnology* 45 (4): 833–842.
185 Sainz, V., Conniot, J., Matos, A.I. et al. (2015). Regulatory aspects on nanomedicines. *Biochemical and Biophysical Research Communications* 468 (3): 504–510.
186 Gaspar, R.S., Florindo, H.F., Silva, L.C. et al. (2014). Regulatory aspects of oncologicals: nanosystems main challenges. In: *Nano-Oncologicals* (ed. M.J. Alonso and M. Garcia-Fuentes), 425–452. Springer.
187 Farjadian, F., Ghasemi, A., Gohari, O. et al. (2019). Nanopharmaceuticals and nanomedicines currently on the market: challenges and opportunities. *Nanomedicine* 14 (1): 93–126.
188 Srivastava, N. and Saxena, S.K. (2020). Opportunities in clinical translation and commercialization of nanomedicine. In: *NanoBioMedicine* (ed. S.K. Saxena and S.M.P. Khurana), 501–517. Springer.

12

Nanorobotics in Nanomedicine

Vaishali Y. Londhe and Rupali S. Bhadale

Shobhaben Pratapbhai Patel School of Pharmacy and Technology Management, SVKM's NMIMS, Mumbai, Maharashtra, India

12.1 Introduction

In 1959, the idea of nanotechnology was initiated by the late Nobel physicist Richard P. Feynman, who worked on the Manhattan Project at Los Alamos during World War II and was also a professor at CalTech. During his lecture, "There's plenty of room at the bottom" (1959) Feynman presented his ideas on the world where the fabrication of material at the nanoscale was possible. He explored the possibility of the applications of his concept in the healthcare industry; thus, offering a proposal to provide therapy for heart disease [1]. He says, "A friend of mine Albert R. Hibbs suggests a very interesting possibility for relatively small machines. He says that, although it is a very wild idea, it would be interesting in surgery if you could swallow the surgeon. You put the mechanical surgeon inside the blood vessel and it goes into the heart and looks around. It finds out which valve is the faulty one and takes a little knife and slices it out. Other small machines might be permanently incorporated in the body to assist some inadequately functioning organ" [1].

In 1974, Norio Taniguchi introduced the term nanotechnology while discussing the possibilities of designing materials at nanoscales [2]. Nanotechnology proposes novel opportunities for designing new materials and for the fabrication of precise nanostructures and ultimately nanodevices for drug delivery and diagnosis. In 2019, BCC's (Business Communications Company) research assessed the global nanotechnology market was \$2.0 billion in 2018 and expected it to reach \$2.1 billion in 2023 at a compound annual growth rate of 19.4% from 2018 to 2023 [3].

Nanomedicine is a most promising application of medicinal nanotechnology with the aid of a molar device and microscopic information of human physiology [4]. Nanomedicine covers several kinds of detailed assemblies at the molecular level: nanobiological nanomaterials, biotechnological material, engineered organisms, and nonbiological devices [5]. In the foreseeable future, molecular nanomedicine will comprise biological nanomaterials with a precise nanoscale. In the interim, the information received from genomics and

Nanomaterials and Nanotechnology in Medicine, First Edition. Edited by P.M. Visakh.

proteomics will enable the design of new personalized therapy, a novel chemical entity that targets pathogens whose genomes being deciphered, and includes stem-cell therapy.

Genomic therapy, tissue engineering, and several side-shoots of biological technology will become regular medical practice. Also, scientists may be able to extract bacteria or other mobile cells with biological robots and then redesign or reschedule them with synthetic biological tools (organic engines or self-assembled deoxyribonucleic acid (DNA)-based constructions) for various applications in the healthcare industry. Nanomedicine has the potential to take massive steps in medical treatment and diagnosis. It has the potential to enable, *in vivo* nanorobots traveling directly to a target cell and complete its restoration at the cellular level without disturbance, pain, or defacement [6]. Nanomedicines and molecular nanotechnology are green technologies that complement one another [7].

In the 2020s and beyond, nanorobotics are becoming the most promising combination of three categories of nanomedicine, yet challenging to convert to actual clinical treatments owing to the complexity of operating inside the human body but initiated to give an impression in the healthcare sector. Animatronics could develop and design parts of nanodevices like gears, bearings, and ratchets. An individual nano part might consist of a several thousand specifically sited atoms. These nano parts might be assembled to produce micron-scale nanosensors, nan manipulators, nano pumps, nanocomputers, and even nanorobots. The external computer system is necessary for analyzing the data, monitoring the activities, and to enable the physician to control the nano device during treatment. Nanorobots may be able to treat chronic diseases with the precise delivery of medicine, which can improve human health and life span.

12.2 What is Nanorobotics?

Nanorobotics is a miniature engine intended to achieve definite functions continually with precision at nanoscale dimensions. Nanorobotics, occasionally called molecular robotics, is an embryonic research area as evinced by current literature. These nanodevices will be utilized in humans to maintain health and defend against pathogens. They will be fabricated of nanoparts and sub-assemblies as bearings and gears, nanomotors and pumps, nanomanipulators, nanosensors, and nanocomputers all having a diameter of around 0.5–3 μm.

Nanorobotics are composed of carbon in the form of diamond/fullerene nanocomposites due to diamond/fullerene's physicochemical properties. Oxygen and nitrogen may be employed for a specific feature in nanorobots. The diamond coating on the exterior parts protects against attack by the human immune system. The energy for the nano-device could be supplied by metabolizing glucose and oxygen entity at the cellular level. Another source of energy may be acoustic waves that are supplied externally. Figure 12.1 is a schematic representation of a nanorobot that can be activated by the cell itself, when it is needed. The external shell comprises of the tunable pore that releases active agents after the activation of nanorobots and targeting agents to operate at a particular destination.

As a result, nanorobotics is a technology that is proficient in translating power sources into kinetic energy:

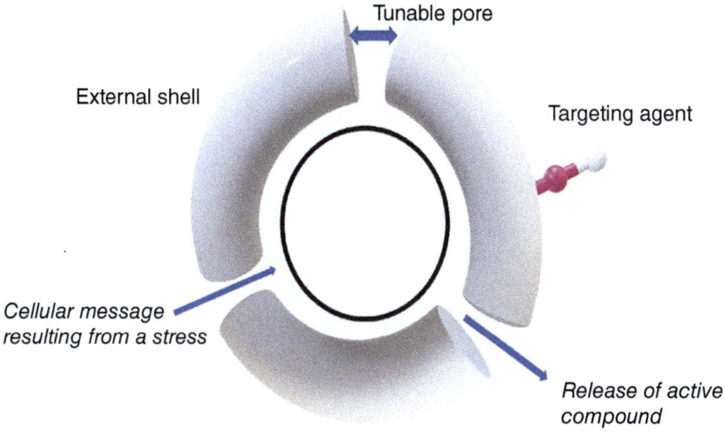

Figure 12.1 Schematic representation of activated nanorobotics at the cellular level.

1) In biologically driven nanorobots artificial nanodevice combine with mobile microorganisms that provide power to the nanorobots [8–10]. In this loop, biological assemblies like proteins may act as motors (myosin, kinesin, dynein, and E-coli), mechanical joints, transmission elements, sensors, etc. Biomolecular motors have a high efficiency, are easily available, and are self-replicating.
2) Chemically driven nanorobots utilize lopsided catalyzing machines to electively translate biochemical energies into propulsion [11–13]. There are three types of chemical species that are employed as a propellant solid, liquid, and gas. Among them, a liquid propellant is most frequently used for a prolonged period. For example, Janus's spherical nanorobots comprise gold and platinum that provide power for the propulsion of the nanodevices. Subsequently, they release oxygen after contacting hydrogen peroxide.
3) Based on the motor configuration and outline, physically motorized nanorobots convert external energy like magnetic waves, ultrasound waves, or light waves into locomotion [14–17].

Broadcast-type acoustic signals may be used to communicate with nanodevices. The natural excretory system will be used to expel nanorobots from the human body. An active scavenger system may be an alternative option for eliminating nanodevices that are design-dependent [18].

Nanorobotics, as a concept, has different definitions in the literature. Thus, nanorobotics can be divided into two key sectors, the first sector looks at the nanoscale structure, reproduction, control, and management of robots. Plenty of research performed in this field remains theoretical due to complications with manufacturing such nanodevices. While artificial nanorobots don't exist, biological nanorobots do exist and offer proof that such a structure is at least conceivable [19] as they utilize the requisite components.

The second sector concerns the manipulation and/or fabrication of nanoscale parts in macroscale robots, i.e. the nanomanipulator. Much research has been carried out and documented. The practical development of this sector is essential due to the advancement in the potential applications of nanotechnology. Similarly, nanomanipulation and nanoassembly might play a vital role in designing artificial nanorobotics. Manipulation at the nanorange is yet to begin and the physical and chemical singularities at the nanoscale are not established [20].

12.3 Nanorobotics in Nanomedicines

Previously, the downsizing of technology was just a scientific invention and elucidation. Firstly, the movie "Fantastic Voyage" displayed a downsized engine named Proteus used to detach a clot present in the scientist's brain. It seems strange to hear about such things in reality. Then, in the twenty-first century, scientists succeed in transmuting that imagination into reality. Research activities of nanorobotics grew exponentially to 2019, which demonstrates the increased curiosity of scientists (Figure 12.2).

In the laboratory, the capability of nanorobots to execute several challenges has been demonstrated, providing proof of claims comprising specific freight distribution [22, 23], liquid blending [24, 25], and corporeal management of microstructures [26, 27]. In 2019 BCC's research evaluated the global market for nanodevices and nanomachines. It was \$736.2 million in 2018 and expected to reach \$1.2 billion in 2023 and projected to reach \$2.7 billion in 2028 at a compound annual growth rate of 11.6% from 2018 to 2023 and 16.0% from 2023 to 2028 [3].

In recent times Professor Ido Bachelet, principal investigator at the Nanotechnology Center, Bar Ilan University proclaimed the therapy of cancer with nanomedicine. He designed DNA nanorobots that could detect dozens of distinct categories of cancer tumors in humans and promote fast recovery in the cancer patient and assist spinal cord damage [28]. In addition to sensing and finding cancer cells, nanorobots can distribute drugs in a controlled manner [29].

In this loop, "Respirocytes" is an artificial mechanical red blood cell (RBC) having the ability to deliver 236 times more oxygen to cells as compared to natural RBCs [30]. Another nanorobot, "Microbivores" can destroy microbiological pathogens, and a serum oxyglucose powered spherical nanorobot, "Clottocytes" can be used to heal a wound using a fiber mesh [31]. A hypothetical motile cell repair nanorobot, "Chromallocytes" can reverse

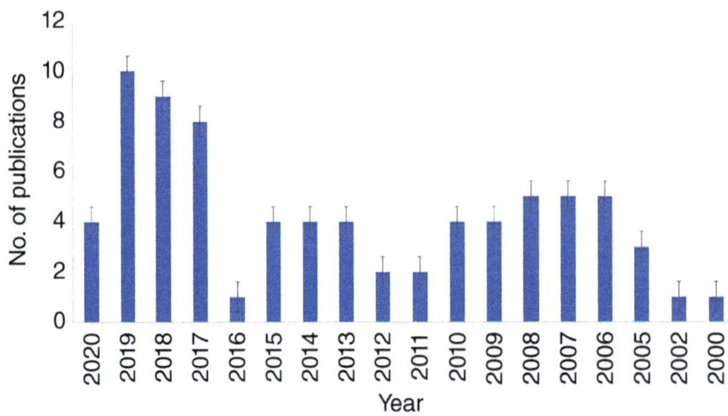

Figure 12.2 Cumulative publications on the nanorobotics in nanomedicines. *Source:* Modified from [21].

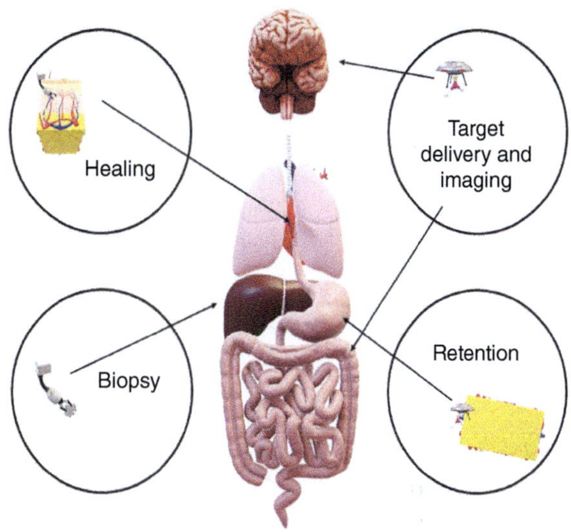

Figure 12.3 Outline of therapeutic perceptions in nanorobotics for *in vivo* human applications.

effects by swapping whole chromosomes in genetic diseases [32]. As a result, the *in vivo* applications of nanorobots comprise performing surgery [33], wound recovery [34], improved retainment in tissue [32], and target delivery [35] (Figure 12.3).

12.4 Nanorobots for Medical Imaging

Previously, limited research has been reported on the detection of nanomotors in *in vitro* models [34, 36, 37] or *in vivo* with a limited application in animals [32, 38]. The main disadvantage is the lack of sensitivity in tracking the location of the nanodevice *in vivo* [39], which is necessary to monitor their circulation, eradication, and the specific site or tissue delivery. A further challenge is to design accurate and dependable micro engine controllers. Some imaging techniques along with their merits and disadvantages are shown in Table 12.1.

Tracking and real-time imaging of the nanorobot a challenge and currently restricts progress, the literature does not describe any chemically driven nanorobots with this facility. Fluorescence is a technique that merges with biologically driven nanorobots to produce real-time imaging [40–42]. Physically motorized nanorobots fuse with different types of imaging techniques. Procedures include: the use of endoscopy and X-rays to sense microgrippers in the GI tract [33]; optical imagining to check motion in the eye [43, 44]; and fluorescence-imaging systems to trail the location of magnetically triggered helical nanorobots in the abdominal cavity [45] or subcutaneous layer [46] of a mouse. Also, dual-imaging techniques were utilized to sense biodegradable magnetic micro-helix nanorobots inside rats.

The fluorescence technique and magnetic-resonance-based imaging [47] were utilized to sense the nanorobots in the subdermal layer, and abdominal cavity of a mouse. In the future, the optical service of a nanodevice needs a symbiotic combination of some systems like imaging technology and image processing, control theory, and contrast-agent functionalization [39, 48, 49]. The application of two-dimensional (2D) and three-dimensional (3D)

Table 12.1 Comparison of imaging technique for detecting nanorobots.

Techniques	Merits	Limitation
Optical tracking	• Real-time computation • A basic device • Affordable price • High resolution (3D)	• Difficulty in 3D tracking of nanorobots • Low signal to noise ratio • Utilized in clear microhabitat like the vitreous humor in the eye
MRI	• A magnetic field with non-ionizing radiation • The capability of 3D tracking • Also utilized to manipulate nanorobots • Accuracy in detection	• Magnetic material will relics • Real-time imaging is difficult • Expensive • Low resolution (3D)
Myocardial Perfuming Imaging (MPI)	• A magnetic field with non-ionizing radiation • The capability of 3D tracking • Accuracy in detection • Quickly scrutinized	• Used at lab-scale only • Restricted to only magnetically triggered nanorobots
X-ray fluoroscopy	• A real-time imaging • Precise while tracking • High resolution • Maximum diffusion in living tissue	• Low resolution (3D) • Radiation exposure
X-ray computed tomography	• Maximum diffusion in living tissue	• Low resolution (3D) • Radiation exposure • Slowly analyzed
Ultrasound	• Real-time imaging • Negligible negative impact • Also utilized to manipulate nanorobots	• Low resolution • Low signal-to-noise ratio • Trickled by bones and air pocket • Low penetration capacity • Restricted area for scanning
Fluorescence	• High planer resolution • Real-time imaging • Highly sensitive	• Low resolution (3D) • Low penetration capacity • Low quality of images • Aid to the cost of nanodevices

optical microscopes can be used to see superficial area organs due to the low ability of visible light that penetrates biological tissue [50]. Real-time tracking of nanorobots can be done using ultrasound and can be more effective in living tissue.

From a different perspective, a bubble-propulsion mechanism can be utilized to locate nanorobots, currently it is only used to remove microbubbles [51]. Further high-precision imaging methods applied to track nanorobots are computed tomography (CT) or magnetic resonance imaging (MRI) [37] depict insufficient sensitivity, which presents enormous challenges [52, 53]. On the other hand, the integration of positron or gamma emitters with

Table 12.2 Summary of *in vivo* application of nanorobots based on a source of energy.

Source of energy	Material or therapeutic agent and structure of nanorobots	Dimension (μm)	Animal model and site	Applications
Biological	Magnetococcus marine	2	Intra-peritoneal cavity of a mouse	Fluorescence imaging of targeted payload delivery for tracking gene expression [40]
	S. Typhimurium	3	Thigh, tail vein, and blood	Fluorescence imaging of tumor [41]
Chemical	Zinc micro rocket	15	Stomach of mouse	Imaging and retention of cargo in the stomach [32]
	Magnesium micro rocket	15	Gastrointestinal (GI) tract of the mouse	Retention of nanorobots at the targeted site and imaging (studied at a different site of the GI tract) [53]
Physical	Helical design	20	Intra-peritoneal cavity of a mouse	Optical imaging used to detect and navigate nanorobot [45]
	Spirulina micro-algae merged ferric oxide	100 nm	Subcutaneous tissue, stomach, intra-peritoneal cavity of a mouse	Optical and magnetic imaging used to detect and navigate nanorobot [47]

nanorobots can be seen as an idyllic approach for locating nanorobots inside the human body through nuclear imaging techniques such as positron emission tomography (PET) or single-photon emission computed tomography (SPECT). Currently, these methods possess inadequate morphological data and often operate with CT. This technique also only analyzes gamma rays either instantly emanated via gamma-emitting isotopes (single-photon emission tomography; SPET) or those produced by PET.

The nuclear-imaging system is marginally intrusive, highly sensitive, spatially dependent, and quantifies the concentration of labeled material inside the human body, which is directly proportional to the emitted gamma rays. Also, gamma rays have the ability to penetrate the biological tissue, aiding PET, and SPET techniques, to enable translation into actual clinical treatment. The early diagnosis and analysis of the efficacy of treatment is determined with the aid of PET and SPECT respectively or blending with CT. Some of the examples of *in vivo* imaging information of nanorobots based on the source of energy are summarized in Table 12.2, and provides data on several internal organs.

12.5 Nanorobots for Targeted Drug Delivery

In 1906 Ehrlich first postulated the term "magic bullet." The strength of this perception is a robust sign of its demand, but the "magic bullet" elusive. The objective being threefold: to locate the target site, to provide an effective therapeutic agent, and create an effective carrier. Nanorobots provide and efficient carrier of the therapeutic agent coupled with targeting ligands partially accomplish the concept of a "magic bullet." In this section, we emphasize targeted drug delivery with nanorobots for a speedy recovery and to improve the efficiency of the therapeutic agent in the area of interest. The annals of nanorobots recommended for chronic disease therapy are demonstrative of their potential for translation into actual clinical therapies.

12.5.1 Bacterial, Viral, and Parasitic Infection

The most familiar disease, to most people, is an infection in the human body due to various viruses, bacteria, protozoa, or microscopic parasites. Microbivore, a white-cell substitute nanorobot composed of diamond and sapphire, will detect blood-borne pathogens and aid as primary therapy for pathogen-related diseases. These nanorobots will be quicker, reliable, and controlled by humans to follow the mechanism of phagocytosis to remove problematic microbes. Nanorobots can securely travel through the thinnest human capillary and tightest junctions in the spleen such as the interendothelial spleno fenestral slits.

Microbivores comprise of two cylinders, an inner cylinder that crushes the bacteria to the nanorange and the remnants are collected in the second outer cylinder. In a prescheduled system, different enzymes may be injected and mined six times in one digestion cycle. With the aid of molecular sorting rotors, the morcellator reduces the tissues to a mono-residue amino acid, mono-nucleotide, glycerol, and simple sugar and releases them into the bloodstream. The nanorobots get rid of the kidney stones after receiving an ultrasound signal at the end of therapy [54]. A singular terabit (1012 nanorobots) dose of microbivores can completely and safely eradicate blood-borne pathogens. Microbivores possess 1000 times the fast-acting capacity compared to antibiotic therapy, they also digest 100 times more than a natural white cell in a given period. Microbivores have a long durability of phagocytosis versus natural white blood cells. Additionally, microbivores can eliminate the respiratory or cerebrospinal infection or infection in a pleural, synovial, or urinary fluid. A slight surface modification of microbivores may cure a site-specific infection.

12.5.2 Cancer

Current phases of medical technologies and treatment may effectively treat cancer. Various factors can determine the survival rate of cancer patients such as early diagnosis and efficient target delivery of chemotherapy. Nanorobots can be a good option for diagnosis and target delivery for the treatment of cancer, which may reduce expected side effects, given the features of nanorobots to navigate as blood-borne devices that can diagnose and deliver an anticancer drug at the target site. Nanorobots with chemical biosensors can detect tumor cells in the early stages of development in the human body. Thus, integrated nanosensors can be used to detect the intensity of E-cadherin signals. The modified microbivore can detect cancerous cells or tumors and then engulf these cells without harming healthy cells. Microbivores can digest 1 μm tumor mass per minute, 4000 nanodevices require one hour to digest a 100 μm tumor mass.

The larger tumor masses can be digested by endoscopic nanosurgery having a special nanoprobe such as nanosyringoscope whose tip is mobile and directed by sensors to sense the tumor region. These nanorobots are inserted into diseased tissue then modify, digest, or vacuum out the infected cells like atherectomies [55] or laser-based tumor debulking methods [56]. Choramallocytes can be used for *in situ* chromosome replacement treatment, which may genetically cure infected cancer cells, but not for larger tumor cells, as most tumor cells comprise surplus tissue which require removal rather than repair.

12.5.3 Cardiovascular Disease

Atherosclerosis is the most common root cause of cardiovascular disease in which is an accumulation of fatty white cells (atheroma) in the inner lining of the arterial wall. In the future, cardiovascular disease might be reduced as atherosclerosis is moderately rescindable by monitoring blood lipids [57], and imminent nanorobotic control of gene expression or stem cell [58] treatment as previously verified in rodents can be more efficient as preventive measures.

The vasculocyte may become the first-choice therapy in the initial stage of arteriosclerosis. The nanodevice, created as a squat, hexagonal curved cornered nanorobot, and telescoping attachment attached on the underside, may be used to manipulate blood clots, for foam cell disassembly, endothelial cell herding, exclusion of sticky glycoprotein, and similar items. The syringe may suck or inject the therapeutic agent via a 10 nm piercing in the cell membrane when required. The nanodevice can travel through the narrowest capillary in the human body. The 400-billion molecule assembly of a nanodevice might be around 8 picograms. The domed top of a nanodevice can be covered with 174 000 rotors to permit the speedy interchange of a specific atom among internal nanorobots and the human bloodstream.

Vasculocytes travelling through the bloodstream will detect and collect sclerotic tissue. The nanorobots combine to form mutual communication with watertight arterial strapping via interlocking with expandable cushions and commentators firmly inserted to initiate repair operations, which may be observed by a physician. The entire computational control intrinsic to each strapping would be equally remarkable as a 1 cm^2 patch of allied vasculocytes can have a 5 MB memory. Also, nanorobots can repair within 24 hours whereas hemangioblast precursor cells derived from human stem cells demonstrate the restorative mechanism of dented rat/mouse vasculature in 24–48 hours [59]. Figure 12.4 shows the eight steps required to repair cells. Additionally, nanorobots can treat nonatheroma lesions of the vasculature caused by viral intruders [60].

12.5.4 Cerebrovascular Disease

A cerebrovascular disease (CBVD) is a group of conditions that includes stroke, transient ischemic attack, aneurysms, and blocked arteries. In industrialized countries, stroke is the most common cause of neurological damage by depositing atheroma in the carotid artery, which may reduce blood supply to the brain. Also, if the atheroma is ruptured in either the carotid artery or heart then clots may be transported by blood and trapped in smaller brain vessels. When blood circulation to the brain is interrupted, brain cells may be damaged due to a lack of oxygen. Nanorobots can be crucial in the treatment of cardiovascular diseases.

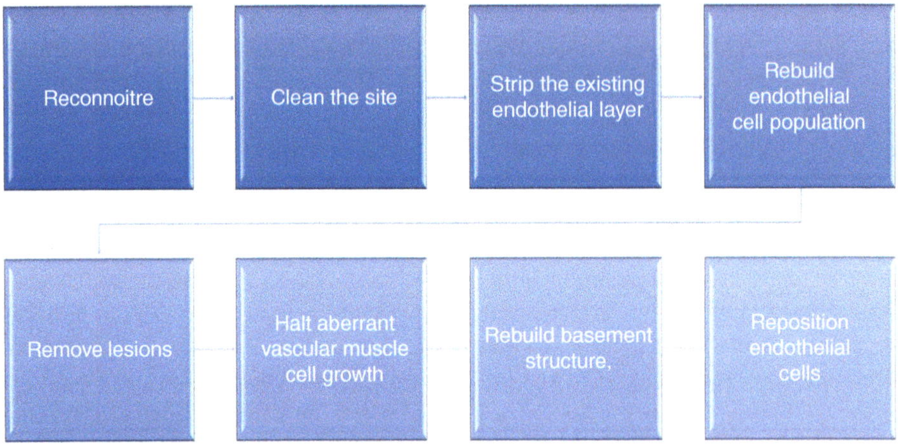

Figure 12.4 Schematic representation of steps involved in repairing cells.

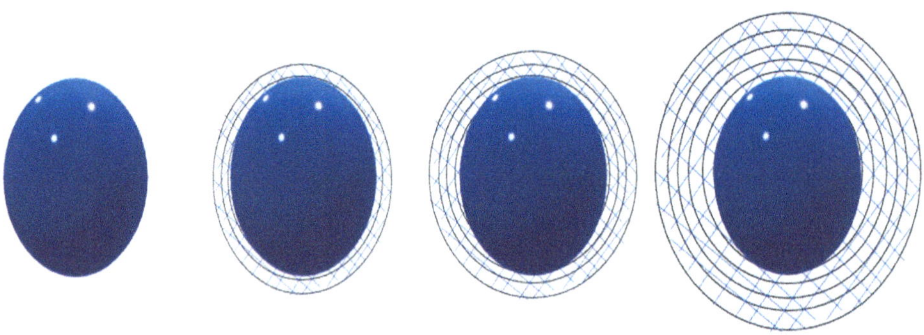

Figure 12.5 Schematic representation of a mechanical platelet, clottocyte, that quickly expands its netting at the intracranial hemorrhage, impeding bleeding in a second.

Vasculocytes, described earlier, can clear a partial occlusion, repair vascular walls, and enlarge vessel lumen to a normal diameter. Another approach is to digest the obstruction – delving tissue-mobile microbivore can engulf the ruptured atheroma in a few minutes. Besides, delving microbivore with respirocytes in a cocktail would maintain oxygen at the affected tissue through the diffusion mechanism to adjacent capillaries. Incorporation of endoscopic nanosurgery, vascular gates, and clottocytes (Figure 12.5) may clear a total occlusion and intracranial hemorrhage.

12.5.5 Hormonal, Metabolic, and Genetic Disease

Diabetic mellitus is a hormonal disorder that causes blindness and problems in the kidneys, nerves, and cardiovascular damage. Immune disorder reasons causing type I diabetes can be eradicated with the help of immune engineering that has been demonstrated effective in animals and immune engineering can cause a fluctuation in the gene expression of age-related Type II diabetes, which makes them both preventable. Chromallocytes can repair any genetic vulnerabilities at their source, for example, a reconstruction of a mislaid pancreatic β-cells by genomic replacement in an existing cell and the production of fresh β-cells that leads to more resistance to autoimmune devastation via erasing pancreas virus, consequently curing diabetes. Also, chromallocytes can remedy other endocrine diseases such as hypopituitarism, hyperthyroidism, adrenal malfunction, and metabolic diseases.

On the other hand, microbivore nanodevices can erase autoreactive immune cells. Cell repair nanorobots can correct abnormal or defective gene expression in targeted cells to eradicate the impact of insulin resistance. In a different application, an artificially entrenched nanorobot namely, pharmacyte, would expansively regulate serum levels of insulin and deliver deglycating enzymes via diffusion, which saturate the targeted tissue to digest cross-linkage molecules by reducing tissue stiffness.

Therapeutic nanorobots can compare the molecular structure of DNA and protein, which is found in cells, to detect reference structures for the treatment of genetic disorders. Also, modifications can made at the same place. Sometimes chromosomal replacement therapy is more effective than cyto-repair. An assembler-built repair vessel performs some genetic maintenance via floating inside the nucleus of a human cell. The nanomachine pulls the unwound strand by elongating a supercoil of DNA between its lower pair of robot arms for analysis. Meanwhile, detached regulatory proteins from the chain are placed in the upper arm. The detected molecular structures of DNA and protein are compared to a database via a nanocomputer situated outside the nucleus and linked to a cell repair unit. Subsequently, correcting the structure of proteins and reattaching them to the DNA chain converts it to its original form.

12.5.6 Kidney Diseases

A kidney stone is the most common problem formed by saturation of a crystallized substance and is extremely painful. A nanorobot can break the kidney stone with an ultrasonic wave or laser and subsequently microbivore engulf the broken stone and it is excreted from the body. Polycystic kidney disease, another genetic disease, produces many cysts. Tissue-mobile nanorobots can be used to remove a cyst from the kidney without interrupting other cells [61].

12.5.7 Retinal Diseases

The intravitreal delivery of a therapeutic agent plays a vital role in ocular drug delivery and is utilized to treat glaucoma, diabetic retinopathy, and macular edema. Novel intravitreal delivery of a nanodevice namely, slippery micro-propellers, are driven by magnetic waves to reach the retina and aim to detect and deliver a therapeutic agent (Figure 12.6). These are helical in shape and coated with slippery agents such as silicon oil and fluorocarbon. These micro-propellers are used to reduce contact between the nanorobot and collagen bundle existing in the vitreous subsequently reaching the retina within 30 minutes [62].

12.6 Enzymatic Nanolithography

Enzymes are a special category of proteins reliable for catalyzing chemical reactions inside the cells and therefore are idyllic drug targets. Enzymes are highly specific in nature, they bind with high efficiency. Nanolithography is the science of etching, writing, or printing to

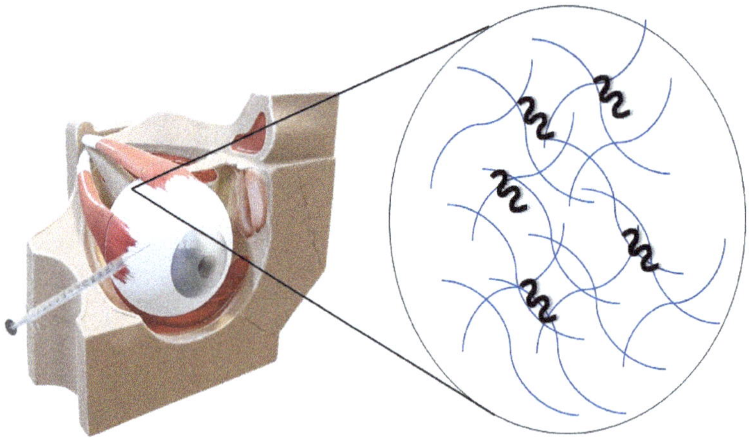

Figure 12.6 Schematic representation of target delivery of magnetically driven slippery micro propellers.

modify a material surface at a nanoscale. Enzymatic nanolithography, a novel approach for writing biological specificity under 100 nm, where a proteolytic enzyme is used to etch a groove in a hardened protein layer.

In nanorobotics, enzymatic nanolithography plays a crucial role in the target delivery of therapeutic agents. The simplest technique for enzymatic nanolithography, is where enzymes were immobilized to an atomic force microscope (AFM) tip on the nano parts of nanorobots, which can bind at the targeted site after insertion into the body. An alternative way to deposit mobile enzymes is with dip-pen technology. As dip-pen technology is performed in the air, there is some risk of denaturation of enzymes with the consequence of decreasing their activity, which can be overcome by using the micropipette technique. Vega et al. patterned and immobilized tobacco mosaic viruses with dip-pen nanolithography (DPN) [63]. Walker et al. fabricated magnetic propellers with surface-immobilized urease that efficiently penetrates mucus, thus delivering the therapeutic agent [64]

12.7 Biomimetic Approach

Progression of nature is effectual behavior in energy/material employment and recital which offers several concepts, ideologies, and strategies that can be applied. Bionanorobotics are designed, fabricated, or driven to mimic biological matters that transpire in nature (Figure 12.7) [19].

Bionanorobotics were fabricated using preprogrammed biologically functioned elements (proteins and DNA) as nanoparts such as motors, sensors, or transmission elements with high efficiency [66] and minimum side effects [67]. Microbivores are a type of biomimetic-derived nanorobot-like white blood cells in the bloodstream that bind to a specific site and destroy pathogens. Another example of a biomimetic-derived nanorobot is Respirocyte, an artificial RBC that provides 200 times more oxygen as compared to natural RBCs. Ali et al.

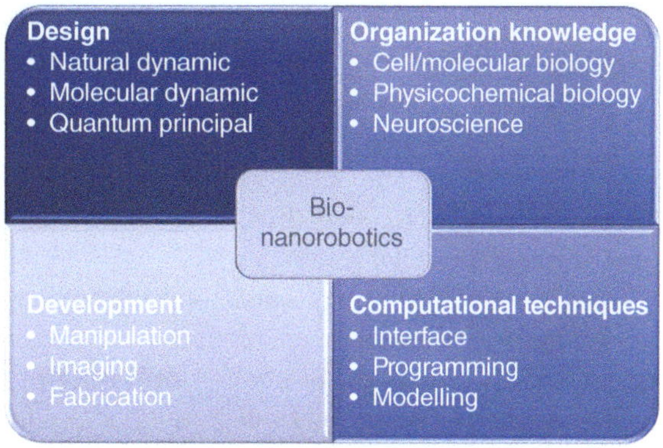

Figure 12.7 Schematic representation of different factors considered in the fabrication of bionanorobotics. *Source:* Based on Stock et al. (2010) [65].

fabricated a magnetically driven self-assembled bacterial flagellar nanorobotic swimmer mimicking the rotary motion of helical bacterial flagella for propulsion [68] to transmit nanorobots.

12.8 Cell Biochips

A micro- or nanosystem cell biochip is fabricated with electronic and microfluidic functions to control or scrutinize living cells. Researchers have formulated an entire series of cell chips for instantaneous or chronological investigation of single cells, cell groups, or cell tissues rebuilt on the chip. For parallel analysis, cells are organized in a square array on a parallel cell chip, whereas in the case of a series cell chip, cells are observed and treated individually in a micro-channel. Unlike the use of biochips for high-output analysis, single-cell chips emphasize the investigation of a single segregated cell. For example, DNA microarrays comprised of several oligonucleotides which were arranged in a matrix array, parallel cell chips match the order of living cells. The cells can be isolated by array, which enables parallel investigation or dispensation by cell type.

Le Pioufle et al. describe a microdevice for the cultivation of single cells or small groups of cells in a micro-pit array that are stimulated by electricity or fluids [69]. Among the proposed theoretical uses are gene transfer, cell sorting, and screening in pharmacology. Bailey et al. combined DNA microarray and cell biochip. The cultured cells transfect on the substrate depending on their position in the array [70]. Cell biochips aim to facilitate the invention of the novel active molecule, definitely targeting ion channels that will interact easily with proteins. Thus, this progress merged with high-output screening will facilitate the detection of targeted cell types earlier in the human body.

12.9 Nanorobots for Precision Surgery

Robotics can reduce the problems encountered with complex surgery expand the range of a medical practitioner. Robot-assisted surgery can achieve high precision, mobility, and control. Nanorobots can navigate through the human body and access hard-to-reach tissue locations, thus targeted delivery can be achieved.

Current developments have revealed the significant potential for addressing the challenges with these microdevices for precision surgery. Unfettered nanorobots, ranging from nanodrillers to micrgrippers and microbullets, provide a key feature for high precision minimally invasive surgery. The mobile nanorobots, supported with diverse energy and the nanoscale surgical component, can pierce or recover cellular tissues for precision surgery. These nanorobots can travel through the narrowest capillaries of the body and conduct trials down to the cellular level.

Unstrapped microgrippers signify a substantial step to create an autonomic robotic tool for microsurgery. These mobile microgrippers can detect and recover hard-to-reach tissues and cells. Generally, mechanically or electrically driven conventional microgrippers produce signals through exterior networks such as wires and tubes that limit their size and flexibility. The gripping action of untethered microgrippers includes an opening/closing of the device, akin to their large tethered complements. Leong et al. have fabricated self-driven responsive microgrippers employed as minimally invasive microsurgical tools [71]. These microgrippers are fabricated with conventional multilayer photolithography designed to affect biological adjuncts [33]. Soft microgrippers with a built-in self-folding actuation response allow for the removal of external tethers. Self-folding microgrippers are triggered by various reactive mechanisms based on pH, or enzyme stimuli, and temperature [72]. Singh et al. demonstrated the capability of a tether-free thermochemically stimulated microgripper to grab live fibroblast cells from a dense cell mass in a capillary tube. With an *in vitro* tissue biopsy, the microgripper may extract cells from a capillary tube with trapped cells thus demonstrating its strength. The ablation has been demonstrated to be an additional biomedical function [73].

In the same way, magnetically activated nanorobots have revealed significant potential in *in vivo* surgical procedures as magnetic fields can penetrate thick biological tissues. Chatzipirpirdis et al. implanted a magnetic tubular nanorobot in a position to execute surgery on the posterior segment of the eye [74]. The electrochemically driven nanorobots were inserted by a 23-gauge needle in the vitreous humor and detected using an integrated camera and ophthalmoscope. The *in vivo* study of ophthalmic magnetic nanorobots has been driven with wireless control in the rabbit eye. Similarly, magnetic microtubes can be designed and utilized as an implantable device for target delivery in a different restricted area of the human body.

Recently, ultrasound was used to produce nanorobots with notable tissue-diffusion abilities. Kagan et al. used ultrasound-driven high-speed propulsion, permitted via rapid evaporation of perfluorocarbon that was used as biocompatible fuel [75]. In response to an external ultrasound stimulus, conical tube-shaped microbullets comprising energy sources show high-speed mobility with a speed of over $6\,\mathrm{m\,s^{-1}}$. This can deliver nanorobots into deep tissue for penetration, excision, and devastation. Similarly, nanorobots can be ultrasound-driven via rapid evaporation of perfluorocarbon applied in tube-shaped micrometric cannon proficient in loading and ejecting nano bullets at notable velocity [76]. This microballistic device is applied to discharge nanobullets and firing a variety of cargos to hard-to-reach contaminated tissues. Current prototype research has confirmed that unstrapped nanoscale cargos can execute a single-cell surgical operation. Solovev et al. designed a powerful nanodevice propelled by InGaAs/GaAs/(Cr)Pt microjets. The self-driven folded tubings with a diameter of 280–600 nm achieved up to $180\,\mathrm{\mu m\,s^{-1}}$ velocity in hydrogen peroxide solution [77]. These tubes are driven via transferring chemical energy, which pierces and places in a single cell of biological tissue. Also, the same group of researchers designed a rolled fuel-free microdriller that can be controlled via a rotational magnetic field as hydrogen peroxide might be inappropriate in a live cell [78].

The self-folded sharp-edged magnetically driven nanotool allowed piercing in pig liver tissues *ex vivo*. Strivastava et al. created magnetically driven micro-daggers that can lead to cellular penetration for drug release [79]. These studies demonstrated the significant performance of precision surgery at molecular levels. The ability of surgical nanorobots can be enhanced based on the choice of propulsion method and the use of real-time localization and mapping with a robust control system, improves penetration, detection, and eradication of infected cells.

12.10 Nanorobots for Detoxification

Self-driven nanorobots have been demonstrated as high-cleaning detoxification devices. Like medical imaging/sensing, detoxification relies on self-driven nanorobots that quickly detect and eradicate the toxin to make the environment's toxicity. The collision and binding of toxins to the motors coated with desired functional materials are due to effective mobility. As an illustration, cell-derived natural material was used to fabricate nanomotors that can mimic the natural features of their source cell on the way to innovative tiny biodetoxification tools. RBCs have shown an exceptional ability to absorb toxins as nanosponges, which counteract and eliminate hazardous "pore-forming toxins" (PFTs) from the circulatory system [80]. Some distinct cell-mimicking nanomotors are driven by biotic features of RBCs constructed for detoxification. Wu et al. demonstrated a cell-mimicking water-powered nanomotor constructed on RBC membrane-coated magnesium microparticles, which can estimably grab and deactivate α-toxin in the bloodstream (Figure 12.8) [82].

Figure 12.8 Schematic representation of RBCs-Mg Janus micromotor mobilized in the bloodstream and detoxifying an α-toxin. *Source:* Based on Wu et al. (2015) [81].

An alternative method for detoxification is through mixing RBC membranes with ultrasound-propelled nanomotors as a biomimetic base that may efficiently absorb and neutralize PFTs [81]. In the same way, a self-propelling 3D-printed microfish comprising polydiacetylene (PDA) nanoparticles that bind to attract, capture, and neutralize toxins [83] can be used. Self-propelling 3D microfishes produced in the toxin solution displayed higher fluorescence intensities concerning stationary microfishes, concluding the status of mobility for enhancement in detoxification processes.

12.11 Fabrication of Nanorobots

The design of the nanorobot device is allied to existing techniques. In the past decade, the development of the microfabrication process has been stable. Nanofabrication methods work actively, and a prototype produced via these methods is in progress. Most micro- and nanofabrication techniques have their roots in standard fabrication procedures established for the semiconductor industry [84–87]. Thus, a sound knowledge of these techniques is essential for somebody introducing a research and development path in the micro and nano fields.

12.11.1 Microfabrication

The process used in typical microfabrication uses lithography, thin-film deposition, chemical etching, and electrodeposition.

12.11.1.1 Lithography

Lithography is a method that transforms a computerized pattern onto a substrate such as silicon, glass, for various purposes. Photolithography is the most widely used lithography technique in microfabrication in which ultraviolet light is used for etching. Also, electron-beam and X-ray lithographies are two alternatives involved in micro-electromechanical system (MEMS) and nanofabrication. The initial step for a specific fabrication sequence is a creation of a photomask with a computer layout. Using the optical or e-beam generator, the desired pattern is fabricated in the form of a thin chromium layer on a glass plate. The photolithography process starts after the spin coating of the desired material (photoresist) on the substrate. Hexamethyldisilazane (HMDS) is used as an adhesion promoter prior to the resist application. The resist thickness (0.5–2.5 μm) depends on the photoresist and the spinning speed. Then, the final fabrication is completed with the help of wet and dry etching. Figure 12.9 displays a schematic representation of the photolithography steps with a positive photoresist.

12.11.1.2 Thin-film Deposition and Doping

In micro- and nanofabrication, thin-film deposition and doping are widely employed. Various deposition techniques are used to fabricate designs that contain materials other than that of the substrate. These methods comprise chemical vapor deposition (CVD), physical vapor deposition (PVD), oxidation, doping, and electroplating.

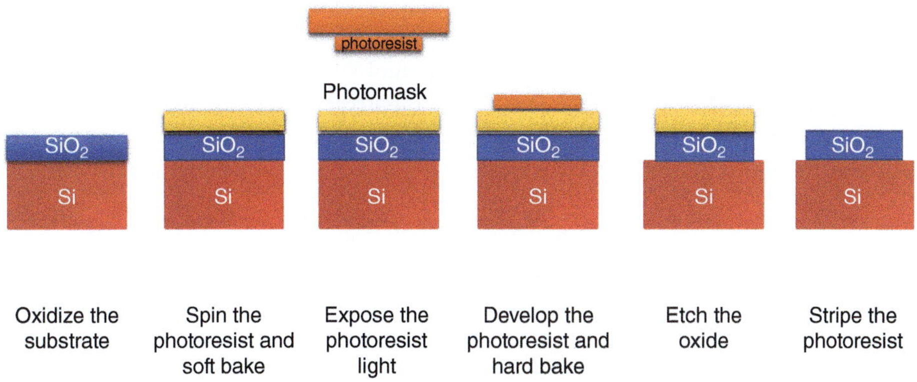

Figure 12.9 Schematic drawing of the photolithographic stages.

Vapor deposition is a complex process, followed by photolithography and etching to accomplish the manufacturing of nanorobots. This method is applied to set a metal film on a silicon substrate. After using an appropriate etching method and photolithographic method, nanorobots are designed. There are two different methods of vapor deposition. The first method is the PVD-based, in which the heated materials and vapors produced from the source are placed on the substrate surface. The second is CVD. In this method, the heated substrate chemically reacts with inert gases to form a thin film.

12.11.1.3 Etching

For micro and nanofabrication, etching is the concluding step after lithography, which is used to fabricate various mechanical structures (beams, plates, etc.). In this method, a thin film that is placed on the substrate is etched. There are two different methods of etching, i.e. dry and wet.

Dry etching can be done by two different methods, i.e. the physical method and the chemical method. In a physical method, passive ions strike the surface to be etched using ion beam milling (IBM). When ions are showered, it is defined as showered IBM (SIBM) whereas when ions are focused, it is called focused IBM (FIBM). In the chemical method, the generated plasma of reactive ions is placed on the material to be etched with reactive ion etching (RIE). In wet etching, the additional substance is eradicated via dipping in the bath, having a chemical etchant. Isotropic and anisotropic are two methods of wet etching. In isotropic etching, ions strike at the same rate in all directions on the materials (i.e. polysilicon, oxides, nitrides, gold or silicon, aluminum). In anisotropic etching, anisotropic ions strike the material at different rates in the same direction.

Mostly, dry-etching methods are plasma-based having benefits in comparison to wet etching. These include minor undercuts, which allow smaller lines to be patterned, and higher anisotropy, which allows high-aspect-ratio vertical structures. The speed and selectivity of dry etching procedures is less compared to wet etchants, and it is essential to consider the predetermined rate of etching of the masking materials. The basic dry-etching methods are specifically high-pressure plasma etching, RIE, and ion milling, which employ other modes to find directivity.

12.11.2 Nanofabrication

Several researchers have an interest in the design and fabrication of the nanoelectromechanical system (NEMS). For nanofabrication, two approaches namely, bottom-up and top-down have been acknowledged via the nanotechnology research community and various researches are being autonomously explored. Top-down strategies are constructed on microfabrication such as nanolithography, nanoimprinting, and chemical etching. Currently, these are 2D fabrication strategies with comparatively nominal resolution. Bottom-up systems are assembly-based techniques that comprise methods namely self-assembly, and dip-pen lithography. These methods may produce wide-ranging systematic patterns in nanoscale. Mostly, DPN, epitaxy and strain engineering, and e-beam and nanoimprint technique are utilized for nanofabrication.

12.11.2.1 E-Beam and Nanoimprint Fabrication

In prior sections, we reviewed some significant lithography methods employed in MEMS and microfabrication. These comprise ultraviolet (UV; usual, intense, and excessive) and X-ray lithography. These techniques are not suitable for nanofabrication as a result of a deficiency of resolution (as to the UV), or the problems encountered with the fabrication mask and radiation sources (X-ray). E-beam lithography is a substitute method for nanofabrication of nanorobots [84, 88]. An electron-sensitive resist namely polymethyl methacrylate (PMMA) dissolved in trichlorobenzene (positive) or poly-chloromethyl styrene (negative) that offers high resolution with low cost and ease of process. The e-beam gun is generally a component of scanning electron microscopy (SEM), while transmission electron microscopy (TEM) may be also utilized. The power of radiation and design are generated with a computer interface.

E-beam lithography is sequential but strictly restricts large-scale nanofabrication. Various nanostructures can be fabricated by e-beam lithography combined with lift-off, etching, and electrodeposition. The nanoimprinting technology is a novel method that evades the periodic and low-efficiency limits of the e-beam lithography for nanofabrication [84, 89]. This method utilizes an e-beam-fabricated hard material mold to stamp and deforms a polymeric resist followed by a RIE to transfer the stamped pattern to the substrate. This novel method is economical as a single stamp can be used continually to manufacture several nanostructures.

12.11.2.2 Epitaxy and Strain Engineering

Atomic-precision deposition methods namely molecular-beam epitaxy (MBE) and metallo-organic chemical vapor deposition (MOCVD) demonstrated their effectiveness as tools in the fabrication of various significant quarantine assemblies and strategies like quantum well lasers, photodetectors, and resonant tunneling diodes [84, 90–92].

12.11.2.3 Dip-pen Nanolithography

In DPN, the end-point of an AFM is painted in a selected chemical and contacted with the surface. Like a fountain pen, the paint flows from the endpoint of an AFM onto the surface.

This technique demonstrated ink thicknesses minimized to 12 nm with a 3D resolution of 5 nm [84, 93]. Types designed with DPN comprise gold, polymers, alkanethiols dendrimers, antibodies, organic dyes, and DNA.

12.12 Toxicity

Many researchers produced qualitative analyses of safety and toxicology based on histological assays using animal models. Precise target delivery can be achieved by considering the accumulation of foreign material in the body and minimizing the circulation of injected nanorobots to nontarget cells. Entire nanorobots have distinct safety concerns. Biohybrids could penetrate and propagate in an undesirable way. Chemically driven nanorobots may alter the chemical environment, which can have an effect on the microbiome of the gastrointestinal tract. Physically driven nanorobots are rigid and made from a nondegradable material, which may pose a danger. Scientists may also reveal additional parameters that detail the initial toxicity. The probable parameters include the number of motors used for treatment (grams) selected as an escalation of units (inadequacy and toxicity limit). Also, the distribution and toxicity of material used to fabricate nanorobots are important factors to consider.

12.13 Administration and Retrieval

Most commonly, administration methods include oral administration, injection, catheter, and topical administration (Figure 12.10). Considering retrieval strategy for nanorobots, researchers claim that biohybrid and chemically driven nanorobots are classified as biodegradable. Sometimes the effect of the synthetic material used was not specified or verified. Few researchers provided a retrieval mechanism for physically driven nanorobots, which are considered biodegradable. The use of a magnetic catheter for disposition and retrieval of nanorobots was suggested as a possible solution for retrieval in clinical practice [94]. The current *in vitro* research demonstrates nanorobot systems that are fully biodegradable [8–10].

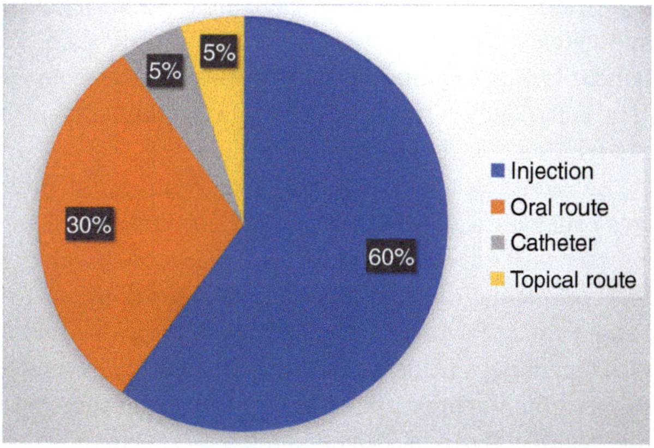

Figure 12.10 Pie chart representing the estimated ratio of the route of administration of nanorobots.

12.14 Clinical Presence of Nanorobots

The fundamental research is a revolutionary step in medical advances and economic growth. Most of the fundamental research requires prolonged time for translation from basic to clinical trials. In conclusion, most trials showed that novel technologies are inadequate compared with current standards of care, which should be considered in the development of nanorobots. It is crucial to recognize pros and cons in the consideration of their significance for future growth. The translational motion and mixing capabilities are the foundation of the most recent generation of nanorobots. The concept behind the smart nanorobot is still beyond the current capabilities and progress needs to be made on the ability to locate them and how they relate to localized stimuli.

The critical steps of environmental manipulation, nanorobot retrieval, and toxicology still require discussion and investigation. Medical technologies require an extended period for clinical translation/commercialization. However, significant development has been achieved in the field, with progress, in the past two decades from initial proof of the concept using chemical-assisted nanowires in peroxide [95] to the recent outbreak of *in vivo* studies. During clinical transformation it is essential to regulate the procedures and the precision of nanorobots. The commercial viewpoint also needs to be taken into account.

12.15 Reproducibility and Standardization

The efficacy of nanorobots *in vitro* is based on their velocity and directivity, or by analyzing the clinical outcome when compared to passive diffusion. For early proof-of-concept studies, these methods were utilized in efficacious clinical translation. The variation between current standards and *in vivo* nanorobotics means that it difficult to deliver a comparative or quantifiable analysis. A standardized protocol for reporting experimental methodology, characterization, and results, is required in order to make an effective transformation of integrative research to enhanced medical effects.

Generally, experimental findings might have a stronger case for translation if the findings are replicated by other researchers. In the same way, guidelines should be followed for reporting safety and toxicology data, thus permitting the comparison of nanorobot efficacy in qualitative or quantitative ways. The specifications for novel technologies need clear standards underpinning them. As a substitute, an equivalence of an investigational report may enable better experimental design, assessment, and enhanced precision between several nanorobotic platforms.

The identification of the energy source utilized to provide the power supply to nanorobots could assist in detailing the exclusive benefits of each method. The specifications of materials used for the fabrication of nanorobots comprises surface charge, stability, the average lifetime of propulsion, and size distribution. Knowing these will be helpful the design, manufacturing, and transportation in therapeutic applications. Based on experiment-dependent parameters, the efficacy of nanorobots can be analyzed. For drug delivery, the important parameter is administration and delivery dose (percent by mass). The number of cargos located in the target region is a crucial parameter in retention studies. In biopsy application, an important parameter is the force applied to a definite part of tissue (Pascal). Also, the efficacy of nanorobots needs to be compared with the state-of-the-art treatment, not with passive diffusion. Lastly, variables are considered to rationalize translation into existing therapies such as imaging, toxicity, route of administration, and retrieval region.

12.16 Regulatory Issues

The nanorobotic industry is growing. The nanorobot market was expected to grow 6.06% in 2016–2020 with their primary growth market in medicine [96]. The US government is constructing efforts to deal with progress in the industry. Regulation may be defined as the intended effort to channel the behavior of individuals. The nanodevice industry is becoming more commercialized, a similar regulatory process exists for nanorobots as with other medical devices. The nanorobots inventors interface with two most significant centralized agencies: the Food and Drug Administration (FDA), which sanctions medical products; and the Patent and Trademark Office (PTO), which deals with the intellectual property of innovation. Though these agencies are making an effort to focus on the innovative feature of nanorobotics in nanomedicine, industries face substantial challenges during commercialization.

12.16.1 The FDA and Nanorobots

To address the issues of nanotechnology, the FDA established the nanotechnology task force in August 2006 [97]. The report issued by this force emphasized the requirement for a regulatory path for nanorobots and related technologies but no proposals for what that should be [98]. In December 2017, the FDA's Center for Drug Evaluation and Research (CDER) and the Center for Biologics Evaluation and Research (CBER) released draft guidance that may provide regulation of nanorobotics. The draft guidance provides information regarding nanomaterials for drug products but what level is functional to nanorobotics is unclear. Also, it details how nanomaterial might be processed in investigational new drug applications (INDs), new drug applications (NDAs), abbreviated new drug applications (ANDAs), and biologics license applications (BLAs) but does not detail how nanorobots may be handled in the premarket approval (PMA). The Center for Devices and Radiological Health (CDRH) controls medical devices including nanorobots [99].

The FDA-approved medical products premarket processes use categories of products such as mechanical, biological, or chemical products. The CDRH regulates mechanically driven medical devices. Chemical action-based drugs are regulated by the CDER while biological products are regulated by the CBER. The draft guidance offers a list of factors for the evaluation of nanoscale products by the CBER and CDER, thus claimants will be required to consult with the FDA about the information needed to acquire the approval of nanorobots. A nanorobot can be driven by different techniques requiring classification on the mode of action for the approval of nanorobot medical devices. The primary mode of action is considered in a combination product that possesses more than one action mode or technique. The growing number of combination products to be submitted for FDA approvals led to the establishing of an Office of Combination Products (OCP), which deals with the designation of the product and picking a lead center for review. Eventually, once nanorobots are designated and allotted to a lead center, their review is completed in the same way as other medical products.

12.16.2 PTO and Nanorobots

The necessity for the procurement of a patent on nanotechnology is the same as for other innovations. Generally, an originator applies to the PTO with details of the new invention, the PTO reviews the application and compares the claims to identify the technologies. While the patent needs for nanorobots are not different from other technologies but the terminology is not yet clearly defined for, for example, nanoparticles, nanocrystal, quantum dots, and nanodots. As nanorobots are novel, the literature search is not as developed as other established fields. The PTO might not be able to search references that recommend innovator claims the lead to the issue but not the immediate issue of excessively extensive patents.

If a patent is presented, which could be considered excessive, then an opponent can challenge the patent. This could result in the patent application being cancelled in post-grant PTO procedures or become involved in a court case to verify its legality. To protect against such occurrences, innovators should present some patents with claims of variable scope to protect their innovations. If the widest claims are approved by the PTO, innovators may file extensive claims to acquire patents on supplementary claims. The innovator must present precise claims for the commercialization of nanorobots that are specific to several features such as claims, nanoparts, fabrication processes, and applications of nanorobots.

Another issue that arises from the undeveloped literature base in nanorobots is that the PTO, when checking this literature finds that claims in the literature are ahead of the actual facts. Thus, an application must contain a background segment relating to the state of the art and the new developments in the unique nanorobot patent being applied for. Lastly, the innovator of the nanorobot may experience refusal from PTO founded on it not being considered a nano-sized product or structure. Altering the size of a structure does not essentially create a new patentable product. Hence, the patent application must designate the method of fabricating nanorobots, the material used, and the advantages of such small devices.

Briefly, the PTO procedures for gaining patent protection for nanorobots appears to be better established compared to FDA approval. Several factors must be reviewed while drafting and submitting patent applications for nanorobots.

12.17 Conclusion

Recently, nanotechnology has begun to alter medical procedures with the aim of improving medical services. In this process, nanotechnology in healthcare comprises a structure, manufacture, regulatory guidelines, and utilization of pharmaceutically active agents and machines being nano-sized (≤100 nm). An emerging and favorable area of nanotechnology is the fabrication of nanorobots that are nanoscale. This nanodevice seems to offer better opportunities, several of which are presently being studied and developed. Nanorobotics can provide personalized treatment to eradicate diseases such as cardiovascular diseases, cancer, neurodegenerative disorders, diabetes, and viral infection with enhanced efficacy and reduced side effects. Also, nanorobots will be able to offer mutual action in delivering drugs and diagnostic agents or imaging actions during surgery.

The arrival of molecular nanotechnology will improve productivity and outcome through reduced risk, cost, and intrusiveness. This technology has a robust capability to transform healthcare in the future and copious research for this promise to be achieved. Upcoming nanorobotic treatment will characteristically include the concoction of various nanorobots, some executing the primary task while others serve in a supportive manner [100]. Within a few decades nanorobots could be prominent in nanomedicine and biotherapies substituting earlier drug-delivery techniques [101].

References

1 Feynman, R. (1996). There's plenty of room at the bottom. *Students Review CalTech* 4 (1): 22–23. https://doi.org/10.1353/imag.2003.0081.

2 Sahoo, S., Parveen, S., and Panda, J. (2007). The present and future of nanotechnology in human health care. *Nanomedicine: Nanotechnology, Biology and Medicine* 3 (1): 20–31. https://doi.org/10.1016/j.nano.2006.11.008.

3 BCC Research (2020). BCC Research: Market Research Reports & Industry Analysis. http://Bccresearch.com. https://www.bccresearch.com (acceessed 1 July 2020).

4 Svenson, S. (2014). What nanomedicine in the clinic right now really forms nanoparticles? *Wiley Interdisciplinary Reviews: Nanomedicine and Nanobiotechnology* 6 (2): 125–135. https://doi.org/10.1002/wnan.1257.

5 Chen, L., Henein, G., and Luciani, V. (2011). Nanofabrication techniques for controlled drug-release devices. *Nanomedicine* 6 (1): 1–6. https://doi.org/10.2217/nnm.10.140.

6 Cavalcanti, A. and Freitas, R. (2005). Nanorobotics control design: a collective behavior approach for medicine. *IEEE Transactions on NanoBioscience* 4 (2): 133–140. https://doi.org/10.1109/tnb.2005.850469.

7 Zor, F., Selek, F., Orlando, G., and Williams, D. (2019). Biocompatibility in regenerative nanomedicine. *Nanomedicine* 14 (20): 2763–2775. https://doi.org/10.2217/nnm-2019-0140.

8 Kim, J. and Tung, S. (2015). Bio-hybrid micro/nanodevices powered by flagellar motor: challenges and strategies. *Frontiers in Bioengineering and Biotechnology* 3. https://doi.org/10.3389/fbioe.2015.00100.

9 Bente, K., Codutti, A., Bachmann, F., and Faivre, D. (2018). Biohybrid and bioinspired magnetic microswimmers. *Small* 14 (29): 1704374. https://doi.org/10.1002/smll.201704374.

10 Palagi, S. and Fischer, P. (2018). Bioinspired microrobots. *Nature Reviews Materials* 3 (6): 113–124. https://doi.org/10.1038/s41578-018-0016-9.

11 Chen, C., Karshalev, E., Li, J. et al. (2016). Transient micromotors that disappear when no longer needed. *ACS Nano* 10 (11): 10389–10396. https://doi.org/10.1021/acsnano.6b06256.

12 Nourhani, A., Brown, D., Pletzer, N., and Gibbs, J. (2017). Engineering contactless particle-particle interactions in active microswimmers. *Advanced Materials* 29 (47): 1703910. https://doi.org/10.1002/adma.201703910.

13 Schattling, P.S., Ramos-Docampo, M.A., Salgueiriño, V., and Städler, B. (2017). Double-fueled Janus swimmers with magnetotactic behavior. *ACS Nano* 11: 3973–3983. https://doi.org/10.1021/acsnano.7b00441.

14 Garcia-Gradilla, V., Orozco, J., Sattayasamitsathit, S. et al. (2013). Functionalized ultrasound-propelled magnetically guided nanomotors: toward practical biomedical applications. *ACS Nano* 7 (10): 9232–9240. https://doi.org/10.1021/nn403851v.

15 Dai, B., Wang, J., Xiong, Z. et al. (2016). Programmable artificial phototactic microswimmer. *Nature Nanotechnology* 11 (12): 1087–1092. https://doi.org/10.1038/nnano.2016.187.

16 Bi, C., Guix, M., Johnson, B. et al. (2018). Design of microscale magnetic tumbling robots for locomotion in multiple environments and complex terrains. *Micromachines* 9 (2): 68. https://doi.org/10.3390/mi9020068.

17 Pal, M., Somalwar, N., Singh, A. et al. (2018). Maneuverability of magnetic nanomotors inside living cells. *Advanced Materials* 30 (22): 1800429. https://doi.org/10.1002/adma.201800429.

18 Patel, G., Patel, G., Patel, R. et al. (2006). Nanorobot: a versatile tool in nanomedicine. *Journal of Drug Targeting* 14 (2): 63–67. https://doi.org/10.1080/10611860600612862.

19 Requicha, A. (2003). Nanorobots, NEMS, and nanoassembly. *Proceedings of the IEEE* 9 (11): 1922–1933. https://doi.org/10.1109/jproc.2003.818333.

20 Li, F., Lu, S., Feng, Q., and Lanza, R. (2009). Large scale generation of functional megakaryocytes from human Embryonic Stem Cells (hESCs) under stromal-free conditions. *Blood* 114 (22): 2540–2540. https://doi.org/10.1182/blood.v114.22.2540.2540.

21 PubMed. (2020). *Nanorobotics in Nanomedicines - Search Results - PubMed*. Retrieved 8 July 2020, from https://pubmed.ncbi.nlm.nih.gov/?term=Nanorobotics+in+Nanomedicines.

22 Solovev, A., Sanchez, S., Pumera, M. et al. (2010). Magnetic control of tubular catalytic microbots for the transport, assembly, and delivery of micro-objects. *Advanced Functional Materials* 20 (15): 2430–2435. https://doi.org/10.1002/adfm.200902376.

23 Srivastava, S., Clergeaud, G., Andresen, T., and Boisen, A. (2019). Micromotors for drug delivery in vivo: the road ahead. *Advanced Drug Delivery Reviews* 138: 41–55. https://doi.org/10.1016/j.addr.2018.09.005.

24 Singh, V., Soto, F., Kaufmann, K., and Wang, J. (2015). Micromotor-based energy generation. *Angewandte Chemie* 127 (23): 7000–7003. https://doi.org/10.1002/ange.201501971.

25 Orozco, J., Cheng, G., Vilela, D. et al. (2013). Micromotor-based high-yielding fast oxidative detoxification of chemical threats. *Angewandte Chemie International Edition* 52 (50): 13276–13279. https://doi.org/10.1002/anie.201308072.

26 Cappelleri, D., Efthymiou, D., Goswami, A. et al. (2014). Towards mobile microrobot swarms for additive micromanufacturing. *International Jouranl of Advanced Robotic Systems* 11: 150. https://doi.org/10.5772/58985.

27 Schuerle, S., Vizcarra, I.A., Moeller, J. et al. (2017). Robotically controlled microprey to resolve initial attack modes preceding phagocytosis. *Science robotics* 2: eaah6094. https://doi.org/10.1126/scirobotics.aah6094.

28 Douglas, S., Bachelet, I., and Church, G. (2012). A logic-gated nanorobot for targeted transport of molecular payloads. *Science* 335 (6070): 831–834. https://doi.org/10.1126/science.1214081.

29 Fu, J. and Yan, H. (2012). Controlled drug release by a nanorobot. *Nature Biotechnology* 30 (5): 407–408. https://doi.org/10.1038/nbt.2206.

30 Foresight (2020). Respirocytes: A Mechanical Artificial Red Cell page 4. http://Foresight.org. https://foresight.org/Nanomedicine/Respirocytes4.html (accessed 7 July 2020).

31 Ahmad, U., Faiyazuddin, M., Hussain, M. et al. (2015). Silymarin: an insight to its formulation and analytical prospects. *Acta Physiologiae Plantarum* 37 (11): 1–17. https://doi.org/10.1007/s11738-015-2008-3.

32 Gao, W., Dong, R., Thamphiwatana, S. et al. (2015). Artificial micromotors in the mouse's stomach: a step toward in vivo use of synthetic motors. *ACS Nano* 9 (1): 117–123. https://doi.org/10.1021/nn507097k.

33 Gultepe, E., Randhawa, J., Kadam, S. et al. (2012). Biopsy with thermally-responsive untethered microtools. *Advanced Materials* 25 (4): 514–519. https://doi.org/10.1002/adma.201203348.

34 He, W., Frueh, J., Hu, N. et al. (2016). Guidable thermophoretic Janus micromotors containing gold nanocolorifiers for infrared laser-assisted tissue welding. *Advanced Science* 3 (12): 1600206. https://doi.org/10.1002/advs.201600206.

35 Felfoul, O., Mohammadi, M., Taherkhani, S. et al. (2016). Magneto-aerotactic bacteria deliver drug-containing nanoliposomes to tumor hypoxic regions. *Nature Nanotechnology* 11 (11): 941–947. https://doi.org/10.1038/nnano.2016.137.

36 Li, J., Angsantikul, P., Liu, W. et al. (2017). Micromotors spontaneously neutralize gastric acid for pH-responsive payload release. *Angewandte Chemie International Edition* 56 (8): 2156–2161. https://doi.org/10.1002/anie.201611774.

37 Sharafi, A., Olamaei, N., and Martel, S. (2015). MRI-based communication for untethered intelligent medical microrobots. *Journal of Micro-Bio Robotics* 10 (1–4): 27–35. https://doi.org/10.1007/s12213-015-0081-8.

38 de Ávila, B., Angsantikul, P., Li, J. et al. (2017). Author correction: micromotor-enabled active drug delivery for in vivo treatment of stomach infection. *Nature Communications* 8 (1): 1299. https://doi.org/10.1038/s41467-017-01616-y.

39 Medina-Sánchez, M. and Schmidt, O. (2017). Medical microbots need better imaging and control. *Nature* 545 (7655): 406–408. https://doi.org/10.1038/545406a.

40 Akin, D., Sturgis, J., Ragheb, K. et al. (2007). Bacteria-mediated delivery of nanoparticles and cargo into cells. *Nature Nanotechnology* 2: 441–449. https://doi.org/10.1038/nnano.2007.149.

41 Park, S.J., Park, S.-H., Cho, S. et al. (2013). New paradigm for tumor theranostic methodology using bacteria-based microrobot. *Scientific Reports* 3: 3394. https://doi.org/10.1038/srep03394.

42 Din, M.O., Danino, T., Prindle, A. et al. (2016). Synchronized cycles of bacterial lysis for in vivo delivery. *Nature* 536: 81–85. https://doi.org/10.1038/nature18930.

43 Ullrich, F., Bergeles, C., Pokki, J. et al. (2013). Mobility experiments with microrobots for minimally invasive intraocular surgery. *Investigative Ophthalmology and Visual Science* 54: 2853. https://doi.org/10.1167/iovs.13-11825.

44 Pokki, J., Ergeneman, O., Chatzipirpiridis, G. et al. (2016). Protective coatings for intraocular wirelessly controlled microrobots for implantation: corrosion, cell culture, and in vivo animal tests. *Journal of Biomedical Materials Research. Part B Applied Biomaterials* 105: 836–845. https://doi.org/10.1002/jbm.b.33618.

45 Servant, A., Qiu, F., Mazza, M. et al. (2015). Controlled in vivo swimming of a swarm of bacteria-like micro-robotic flagella. *Advanced Materials* 27: 2981–2988. https://doi.org/10.1002/adma.201404444.

46 Li, J., Li, X., Luo, T. et al. (2018). Development of a magnetic microrobot for carrying and delivering targeted cells. *Science Robotics* 3: eaat8829. https://doi.org/10.1126/scirobotics.aat8829.

47 Yan, X., Zhou, Q., Vincent, M. et al. (2017). Multifunctional biohybrid magnetite microrobots for imaging-guided therapy. *Science Robotics* 2: eaaq1155. https://doi.org/10.1126/scirobotics.aaq1155.

48 Vilela, D., Cossío, U., Parmar, J. et al. (2018). Medical imaging for the tracking of micromotors. *ACS Nano* 12: 1220–1227. https://doi.org/10.1021/acsnano.7b07220.

49 Wang, B., Zhang, Y., and Zhang, L. (2018a). Recent progress on micro- and nanorobots: towards in vivo tracking and localization. *Quantiative Imaging in Medicine Surgery* 8: 461–479. https://doi.org/10.21037/qims.2018.06.07.

50 Glückstad, J., Villangca, M., Palima, D., & Bañas, A. (2017). Light-actuated microrobots for biomedical science. SPIE Newsroom. http://dx.doi.org/10.1117/2.1201701.006860

51 Khalil, I., Magdanz, V., Sanchez, S. et al. (2014). Wireless magnetic-based closed-loop control of self-propelled microjets. *PLoS One* 9 (2): e83053. https://doi.org/10.1371/journal.pone.0083053.

52 Cunha, L., Horvath, I., Ferreira, S. et al. (2013). Preclinical imaging: an essential ally in modern biosciences. *Molecular Diagnosis & Therapy* 18 (2): 153–173. https://doi.org/10.1007/s40291-013-0062-3.

53 Li, J., Thamphiwatana, S., Liu, W. et al. (2016). Enteric micromotors can selectively position and spontaneously propel in the gastrointestinal tract. *ACS Nano* 10: 9536–9542. https://doi.org/10.1021/acsnano.6b04795.

54 Qian, K., Ma, X., Dai, X., and Fang, F. (2010). POMDP navigation of service robots with human motion prediction. *ROBOT* 32 (1): 18–24. https://doi.org/10.3724/sp.j.1218.2010.00018.

55 US National Library of Medicine (2020). Post-Market Registry of AURYON™ Atherectomy Device in Subjects Affected With Infrainguinal Peripheral Artery Disease https://clinicaltrials.gov/ct2/show/NCT04229563 (accessed 16 February 2022). https://doi.org/10.31525/ct1-nct04229563

56 Rapoport, S., Deeb, Z., and Scriven, K. (2017). Treatment of acute upper airway obstruction. *JAMA Otolaryngology. Head & Neck Surgery* 144 (1): 90–91. https://doi.org/10.1001/jamaoto.2017.1922.

57 Bertrand, M., Abran, M., Merlet, N. et al. (2016). Bimodal intravascular ultrasound (IVUS)/near-infrared fluorescence (NIRF) molecular imaging of atherosclerotic rabbits. *Canadian Journal of Cardiology* 32 (10): S198–S199. https://doi.org/10.1016/j.cjca.2016.07.311.

58 Li, W. (2009). Nanomanipulation technology [the editor's desk]. *IEEE Nanotechnology Magazine* 3 (3): 4–4. https://doi.org/10.1109/mnano.2009.934210.

59 Lu, S., Feng, Q., Caballero, S. et al. (2007). Generation of functional hemangioblasts from human embryonic stem cells. *Nature Methods* 4 (6): 501–509. https://doi.org/10.1038/nmeth1041.

60 Sahni, S. (2007). Endothelial cell infection and hemostasis. *Thrombosis Research* 119 (5): 531–549. https://doi.org/10.1016/j.thromres.2006.06.006.

61 Shop, I., Lang, F., Papadopoulos, L., & Yilmaz, D. (2020). Ultrasonic Powered Nanobots Remove Toxins and Bacteria from Blood. Interesting Engineering https://interestingengineering.com/ultrasonic-powered-nanobots-remove-toxins-and-bacteria-from-blood (accessed 26 July 2020).

62 Wu, Z., Troll, J., Jeong, H. et al. (2018). A swarm of slippery micro propellers penetrates the vitreous body of the eye. *Science Advances* 4 (11): eaat4388. https://doi.org/10.1126/sciadv.aat4388.

63 Vega, R., Maspoch, D., Salaita, K., and Mirkin, C. (2005). Nanoarrays of single virus particles. *Angewandte Chemie International Edition* 44 (37): 6013–6015. https://doi.org/10.1002/anie.200501978.

64 Walker, D., Käsdorf, B., Jeong, H. et al. (2015). Enzymatically active biomimetic micro propellers for the penetration of mucin gels. *Science Advances* 1 (11): e1500501. https://doi.org/10.1126/sciadv.1500501.

65 Stock, D., W. Leslie, A., and Walker, J. (2010). Refined structure of yeast F1c10 ATPase complex to 3 A resolution RCSB Protein Data Bank. https://www.rcsb.org/structure/2XOK (accessed date 16 February 20200).

66 Kinosita, K., Yasuda, R., Noji, H., and Adachi, K. (2000). A rotary molecular motor that can work at near 100% efficiency. *Philosophical Transactions of the Royal Society of London. Series B: Biological Sciences* 355 (1396): 473–489. https://doi.org/10.1098/rstb.2000.0589.

67 Vijayakumar, R., Jagannathan, S., Gandhi, P.R., and Chaansha, S. (2011). Nanorobotics: a newer platform for molecular diagnose. *Nano Biomedicine and Engineering* 3 (3): 192–201. https://doi.org/10.5101/nbe.v3i3.p192-201.

68 Ali, J., Chcang, U., Martindale, J. et al. (2017). Bacteria-inspired nanorobots with flagellar polymorphic transformations and bundling. *Scientific Reports* 7 (1): 14098. https://doi.org/10.1038/s41598-017-14457-y.

69 Le Pioufle, B., Surbled, P., Nagai, H. et al. (2000). Living cells captured on a bio-microsystem devoted to DNA injection. *Materials Science and Engineering: C* 12 (1–2): 77–81. https://doi.org/10.1016/s0928-4931(00)00162-4.

70 Bailey, S., Wu, R., and Sabatini, D. (2002). Applications of transfected cell microarrays in high-throughput drug discovery. *Drug Discovery Today* 7 (18): S113–S118. https://doi.org/10.1016/s1359-6446(02)02386-3.

71 Leong, T., Randall, C., Benson, B. et al. (2009). Tetherless thermobiochemically actuated microgrippers. *Proceedings of the National Academy of Sciences of the United States of America* 106 (3): 703–708. https://doi.org/10.1073/pnas.0807698106.

72 Breger, J., Yoon, C., Xiao, R. et al. (2015). Self-folding thermo-magnetically responsive soft microgrippers. *ACS Applied Materials & Interfaces* 7 (5): 3398–3405. https://doi.org/10.1021/am508621s.

73 Vikram Singh, A. and Sitti, M. (2016). Targeted drug delivery and imaging using mobile milli/microrobots: a promising future towards theranostic pharmaceutical design. *Current Pharmaceutical Design* 22 (11): 1418–1428. https://doi.org/10.2174/1381612822666151210124326.

74 Chatzipirpiridis, G., Ergeneman, O., Pokki, J. et al. (2014). Electroforming of implantable tubular magnetic microrobots for wireless ophthalmologic applications. *Advanced Healthcare Materials* 4 (2): 209–214. https://doi.org/10.1002/adhm.201400256.

75 Kagan, D., Benchimol, M., Claussen, J. et al. (2012). Acoustic droplet vaporization and propulsion of perfluorocarbon-loaded microbullets for targeted tissue penetration and deformation. *Angewandte Chemie International Edition* 51 (30): 7519–7522. https://doi.org/10.1002/anie.201201902.

76 Soto, F., Martin, A., Ibsen, S. et al. (2015). Acoustic microcannons: toward advanced microballistics. *ACS Nano* 10 (1): 1522–1528. https://doi.org/10.1021/acsnano.5b07080.

77 Solovev, A., Xi, W., Gracias, D. et al. (2012). Self-propelled nanotools. *ACS Nano* 6 (2): 1751–1756. https://doi.org/10.1021/nn204762w.

78 Xi, W., Solovev, A., Ananth, A. et al. (2013). Rolled-up magnetic microdrillers: towards remotely controlled minimally invasive surgery. *Nanoscale* 5 (4): 1294–1297. https://doi.org/10.1039/c2nr32798h.

79 Srivastava, S., Medina-Sánchez, M., Koch, B., and Schmidt, O. (2015). Medibots: dual-action biogenic microdaggers for single-cell surgery and drug release. *Advanced Materials* 28 (5): 832–837. https://doi.org/10.1002/adma.201504327.

80 Hu, C., Fang, R., Copp, J. et al. (2013). A biomimetic nanosponge that absorbs pore-forming toxins. *Nature Nanotechnology* 8 (5): 336–340. https://doi.org/10.1038/nnano.2013.54.

81 Wu, Z., Li, T., Gao, W. et al. (2015). Cell-membrane-coated synthetic nanomotors for effective biodetoxification. *Advanced Functional Materials* 25 (25): 3881–3887. https://doi.org/10.1002/adfm.201501050.

82 Wu, Z., Li, J., de Ávila, B. et al. (2015). Water-powered cell-mimicking Janus micromotor. *Advanced Functional Materials* 25 (48): 7497–7501. https://doi.org/10.1002/adfm.201503441.

83 Zhu, W., Li, J., Leong, Y. et al. (2015). 3D-printed artificial microfish. *Advanced Materials* 27 (30): 4411–4417. https://doi.org/10.1002/adma.201501372.

84 Drexler, K. (1992). *Nanosystems: Molecular Machinery, Manufacturing and Computation*. New York: Wiley.

85 Campbell, S.A. (2001). *The Science and Engineering of Microelectronic Fabrication*. New York: Oxford University Press.

86 Jaeger, C.J. (2002). *Introduction to Microelectronic Fabrication*. Upper Saddle River: Prentice Hall.

87 Plummer, J.D., Deal, M.D., and Griffin, P.B. (2000). *Silicon VLSI Technology*. Upper Saddle River: Prentice Hall.

88 Rai-Choudhury, P. (ed.) (1997). *Handbook of Microlithography, Micromachining and Microfabrication*. Bellingham: SPIE.

89 Chou, S.Y. (2001). Nano-imprint lithography and lithographically induced self-assembly. *MRS Bulletin* 26: 512–517.

90 Herman, M.A. (1996). *Molecular Beam Epitaxy: Fundamentals and Current Status*. New York: Springer.

91 Frood, J.S., Davis, G.J., and Tsang, W.T. (1997). *Chemical Beam Epitaxy and Related Techniques*. New York: Wiley.

92 Mahajan, S. and Harsha, K.S.S. (1999). *Principles of Growth and Processing of Semiconductors*. New York: McGraw-Hill.

93 Mirkin, C.A. (2001). Dip-pen nanolithography: automated fabrication of custom multicomponent, sub-100 nanometer surface architectures. *MRS Bulletin* 26: 535–538.

94 Iacovacci, V., Ricotti, L., Sinibaldi, E. et al. (2018). An intravascular magnetic catheter enables the retrieval of nanoagents from the bloodstream. *Advanced Science* 5: 1800807. https://doi.org/10.1002/advs.201800807.

95 Paxton, W.F., Kistler, K.C., Olmeda, C.C. et al. (2004). Catalytic nanomotors: autonomous movement of striped nanorods. *Journal of the American Chemical Society* 126: 13424–13431. https://doi.org/10.1021/ja047697z.

96 Research and Markets (2020). Global nanorobotics systems market growth at 6% CAGR - analysis, technologies & forecasts to 2020 - Research and Markets https://www.prnewswire.com/news-releases/global-nanorobotics-systems-market-growth-at-6-cagr-analysis-technologies-forecasts-to-2020-research-and-markets-300375278.html (acceessed 3 August 2020).

97 US FDA Nanotechnology Task Force. (2020). https://www.fda.gov/science-research/nanotechnology-programs-fda/nanotechnology-task-force (accessed 8 August 2020).

98 US FDA Nanotechnology Task Force (2007). Nanotechnology. A Report of the U.S. Food and Drug Administration Nanotechnology Task Force July 25, 2007 https://www.fda.gov/media/74257/download (accessed 8 August 2020).

99 US FDA Nanotechnology Task Force (2017). Drug Products, Including Biological Products, that Contain Nanomaterials Guidance for Industry https://www.fda.gov/media/109910/download (accessed 9 August 2020).

100 Karshalev, E., Ávila, B.E.-F.D., Beltrán-Gastélum, M. et al. (2018). Micromotor pills as a dynamic oral delivery platform. *ACS Nano* 12: 8397–8405. https://doi.org/10.1021/acsnano.8b03760.

101 Li, J., Esteban-Fernández de Ávila, B., Gao, W. et al. (2017). Micro/nano robots for biomedicine: delivery, surgery, sensing, and detoxification. *Science Robotics* 2: eaam6431. https://doi.org/10.1126/scirobotics.aam6431.

13

Nanosensors

Asit Behera[1], A.K. Sahoo[1], and S.S. Mohapatra[2]

[1] *School of Mechanical Engineering, KIIT, Bhubaneswar, India*
[2] *Department of Mechanical Engineering, NIT Rourkela, Odisha, India*

13.1 Introduction and Recent Advances

A sensor is a device that can measure a quantity. Generally, the quantity is detected electronically and converted to a signal in order to measure it properly. A sensor must have good sensitivity, stability, accuracy, and diversity. The chapter discusses nanosensors and their applications especially in the field of healthcare.

13.1.1 What are Nanosensors?

A nanosensor is a detecting device that has at least one dimension smaller than 100 nm and senses a variable quantity in the nanoscale range [1]. A nanosensor can be used for the measurement of physical variables, the detection of chemicals, gases, etc. A nanosensor may be in the nanoscale range or may measure one dimension in the nanorange. It is not mandatory that a nanosensor must always be in the nanorange. Each nanosensor has a generalized workflow while detecting a variable: binding to an analyte, signal generation while interacting with molecules, and processing of the signal. Nanosensors have wide applications in the fields of biomedicine, defense, the cosmetics industry, etc. [2]. Figure 13.1 shows a nanosensor embedded in a transparent material.

13.1.2 Why are they Important?

Nanosensors are presenting opportunities for building new generation devices They are unique and can interact easily at the nanorange making them very different to the macro level. Nanosensors are small, durable, and portable. They have more accuracy and high sensitivity. They consume less power, have low response times, and more speed. They can detect multiple quantities at the same time making them convenient for real-time analysis.

Nanomaterials and Nanotechnology in Medicine, First Edition. Edited by P.M. Visakh.

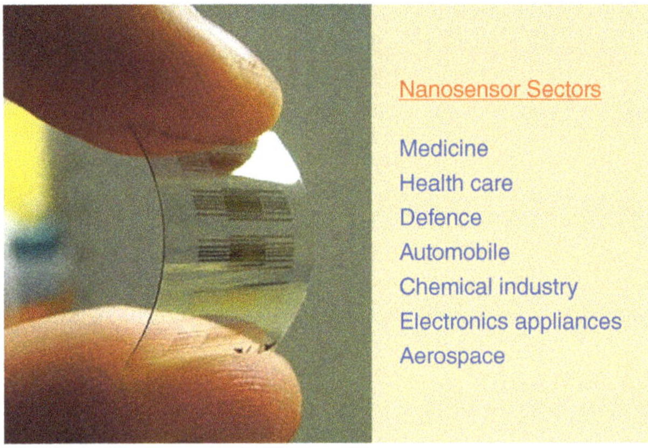

Figure 13.1 A schematic view of nanosensors.

13.1.3 How Nanosensors Work

Nanosensors work like a transducer. They convert one form of energy into another form of energy. Generally, as a first step the nanosensor senses a variable and converts it to an electrical signal. Most nanosensors can convert the electrical signal to a readable form. Nanosensors detect changes from external excitations and convey the changes to other nanoparticles [3].

13.1.4 Some Recent Advancements in Nanosensors

In the past decade many more advancements have been achieved in the fabrication of nanosensors that provide greater flexibility, versatility, and sensitivity. Recent advancements in nanosensors have applications in: the finding the pathology of tissues such as tumors; metabolite monitoring within body fluids; and detecting important molecules such as proteins and pathogens, etc. [4]. The advancements are based on a continuous evaluation of nanosensors. Some of the past inventions, on which recent advancements are based, are depicted in Table 13.1.

A recent advancement in nanosensor technology incorporates molecular imprinting technology that produces molecular matrices from monomers these then may act as a receptor in the detection of foreign molecules [33]. The nanosensor uses functional monomers with template molecules to form polymer matrices thus enhancing the selectivity and affinity of matrices. This specific technology helps nanosensors to detect chemical entities. In order to overcome the limitations of a single nanosensor, in terms of power and effectiveness, in recent past a nanonetwork has been used. Nanonetworks are usually more effective and work through electromagnetic communication [34]. Field-effect transistors have good sensitivity. So, many carbon-based nanotubes are designed as field-effect transistors. Due to chemical doping and charge transfer the electrical conductivity changes in the nanosensor, which makes it convenient to use as a detector [35]. Using this technique carbon nanotubes (CNTs) are used to sense ionization of gas.

Biological materials are being identified by quantum dots (QDs) in the applications of nanophotonics [36]. In some organizations nanocantilevers are used to detect food-borne pathogens. Tumors in the body can be discovered by using cadmium selenide QDs [37].

Table 13.1 List of patents filed on nanosensor evolution.

Year of filing	Gradual development of Nanosensor application in biomedical [patents]
2002	Surface-enhanced Raman nanobiosensor [5]
2002	Nanoelectronic measurement system for physiologic gases and improved nanosensor for carbon dioxide [6]
2002	Carbon dioxide nanosensor, and respiratory CO_2 monitors [7]
2002	Anesthesia monitor, capacitance nanosensors and dynamic sensor sampling method [8]
2003	Microsensors and nanosensors for chemical and biological species with surface plasmons [9]
2003	Sensor for detecting an analyte using silver nanoparticles [10]
2004	Optical sensor with layered plasmon structure for enhanced detection of chemical groups by sers [11]
2005	Diagnostic nanosensor and its use in medicine [12]
2005	Anesthesia monitor, capacitance nanosensors and dynamic sensor sampling method [13]
2005	Ammonia nanosensors, and environmental control system [14]
2008	Magnetic nanosensor compositions and bioanalytical assays therefor [15]
2009	Surface modification of nanosensor platforms to increase sensitivity and reproducibility [16]
2009	Nanosensor and breath analyzer for ammonia detection in exhaled human breath [17]
2009	Optical nanosensors comprising photoluminescent nanostructures [18]
2010	Chemical and biomedical nanosensors [19]
2010	A molecular-imprint nanosensor for ultrasensitive detection of proteins [20]
2011	Microbe detection via hybridizing magnetic relaxation nanosensors [21]
2011	Use and making of biosensors utilizing antimicrobial peptides for highly sensitive biological monitoring [22]
2012	Implantable nanosensor [23]
2012	Enzymatic nanosensor compositions and methods [24]
2013	Nanosensor for detecting the activity of glycosaminoglycan-cleaving enzymes and uses thereof [25]
2014	Analyte detection system and method for intradermal implantation of biocompatible optode nanosensors [26]
2014	Saccharide-responsive optical nanosensors [27]
2014	Graphene-based nanosensor for identifying target analytes [28]
2015	Microfluidic aptasensor including a graphene nanosensor [29]
2015	Porous microneedles through sacrificial sugar incorporation, analyte detection system, and method for intradermal optode nanosensor implantation [30]
2016	Noninvasive electroactive photonic protein nanosensor with polymer photovoltaic optics for memory transduction using organic and inorganic elements as platforms [31]
2018	Field-effect nanosensor for detecting simple metabolites in living organisms [32]

By linking artificial intelligence (AI) with nanosensors, healthcare practices are enhanced. The temperature inside the body of a living being can be measured using thermosensors equipped with AI [38]. Printed electronics, silicon photonics, for example, also contribute toward the advancement of nanosensors. Raman spectroscopy is a powerful tool that, when incorporated with a nanosensor, can detect a biological specimen [39]. Surface-enhanced Raman spectroscopy (SERS) is the most advanced technique in this field. Localized surface plasmon resonance (LSPR) depends on various factors of nanoparticles. LSPR's high sensitivity can detect large biomolecules [40]. One-dimensional inorganic semiconductor nanostructure-based photodetectors also currently find important applications. This technique enables optical input to be processed efficiently producing electrical outputs [41].

13.2 Classification of Nanosensors

Nanosensors may be classified as physical nanosensors, chemical nanosensors, and biological nanosensors, according to the signal they detect. They may also be classified according to applications, structure, and energy sources. Some of the nanosensors are described.

13.2.1 Mechanical Nanosensors

Mechanical nanosensors can measure pressure, stress, strain, etc. These are essential sensors in nanodevice components in micro-electronic devices. Mechanical nanosensors have various advantages over electromagnetic nanosensors and optical nanosensors. Nano mechanical cantilever sensors and CNT-based fluidic shear–stress sensors are the two most advanced mechanical nanosensors. Binh et al. discovered a mechanical nanosensor that can monitor the elastic and vibrational properties of a nanosphere [42].

13.2.2 Gas Nanosensors

Gas nanosensors are often categorized under chemical nanosensors. Chemical nanosensors are mostly used in medicine, environmental monitoring, food safety, and homeland security. Pearton and colleagues' review identified chemical sensors that are able to analyze a wide variety of environmental and biological gases and liquids and are able selectively target a specific analyte with high sensitivity [43]. Gas sensors measure the volume of a desired gas by oxidizing or reducing the gas and in turn by the resulting current change. A zinc oxide (ZnO) nanobelt sensor can be used to detect ammonia (NH_3) with high accuracy [44].

13.2.3 Optical Nanosensors

Optical nanosensors are used to monitor chemical make-up in an environment. The chemical analysis can be carried out by measuring changes in the optical characteristics of materials. Optical nanosensors are used in many areas like biotechnology, environmental science, and the chemical industry. Optical nanosensors usually employ photovoltaic units and one binding component. In the luminescence process light is absorbed by a fluorophore that is followed by the emission of light [45]. The first optical nanosensor based on fluorescein and was used for the measurement of pH.

13.2.4 Fiber Optic Nanosensors

Fiber optic sensors are very efficient in monitoring cell processes *in vivo*. Vo-Dinh et al. have studied fiber optic nanosensors. It is evident that when an interaction between a receptor and a target takes place there is a generation of a physiochemical behavior. This change is then converted into an electrical signal, which can be easily measured. This electrical signal can be collected by a probe and stored in a database [46].

13.2.5 Nanosensors for Electrical Current Measurement

Geng et al. investigated the interaction between gold (Au) nanoparticles and hydrogen sulfide gas molecules [47]. The generation of a sulfide shell promotes the charge transfer from one nanoparticle to another nanoparticle. This phenomenon is called the hopping phenomenon and is depicted in Figure 13.2. The hopping of electrons can be created by varying voltage and current.

13.2.6 Magnetic Nanosensors

Magnetic nanosensors are made up of magnetic nanoparticles such as iron oxide. When these nanoparticles combine with target materials they generate stable nanoassemblies. This may cause a change in the spin–spin relaxation time, which can be detected by magnetic resonance imaging [48]. Most often magnetic nanosensors are used to detect biomolecules.

13.2.7 Biosensors

Biosensors can be considered as a subset of chemical nanosensors. These sensors can be used for cancer detection, specific deoxyribonucleic acid (DNA) detection, and various diseases detection. Among the various biosensors field-effect transistors are the most efficient

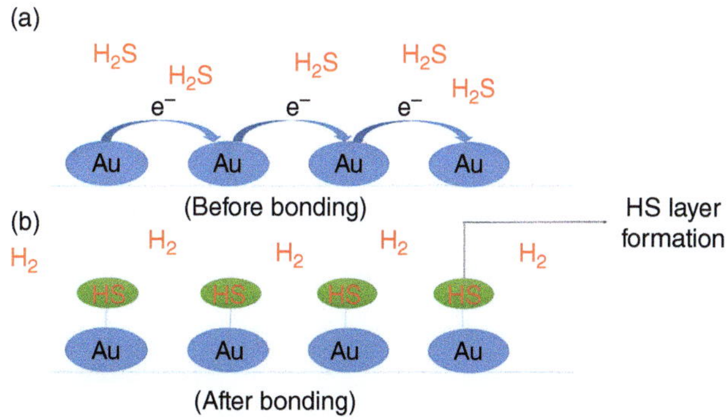

Figure 13.2 Electron hopping phenomenon (a) before bonding, (b) after bonding.

due to mass production capability, low-cost manufacturing, and ultrasensitive detection. Biosensors-on-chip are used frequently in diagnosis as they are robust and effective in the human body [49]. Chemical and biological components are merged into a single platform by the integration of biosensor and microfluidic technologies, which offer a new direction in biosensing. Again, nanosensors can be classified based on the materials used.

13.2.8 Nanosensors Based on Nanoparticles and Nanoclusters

Optical nanosensors are built by using nanoparticles such as noble metals and metal oxides. Apart from nanoparticles the nanosensors can be built with semiconductor QDs. In some nanosensors nanoprobes containing dyes are used whose fluorescence is quenched in the presence of the analyte to be determined. For illustration, an Au nanoparticle-based enzyme biomarker has been developed that can detect proteases disease in humans [50]. The nanosensor indicates when proteases are present. The process is shown in Figure 13.3.

13.2.9 Nanosensors Based on Nanowires, Nanofibers, and CNTs

CNTs are robust and inert making them good to use in chemical doping. The functionalization of CNT is excellent at selecting the target analyte. Most CNT nanosensors are basically field-effect transistors. Nanofibers and nanowires are also used to diagnose disease by forming chemi-resistive sensors. In exhaled breath analysis, as shown in Figure 13.4, they can be used to detect volatile organic compounds [51].

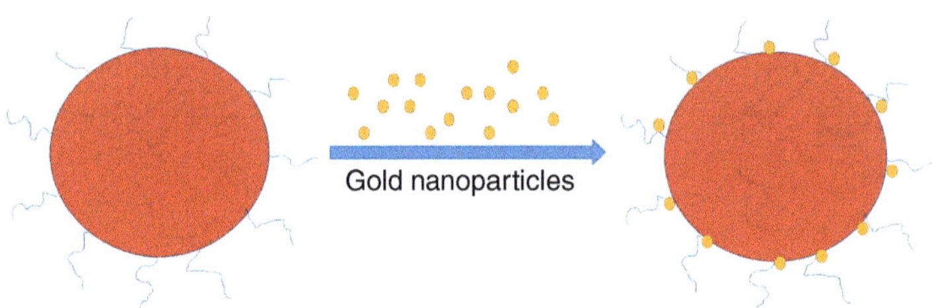

Figure 13.3 Gold nanoparticles covered in casein. Proteases eat the protective casein barrier and expose the surface of the gold nanoparticles.

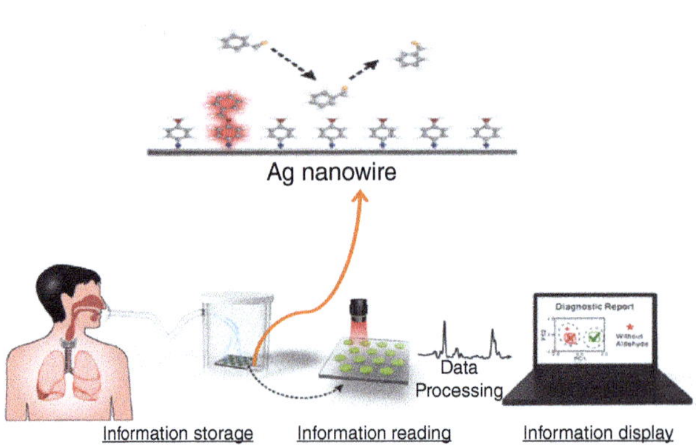

Figure 13.4 Exhaled breath analysis.

13.2.10 Nanosensors Based on Graphene

Graphene-based nanosensors have good application in the field of biomedical and chemical sensing. Chemical vapors change the noise spectra of the graphene transistors allowing them to sense selective gases. In some sectors three-dimensional graphene sheets are used, which have high conductivity [52]. Graphene foams are also finding promising applications in various fields.

13.2.11 Nanosensors Based on Bulk Nanostructured Materials

Some nanosensors based on bulk nanostructured materials have superior properties. The high surface area of nanoparticles makes them suitable for biomaterial coating, immobilizing molecules. The catalytic behavior of nanoparticles helps nanosensors by being electrochemical sensing devices. In the design of gas diffusion electrodes, platinum (Pt) nanoparticles with porous carbon have higher effectiveness [53].

13.2.12 Nanosensors Based on Metal–organic Frameworks

Metal-organic frameworks (MOFs) are inorganic–organic hybrid porous materials that consist of a series of positive metallic ions surrounded by organic "linker" molecules. The metal ions bind the linkages to form a structure. Due to the large surface area and hollow structure MOFs are extensively used for gas-sensing applications [54]. The application of MOFs in gas sensing is illustrated in Figure 13.5.

13.3 Nanosensor Fabrication

Nanosensors are manufactured using different processes. Top-down lithography, bottom-up fabrication, and self-assembled nanostructures are the most common methods for producing nanosensors. The basic difference between top-down and bottom-up methods is shown in Figure 13.6.

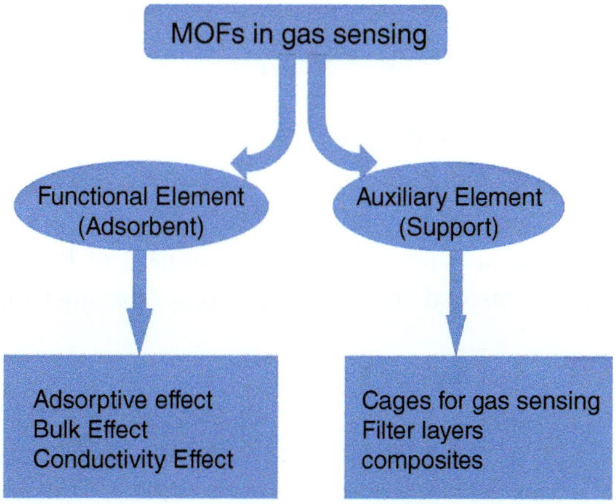

Figure 13.5 MOFs in gas sensing.

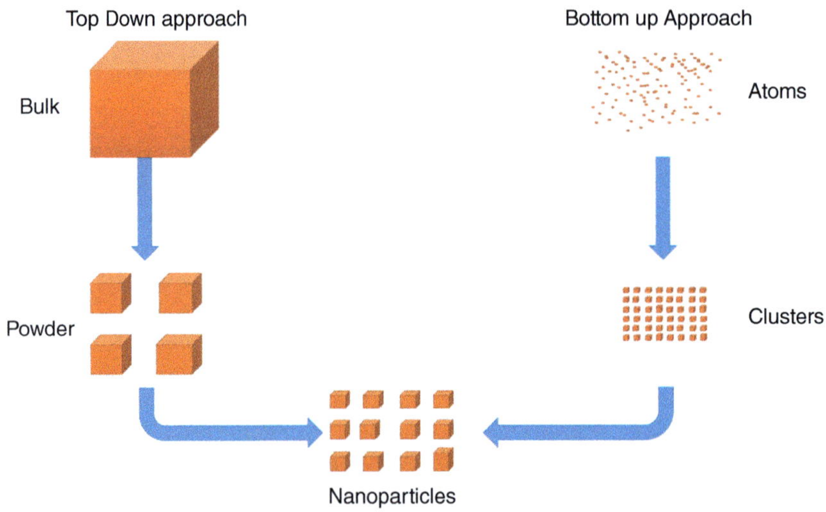

Figure 13.6 Comparison of top-down and bottom-up approach.

13.3.1 Top-down Methods

This method is usually employed in the micro-electronics industry. It is a subtractive process in which material is removed from the parental deposited material to produce the desired shape in the nanorange. Using this method items the size of a few tens of nanometers can be produced. It is such a controlled process that the desired size and shape can be produced. Some of the top-down methods for manufacturing nanosensors are expanded next.

13.3.1.1 Lithography

This is a popular top-down method in which larger material is continuously carved to achieve the desired shape. The carved products are generally in the microrange, these are used extensively in microelectromechanical systems. It is now possible to get this in the nanorange. Electron beam lithography is a well-known process of producing nanosensors but is quite expensive. This method is proving to be very effective.

13.3.1.2 Fiber Pulling

The fiber pulling process is usually employed in fiber optics technology. In this process a fiber is pulled to produce tensile force along the major axis of the fiber so as to achieve nano-sized scale. For easy achievement of the elongation the fiber is heated. The fiber-pulling method and lithography processes are shown in Figure 13.7.

13.3.1.3 Chemical Etching

There are two main types of chemical etching: (i) tube etching and (ii) the Turner method. In tube etching hydrofluoric acid is used in which optical fiber is laid to be etched. In this method the silica fiber is surrounded with a cladding material is exposed to hydrofluoric

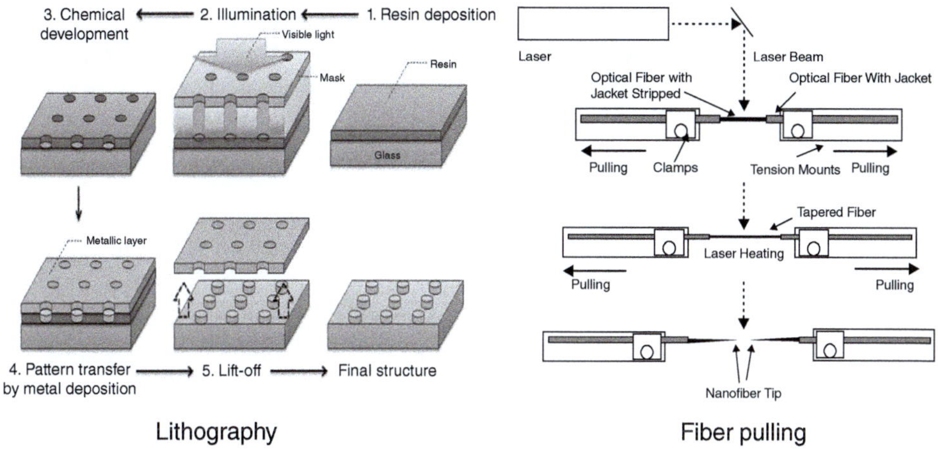

Figure 13.7 Lithography and fiber pulling process.

acid. The acid etches away the tip of the fiber without disturbing cladding material. After etching, a cone with smooth taper is produced [55].

In the Turner method the point of the fiber is placed at the meniscus of hydrofluoric acid and organic overlayer. This method produces a large taper angle and tip diameter. This etching process is more susceptible to the environment [56].

13.3.2 Bottom-up Methods

Bottom-up fabrication process is an additive process in which the desired shape of the nanosensor can be formed by adding and rearranging atoms and molecules. The rearrangement of atoms and molecules is achieved using atomic force microscopy. Basically, by assembling the smaller components we get the desired shape of the nanosensor. Liquid-phase formation and gas-phase synthesis are the two well-known bottom-up methods through which nanosensors are manufactured. This method may include metal organic decomposition, chemical vapor deposition (CVD) process, solgel method, laser pyrolysis, plasma arcing, molecular beam epitaxy, self-assembly processes, or wet synthesis.

13.3.2.1 Plasma Arcing

Plasma arcing is a popular method of forming nanoparticles and in turn nanosensors. It produces positively charged ions to be deposited as nanoparticles. It requires highly ionized gas atoms that possess very high energy. The ample amount of energy leads to electrons being thrown out of the valence shell of the material. The end product is positively charged atoms. Due to high potential difference electron separation from the atoms is possible. It uses inert gas as the heat source. The avalanche of electrons results in an arc formation between electrodes. The positively charged ions formed in these processes possess a great amount of kinetic energy. The ions are driven by voltage toward the cathode and are deposited as nanoparticles [57].

13.3.2.2 Chemical Vapor Deposition

The CVD process includes a substrate and volatile precursors. At a high temperature the precursors are absorbed on to the substrate, react with it, and decompose it to deposit the byproduct as crystals. The process is commonly used in the semiconductor industry [58].

13.3.2.3 Molecular Beam Epitaxy

Molecular beam epitaxy is a physical evaporation process in which no chemical reaction is involved. It is performed in a low temperature environment. In this process vacuum evaporation takes place where under high vacuum conditions atomic beams are directly bombarded on a heated substrate [59]. The process is illustrated in Figure 13.8.

13.3.2.4 Solgel Synthesis

Solgel is the combination of sol and gel. Sol refers to mixture where solid colloidal particles are dispersed in a liquid. Gel is considered to be a rigid network in which nanostructures are interconnected in a liquid phase. In this process, solid nanoparticles are mixed in a liquid medium and form a gel with a pore diameter in nanorange. The properties of solgels depend on parameters like temperature, catalysts, pH, time, and type of solvent. Mixing, of sol, gelation, aging, drying, dehydration, or chemical stabilization and densification are the processes that are involved in this method.

13.3.3 Molecular Self-assembly

It is one of the most effective processes for manufacturing nanosensors. It can fabricate nanostructures in the range of 1–100 nm [60]. Natural materials production, like DNA production, takes place through molecular self-assembly. In this method atoms and molecules are assembled together to form a well-defined nanostructure joined by noncovalent bonds.

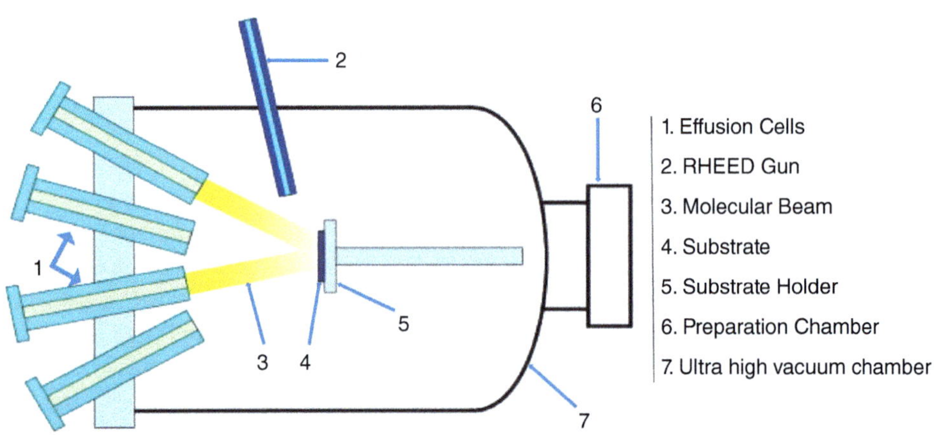

Figure 13.8 Molecular beam epitaxy process.

13.4 Inorganic Nanosensors

13.4.1 Materials Used

13.4.1.1 Carbon-based Nanomaterials

Carbon-based nanomaterials provide exposure of functional surface to analytes making an effective application for the detection of a foreign material. These materials have high stability, good conductivity, easy surface functionalization, etc. For various electro-analytical applications grapheme, CNTs, etc. are used which are outlined here.

Carbon Nanotubes CNTs have high stability in aqueous and nonaqueous solutions. They have high electronic conductivity. Apart from these properties, CNTs possess many other electronic properties making these perfect nanosensors. CNTs can be manufactured by laser ablation, CVD, or electrical discharge machining. The sensing mechanism of CNT-based gas sensors is based on their p-type CNT semiconducting property. They have large surface area and have capacity of fast charge transfer [61].

Graphene The electro-chemical property makes graphene useful for many applications. By enlarging the surface area of graphene the electro-chemical activity of the materials is increased. Graphene can be manufactured by CVD, electro-chemical reduction, the Hummers method, etc. Graphene has a two-dimensional nanostructure that allows fast electron transfer to be used in electro-chemical sensors and biosensors.

Porous Carbon Due to the larger surface area, narrow electron transfer passage porous carbon is used in many applications. Porous carbon nanomaterials are divided into microporous or mesoporous materials depending on their size. Mesoporous nanomaterials are categorized into three categories: microporous, mesoporous, and macroporous. Particles that have pore sizes < 2 nm are called microporous, between 2 and 50 nm are classified as mesoporous, and > 50 nm is considered macroporous nanomaterials [62]. Mesoporous silica nanoparticles popularly known as MSNs are very familiar in the nano world and are a type of mesoporous material. These particles possess good surface properties such as a large surface area and suitable pore size.

Mesoporous Nanomaterials In the recent past mesoporous nanomaterials have provided sustainable solutions in the healthcare sector. Mesoporous nanomaterials are very useful in diagnostic applications. The large pore size and surface area of MSNs protect the body from premature release and degradation of drugs before reaching the target. Surface modification in the case of MSNs leads to the enhancement of biocompatibility, providing functional groups and inhibiting adsorption. Physically, through self-assembly, and chemically, through surface fictionalization, the layer-by-layer self-assembly (LSA) surface is modified. Nanoparticles with a positive charge have a more active nature compared to any other charge. Positively charged nanoparticles possess more cytotoxicity and immune response compared to any other charged particle. MSNs having negative zeta potential associate with serum opsonin [63]. But there is a major limitation in the use of MSNs in biomedical applications. MSNs possess (-SiOH) groups on their surface. These functional groups can deteriorate plasma proteins and lipids present in the cellular membrane. So, in order to be biocompatible MSNs have to be modified for reactivity alteration [64].

Silica nanoparticles can be considered mesoporous nanomaterials. These particles carry silica adduct, which have a high number of pores. The pores can accommodate various molecules. In addition, silica has an external surface between anionic molecules and pore openings. The anionic molecules are usually attached to the outer surface of silica. The anionic molecules provide hydrophilicity to the submicron level. The surfaces have a higher surface-to-volume ratio than simple particles. Hence, these are used for cargo applications [65]. Moreover, these mesoporous materials possess easy surface modification, biocompatibility, etc., which makes them convenient for use in biomedical applications [66, 67].

Carbon Dots Carbon dots can be produced by electro-chemical oxidation, laser irradiation, hydrothermal treatment, microwave treatment, etc. The process of production of carbon dots is easy and cost effective. It has high biocompatibility. It is mostly used in temperature nanosensors for physiological applications.

13.4.1.2 Metal- and Noble Metals-Based Nanomaterials

Metal-based nanoparticles have unique chemical and physical properties that make these nanoparticles applicable in various fields. Metal-based nanomaterials have a strong potential with good selectivity and stability. Metals like Au, Pt, Ag, cobalt (Co), copper (Cu) and rare earth metals have a good sensitivity that can be employed in nanosensors. These metal-based nanomaterials have a high surface-to-volume ratio making them useful for application in various fields like medicine, electrodes, catalysis, fuel cells, and mechanical actuators. Some of the metal-based nanomaterials are elaborated here.

Gold-based Nanoparticles Au-based nanoparticles have a large surface area, improved optical properties, and a high capacity for surface modification. Au-based nanosensors are mostly used in environmental applications. Au-based electrodes have high sensitivity, higher signal-to-noise ratio, enhanced catalytic action, etc. Au nanoparticles present in electrodes lead to electron transfer between biomolecules and the transducer. Au has a better bioanalytical property. Because of its higher stability and complete recovery in chemical redox reaction Au-based nanoparticles are considered to be effective catalysts [68].

Due to their enhanced catalytic, chemical, and optical properties Au nanoparticles find application in chemical and biological sensors.

Silver-based Nanoparticles Ag-based nanoparticles are suitable for use in biosensors due to their improved catalytic activity and good surface-enhanced Raman scattering. From research [69] it is evident that Ag nanoparticles combined with various matrixes like metals, polymers, and fibers to provide high stability and good sensing efficiency [69].

Platinum-based Nanoparticles Pt-based nanoparticles are extensively used for the fabrication of electrodes. The nature of electrodes depends upon the manufacturing process of the electrodes. For the fabrication of Pt-based nanoparticles various techniques such as chemical reduction, photo-chemical deposition, metal-vapor synthesis, electro-chemical deposition, etc. are used. The fabrication method determines the chemical inertness and the sensing capability. On the whole Pt nanoparticle possess excellent catalytic properties and are extensively used in electro-chemical analysis. The function of Pt nanoparticles depends upon the bond length, optical property, crystallographic orientation, melting point, etc.

Palladium-based Nanoparticles Palladium (Pd) is easily available. The abundance and cheapness of Pd makes it more useful substitute over Au and Pt. Pd nanoparticles have high electrocatalytic potential toward analytes. So, its main use is in sensor applications for gases, toxic molecules, and biomolecules [70]. The Pd nanoparticles offer an electro-tunneling effect, which facilitates the electron transfer between electrode and target thus providing good electro-chemical sensing.

Copper Cu has good stability, good conductivity, and a low cost. Its low cost makes it attractive to use in a variety of industries. Cu nanoparticles have an improved signal-to-noise ratio in electroanalytical measurements and they also have a high surface-to-volume ratio and a high transport rate. These unique properties of Cu nanostructures lead to its extensive use.

Metal Oxide Nanoparticles Metal oxides have enhanced electrical, chemical, electronics, and physical properties that sense environmental conditions very effectively. Changes in electrical properties in metal oxides makes it suitable for use as a sensing element. Metal oxides have a low cost and a very large surface area.

Tin Oxides Generally, in gas sensors, tin (Sn) oxides are used. Hierarchical nanostructures have higher sensitivity for ethanol and good selectivity for NH_3, CO_2, H_2, etc. Khaong et al. studied hierarchical SnO_2 nanostructures for ethanol [71]. Pan et al. also studied electro-deposited SnO_2 with the addition of polyethylene glycol. It was used as a surfactant. The surfactant produced spherical SnO_2, which has better sensitivity [72].

Nickel Oxides Nickel oxide (NiO) nanomaterials have p-type conductivity. They are used in gas sensors, catalysis, battery electrodes, etc. They enhance the electro-chemical activity of the electrode. They also provide a large contact area between the target and the electrode. From the literature it is evident that NiO shows great sensitivity for formaldehyde gas sensing [73].

Titanium Oxides Titanium dioxide (TiO2) nanomaterials are generally used in pharmaceutical and medical applications for electro-chemical sensing. A range of proteins can be immobilized into nanoporous TiO_2 film electrodes. This process is used to make optical and electro-chemical biosensors. Topoglidis et al. investigated hemoglobin/TiO_2 films and found that these films allow optical sensing of nitric oxide. In particular, this concept is used to measure nitric oxide at micromolar levels [74].

Zinc oxide-based Nanostructures Zinc-based nanostructures have diverse applications in various fields. Methods of manufacture of ZnO nanostructures include laser ablation, vapor phase transport, wet chemical methods, magnetron sputtering, etc. When these methods are assisted by microwaves the process becomes fast. The sensitivity of ZnO is highly affected by additives.

Quantum Dots QDs have bright luminescence and a wide excitation spectra. They are used for imaging and fluorescent labelling. The luminescence intensity of QDs decreases with an increase in temperature, which is indicated by a widening of the emission spectra. QDs can generally be used in temperature sensors to detect variations in temperature. Normally QDs are manufactured by using highly toxic elements such as tellurium, cadmium, and selenium [75].

Photon Upconverting Nanoparticles Photon upconverting nanoparticles (UCNPs) are used for sensing temperature [76]. UCNPs exhibit multiple and sharp emission peaks that depend on lanthanide dopants. This property makes them suitable for use in *in vivo* imaging. They have an excellent signal-to-noise ratio, a lack of photobleaching, good tissue penetration, etc. UCNPs are frequently prepared from a lanthanide-doped host material. UCNPs can be photoexcited using near infrared (NIR) continuous-wave diode lasers. These are mostly used for detection of temperature. Figure 13.9 shows the fundamental action of both QDs and UCNPs.

13.5 Biopolymer-derived Nanosensors

13.5.1 Materials Used

Biopolymer-based nanomaterials show greater selectivity and sensitivity with a high performance. The enhanced properties of the nanomaterials enable them to be used in nanosensors. The properties of the nanomaterials add value to the nanosensors for enhancement in physical, chemical, thermal, and electrical properties.

13.5.1.1 Polymer Nanomaterials

Polymeric nanomaterials are generally used to determine toxic and chemical contamination in gas and liquid. They are used for the purpose of health applications. Nanocomposites made of CNT, grapheme, metal nanoparticles, or a combination of all these lead to further enhanced electro-chemical detection properties. In these composites, the combination of filler material and matrix plays a great role in determining the properties of the

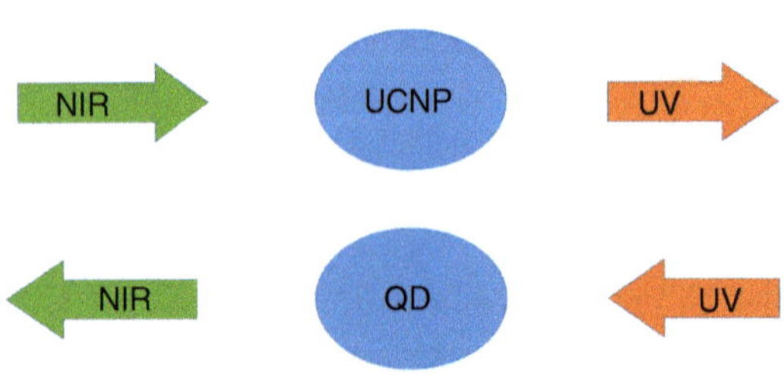

Figure 13.9 Emission of a photon (UV) is higher than that of the incident photon (NIR) for UCNPs whereas QDs emit photons with a lower energy than the incident photons.

nanocomposites or nanosensors. Selectivity, sensitivity, and biocompatibility are also affected by the combinations. In the healthcare industry polymeric nanomaterials are extensively used in medical coatings, prosthesis, implantation, medical devices, etc. [77]. These materials are very much compatible with human tissues and cells. Polymer materials are basically of two types: natural and synthetic. Natural polymers are very complex and difficult to convert to any other form but synthetic polymers are convenient to use as nanosensors. Wool protein and silk are examples of natural occurring polymers that are derived from mother nature. There are four types of synthetic polymers: elastomers, thermoplastics, thermosets, and synthetic fibers, nylon, Teflon, and polypropylene, etc. are some popular synthetic fibers.

Polymer nanostructures can be formed through two different methods: (i) the template free method and (ii) the template-based method. In template-free methods no template is used while fabricating the nanostructures. These are basically self-assembly techniques in which end product is formed by, for example, Vander Waals forces. Electrospinning is also a template-free method in which a high electric field is generated. In template-based methods nanostructure materials are set as a template over which one-dimensional polymers are grown.

13.5.1.2 Bionanomaterials

In bionanomaterials, biomolecules are formed as nanomaterials. The analytical property of biomolecules with the combination of nanostructures enhances the properties to be used as nanosensors. For electro-chemical detection of capsaicin Sabela et al. prepared multiwalled carbon nanotubes (MWNTs) using a nanobiocomposite of L-phenylalanine ammonia-lyase enzyme [78]. Li et al. also prepared nanosensors for the detection of *Escherichia coli* [79]. This was done by developing a self-assembled monolayer (SAM).

13.5.1.3 DNA-Based Nanostructures

DNA is a genetic molecule for storing hereditary information. It contains a phosphate group, a sugar molecule, and a nitrogen containing a nucleobase. In a DNA molecule a double stranded helical structure is present which makes it convenient for use in self-assembly processes in nanosensors. It has several major effective characteristics. Firstly, it acts as a rigid polymer below the size of 50 nm. Secondly, DNA molecules can be segregated by a self-assembly process leading to possibilities of fabrication in the nanorange. Thirdly, by chemical coding processes genetic information is coded. So, intermolecular interaction of molecules can be programmed easily. These advantages makes it possible for DNA to be used for nanostructures in nanomaterials.

13.5.1.4 Dye-doped Nanoparticles

For the preparation of temperature-sensitive nanomaterials the encapsulation of dyes into nanoparticles is carried out. The precipitation process is an easy way to prepare nano-sized temperature sensors. In this process a homogeneous solution is prepared by dissolving a host material and a probe into an organic solvent. Another solvent, in which the host material and probe cannot be dissolved, is added to the existing solution. The probe is encapsulated and the host material is precipitated afterwards. This method is applicable for a large variety of temperature sensors.

13.5.1.5 Thermo-responsive Organic Gels

Some organic gels can also be used for temperature sensing *in vivo*. Here phase transition plays a great role to sense temperature. For example, poly(N-isopropylacrylamide) is a thermo-responsive polymer that undergoes phase change in an aqueous solution. Below 305 K the polymer is in a swollen state but as the temperature rises the polymer is coiled and forms nanoparticles in the range 100 nm. NIR probes, a particular form of probe with special features, can be attached to polymer. This can extend the applicability of the device for use *in vivo* for sensing temperature [80].

13.6 Applications

Nanosensors have applications in detecting biomolecules in cellular organelles, monitoring chemical and physical phenomena in regions that are difficult to reach, and measuring nanoscopic particles in environment and the industry. The ability of nanomaterials to detect nucleic acid, proteins, etc. finds application in the health sector. Carbon-based nanomaterials are used for bioimaging, antibody detection, etc. Noble metals and their metal oxides are extensively used nanomaterials for their chemical stability and easy fabrication techniques. Nanosensors are greatly used in diagnostic medicines enabling early identification of diseases without relying on symptoms. Nanosensors, currently, are equipped with AI for real-time analysis. Using nanosensors and AI a patient's breath can be analyzed. Thermosensors and QDs, etc. are being used in the analysis of human health. This section illustrates the applications of nanosensors in healthcare.

13.6.1 Drug Discovery

The development and detection of pharmaceuticals becomes easier if molecules in it can hold proteins and organic molecules together. One example shows the diagnosis of molecular inhibitors to tyrosine kinases. Proteins that intervene signal conveyance in mammalian cells via phosphorylation of a tyrosine residue from a substratum protein through ATP are called tyrosine kinases as shown in Figure 13.10 [81].

The deregulation of phosphorylation in the body is a cause of illness, potentially leading to cancer. The kinase Abl (protein) adsorbed in silicon nanosensors tested the bonding of adenosine triphosphate (ATP) in order to control and fix tiny molecule inhibitors to tyrosine kinases ATP bonds with biological atoms such as the drug Gleevec. ATP is negatively charged, which binds with Abl due to the drop or rise of performance of a p-type nanosensor. When negatively charged ATP bonds with Abl the conductivity increases extensively. This provides excellent potential for use in therapeutic testing. In low molecular concentrations the conductivity may decrease indicating the dependence of degree of inhibition on the molecular structure [82, 83].

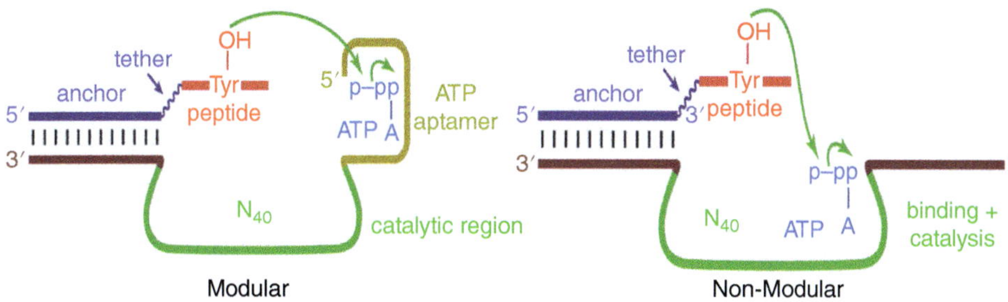

Figure 13.10 Tyrosine Kinase activity for DNA.

13.6.2 Monitoring the Effect of Drugs in Mammalian Cell

In the past decades it has been interesting to develop nanosensors that evaluate the effect of anticancer drugs in human cells. In the treatment of cancer, during the application of anticancer drugs, a very highly organized process called apoptosis takes place in which a series of cells dies. Caspases are activated in a self-amplifying cascade during apoptosis. Though all cell death does not occur by activation of caspase activation it is suggested that caspase activation is needed for the apoptotic process. For the initialization of apoptosis, caspases have an important role. Caspases also have an essential role in downstream processes that produce various hallmark signs of apoptosis. For example, activation of 2, 8, 9, and 10 caspases leads to proteolytic activation of 3, 6, and 7 caspases [84].

Nanobiosensors were used to control the initiation of the mitochondrial apoptosis pathway in a living cell. This is done by determining enzyme activity by caspase 9. A renascent field excites the cleaved AMC and the resulting fluorescent signal is detected. Quantitative monitoring of changes in fluorescent signals revealed caspase-9 activity in single living MCF-7 cells. Photodynamic therapy (PDT) protocols with δ-aminolevulinic acid (5-ALA) were used to induce apoptosis in MCF-7 cells. The LEHD-AMC substrate is knocked down by caspase-9 and the released AMC molecules are excited and emit a fluorescent signal. Using comparative apoptotic cell fluorescence and uninduced control, Kasili et al. found and identified caspase-9 activity. Comparing the fluorescent signals of apoptotic cells induced by photodynamic treatment against native activity with non-apoptotic cells, the researchers found successful caspase-9 activity. Detection of caspase-9 indicates the onset of apoptosis in cells [85].

The ability to examine mammalian cells without altering their physiological condition avoids the need for sampling and mechanical manipulation. Mechanical manipulations, such as cell lysis, can cause cell destruction before prior to the sample being taken and can interfere with cell biochemistry measurements. In optical nanobiosensors, the lack of cell manipulation prior to measurement limits the significant effects on the cell at the time of sampling. Based on the results of Poreba et al's intracellular study, nanobiosensors were shown to be extremely sensitive to biomedically significant activities (apoptosis) [86]. This high sensitivity is largely due to the fact that a single enzyme molecule can catalyze the cleavage of many biomolecules, so that it produces large quantities of highly fluorescent R110. Another useful feature of the nanobiosensor is its selectivity, which depends on the high accuracy with which the enzymes act. Just as caspase-9 specifically encapsulates the

LEHD sequence to produce free AMC for detection in the previous example, the active site of caspase-7 specifically interacts with DEVD-R110 to generate highly fluorescent R110. Together, these two fluorescent probes illustrate that fluorescent fibro-optic nanoprobes can effectively provide biomedical information of individual living mammalian cells. This work demonstrates the practical application of optical nanobiosensors for molecular recognition of the level of individual cells with the substrate derived from R110, which contains the recognition site of caspase-7, DEVD-R110. Optical nanobiosensors for the study of intact living cells promise not only minimally invasive dynamic analysis of individual living cells, but also to offer a new concept in cellular analysis by offering highly miniaturized and practical nanoscale devices that perform analysis based on an affordable level in a single cell [86].

13.6.3 Cancer Diagnosis

Diagnostic devices that use nanosensors have greater speed and sensitivity compared to conventional diagnostic techniques. In the treatment of medical ailments nanosensors are extensively used in a variety of situations. To increase the antimicrobial property and mechanical properties of dental filler material ZnO nanoparticles are used. Through ultrasonic processes ZnO nanoparticles and hydroxyapatite (HAP) are mixed to form a ZnO–HAP nanocomposite. This can improve antimicrobial activity and can detect *E. coli* DH5a. ZnO nanoparticles have the capability to destroy cancer cells. It is also used against human brain tumors and in anticancer studies [87].

Au nanoparticles are also used for cancer diagnosis. They detect volatile organic compounds in breath. To detect lung cancer, peptide-coated nanoparticles used as activity-based sensors are used in human beings. This is in early development and the research for advancement continues. Au nanoparticles can detect a tumor in the size order of some millimeters. It is a portable, noninvasive, and cost-effective diagnostic tool. Raman spectroscopy is also implemented for biological and histological diagnosis. It is specially used to target biochemical compounds and cells for identification and tracking. DNA-based nanostructures are also being used for diagnostic applications. DNA-based nanostructures have greater uniformity, diversity, and biocompatibility. Electroporation is used for gene delivery that demands high voltage but, in this process, many living cells die due to the high voltage. In order to overcome this limitation, molecular delivery assisted by shock waves is employed. These shock waves are generated using nanoenergetic materials. This is used for gene transfection purposes. The small size of the nanostructures enables easy access to the internal parts of the body without affecting other bodily functions.

QDs are highly luminescent and stable. These provide new opportunities to study proteins, drug targets, and genes in a living cell. Thus, it facilitates the visualization and study of cancer cells in human body. For example, in the case of prostate cancer, a method is developed for the detection of protein analytes. In this method magnetic microparticle probes and antibodies bind a target of interest like a prostate-specific antigen. QDs incorporated with antibodies and polyacrylate cap are used for immunofluorescent labelling of the breast cancer marker Her2 [88]. Labelling is more stable, specific, and brighter than fluorescent markers. QDs with fluorescence microscopy follow cells at high resolution in living animals as shown in Figure 13.11.

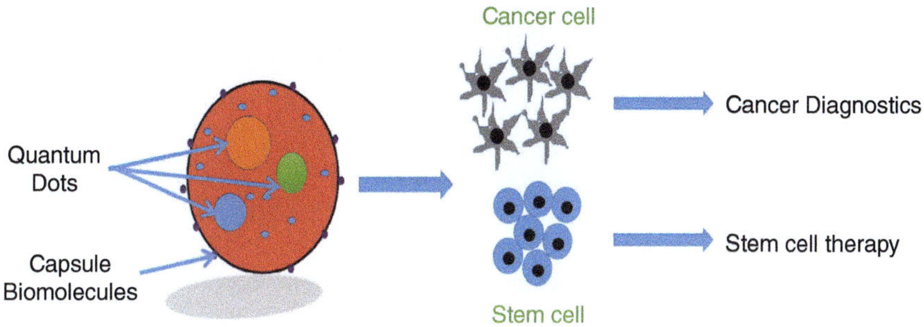

Figure 13.11 Quantum dots used in cancer diagnostic and stem cell therapy.

13.6.4 Tumor Detection

Tumor detection utilizes the fluorescence property of the material. Cadmium selenide is one of the materials used for the detection of tumors in the body but it is very toxic to the body. So, there is a need to use a less toxic fluorescent material for the detection of a tumor. Recently it was found that zinc sulfide dots are very useful for the purpose but they have less fluorescence than cadmium selenide. But when zinc sulfide is mixed with other materials, like lanthanide elements or manganese, their fluorescence property increases, which helps to identify human tumor cells in the body. Raman spectroscopy is an efficient technique for biological specimen identification. SERS increases the identification accuracy of the specimen. In order to detect tumors SERS is equipped with Au and Ag nanoparticles. By applying a bundled assembly of helical polymeric nanostructures laden with Pt drugs the therapeutic efficiency against pancreatic tumor has been increased.

13.6.5 Glucose Monitoring

For glucose sensing *in vivo* nanosensors can play a great role. For the detection of hypoglycemia in people with insulin-dependent diabetes glucose-sensing nanosensors have been created. For this purpose, fluorescent micro/nanoscale devices are prepared for glucose sensing. Achieving sensitive response nanotechnology with microcapsules permits control over catalytic properties. The presence of nanoparticles in the dermis allows transdermal monitoring of glucose changes in interstitial fluid. Noninvasive glucose sensing is an effective way to overcome biocompatibility problems of implants. Single-walled CNT can be used as glucose sensor [89]. Emission is modulated by it in response to adsorption of biomolecules. It works on signal transduction-fluorescence quenching and charge transfer. It is coated with glucose oxidase. It is the enzyme that breaks down the glucose molecule. An electron-deficient molecule called ferricyanide is then sprinkled over it. Ferricyanide makes the material to glow when excited by infrared light by moving electrons from the nanotubes. Higher levels of glucose lead to higher level of infrared fluorescence released by the nanotube. A tube containing oxidase and ferricyanide-coated nanotubes is implanted in human skin, (Figure 13.12) and the sensor can be excited with infrared light to detect fluorescence [90].

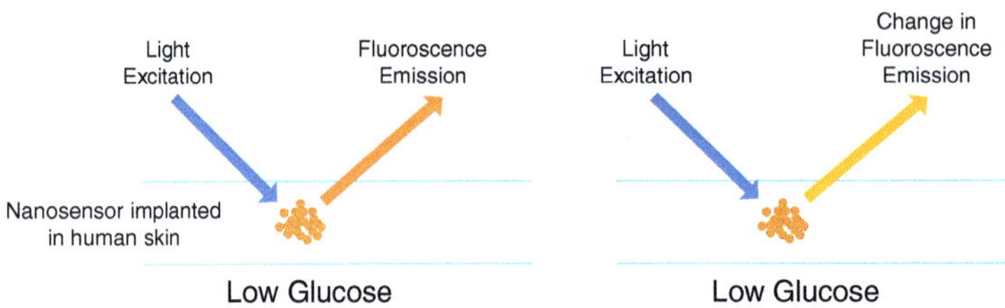

Figure 13.12 Behavior of nanosensor toward fluorescence, i.e. implanted in skin.

By immobilizing glucose oxidase onto a microcantilever surface, glucose concentrations can be also detected. In this method the microcantilever bends due to the variation of surface stress impacted by the reaction between glucose in solution and the glucose oxidase immobilized on the surface.

13.6.6 pH Sensing

Nanowire made devices can detect hydrogen-ion concentration or pH in a liquid. For this purpose, a basic p-type silicon nanowire device can be used. It is improved with 3-aminopropyltriethoxy silane that yields amino groups at the nanowire surface along with the naturally occurring silanol. The amino and silanol acts as receptors for hydrogen ions. It undergoes protonation/deprotonation reactions that change net nanowire surface. P-type silicon nanowire devices exhibit stepwise increase in conductance as the pH of the solution increases [91]. The presence of two distinct receptor groups leads to an increase in conductance with pH. This helps in the detection of pH. The response of a nanowire can also be tested by pH response. In this case a silanol group can function as a receptor for hydrogen ions. For chemically specific biosensing SERS has become a potential solution. This method is applied after cell lysis. It may rely upon endocytotic/phagocytotic uptake of noble metal nanoparticles. But phagocytosis and endocytosis can move SERS-activated colloids directly into the lysosomal pathway thus decreasing the biochemical information for the cell [92].

In addition, basic cell resolution and diffusion constraints similar to those that cause heterogeneous intracellular distribution of more traditional fluorescent dyes may limit cell areas in which nanoparticles or QD-based methods allow biochemically useful data to be obtained. In particular, heterogeneous intracellular nanoparticle distributions (e.g., heterogeneous fluorine distribution) can lead to severe results associated with nanoparticle concentration or nanoparticle aggregation rather than analyzer concentration. As an alternative to the dye and nanoparticle-based approach submicron SERS-active fiber-optic nanoprobes can be used for real-time intracellular biosensitivity. It was demonstrated previously the usefulness of glass-based fluorescence sensors of glass optics for intracellular chemical sensation. A SERS-based fiber-optic nanoprobe can be used for any chemical measurement in a single live mammal [93].

Among the existing fluorescence-based nanoprobes, SERS-based nanoprobes can be used to monitor intracellular conditions during biochemical/biomedical treatment, environmental exposure, apoptosis, and many other important biological processes. Unlike existing color-based and nanoparticle-based methods, the fiber-optic SERS nanoprobe allows simultaneous determination of intracellular and extracellular pH without major physical limitations imposed on cellular juxtaposition, intracellular diffusion, and lysosomal pathway. In addition, the spectral range of subnanometer novels and SERS peaks has the potential to allow strongly multiplexed *in vivo* analysis of unicellular nanobio [94].

13.6.7 Asthma Detection

By testing regularly asthma can be controlled in a similar way to testing blood sugar for diabetes, thus saving life. By using a handheld device to test the nitric oxide level in the breath an asthma attack can be detected early. The asthma patient can take appropriate action if they know level of nitric oxide in their breath [95], whether it be increasing or too high. The level will indicate the risk of asthma in the patient. For this purpose, a nanobiosensor can be used to detect asthma three weeks before an attack. The nanosensor is a polymer-coated nanotube field-effect transistor (NTFET) that contains a network of single-walled CNTs on a silicon oxide substrate between source and drain Au electrodes.

13.6.8 Cell Monitoring

Intracellular species with fluorescent labels are monitored by a cell monitoring process. It is an optical method for cell monitoring. Fiberoptics as well as nanosensors can be tested for sodium, potassium, glucose detection, etc. for use in mammalian cells [96]. The selectivity and signal-to-noise ratio is adequate for intracellular work, which overcomes the problem of photobleaching, leaching, fragility, and invasiveness. Cell monitoring is a potential application for nanosensors. For chemical analysis of single cells optical imaging fibers are implemented. The bunch of fibers consists of a number of single fibers. These fibers are drawn together in a coherent manner so that each fiber has an isolated optical signal from one end to another. Through a wet-etching process individual femtoliter-sized wells are formed at the distal tip of an optical-imaging fiber. The wells are fabricated in such a way that depth and diameter allow one cell to be accommodated in a well. NIH 3T3 mouse fibroblast cells, which are given fluorescence, are well equipped in an array of wells by the suspension of cells in the well and adhering to the bottom [97]. By exciting cell membranes at their required wavelength the pattern of the cell in the well can be found. Fluorescence measurements can be taken easily once the location of cell array is determined. By release or consumption of an analyte with a change in fluorescent intensity, the chemical environment of cell can be determined and monitored.

13.6.9 Microorganism Detection

Nanowire and nanotube materials can be used to sense viruses and bacteria. The nanotube has to be excited by attaching an antibody to it. Whenever a bacterium comes in contact with the nanotube and bonds with the antibody the conductance changes. Specifically, here the nanotube is attached to a metal contact detector. Then voltage is applied to the nanotube. When the bacteria or virus binds to the nanotube the current flow in the nanotube changes. It generates a signal. From the signal the presence of microorganism can be detected. This is believed to be a fast process for detecting microorganisms.

13.6.9.1 Detection of Bacteria

Detection of bacteria in the human body should be fast and more sensitive. This will lead to better and more rapid diagnosis. Most of the diagnosis processes are time consuming and less sensitive. Nanotechnology adds value to the diagnostic applications. Nanotechnology-based methods like use of ceramic nanospheres and magnetic nanoparticles are very effective for diagnostic purposes. A bacteria can be detected within 20 minutes if a bioconjugated nanoparticle-based bioassay is used. Due to high fluorescence the nanoparticle can be used for biorecognition of certain particles. It is been found that in comparison to absorbance-based method, the nanoparticle-based colorimetric assay enhances detection sensitivity fourfold [98]. Upon binding a bacterium to a silicon nitride cantilever surface the detection of a few *Salmonella* bacteria is possible as the surface stress changes. Using a scanning electron microscope it is clear that about 25 microorganisms suffice for the process of detection. Another nanotechnology-based process, sensing of phage-triggered ion cascade (SEPTIC), is used for detection of bacterium. A nanowell device with two antenna-like electrodes is employed to sense the electric-field fluctuations and in turn identifies the bacteria [99].

13.6.9.2 Detection of Viruses

Detection of virus should also be very fast and reliable. Transmission electron microscopy, plaque analysis, immunological assay, polymerase chain reaction (PCR)-based testing on viral nucleic acid etc. are some of the established methods for the detection of a single virus. But, for the detection of virus cells generally a larger sample is required. In addition, these methods are not fast in case of a single virus. This also requires high level of sample manipulation. Thus, these methods are not convenient. However, the use of a nanowire field-effect transistor for virus detection is convenient using real-time electrical detection of viruses.

For the detection of Influenza A nanowire arrays are used in which the virus binds or unbinds as the electrical conductivity changes. By using fluorescently labeled Influenza A the optical and electrical measurements can detect a single virus [100]. Further from pH-dependent study it is evident that nanowire sensors are used to determine the kinetics of virus–virus binding. When a virus binds to a receptor body on a nanowire sensor, the conductance of the sensor is changed from baseline value and again when virus unbinds the conductance returns to the baseline value. A high conductance change is observed when highly dilute Influenza A virus solution is fed to a p-type silicon nanowire. This indicates the binding and unbinding of virus to the receptor cell. As it is clear from electrical and optical data the trend is such that when the virus touches the nanowire the conductance is at the baseline value. As the virus starts diffusing into the nanowire, the conductance starts increasing rapidly changing from the baseline value. When virus starts diffusing from the nanowire again the conductance decreases to baseline value. This indicates that for the detection of virus it must be embedded on the nanowire surface to yield an electrical response. Development is in progress to develop ultra-dense nanowire device arrays to detect virus as small as possible. Recently, nanomechanical microcantilevers coated with antibodies and polycrystalline silicon were used to detect *Baculovirus* of varying concentrations [101].

13.6.10 Protein and DNA Detection

When specific receptors are linked, biological macromolecules such as nucleic acid and protein can be sensed. For example, the molecule biotin present in the p-type silicon nanosensor binds selectively with streptavidin molecule. The binding takes place at the oxide surface of nanosensor. When the molecule streptavidin comes in contact with the nanosensor surface and binds with it the conductance increases abruptly to a constant value. This conductance can be maintained by the addition of a buffer solution. This results in a small negative charge on streptavidin and the dissociation rate is very slow. The biotin surface on the nanosensor plays a role in determining the sensation of streptavidin. It is evident by blocking the streptavidin binding sites or by the addition of streptavidin to the nanosensor surface yields no result. It is evident that the detection of electrical entity is up to 10 pM, which is the level required for disease marker proteins [102]. From a number of research findings it is found that the binding method is highly specific and can detect macromolecules at a desired level. Hence, this method can be used as an efficient detection method. The detection for various bioproducts is illustrated in Figure 13.13.

The detection of single-stranded DNA is possible using a nanosensor containing a silicon nanowire. Single-stranded DNA is a negatively charged macromolecule. It can bind to p-type nanowires resulting in an increase in conductance. Peptide nucleic acids (PNA) are used as a receptor for DNA detection. So PNA act as the target material for DNA detection. PNA has a great affinity to DNA. P-type nanosensors embedded with PNA can sense DF508 mutation site in the cystic fibrosis transmembrane receptor gene. By the addition of a 60 fM wild-type DNA sample solution, the conductance in the nanosensor increases. As the negatively charged DNA continues binding with the receptor the negative surface charge density increases. Hence the conductance of the silicon nanowire keeps on

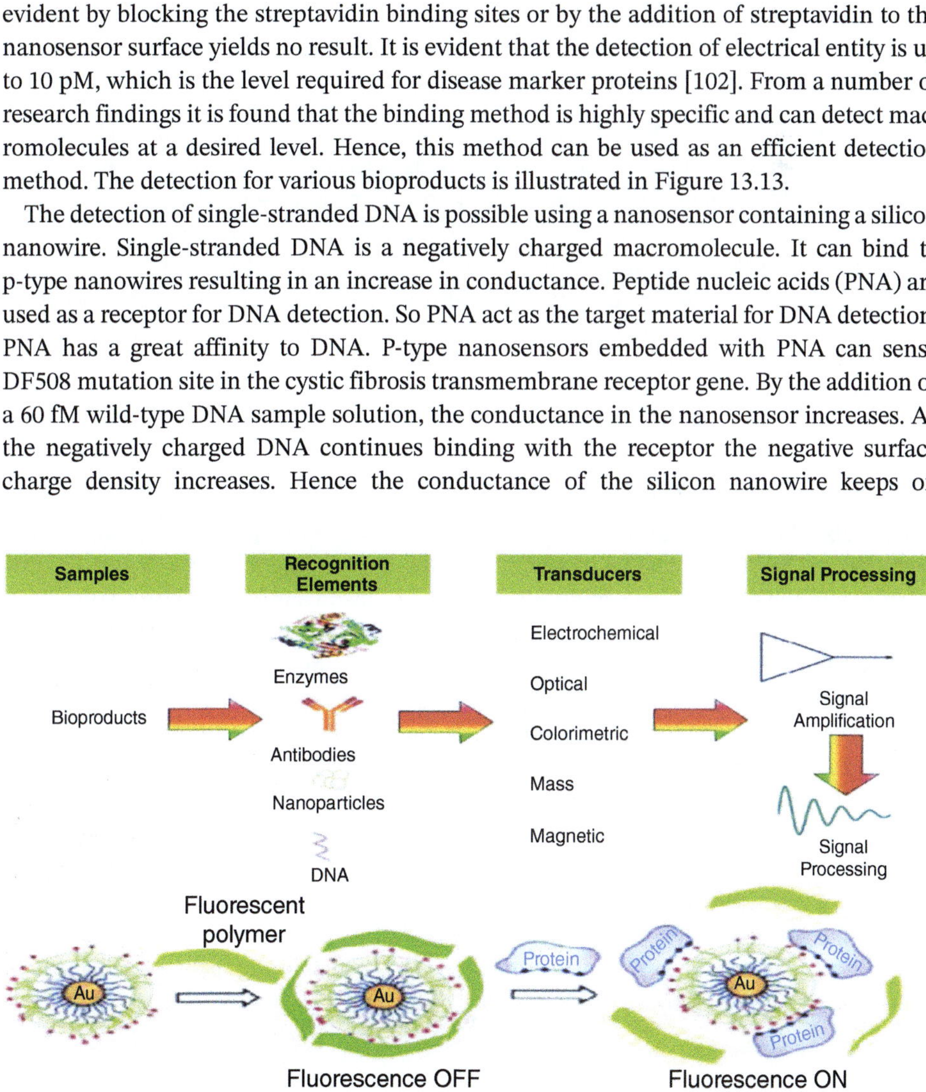

Figure 13.13 Protein and DNA detection in nanosensors.

increasing. The specificity of the sequence leads to the development of nanowire sensors for disease detection [103].

The concept of mechanical resonance on a microcantilever is also used for detection of a DNA molecule. In this method Au nanoparticles are used for detection. In this innovative process the concept is that a change in the mass of the cantilever by DNA molecules changes the resonance frequency of the microcantilever beam. This alteration is measured and calibrated to identify the quantity of DNA. Through the use of Au nanoparticles chemical amplification occurs on the microcantilever surface due to the nucleation of Ag. The process of sensation is limited to a concentration of 0.05 nM. This method is being used extensively for diagnostic purpose [104].

13.6.11 Contamination in Organic Implant

Nanosensors are used for detecting contamination in organic implants. Nanosensors can identify whether the implant is healthy and working perfectly. The sensor is embedded into the implant and is meant for the detection of contamination. Generally, the nanosensor works with piezo-electric effect to self-power. The sensor continuously sends an electrical signal to the outside world so that the provider can track whether the implant is okay or not. Analyzing on the electrical signal the healthcare service provider can know whether the cells are healthy or contaminated with bacteria [105]. However, in long-term use tissue grows on the implant, which restricts the sending of the electrical signal to the service provider. Thus, the effectiveness of the implant may deteriorate.

13.6.12 Astronaut's Diagnosis

Astronauts who work in an extra planetary environment in a spacecraft, as well as the ground-based personnel who work around a spacecraft, are potentially exposed to hazardous gases. In the spacecraft generally a hazardous gas, hydrazine, is used [106]. Spaceships often use hydrazine and other dangerous gases. So, it is essential to detect this kind of gas in the spaceship as the concentration of about tens of parts per billion is hazardous to humans. It, therefore, needs to be detected, identified, and quantified. Here nanosensors play a great role in detecting, identifying, and quantifying such gases in the working environment.

Another type of biosensor is derived from synthetic polymers called dendrines. They are less than 5 nm and are created layer by layer. These nanosensors can be passed through membranes to the white blood cell lymphocytes. Through biochemical changes the sensors can detect early infection or radiation damage in astronauts. Astronauts are at high risk of cancer due to cell damage when on space missions. Due to their very small size, bionanosensors can be introduced to the body through the skin. The use of nanosensors eliminates the need of blood samples for testing, intravenous drips would be avoided during space missions [107].

13.6.13 Other Diagnosis

For the detection of biomarkers in human blood serum a hybrid ZnO nanorod is used. ZnO nanorods enhance the fluorescence property of the material. The nanorod is grown on a glass substrate over which a polymer brush can grow. A nanobrush like "POEGMA-co-GMA" retains the higher intensity of the antibodies in the body. The range of detection of the biomarker is limited to $100\,fg\,ml^{-1}$ in the serum [108]. The process of biomarker detection is shown in Figure 13.14. Graphene oxide-based Au nanoparticles are used for the colorimetric detection of uric acid in human serum. This is carried out using a paper-based platform. It has a high sensitivity of 4 ppm and provides quick results within five minutes [109].

Nanocantilevers are, currently, extensively used in the diagnostic applications. Nanocantilevers require a less concentrated analytes and have higher sensitivity. They have faster response times and consume less power. The nanocantilevers detect analytes with higher sensitivity when surface functionalization of these cantilevers is incorporated using a bottom-up approach. This case ensures the use of nanocantilevers for diagnostic applications in the future. DNA machines are being developed for the diagnosis of antibodies. When an antibody embeds with the DNA machine structural change occurs. The quencher and light-emitting fluorophore are attached to the single-stranded DNA stems. Recognition elements like peptides and oligonucleotides capable of binding the molecules are attached to the DNA stem [110]. As the recognition elements bind the molecule the DNA structure breaks and separates into two parts. Light-emitting fluorophore and quencher are separated by generating a light signal. DNA nanomachines are very efficient and comparatively more accurate than micromachines.

During blood transfusion in patients some safety measures are taken. According to the guidelines it is mandatory to determine the blood products and blood type of the patient. Virus contamination of the blood must also to be checked – all these things can be executed by diagnostic nanosensors. The diagnostic nanosensors are equipped with bioorganic interactive materials that can determine blood type and contamination in the blood. If bioorganic interactive nanomaterials are incorporated with some self luminescence nanoparticles the interaction can be visualized by the naked eye. The diagnostic nanosensors can be incorporated in a pipeline through which blood is collected. Also, the nanosensor can be dipped into the blood bag, thus blood type and contamination can be detected, thereby, any extra equipment for detection of contamination of virus is eliminated. To determine the glucose homeostasis or coagulation in the patient the serum component in the blood is

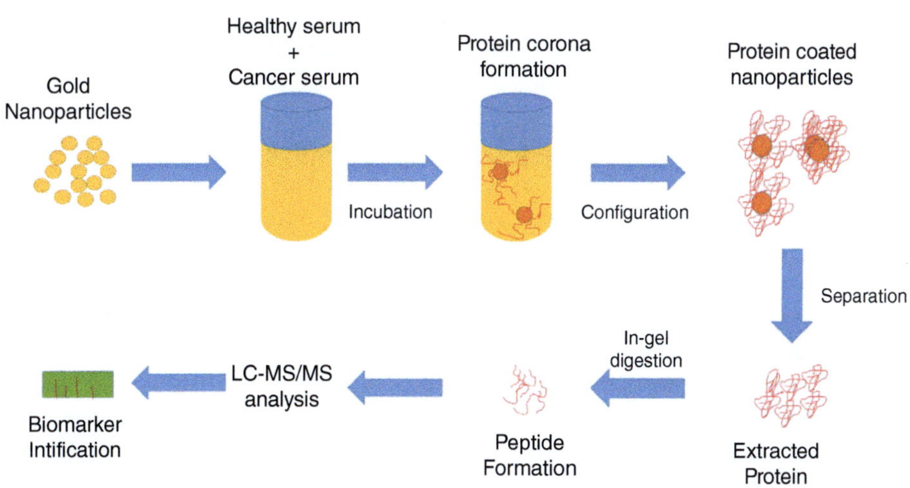

Figure 13.14 Process of biomarker identification.

determined quantitatively. In the same way the diagnostic nanosensors are equipped with the bioorganic interactive material, which enables identification of glucose or components of the coagulation system. When glucose or components of the coagulation system are embedded in to the nanostructured array the plasmon resonance frequencies on the surface of nanosensor change. Thus, the effect enables the measurement of glucose or components of the coagulation system quantitatively. This process can be conducted on a hand-held photometer or spectral photometer.

13.6.14 Other Examples of Nanosensor Applications

Potential difference in a biological cell wall membrane can be detected by a nanoscale sensor and fluorescence. This type of nanosensor is known as an optical voltage nanosensor, which is a noncontact sensing process. It is based on fast resonance energy transfer. A gel-based nanosensor can be used for radiotherapy. This is quite helpful for oncologist and physicists. This also helps in reducing overexposures and also reduces radiation-induced toxicity. These kind of plasmonic nanosensor gels, compared to conventional sensors, indicate the exposure level of radiation when radiation to the hydrogel increases. Wireless graphene nanosensors can be interfaced with biomaterials by silk bioresorption. In order to monitor harmful oral bacteria graphene nanosensor tattoos on teeth can be utilized too [111].

13.7 Regulatory Issues

Nanosciences and nanotechnologies cover a wide range of research institutions, industrial sectors, and applications. Therefore, it is likely that several regulators will need to consider the impact that nanosensors have in each of their areas. It is believed that flexible and proportionate regulatory measures data are beneficial for all: the public, consumers, and employees are protected from harm, while industry can participate in the development of standards and make recommendations to ensure a level playing field and reduce the risk of liability. Regulation requires an assessment of the hazards and an assessment of the probability or duration of exposure. The overall objective is to identify risk management measures necessary to eliminate risks or reduce them to an acceptable level. When possible, evidence is reported, usually derived from toxicological, environmental, or epidemiological studies. The precautionary principle comes into force when there is no complete scientific certainty about the risk of harm of the substance. The potential hazard must then be assumed on the basis of available evidence and best judgments about the hazardous properties of the substance should be drawn. An exposure risk assessment is then required, for example, in the workplace or for the general public from the use of the products.

The need to control the use of hazardous substances to prevent harm to humans is essential. Only those substances that pose the most serious risks to health or the environment, such as certain carcinogens, are prohibited. Extensive national and European legislation already covers various aspects of the use of hazardous substances. In addition, various international agreements have been developed to control the global aspects of the problem. There are several regulatory options available:

- workplace controls;
- control of emissions to air, water, and land;
- waste disposal restrictions;
- classification and labelling measures;
- marketing and use restrictions; and
- prohibition.

Research is underway to introduce nanosensors into medical diagnosis and treatment. Although such materials would be subject to a strict regulatory regime that regulates all new interventions in medicine, the special properties of nanosensors suggest the possibility of unforeseen toxicity if they are ingested in large quantities. So it is imperative that particle size and chemistry are taken into account when examining potential adverse drug side effects. Regulations that are particularly important for nanosensors are regulations on the control of substances hazardous to health (COSHH), which establishes general requirements for the reduction of occupational ill health by establishing a simple framework for the control of hazardous substances in the workplace. Concerns have been expressed about the manufacturing of nanotubes and nanoparticles and potential risk for laborers involved in the production process. For individual substances personal exposure limits are regulated by compliance with occupational exposure limits (OELs). These are determined separately, reviewed, and adjusted in the light of new knowledge.

13.8 Conclusions

This chapter discussed nanosensors and their application in medicine and healthcare. By incorporating nanotechnology in nanosensors it was found that the effectiveness and convenience increases. In turn when they are used in the healthcare sector they provide easy access and diagnosis in humans. As discussed, nanosensors can be used for glucose monitoring, cancer detection, tumor detection, etc. The global-market share of different nanosensor applications is given in Figure 13.15.

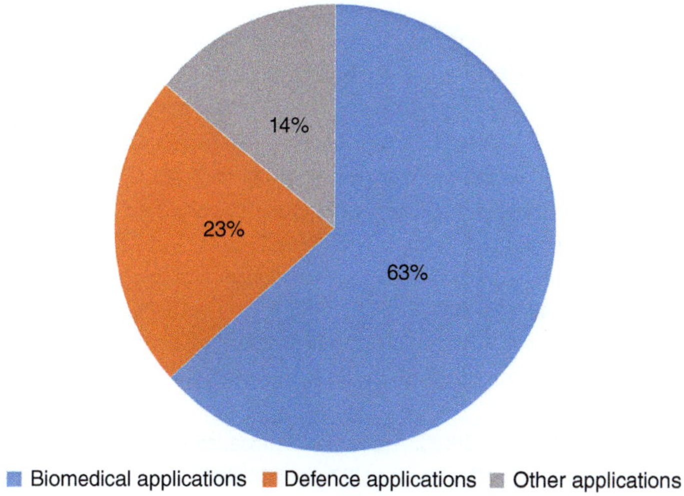

Figure 13.15 Share of nanosensor applications in global market.

Many advancements are also incorporated in nanosensors for broad applications in the health sector. The chapter discussed various advancements in this sector. There is currently much focus and extensive investigation into nanosensors and their uses. It is clear that the world will witness much more advancement in field of nanosensors and an increase in applications is also very likely.

References

1 Munawar, A., Ong, Y., Schirhagl, R. et al. (2019). Nanosensors for diagnosis with optical, electric and mechanical transducers. *RSC Advances* (12): 6793–6803. https://doi.org/10.1039/C8RA10144B.

2 Gnanamani, S., Chidambaram, S., and Prabaharan, M. (2011). Prospects of nanosensors in environmental and biomedical fields in smart nanomaterials for sensor applications. In: *Smart Nanomaterials for Sensor Application* (ed. L. Songiun, Y. Li and H. Li), 82–92. Bentham Science.

3 Lu, J. and Bowles, M. (2013). How will Nano-sensor technology affect logistics management. *Journal of Convergence Information Technology* 8 (2): 820–828. https://doi.org/10.4156/jcit.vol8.issue2.98.

4 Patra, J.K., Das, G., Fraceto, L.F. et al. (2018). Nano based drug delivery systems: recent developments and future prospects. *Journal of Nanobiotechnology* 16: 71. http://doi.org/10.1186/s12951-018-0392-8.

5 Van Duyne, R.P., Glucksberg, M.R., Peltier, K.E. et al. (2002). Surface-enhanced raman nanobiosensor, Patent application no. US8592226B2US8592226B2.

6 Ben-Barak, N., Chang, D.M., Gabriel, J.-C.P. et al. (2002). Nanoelectronic measurement system for physiologic gases and improved nanosensor for carbon dioxide. Patent application no. US7714398B2.

7 Chang, D., Chang, Y.-L., Gabriel, J.-C., et al. (2002). Carbon dioxide nanosensor, and respiratory CO2 monitors. Patent application no. US20070048181A1.

8 Gabriel, J.-C.P., Joshi, V., Passmore, J.L. et al. (2002). Anesthesia monitor, capacitance nanosensors and dynamic sensor sampling method. Patent application no. US20100085067A1.

9 Adam, P.M., Bijeon, J.-L., Chaton, P. and (2003). Microsensors and nanosensors for chemical and biological species with surface plasmons. Patent application WO2005033335A3.

10 Brennan, M.E., Armstrong, G.J., Kelly, J.M. et al. (2003). Sensor for detecting an analyte using silver nanoparticles. Patent application no. WO2004086044A1 086044A1.

11 Poponin, V. (2004). Optical sensor with layered plasmon structure for enhanced detection of chemical groups by SERS. Patent application no. US7351588B2.

12 Pison, U., Giersig, M., and Schaefer, A., (2005). Diagnostic nanosensor and its use in medicine. Patent application no. US8846580B2.

13 Gabriel, J.-C., Joshi, V., Passmore, J., et al. (2005). Anesthesia monitor, capacitance nanosensors and dynamic sensor sampling method. Patent application no. US20080021339A1.

14 Briman, M.l, Bryant, C., Chang, Y.-L. et al. (2005). Ammonia Nanosensors, and Environmental Control System. Patent application no.US20160123947A1.

15 Lowery, T.J., Koziarz, J.J., Levinson, D.A. et al. (2008). Magnetic nanosensor compositions and bioanalytical assays therefor. Patent application no WO2010002479A1.

16 Thompson, M.E., Zhou, C., Cote, R.J., et al. (2009). Surface modification of nanosensor platforms to increase sensitivity and reproducibility. Patent application no. WO2010115143A1.

17 Gouma, P., Kalyanasundaram, K., Yun, X. et al. (Jan. 2010). Nanosensor and breath analyzer for ammonia detection in exhaled human breath. *IEEE Sensors Journal* 10 (1): 49–53. https://doi.org/10.1109/JSEN.2009.2036050.

18 Strano, M.S., Kim, J.-H., Zhang, J.g. and Heller, D.A. (2009). Optical nanosensors comprising photoluminescent nanostructures. Patent application no. US8486709B2.

19 Novotny, V.J. (2010). Chemical and Biomedical NanoSensors. Patent application no. US20120304776A1.

20 Cai, D., Ren, L., Zhao, H. et al. (2010). A molecular-imprint nanosensor for ultrasensitive detection of proteins. *Nature Nanotechnology* 5: 597–601. https://doi.org/10.1038/nnano.2010.114.

21 Naser, S., Perez, J.M., and Kaittanis, C. (2011). Microbe detection via hybridizing magnetic relaxation nanosensors. Patent application no. CA2839865A1.

22 McAlpine, M.C. and Mannoor, M.S. (2011). Use and making of biosensors utilizing antimicrobial peptides for highly sensitive biological monitoring. Patent application no. US20160223538A1.

23 Barcelo, S.J., Lee, Z., Gold, A., and Gibson, G. (2012). Implantable nanosensor. Patent application no. CN104487824A.

24 Clark, H.A. and Cash, K.J. (2012). Enzymatic nanosensor compositions and methods. Patent application no. EP2823051A4.

25 Desai, U.R., Kalita, M., and Balagurunathan, K. (2013). Nanosensor for detecting the activity of glycosaminoglycan-cleaving enzymes and uses thereof. Patent application no. US20160201112A1.

26 Kam, K., Schwartz, J.J., Demas, V., and Bajaj, V.S. (2014). Analyte detection system and method for intradermal implantation of biocompatible optode nanosensors. Patent application no. US9974471B1.

27 McNicholas, T.P., Ahn, J., and Strano, M.S. (2014). Saccharide responsive optical nanosensors. Patent application no. US20170131287A1.

28 Lin, Q., Zhu, Y., Shang, J. et al. (2014). Graphene-based nanosensor for identifying target analytes. Patent application no. US20170181669A1.

29 Lin, Q., Wang, X., Yang, J., and Cheng, W. (2015). Microfluidic aptasensor including a graphene nanosensor. Patent application no. WO2016112079A1.

30 National Center for Biotechnology Information (2015). Porous microneedles through sacrificial sugar incorporation, analyte detection system, and method for intradermal optode nanosensor implantation PubChem Patent, US10098574. https://pubchem.ncbi.nlm.nih.gov/patent/US10098574.

31 Broderick, P.A. (2016). Noninvasive electroactive protonic protein nanosensor with polymer photovoltaic optics for memory transduction using organic and inorganic elements. Patent application no. WO2017117223A1.

32 Botero, J. and Tejada, M. (2018). Field-effect nanosensor for detecting simple metabolites in living organisms. Patent application no. WO/2020/039377.

33 Crapnell, R.D., Hudson, A., Foster, C.W. et al. (2019). Recent advances in electrosynthesized molecularly imprinted polymer sensing platforms for bioanalyte detection. *Sensors (Basel)* 19 (5): 1204. https://doi.org/10.3390/s19051204.

34 Galal, A. and Hesselbach, X. (2018). Nano-networks communication architecture: modeling and functions. *Nano Communication Networks* 17: 45–62. https://doi.org/10.1016/j.nancom.2018.07.001.

35 Randa, A.-K., Reda, Y., and Abdel-Fattah, A. (2020) Review – nanostructured materials-based nanosensors. *Journal of the Electrochemical Society*, 167: 037554. https://doi.org/10.1149/1945-7111/ab67aa.

36 Conde, J., Rosa, J., Lima, J.C., and Baptista, P.V. Nanophotonics for molecular diagnostics and therapy applications. *Hindawi Publishing Corporation International Journal of Photoenergy* 2012: 619530, 11 pages https://doi.org/10.1155/2012/619530.

37 Fang, M., Peng, C.W., Pang, D.W., and Li, Y. (2012). Quantum dots for cancer research: current status, remaining issues, and future perspectives. *Cancer Biology & Medicine* 9 (3): 151–163. https://doi.org/10.7497/j.issn.2095-3941.2012.03.001.

38 Angelov, G.V., Nikolakov, D.P., Ruskova, I.N. et al. (2019). Healthcare sensing and monitoring. In: *Enhanced Living Environments. Lecture Notes in Computer Science*, vol. 11369 (ed. I. Ganchev, N. Garcia, C. Dobre, et al.), 226–262. Cham: Springer https://doi.org/10.1007/978-3-030-10752-9_10.

39 Vo-Dinh, T., Cullum, B., and Stokes, D. (2001). Nanosensors and biochips: Frontiers in biomolecular diagnostics. *Sensors and Actuators B: Chemical* 74: 2–11. https://doi.org/10.1016/S0925-4005(00)00705-X.

40 Hong, Y., Huh, Y.-M., Yoon, D.S., and Yang, J. (2012). Nanobiosensors based on localized surface plasmon resonance for biomarker detection. 2012: 759830, 13 pages, https://doi.org/10.1155/2012/759830.

41 Chen, H., Liu, H., Zhang, Z. et al. (2015). Nanostructured photodetectors: from ultraviolet to terahertz. *Advanced Materials* https://doi.org/10.1002/adma.201503534.

42 Binh, V.T., Garcia, N., and Levanuyk, A.L. (1994). A mechanical nanosensor in the gigahertz range: where mechanics meets electronics. *Surface Science* 301: L224–L228.

43 Sippel-Oakley, J., Wang, H.-T., Kang, B.S. et al. (2005). Carbon nanotube films for room temperature hydrogen sensing. *Nanotechnology* 16: 2218–2221.

44 Mhlongo, G., Motaung, D., Cummings, F. et al. (2019). A highly responsive NH_3 sensor based on Pd-loaded ZnO nanoparticles prepared via a chemical precipitation approach. *Scientific Reports* 9: 9881. https://doi.org/10.1038/s41598-019-46247-z.

45 Quaranta, M., Borisov, S.M., and Klimant, I. (2012). Indicators for optical oxygen sensors. *Bioanalytical Reviews* 4 (2–4): 115–157. https://doi.org/10.1007/s12566-012-0032-y.

46 Vo-Dinh, T., Scaffidi, J., Gregas, M. et al. (2009). Applications of fiber-optics-based nanosensors to drug discovery. *Expert Opinion on Drug Discovery* 4 (8): 889–900. https://doi.org/10.1517/17460440903085112.

47 Geng, J., Thomas, M., Shephard, D., and Johnson, B. (2005). Suppressed electron hopping in a Au nanoparticle/H_2S system: development towards a H_2S nanosensor. *Chemical Communications (Cambridge, England)* 1895–1897. https://doi.org/10.1039/b418559e.

48 Perez, J.M., Josephson, L., and Weissleder, R. (2004). Use of magnetic nanoparticles as nanosensors to probe for molecular interactions. *ChemBioChem* 5: 261–264. https://doi.org/10.1002/cbic.200300730.

49 Bhalla, N., Jolly, P., Formisano, N., and Estrela, P. (2016). Introduction to biosensors. *Essays in Biochemistry* 60 (1): 1–8. https://doi.org/10.1042/EBC20150001.

50 Zhou, W., Gao, X., Liu, D., and Chen, X. (2015). Gold nanoparticles for in vitro diagnostics. *Chemical Reviews* 115 (19): 10575–10636. https://doi.org/10.1021/acs.chemrev.5b00100.

51 Konvalina, G. and Haick, H. (2014). Sensors for breath testing: from nanomaterials to comprehensive disease detection. *Accounts of Chemical Research* 47 (1): 66–76. https://doi.org/10.1021/ar400070m.

52 Cao, X., Yin, Z., and Zhang, H. (2014). Three-dimensional graphene materials: preparation, structures and application in supercapacitors. *Energy and Environmental Science* 7: 1850–1865.

53 Rassaei, L., Amiri, M., Cirtiu, C.M. et al. (2011). Nanoparticles in electrochemical sensors for environmental monitoring. *TrAC Trends in Analytical Chemistry* 30: 1704–1715.

54 Yap, M.H., Fow, K.L., and Chen, G.Z. (2017). Synthesis and applications of MOF-derived porous nanostructures. *Green Energy & Environment* 2 (3): 218–245, ISSN 2468-0257, https://doi.org/10.1016/j.gee.2017.05.003.

55 Haddock, H., Shankar, P.M., and Mutharasan, R. (2003). Fabrication of biconical tapered optical fibers using hydrofluoric acid. *Materials Science and Engineering B* 97: 87–93. https://doi.org/10.1016/S0921-5107(02)00434-8.

56 Stöckle, R., Fokas, C., Deckert, V. et al. (1999). High-quality near-field optical probes by tube etching. *Applied Physics Letters* 75: 160–162. https://doi.org/10.1063/1.124305.

57 Kumar, S., Bhushan, P., and Bhattacharya, S. (2018). Fabrication of nanostructures with bottom-up approach and their utility in diagnostics, therapeutics, and others. In: *Environmental, Chemical and Medical Sensors. Energy, Environment, and Sustainability* (ed. S. Bhattacharya, A. Agarwal, N. Chanda, et al.), 167–198. Singapore: Springer https://doi.org/10.1007/978-981-10-7751-7_8.

58 Doll, G.L., Mensah, B.A., Mohseni, H. et al. (2010). Chemical vapor deposition and atomic layer deposition of coatings for mechanical applications. *Journal of Thermal Spray Technology* 19: 510–516. https://doi.org/10.1007/s11666-009-9364-8.

59 Arthur, J. (2002). Molecular beam epitaxy. *Surface Science* 500: 189–217. https://doi.org/10.1016/S0039-6028(01)01525-4.

60 Liyanage, T., Rael, A., Shaffer, S. et al. (2018). Fabrication of a self-assembled and flexible SERS nanosensor for explosive detection at parts-per-quadrillion levels from fingerprints. *Analyst* 143 (9): 2012–2022.

61 Wang, Y. and Yeow, J.T.W. A review of carbon nanotubes-based gas sensors, nanomaterials for chemical sensing technologies. *Journal of Sensors, Hindawi Publication* 2009: 493904, 24 pages, https://doi.org/10.1155/2009/493904.

62 Mabrouk, M., Rajendran, R., Soliman, I.E. et al. (2019). Nanoparticle- and nanoporous-membrane-mediated delivery of therapeutics. *Pharmaceutics* 11 (6): 294. Published 2019 Jun 21. https://doi.org/10.3390/pharmaceutics11060294.

63 Parviz, M. (2018). Mesoporous silica nanoparticles: synthesis, modification and applications. *Nanomedicine Nanotechnology* 3 (2): 000136.

64 Casares, D., Escribá, P.V., and Rosselló, C.A. (2019). Membrane lipid composition: effect on membrane and organelle structure, function and compartmentalization and therapeutic avenues. *International Journal of Molecular Sciences* 20 (9): 2167. https://doi.org/10.3390/ijms20092167.

65 Jafari, S., Derakhshankhah, H., Alaei, L. et al. (2019). Mesoporous silica nanoparticles for therapeutic/diagnostic applications. *Biomedicine & Pharmacotherapy* 109: 1100–1111. ISSN 0753-3322, https://doi.org/10.1016/j.biopha.2018.10.167.

66 Bharti, C., Nagaich, U., Pal, A.K., and Gulati, N. (2015). Mesoporous silica nanoparticles in target drug delivery system: a review. *International Journal of Pharmaceutical Investigation* 5 (3): 124–133. https://doi.org/10.4103/2230-973X.160844.

67 Narayan, R., Nayak, U.Y., Raichur, A.M., and Garg, S. (2018). Mesoporous silica nanoparticles: a comprehensive review on synthesis and recent advances. *Pharmaceutics* 10 (3): 118. https://doi.org/10.3390/pharmaceutics10030118.

68 Ansari, A.A., Alhoshan, M., Alsalhi, M.S., and Aldwayyan, A.S. (2010). Prospects of nanotechnology in clinical immunodiagnostics. *Sensors* 10 (7): 6535–6581. https://doi.org/10.3390/s100706535.

69 Lee, S.H. and Jun, B.H. (2019). Silver nanoparticles: synthesis and application for nanomedicine. *International Journal of Molecular Sciences* 20 (4): 865. https://doi.org/10.3390/ijms20040865.

70 Maduraiveeran, G. and Jin, W. (2017). Nanomaterials based electrochemical sensor and biosensor platforms for environmental applications. *Trends in Environmental Analytical Chemistry* 13: 10–23. ISSN 2214-1588, https://doi.org/10.1016/j.teac.2017.02.001.

71 Khoang, N.D., Trung, D., Duy, N. et al. (2012). Design of SnO_2/ZnO hierarchical nanostructures for enhanced ethanol gas-sensing performance. *Sensors and Actuators B: Chemical* 174: 594–601. https://doi.org/10.1016/j.snb.2012.07.118.

72 Pan, S., Lu, W., Chu, Z. et al. (2015). Deep ultraviolet emission from water-soluble SnO_2 quantum dots grown via a facile 'top-down' strategy. *Journal of Materials Science and Technology* 31 (6): 670–673.

73 Dirksen, J. and Duval, K. (2001). NiO thin-film formaldehyde gas sensor. *Sensors and Actuators B: Chemical* 80: 106–115. https://doi.org/10.1016/S0925-4005(01)00898-X.

74 Topoglidis, E., Campbell, C., Cass, T., and Durrant, J. .R. (2006). Nitric oxide biosensors based on the immobilization of hemoglobin on mesoporous titania electrodes. *Electroanalysis* 18: 882–887. https://doi.org/10.1002/elan.200503473.

75 Zhang, Y. and Wang, T.H. (2012). Quantum dot enabled molecular sensing and diagnostics. *Theranostics* 2 (7): 631–654. https://doi.org/10.7150/thno.4308.

76 Sedlmeier, A., Achatz, D., Fischer, L. et al. (2012). Photon upconverting nanoparticles for luminescent sensing of temperature. *Nanoscale* 4: 7090–7096. https://doi.org/10.1039/c2nr32314a.

77 Polívková, M., Hubáček, T., Staszek, M. et al. (2017). Antimicrobial treatment of polymeric medical devices by silver nanomaterials and related technology. *International Journal of Molecular Sciences* 18 (2): 419. http://doi.org/10.3390/ijms18020419.

78 Sabela, M., Mpanza, T., Kanchi, S. et al. (2016). Electrochemical sensing platform amplified with a nanobiocomposite of L-phenylalanine ammonia-lyase enzyme for the detection of capsaicin. *Biosensors and Bioelectronics* 83: 45–53. https://doi.org/10.1016/j.bios.2016.04.037.

79 Li, Z., Fu, Y., Fang, W., and Li, Y. (2015). Electrochemical impedance Immunosensor based on self-assembled monolayers for rapid detection of *Escherichia coli* O157:H7 with signal amplification using lectin. *Sensors (Switzerland)* 15: 19212–19224. https://doi.org/10.3390/s150819212.

80 Xu, X., Liu, Y., Fu, W. et al. (2020). Poly(N-isopropylacrylamide)-based thermoresponsive composite hydrogels for biomedical applications. *Polymers* 12: 580.

81 Paul, M.K. and Mukhopadhyay, A.K. (2004). Tyrosine kinase – role and significance in cancer. *International Journal of Medical Sciences* 1 (2): 101–115. https://doi.org/10.7150/ijms.1.101.

82 Wang, L., Hu, C., and Shao, L. (2017). The antimicrobial activity of nanoparticles: present situation and prospects for the future. *International Journal of Nanomedicine* 12: 1227–1249. http://doi.org/10.2147/IJN.S121956.

83 Egorova, K.S., Gordeev, E.G., and Ananikov, V.P. (2017, 2017). Biological activity of ionic liquids and their application in pharmaceutics and medicine. *Chemical Reviews* 117 (10): 7132–7189.

84 Parrish, A.B., Freel, C.D., and Kornbluth, S. (2013). Cellular mechanisms controlling caspase activation and function. *Cold Spring Harbor Perspectives in Biology* 5 (6): a008672. https://doi.org/10.1101/cshperspect.a008672.

85 Kasili, P.M., Song, J.M., and Vo-Dinh, T. (2004). Optical sensor for the detection of caspase-9 activity in a single cell. *Journal of the American Chemical Society* 126 (9): 2799–2806. https://doi.org/10.1021/ja037388t.

86 Poreba, M., Strózyk, A., Salvesen, G.S., and Drag, M. (2013). Caspase substrates and inhibitors. *Cold Spring Harbor Perspectives in Biology* 5 (8): a008680. Published 2013 Aug 1. https://doi.org/10.1101/cshperspect.a008680.

87 Siddiqi, K.S., Ur Rahman, A., and Tajuddin, H.A. (2018). Properties of zinc oxide nanoparticles and their activity against microbes. *Nanoscale Research Letters* 13 (1): 141. https://doi.org/10.1186/s11671-018-2532-3.

88 Miyashita, M., Gonda, K., Tada, H. et al. (2016). Quantitative diagnosis of HER2 protein expressing breast cancer by single-particle quantum dot imaging. *Cancer Medicine* 5 (10): 2813–2824. https://doi.org/10.1002/cam4.898.

89 Barone, P.W. and Strano, M.S. (2009). Single walled carbon nanotubes as reporters for the optical detection of glucose. *Journal of Diabetes Science and Technology* 3 (2): 242–252. Published 2009 Mar 1. https://doi.org/10.1177/193229680900300204.

90 Klonoff, D.C. (2012). Overview of fluorescence glucose sensing: a technology with a bright future. *Journal of Diabetes Science and Technology* 6 (6): 1242–1250. Published 2012 Nov 1. https://doi.org/10.1177/193229681200600602.

91 Patolsky, F. and Lieber, C.M. (2005). Nanowire nanosensors. *Materials Today* 8 (4): 20–28. ISSN 1369-7021, https://doi.org/10.1016/S1369-7021(05)00791-1.

92 Behzadi, S., Serpooshan, V., Tao, W. et al. (2017). Cellular uptake of nanoparticles: journey inside the cell. *Chemical Society Reviews* 46 (14): 4218–4244. https://doi.org/10.1039/c6cs00636a.

93 Lussier, F., Brule, T., Vishwakarma, M. et al. (2016). Dynamic-SERS Optophysiology: A nanosensor for monitoring cell secretion events. *Nano Letters* 16 (6): 3866–3871.

94 Scaffidi, J.P., Gregas, M.K., Seewaldt, V., and Vo-Dinh, T. (2009). SERS-based plasmonic nanobiosensing in single living cells. *Analytical and Bioanalytical Chemistry* 393 (4): 1135–1141. https://doi.org/10.1007/s00216-008-2521-y.

95 Aytekin, M. and Dweik, R.A. (2012). Nitric oxide and asthma severity: towards a better understanding of asthma phenotypes. *Clinical and Experimental Allergy* 42 (5): 614–616. https://doi.org/10.1111/j.1365-2222.2012.03976.x.

96 Cash, K.J. and Clark, H.A. (2010). Nanosensors and nanomaterials for monitoring glucose in diabetes. *Trends in Molecular Medicine* 16 (12): 584–593. https://doi.org/10.1016/j.molmed.2010.08.002.

97 Rejmontová, P., Capáková, Z., Mikušová, N. et al. (2016). Adhesion, proliferation and migration of NIH/3T3 cells on modified polyaniline surfaces. *International Journal of Molecular Sciences* 17 (9): 1439. https://doi.org/10.3390/ijms17091439.

98 Chang, C.C., Chen, C.P., Wu, T.H. et al. (2019). Gold nanoparticle-based colorimetric strategies for chemical and biological sensing applications. *Nanomaterials (Basel)* 9 (6): 861. doi: 10.3390/nano9060861.

99 Kish, L., Dobozi-King, M., Young, R. et al. (2006). Sensing phage-triggered ion cascade (SEPTIC), Patent application no. WO / 2006/058263.

100 Patolsky, F., Zheng, G., Hayden, O. et al. (2004). Electrical detection of single viruses. *Proceedings of the National Academy of Sciences of the United States of America* 101: 14017–14022. https://doi.org/10.1073/pnas.0406159101.

101 Ilic, B., Yang, Y., and Craighead, H.G. (2004). Virus detection using nanoelectromechanical devices. *Applied Physics Letters* 85: 2604–2606. https://doi.org/10.1063/1.1794378.

102 Rusling, J.F., Kumar, C.V., Gutkind, J.S., and Patel, V. (2010). Measurement of biomarker proteins for point-of-care early detection and monitoring of cancer. *Analyst* 135 (10): 2496–2511. https://doi.org/10.1039/c0an00204f.

103 Rashid, J.I.A., Abdullah, J., Yusof, N.A., and Hajian, R. The development of silicon nanowire as sensing material and its applications. 2013: 16. http://doi.org/10.1155/2013/328093.

104 Cordeiro, M., Ferreira Carlos, F., Pedrosa, P. et al. (2016). Gold nanoparticles for diagnostics: advances towards points of care. *Diagnostics (Basel)* 6 (4): 43. doi: 10.3390/diagnostics6040043.

105 Majumder, S., Mondal, T., and Deen, M.J. (2017). Wearable sensors for remote health monitoring. *Sensors (Basel)* 17, 130 (1): https://doi.org/10.3390/s17010130.

106 Wikipedia (2022). Hydrazine https://en.wikipedia.org/wiki/Hydrazine (access date 3 February 2022)

107 Swierczewska, M., Liu, G., Lee, S., and Chen, X. (2012). High-sensitivity nanosensors for biomarker detection. *Chemical Society Reviews* 41 (7): 2641–2655. https://doi.org/10.1039/c1cs15238f.

108 Kumar, S., Bhushan, P., and Bhattacharya, S. (2017). Fabrication of nanostructures with bottom-up approach and their utility in diagnostics, therapeutics, and others. In: *Environmental, Chemical and Medical Sensors* (ed. S. Bhattacharya, A.K. Agarwal, N. Chanden, et al.), 167–198. Springer http://doi.org/10.1007/978-981-10-7751-7_8.

109 Kumar, S., Bhushan, P., and Bhattacharya, S. (2016). Development of a paper-based analytical device for colorimetric detection of uric acid using gold nanoparticles–graphene oxide (AuNPs–GO) conjugates. *Analytical Methods* 8: 6965–6973. https://doi.org/10.1039/C6AY01926A.

110 Ranallo, S., Rossetti, M., Plaxco, K.W. et al. (2015). A modular, DNA-based beacon for single-step fluorescence detection of antibodies and other proteins. *Angewandte Chemie International Edition* 54 (45): 13214–13218. https://doi.org/10.1002/anie.201505179.

111 Mannoor, M.S., Tao, H., Clayton, J. et al. (2012). Corrigendum: graphene-based wireless bacteria detection on tooth enamel. *Nature Communications* 3: 763. https://doi.org/10.1038/ncomms1767.

14

Inorganic Nanoparticles for Drug-delivery Applications

Chinnu Sabu, V.K. Ameena Shirin, Renu Sankar, and K. Pramod

College of Pharmaceutical Sciences, Government Medical College, Kozhikode, Kerala, India

14.1 Introduction

Recently we have witnessed interesting drug-delivery applications of nanoparticles in nanomedicine. The characteristic nanoparticle size, shape, and composition make them unique. The amount of drugs released from the nanoparticles can be controlled, which helps to reduce the dose of drug administration and it is already evident that the delivery of drugs using nanoparticles results in increased bioavailability and lower toxicity [1]. Among the different types of nanomaterials, the applications of inorganic nanoparticles (INPs) have levitated due to their characteristic properties, which include rigidity, uniformity, ease of synthesis, and ease of functionalization [2].

Metal oxide particles, or particles with metallic composition, are known as INPs [1]. The nanoscale size range of INPs enables their easy permeation through the biological membranes and more interaction with the cells. Due to unusual mechanical, optical, thermal, physicochemical, and electrical properties their applications in drug delivery, diagnostic, and theranostic fields are increasing. Compared to organic particles INPs are more stable [3]. gold nanoparticles (AuNPs), silver nanoparticles (AgNPs), mesoporous silica nanoparticles (MSNPs), etc. are the major INPs currently in use. These nanoparticles can be easily prepared and functionalized with various moieties, which helps them to interact with biomolecules easily. The surface modification of an INP with suitable materials will enhance its diagnostic, drug-delivery application [4, 5].

There are several methods available for the synthesis of INPs. This c describes in detail the recently used methods of synthesis which include sonochemical synthesis, solvothermal process, nucleic acid-mediated synthesis, etc. By employing newer strategies in the synthesis, different types of INP with diverse properties can be prepared [6]. The knowledge of the nature of particle formation during the synthesis of the INP will help to control the size of the particle formed [4]. Due to the limited availability of data regarding safety and toxicities of INPs, their use in drug-delivery applications faces some challenges [5].

Nanomaterials and Nanotechnology in Medicine, First Edition. Edited by P.M. Visakh.

This chapter gives an in-depth insight into the methods of synthesis, properties, and functionalization of INPs. Besides this, the novel applications of AgNPs, silica nanoparticles, mesoporous nanoparticles, AuNPs, etc. are discussed in detail.

14.2 Synthesis of Inorganic Nanoparticles

14.2.1 Sonochemical Synthesis

Sonochemical synthesis can be used for the preparation of various INPs. Here, the huge amount of energy released by the collapse of hot spots is responsible for the conduction of chemical reactions between the reactants. This method is mainly used for the synthesis of magnetic nanoparticles. For example, $FeCl_2$ and $FeCl_3$ solutions in distilled water were sonicated and oleic acid was added during sonication. Finally, the obtained black colored solution is filtered, washed, and then particles are separated by centrifugation. Similarly, the nanoparticles of silver, copper, and cobalt (Co) oxide can be prepared by this method. By maintaining the reaction property, the nanoparticles can be prepared within the narrow size range using this method [7].

AgNPs were prepared by this method using starch. Interestingly, this nanoparticle showed an increase in absorption from 420 to 440 nm and can be used for the preparation of 2-aryl benzimidazole [8]. Zinc oxide (ZnO) nanoparticles have been synthesized by an environmentally friendly approach using a mini submersible pump assisted sonochemical reactor [9]. The preparation of rhenium nanoparticles using the sonication method revealed that nanoparticle size, distribution, and morphology are influenced by the intensity and the properties of the solvents used [10].

14.2.2 Solvothermal Process

This method, also known as the hydrothermal process, involves the crystallization of solutions and has two main steps: (i) nucleation and (ii) further growth. In this method, the size and morphology of the nanoparticle can be controlled by maintaining the optimal conditions of temperature, pressure, and other processing variables. The size of the metal oxide nanoparticle can be controlled with a flow-through supercritical water method. The conduction of the solvothermal process in supercritical water will enhance the reaction rate more than 10 000 times [11]. The fact that the variations in conditions of the hydrothermal process influence the particle size and morphology of the nanoparticles was proven by the synthesis of nanoparticles of hematite [12]. Using a spiderweb as a reducing agent and stabilizing agent, AuNPs have been synthesized by this approach. The hydrolyzed spiderwebs were taken in a beaker and chloroauric acid was mixed in. Then the mixture was stirred continuously for 30 minutes. The homogenous mixture is then placed in a Teflon vessel and sealed using stainless steel. This sealed container is placed in an autoclave for three hours at 180 °C. After the reaction is complete, the autoclave is cooled to room temperature. The final product obtained was subject to centrifugation and washed several times with ethanol and water and finally dried to get AuNPs [13].

14.2.3 Nucleic Acid-mediated Synthesis

Nucleic acids can be used as the templates for the synthesis of INPs. Nucleic acids can control the morphology and growth of the INP and they can also improve the physicochemical properties of the obtained INPs. The mechanism involved in the synthesis of INP by the use of nucleic acid consists of: (i) initiation, (ii) growth, (iii) termination, (iv) passivation, and (v) solubilization. The initiation step involves the binding of the nucleic acid ligand to the metal and results in the formation of the precursor for nanoparticles. The growth step involves the formation of a negatively charged layer as a coating to the clusters formed during the initiation step. This process is called the capping of nucleic acid. At the end of this step, the small clusters tend to form larger aggregates. The termination step involves the termination of further aggregation by providing an electrostatic barrier upon reaching a suitable size. Passivation involves the suppression of defect states on the nanoparticles and results in the improvement of its physicochemical properties. In the solubilization step, nucleic acid ensures that the nanoparticles remain dispersed in the solution [14].

Peptide nucleic acid (PNA) is a DNA (deoxyribonucleic acid)- or RNA (ribonucleic acid)-related synthetic polymer. Platinum (Pt) nanoparticles can be synthesized by a reduction process of the precursor complex of Pt prebound on a PNA template. By adjusting the processing time, the dimensions of the resultant Pt nanoparticles can be controlled. Synthesis of Pt nanoparticles is due to the development of Pt—Pt bond between the Pt that binds to the biopolymer and the one free in solution [15]. Palladium (Pd) nanoparticles can be synthesized by exposing RNA sequences to Pd precursors. A single RNA sequence is capable of catalyzing the formation of Pd nanoparticles and the RNA sequence is related to the final shape of the formed nanoparticle and the rate of formation of the nanoparticles. At the end of eight cycles, most of the nanoparticles are hexagonal shaped, and fewer rod and cube-shaped nanoparticles are found [16].

14.2.4 Synthesis by Ionizing Radiations

The main principle involved in the synthesis of INPs is the ionization radiation radiolysis of water and the formation of intermediate products with the compounds dissolved. Varieties of INPs with enhanced properties can be prepared by this method. By changing the composition and dose of the irradiated solution, nanoparticles' size varies from sub-nanometric to tens of nm. This radiation-based method is also responsible for the modification of the nanoparticle structure. This method depends on the irradiation of solution containing nanoparticle precursors with either accelerated electrons or gamma rays at room temperature. There exists a small difference in the quality of the obtained nanoparticles depending upon the use of accelerated electrons or gamma rays. This method has several advantages: it does not require any strict experimental conditions; it is temperature independent; and it has a high potential for producing a narrower size. AgNPs having a size range 10–60 nm were prepared by irradiating an aqueous solution of Ag precursor with accelerated electrons using polyvinyl alcohol (PVA) as a stabilizer [17]. Cuprous oxide nanocrystals can be developed by the reduction of copper salts present in the solution induced by radiation. These cuprous oxide nanocrystals, having particle size 14 nm, were obtained by the irradiation of the copper sulfate present in the deoxygenated aqueous solution by gamma radiation. Apart from copper sulfate, the aqueous solution also contains sodium dodecyl sulfate (SDS), propanol, and a 4–4.5 pH acetate buffer. The surfactant SDS was used to control the formed nanoparticles size [18]. Uranium dioxide nanocrystals can be prepared by irradiation of uranyl nitrate by using gamma irradiation or accelerated electron beam in the presence of propanol [19]. ZnO nanostructures can be synthesized by the radiolytic method. Hydrated electrons, produced by water hydrolysis, were used for the synthesis of ZnO nanostructures from zinc salt. The resulting ZnO nanostructure exhibits very strong fluorescence at room temperature. The dynamic light scattering (DLS) experiment elicited nanostructures in the size range of 6 to 8 nm [20].

14.2.5 Biosynthesis

INPs can be synthesized by biological processes. This method is based on the exploitation of the biomass of microorganisms like fungi, bacteria, algae, and plant extracts in the form of aqueous extract or whole cell. By adjusting the parameters like temperature, pH, amount of the metal precursor, and the biological extract, nanoparticle size and shape can be adjusted [21]. Flower-like AuNPs can be prepared by using an aqueous extract of *Rhizopus oryzae*. Different shaped (hexagonal, rod, and triangle) AuNPs were easily prepared by adjusting the time for reaction, the amount of gold precursor, and the pH. Here, the aqueous extract of the fungus acts as a pool of stabilizing and reducing agent. The protein content in the extract of fungus is responsible for the reduction of Au (III) to Au (0) and results in the production of AuNPs [22]. Polydisperse spherical-shaped AgNPs, having a size range of 2 to 15 nm, can be prepared by using biomass of microalga *Acutodesmus dimorphus* as a reducing agent [23]. Spherical shaped PdNPs having a size range of 2 to 4 nm with effective catalytic properties can be prepared by using leaf extract of *Delonix regia* [24]. Biosynthesis of very tiny spherical PtNPs of size 2.2 nm was prepared by mixing extract of orange peel and hydrogen hexachloroplatinate [25].

14.2.6 Precipitation of Salts in an Aqueous Medium

This method is widely used for the production of magnetic nanoparticles. Mainly iron oxide nanoparticles (IONPs) are synthesized by this method. Nanoparticles with size 2–20 nm were obtained by this and the size of the nanoparticles can be tuned by modulating the experimental conditions. This method can result in the production of large amounts of magnetic nanoparticles and this method is also very simple and not expensive. Two approaches are mainly used in this method. The first involves the oxidation of a suspension of ferrous hydroxide using oxidizing agents like hydrogen peroxide or potassium nitrate [26]. The second approach involves the co-precipitation of ferric and ferrous salts mixtures [27].

14.2.7 Decomposition in Organic Media

This method involves the decomposition of metal precursors in organic solvents thermally. This method results in the formation of INPs with a high degree of crystalline nature and narrow size distribution. Generally, this method consists of a metal precursor, which may be inorganic salt of a coordination complex, one or more stabilizers like amines, carboxylic acids, phosphines, etc., and an oxidizing or reducing agent if necessary. A highly elevated temperature is required for the precursor decomposition, so the process is carried out in solvents having high boiling points like benzyl and phenyl ether, and octadecane. By adjusting the parameters like temperature, stirring, the concentration of the oxidizing and reducing agent, the metal precursor, and the solvent boiling point, the nanoparticles size and morphology can be controlled [28].

14.2.8 Synthesis Using Water-in-oil (W/O) Microemulsions

This method consists of a metal salt solution as microdroplets in water, dispersed in a nonpolar solvent stabilized by a surfactant. The cavities developed into these systems have a diameter in the sub-micrometer range, so they are the best method for the formation of nanoparticles. The most commonly used surfactants are SDS, cetyltrimethylammonium bromide (CTAB), and sodium bis(2-ethylhexyl) sulphosuccinate (AOT). IONPs are first synthesized by this method. Here, iron sulfate air oxidation occurs in an organic solvent using AOT as a surfactant [29]. Co and CoPt alloy nanoparticles can be synthesized by this method using CTAB as a surfactant, butanol as a co-surfactant, and octane as the oil phase [30]. Fe_2O_3 and Fe_3O_4 nanoparticles can be synthesized by this method using SDS as a surfactant [31].

14.2.9 Polyol Process

This method involves the synthesis of INPs in polyalcohols by metal precursor reduction. These polyalcohols have a high boiling point and dielectric constants, so the metal precursors can partially or completely dissolve and the reaction can be carried out at different temperatures. The polyol simultaneously acts as both the reducing agent and solvent. It can also prevent aggregation by acting as a stabilizer and act along with other compounds like polyvinylpyrrolidone. This method is done in the presence of alkalis like NaOH, which helps in the process of reduction. This was first used for the synthesis of Co, Cu, and Ni nanoparticles by reduction of their hydroxides, acetates, or oxides in refluxing ethylene glycol [32]. Nonmagnetic IONPs with size 80–100 nm were synthesized in ethylene glycol with a small amount of water [33].

14.3 Properties of Inorganic Nanoparticles

The silica nanoparticle, AgNPs, MSNPs, etc. are the major INPs used in nanomedicine. They are suitable for drug-delivery applications due to their unique properties such as biocompatibility, easy synthesis, and functionalization. Among all types of silica nanoparticles, the nonporous silica nanoparticles are the most widely used. The properties like the ease of large-scale preparation, the hydrophobic nature of the surface, the flexibility of size and shape, mechanical stiffness, etc. make them unique. Similarly, the high drug-loading capacity, controllable size, and shape properties of MSNPs increase their biomedical applications. The specialized characteristics of the nanoparticles of gold will results in easy functionalization with biomolecules, which further helps to expand the applications in drug delivery and diagnosis. Further developments in this area to determine the interaction between INPs and biological membranes will expand their applications [34].

14.3.1 Interfacial Properties

Nanoparticles effectively mediate many molecular processes essential for the regulation of cell functions. The interfacial properties of INPs like redox state change, adsorption capacity, the surface-mediated reaction rate will affect the fate of nanoparticles in the environment, and the toxicity in organisms. For smaller nanoparticles of a size less than 20 nm, the interfacial reactivity and reactivity within the environment can easily be improved. Electron uptake affinity, aggregation, and interfacial phenomena like adsorption of macromolecules or pollutants can result in nanoparticle hazards. These factors play a crucial role in the INPs' toxic properties by the transformation of chemical species, development of reactive oxygen species (ROS), and the release of toxic substances into solution. Active sites present on the nanoparticles produce ROS and result in toxicity [35]. Toxicity of CeO_2 and TiO_2 nanoparticles occurs due to the generation of ROS [36].

14.3.2 Optical and Electronic Properties

Both optical and electronic properties are interdependent. All noble metal nanoparticles have optical properties reliant on size and result in the exhibition of a strong UV-visible extinction band in the spectrum that is absent in the spectrum of bulk material. This band results when the frequency of the incident photon has become constant with excitation of the conductive electrons and this is called localized surface plasma resonance (LSPR) [37]. Free electrons present on the surface of Ag and AuNPs are very freely transportable. The mean path for Au and Ag is 50 nm, this is greater than the size of nanoparticles of these materials. So, scattering from the bulk does not occur upon interaction with light. But they get set into standing resonance conditions, which is responsible for the LSPR in these nanoparticles [38].

14.4 Functionalization of Inorganic Nanoparticles

INPs have tremendous applications in nanomedicine. Recent studies have shown that the surface modification or functionalization of nanoparticles improves the properties and thereby increase the therapeutic applications. Several studies prove that functionalization has a great impact on altering the properties of the INPs.

The INPs administered are cleared rapidly by the human body, which reduces their applicability. The therapeutic and diagnostic applications are restricted due to this rapid clearance from the body. There are many strategies adopted to solve this issue. A polymer coating is one such strategy that can be achieved by two methods, i.e. the "one-pot" method and the "two-step" process. In the one-pot method, the synthesis of nanoparticles and the polymer coating occurs in a single step whereas the two-step process, as the name suggests, occurs in two separate steps. The polymer coating of the nanoparticle prevents the adsorption by proteins and uptake by macrophages. There are various types of polymer coating depending upon the intended purpose of the nanoparticles. For example, if we want to target the nanoparticles toward tumor cells then the targeting ligands are incorporated into the polymer coating [3].

Another area of emerging research is the functionalization of INPs with pillar[n]arenes. The advantages of the pillar[n]arenes over other hosts include ease of synthesis, flexible nature, pillar-like structure, rigidity, and π-rich cavities [39]. The AuNPs functionalized with carboxylated pillar[n]arenes were synthesized by a one-pot method. Such nanoparticles are found to be more stable in conditions where the conventional ones are not. Also under certain conditions, this platform shows enzyme-like activity [40]. Similarly, an MSNP-delivery system doped with carboxylatopillar[5] arene modified AuNPs has been developed. This delivery system exhibits advantages like exceptional responsiveness to competitive binding and temperature changes, and a high drug-loading capacity [41]. The AgNPs modified by using a water-soluble pillar[5]arene shows a high degree of stability and has excellent power in the detection of even a small amount of spermine. The presence of spermine causes the aggregation of these nanoparticles and changes the color from yellow to brown [42]. The INPs modified by using pillar[n]arenes open a wide range of opportunities for researchers and many more structures and properties can be developed by using this platform.

The INP offers a wide variety of bioimaging applications. The binding of the desired biomolecule with these nanoparticles results in targeted delivery, reduced toxic effects, and increased efficiency. There are several approaches adopted like silica coating and synthetic polymer coating [43]. Also, the nanoparticles of silver, copper, iron, etc. have antimicrobial activity. But the frequent administration of these nanoparticles results in the development of resistance. Studies revealed that the functionalization of the nanoparticles tops up the antimicrobial activity by promoting the interaction with cells [44].

In a study, the magnetic MSNPs modified with folic acid (FA)-linked polyethyleneimine were fabricated as a carrier for disulfiram. The solubility and the selectivity of the nanoparticles toward the tumor cells were enhanced by this modification. Also, the combination with copper and sodium nitroprusside increased the cytotoxicity of the carrier toward cancer cells [45]. The poor loading capacity of the carriers results in their requirement in high doses, which substantially increases their side effects. Interestingly, MSNPs functionalized with cyclodextrin-grafted polyethyleneimine offers high DOX (doxorubicin) and siRNA (small interfering RNA) loading capacity. This platform acts as an effective drug-delivery system for drugs and siRNA to tumor cells [46].

The hybridization of INPs with biomolecules enhances biomedical applications. The MSNP surface functionalized with poly(2-dimethyl aminoethyl methacrylate) has been developed, which enables the highly efficient delivery of siRNA to the tumor cells. Furthermore, it shows enhanced cellular uptake and enhanced binding capacity toward siRNA. The advantages of the system include lower cost and lower cytotoxicity [47].

The modified nanoparticles of ZnO were synthesized with help of a silane coupling agent and are incorporated into the matrix of HBPU (hyperbranched polyurethane) for coating. This modification forms a highly cross-linked network within it and thereby improves the storage and tensile characteristics [48]. The N-(6-aminohexyl)-3-aminopropyl tri methoxy silane (AHAPS) functionalized silica nanoparticles are positively charged and have high colloidal stability. Their positive charge enables the rapid uptake by the HeLa cells. This study proves that high colloidal stability has a great influence on cellular uptake [49].

14.5 Quantum Dots for Drug Delivery

14.5.1 As Nanocarriers for Drugs in Targeted Drug Delivery

Quantum dots (QDs) act as a good candidate for drug delivery [50]. The ideal properties of a QD nanocarrier include high drug loading and encapsulation efficiency, good biocompatibility, nontoxicity, great stability, and prolonged residence time *in vivo* [51]. The hybridization of QDs with protein takes place by chemical coupling or physical entrapment resulting in high stability [52]. Zein-ZnS QD nanoparticles were employed for the drug delivery of 5-fluorouracil, which shows efficient penetration into tumor cells [53].

FA can be used as a targeting agent because of its high binding affinity. Hydrophobic ZnSe: Mn/ZnS core-shell fluorescence QDs and anticancer drug paclitaxel (PTX) have been coloaded through F127 micelles onto folate conjugated silica nanocapsules that prove to be useful in chemotherapy [54]. DOX-loaded ZnO QD drug release based on pH can be used for the controlled delivery of drugs. Targeting ligand, hyaluronic acid conjugated to this system binds to CD44 cells and releases DOX at an acidic pH due to the dissociation of the metal-DOX complex. A synergic drug effect was observed due to the presence of Zn^{2+} and DOX [55] (Figure 14.1a). The 5-fluorouracil drug loaded into chitosan (CS) biopolymer embedded in Mn-ZnS QDs and conjugated to FA shows high apoptosis against breast cancer cells [57].

Dual enzyme-sensitive gemcitabine was prepared by functionalizing CdSe/ZnS QDs with detachable matrix metalloproteinase-9 (MMP-9), polyethylene glycol (PEG), cathepsin B cleavable gemcitabine, and targeting CycloRGD ligand. PEG ensures enhanced circulation time and avoids non-specific interactions. When QDs are accumulated inside the tumor, the outer covering is removed for cellular internalization. The increased level of cathepsin B promotes the release of gemcitabine in pancreatic cancer cells [56] (Figure 14.1b). QDs and PTX cocharged lipid nanoparticles serve as a promising parenteral drug-delivery platform in cancer theranostics [58].

Small QDs can accumulate in tumor tissues; thereby avoiding uptake by the reticuloendothelial system. The impact of the size of particles on water-soluble QDs on *in vivo* distribution indicated that smaller QDs are capable of extravasating and diffusing themselves away from the blood vessels [59]. Various delivery vehicles like liposomes, micelles, and polymeric nanoparticles show high drug loading [60]. A polysaccharide-based hybrid nanogel obtained by immobilizing QDs within a dual controllable polysaccharide copolymer was used for the controlled delivery of temozolomide by a pH-dependent release mechanism. [61]. Near-infrared (NIR) emission of micelles loaded with QDs comprising the antitumor drug Herceptin for tumor treatment and monitoring shows high antitumor efficiency [62]. DOX-loaded immunoliposomes with QDs and single-chain anti-HER2 Fv fragments bonded to the liposome surface through PEG chains can be used in a xenograft mouse model for tumor-targeted imaging. [63].

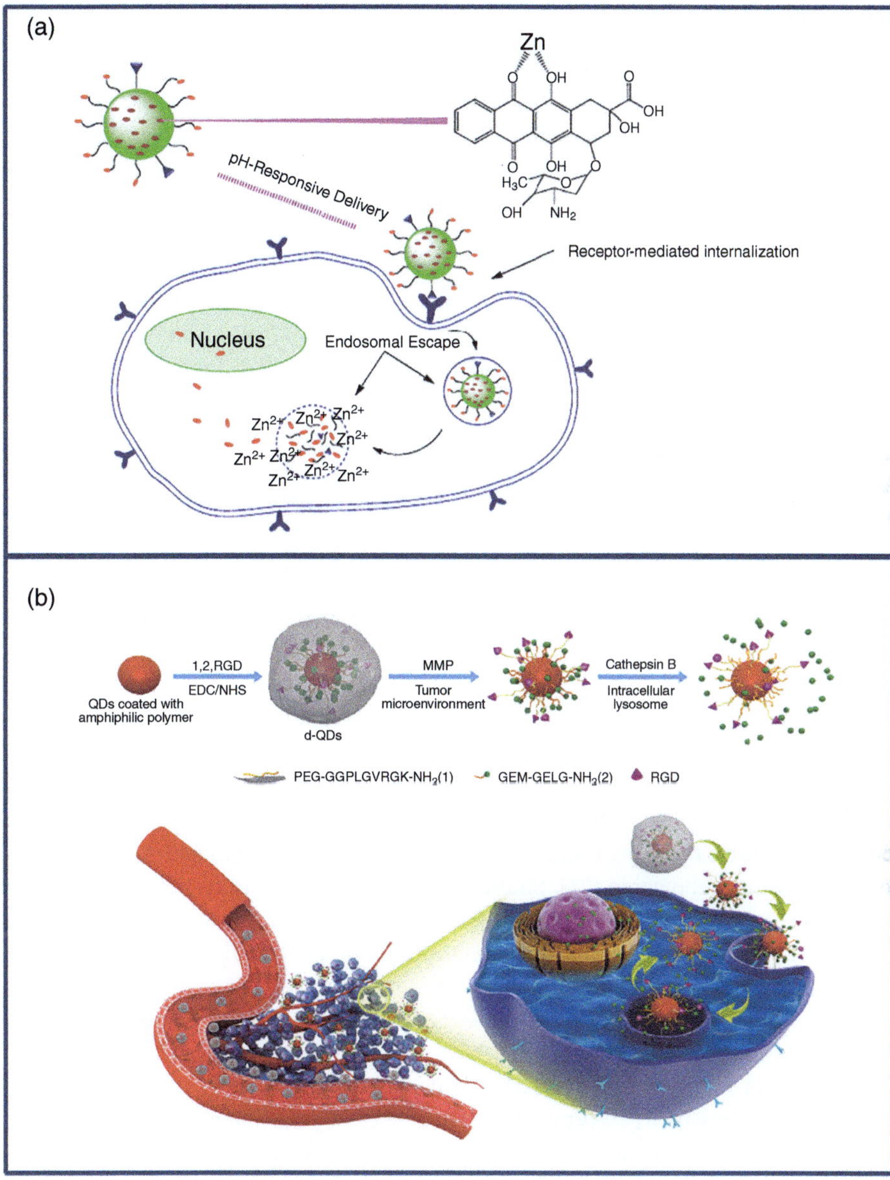

Figure 14.1 (a) Schematic representation of pH-responsive DOX-loaded zinc oxide QDs for cancer therapy. *Source:* Reprinted with permission from [55]. Copyright (2016) American Chemical Society. (b) Schematic illustration of CdSe/ZnS QDs for gemcitabine delivery for pancreatic cancer. *Source:* Reprinted with permission from [56]. Copyright (2017) American Chemical Society.

14.5.2 Topical Application

The colloidal core of QDs coated with various chemical agents can aid in absorption and toxicity [64]. Studies have indicated relatively low QD penetration takes place through an intact skin barrier. The penetration ability of QDs is determined by surface coating with different charges. Studies were conducted with negative and neutral charged QDs on derma-braided SKH hairless mice skin. Reports of the elemental analysis show that neutrally charged QDs have better penetration than negatively charged QDs [65, 66]. Another research was conducted to study the impact of positive, neutral, and negative QDs. The neutrally charged QDs show a better effect than positive and negative QDs. All studies report that negatively charged QDs are highly toxic at a >20 nM concentration [67]. Diphtheria toxin vaccine incorporated into QDs, administered transdermally to mice, showed a strong immune response [68].

14.6 Drug Delivery by Mesoporous Silica Nanoparticles

14.6.1 As Selective Targeting Agent for Localized Therapy

MSNPs serve as a novel therapeutic armamentarium that hosts, protects, and transports drugs to targeted sites. Targeting agents are conjugated with the exterior surface of MSNPs to enable them to deliver the drug directly to the target site and prevent unintended adverse effects [69]. The outermost surface of MSNPs can be decorated with agents that react with membrane receptors overexpressed in diseased cells. Also, MSNPs can be decorated with a targeting ligand that can interact with blood vessels resulting in tumor destruction. The above approach of particle uptake by tumor cells or tumor blood vessels is known as active targeting [70]. Passive targeting is defined as the accumulation of nanoparticles inside the tumor cells to release the cargo [71]. Studies have shown that merging active and passive targeting achieves greater therapeutic effect. A novel targeting agent was prepared by the combined approach for the treatment of neuroblastoma using drug-loaded MSNPs [72]. Another approach is dual targeting, which consists of two targeting agents directed toward the same nanocarrier, can detect two different organelles [73]. Recent studies elicit that asymmetrically functionalized Janus-type MSNPs with targeting agents FA and triphenylphosphine find a potential sequential cell-organelle-targeting process. FA increases the concentration of Janus-type MSNPs inside the cancer cell by the folate membrane receptor followed by an interaction with the mitochondria by triphenylphosphine leading to increased efficiency of MSNPs in cancer treatment [74] (Figure 14.2).

One of the disadvantages associated with this nanosystem is its low-penetration potential of the tumor cell mass owing to the presence of a collagen-rich intracellular matrix that hinders the diffusion. Two approaches have been devised to improve the penetration of MSNPs into tumor cells. The initial approach is the preparation of pH-sensitive collagenase nanocapsules attached to the MSNPs' surface [75]. The mechanism involved is that collagenase, being a proteolytic enzyme, degrades the extracellular matrix and enhances the penetration ability. However, the degradation of the enzyme under various conditions is prevented by using nanocapsules prepared by various monomers such as acrylamide, 2-aminoethyl methacrylate, and ethylene glycol dimethacrylate. The second approach

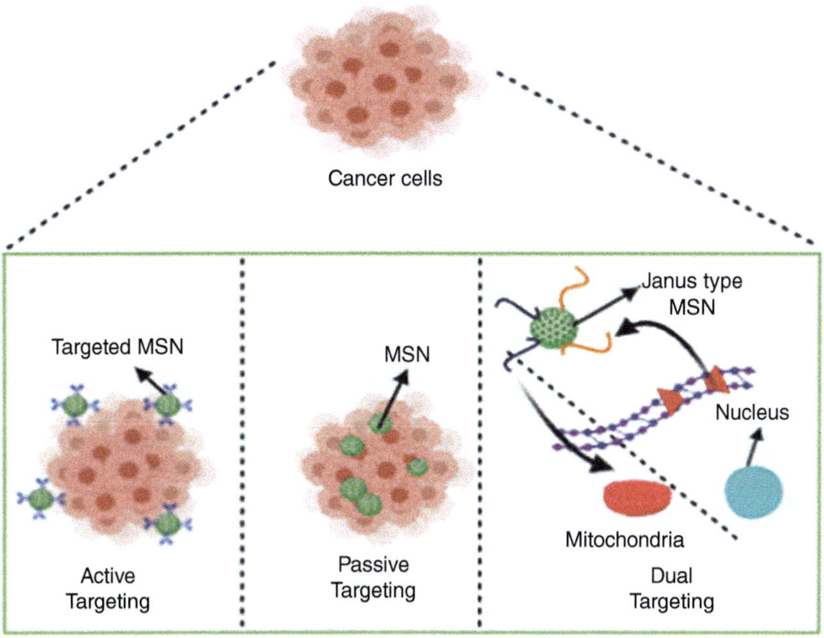

Figure 14.2 Various targeting approaches to targeted cancer therapy.

involves the ability of human mesenchymal stem cells to migrate toward deeper areas of tumor [76]. A new system comprising of the decidua of human placenta mesenchymal stem cells and DOX-loaded MSNPs elicit good anticancer efficiency [77].

14.6.2 As Stimuli-responsive Drug-delivery System

It is simple to accumulate a therapeutic cargo inside MSNPs and diffuse it out of MSNPs. An approach devised to block the early release of drugs is the capping of the pores of MSNPs using stimuli-responsive gatekeepers [69]. The detachment of the capping agent on various stimuli application would trigger the release of the cargo, on demand. Various internal stimuli such as pH, redox potential, enzymes, or external stimuli magnetic fields, ultrasounds, or light open the pores and release the cargo [78]. In general, the stimuli-responsive MSNP drug delivery consists of a capping agent, a targeting ligand, cargo, a tracking marker, and a sensitive linker [79] (Figure 14.3a). The three major capping agents used are: reusable caps, which use a bulky capping molecule that attach reversibly; completely reversible caps that have the ability to connect with two or more binding sites; and irreversible caps, dependent on the capping molecule's chemical bond cleavage, which contributes toward a permanent detachment from the MSNP's pore entrance [71]. In the presence of the given stimulus, the sensitive linker can split, degrade, or experience a conformation transition [69].

Internal stimulation is characteristic of treated pathology and hence the resulting MSNPs are capable of reacting to a chemical change in the body. pH-sensitive MSNPs release drugs in response to pH changes whereas in enzyme-sensitive MSNPs overexpression of certain

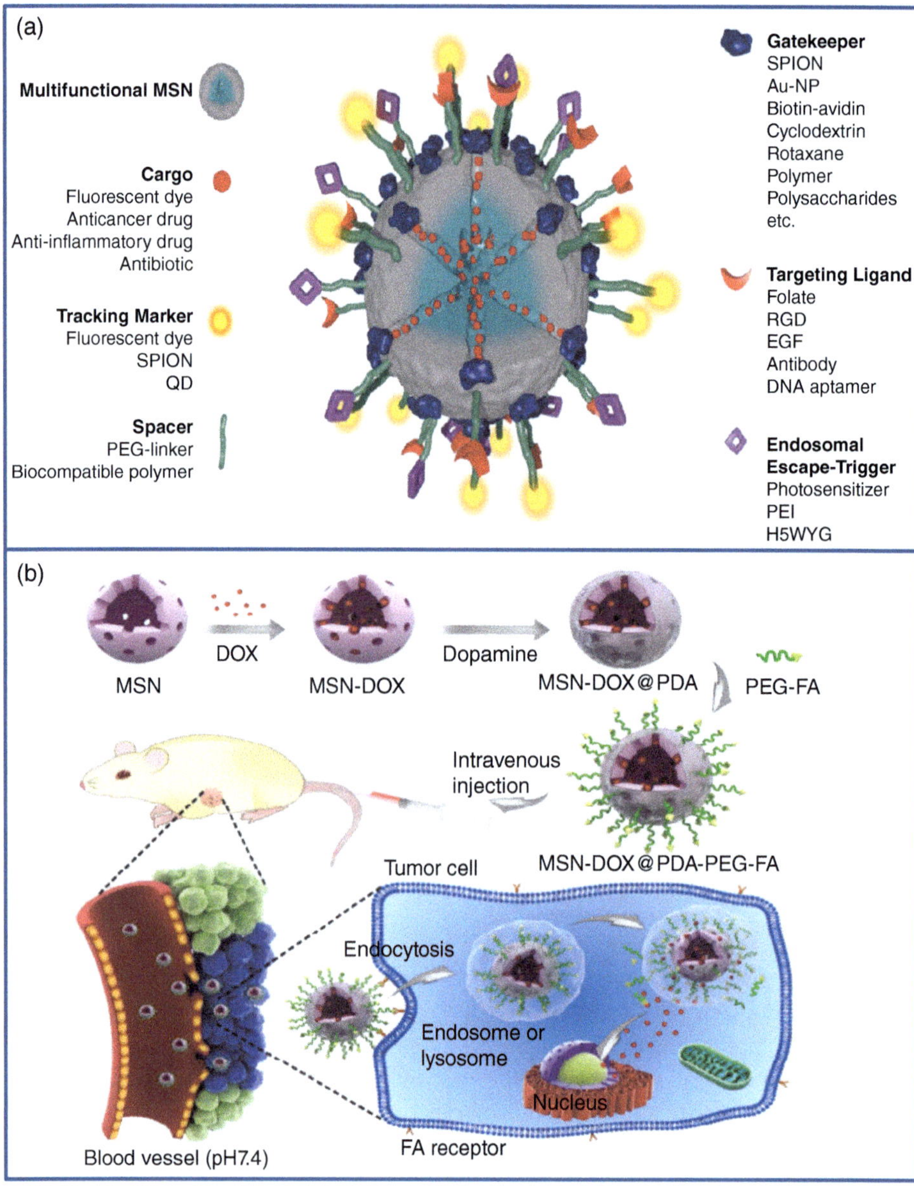

Figure 14.3 (a) Schematic representation of necessary features MSNPs for stimuli-responsive delivery. *Source:* Reprinted with permission from [79]. Copyright (2014) American Chemical Society. (b) Mechanism of poly(ethylene glycol) FA functionalized polydopamine-modified MSNPs for delivery of DOX. *Source:* Reprinted with permission from [80] Copyright (2017) American Chemical Society.

enzymes aids in drug release. In redox-potential-responsive MSNPs, an imbalance of the reductive species inside and outside the cells trigger the release of the therapeutic cargo [81]. Poly(ethylene glycol) FA-functionalized polydopamine-modified MSNPs were designed for the controlled delivery of DOX. Under acidic condition, the polydopamine films get destroyed, and loaded DOX get released [80, 82] (Figure 14.3b). External stimuli can be triggered manually by a clinician who has full influence all the time. The various advantages associated with external stimuli are pulsatile response and increased precision [71]. The application of the magnetic field is popular since they produce magnetic control in the presence of a permanent magnetic field or a rise in temperature in the presence of an alternating magnetic field. Ultrasound-triggered MSNPs can attain a spatiotemporal control over drug delivery by thermal effect resulting in noninvasive penetration. Light-triggered MSNPs achieve drug release on-demand via illumination with a particular wavelength such as ultraviolet, visible, and NIR leading to low toxicity and precise focalization [83].

14.6.3 As Immediate/Sustained Drug-delivery Systems

The large pore volume as well as the high surface area of MSNPs helps to release a high payload of drugs and to increase the solubility of the insoluble drug [84]. As MSNPs' structural integrity is maintained in inorganic solvents, they are ideal for the delivery of hydrophobic drugs. Therefore, solvent selection for the loading of the drug depends upon the characteristics of the drug. However, in the case of hydrophilic drug charge-matching-based adsorption in aqueous solvent attains high drug loading where the release mechanism is highly diffusion controlled [85]. Various types of drugs such as anti-inflammatory (ibuprofen, naproxen); chemotherapeutic drugs (DOX, camptothecin); and antibiotics (erythromycin, vancomycin) can be delivered through MSNPs [86].

Resveratrol is a promising drug with poor aqueous solubility. Resveratrol incorporated MSNPs increases the solubility and *in vitro* drug release [87]. Furthermore, oral administration of fenofibrate-loaded MSNPs showed increased absorption [88]. Moreover, MSNPs find great potential in the delivery of therapeutic protein due to their biocompatibility, high stability, and tunable surface chemistry. Various MSNPs glucose-responsive insulin release systems are being developed for diabetes therapy [89].

14.7 Silver Nanoparticles for Drug Delivery

14.7.1 As an Antibacterial Agent

Solvent selection for the loading of the drug depends upon the characteristics of the drug. The antibacterial activity of AgNPs is exerted through several mechanisms [90–92]. The first mode of action involves the attachment of AgNPs to the walls of the cell surface by electrostatic interaction of the positively charged AgNPs with a negatively charged microbial surface [93]. The second mode of action involves the penetration of AgNPs within the cell followed by the destruction of intracellular contents. The bacterial cell wall is disrupted irreversibly by the interaction of Ag^+ ions with cell wall proteins [94]. Moreover, Ag^+ ions can bind with sulfur and phosphorous of DNA causing damage in DNA responsible for the replication of bacteria and leads to cell death [95]. Another mechanism involves ROS and free radicals resulting in cellular toxicity as well as increased oxidative stress [96]. A different mechanism of action for the antimicrobial activity of AgNPs is the modulation of microbial signal transduction pathways [91, 95] (Figure 14.4a).

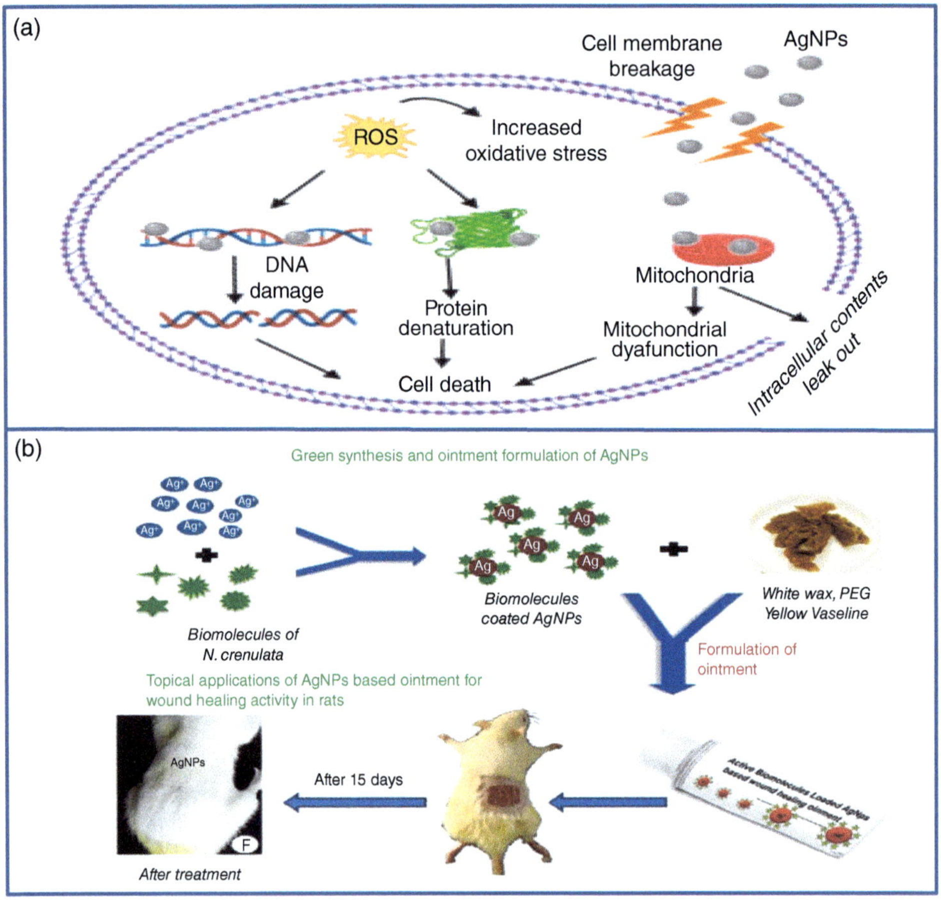

Figure 14.4 (a) Schematic representation of the mechanism of antimicrobial action of AgNPs (b) Wound-healing mechanism of AgNPs from the leaves of *Naringicrenulata*. *Source:* Reprinted with permission from [97]. Copyright (2014) Elsevier.

The small controllable size, large surface area, and enhanced reactivity make AgNPs a potent antimicrobial agent against infections. The AgNPs' surface ligand coating integrated with the drug serves as a medium for drug delivery due to the sustained release and minimized side effects [98]. AgNP-decorated Co nanocomposites exhibiting AgNP-dependent antibacterial activity is considered an effective drug carrier. The antibacterial activity is directly related to AgNP concentration [99]. AgNPs with proline-rich peptide C-Bac3.4 elicits pronounced antibacterial activity against standard and multidrug-resistant bacteria [100]. Curcumin-loaded AgNPs hydrogel releases curcumin in a controlled manner showing that polymer decorated AgNPs are prominent drug-delivery systems with great antibacterial effect against *E. coli* [101, 102]. Laser-receptive polyelectrolyte thin films doped with AgNPs incorporated with moxifloxacin hydrochloride shows about 85% drug release [103].

14.7.2 As Anticancer Agents

Drug-mediated toxicity of healthy cells is one of the major disadvantages associated with conventional cancer therapy. The efficiency of cancer therapy can be improved and side effects can be reduced with AgNPs. As AgNPs process the ability to cross various biological barriers, they can be thought of as a targeted delivery system. The cytotoxicity of AgNPs is affected by various factors such as particle size, shape, and surface chemistry. It is noted that the tumor cell viability reduces as the concentration of AgNPs increases [104, 105].

AgNPs prepared from the various concentration of seed extracts of *Setariaverticillata* loaded with hydrophilic antitumor drug DOX and daunorubicin is considered a novel anticancer system for the delivery of drugs [105]. The concurrent intracellular delivery of DOX and alendronate can be performed using an AgNP-based delivery system, which shows a higher anticancer activity than individual drugs [106]. AgNPs prepared from *Moringa olifera* stem bark extract was effective against human cervical carcinoma cells by apoptosis-induced ROS generation [107]. AgNPs conjugated with PEG act as a platform for the delivery of the anticancer drug in tumor treatment [108].

14.7.3 As Antiviral Agents

AgNPs represent an emerging class of antiviral agents because of their inhibitory activity against various viruses. The viral inactivation occurs due to the reaction of AgNPs with sulfhydryl, amino, carboxyl, phosphate, and imidazole groups [109]. AgNPs conjugated with PEG act as a platform for the delivery of an anticancer drug in tumor treatment [110]. AgNPs prepared by the oxidation–reduction method are active against H3N2 in MDCK cells leading to the destruction of viral structure and reduction of H3N2 associated apoptosis [111]. AgNPs were found to inhibit hepatitis B virus replication [112]. Also, decorated AgNPs were found to be strong virucidal agents. Studies have shown that polysaccharide-coated AgNPs are effective against TCRV (tissue culture rinderpest vaccine) and HPV (human papilloma virus) whereas poly(N-vinyl-2-pyrrolidone)- coated AgNPs are effective against HIV-1 and RSV (respiratory syncytial virus). Mercaptoethane sulfonate-coated AgNPs are active against the herpes simplex virus [100].

14.7.4 As an Antifungal Agent

Miconazole-conjugated AgNPs exhibit antifungal activity against pathogenic *Candida* strains and as drug-delivery vectors. It was observed that miconazole remains on the nanoparticle surface for a long period, thereby side effects to normal tissue are reduced [113]. Another innovative green strategy for treating the fungal infection is amphotericin B-conjugated AgNPs. The conjugated system serves efficient drug-delivery vehicles in the treatment of *Candida* infections [114]. Additionally, a combination therapy of nystatin and fluconazole with AgNPs has been observed to enhance antifungal efficiency [115].

14.7.5 As Drug-targeted Delivery System

AgNPs are considered as a targeted drug-delivery system because of their ability for passive and active targeting of a specific site. They can be targeted at a particular site by surface modification with a particular targeting agent or by encasing with a biomimetic polymer [116, 117]. It was observed that the delivery of the drug into cells by endocytosis relies on the shape of the nanoparticles. Spherical AgNPs from *Aervajavanica* combined with the anticancer agent gefitinib reduce the cell viability of cancer cells and have fewer side effects [118]. The biofunctionalization of AgNPs conjugated with biomacromolecules serves as an antibiotic vehicle where the drug is encapsulated in the macromolecule cavity [119]. Attachment of antimicrobial peptide to the surface of AgNPs, with the help of a linker molecule, possesses the ability to target multidrug-resistant microbes [120, 121].

14.7.6 As Photoactivated Drug-delivery Vectors

The photophysical properties such as the large value of extinction coefficient and blue-shifted Plasmon resonance peaks of AgNPs makes them promising photoactivated drug-delivery vectors [122]. They are mainly used as biological sensors because of their ability to detect a specific target depending on surface plasmon resonance (SPR). The characteristics of AgNPs are that the SPR spectra can be tuned and mainly depend on particle shape, size, and environment. AgNPs decorated with thiol-terminated photolabile – DNA oligonucleotides find potential as photo-activated vectors for the delivery of drugs that increase hybridization activity and cellular uptake [123].

14.7.7 As Topical Wound Healing Agents

AgNPs possess the capability to release ROS resulting in the destruction of the microbial cell, which endows specific antibacterial and antifungal effects [124]. Dependent on this fact, AgNPs have a great promise in treating wound inflammation and wound healing. Various parameters such as penetration ability and safety need to be studied for AgNPs' topical application. Studies report that the skin penetration ability of AgNPs is considerably lower compared to other INPs due to the presence of large numbers of free ions precipitated on the outer layers of the skin [64]. The skin penetration ability of polyvinylpyrrolidone-coated AgNPs was very low [125, 126]. Toxicity of AgNPs is conducted *in vivo* on three types of AgNPs; washed, unwashed, and carbon-coated. The toxicity of unwashed AgNPs was observed to be higher in comparison with the other two due to the presence of contaminants like formaldehyde [127].

AgNPs are considered an effective agent to promote wound healing. The impact of AgNPs on wound healing was carried out in an animal model using silver sulfadiazine as a control. Results showed that AgNPs exhibit a favorable effect on wound healing through their antimicrobial properties as opposed to silver sulfadiazine, which has a poorer effect [128]. Moreover, a positive effect was observed for post-surgical wounds treated with AgNPs. However, local tissue edema occurs with AgNPs which limits their application and needs resolution [129]. Biosynthesized AgNPs from the leaves of *Naringicrenulata* extracted by microwave irradiation is formulated as an ointment and showed potential wound-healing power in experimental rats [97] (Figure 14.4b).

14.8 Gold Nanoparticles for Drug Delivery

14.8.1 For Anticancer Drug Delivery

AuNPs are mostly used for the treatment of malignant tumors due to their characteristics like, unique quenching efficiency, hydrophilicity, biocompatibility, and surface modifiability. The drug efficiency of the anticancer drug is lowered by poor bioavailability and side effects. Hence a combination strategy was developed using dual delivery of anticancer drugs to resolve these limitations. AuNPs loaded with bleomycin and DOX shows high drug loading and increased uptake against HeLa cells [130]. DOX, being an intercalating drug, blocks the process of replication and results in inhibition of macromolecule synthesis [131] whereas bleomycin produces ROS and causes DNA damage [132].

Also, conjugation of AuNPs with drugs by ionic or covalent interaction or by physical absorption can increase drug efficiency. Methotrexate, an anticancer drug can hinder tumor cell growth. The concentration of methotrexate delivered to diseased tissue is higher when AuNP is conjugated with methotrexate rather than free methotrexate [133]. DOX-tethered responsive AuNPs cause a rapid increase in intracellular drug concentration enhancing the efficiency against cancer cells. Attachment of DOX to AuNPs with spacer by acid-labile linkage triggered the release of DOX under acidic conditions following endocytosis [134] (Figure 14.5a). In the active form, AuNPs conjugated to oxaliplatin have greater effectiveness over free oxaliplatin in cancer cell destruction [136]. AuNPs oligonucleotide was utilized for the delivery of Pt compounds. The Pt compounds can be delivered using AuNPs oligonucleotide. The DNA-AuNPs decorated with Pt were internalized into the cell and caused the reduction of intracellular pH leading to decreased cisplatin release from the prodrug that aids the death of cancer cells [137].

The surface of AuNP can be decorated using PEG. Pt-tethered nanoparticles functionalized with a PEG monolayer exhibits good cytotoxicity [136]. The amphiphilic properties of the polymer can provide excellent stability under different physiological conditions [138]. PEG-modified mercaptopropionic acid-linked AuNPs serve as a promising nuclear-targeted drug-delivery carrier [139].

14.8.2 For Targeted Drug Delivery

Site-specific drug delivery can be accomplished by combining drug-delivery vehicles with a biomolecule that specifically interacts with the ligand [140]. The effectiveness of AuNPs as a targeted system can be assessed by analyzing the effects of the galactose targeting ligand, PEGylation, and nanoparticle size on the targeting of hepatocytes [141]. The choice of a targeting ligand is an essential parameter in targeted delivery system formation [142]. Ligand targeting depends on different factors such as biocompatibility of ligand, cell specificity, affinity to binding, ligand purity, ligand size and ligand molecule charge, and their ease of nanoparticle modification and conjugation. Conjugation of thiol derivatized PEG linked to protein TNF adsorbed on AuNPs is used for cancer therapy [141].

Aptamers are now considered as potential ligands for targeting cancer. Studies have shown that aptamer-conjugated AuNPs help in the detection of cancer cells [143]. As they are upregulated in tumor cells, FA is often considered as another targeting agent

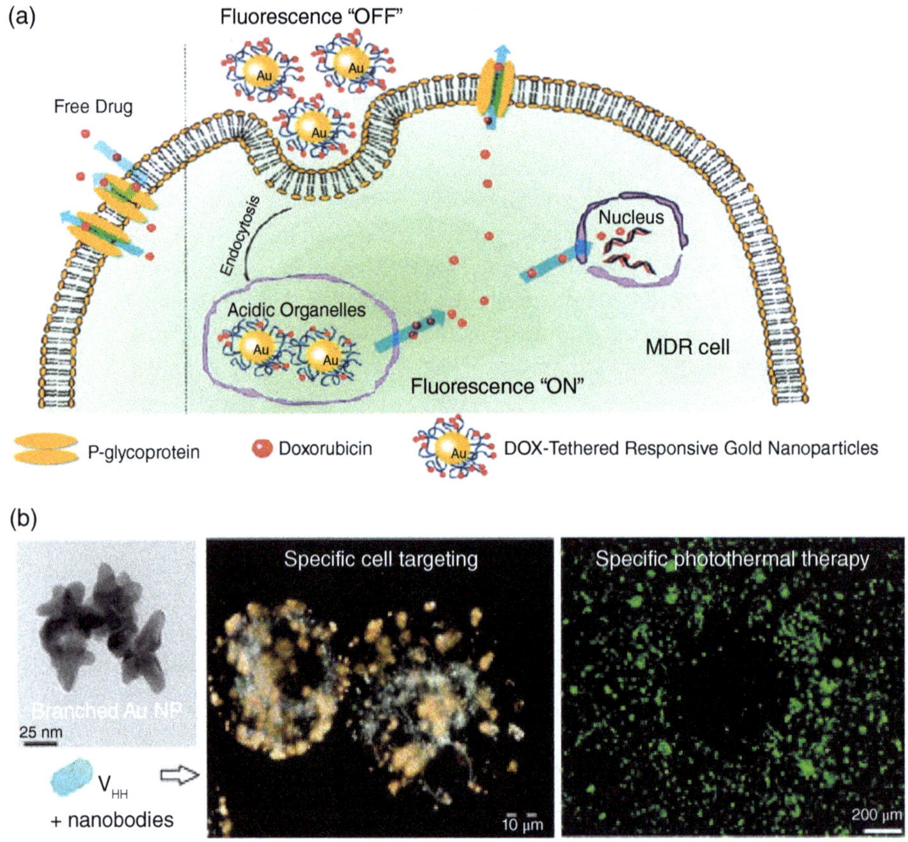

Figure 14.5 (a) Mechanism of DOX-tethered AuNPs' drug delivery. *Source:* Reprinted with permission from [134]. Copyright (2011) American Chemical Society. (b) AuNPs functionalized with nanobodies for photothermal therapy. *Source:* Reprinted with permission from [135]. Copyright (2011) American Chemical Society.

for drug-delivery applications [144]. Thioctic acid anchored to PEG and FA conjugated with AuNPs is used for targeting of tumor cells having excellent stability and biocompatibility [145]. Studies showed that tumor necrosis factor conjugated to AuNPs was found to accumulate in tumor cells indicating its potential for therapeutic bound particles [140].

14.8.3 For Glutathione-mediated Drug Release

Glutathione sensitive release of drugs is a novel method for intracellular nonenzymatic drug delivery [146]. The efficient release of the payload happens as a result of disulfide bond breakage in response to elevated cellular glutathione levels. Tiopronin-conjugated AuNPs for the therapy of acute liver injury is achieved by glutathione-mediated drug release [147].

14.8.4 For Photothermal Therapy

Alteration of the shape and size of AuNPs can tune their energy absorption profile, which finds application in photothermal therapy (PTT) [148]. The ability of AuNPs to convert visible light into heat energy is used in PTT. PTT therapy involves exposure to the NIR region, which induces cytotoxic damage to the infected tissues. NIR light is used because it causes reduced absorption by tissues and physiological fluid at this wavelength [149]. AuNPs such as nanorods, nanoshells, and nanocages, having optimum absorption in the NIR, are suitable for PTT since they are capable of absorbing NIR light and releasing the absorbed energy as heat at their localized site of accumulation [148]. The initial study was carried out on Au nanospheres conjugated to IgG, which selectively targets CD8 receptors on lymphocytic cells, and caused by about 95% death of cells [150]. Later studies showed that maximum absorption at 785 nm can be used as a PTT agent, which is active against cancer cells leaving the healthy cells intact [151]. PEGylated AuNPs causes inhibition of tumor cells without causing damage to surrounding tissues [148]. Recent studies showed that branched AuNPs' surface modified with nanobodies finds application as an antigen targeted PTT agent. These nanobodies attach to the HER2 antigen that is strongly expressed in breast and ovarian cancer cells [135] (Figure 14.5b).

AuNPs were found to bolster radiation efficiency on tumors. Tumors loaded with AuNPs received a high dose of radiation in comparison to the dose of radiation received without the presence of AuNPs [152]. Studies showed that improved efficiency was observed in mammary carcinoma of mice and on a mouse model of melanoma [153]. Another approach for the destruction of the cancer cell is radiofrequency ablation involving probe insertion [154].

14.9 Superparamagnetic Iron Oxide Nanoparticles for Drug Delivery

14.9.1 For Magnetic Target-based Drug Delivery

Magnetic drug delivery involves the targeted delivery of drugs with the aid of a permanent magnet. The magnetic targeting decreases the systemic distribution of drugs and delivers drugs specifically to the targeted sites [155]. The various factors that determine the effectiveness of this method involves the strength of the magnetic field, its geometry, depth of the target tissue, rate of blood flow, and vascular supply [156]. Epirubicin-loaded SPIONs (super-paramagnetic iron oxide nanoparticles) were used to target solid tumors but this caused embolization of blood vessels and toxic responses to the nanosystem [157]. A clinical trial of DOX-like magnetic nanoparticles showed that 75% of the tumor remained stable [158].

Nanomagnetosol inhalation is beneficial for different lung disorders [159]. An aerosol composed of SPIONs, PEG, and hydroxypropyl-b-cyclodextrin was developed using the spray-drying method. The aerodynamic properties showed that they showed increased deposition in the presence of a magnetic field [160]. Another approach of aerosol using SPIONs, model drug fluorescein, and plasmid DNA in the vicinity of the external magnetic field in mice lungs was studied. The presence of an external magnetic field caused a higher accumulation of drug fluorescein than the absence of an external magnetic field [161].

Repeated intra-articular injection of corticosteroids for the treatment of arthritis can cause severe side effects. SPION-corticosteroid conjugates were developed to solve this issue [162]. Long-term effects at the desired site can be achieved by an external magnetic field, which delays uptake by macrophage [163]. Intra-articular injections of amino-polyvinyl alcohol-Cy2.5 SPIONs targeted with a permanent magnet on carpometaphalangeal joints of sheep indicate their ability to extend the drug action [164]. SPIONs and dexamethasone acetate coencapsulated into microparticles of polylactic-co-glycolic acid also act as a targeted-delivery platform for joint therapies [165].

14.9.2 For Active and Passive Targeting in Drug Delivery

Active targeting using SPION is receiving more attention since passive targeting can't distinguish cancer cells from noncancerous cells. Their targeting ability depends on the ligand used for targeting and the method of conjugation to SPIONs [158]. The loading of the drug into SPIONs can increase the stability of hydrophobic drugs. Better regulation of drug loading and drug release is achieved when the drug is directly conjugated to SPIONs. However, the presence of functional moieties can decrease drug loading [166].

In reports focused on current researches, SPIONs seem to very promising in oncological therapy mainly for breast, prostate, and pancreatic tumors [167]. Some agents such as chlorotoxin serve as a targeting agent and a therapeutic for a brain tumor. Hsp70SPION conjugate combines with the CD40 receptor of glioma cells [168]. Newer docetaxel loaded, prostate-specific membrane antigen targeted SPION formulation was developed for prostate cancer therapy. A potent anticancer effect is caused by a significant increase in pro-apoptotic proteins and by downregulation of anti-apoptotic proteins in the prostate cancer cell line [169]. Also, fluorescently mantled carbon-coated core-shell SPIONS find application in neuronal therapeutics in neuro-engineering applications [170]. Oleic acid-coated iron oxide and Pluronic-stabilized nanoparticles loaded with DOX and PTX show synergistic effects against breast tumor [171].

A folate-functionalized SPION-nanosystem was self-assembled to form nanoscaled micelles incorporated with fluorescein diacetate. The human hepatic Bel-7402 cancer cells were internalized following a folate receptor-mediated targeting mechanism and finally released the drug inside lysosomal compartments [158]. SPIONs coated with PEG and PEI (poly[ethylenimine]) and further conjugated with FA are used for the treatment of tumor cells. FA attaches with the receptor on the tumor cells and undergoes endocytosis to release the DOX drug [172] (Figure 14.6).

14.10 Hybrid Systems of Inorganic Nanoparticles

The design of a hybrid system is a breakthrough that offers various applications in drug delivery. Hybrid systems offer great advantages because they have the properties of both. Thiolated cell-penetrating peptide-functionalized Ag-Fe_3O_4 nanocomposites act as a dual system for thiolated drug delivery. Once the nanocomposites enter the systemic circulation, the nano-vehicle is directed to the tumor site by the magnetic field and the cell-penetrating peptide facilitates efficient tumor intracellular uptake [173]. The bimetallic system is

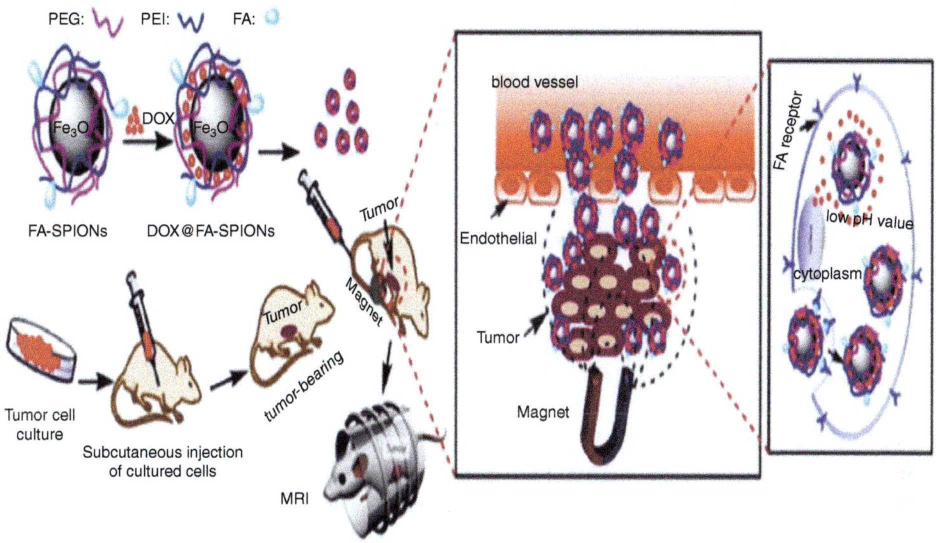

Figure 14.6 Schematic representation of FA conjugated SPIONs for tumor therapy. *Source:* Reprinted with permission from [172]. Copyright (2016) Elsevier.

usually preferred as they incorporate the characteristics of two metals in a single entity. A combined system of AuNPs and AgNPs with chitosan as a template makes it a nanocomposite for cancer therapy [174]. Another study showed that cross-linked chitosan composites modified with nanoparticles of silver and gold exhibit antimicrobial properties to *E. coli* [175]. A novel Janus silver-mesoporous silica nanoparticle acts as a pH-responsive delivery system for DOX delivery in cancer treatment [176].

A stimuli-responsive drug-delivery system was prepared by end-capping MSNPs with functional peptide-coated AuNP [177]. The end-capped MSNPs offer improved biocompatibility and controlled release of drugs. MSNPs are functionalized with carboxyl groups sensitive to pH that carries a negative charge whereas AgNPs are coated with peptide-containing oligo-lysine to maintain a positive charge [178]. Under neutral condition, a layer of AuNPs@peptide can coat the MSNPs by interaction to cap the pores and prevent premature leakage. Upon endocytosis, the charge reversal of the MSNPs takes place leading to the release of the entrapped drug [179]. Mesoporous silica coated with AuNPs incorporated with DOX shows high therapeutic efficiency in the treatment of tumor cells [180]. Another study showed that DOX-loaded mesoporous silica-encapsulated gold nanorods cause light-triggered DOX release resulting in high therapeutic efficiency and low systemic toxicity [181]. Another new approach involves encapsulation of NIR-responsive gold nanorods inside MSNPs conjugated with DOX, which is further coated with a thermo-responsive polymer, poly(N-isopropyl acrylamide). Below the lower critical solution temperature of poly(N-isopropyl acrylamide) the coating on the voids of the MSNPs is removed causing a rapid release of an incorporated drug. To avoid the unwanted interactions of proteins in the blood circulation, the lower critical solution temperature of poly(N-isopropyl acrylamide) should be modified by copolymerization with a hydrophilic monomer. The proposed system shows enhanced drug-loading efficiency where the release can be controlled by various external and internal triggers [182]. A SPION-MSNP composite prepared by the sol-gel method can act as a potential drug-delivery vector [183].

14.11 Prospects of Inorganic Nanoparticles

INP science is the most interesting area in research fields. By using INPs for delivering the exact amount of medication to the infected cells, INP's application is definitely a concept that will continue to be studied and developed for decades to come. The current research focuses more on aspects like biodistribution and toxicity, which remains a challenge. The application of INPs in both diagnosis and treatment is a clinical field that may theoretically open up as a research avenue.

The optimization of the surface ligand is necessary for the preparation of AgNPs for commercial applications against microbial infections. INPs, as such, impart low biocompatibility as drug carriers. Therefore, conjugation of INPs with organic material will help to improve the biocompatibility without impairing their intrinsic properties. Personalized medicine is now attracting interest and the application of nanotechnology is anticipated to contribute to the overall success in this field. Multifunctional IONPs will be an encouraging material for biomedical applications, which might alter the current pharmaceutical industry business model.

14.12 Conclusion

Nano-enabled drug delivery enables the permeation of drugs across cell walls, which is essential for the predicted advances in genetic medicine. In this chapter, we reviewed the important application of INPs in drug delivery and targeted therapy. We addressed some aspects like properties, synthesis, and functionalization of INPs such as MSNPs, QDs, SPIONs, AuNPs, AgNPs, and a hybrid system of INPs. INPs play a crucial role in drug delivery and imaging. They improve the drug efficacy by targeting specific sites rendering them biocompatible and less toxic. SPIONs are considered to be highly attractive for active targeting. However, AuNPs show unique properties in the controlled release of drugs. Moreover, QDs and AgNPs play a crucial part in drug-delivery applications. However, it is crucial to modify these INPs by combining them with organic materials to be used efficiently in the respective applications, rendering them highly biocompatible. Additionally, INPs require an interdisciplinary approach, including biopolymers, biopharmaceutics, and materials science, to promote their future applications in drug delivery.

References

1 Fadeel, B. and Garcia-Bennett, A.E. (2010). Better safe than sorry: understanding the toxicological properties of inorganic nanoparticles manufactured for biomedical applications. *Advanced Drug Delivery Reviews* 62 (3): 362–374. http://www.sciencedirect.com/science/article/pii/S0169409X09003469.

2 Chen, J., Wang, Y., Wang, C. et al. (2019). Functionalization of inorganic nanomaterials with pillar[n]arenes. *Chemical Communications* 55 (48): 6817–6826. http://dx.doi.org/10.1039/C9CC03165K.

3 Neoh, K.G. and Kang, E.T. (2011). Functionalization of inorganic nanoparticles with polymers for stealth biomedical applications. *Polymer Chemistry* 2 (4): 747–759.

4 Salas, G., Costo, R., and Morales, M.d.P. (2012). Synthesis of inorganic nanoparticles. In: *Nanobiotechnology* (ed. J.M. de la Fuente and V. Grazu), 35–79. Elsevier http://www.sciencedirect.com/science/article/pii/B9780124157699000029.

5 Pandey, P. and Dahiya, M. (2016). A brief review on inorganic nanoparticles. *Journal of Critical Reviews* 3: 18–26.

6 Rao, C., Vivekchand, S., Biswas, K., and Govindaraj, A. (2007). Synthesis of inorganic nanomaterials. *Dalton Transactions* 34: 3728–3749.

7 Kis-Csitari, J., Konya, Z., and Kiricsi, I. (2008). Sonochemical synthesis of inorganic nanoparticles BT – functionalized nanoscale materials, devices and systems. In: *Functionalized Nanoscale Materials, Devices and Systems* (ed. A. Vaseashta and I.N. Mihailescu), 369–372. Dordrecht: Springer Netherlands.

8 Kumar, B., Smita, K., Cumbal, L. et al. (2014). Sonochemical synthesis of silver nanoparticles using starch: a comparison. *Bioinorganic Chemistry and Applications* 2014: 784268.

9 Gupta, A. and Srivastava, R. (2019). Mini submersible pump assisted sonochemical reactors: large-scale synthesis of zinc oxide nanoparticles and nanoleaves for antibacterial and anticounterfeiting applications. *Ultrasonics Sonochemistry* 52: 414–427.

10 Sakkas, P.M., Argirusi, M., Sourkouni, G., and Argirusis, C. (2020). Rhenium oxide nanoparticles – sonochemical synthesis and integration on anode powders for solid oxide fuel cells. *Ultrasonics Sonochemistry* 69: 105250.

11 Hayashi, H. and Hakuta, Y. (2010). Hydrothermal synthesis of metal oxide nanoparticles in supercritical water. *Materials* 3 (7): 3794–3817.

12 Tadic, M., Trpkov, D., Kopanja, L. et al. (2019). Hydrothermal synthesis of hematite (α-Fe2O_3) nanoparticle forms: synthesis conditions, structure, particle shape analysis, cytotoxicity and magnetic properties. *Journal of Alloys and Compounds* 792: 599–609.

13 Vinay, S.P., Udayabhanu, N.G., Chandrappa, C.P., and Chandrasekhar, N. (2020). Hydrothermal synthesis of gold nanoparticles using spider cobweb as novel biomaterial: application to photocatalytic. *Chemical Physics Letters* 748: 137402.

14 Hinds, S., Taft, B.J., Levina, L. et al. (2006). Nucleotide-directed growth of semiconductor nanocrystals. *Journal of the American Chemical Society* 128 (1): 64–65.

15 Wang, X., Pandey, R.R., Singh, K.V. et al. (2006). Synthesis and characterization of peptide nucleic acid-platinum nanoclusters. *Nanotechnology* 17 (5): 1177–1183.

16 Gugliotti, L.A., Feldheim, D.L., and Eaton, B.E. (2004). RNA-mediated metal-metal bond formation in the synthesis of hexagonal palladium nanoparticles. *Science* 304 (5672): 850–852.

17 Bogle, K.A., Dhole, S.D., and Bhoraskar, V.N. (2006). Silver nanoparticles: synthesis and size control by electron irradiation. *Nanotechnology* 17 (13): 3204–3208.

18 Zhu, Y., Qian, Y., Zhang, M. et al. (1994). Preparation and characterization of nanocrystalline powders of cuprous oxide by using Y-radiation. *Materials Research Bulletin* 29 (4): 377–383.

19 Roth, O., Hasselberg, H., and Jonsson, M. (2009). Radiation chemical synthesis and characterization of UO_2 nanoparticles. *Journal of Nuclear Materials* 383 (3): 231–236.

20 Rath, M.C., Sunitha, Y., Ghosh, H.N. et al. (2009). Investigation of the dynamics of radiolytic formation of ZnO nanostructured materials by pulse radiolysis. *Radiation Physics and Chemistry* 78 (2): 77–80.

21 Dahoumane, S.A., Jeffryes, C., Mechouet, M., and Agathos, S.N. (2017). Biosynthesis of inorganic nanoparticles: a fresh look at the control of shape, size and composition. *Bioengineering* 4, 4 (19): PMC5590428.

22 Das, S.K., Das, A.R., and Guha, A.K. (2010). Microbial synthesis of multishaped gold nanostructures. *Small* 6 (9): 1012–1021. https://doi.org/10.1002/smll.200902011.

23 Chokshi, K., Pancha, I., Ghosh, T. et al. (2016). Green synthesis, characterization and antioxidant potential of silver nanoparticles biosynthesized from de-oiled biomass of thermotolerant oleaginous microalgae *Acutodesmus dimorphus*. *RSC Advances* 6 (76): 72269–72274.

24 Dauthal, P. and Mukhopadhyay, M. (2013). Biosynthesis of palladium nanoparticles using *Delonix regia* leaf extract and its catalytic activity for nitro-aromatics hydrogenation. *Industrial and Engineering Chemistry Research* 52 (51): 18131–18139.

25 Castro, L., Blazquez, M.L., Gonzalez, F. et al. (2015). Biosynthesis of silver and platinum nanoparticles using orange peel extract: characterisation and applications. *IET Nanobiotechnology* 9 (5): 252–258.

26 Tada, M., Hatanaka, S., Sanbonsugi, H. et al. (2003). Method for synthesizing ferrite nanoparticles ~30 nm in diameter on neutral pH condition for biomedical applications. *Journal of Applied Physics* 93 (10): 7566–7568.

27 Jolivet, J.P., Tronc, E., and Chaneac, C. (2000). Synthesis of iron oxide- and metal-based nanomaterials. *European Physical Journal Applied Physics* 10 (3): 167–172.

28 Levy, M., Quarta, A., Espinosa, A. et al. (2011). Correlating magneto-structural properties to hyperthermia performance of highly monodisperse iron oxide nanoparticles prepared by a seeded-growth route. *Chemistry of Materials* 23 (18): 4170–4180.

29 Inouye, K., Endo, R., Otsuka, Y. et al. (1982). Oxygenation of ferrous ions in reversed micelle and reversed microemulsion. *The Journal of Physical Chemistry* 86 (8): 1465–1469.

30 Carpenter, E.E., Seip, C.T., and O'Connor, C.J. (1999). Magnetism of nanophase metal and metal alloy particles formed in ordered phases. *Journal of Applied Physics* 85 (8): 5184–5186.

31 Feltin, N. and Pileni, M.P. (1997). New technique for synthesizing iron ferrite magnetic nanosized particles. *Langmuir* 13 (15): 3927–3933.

32 Fievet, F., Lagier, J.P., Blin, B. et al. (1989). Homogeneous and heterogeneous nucleations in the polyol process for the preparation of micron and submicron size metal particles. *Solid State Ionics* 32–33: 198–205.

33 Jungk, H.-O. and Feldmann, C. (2000). Nonagglomerated, submicron α-Fe_2O_3 particles: preparation and application. *Journal of Materials Research* 15: 2244–2248.

34 Nasirzadeh, K., Nazarian, S., and Gheibi Hayat, S.M. (2016). Inorganic nanomaterials: a brief overview of the applications and developments in sensing and drug delivery. *Journal of Applied Biotechnology Reports* 3: 395–402.

35 Jiang, W., Kim, B.Y.S., Rutka, J.T., and Chan, W.C.W. (2008). Nanoparticle-mediated cellular response is size-dependent. *Nature Nanotechnology* 3 (3): 145–150.

36 Nel, A., Xia, T., Madler, L., and Li, N. (2006). Toxic potential of materials at the nanolevel. *Science* 311 (5761): 622–627.

37 Eustis, S. and El-Sayed, M.A. (2006). Why gold nanoparticles are more precious than pretty gold: noble metal surface plasmon resonance and its enhancement of the radiative and nonradiative properties of nanocrystals of different shapes. *Chemical Society Reviews* 35 (3): 209–217.

38 Khlebtsov, N.G. and Dykman, L.A. (2010). Optical properties and biomedical applications of plasmonic nanoparticles. *Journal of Quantitative Spectroscopy and Radiation Transfer* 111 (1): 1–35.

39 Shi, B., Xia, D., and Yao, Y. (2014). A water-soluble supramolecular polymer constructed by pillar[5]arene-based molecular recognition. *Chemical Communications* 50 (90): 13932–13935.

40 Park, C., Jeong, E.S., Lee, K.J. et al. (2014). Carboxylated pillar[5]arene-coated gold nanoparticles with chemical stability and enzyme-like activity. *Chemistry: An Asian Journal* 9 (10): 2761–2764.

41 Gao, B., Tan, L.-L., Song, N. et al. (2016). A high-yield synthesis of [m]biphenyl-extended pillar[n]arenes for an efficient selective inclusion of toluene and m-xylene in the solid state. *Chemical Communications* 52 (34): 5804–5807.

42 Yao, Y., Zhou, Y., Dai, J. et al. (2014). Host-guest recognition-induced color change of water-soluble pillar[5]arene modified silver nanoparticles for visual detection of spermine analogues. *Chemical Communications* 50 (7): 869–871.

43 Erathodiyil, N. and Ying, J.Y. (2011). Functionalization of inorganic nanoparticles for bioimaging applications. *Accounts of Chemical Research* 44 (10): 925–935.

44 Khan, K. and Javed, S. (2018). Functionalization of inorganic nanoparticles to augment antimicrobial efficiency: a critical analysis. *Current Pharmaceutical Biotechnology* 19 (7): 523–536.

45 Solak, K., Mavi, A., and Yilmaz, B. (2021). Disulfiram-loaded functionalized magnetic nanoparticles combined with copper and sodium nitroprusside in breast cancer cells. *Materials Science and Engineering: C* 119: 111452.

46 Shen, J., Liu, H., Mu, C. et al. (2017). Multi-step encapsulation of chemotherapy and gene silencing agents in functionalized mesoporous silica nanoparticles. *Nanoscale* 9 (16): 5329–5341.

47 Lin, D., Cheng, Q., Jiang, Q. et al. (2013). Intracellular cleavable poly(2-dimethylaminoethyl methacrylate) functionalized mesoporous silica nanoparticles for efficient siRNA delivery in vitro and in vivo. *Nanoscale* 5 (10): 4291–4301.

48 Jena, K.K., Narayan, R., and Raju, K.V.S.N. (2015). Surface functionalized zinc oxide (ZnO) nanoparticle filled organic-inorganic hybrid materials with enhanced thermomechanical properties. *Progress in Organic Coatings* 89: 82–90.

49 Graf, C., Gao, Q., Schutz, I. et al. (2012). Surface functionalization of silica nanoparticles supports colloidal stability in physiological media and facilitates internalization in cells. *Langmuir* 28 (20): 7598–7613.

50 Matea, C.T., Mocan, T., Tabaran, F. et al. (2017). Quantum dots in imaging, drug delivery and sensor applications. *International Journal of Nanomedicine* 12: 5421–5431.

51 Zhao, M. and Zhu, B. (2016). The research and applications of quantum dots as nano-carriers for targeted drug delivery and cancer therapy. *Nanoscale Research Letters* 11: 207.

52 Zayed, D.G., AbdElhamid, A.S., Freag, M.S., and Elzoghby, A.O. (2019). Hybrid quantum dot-based theranostic nanomedicines for tumor-targeted drug delivery and cancer imaging. *Nanomedicine* 14 (3): 225–228.

53 Girija Aswathy, R., Sivakumar, B., Brahatheeswaran, D. et al. (2012). Biocompatible fluorescent zein nanoparticles for simultaneous bioimaging and drug delivery application. *Advances in Natural Sciences: Nanoscience and Nanotechnology* 3 (2): 25006.

54 Zhao, T., Liu, X., Li, Y. et al. (2017). Fluorescence and drug loading properties of ZnSe:Mn/ZnS-paclitaxel/SiO_2 nanocapsules templated by F127 micelles. *Journal of Colloid and Interface Science* 490: 436–443.

55 Cai, X., Luo, Y., Zhang, W. et al. (2016). pH-Sensitive ZnO quantum dots-doxorubicin nanoparticles for lung cancer targeted drug delivery. *ACS Applied Materials & Interfaces* 8 (34): 22442–22450.

56 Han, H., Valdeperez, D., Jin, Q. et al. (2017). Dual enzymatic reaction-assisted gemcitabine delivery systems for programmed pancreatic cancer therapy. *ACS Nano* 11 (2): 1281–1291.

57 Bwatanglang, I.B., Mohammad, F., Yusof, N.A. et al. (2016). In vivo tumor targeting and anti-tumor effects of 5-fluororacil loaded, folic acid targeted quantum dot system. *Journal of Colloid and Interface Science* 480: 146–158.

58 Olerile, L.D., Liu, Y., Zhang, B. et al. (2017). Near-infrared mediated quantum dots and paclitaxel co-loaded nanostructured lipid carriers for cancer theragnostic. *Colloids and Surfaces, B: Biointerfaces* 150: 121–130.

59 Popovic, Z., Liu, W., Chauhan, V.P. et al. (2010). A nanoparticle size series for in vivo fluorescence imaging. *Angewandte Chemie, International Edition* 49 (46): 8649–8652.

60 Bilan, R., Nabiev, I., and Sukhanova, A. (2016). Quantum dot-based nanotools for bioimaging, diagnostics, and drug delivery. *ChemBioChem* 17 (22): 2103–2114.

61 Wu, W., Aiello, M., Zhou, T. et al. (2010). In-situ immobilization of quantum dots in polysaccharide-based nanogels for integration of optical pH-sensing, tumor cell imaging, and drug delivery. *Biomaterials* 31 (11): 3023–3031.

62 Nurunnabi, M., Cho, K.J., Choi, J.S. et al. (2010). Targeted near-IR QDs-loaded micelles for cancer therapy and imaging. *Biomaterials* 31 (20): 5436–5444.

63 Weng, K.C., Noble, C.O., Papahadjopoulos-Sternberg, B. et al. (2008). Targeted tumor cell internalization and imaging of multifunctional quantum dot-conjugated immunoliposomes in vitro and in vivo. *Nano Letters* 8 (9): 2851–2857.

64 Wang, M., Marepally, S.K., Vemula, P.K., and Xu, C. (2016). Inorganic nanoparticles for transdermal drug delivery and topical application. In: *Nanoscience in Dermatology* (ed. M.R. Hamblin, P. Avci and T.W. Prow), 57–72. Boston: Academic Press.

65 DeLouise, L.A. (2012). Applications of nanotechnology in dermatology. *The Journal of Investigative Dermatology* 132 (3 Pt 2): 964–975.

66 Gopee, N.V., Roberts, D.W., Webb, P. et al. (2009). Quantitative determination of skin penetration of PEG-coated CdSe quantum dots in dermabraded but not intact SKH-1 hairless mouse skin. *Toxicological Sciences* 111 (1): 37–48.

67 Pedata, P., Boccellino, M., Porta, R. et al. (2011). Interaction between combustion-generated organic nanoparticles and biological systems: in vitro study of cell toxicity and apoptosis in human keratinocytes. *Nanotoxicology* 6: 338–352.

68 Upadhyay, P. (2006). Enhanced transdermal-immunization with diptheria-toxoid using local hyperthermia. *Vaccine* 24 (27–28): 5593–5598.

69 Vallet-Regí, M., Colilla, M., Izquierdo-Barba, I., and Manzano, M. (2017). Mesoporous silica nanoparticles for drug delivery: current insights. *Molecules* 23 (1): 47.

70 Ruoslahti, E., Bhatia, S.N., and Sailor, M.J. (2010). Targeting of drugs and nanoparticles to tumors. *The Journal of Cell Biology* 188 (6): 759–768.

71 Manzano, M. and Vallet-Regí, M. (2020). Mesoporous silica nanoparticles for drug delivery. *Advanced Functional Materials* 30 (2): 1902634.

72 Villaverde, G., Baeza, A., Melen, G.J. et al. (2015). A new targeting agent for the selective drug delivery of nanocarriers for treating neuroblastoma. *Journal of Materials Chemistry B* 3 (24): 4831–4842.

73 Zhao, Y., Jiang, Y., Lv, W. et al. (2016). Dual targeted nanocarrier for brain ischemic stroke treatment. *Journal of Controlled Release* 233: 64–71.

74 Lopez, V., Villegas, M.R., Rodriguez, V. et al. (2017). Janus mesoporous silica nanoparticles for dual targeting of tumor cells and mitochondria. *ACS Applied Materials & Interfaces* 9 (32): 26697–26706.

75 Villegas, M.R., Baeza, A., and Vallet-Regí, M. (2015). Hybrid collagenase nanocapsules for enhanced nanocarrier penetration in tumoral tissues. *ACS Applied Materials & Interfaces* 7 (43): 24075–24081.

76 Vegh, I., Grau, M., Gracia, M. et al. (2013). Decidua mesenchymal stem cells migrated toward mammary tumors in vitro and in vivo affecting tumor growth and tumor development. *Cancer Gene Therapy* 20 (1): 8–16.

77 Paris, J.L., de la Torre, P., Manzano, M. et al. (2016). Decidua-derived mesenchymal stem cells as carriers of mesoporous silica nanoparticles. In vitro and in vivo evaluation on mammary tumors. *Acta Biomaterialia* 33: 275–282.

78 Karimi, M., Mirshekari, H., Aliakbari, M. et al. (2016). Smart mesoporous silica nanoparticles for controlled-release drug delivery. *Nanotechnology Reviews* 5 (2): 195–207.

79 Argyo, C., Weiss, V., Brauchle, C., and Bein, T. (2014). Multifunctional mesoporous silica nanoparticles as a universal platform for drug delivery. *Chemistry of Materials* 26 (1): 435–451.

80 Cheng, W., Nie, J., Xu, L. et al. (2017). pH-sensitive delivery vehicle based on folic acid-conjugated polydopamine-modified mesoporous silica nanoparticles for targeted cancer therapy. *ACS Applied Materials & Interfaces* 9 (22): 18462–18473.

81 Baeza, A., Colilla, M., and Vallet-Regí, M. (2015). Advances in mesoporous silica nanoparticles for targeted stimuli-responsive drug delivery. *Expert Opinion on Drug Delivery* 12 (2): 319–337.

82 Zhou, Y., Quan, G., Wu, Q. et al. (2018). Mesoporous silica nanoparticles for drug and gene delivery. *Acta Pharmaceutica Sinica B* 8 (2): 165–177.

83 Mura, S., Nicolas, J., and Couvreur, P. (2013). Stimuli-responsive nanocarriers for drug delivery. *Nature Materials* 12 (11): 991–1003.

84 Li, Z., Zhang, Y., and Feng, N. (2019). Mesoporous silica nanoparticles: synthesis, classification, drug loading, pharmacokinetics, biocompatibility, and application in drug delivery. *Expert Opinion on Drug Delivery* 16 (3): 219–237.

85 Zhang, J. and Rosenholm, J.M. (2015). The viability of mesoporous silica nanoparticles for drug delivery. *Therapeutic Delivery* 6 (8): 891–893.

86 Bharti, C., Nagaich, U., Pal, A.K., and Gulati, N. (2015). Mesoporous silica nanoparticles in target drug delivery system: a review. *International Journal of Pharmaceutical Investigation* 5 (3): 124–133.

87 Summerlin, N., Qu, Z., Pujara, N. et al. (2016). Colloidal mesoporous silica nanoparticles enhance the biological activity of resveratrol. *Colloids and Surfaces, B: Biointerfaces* 144: 1–7.
88 Bukara, K., Schueller, L., Rosier, J. et al. (2016). Ordered mesoporous silica to enhance the bioavailability of poorly water-soluble drugs: proof of concept in man. *European Journal of Pharmaceutics and Biopharmaceutics* 108: 220–225.
89 Xu, C., Lei, C., and Yu, C. (2019). Mesoporous silica nanoparticles for protein protection and delivery. *Frontiers in Chemistry* 7: 290.
90 Ahmed, S., Ahmad, M., Swami, B.L., and Ikram, S. (2016). A review on plants extract mediated synthesis of silver nanoparticles for antimicrobial applications: a green expertise. *Journal of Advanced Research* 7 (1): 17–28.
91 Dakal, T.C., Kumar, A., Majumdar, R.S., and Yadav, V. (2016). Mechanistic basis of antimicrobial actions of silver nanoparticles. *Frontiers in Microbiology* 7: 1831.
92 Duran, N., Duran, M., de Jesus, M.B. et al. (2016). Silver nanoparticles: a new view on mechanistic aspects on antimicrobial activity. *Nanomedicine: Nanotechnology, Biology and Medicine* 12 (3): 789–799.
93 Abbaszadegan, A., Ghahramani, Y., Gholami, A. et al. (2014). The effect of charge at the surface of silver nanoparticles on antimicrobial activity against gram-positive and gram-negative bacteria: a preliminary study. *Journal of Nanomaterials* 2015: 8.
94 Keat, C.L., Aziz, A., Eid, A.M., and Elmarzugi, N.A. (2015). Biosynthesis of nanoparticles and silver nanoparticles. *Bioresources and Bioprocessing* 2 (1): 47.
95 Prabhu, S. and Poulose, E.K. (2012). Silver nanoparticles: mechanism of antimicrobial action, synthesis, medical applications, and toxicity effects. *International Nano Letters* 2 (1): 32.
96 Huang, C.-C., Aronstam, R.S., Chen, D.-R., and Huang, Y.-W. (2010). Oxidative stress, calcium homeostasis, and altered gene expression in human lung epithelial cells exposed to ZnO nanoparticles. *Toxicology in vitro* 24 (1): 45–55.
97 Bhuvaneswari, T., Thiyagarajan, M., Geetha, N., and Venkatachalam, P. (2014). Bioactive compound loaded stable silver nanoparticle synthesis from microwave irradiated aqueous extracellular leaf extracts of *Naringi crenulata* and its wound healing activity in experimental rat model. *Acta Tropica* 135: 55–61.
98 Mandal, A.K. (2017). Silver nanoparticles as drug delivery vehicle against infections. *Global Journal of Nanomedicine* 3 (2): 31–34.
99 Kanwal, Z., Raza, M.A., Riaz, S. et al. (2020). Synthesis and characterization of silver nanoparticle-decorated cobalt nanocomposites (Co@AgNPs) and their density-dependent antibacterial activity. *Royal Society Open Science* 6 (5): 182135.
100 Rai, M., Ingle, A.P., Gupta, I., and Brandelli, A. (2015). Bioactivity of noble metal nanoparticles decorated with biopolymers and their application in drug delivery. *International Journal of Pharmaceutics* 496 (2): 159–172.
101 Ravindra, S., Mulaba-Bafubiandi, A., Vanarchi, R. et al. (2012). Development and characterization of curcumin loaded silver nanoparticle hydrogels for antibacterial and drug delivery applications. *Journal of Inorganic and Organometallic Polymers and Materials* 22: 1254–1262.
102 Vimala, K., Varaprasad, K., Sadiku, R. et al. (2014). Development of novel protein-Ag nanocomposite for drug delivery and inactivation of bacterial applications. *International Journal of Biological Macromolecules* 63: 75–82.

103 Sripriya, J., Anandhakumar, S., Achiraman, S. et al. (2013). Laser receptive polyelectrolyte thin films doped with biosynthesized silver nanoparticles for antibacterial coatings and drug delivery applications. *International Journal of Pharmaceutics* 457 (1): 206–213.

104 Rao, P.V., Nallappan, D., Madhavi, K. et al. (2016). Phytochemicals and biogenic metallic nanoparticles as anticancer agents. *Oxidative Medicine and Cellular Longevity* 2016: 3685671.

105 Naz, M., Nasiri, N., Ikram, M. et al. (2017). Eco-friendly biosynthesis, anticancer drug loading and cytotoxic effect of capped Ag-nanoparticles against breast cancer. *Applied Nanoscience* 7 (8): 793–802.

106 Benyettou, F., Rezgui, R., Ravaux, F. et al. (2015). Synthesis of silver nanoparticles for the dual delivery of doxorubicin and alendronate to cancer cells. *Journal of Materials Chemistry B* 3 (36): 7237–7245.

107 Vasanth, K., Ilango, K., and MohanKumar, R. (2014). Anticancer activity of *Moringa oleifera* mediated silver nanoparticles on human cervical carcinoma cells by apoptosis induction. *Colloids and Surfaces, B: Biointerfaces* 117: 354–359.

108 Soumya, R.S. and Hela, P.G. (2013). Nano silver based targeted drug delivery for treatment of cancer. *Der Pharmacia Lettre* 5 (4): 189–197.

109 Lara, H.H., Garza-Trevino, E.N., Ixtepan-Turrent, L. et al. (2011). Silver nanoparticles are broad-spectrum bactericidal and virucidal compounds. *Journal of Nanobiotechnology* 9: 30. https://doi.org/10.1186/1477-3155-9-30.

110 Sun, R.W.-Y., Chen, R., Chung, N.P.-Y. et al. (2005). Silver nanoparticles fabricated in Hepes buffer exhibit cytoprotective activities toward HIV-1 infected cells. *Chemical Communications* (40): 5059–5061.

111 Xiang, D., Zheng, Y., Duan, W. et al. (2013). Inhibition of A/Human/Hubei/3/2005 (H_3N_2) influenza virus infection by silver nanoparticles in vitro and in vivo. *International Journal of Nanomedicine* 8: 4103–4113.

112 Galdiero, S., Falanga, A., Vitiello, M. et al. (2011). Silver nanoparticles as potential antiviral agents. *Molecules* 16 (10): 8894–8918.

113 Kumar, C.G. and Poornachandra, Y. (2015). Biodirected synthesis of miconazole-conjugated bacterial silver nanoparticles and their application as antifungal agents and drug delivery vehicles. *Colloids and Surfaces, B: Biointerfaces* 125: 110–119.

114 Ahmad, A., Wei, Y., Syed, F. et al. (2016). Amphotericin B-conjugated biogenic silver nanoparticles as an innovative strategy for fungal infections. *Microbial Pathogenesis* 99: 271–281.

115 Hussain, M., Ahmed, D., Anwar, A. et al. (2018). Combination therapy of clinically approved antifungal drugs is enhanced by conjugation with silver nanoparticles. *International Microbiology* 22 (2): –239, 246.

116 Jeyaraj, M., Rajesh, M., Arun, R. et al. (2013). An investigation on the cytotoxicity and caspase-mediated apoptotic effect of biologically synthesized silver nanoparticles using *Podophyllum hexandrum* on human cervical carcinoma cells. *Colloids and Surfaces, B: Biointerfaces* 102: 708–717.

117 Blanco, M.D., Teijón, C., Olmo, R.M. et al. (2012). Targeted nanoparticles for cancer therapy. In: *Recent Advances in Novel Drug Carrier Systems* (ed. A.D. Sezer), 241–278. Rijeka: IntechOpen.

118 Khalid, S. and Hanif, R. (2017). Green biosynthesis of silver nanoparticles. *International Journal of Advanced Science Engineering and Technology* 5 (2): 59–63.

119 Singh, A., Dar, M.Y., Joshi, B. et al. (2018). Phytofabrication of silver nanoparticles: novel drug to overcome hepatocellular ailments. *Toxicology Reports* 5: 333–342.

120 Jelinkova, P., Mazumdar, A., Sur, V.P. et al. (2019). Nanoparticle-drug conjugates treating bacterial infections. *Journal of Controlled Release* 307: 166–185.

121 Prasher, P., Sharma, M., Mudila, H. et al. (2020). Emerging trends in clinical implications of bio-conjugated silver nanoparticles in drug delivery. *Colloid Interface Science Communication* 35: 100244.

122 Ivanova, N., Gugleva, V., Dobreva, M. et al. (2019). Silver nanoparticles as multi-functional drug delivery systens. In: *Nanomedicines* (ed. M.A. Farrukh), 71–92. Rijeka: IntechOpen.

123 Brown, P.K., Qureshi, A.T., Moll, A.N. et al. (2013). Silver nanoscale antisense drug delivery system for photoactivated gene silencing. *ACS Nano* 7 (4): 2948–2959.

124 Levard, C., Hotze, E.M., Lowry, G.V., and Brown, G.E. (2012). Environmental transformations of silver nanoparticles: impact on stability and toxicity. *Environmental Science & Technology* 46 (13): 6900–6914.

125 Larese, F.F., D'Agostin, F., Crosera, M. et al. (2009). Human skin penetration of silver nanoparticles through intact and damaged skin. *Toxicology* 255 (1): 33–37.

126 George, R., Merten, S., Wang, T.T. et al. (2014). In vivo analysis of dermal and systemic absorption of silver nanoparticles through healthy human skin. *The Australasian Journal of Dermatology* 55 (3): 185–190.

127 Samberg, M.E., Oldenburg, S.J., and Monteiro-Riviere, N.A. (2009, 2010). Evaluation of silver nanoparticle toxicity in skin in vivo and keratinocytes in vitro. *Environmental Health Perspectives* 118 (3): 407–413.

128 Tian, J., Wong, K.K.Y., Ho, C.-M. et al. (2007). Topical delivery of silver nanoparticles promotes wound healing. *ChemMedChem* 2 (1): 129–136.

129 Chowdhury, S., De, M., Guha, R. et al. (2014). Influence of silver nanoparticles on post-surgical wound healing following topical application. *European Journal of Nanomedicine* 6 (4): 237–247.

130 Farooq, M.U., Novosad, V., Rozhkova, E.A. et al. (2018). Gold nanoparticles-enabled efficient dual delivery of anticancer therapeutics to HeLa cells. *Scientific Reports* 8 (1): 2907.

131 Tacar, O., Sriamornsak, P., and Dass, C.R. (2013). Doxorubicin: an update on anticancer molecular action, toxicity and novel drug delivery systems. *The Journal of Pharmacy and Pharmacology* 65 (2): 157–170.

132 Stubbe, J. and Kozarich, J.W. (1987). Mechanisms of bleomycin-induced DNA degradation. *Chemical Reviews* 87 (5): 1107–1136.

133 Chen, Y.-H., Tsai, C.-Y., Huang, P.-Y. et al. (2007). Methotrexate conjugated to gold nanoparticles inhibits tumor growth in a syngeneic lung tumor model. *Molecular Pharmaceutics* 4 (5): 713–722.

134 Wang, F., Wang, Y.-C., Dou, S. et al. (2011). Doxorubicin-tethered responsive gold nanoparticles facilitate intracellular drug delivery for overcoming multidrug resistance in cancer cells. *ACS Nano* 5 (5): 3679–3692.

135 Van de Broek, B., Devoogdt, N., D'Hollander, A. et al. (2011). Specific cell targeting with nanobody conjugated branched gold nanoparticles for photothermal therapy. *ACS Nano* 5 (6): 4319–4328.

136 Brown, S.D., Nativo, P., Smith, J.-A. et al. (2010). Gold nanoparticles for the improved anticancer drug delivery of the active component of oxaliplatin. *Journal of the American Chemical Society* 132 (13): 4678–4684.

137 Dhar, S., Daniel, W.L., Giljohann, D.A. et al. (2009). Polyvalent oligonucleotide gold nanoparticle conjugates as delivery vehicles for platinum(IV) warheads. *Journal of the American Chemical Society* 131 (41): 14652–14653.

138 Kong, F.-Y., Zhang, J.-W., Li, R.-F. et al. (2017). Unique roles of gold nanoparticles in drug delivery, targeting and imaging applications. *Molecules* 22: 1445.

139 Gu, Y.-J., Cheng, J., Lin, C.-C. et al. (2009). Nuclear penetration of surface functionalized gold nanoparticles. *Toxicology and Applied Pharmacology* 237 (2): 196–204.

140 Han, G., Ghosh, P., and Rotello, V.M. (2007). Functionalized gold nanoparticles for drug delivery. *Nanomedicine* 2 (1): 113–123.

141 Chandran, P.R. and Thomas, R.T. (2015). Gold nanoparticles in cancer drug delivery. In: *Nanotechnology Applications for Tissue Engineering* (ed. S. Thomas, Y. Grohens and N. Ninan), 221–237. Oxford: William Andrew Publishing.

142 Allen, T.M. (2002). Ligand-targeted therapeutics in anticancer therapy. *Nature Reviews Cancer* 2 (10): 750–763.

143 Medley, C.D., Smith, J.E., Tang, Z. et al. (2008). Gold nanoparticle-based colorimetric assay for the direct detection of cancerous cells. *Analytical Chemistry* 80 (4): 1067–1072.

144 Lin, J.-J., Chen, J.-S., Huang, S.-J. et al. (2009). Folic acid-Pluronic F127 magnetic nanoparticle clusters for combined targeting, diagnosis, and therapy applications. *Biomaterials* 30 (28): 5114–5124.

145 Dixit, V., Van den Bossche, J., Sherman, D.M. et al. (2006). Synthesis and grafting of thioctic acid-PEG-folate conjugates onto Au nanoparticles for selective targeting of folate receptor-positive tumor cells. *Bioconjugate Chemistry* 17 (3): 603–609.

146 Anderson, M.E. (1998). Glutathione: an overview of biosynthesis and modulation. *Chemico-Biological Interactions* 111–112: 1–14.

147 Bao, Q.-Y., Geng, D.-D., Xue, J.-W. et al. (2013). Glutathione-mediated drug release from Tiopronin-conjugated gold nanoparticles for acute liver injury therapy. *International Journal of Pharmaceutics* 446 (1): 112–118.

148 Dreaden, E.C., Austin, L.A., Mackey, M.A., and El-Sayed, M.A. (2012). Size matters: gold nanoparticles in targeted cancer drug delivery. *Therapeutic Delivery* 3 (4): 457–478.

149 Weissleder, R. (2001). A clearer vision for in vivo imaging. *Nature Biotechnology* 19 (4): 316–317.

150 Pitsillides, C.M., Joe, E.K., Wei, X. et al. (2003). Selective cell targeting with light-absorbing microparticles and nanoparticles. *Biophysical Journal* 84 (6): 4023–4032.

151 El-Sayed, I.H., Huang, X., and El-Sayed, M.A. (2006). Selective laser photo-thermal therapy of epithelial carcinoma using anti-EGFR antibody conjugated gold nanoparticles. *Cancer Letters* 239 (1): 129–135.

152 Hainfeld, J.F., Dilmanian, F.A., Slatkin, D.N., and Smilowitz, H.M. (2008). Radiotherapy enhancement with gold nanoparticles. *The Journal of Pharmacy and Pharmacology* 60 (8): 977–985.

153 Chang, M.-Y., Shiau, A.-L., Chen, Y.-H. et al. (2008). Increased apoptotic potential and dose-enhancing effect of gold nanoparticles in combination with single-dose clinical electron beams on tumor-bearing mice. *Cancer Science* 99 (7): 1479–1484.

154 Cardinal, J., Klune, J.R., Chory, E. et al. (2008). Noninvasive radiofrequency ablation of cancer targeted by gold nanoparticles. *Surgery* 144 (2): 125–132.

155 Lee, N., Yoo, D., Ling, D. et al. (2015). Iron oxide based nanoparticles for multimodal imaging and magnetoresponsive therapy. *Chemical Reviews* 115 (19): 10637–10689.

156 Neuberger, T., Schopf, B., Hofmann, H. et al. (2005). Superparamagnetic nanoparticles for biomedical applications: possibilities and limitations of a new drug delivery system. *Journal of Magnetism and Magnetic Materials* 293 (1): 483–496.

157 Sun, C., Lee, J.S.H., and Zhang, M. (2008). Magnetic nanoparticles in MR imaging and drug delivery. *Advanced Drug Delivery Reviews* 60 (11): 1252–1265.

158 Chee, C.F., Leo, B.F., and Lai, C.W. (2018). Superparamagnetic iron oxide nanoparticles for drug delivery. In: *Applications of Nanocomposite Materials in Drug Delivery*, Woodhead Publishing Series in Biomaterials (ed. Inamuddin, A.M. Asiri and A. Mohammad), 861–903. Woodhead Publishing.

159 Dames, P., Gleich, B., Flemmer, A. et al. (2007). Targeted delivery of magnetic aerosol droplets to the lung. *Nature Nanotechnology* 2 (8): 495–499.

160 Tewes, F., Ehrhardt, C., and Healy, A.M. (2014). Superparamagnetic iron oxide nanoparticles (SPIONs)-loaded trojan microparticles for targeted aerosol delivery to the lung. *European Journal of Pharmaceutics and Biopharmaceutics* 86 (1): 98–104.

161 Hasenpusch, G., Geiger, J., Wagner, K. et al. (2012). Magnetized aerosols comprising superparamagnetic iron oxide nanoparticles improve targeted drug and gene delivery to the lung. *Pharmaceutical Research* 29 (5): 1308–1318.

162 Wahajuddin, A.S. (2012). Superparamagnetic iron oxide nanoparticles: magnetic nanoplatforms as drug carriers. *International Journal of Nanomedicine* 7: 3445–3471.

163 Butoescu, N., Jordan, O., Burdet, P. et al. (2009). Dexamethasone-containing biodegradable superparamagnetic microparticles for intra-articular administration: physicochemical and magnetic properties, in vitro and in vivo drug release. *European Journal of Pharmaceutics and Biopharmaceutics* 72 (3): 529–538.

164 Schulze, K., Koch, A., Schopf, B. et al. Intraarticular application of superparamagnetic nanoparticles and their uptake by synovial membrane – an experimental study in sheep. *Journal of Magnetism and Magnetic Materials* 293: 419–432.

165 Butoescu, N., Seemayer, C.A., Foti, M. et al. (2009). Dexamethasone-containing PLGA superparamagnetic microparticles as carriers for the local treatment of arthritis. *Biomaterials* 30 (9): 1772–1780.

166 Palanisamy, S. and Wang, Y.-M. (2019). Superparamagnetic iron oxide nanoparticulate system: synthesis, targeting, drug delivery and therapy in cancer. *Dalton Transactions* 48 (26): 9490–9515.

167 Dulinska-Litewka, J., Tazarczyk, A., Halubiec, P. et al. (2019). Superparamagnetic iron oxide nanoparticles – current and prospective medical applications. *Materials (Basel, Switzerland)* 12 (4): 617.

168 Shevtsov, M.A., Yakovleva, L.Y., Nikolaev, B.P. et al. (2013). Tumor targeting using magnetic nanoparticle Hsp70 conjugate in a model of C6 glioma. *Neuro-Oncology* 16 (1): 38–49.

169 Nagesh, P.K.B., Johnson, N.R., Boya, V.K.N. et al. (2016). PSMA targeted docetaxel-loaded superparamagnetic iron oxide nanoparticles for prostate cancer. *Colloids and Surfaces, B: Biointerfaces* 144: 8–20.

170 Tiwari, A., Kumar, R., Shefi, O., and Randhawa, J.K. (2020). Fluorescent mantle carbon coated core-shell SPIONs for neuroengineering applications. *ACS Applied Bio Materials* 3 (7): 4665–4673.

171 Paul, W. and Sharma, C.P. (2020). Inorganic nanoparticles for targeted drug delivery. In: *Biointegration of Medical Implant Materials*, Woodhead Publishing Series in Biomaterials, 2e (ed. C. Sharma), 333–373. Woodhead Publishing.

172 Huang, Y., Mao, K., Zhang, B., and Zhao, Y. (2017). Superparamagnetic iron oxide nanoparticles conjugated with folic acid for dual target-specific drug delivery and MRI in cancer theranostics. *Materials Science and Engineering: C* 70: 763–771.

173 Liu, E., Zhang, M., Cui, H. et al. (2018). Tat-functionalized Ag-Fe_3O_4 nano-composites as tissue-penetrating vehicles for tumor magnetic targeting and drug delivery. *Acta Pharmaceutica Sinica B* 8 (6): 956–968.

174 Dutta, D., Sahoo, A., Chattopadhyay, A., and Ghosh, S. (2015). Bimetallic silver nanoparticle–gold nanocluster embedded composite nanoparticles for cancer theranostics. *Journal of Materials Chemistry B* 4: 793–800.

175 Ryan, C., Alcock, E., Buttimer, F. et al. (2017). Synthesis and characterisation of cross-linked chitosan composites functionalised with silver and gold nanoparticles for antimicrobial applications. *Science and Technology of Advanced Materials* 18 (1): 528–540.

176 Shao, D., Zhang, X., Liu, W. et al. (2016). Janus silver-mesoporous silica nanocarriers for SERS traceable and pH-sensitive drug delivery in cancer therapy. *ACS Applied Materials & Interfaces* 8 (7): 4303–4308.

177 Chen, G., Xie, Y., Peltier, R. et al. (2016). Peptide-decorated gold nanoparticles as functional nano-capping agent of mesoporous silica container for targeting drug delivery. *ACS Applied Materials & Interfaces* 8 (18): 11204–11209.

178 Luo, G.-F., Chen, W.-H., Liu, Y. et al. (2013). Charge-reversal plug gate nanovalves on peptide-functionalized mesoporous silica nanoparticles for targeted drug delivery. *Journal of Materials Chemistry B* 1 (41): 5723–5732.

179 Du, J.-Z., Du, X.-J., Mao, C.-Q., and Wang, J. (2011). Tailor-made dual pH-sensitive polymer–doxorubicin nanoparticles for efficient anticancer drug delivery. *Journal of the American Chemical Society* 133 (44): 17560–17563.

180 Monem, A.S., Elbialy, N., and Mohamed, N. (2014). Mesoporous silica coated gold nanorods loaded doxorubicin for combined chemo-photothermal therapy. *International Journal of Pharmaceutics* 470 (1): 1–7.

181 Shen, S., Tang, H., Zhang, X. et al. (2013). Targeting mesoporous silica-encapsulated gold nanorods for chemo-photothermal therapy with near-infrared radiation. *Biomaterials* 34 (12): 3150–3158.

182 Baek, S., Singh, R.K., Kim, T.-H. et al. (2016). Triple hit with drug carriers: pH- and temperature-responsive theranostics for multimodal chemo-and photothermal therapy and diagnostic applications. *ACS Applied Materials & Interfaces* 8 (14): 8967–8979.

183 Huang, S., Li, C., Cheng, Z. et al. (2012). Magnetic Fe_3O_4@mesoporous silica composites for drug delivery and bioadsorption. *Journal of Colloid and Interface Science* 376: 312–321.

15

Intelligent Nanomaterials for Medicine

Ajit Behera[1] *and Ranjan K. Mohapatra*[2]

[1] *Department of Metallurgical & Materials Engineering, National Institute of Technology, Rourkela, Odisha, India*
[2] *Department of Chemistry, Government College of Enginnering, Keonjhar, Odisha, India*

15.1 Introduction to Intelligent Nanomaterials

Currently, researchers are focused on finding suitable materials to detect diseases in disease-prone areas and to deliver the required treatment preferably avoiding surgery. These biological and medical applications are possible due to the development of intelligent nanomaterials. Intelligent nanomaterials have caused a revolution in the area of multiple disciplines, including micro- and nano-engineering, bionanotechnology, biomedical engineering, and nanomedicine, as well as pharmaceutical and biomedical industries [1–4]. Over the past decade, nanotechnology has had a steadily increasing impact on preclinical development in medicine, shaping the emerging scientific field of nanomedicine. An important topic is the development of composite nanosystems for diagnosis and therapy within the body. The size of the nanomaterials particles, as used here, is around same as that of biological molecules and structures. Intelligent stimuli-responsive nanostructures generally act as carrier for drug-targeting, controlled releasing at specific tissues and other therapeutic genes that are included in the term as *smart drug delivery* [5, 6]. Smart drug delivery primarily focused on: carrying a drug to release at the selected site, reducing as much as possible the side effects on the tissue/organ, optimizing the efficacy of curative drugs, avoiding the enzymatic/proteolytic degradations of carriers and drug, and minimizing the dosing frequency [7].

Various nanomaterials such as dendrimers, liposomes, nanoemulsions, micelles, and metallic nanoparticles (NPs) can be successfully used as nanodrug-delivery systems. With innovative nanoscience- and nanotechnology-based approaches, nanomaterials that have different functionalities, shapes, and distinct physical, chemical, and biological features have been designed and developed for the successful applications in the medicine industry [8]. Some intelligent nanomaterials can change their unique features and this enables them to be applied in drug-delivery materials. For example, intelligent polymeric nanomaterials

Nanomaterials and Nanotechnology in Medicine, First Edition. Edited by P.M. Visakh.

can open at a cancer cell to release the drugs imbedded in it, operated by the change in pH. One of latest interesting and promising research fields has developed from intelligent nanomaterials whose structure and chemical features are responsive to the biocatalytic action of natural enzymes [9]. Various examples of enzyme-responsive intelligent nanomaterials are polymeric nanomaterials, phospholipids, and inorganic nanomaterials. Functional mesoporous SiNPs (silicon NPs) were successfully made by applying alkoxysilane tether, a-cyclodextrin, and multifunctional peptides to targeted tumor cells to decrease the side effects of doxorubicin (DOX), which is an antitumor drug. This effective enzyme-activated NP-based drug-delivery system exhibits excellent potential in the area of nanomedicine [10].

The worldwide nanomedicine market is underpinned by cutting-edge technology for drug release, different applications in healthcare, and low-cost therapies. The global nanomedicine market is categorized into: diagnostic imaging, drug release, vaccines, regenerative medicine, and implants, etc. The global nanomedicine market accounted for US$111 912 million in 2016, and is expected to reach US$261 063 million by 2023 [11]. With the rapid progress in nanotechnology and nanoscience, the National Institute of Health (NIH) promoted a program – NNI (National Nanotechnology Initiative) – in 2000 to accelerate nano-medicinal studies. With novel approaches in nanomedicine, the area of nanoscience was rapidly accepted by researchers in the pharmaceuticals industry to create nanopharmaceuticals [12].

Nanocarriers are nano-sized materials (diameter = 1–100 nm) that can transfer various things like small drug molecules, imaging agents, proteins, and nucleic acids. Nanocarriers are designed to improve the efficiency and safety of drug delivery in general and the delivery of nonviral drugs that are specific to a particular target. The design of these nanomaterials requires the ability to control particle size, provide biocompatibility and stealth properties, optimize specificity, and achieve controlled release and functionality [13]. Polymers, carbon nanotubes (CNTs), metal NPs, organic NPs, quantum dots (QDs), nanogels, and peptide NPs were used as basic materials as possible platforms for supporting nanotechnology, as depicted in Figure 15.1.

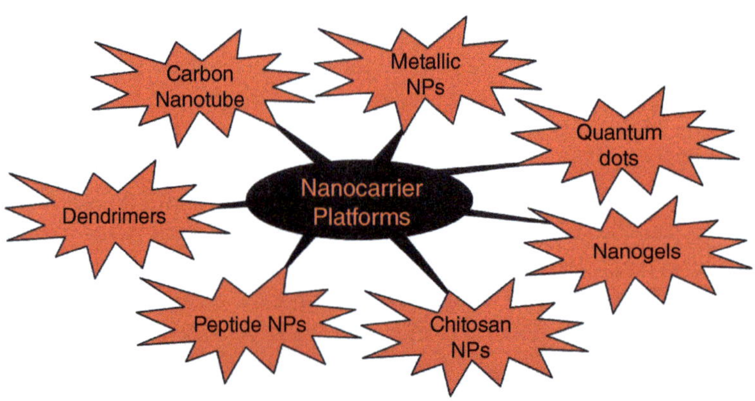

Figure 15.1 Various nanocarrier platforms in medicine industries.

15.2 Design and Function of Intelligent Nanoparticles

Stimuli-responsive targeting is currently receiving attention around the globe as it allows triggered drug release. Major efforts have been devoted to develop suitable nanotechnological platforms to improve drug delivery to a particular target tissue area. For the development of such platforms, several factors need to be considered [14, 15]: (i) the size of NPs influence the distribution, clearance, and payload uptake; (ii) a high target specificity for drug delivery or advanced functionality, (iii) controlled release mechanisms triggered by particular stimuli, (iv) biocompatibility, (v) immunological recognition and serum protein interactions in NPs stealth properties. Multiple nanoscale platforms have been developed for this purpose. In this, the nanoparticulate system can undergo physicochemical changes in its structure resulting in the release of the drug at a particular location and time when subjected to endogenous stimuli such as low pH, presence of redox gradients or certain enzymes in tumor microenvironment, or exogenous stimuli such as light, heat, electric/magnetic fields, and ultrasound. For example, Thermodox (thermosensitive liposomaldoxorubicin) has temperature-sensitive lipids and polymers. Such a combination allows the nanomaterial to remain stable at physiologic temperature and later undergo phase change increasing their permeability upon heating hence triggering the release of the drug [16].

Currently, the design of intelligent NPs for cancer therapy is the most desired field in medicine. For cancer therapy, the typical intelligent nanomaterials used are polymers, inorganic NPs such as gold (Au), iron oxide (IO), and mesoporous silica, and carbon-based materials such as CNTs or graphene. Polymer NPs can be synthesized by solvent evaporation, nanoprecipitation, emulsion polymerization, and controlled/living radical polymerizations. The most commonly employed polymeric material is poly(lactic acid-co-glycolic acid) (PLGA) due to its relatively nontoxic degradation products. Metal and metal-oxide NPs can be fabricated by mechanical attrition, and physical and chemical vapor-deposition techniques [17]. Au and IO NPs have been of great interest because of their ability to be remotely heated by infrared (IR) light and magnetic fields respectively. Mesoporous silicon (Si)-based nanotherapeutics have increased in popularity because of the ability to tailor surface functionality and load drugs into pores. Carbon-based nanomaterials such as CNTs and graphene have also emerged as promising candidates for cancer therapy due to their high surface area and ability to be used in photothermal therapies[18]. It is found that 100–200 nm size nanomaterials have the highest ability to reach cancerous tissues through passive targeting methods. In addition to size, NP shape (spherical, rod, etc.) plays an important role in the ability of a drug-delivery system to reach cancerous tissues. A single-step assembly of PLGA-lecithin-PEG (polyethylene glycol) NPs loaded with DOX and indocyanine green can be used for a single-step chemotherapy photothermal combination therapy for cancer. Surface functionalization of NPs has traditionally been achieved using PEG for biocompatibility and targeting groups such as peptides or antibodies for specificity. One such carrier is a nano-sized chitosan-functionalized graphene oxide (GO), designed as a drug and gene carrier, which serves to enhance biocompatibility and solubility. Loss or inability of active pharmaceutical ingredients to reach their final destination results in ineffective treatment, or sometimes worse, off-target side effects. The ability to encapsulate hydrophobic drug molecules has made liposomes, vesicles, and micelles attractive drug-delivery vehicles. Such nanocarriers are constructed with precise control over molecular architecture, permitting loading of a variety of anticancer agents such as hydrophobic small molecules, proteins, siRNAs (small interfering RNA), and even plasmids. Block copolymers have commonly been used as the synthetic building blocks of these composite nanocarriers [19].

On the basis of surrounding stimulus, intelligent nano-platforms can be categorized into internally (locally) and externally triggered NPs [20]. Locally stimuli are pH, redox potential, enzymatic activity, and temperature. They have the advantage of high specificity due to possible control by a molecular mechanism, which depends on the particular disease. It is expected that internally switchable nanosystems damage healthy tissue less than unswitched materials. Since their control signal is tissue specific, neighboring organs are also spared. Nanomaterials switched by internal stimuli have entered clinical studies, but far-reaching clinical benefits for patients have not yet been achieved. An important challenge is the fact that not all diseases have a specific internal molecular trigger that can be used as a trigger for switchable nanosystems [21]. Externally controlled nanosystems can be divided into temperature-, light-, electromagnetic-, radiation-, or ultrasound-sensitive delivery systems and have the advantage of external regulatory application properties in relation to the area of interest. A limiting factor for externally controlled nanomaterials is the penetration of the control signal into the tissue and the potential for tissue damage between the energy source and the target organ [22]. Tissue damage can occur when certain forms of radiation (heat, light, electromagnetic fields) interact with biomacromolecules (e.g. membrane proteins), which leads to tissue disorders and damage even in the viewer's healthy tissues. Certain externally controlled systems have already proven to be a clinical advantage, as shown by the Nanotherm NPs based on IO, which were developed by MagForce for the treatment of glioblastomas through magnetic hyperthermia or compounds based on the photodynamic Visudyne® therapy, Photofrin® and Levulan®, and Kerastick® for the treatment of aging macular degeneration, esophageal cancer, and actinic keratosis, respectively.

Liposomes and other NPs have found wide application for the delivery of various payloads. Due to their small size, biocompatibility, stability, and stealth properties, they can extend the half-life of the circulatory system, targeting tumor tissue and uptake by cells. Successful targeting requires sufficient stability to prevent premature drug release into the bloodstream and to avoid degradation of the circulating payload before the target is reached. Equally important are adequate stealth properties that allow NPs to avoid interactions with plasma proteins or with the immune system, which could lead to early removal of the material [23].

Passive targeting depends on the EPR (enhanced permeability and retention) effect, which is characterized by a greater accumulation of NPs in the tumor tissue. Rapidly growing tumor tissue is characterized by rapid angiogenesis triggered by tissue anoxia and the expression of growth factors such as vascular endothelial growth factor (VEGF). Newly formed blood vessels show a higher rate of vascular fenestration and, therefore, patency in various tumors. NPs extend through the walls of these blood vessels and tend to show prolonged retention in tumor tissue. NP extravasation is size dependent. In passive targeting strategies, a long enough NP perfusion time is important to allow high contrast between tumor accumulation compared to that of the liver, spleen, and kidneys, in which the particles tend to accumulate in the absence of a suitable target tissue. This goal can be achieved through optimal stealth properties, generally by incorporating polymers like PEG into the nanomaterial layer. Furthermore, premature release of the payload from nanocarriers (e.g. replacement of hydrophobic agents by albumin) is an important issue to consider.

Active targeting was conceptually introduced in 1906 by Paul Ehrlich, who coined the term "wonder drug" when he described the need for a system that could target the delivery of specific drugs to the body. For active targeting, the nanocarriers must first reach the target organ. Therefore, the initial design considerations are similar to those for passive targeting. Active targeting then uses specific mechanisms that, thanks to suitable ligands, increase the interaction of target cells with specific particles. These ligands (be they small molecules such as folic acid and carbohydrates or macromolecules such as aptamers, antibodies, proteins, peptides, or oligonucleotides) must be carefully selected to show maximum binding to the target [24].

15.3 Various Intelligent Materials for Medicines

Various materials are used as intelligent nanomaterials such as polymeric, lipid-based, carbon-based, metallic, hybrid, and composite nanostructures are discussed.

15.3.1 Carbon-based Materials

CNTs are a distinct molecular form of carbon atoms having hexagonal moiety. CNTs can be found as both single-walled and multiwalled forms. They yield a 10-times higher mechanical strength than that of steel. Their properties are mainly dependent on graphene folding to form the wall. Recently, CNTs have attracted attention for their use in the uncontrolled release of drugs, as well as the release of nucleic acids, peptides, and antibodies. Their inner core and outer surface allow a certain payload to be introduced into the small inner heart, while the outer surface can be modified to achieve the required biocompatibility in the body or to bind targeting ligands or payloads. The functionalization of the surface of CNTs through the union of suitable molecules has shown a spectacular decrease in their toxicity [25]. Carbon nanostructures can be dimensionally divided into four main categories: (i) 0D (zero-dimensional) carbon nanostructures (fullerenes, carbon dots (CDs), and graphene QD), (ii) 1D (one-dimensional) carbon nanostructures (CNTs and carbon nanofibers (CNFs)), (iii) 2D (two-dimensional) carbon nanostructures (GO, reduced graphene, and multilayer graphite sheets), (iv) 3D (three-dimensional) carbon nanostructures (CNT and graphene hybrids, 3D carbon microspheres, and 3D graphene superstructures).

Fullerenes can be thought of as the smallest carbon nanostructure that can be synthesized from graphite using various methods such as laser irradiation, plasma discharges, chemical vapor deposition, and arc discharge. Carbon spots were discovered in the single-walled CNT (SWCNT) synthesis using the arc-discharge method. Carbon QDs and polymer QDs as further types of 0D carbon nanostructures have been extensively studied recently. These nanostructures have been shown to be usable in a variety of fields due to the strong fluorescence and photoluminescence that can be achieved with them, particularly in relation to optical properties such as bioimaging and photocatalysis and biological labeling. For example, the superior fluorescent and optical properties of CDs make them attractive for creating TiO_2/C dots and SiO_2/C dots to convert the spectrum of sunlight to a desired wavelength for the degradation of pollutants like polyaromatic hydrocarbons. Another important application of the drug-classified (DC) family is the administration of medications. Due to their different chemical functional groups, the combination of CD nanostructures with magnetic nanostructures like Fe_3O_4 and Fe_2O_3 can be used as an structure to improve drug release, control, and delivery [26]. CNTs, discovered in 1991, are one of the best known 1D carbon nanostructures. CNTs consist of layers of graphene that are combined into a tubular or cylindrical shape with diameters of around 0.5–10 nm and a micrometer length. When it comes to the number of graphene layers in CNTs, they fall into three main categories: SWCNT, double-walled CNT (DWCNT), and multiwalled CNT (MWCNT). CNFs are another group of carbon nanostructures. CNTs have a hollow composition and are empty cylinders. In contrast, CNFs are not hollow. Regarding the high aspect ratio of CNTs, they are practical, usable nanostructures for use as polymeric additives for the manufacture of nanocomposites that improve physical and chemical properties. Graphene is known to be the thinnest and strongest 2D carbon nanostructure. Graphene layers consist of a 2D layer in which the sp^2 carbon atoms are connected to form a honeycomb network. Graphene can be synthesized using: chemical techniques (one of the most popular approaches is the Hammer method) and chemical vapor deposition, where graphene is typically synthesized with a graphite target, a "peeling" technique. Peeling separates the graphene layers by breaking up these weak Vander-Waals interactions. From a medical and biomedical point of view, graphene, GO, functionalized graphene and other derivatives of graphene have shown practical applications including disease diagnosis, bioimaging, cancer control, tissue engineering, drug delivery, cancer therapy, gene delivery, antibacterial and antiviral materials, etc. For example, GO and functionalized GO have shown promising results for the delivery of water-insoluble anticancer drugs due to their high surface area and excellent chemical functional groups. Furthermore, the extraordinary optical properties of graphene and its relatives open up a new method for treating cancer. 3D carbon nanostructures are generally synthesized using graphene sheets like 2D sheets and CNT, CNF, carbon nanohorn, and other 1D carbon nanostructures [27]. Due to the fascinating chemical and physical properties of these nanostructures, CNTs, and hybrid graphenes have been widely reported as 3D carbon nanostructures. 3D carbon nanostructures have a larger specific surface area due to an additional dimension and chemical vapor is commonly used for their production. It is believed that if 1D carbon nanostructures were arranged vertically on graphene sheets, they would be more desirable for specific applications such as drug delivery and catalysis. 3D carbon nanostructures are based on the hybridization of 1D and 2D carbon nanostructures, which represent a new structure with the properties of its components [28].

15.3.2 Metal-based Nanomaterials

Metallic NPs such as Au, silver (Ag), and IO have been developed and modified for use in drug delivery, magnetic separation, and diagnostic imaging. Super paramagnetic NPs built from oxide NPs, such as magnetite (Fe_3O_4) and maghemite (Fe_2O_3), exhibit particular features like ultrafine size, biocompatibility, and magnetic properties [29]. Superparamagnetic properties are manifested when a magnetic moment is induced by the application of a magnetic field. The large magnetic moment generates a strong signal change in magnetic resonance imaging (MRI) and thus enables sensitive detection with high resolution. Another application of IO NPs is the treatment of tumors by magnetically induced hyperthermia. Due to its chemical inertness and adapted mechanical properties, Au was used medicinally for dental implants and is also used in cancer radiation. AuNPs can be formed with core sizes ranging from 1 to 100 nm. The initial claim that there was no cytotoxicity provided an excellent drug-delivery system, although further recognition of size-dependent cytotoxicity should be considered prior to use. Au NPs can deliver peptides, proteins, nucleic acids, or small molecules. If functionalized with quaternary ammonium groups, they can bind to negatively charged DNA (deoxyribonucleic acid) or RNA (ribonucleic acid) and protect nucleic acids from enzymatic degradation [30]. Ag has long been used as a stain for glass and as a bacteriostatic agent. Elemental Ag and Ag salts are known to be relatively nontoxic to human cells and have been used in the treatment of burns, diabetic skin ulcers, and eye infections in newborns. Therefore, Ag NPs have been developed for antibacterial applications, particularly for coating medical catheters to prevent contamination and infection. The electromagnetic properties of various metallic NPs that are used in an interesting way for electromagnetic control, in particular induced hyperthermia. The effects of plasmon on metallic NPs are another physical effect with which nanoscale switches can be incorporated. Metal NPs can be surface functionalized with relative ease, so the particles can be endowed with a variety of advanced functions, including pH sensitivity and redox reactivity. One possible disadvantage is the weight-to-weight ratio between the functional payload and the inert platform, which tends to be low in metal NPs. The lack of an internal cavity is a disadvantage and the problem of long-term behavior (mass balance, bioaccumulation, elimination pathways) after injection into an organism and cellular toxicity must be carefully considered when applied to testing. Clinical applications are being considered. A potential problem with metal NPs that use heavy metals is their release into the environment in a form that may not be recyclable. A body mass balance and an estimate of the environmental mass balance of these materials appears prudent in the development of industrial products [31].

QD are small semiconductor nanocrystals (1–10 nm) consisting of an inorganic elemental nucleus (for example, cadmium and selenium) surrounded by a metal layer (ZnS) that forms a barrier between them and the nucleus is optically active. QDs can be modified by ligand binding or encapsulated with amphiphilic polymers to improve solubility, specificity, size, and display properties in tissues. Due to their resistance to photobleaching and superior quantum efficiency compared to organic dyes, they are widely used as a fluorescence imaging tools in biological research. QDs can also be used as drug carriers or simply as fluorescent markers for other drug carriers. Zintchenko et al. introduced a type of nanocarrier of drugs called "Quantoplex." Quantoplex is a polyethyleneimine nanocarrier that contains plasmid DNA and QDs and allows real-time monitoring of NPs in live animals. QDs are particularly interesting when combined with other nanoplatforms, even as a building blocks. For clinical application, its ability to combine molecular imaging and cancer therapy may open new doors in the field of cancer therapy, but the toxicity of certain materials used in QD synthesis is of considerable concern [32]. Smart nanostructured metals in the forms of powder, wire, bulk, and porous have received a lot of attention due to the combination of shape memory and superelastic properties. Nanostructured smart metals showed better interaction with different cell lines, higher mechanical properties, and smart therapeutic behavior. Shape-memory alloys as smart materials in reconstructive surgery are attracting increasing interest. Among these, the NiTi (nickel titanium) shape-memory alloy exhibits exceptional recognition and performance behavior. Good biocompatibility, fine corrosion resistance, and promising osteoconductivity have been reported for porous NiTi alloys in particular. NiTi alloys have a combination of shape memory and superelasticity, which makes this alloy particularly promising for cardiovascular stents. After complete crushing, NiTi stents can elastically return to their original shape without changing the diameter of the lumen, as the elasticity of NiTi is approximately 10 times that of the stainless steel currently used in devices. This alloy is used in the form of powder, wire, ground, mass with surface treatment and porous structure. Here nanotechnology plays an important role in improving its biological properties. The micro- and nanostructured properties of the surface of dental implants and stents generally show special behavior. Synthesized NiTi alloys with different surface roughness, from microns to nanometers, use different sizes of NiTi powder particles. The endothelial cell culture results showed that endothelial cells adhere much better to fine-grained NiTi compacts. Furthermore, cells in NiTi with higher surface roughness in the submicron- and nanorange increased more than in coarse-grained NiTi. Certain face-centered cubic (FCC) monocrystalline metallic nanowires can exhibit both shape memory and pseudoelastic properties. The mentioned properties refer to nanoscale phenomena that have not been demonstrated for bulk metals. This shows that metallic nanowires can be used as self-healing materials in future structures in the nanoscale and micrometer range. Shape-memory nanowires could include improvements over bulk alloys in a number of ways, including the ability to withstand higher tensile stresses with more than 40% reversible strains, compared to about 10% for bulk alloys. Ultrapure, highly crystalline NPs developed by mechanical alloy combustion synthesis offer a high specific surface area, which is desirable for improving osteoconductivity. High-energy milling is an innovative approach to producing nanocrystalline NiTi materials. During the milling process, nanoscale component interactions take place at atomic levels. Biological properties, including biocompatibility, bioactivity, and biodegradability appear to be controlled by grinding speed/time, compression pressure, and sintering temperature/times. Shape-memory NiTi alloys can be developed from compact aggregates that are alloyed both elementally and mechanically [33]. Furthermore, liquid laser ablation is known as a technique to generate NPs from colloidal shape memory alloys. The creation of NPs from shape-memory NiTi alloys for surface coatings has proven to be a powerful tool for providing access to nanomaterials for biomedical purposes. Cell adhesion was observed on a surface coated with laser generated NiTi NPs. Due to the effect of shape memory, superelasticity, biocompatibility, and high buffering capacity, smart metals are used in a variety of applications, particularly in biomedical engineering and MEMS (micro-electromechanical system). It has drawn a lot of attention for applications in orthodontic wires, vascular stents, orthopedic implants, and eyeglass frames. In addition, the application of nanotechnology can improve the success rates of implants for orthopedic and dental applications. NiTi has both mechanical and thermal shape-memory behavior. Thus, NiTi vascular stents can deploy when exposed to convective heat flow from the blood. Furthermore, the elasticity of NiTi shows a high degree of flexibility, which is advantageous for stenting at the vascular site [34].

The corrosion resistance of an NiTi alloy is based on the titanium dioxide (TiO_2) that forms naturally on its surface. However, this layer is too thin (typically 5–10 nm thick) and easily deteriorates. Pitting corrosion is the main mechanism behind NiTi implants in the human body. Chloride ions (Cl-) in body fluid have high diffusibility and break down the passive layer, leading to the breakdown of the oxide layer and ultimately pitting corrosion. Severe pitting and fracture corrosion was found in NiTi stent wires from explanted endovascular aortic grafts. The appearance of corrosion and breakage in NiTi stents can induce the release of toxic Ni, subsequent inflammation, and possible carcinogenic effects. Regarding the NiTi alloy, cytotoxicity of Ni ions released in endothelial cells was observed. Furthermore, the inert surface of NiTi does not allow easy attachment of endothelial cells and the formation of a monolayer. In long-term cell culture tests, cells did not grow well in NiTi with a larger grain size because the coarse particles had large surface voids, while the presence of a greater number of homogeneous interparticle voids in fine-grained NiTi appears to be most useful for endothelial cell attachment and proliferation resulting in the formation of an endothelial cell monolayer. Furthermore, data from laboratory experiments confirmed that endothelialization of an NiTi alloy can currently be enhanced by micro-/nano-surface structuring [35].

Various types of surface treatments have been reported, such as chemical or electrochemical processes, heat treatment, and ion implantation. Anodization has also been found to form a compact oxide layer enriched in titanium, particularly reducing the release of nickel from the surface of the material. Another problem with the use of NiTi is the lack of superficial osteoconductivity, which is a characteristic of metallic biomaterials but which has been less considered for NiTi and can be improved by the presence of a bioactive layer on the surface. For example, hydroxyapatite and other calcium phosphate coatings have been developed to promote adhesion of bone tissue and increase the interfacial bonding of nitinol surfaces for orthopedic applications. While these coatings can improve surface bioactivity, poor substrate bond strength and malleability are major long-term problems. If production techniques and corrosion rates are well controlled, these metallic nanostructures may be promising options for cardiovascular, orthopedic, dental, and drug-delivery system applications [36].

Metallic NPs, such as Au or silicon dioxide-gold (SiO_2-Au) NPs, have been incorporated into temperature-sensitive interpenetrating polymer matrices as novel and intelligent therapeutic nanocomposite systems. NPs or nanococci absorb light, then convert it to heat and transfer it locally to the surrounding smart therapeutic agent. Au NPs were incorporated into a polyacrylamide/polyacrylic acid coating by *in situ* inverse emulsion polymerization and then their surface was treated with PEG by covalent grafting. Depending on the type and size, the metal NP core can be adjusted to absorb visible near infrared (NIR) at wavelengths of 530 to 1200 nm. As the temperature increases, the polymer begins to swell, releasing all the beneficial agents encapsulated in the polymer network. Intelligent therapy systems using metals can revolutionize drug delivery and disease treatment based on heat-sensitive metal-polymer nanocomposites. Metallic nanowires have been studied extensively in the past decade due to their better mechanical, electrical, and optical properties as a result of their nanoscale size [37].

15.3.3 Polymer-based Nanomaterials

A few decades ago, drug-delivery mechanisms were improved through NP encapsulation and delivery, resulting in delayed release and effectiveness. Due to primary postpartum complications such as surface modification was required to improve clinical outcomes.

Polymer-based nanomaterials are known as dendrimers, large, complex molecules with a very regular chemical structure, were developed in the early 1980s. They are almost perfect monodisperse macromolecules with a regular and highly branched architecture. Dendrimers are built through an iterative series of chemical reaction steps, resulting in predictable changes in their size that are determined by each generation. In commonly used chemical syntheses, dendrimers are structures with a size of 1–10 nm and a hydrophobic interior, which allows the delivery of hydrophobic compounds such as anticancer drugs. Poly(amidoamine) (PAMAM) is the best-known dendrimer for biological applications [38].

Surface modification is particularly advantageous in cancer treatments because it can improve the selective absorption capacity of the delivery system in fighting tumors. In this regard, the polymeric compositions help to reduce undesirable physiological effects. The ease of surface modification gives polymer-based NPs a distinct advantage. Retention and permeability improvement (RPE) is a property of nanocarriers that has made them more suitable for certain medical fields. A targeted accumulation can manifest a high concentration of therapeutic agents in hydrophobic polymer NPs due to the location in the vicinity of a certain point of pathology. Water-insoluble agents can be encapsulated in the polymeric micelle, which is surrounded by its hydrophilic layer, allowing the drug-delivery system to migrate through blood and other aqueous media. Swelling also has a significant impact on the rate of drug release. Mechanisms of drug release include swelling due to the osmotic effects of the aqueous environment followed by an enzyme-induced disorder leading to drug release. However, the end result of targeting NPs depends on certain physical properties. The loading and releasing capacity of the drug depends on the chemical nature of the drug and polymer. Scientists can deduce the binding rate from the adsorption isotherms. Furthermore, manipulation of polymeric matrices can change the drug-release effect and there is a direct ratio between loading and release. The copolymer blend can also be used to control release. The intelligent nature of these NPs promotes structural self-organization in micellar form with polymers with varying degrees of hydrophobicity. The hydrophobic component of the micellar functions in pharmaceutical administration stabilizes the structure in the blood through steric protection. Smart polymer nanocomposites can also be activated by external stimuli such as a change in pH or temperature can be stimulated to release the drug. Biodegradable polymer systems have a controlled and delayed release, are safe, nonimmunogenic, and can be encapsulated for nano delivery [39].

A variety of polymeric NPs, comprising dispersion of preformed polymers or polymerization of monomers, can be prepared using techniques such as solvent evaporation, salting, dialysis, supercritical fluid technology, microemulsion, miniemulsion, surfactant-free emulsion, and interfacial polymerization. The selection of one of these depends on several factors, such as particle size, particle size distribution, and scope. Research has already correlated the physicochemical properties of polymeric NPs with their biological reactions, in which the morphology and surface charges of biodegradable entities were examined in a recent review highlighting the challenges posed by biodegradable entities. Polymeric NPs are known to offer not only protection against environmental stimuli, but also site-specific deliveries, especially for loaded NPs that are well protected. Parameters such as absorption, bioavailability, and long-term therapeutic efficacy can be optimized by controlling their electrostatic interactions. Investigation of the drug-release properties of diazepam-loaded

PLGA-NPs confirmed that parameters such as ultrasound time, polymer content, surfactant, the relationship between organic phase and aqueous phase, and the amount of drug influence its size [40].

These NPs have a core-shell structure and are composed of biodegradable polymers and copolymers to increase the circulation time in the body and decrease the absorption of the mononuclear phagocyte system (MPS). For example, some of the well-studied NPs are poly (glycolic acid), poly(lactic acid), poly(methyl methacrylate), and poly(lactic-co-glycolic acid). These are FDA approved and biodegradable. Furthermore, the crown is made up of hydrophilic polymers such as PEG and poly(vinylpyrrolidone), and the core is a polymeric matrix to encapsulate hydrophobic drugs. The active substance molecule can be physically adsorbed or chemically attached to the surface or encased in the NP.

Poly-d, 1-lactide-co-glycolide (PLGA) is an efficient synthetic NP that undergoes enzymatic hydrolysis to produce nontoxic metabolic products. Synthetic polymer NPs have a dramatic effect on improving sustained release compared to natural polymers. They have been used in the development of bone implants and screws. Nanoprecipitation is the most common synthesis method for this type of polymer NP. This polymer is often mixed with alginate or chitosan to form a compound that is useful for nanomedical and marketable purposes. Furthermore, bacterial infections such as tetanus and postmenopausal hormone therapy are relevant examples of PLGA-NP encapsulation and drug delivery [41]. Polylactic acid (PLA) is a biocompatible polymeric NP that can be produced by evaporating the solvent. Of note is its role in the encapsulation efficiency of antipsychotics, which is increased by approximately 95% and is supported by neuroleptics *in vitro* and *in vivo*. PLA can dramatically improve the blood flow time of Oridonin, which targets malignant lymphoid cells for apoptosis [42]. Polycaprolactone (PCL) is generally synthesized by nanoprecipitation. Due to its low rate of degradation, it has the strongest case for use in long-lasting implantable instruments. Similar to the aforementioned polymer NPs, PCL is effective in cancer treatments and, in addition to improving tamoxifen encapsulation efficacy, both drug release time and shelf life. It has also been used as a carrier vehicle for insulin with a positive glycemic response that preserves the bioactivity of the hormone and ultimately proves to be of great use to diabetics. The encapsulation of PCL NPs with amphotericin-B not only increases drug efficacy through delayed release, but also reduces associated toxic side effects [43].

Chitosan is a natural polymer derived from exoskeletal chitin found in arthropods. Its synthetic methods include ionotropic gelling and microemulsion. Chitosan significantly increases insulin absorption from the gastrointestinal tract after encapsulation with a hormone. Nasal insulin absorption was also improved in Arat's model when it was administered in encapsulated chitosan NPs. Gastrointestinal abnormalities such as ulcerative colitis and Crohn's disease can be combated by inhibiting the activation of T cells by cyclosporine A (Cy A) and the subsequent autoimmune reaction. Chitosan has been used effectively in combination with Cy A. In addition, systemic immune responses directed to diphtheria toxoid (DT) have been greatly enhanced during *in vivo* applications. Applications of chitosan-based NPs in the literature are widespread and indicate proven results in the treatment of various pathologies, including ocular administration and cancer therapy [44].

The bioactive and nontoxic behavior of gelatin, in addition to its widespread use in food, makes it attractive for nanomedical applications. It is derived from the breakdown of collagen protein. Gelatin is a hydrophilic polymer that can swell in an aqueous environment and is manufactured by a variety of techniques including desolvation and emulsification. The drug is loaded by incubation in the gelatin solution for a certain period of time. Two-phase drug release is characterized by rapid initial release followed by delayed release as a result of slow diffusion. Encapsulation of bovine serum albumin (BSA) maximizes the water absorbency of the gelatin NPs up to approximately 70%. Cancer and HIV drugs have also been encapsulated in gelatin NPs. In particular, these polymeric NPs can bind tumorigenic cells in a specific ligand-receptor interaction. Due to the abundance of collagen present in the eye, ocular drug delivery is another application of gelatin NPs that improves both corneal stability and penetration due to the electrostatic exchange of the polymer. Mucin also shows promise in gene and peptide drug delivery [45].

Dendrimers differ significantly in the size of micelles, liposomes, and polymer systems, leading to different distribution behaviors after injection into an organism. They form a single covalently bound structure and are mechanically more stable than liposomes. They provide a hydrophobic internal void space for encapsulation and change of payloads, but the payload-to-carrier ratio measured as the weight-to-weight ratio is lower compared to micelles and lipid/polymer vesicles. This is a limitation if the smart device consumes storage space. Nanoscale sensors or switches must be integrated into a nanosystem. A challenge for developing smart dendrimer-based nanoscale systems as commercial products today is the fact that clinical experience with such systems is still limited [45].

15.3.4 Micelles and Liposomes Nanomaterials

Micelles are nanometric structures that are characterized by a hydrophobic core and a hydrophilic layer and that form spontaneously from amphiphilic molecules in aqueous environments. Liposomes are attractive carrier candidates for the delivery of smart switches at the nanoscale because the inner aqueous core provides a "nanocompartment" in which processes that require protection from surrounding body fluids can take place when injected into an organism. Liposomes are already included in smart nanosystems by incorporating a variety of stimulus response characteristics such as temperature, light, pH, ultrasound, enzyme response, or even as a drug-delivery system for them. Radiation-sensitive NPs trigger liposomes as simple and effective multimodal vectors[46]. Liposomes are self-assembling spherical structures formed by a lipid bilayer that completely surrounds an aqueous core and can release various types of biomolecules. Depending on the assembly technology used, the size of the vesicles can vary from a few tens of nanometers to micrometers. Under certain conditions, liposomes of approximately 100 nm in diameter have been used successfully to deliver chemotherapeutic agents to tumors. The release of the drug from poorly soluble molecules can be achieved by micelles that use lipid units as hydrophobic building blocks that bind to hydrophilic polymers. Different lipids have different fatty acid chain lengths and different head groups, leading to a wide range of achievable physicochemical properties, such as minimum micelle concentration or melting temperature. They are sensitive to these environmental liposomes (e.g. temperature sensitive, pH-sensitive, mechanosensitive) by selecting the specific configuration. Several liposome-based anticancer drugs have entered the clinical field, e.g. the anthracycline DOX (Doxil, Myocet, Caelyx) that is used to treat ovarian cancer, multiple myeloma, metastatic breast cancer, or daunorubicin (DaunoXome), which are used to treat Kaposi's sarcoma. In addition to the approved active ingredients, liposomal chemotherapy is also used in clinical studies. An important future direction is the development of receptor-specific targeting for cell-specific delivery, which may make liposomes suitable for purposes such as siRNA delivery [47].

15.3.5 Nanogels

The field of nanogels is a new and rapidly developing area of research for nanomedical drug delivery. Nanogels are a 3D network that is formed chemically (covalent bonds) or physically (hydrogen bonds, Van der Waals, and electrostatic interactions) from cross-linked hydrophilic polymers with a size of 200 nm. Due to their hydrophilic properties, nanogels can swell when added to an aqueous solution and encapsulate a large volume of water. Due to its polymeric nature, a large number of chemical modifications are possible. Such hydrogels can by spontaneously process a large number of such biological molecules, such as DNA, RNA, and proteins, which make them good for drug delivery. For site-specific tissue selection in the body, the surface structure of nanogels can be chemically modified with various ligands. Nanogels have demonstrated their efficacy as siRNA carrier platforms and reactivity to stimuli (pH, temperature). Yin et al. [48] feature multilayered multifunctional nanogels that combine magnetic regulation with biochemical detection. They formed a core of magnetic Ni and fluorescent Ag, surrounded by a layer of pH-sensitive p nanogel (EG-MAA). The combination of a pH-dependent magnetic reaction with fluorescence pH detection enables the creation of an intelligent support system for diagnostic and therapeutic applications. Nanogels exhibit high *in vivo* stability, size control ability, biodegradability, surface functionality, and specific drug delivery. Nanogels have found their place in clinical applications.

Peptide-based NPs can be used as a delivery system. Peptides can have lipophilic or hydrophilic properties due to their amino acid composition and can therefore be used to build amphiphilic molecules that self-assemble into nanostructures. The design considerations for these drug-bearing biopolymers are similar to those for other biocompatible polymers, but should take into account that peptides can act as potent immunogens and that the body already contains several lines of defense against proteins and foreign and peptide structures such as viral capsids. There is also a variety of peptidase in the body, when a system is developed preclinically for subsequent clinical use in humans, species differences in peptidase expression must be carefully considered [49].

15.3.6 Hybrid Nanostructures

The development of smart hybrid and composite nanomaterials has attracted a great deal of attention in recent years due to their extraordinary unique physicochemical properties. These nanomaterials are developed by hybridizing organic and inorganic components to provide enhanced functions and properties. One of their general characteristics is its ability to respond to various internal or external stimuli, including, pH, temperature, redox potential, light, and magnetic fields. As a result, these intelligent stimulus-sensitive systems, as

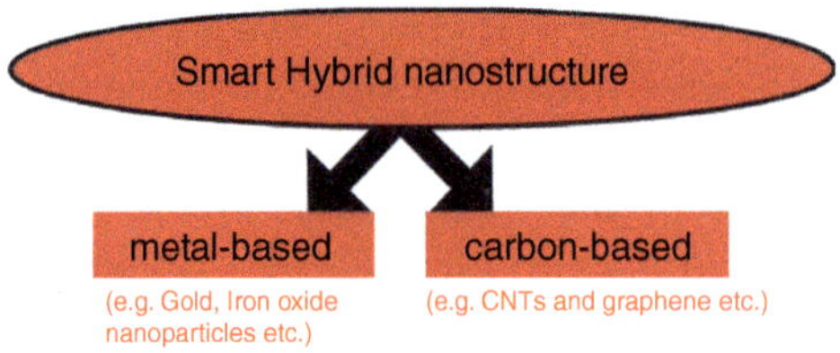

Figure 15.2 Categorization of the drug-delivery platform of smart nanostructured system.

promising platforms, offer many opportunities in many different areas, including nanomedicine as a means of drug transport, theranostics, and imaging [50]. Smart nanostructured drug-delivery platforms can be divided into categories such as metal-based smart nanostructures and carbon-based smart nanostructures (Figure 15.2).

The intrinsic properties of metals (such as magnetic NPs made of Au and IO) have led to the development of a wide variety of smart hybrid nanostructures for theranostic applications. Au NP-based drug and nucleic acid transfer vehicles have received a great deal of attention due to their advantages such as ease of manufacture, ease of surface functionalization, ease of bioconjugation, and biocompatibility. To achieve tumor-directed delivery, Au NPs would have to be modified with molecules that have a high affinity to cancer cells and interact with overexpressed cell surface receptors [51]. Au NPs can be functionalized by thiolated ligands using an Au—S bond formation, and many molecules can be covalently attached to Au using this approach. For example, pH-sensitive composite nanostructures with an Au core, an inner layer made of hydrophobic poly (L-aspartate-doxorubicin), and an outer layer made of poly(ethylene glycol) and folate-conjugated PEG-.OH/FA were produced successfully. Binding of the acid-cleavable hydrazone binds the cancer drug DOX covalently to the hydrophobic inner layer. The vehicle showed a greater cytotoxic effect on the 4T1 mouse mammary carcinoma cell line with pH-activated drug release and targeting ability.

In recent years, carbon-based nanostructured materials, such as CNTs and graphene with sp^2 hybrid carbon atoms, have made significant advances in clinical therapy. Functionalized CNTs have unique properties that enable a wide range of biomedical fields and can be used as drug carriers for *in vitro* and *in vivo* administration [52]. Covalent and noncovalent strategies can be used to treat the surface of CNTs, affecting toxicity and water solubility. The high surface area and high loading capacity, cell internalization, and biocompatibility of NTCs support applications in drug delivery. Biomolecules (e.g. DNA and proteins) can also be attached to functional groups on modified CNTs by electrostatic bonding. In FA-MWCNT, Fe nanocomposites were manufactured from MWCNT difunctionalized by folate and iron for biological (active) or magnetic (passive) targeting. Yun et al. use SWCNT-targeted drug-delivery systems that are activated by changing the pH but use poly (acrylic acid) and poly (vinyl alcohol) for MWCNT [53]. MWCNTs have been used to improve electro-reactive properties. Yun et al. developed a transdermal electro-reactive drug-delivery system by electrospinning polyvinyl alcohol/polyacrylic acid/MWCNT nanocomposites [54]. Other intelligent carbon-based hybrid nanomaterials are electrosensitive gelatin/CNT hybrid microgels, pH-sensitive compounds, decorated with hyaluronic acid (HA), heat-sensitive GO-poly (N-isopropylacrylamide), Arg-Gly, with the peptide support SWCNT-Asp (RGD) [54].

15.4 Type of Stimuli in Intelligent Nanomaterials in Medicine

The mechanism of triggering the stimuli-sensitive nanomaterials, mainly due to the pathological changes occurs in tissues. The stimuli for triggering are: intrinsic stimuli and extrinsic stimuli (Figure 15.3).

Tumorigenic regions have a lower pH value, a higher temperature, a higher redox potential, and certain enzymes. These parameters are called "intrinsic stimuli" because the local stimuli are present in the tumor tissue. The importance of pH-sensitive NPs includes their ability to deliver loaded drugs to the site of pathology. Diseases in which pH-sensitive drug-delivery systems show promise include gastric ulcers, asthma, and cancer. This can be achieved by two strategies: i) by using polymers containing at least one ionizable group that undergo conformational or solubility changes in response to changes in pH in the environment, and ii) by designing bond-containing polymer systems. The acid-sensitive ionizable group is influenced by the change in pH of various regions. For clinical applications, intelligent stimulus-sensitive drug-delivery systems deliver drugs to specific tissues during routine administration. Other pathological changes in tumorigenic regions alter the expression of certain specific enzymes in the extracellular environment [55]. Enzyme-mediated drug delivery is primarily related to changes in the expression profile of specific enzymes under pathological conditions. The expression of specific enzymes such as phospholipases, proteases, or glycosidases is usually more pronounced in pathological tissues such as foci of inflammation or cancer. This uses the constructed NPs that react to modified enzyme gradients in the pathological tissue. Enzyme-activated systems exist in the tumor microenvironment. Matrix metalloproteinases (MMPs), especially MMP-2 and MMP-9, are the best-studied enzymes for on-demand delivery of enzymatic drugs. Recent studies have reported the use of the MMP cleavable linker between PEG chains or the transcription-functionalized liposome transactivator. Furthermore, redox sensitive systems refer to the breaking of disulfide bonds in the presence of glutathione. The release of the triggering drug depends on the amount of glutathione in the intracellular and extracellular compartments. This can be achieved by designing disulfide bonds in the backbone of the hydrophobic polymer component. The other approach could be related to active substances that are sensitive to glutathione in the nucleus or in the NP layer such as micelles [56].

Drug-delivery systems, on the other hand, use induction in the target region by externally applied stimuli, including changes in temperature, magnetic fields, ultrasound, or NIR light termed "extrinsic stimuli." The administration of thermoreactive drugs is one of the best-studied exogenous stimulus-sensitive strategies. The thermal reactivity of NPs generally leads

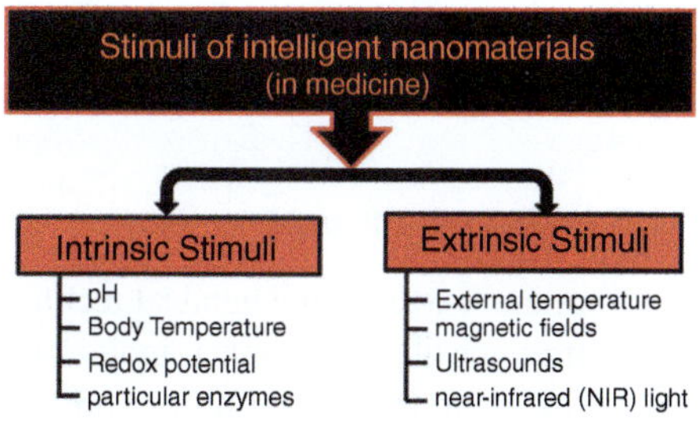

Figure 15.3 Triggering media in intelligent nanomaterials in medicine.

to a sudden change in at least one component of the NP with changes in temperature, leading to triggered drug release. Ideally, NPs should be stable throughout the bloodstream, but they also release a large number of drugs in a slightly hyperthermic state (~40–42 °C). Heat-activated systems generally include liposomes, micelles, or polymers. The heat sensitivity of liposomes results primarily from a gel–liquid phase transition of compound lipids such as dipalmitoylphosphatidylcholine or lysolipids. Another approach to temperature sensitization of liposomal vehicles is the incorporation of synthetic polymers into the membrane where they act as interfering agents at a critical solution temperature below 40 °C. The following polymers have been added to sensitize to liposomes at temperature: poly(N-substituted acrylamides), poly(N-vinyl ethers), and poloxamers. Heat-sensitive polymeric micelles consist of self-assembled heat-sensitive amphiphilic block copolymers. Furthermore, IO NPs associated with alternating magnetic fields can induce hyperthermia-induced drug release [57]. A magnetosensitive NP consists of a magnetically active component in a stable pharmaceutical formulation. The magnetic nanosystem can serve as a magnetic guide under an externally applied permanent magnet and/or a nanothermic generator to control drug release when an alternating magnetic field is applied. Sensitized magnetic fields are most often transmitted by incorporating magnetite or maghemite into NPs. The biomedical application of magnetosensitive NPs includes bioseparations, magnetofection, drug targeting, MRI, and magnetic hyperthermia. Magnetically reactive lipid NPs consist of nano-sized magnetic cores that are encased in lipid layers and contain phospholipids, surfactants, and polymers [58]. Photosensitive systems or photosensitive drug-delivery systems can be mainly divided into two categories. The first is to provide photosensitizers using traditional NPs and the second is to design a light-assisted release of demand control in response to a particular wavelength (in the visible, ultraviolet, or NIR). Liposomal NPs offer the possibility of encapsulating hydrophobic photosensitizers. Photosensitive lipid molecules provide photosensitive liposomal drug-delivery systems with stable circulatory sustained release and light-activated drug delivery at desired sites [59].

15.5 Clinical Applications of Intelligent Nanomaterials

Among the various applications of smart nanomaterials, the most promising applications are in the field of biomedical sciences. Intelligent nanomaterial-based drug-delivery systems aggregate and specifically target diseases with controlled release behavior. The careful design and manufacture of these nanomaterials as effective drug carriers must address several key issues such as biodegradability and biocompatibility, high stability at physiological pH values, excellent drug-loading capacity without toxicity, and large-scale industrial production. For clinical applications. Intelligent stimulus-sensitive drug-delivery systems deliver drugs to specific tissues during routine administration.

15.5.1 In Immunoassay

An immunoassay is a biochemical technique that is effectively applied for the detection of concentrations of a small macromolecule in a sample solution using an antigen or an antibody. The main problem with this approach is obtaining a measurable signal in response to a binding effect. Screen-printed electrodes have attracted much attention as immunosensors due to their unique properties such as miniaturized size, low cost, and ease of manufacture and large-scale production. Only a few microliters of sample solution are used to perform all immunological steps, reducing reagent consumption. Screen-printed electrodes can be successfully coated with NPs, which increase the sensitivity of these electrodes. In an interesting study, a silk-screened electrode with Ag NPs was modified and used effectively for the preliminary detection of cystic fibrosis, a common genetic disease caused by the autosomal recessive gene known as the transmembrane conductivity regulator gene. It affects various organs such as the lungs and intestines, causing irregular transport of chloride and sodium ions through epithelial cells. Ag oxidation has been reported to be sensitive to chloride ion concentration. Chloride levels in sweat are important for an effective diagnosis of cystic fibrosis. A layer of Ag NPs is deposited on the working electrode and anodically removed to create Ag cations. Anodic entrainment voltammetry of Ag NPs gives a single Ag-oxidation peak in the absence of chloride ions, while two voltammetry entrainment peaks (one for AgCl and one for the oxidation of Ag to Ag [I] ions) are obtained in the presence of observed chloride ions. The Ag-chloride peak is used for the quantitative determination of chloride concentration [60].

15.5.2 In Parkinson's Disease and Alzheimer's Disease

These are common neurodegenerative diseases. NP-based sensor systems have been used successfully to distinguish groups of Parkinson's and Alzheimer's disease patients from a healthy control group. The diagnostic approach is based on the identification of patterns of volatile organic compounds in exhaled breath. Disease related changes in blood chemistry can be transferred to the alveolar breath exhaled from the lungs even in the early stages of the disease. The NP-based sensor is subject to rapid and reversible changes in electrical resistance when exposed to characteristic volatile organic compounds. Thus, respiratory pressures can be the basis for the design and production of simple, inexpensive, and noninvasive biomarkers using NP-based sensor systems [61].

15.5.3 In Biological Imaging

Ag NPs can be used for spectroscopic applications. Ag NPs show excellent surface plasmon resonance (SPR) behavior and therefore can perform ultrasensitive analysis down to the single molecule level in the case of higher surface Raman scattering (SERS). However, when using Ag NPs, there are several critical parameters to consider. These factors are the analytes/probes in direct contact with the surface of the Ag NPs that can interfere with the spectroscopic analysis. The mechanism of SPR-mediated chemical reactions is complex, as Ag NPs serve the dual role of Ag being easily oxidized under ambient conditions, reducing plasmon amplification. Ag NPs can be efficiently integrated into graphene to increase its plasmon yield. Shell-isolated Ag has been reported to exhibit excellent plasmonic properties with high stability even after 16 months of storage. Ag NPs also have the advantage of obtaining high-quality Raman signals and can be extended to enhance surface fluorescence, which is used effectively for biological imaging and detection applications [62].

15.5.4 In Electrochemical Sensors

Electrochemical techniques have found many applications in various fields due to their cost effectiveness and the ability to conduct real-time, on-site studies. Modifying the electrode surface in electrochemical sensors by immobilizing the recognition components is an effective approach to achieve strong binding of the target compound with good selectivity and response. Graphene-based electrochemical sensors have received a lot of attention due to their great ability to detect organic compounds with sensitivity. The large specific surface area improves the electron transfer process and allows a very attractive carrier for the incorporation of NPs [63]. In an interesting work by Biris et al. Au NPs were successfully integrated into graphene sheets. The platinum electrode was then modified with the graphene-based nanocomposite, which was manufactured for the sensitive detection of adenine, one of the bases of DNA. In their study, the authors concluded that the kinetics of interfacial charge transfer at the surface of the graphene-modified platinum electrode led to increased electrochemical oxidation of the target compound adenine [64]. In another study, Zhang and colleagues reported the fabrication of a graphene-based electrochemical sensor for efficient detection of paracetamol in pharmaceutical tablet samples. In their study, the reduced GO was doped with phosphorus and then applied to the surface of the glassy carbon electrode. The manufactured electrochemical sensor showed excellent sensitivity and selectivity for paracetamol in pharmaceutical tablet samples. The sensor response was linear in the concentration range of 1.5–120 μM. The detection limit was reached at 0.36 μM [65].

15.5.5 In Separation

Au NPs have been used successfully in capillary electrophoresis-based assays due to improved separation resolutions. These NPs also behave as a pseudo-stationary phase in the existence of polyethylene oxide for efficient separation of double-stranded DNA. The interaction between the polyethylene oxide and the DNA molecule increases in the presence of NPs, which improves the sieving capacity of the polyethylene oxide without changing its viscosity. The separation of acidic and basic proteins using a capillary filled with a NP capillary surfactant has also been reported with high separation efficiency against various target compounds. A simple approach was developed and demonstrated for the production of highly efficient octadecylamine-coated Au-coated NP separation columns for open tubular capillary gas chromatography. Monodisperse silica nanotubes of the desired size and shape have been successfully designed and manufactured using a model-assisted synthesis approach. These developed nanotubes have internal cavities that can be filled with various compounds ranging from large protein molecules to small molecules, and the internal and external surfaces of the nanotubes can be efficiently functionalized. One of the interesting applications of nanotubes is the smart nanophase extractor for the efficient removal of small compounds from solution. Since the outer surface of nanotubes has a hydrophilic property and the inner surface exhibits hydrophobic behavior, these nanomaterials are promising candidates for the efficient extraction of lipophilic compounds from aqueous samples. The antibody functionalized nanotube allows maximum extraction selectivity. The Fab fragments of an antibody raised against the drug 4-[3-(4-fluorophenyl)-2- hydroxy-1-[1,2,4]-triazol-1-yl-propyl]-benzonitrile were immobilized to both the inner and outer surfaces of the silica nanotubes [66].

15.5.6 In Wound Healing

Smart nanomaterials made from glucuronoxylan and Ag NP are also used successfully in wound healing dressings. Polysaccharide hydrogels are efficient and smart materials that are widely used in pharmaceuticals, cosmetics, and in daily life. Hydrogel-forming polysaccharides have attracted great interest in the design and manufacture of Ag NP. The hydrogel layer prevents the dressing from sticking to the wound area due to moisture. Therefore, patient compliance can more easily be achieved by developing these types of bandages. The mechanism of wounds is based on the collagen content and the tensile strength of the epithelial tissues. The tendency for rapid wound healing is based on the crosslinking of collagen fibers. About 25% of diabetics are at high risk of developing foot complications, also known as "diabetic foot ulcers." The wound healing process is impaired in these patients. Various methods have been used to treat diabetic foot ulcers. However, these methods have certain limitations. With the rapid advancements in the field of nanomedicine, NPs have shown remarkable results in wound healing processes. The results obtained show that the diabetic foot ulcer shows a good reaction when using Fe_2O_3 NPs [67].

15.5.7 In Biocide Resistive Surfaces

Biocide resistive surfaces should not be confused with active packaging. In the case of surface biocides, the biocidal agent is intended to help maintain the hygienic state of the food contact surface by preventing or reducing the growth of microbes and contributing to "cleanability." It should not have a preservative effect on food. Surface biocides can have a useful function in food processing equipment (e.g. poultry production lines) and food handling equipment (e.g. conveyor belts) that are difficult to clean. They can also play a role in reusable food containers such as boxes and drawers, as well as the interior liners of refrigerators and freezers. Nanosilver, zinc oxide, or magnesium oxide can be used as an effective surface biocide on plastics that come into contact with food, rubber, silicone, etc. [68].

15.6 Potential Risk Factors in Nanomaterial Application

With the rapid advancement of nanotechnology, some potential risks remain. Most nanomaterials end up in the environment (air, water, and soil). Since many nanomaterials are not biodegradable, they can remain in the environment for a long time. Biological organisms such as microorganisms, plants, and animals could absorb nanomaterials in ecosystems and carry them up the food chain [69]. For example, NPs can easily enter the human body through the respiratory tract. As soon as the NPs enter the bloodstream after passing through the respiratory tract, they can accumulate in certain regions of various organs. Therefore, adverse effects due to nanomaterials or nanoecotoxicology behavior must be carefully evaluated. Pharmaceutical companies are placing increasing emphasis on sustainability by developing pharmaceutical compounds that have the same medicinal value but have a less negative impact on the environment. Even smart research and nanotechnology research is a mature position, but some industries are still operating using traditional manufacturing processes. For the synthesis of smart nanomaterials, advanced technologies are expected in various branches of industry [70, 71].

15.7 Summary and Future Prospects

Smart nanomaterials have attracted a lot of attention from many researchers because they have unique properties such as high physical and chemical stability, versatile functionalization chemistry, and a large specific surface area. The use of smart nanomaterials in industrial-scale pharmaceutical applications is currently expected to play a decisive role in the next generation. Most of the research on the design and development of new smart nanomaterials in pharmaceutical analysis is carried out in the laboratory and reported in the literature. The main goal of modern research is to bridge the gap between scientific research in the laboratory and practical applications in industry. This chapter described information on different types of intelligent material design, the materials used, and the applications. Smart materials nanotechnology offers great added value to existing commercial products and therefore offers new possibilities in various research and development segments of pharmaceutical companies.

References

1 Patra, J.K., Das, G., Fraceto, L.F. et al. (2018). Nano based drug delivery systems: recent developments and future prospects. *Journal of Nanobiotechnology* 16 (1): 71. https://doi.org/10.1186/s12951-018-0392-8.
2 Ventola, C.L. (2012). The nanomedicine revolution: part 1: emerging concepts. *Pharmacy and Therapeutics* 37 (9): 512–525.
3 National Nanotechnology Initiative (2020). The benefits of nanotechnology. https://www.nano.gov/you/nanotechnology-benefits (accessed 19 February 2020)
4 Navya, P.N. and Daima, H.K. (2016). Rational engineering of physicochemical properties of nanomaterials for biomedical applications with nanotoxicological perspectives. *Nano Convergence* 3: 1. https://doi.org/10.1186/s40580-016-0064-z.
5 Karimi, M., Ghasemi, A., Sahandi Zangabad, P. et al. (2016). Smart micro/nanoparticles in stimulus-responsive drug/gene delivery systems. *Chemical Society Reviews* 45 (5): 1457–1501. https://doi.org/10.1039/c5cs00798d.
6 Lombardo, D., Kiselev, M.A., and Caccamo, M.T. (2019). Smart nanoparticles for drug delivery application: development of versatile Nanocarrier platforms in biotechnology and nanomedicine. *Journal of Nanomaterials* 2019: 3702518, 26 pages. https://doi.org/10.1155/2019/3702518.
7 Liu, D., Yang, F., Xiong, F. et al. (2016). The smart drug delivery system and its clinical potential. *Theranostics* 6 (9): 1306–1323. https://doi.org/10.7150/thno.14858.
8 Sharma, D. and Hussain, C.M. (2020). Smart nanomaterials in pharmaceutical analysis. *Arabian Journal of Chemistry* 13 (1): 3319–3343. ISSN 1878-5352, https://doi.org/10.1016/j.arabjc.2018.11.007.

9 Mai, B.T., Fernandes, S., Balakrishnanand, P.B., and Pellegrino, T. (2018). Nanosystems based on magnetic nanoparticles and thermo- or pH-responsive polymers: an update and future perspectives. *Accounts of Chemical Research* 51 (5): 999–1013. https://doi.org/10.1021/acs.accounts.7b00549.

10 Song, Y., Li, Y., Xu, Q., and Liu, Z. (2016). Mesoporous silica nanoparticles for stimuli-responsive controlled drug delivery: advances, challenges, and outlook. *International Journal of Nanomedicine* 12: 87–110. https://doi.org/10.2147/IJN.S117495.

11 Keçili, R., Hussain, C.G., and Hussain, C.M. (2020). Intelligent nanomaterials in pharmaceutical analysis. *Pharma Focus Asia* (39): https://www.pharmafocusasia.com/research-development/intelligent-nanomaterials-pharmaceutical.

12 Kumar, A., Chen, F., Mozhi, A. et al. (2013). Innovative pharmaceutical development based on unique properties of nanoscale delivery formulation. *Nanoscale* 5 (18): 8307–8325. https://doi.org/10.1039/c3nr01525d.

13 Rizvi, S.A.A. and Saleh, A.M. (2018). Applications of nanoparticle systems in drug delivery technology. *Saudi Pharmaceutical Journal* 26 (1): 64–70. https://doi.org/10.1016/j.jsps.2017.10.012.

14 Abdo, G.G., Zagho, M.M., and Khalil, A. (2020). Recent advances in stimuli-responsive drug release and targeting concepts using mesoporous silica nanoparticles. *Emergent Materials* 3: 407–425. https://doi.org/10.1007/s42247-020-00109-x.

15 Yao, J., Feng, J., and Chen, J. (2016). External-stimuli responsive systems for cancer theranostic. *Asian Journal of Pharmaceutical Sciences* 11 (5): 585–595. ISSN 1818-0876, https://doi.org/10.1016/j.ajps.2016.06.001.

16 Bozzuto, G. and Molinari, A. (2015). Liposomes as nanomedical devices. *International Journal of Nanomedicine* 10: 975–999. https://doi.org/10.2147/IJN.S68861.

17 Schubert, S., Heinze, T., Becer, R., and Schubert, U. (2009). Synthetic polymer nanoparticles by nanoprecipitation. *Journal of Materials Chemistry* 19: 3838–3840. https://doi.org/10.1039/b906556n.

18 Navya, P.N., Kaphle, A., Srinivas, S.P. et al. (2019). Current trends and challenges in cancer management and therapy using designer nanomaterials. *Nano Convergence* 6 (1): 23. https://doi.org/10.1186/s40580-019-0193-2.

19 Akbarzadeh, A., Rezaei-Sadabady, R., Davaran, S. et al. (2013). Liposome: classification, preparation, and applications. *Nanoscale Research Letters* 8 (1): 102. https://doi.org/10.1186/1556-276X-8-102.

20 Karimi, M., Sahandi, P., Ghasemi, A., and Hamblin, M. (2015). *Smart External Stimulus-Responsive Nanocarriers for Drug and Gene Delivery*. Morgan and Claypool Publishers https://doi.org/10.1088/978-1-6817-4202-1.

21 Farjadian, F., Ghasemi, A., Gohari, O. et al. (2019). Nanopharmaceuticals and nanomedicines currently on the market: challenges and opportunities. *Nanomedicine (London)* 14 (1): 93–126. https://doi.org/10.2217/nnm-2018-0120.

22 Barua, S. and Mitragotri, S. (2014). Challenges associated with penetration of nanoparticles across cell and tissue barriers: a review of current status and future prospects. *Nano Today* 9 (2): 223–243. https://doi.org/10.1016/j.nantod.2014.04.008.

23 Aggarwal, P., Hall, J.B., McLeland, C.B. et al. (2009). Nanoparticle interaction with plasma proteins as it relates to particle biodistribution, biocompatibility and therapeutic efficacy. *Advanced Drug Delivery Reviews* 61 (6): 428–437. https://doi.org/10.1016/j.addr.2009.03.009.

24 Ruscito, A. and DeRosa, M.C. (2016). Small-molecule binding aptamers: selection strategies, characterization, and applications. *Frontiers in Chemistry* 4: 14. https://doi.org/10.3389/fchem.2016.00014.

25 Vardharajula, S., Ali, S.Z., Tiwari, P.M. et al. (2012). Functionalized carbon nanotubes: biomedical applications. *International Journal of Nanomedicine* 7: 5361–5374. https://doi.org/10.2147/IJN.S35832.

26 Wu, W., Wu, Z., Yu, T. et al. (2015). Recent progress on magnetic iron oxide nanoparticles: synthesis, surface functional strategies and biomedical applications. *Science and Technology of Advanced Materials* 16 (2): 023501. Published 2015 Apr 28. https://doi.org/10.1088/1468-6996/16/2/023501.

27 Jang, L.W., Shim, J., Son, D.I. et al. (2019). Simultaneous growth of three-dimensional carbon nanotubes and ultrathin graphite networks on copper. *Scientific Reports* 9 (1): 12344. Published 2019 Aug 28. https://doi.org/10.1038/s41598-019-48725-w.

28 Wu, Y., Yu, T., and Shen, Z. (2010). Two-dimensional carbon nanostructures: fundamental properties, synthesis, characterization, and potential applications. *Journal of Applied Physics* 108: 071301. https://doi.org/10.1063/1.3460809.

29 Arias, L.S., Pessan, J.P., Vieira, A.P.M. et al. (2018). Iron oxide nanoparticles for biomedical applications: a perspective on synthesis, drugs, antimicrobial activity, and toxicity. *Antibiotics (Basel)* 7 (2): 46. Published 2018 Jun 9. https://doi.org/10.3390/antibiotics 7020046.

30 Delong, R.K., Reynolds, C.M., Malcolm, Y. et al. (2010). Functionalized gold nanoparticles for the binding, stabilization, and delivery of therapeutic DNA, RNA, and other biological macromolecules. *Nanotechnology, Science and Applications* 3: 53–63. https://doi.org/10.2147/NSA.S8984.

31 Subramaniam, M.N., Goh, P.S., Lau, W.J., and Ismail, A.F. (2019) The roles of nanomaterials in conventional and emerging technologies for heavy metal removal: a state-of-the-art review. *Nanomaterials (Basel)* 9 (4): 625. https://doi.org/10.3390/nano9040625.

32 Matea, C.T., Mocan, T., Tabaran, F. et al. (2017). Quantum dots in imaging, drug delivery and sensor applications. *International Journal of Nanomedicine* 12: 5421–5431. https://doi.org/10.2147/IJN.S138624.

33 Mousavi, T., Karimzadeh, F., and Abbasi, M. (2008). Synthesis and characterization of nanocrystalline NiTi intermetallic by mechanical alloying. *Materials Science and Engineering A-structural Materials Properties Microstructure and Processing* 487: 46–51. https://doi.org/10.1016/j.msea.2007.09.051.

34 Jani, J.M., Leary, M., Subic, A., and Gibson, M. (2014). A review of shape memory alloy research, applications and opportunities. *Materials & Design* 56: 1078–1113. https://doi.org/10.1016/j.matdes.2013.11.084.

35 Shen, Y., Wang, G., Chen, L. et al. (2009). Investigation of surface endothelialization on biomedical nitinol (NiTi) alloy: effects of surface micropatterning combined with plasma nanocoatings. *Acta Biomaterialia* 5: 3593–3604. https://doi.org/10.1016/j.actbio.2009.05.021.

36 Jeong, J., Kim, J.H., Shim, J.H. et al. (2019). Bioactive calcium phosphate materials and applications in bone regeneration. *Biomaterials Research* 23: 4. https://doi.org/10.1186/s40824-018-0149-3.

37 Athar, M. and Das, A. (2014). Therapeutic nanoparticles: state-of-the-art of nanomedicine. *Advanced Materials Review* 1: 25–37.

38 Madaan, K., Kumar, S., Poonia, N. et al. (2014). Dendrimers in drug delivery and targeting: drug-dendrimer interactions and toxicity issues. *Journal of Pharmacy & Bioallied Sciences* 6 (3): 139–150. https://doi.org/10.4103/0975-7406.130965.

39 Li, J. and Mooney, D.J. (2016). Designing hydrogels for controlled drug delivery. *Nature Reviews Materials* 1 (12): 16071. https://doi.org/10.1038/natrevmats.2016.71.

40 Bohrey, S., Chourasiya, V., and Pandey, A. (2016). Polymeric nanoparticles containing diazepam: preparation, optimization, characterization, in-vitro drug release and release kinetic study. *Nano Convergence* 3 (1): 3. https://doi.org/10.1186/s40580-016-0061-2.

41 Kumari, A., Sudesh, K.Y., and Subhash, C. (2010). Biodegradable polymeric nanoparticles based drug delivery systems. *Colloids and Surfaces B: Biointerfaces* 75: 1–18.

42 Lee, B.K., Yun, Y., and Park, K. (2016). PLA micro- and nano-particles. *Advanced Drug Delivery Reviews* 107: 176–191. https://doi.org/10.1016/j.addr.2016.05.020.

43 Kumari, A., Yadav, S., and Yadav, S. (2009). Biodegradable polymeric nanoparticles based drug delivery systems. *Colloids and Surfaces B: Biointerfaces* 75: 1–18. https://doi.org/10.1016/j.colsurfb.2009.09.001.

44 Mohammed, M.A., Syeda, J.T.M., Wasan, K.M., and Wasan, E.K. (2017). An overview of chitosan nanoparticles and its application in non-parenteral drug delivery. *Pharmaceutics* 9 (4): 53. https://doi.org/10.3390/pharmaceutics9040053.

45 Tsai, C.H., Wang, P.Y., Lin, I.C. et al. (2018). Ocular drug delivery: role of degradable polymeric nanocarriers for ophthalmic application. *International Journal of Molecular Sciences* 19 (9): 2830. https://doi.org/10.3390/ijms19092830.

46 Zangabad, P.S., Mirkiani, S., Shahsavari, S. et al. (2018). Stimulus-responsive liposomes as smart nanoplatforms for drug delivery applications. *Nanotechnology Reviews* 7 (1): 95–122. https://doi.org/10.1515/ntrev-2017-0154.

47 Yingchoncharoen, P., Kalinowski, D.S., and Richardson, D.R. (2016). Lipid-based drug delivery systems in cancer therapy: what is available and what is yet to come. *Pharmacological Reviews* 68 (3): 701–787. https://doi.org/10.1124/pr.115.012070.

48 Yin, Y., Hu, B., Yuan, X. et al. (2020). Nanogel: a versatile nano-delivery system for biomedical applications. *Pharmaceutics* 12 (3): 290. https://doi.org/10.3390/pharmaceutics12030290.

49 MEROPS (company so no initials) (2017). Classification: three orthogonal approaches. www.ebi.ac.uk/merops/about/classification.shtml (accessed 19 February 2022)

50 Vivero-Escoto, J.L. and Huang, Y.-T. (2011). Inorganic-organic hybrid nanomaterials for therapeutic and diagnostic imaging applications. *International Journal of Molecular Sciences* 12: 3888–3927. https://doi.org/10.3390/ijms12063888.

51 Yoo, J., Park, C., Yi, G. et al. (2019). Active targeting strategies using biological ligands for nanoparticle drug delivery systems. *Cancers (Basel)* 11 (5): 640. https://doi.org/10.3390/cancers11050640.

52 Maiti, D., Tong, X., Mou, X., and Yang, K. (2019). Carbon-based nanomaterials for biomedical applications: a recent study. *Frontiers in Pharmacology* 9: 1401. https://doi.org/10.3389/fphar.2018.01401.

53 Yun, J., Im, J., Lee, Y.-S., and Kim, H.-I. (2011). Electro-responsive transdermal drug delivery behavior of PVA/PAA/MWCNT nanofibers. *European Polymer Journal* 47: 1893–1902. https://doi.org/10.1016/j.eurpolymj.2011.07.024.

54 Baby, R., Saifullah, B., and Hussein, M.Z. (2019). Carbon nanomaterials for the treatment of heavy metal-contaminated water and environmental remediation. *Nanoscale Research Letters* 14: 341. https://doi.org/10.1186/s11671-019-3167-8.

55 Yoshida, T., Lai, T.C., Kwon, G.S., and Sako, K. (2013). pH- and ion-sensitive polymers for drug delivery. *Expert Opinion on Drug Delivery* 10 (11): 1497–1513. https://doi.org/10.1517/17425247.2013.821978.

56 Liechty, W.B., Kryscio, D.R., Slaughter, B.V., and Peppas, N.A. (2010). Polymers for drug delivery systems. *Annual Review of Chemical and Biomolecular Engineering* 1: 149–173. https://doi.org/10.1146/annurev-chembioeng-073009-100847.

57 Pereira Gomes, I., Aparecida Duarte, J., Chaves Maia, A.L. et al. (2019). Thermosensitive nanosystems associated with hyperthermia for cancer treatment. *Pharmaceuticals (Basel)* 12 (4): 171. https://doi.org/10.3390/ph12040171.

58 Puri, A., Loomis, K., Smith, B. et al. (2009). Lipid-based nanoparticles as pharmaceutical drug carriers: from concepts to clinic. *Critical Reviews in Therapeutic Drug Carrier Systems* 26 (6): 523–580. https://doi.org/10.1615/critrevtherdrugcarriersyst.v26.i6.10.

59 Yavlovich, A., Smith, B., Gupta, K. et al. (2010). Light-sensitive lipid-based nanoparticles for drug delivery: design principles and future considerations for biological applications. *Molecular Membrane Biology* 27 (7): 364–381. https://doi.org/10.3109/09687688.2010.507788.

60 Toh, H.S., Batchelor-McAuley, C., Tschulik, K. et al. (2013). The anodic stripping voltammetry of nanoparticles: electrochemical evidence for the surface agglomeration of silver nanoparticles. *Nanoscale* 5 (11): 4884–4893. https://doi.org/10.1039/c3nr00898c.

61 Konvalina, G. and Haick, H. (2013). Sensors for breath testing: from nanomaterials to comprehensive disease detection. *Accounts of Chemical Research* 47: https://doi.org/10.1021/ar400070m.

62 Krajczewski, J. and Kudelski, A. (2019). Shell-isolated nanoparticle-enhanced Raman spectroscopy. *Frontiers in Chemistry* 7, 4: 410. https://doi.org/10.3389/fchem.2019.00410.

63 Yin, P.T., Kim, T.H., Choi, J.W., and Lee, K.B. (2013). Prospects for graphene-nanoparticle-based hybrid sensors. *Physical Chemistry Chemical Physics* 15 (31): 12785–12799. https://doi.org/10.1039/c3cp51901e.

64 Biris, A., Pruneanu, S., Pogacean, F. et al. (2013). Few-layer graphene sheets with embedded gold nanoparticles for electrochemical analysis of adenine. *International Journal of Nanomedicine* 8: 1429–1438. https://doi.org/10.2147/IJN.S42613.

65 Zhang, X., Wang, K.-P., Zhang, L.-N. et al. (2018, 2018). Phosphorus-doped graphene-based electrochemical sensor for sensitive detection of acetaminophen. *Analytica Chimica Acta* 1036: 26–32. https://doi.org/10.1016/j.aca.2018.06.079.

66 Södervall, M. and Mutikainen, I. (2002). Crystal structure of 4-[3-(4-fluorophenyl)-2-hydroxy-1-(1,2,4-triazol-1-yl)-propyl]benzonitrile, C18H17FN4O. *Zeitschrift für Kristallographie – New Crystal Structures.* 217: https://doi.org/10.1524/ncrs.2002.217.jg.38.

67 Naskar, A. and Kim, K.S. (2020). Recent advances in nanomaterial-based wound-healing therapeutics. *Pharmaceutics* 12 (6): 499. https://doi.org/10.3390/pharmaceutics12060499.

68 Brandelli, A., Ritter, A., and Veras, F. (2017). Antimicrobial activities of metal nanoparticles. In: *Metal Nanoparticles in Pharma* (ed. M. Rai and R. Shegokar), 337–363. Springer http://doi.org/10.1007/978-3-319-63790-7_15.

69 Ray, P.C., Yu, H., and Fu, P.P. (2009). Toxicity and environmental risks of nanomaterials: challenges and future needs. *Journal of Environmental Science and Health. Part C Environmental Carcinogenesis and Ecotoxicology Reviews* 27 (1): 1–35. https://doi.org/10.1080/10590500802708267.

70 Wikipedia Nanotechnology https://en.wikipedia.org/wiki/Nanotechnology (acceseed: 28 July 2020)

71 Kumar, R., Singh, R., and Singh, D. (2016). Natural and waste hydrocarbon precursors for the synthesis of carbon based nanomaterials: graphene and CNTs. *Renewable and Sustainable Energy Reviews* 58: 976. https://doi.org/10.1016/j.rser.2015.12.120.

16

Polymer-based Nanocomposites for Biomedical Applications

Chander Amgoth, Kaxi Yu, Shuai Chen, Hongzhen Bai, and Guping Tang

Department of Chemistry, Zhejiang University, Hangzhou, China

16.1 Introduction and Recent Advances

This chapter starts with a review of research on polymer-based nanocomposites followed by their synthesis, preparation, materials, methods, characterizations, and biomedical applications [1]. Further to this there is a detailed introduction to polymer nanocomposites (PNCs), the classification of nanocomposites, recent nanotechnology advances to cure diseases like cancers and other biomedical applications. PNCs are comprised of a wide range of materials consisting of two or more components where at least one component should have dimensions in the range of nanoscale, i.e. from 1 to 100 nm (in diameter) [1, 2]. However, a polymer or its copolymers are micro and nanostructured particles filled with nanofillers, additives, and dopants that are combined, mixed, entrapped, and dispersed in a polymer matrix or host system. Typically, this consists of a macroscopic matrix as a host material, which can be further added and doped with nanoscale particulate, which is termed as filler materials [3]. Interestingly, the filler material can be classified as: zero-dimensional (0D) nanoparticles or capsules; one-dimensional (1D) nanofibers, wires, rods, and tubes; two-dimensional (2D) thin films, QDs (quantum dots); and three-dimensional (3D) cross-linked polymers, branched polymers, network polymers, block copolymers, CNTs (carbon nanotubes) in polymer matrix can all be considered as nanocomposites [4]. Moreover, rubber, engineering plastics, polyolefins, and vulcanized rubber can be used as a host material as a filler in the nanoscale particulates. Typically, less than 5% of nanomaterials are used as filler materials to improve the thermal and mechanical properties of polymer-based nanocomposites. The usage of nanocomposites for biomedical and pharmaceuticals is limited due to biodegradability, biocompatibility, biosafety, and other toxic issues. Polymer-based nanocomposite material can be termed as a host–guest material based on its matrix and filler particulates. There are different methods for creating polymer-based nanocomposites such as *in situ* polymerization and melt-blending compounding process. Basically, various types of nanomaterials are melt-blended with plastics such as nanoclay, nano-inorganic particles (SiO_2), metal and organic

Nanomaterials and Nanotechnology in Medicine, First Edition. Edited by P.M. Visakh.

particles (ZrO_2, ZnO), metallic nanoparticles as dopants (AgNPs (silver nanoparticles), AuNPs (gold nanoparticles)), and nanotubes with the desired dimensions [3–5].

However, the transitions from micro- to meso-, meso- to nano-, and nano- to macro-scale sizes of particles lead to a change in material properties such as physicochemical and thermo-mechanically. In addition, transitions in materials properties can further lead to change in color properties, melting point, stress, strain, brittleness, ductility, tensile strength, elastic nature, creep, fatigue, etc. As the particle size reduces the specific surface area increases and these nanoscale particles have huge volume-to-surface ratio for consideration for use in real-time applications [5]. The synthesis and preparation of polymer-based nanocomposites usually need special care due to the composite structure and the resulting composite should be in the size range of micro- to nanoscale. However, polymer-based nanocomposites show improved properties, which is a good sign to use them for various fields. Various characterization techniques need to be used to check the improved properties of PNCs [4, 5]. XRD (X-ray diffraction) and PXRD (powder X-ray diffraction) are the most common characterization tools for finding the amorphous and crystalline nature of materials. Further, it can be used to determine the internal structural pattern followed by the lattice parameters and crystal structure of materials. Interestingly, TEMs (transmission electron microscopes) offer direct evidence for lamination and delamination of filler particulates and their nanostructured patterns. From the biomedical perspective, there is a need to develop polymer-based nanocomposites with defined characteristic properties and features [6]. Though, there are many PNC materials in use in a wide range of applications in the for biopharmaceutical field like tissue-engineering (TE), bone implantations, biosensor, biomarker, drug delivery, and so forth, but there is still a need to address many issues such as cytotoxicity, side effects, biodegradability, biocompatibility, biosafety, targeted drug delivery, organ and tissue-specific implantation, harmless diagnosis and other various therapeutics. To address all these limitations, there is a need to focus and cover a broad range of ongoing research and biomedical applications on polymer-based nanocomposites [5, 6].

Concurrent demonstration about the polymer-based nanocomposites and recent trends for nanobiotechnological and biomedical applications has been deliberated. It demonstrates the systematic synthesis methods, methodologies, and investigation from the various material property characterizations of PNCs [4–6]. Predominantly, this chapter will provide a good reference to studies about polymer-based nanocomposites for biomedical applications with good scientific rigor.

16.2 Polymers for Biomedical Fields

Based on their properties, polymers are categorized into several types.

i) Thermo-responsive polymers, which can easily mold or changed into the desired size, shape, morphology, and other characteristic features by changing the temperature, pH, concentration, and other external stimuli factors. For example, PNIPAM (poly-N-Isopropyl acryl amide) is a thermo-sensitive polymer that shows LCST (lower critical solution temperature) at ~32 °C.

ii) Biodegradable polymers like TGA (thioglycolic acid), terepthalic acid (TA), and lactic acid (LA).
iii) pH-responsive polymers includes PLGA (poly lactic co-glycolic acid), PEG (poly ethylene glycol), PMMA (poly methyl methacrylate) and amino-acid-based polymers like L-glutamic acid-5-benzyl ester, and L-Aspartic acid-4-benzyl ester.
iv) Electrically responsive polymers are polymers that possess piezoelectric effect, which leads to the change in size, shape, and morphology of polymer nanoparticles, for example PVDF (poly vinylidine fluoride).
v) Chemically responsive polymers that depend on their anionic and cationic nature followed by the oxidation, reduction, and redox properties along with concentration levels and affinity toward other materials, for example PEG shows significant changes with interaction with other substances.
vi) Nonbiodegradable polymers that do not degrade with biological systems, for example cellulose acetate [5–7]. As shown in Figure 16.1 (a) CS (chitosan), (b) PEG, (c) PEI (polyethyleneimine), (d) PLGA, (e) PNIPAM, and (f) sodium conjugated alginate are biocompatible, biodegradable, and most widely used polymers in biomedical applications.

16.2.1 Polymer Properties

To date, polymers and polymer-based nanocomposites have had a remarkable impact and show predominant use in biomedical and pharmaceutical fields: drug-delivery systems, gene-delivery systems (GDSs), cell-delivery systems (CDSs), TE, bone-tissue implantation, tissue grafting, injury healing, biosensor, wound healing, and biomarker applications [7]. The synthesis of polymers and polymer based nanocomposites needs consideration of several factors such as surface tension, density, viscosity, molecular weight, melting point, boiling point, glass transition temperature, rubbery nature,

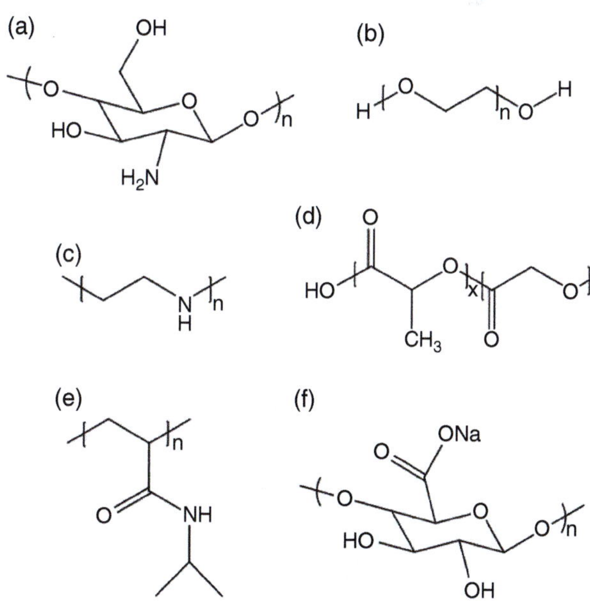

Figure 16.1 Some of the most useful polymers' chemical structure for biomedical applications. (a) CS (chitosan), (b) PEG, (c) PEI (polyethyleneimine), (d) PLGA, (e) PNIPAM, and (f) sodium conjugated alginate.

crystalline and amorphous nature, absorption and adsorption capacity, cross-linking nature, hyper-branching capacity, resilience property, hydrophobic and hydrophilic nature, presence of active functional groups, self-assembly, physiochemical properties, hygroscopic nature, solubility, chemical structure, surface profile, surface charge and charge density, biocompatibility, adhesive nature, thermal properties (heat flow and degradation temperature), mechanical properties such as strength, stress, strain, elongation, compressive nature, brittle, ductility, creep, fatigue, etc. [6–8].

These mentioned characteristic properties of polymers need to be considered for the selection of polymers for synthesis and preparation of polymer-based nanocomposites [8]. Initially, a polymer host (matrix) need to be designed to be compatible with nanoparticulates like nanorods, nanofillers, nanofibers, nanopowders, metal nanoparticles, inorganic nanoparticles, organic nanoparticles, ceramic nanoparticles, iron oxide nanoparticles, etc. to obtain a composite with unique and novel characteristic features for a biomedical application [8, 9].

16.2.2 Classification of Polymer-based Nanocomposites

Figure 16.2 shows the types of nanocomposites classified based on their base materials like the host (matrix) and filler nanoparticles. The final nanocomposite could be homogeneous or heterogeneous but the host matrix material has to be embedded, doped, added, or filled with a nanoscale objective to develop polymer-based nanocomposites.

Figure 16.2a show a metallic nanocomposite, which means the polymer matrix is filled with nano-sized metal nanoparticles and ultimately it is considered to be a type of polymer-metal nanocomposite. Figure 16.2b illustrates a ceramic nanofiller inside the polymer. It has an organic and inorganic host matrix and a scanning electron microscope (SEM) micrograph depicts the polymer matrix doped with mesoporous zinc oxide (ZnO) nanoparticles. Similarly, Figure 16.2c has a polymer matrix filled with nanoscale polymer particles and its nanocomposite has a fish-net type morphology over its surface [6–10].

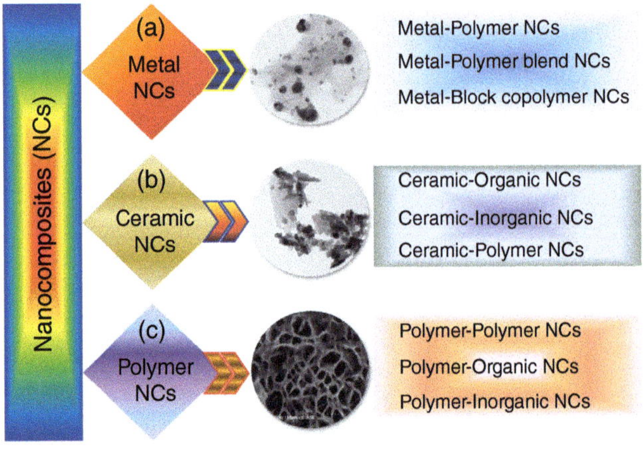

Figure 16.2 Types and classification of nanocomposite materials [2, 6, 10].

16.2.3 Host-guest Concept of Polymer Nanocomposites

As a result of the polymer host matrix, guest filler materials (nano-sized particles like nanorods, nanofibers, metal-nanoparticles) can be doped to improve the properties of composite with compressive and flexural nature. Such composite materials can be termed as hybrid materials or hybrid nanocomposites for various applications. The host-guest concept itself is a novel approach to designing and developing materials with unique properties with specific applications – in particular, properties like melting point, crystalline nature, surface area, surface tension of polymer-based nanocomposites will be different from the source's pure form [10]. These nano-sized filler materials can be completely surrounded with a polymer host matrix and such composites are suitable for biomedical applications. However, properties of nanocomposites can differ from the composites' individual properties. In the domain of polymer-based nanocomposites there are plenty of unmet challenges to focus and address associated with a poor understanding of physiochemical properties and the hybrid host–guest concept of composites for biomedical and other applications [11]. Tailoring the interface between host matrix and filler particulates is an essential task in understanding the mechanism between host–guest molecules. A systemic and mechanistic approach can address the impact and implications of PNCs on biomedical applications. Tethering the polymer properties can help to improve the hyper branching of chains through the network formation that occur as a result of pluronic affinity between host matrix and filler particulates. The charge effect, Vander Waals forces, and electrostatic interactions lead to form a good aggregate of nanocomposite. Figure 16.3a schematic shows a complex nanocomposite, Figure 16.3b illustrates the core-shell concept with the entrapped drug molecules for targeted drug-delivery systems, and Figure 16.3c represents the liposomal micelle and its biomedical use [13, 14]. Figure 16.3d shows a liposomal micelle loaded, dispersed,

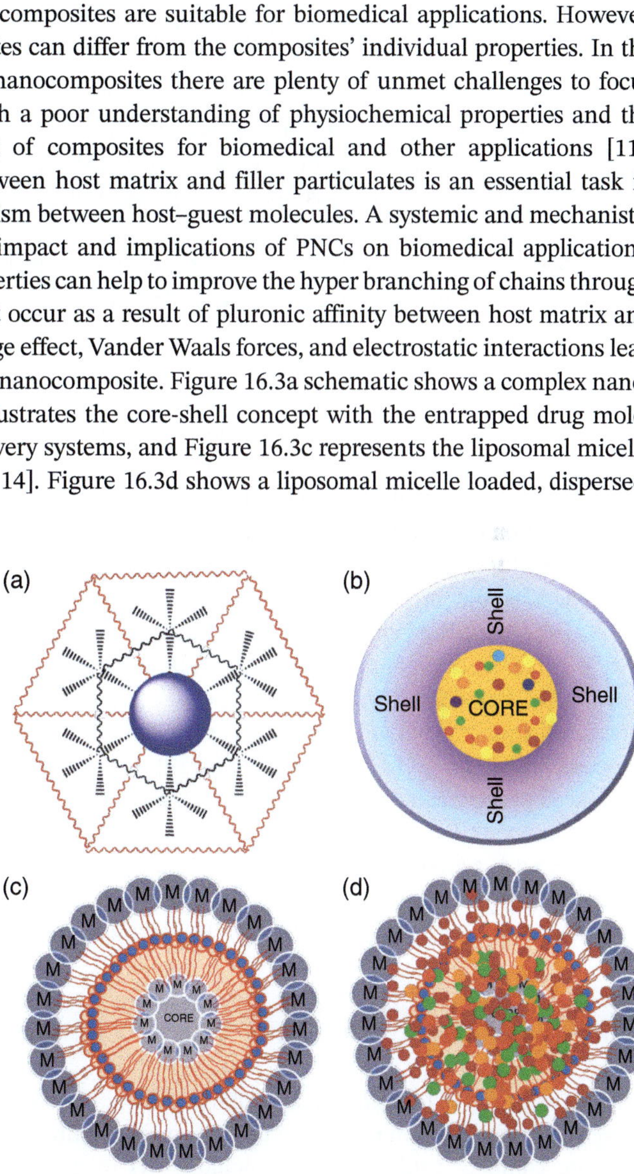

Figure 16.3 Types of nanocomposite materials used for drug delivery and other biomedical applications. Schematic (a) for dendritic capsules, (b) for core-shell concept, (c) for liposomal micelle and (d) for drug encapsulated naocapsule [6, 12].

entrapped, encapsulated, and incorporated with drug molecules for various therapeutic applications. Composite properties like compatibility, thermal stability, and polymer processing conditions, self-assembly, wettability, and interactions with organic and inorganic particulates play a key role in understanding the surface functionalization of interface interactions between the matrix and the filler materials. However, the use of organic and inorganic nanoscale fillers in combination with polymer host matrixes can help to modify the active surface functionalities of composite materials for biomedical uses [8–14].

16.3 Synthesis of Nanocomposites

As per the definition of a nanocomposite, which is discussed in the introduction, a worthwhile synthesis of polymer and filler nanoparticles needs special care concerning their size, shape, morphology, and other properties [11–15]. The polymer host matrix should have an affinity to the filler materials to incorporate and accommodate the foreign objects. The matrix surface should have specific features, for example porosity, functionalized surface, adhesive, and viscous nature that can help to localize the foreign objects for PNCs [16]. Furthermore, foreign objects or filler materials should possess 1D, 2D, and 3D characteristics for better compartmentalization between the host matrix and filler materials, which give a greater range of potential applications for the developed nanocomposite. To obtain a composite with novel and aesthetic features, there is a need to follow specific synthesis methods and methodologies to design and develop the nanocomposites for biomedical applications. Condensation polymerization (CP) and addition polymerization (AP) along with ring-opening polymerization (ROP), precipitation polymerization (PP), and emulsion polymerization (EP) and atom-transfer radical polymerization (ATRP) have a wide range of followers [2, 6, 12, 17]. CP produces a byproduct or subproduct that other polymerization reactions may or may not produce. Moreover, a significant amount of research on PNCs is under progress and has the potential to produce novel and unique outcomes with a wide range of societal applications [18]. The polyamides, polystyrene, vinylidenes, carbonates, polycarbonates, polyethylene, epoxies, acryl, acryl amides, butylenes, terephthalates, urethanes, and polyurethanes are reported as the best host matrix materials for novel nanocomposites for real-time applications [19]. However, PNCs with the desired dimensions need to approach conventional methods as well as modified facile synthesis and fabrication methods. For example, modified Stober's "Sol-Gel" method is the best preferential method used to synthesize spherical nano-sized particles with meso- and nanoscale porosity and even without porosity. PNCs for energy and environmental usage provide a comprehensive and advanced approach through its modern technology [20].

16.4 Characterization Tools for Nanocomposites

The following list of instruments and characterization tools are required for the synthesis, design, and development of PNCs. The equipment required ranges very simple equipment to more complex equipment and includes: (i) microbalance; (ii) homogenizer; (iii) centrifuge; (iv) micropipette; (v) sonicator bath; (vi) EDAX (energy dispersive X-ray spectroscopy) to analyze the elemental composition present in the synthesized PNCs; (vii) SEM; (viii) FE-SEM (field emission SEM) for morphology of nanocomposite; (ix) TEM for size, shape, and morphology of synthesized nanoparticles and capsules and composites; (x) BET (Brunaur, Emmet, Teller) for surface area and porosity; (xi) Thermogravimetric analysis for weight loss measurements; (xii) DSC (differential scanning calorimetry) for heat flow and thermal stability of designed materials; (xiii) PXRD to find out the amorphous, crystalline, and semi-crystalline nature; (xiv) ultraviolet–visible spectrophotometer for absorption measurements of drug released from PNCs with the help of 10 mm optical path length quartz or glass cuvettes; (xv) PL (photo luminescence) spectra for nanocomposites; (xvi) FTIR (Fourier transform infrared) for functional group identification; (xvii) Raman spectrometer for band gap measurements; (xviii) Zeta potential to measure the surface charge density of nanocomposite; (xix) DLS (dynamic light scattering) for particle size analysis; (xx) NMR (nuclear magnetic resonance) for nature of protons and carbons present in the synthesized composite; (xxi) HRMS (high-resolution mass spectrometry) for mass; (xxii) CHNS (carbon, hydrogen, nitrogen, and sulfur) analyzer; and (xxiii) MALDI-TOF (matrix-assisted laser desorption/ionization-time-of-flight) spectrometry for mass and molecular weight of polymer samples.

16.5 Size, Shape, and Morphology of Nanocomposites

Tailoring the interface reactions between the polymer host matrix and filler nanoparticles through the tethering of polymer chains and branches leads to the design of nanocomposites with good characteristic features provides further size, shape, and morphological characterizations. Interestingly, polycarbonates, clay, polymer-silica, polymer-organic, and polymer-ceramic host–guest materials have a significant impact on composite materials [12, 20].

Figure 16.4 depicts the size, shape, and morphology of some spherical capsules with complexes and surface functionalized characteristic features. In Figure 16.4 (a) and (c) are images of complex nanocarrier composites with surface functionalized features acquired through a normal optical microscopic. Figure 16.4 (b) and (d) are SEM micrographs of spherical capsules prepared from the block copolymers. Such surface functionalized polymer-based nanocomposite capsules are promising tools for drug delivery and cancer therapy [20, 21]. The images depict the smooth as well rough-surface nanocapsules with nanoscale porosity throughout the surface of the particles/capsule. This type of spherical-shaped capsules has been developed from the (PEG-MeO-NH)-*b*-(L-GluA) and those capsules have a smooth surface without any porosity. The hydrophilic (PEG-MeO-NH)-*b*-(L-GluA) has been molecular imprinted with hydrophobic polycaprolactone (PCL) solution to obtain porosity throughout the surface of the capsule [22]. Such nanoporous spherical capsules are promising carrier vehicles for targeted drug delivery used in cancer therapies. The most common filler material is montmorillonite clay that enhances the compressive and flexural properties of nanocomposites. Significantly, a nanocomposite for biomedical application should have a good compressive nature with flexural strength, temperature sustainability, heat distortion, etc. The combination of amino-acid-based polymers (L-Histidine, L-Lysine) with biodegradable and biocompatible polymers (PCL, PNIPAM, PEG) can reinforce

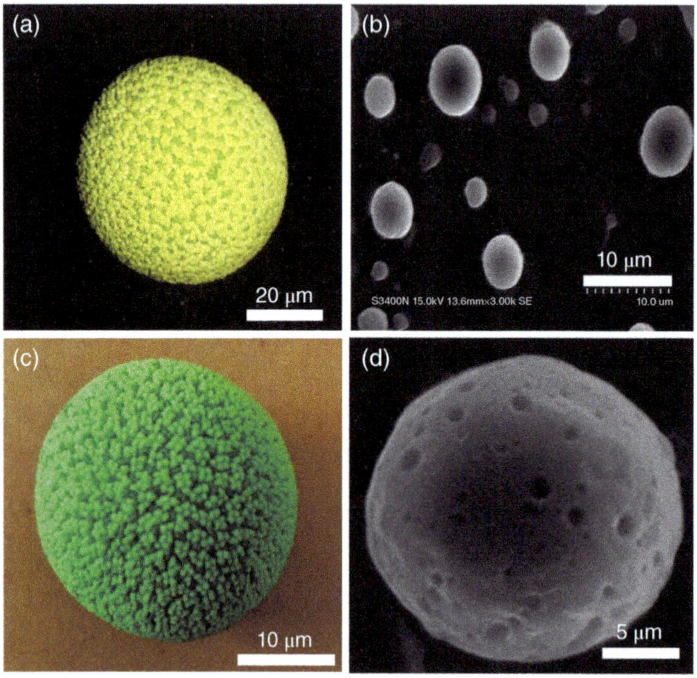

Figure 16.4 (a) and (c) are images of complex nanocarrier composite particles with surface functionalized features acquired through a normal microscopic imaging. (b) A SEM micrograph showing a smooth-surfaced nanocomposite capsule without pores. (d) PNCs with rough-surface and porosity throughout the capsule [6, 12, 20].

nanocomposite properties for biomedical applications. Nowadays, polymer-based nanocomposites are the most predominant and interesting research area for arriving at novel and innovative technologies for real-time applications in diversified fields. A composite that has considerable enhancement in the polymer matrix properties and nano-sized filler properties with a higher surface-to-volume ratio can be an outstanding composite for many applications [23].

For instance, numerous studies deliberate the improvement in nanocomposite properties such as physicochemical and electromechanical properties for wide range utilization. However, Figure 16.5 shows TEM images of various nanocomposite materials [24]. Image (a) in Figure 16.5 corresponds to the polymer-PNC prepared from (PS)-b-(PMMA)] block copolymer depicts nanoscale morphology. Similarly, image (b) in Figure 16.5 shows a polymer matrix doped with ZnO nanoparticles as a filler material and its appearance, image (c) in Figure 16.5 correspond to AgNPs as synthesized and image (d) in Figure 16.5 shows a polymer matrix doped with AgNPs as filler particulates for the development of composite. Most importantly, a key factor for achieving the aforementioned properties of nanocomposites is through the optimal dispersion of nanofiller materials inside the host polymer matrix. However, the thermal stability of polymer-based nanocomposites can be increased by the addition of nanofillers such as graphite, CNTs, and nanorods and this barrier effect is strongly related to the characteristic properties of filler

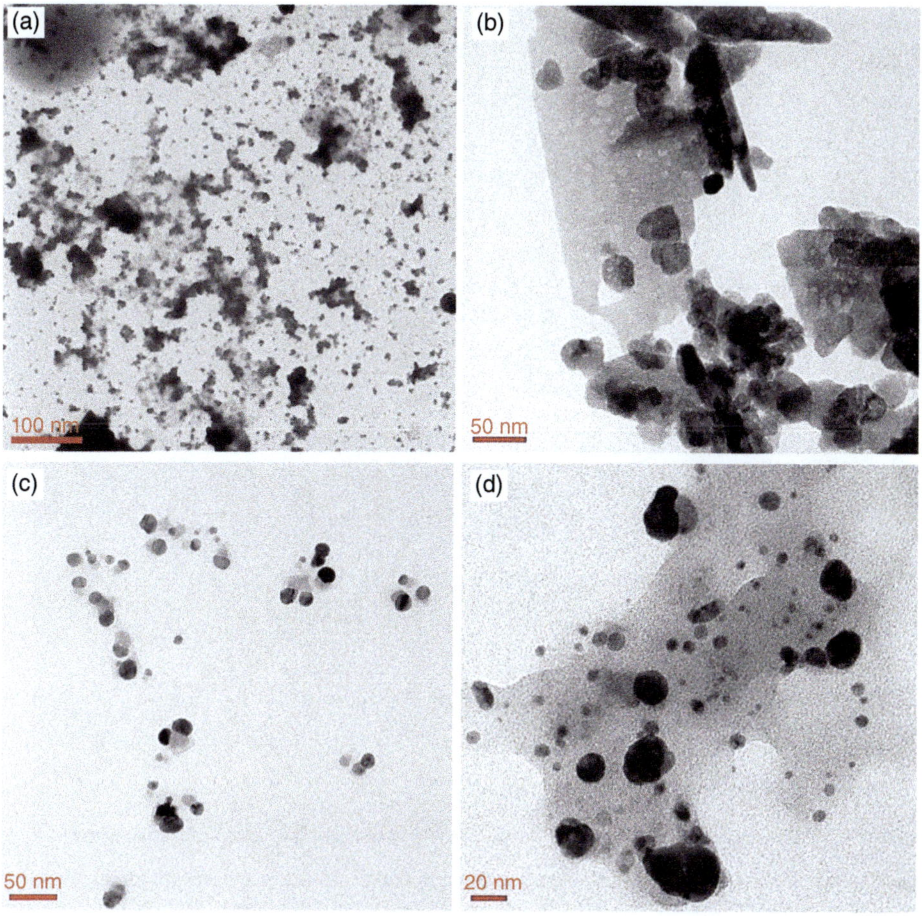

Figure 16.5 TEM micrograph (a) corresponds to a polymer–polymer nanocomposite prepared from a (PS)-*b*-(PMMA) block copolymer. (b) Polymer matrix doped with ZnO nanoparticles as a filler material. (c) AgNPs. (d) Polymer matrix doped with AgNPs as filler particulates followed by the composite development [20–23].

materials [25]. The nanofiller-incorporated composite materials show better performance when used as elastomeric materials; in extensive use for rubber industry, defense systems, and aerospace applications; and in electromechanical and biomedical applications [26].

Significantly, Figure 16.6 (a–d), also shows the morphology of polymer-based nanocomposites developed with different technologies with various filler materials. Such nanocomposites have enhanced improvement in their properties like melting point, surface tension, rubbery nature, glass transition temperature, surface texture, and adhesion properties. Nano-sized filler materials play a crucial role in improving the properties of nanocomposites. From the literature and recent reports, PNCs are the best nanocarrier materials for nanomedicines and anticancer drugs [26, 27].

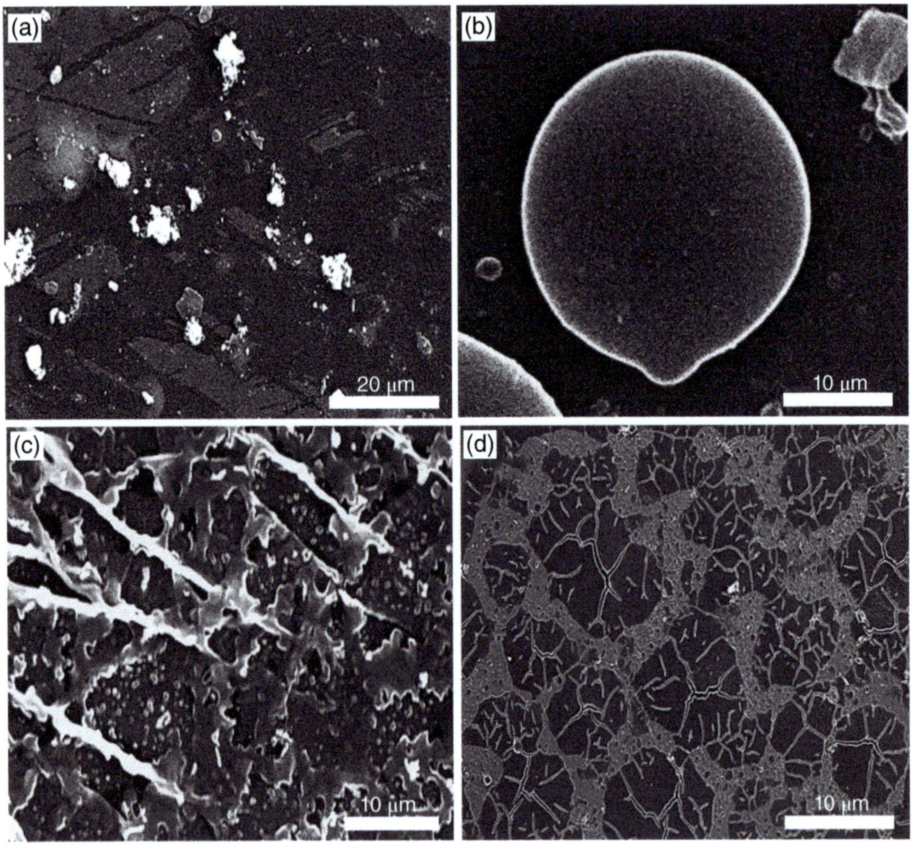

Figure 16.6 SEM micrograph (a) Corresponds to polymer (PCL)-*b*-(Gly) aggregate doped with AgNPs as filler material. (b) Spherical nanocapsule designed from (PCL)-*b*-(L-GluA) corroborates the smooth-surface of the capsule. (c) Rough-surface of polymer matrix prepared from (PCL)-*b*-(NIPAM) block copolymer and localized with ZnO nanorods as filler material. (d) Dendritic nanochannels fabricated from (PCL)-*b*-(L-AspA) [12, 20, 22–27].

16.6 Polymer Nanocomposites for Various Applications

The schematic in Figure 16.7 illustrates the use of polymer-based nanocomposites in various fields such as catalysis, transportation, energy, environmental, instrumentation, defense, biomedical, nanobiotech, sensor, electrical, and electronics. Apart from these, there are many more applications of PNCs, particularly in the automobile sector, pharmaceuticals, oil corporations, industries, for example. These nanocomposites can be used for gas barrier performance with flame retardant properties, for solar cells applications, defense, and aerospace uses as they have high mechanical stability such as stiffness, strength, toughness, and are gas separation permeable materials [28]. Interestingly, thermo-responsive hydrophobic PNCs are preferable for loading with hydrophilic anticancer drugs and nanomedicines. However, nanocomposites designed for defense applications should have greater mechanical properties like strength,

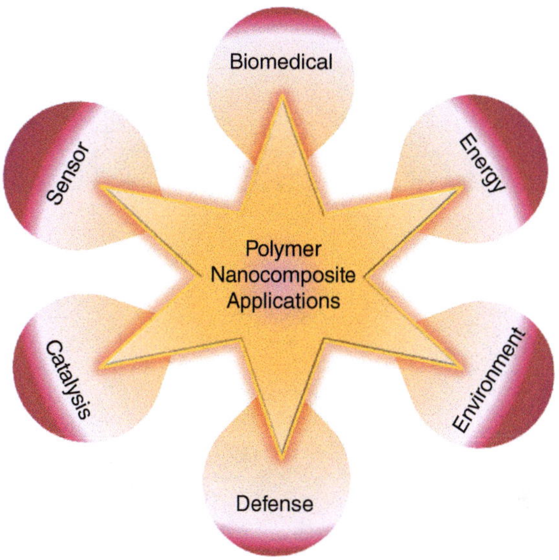

Figure 16.7 Various applications of polymer-based nanocomposites in broader perspective [25–29].

compressive nature, hardness, stress, strain, creep, fatigue, ductile, brittleness, whereas nanocomposites for drug delivery and other biomedical applications should possess biocompatibility, biodegradability, biosafety, with the desired size, shape, and morphology with considerable molecular weight and density [29].

The nanocomposite with good engineering application properties is not suitable for biomedical applications rather these are much fit for other industrial and defense applications (Figure 16.7). As well as drug delivery and biomedical applications, these polymer-based nanocomposites are recommended for many industrial and defense applications to prepare high thermal-resistive materials, bullet-proof jackets, corrosion-free machinery, insulation materials for electrical applications, etc. However, the thermally unstable composites such as ammonium and lactums are not recommended for defense and other applications. Predominantly, PNCs have their own unique characteristics making them suitable for biomedical and pharmaceutical uses. Understanding of the internal structure of nanocomposites can inform the development and desired properties of materials required for specific applications [30].

16.7 Nanocomposites for Molecular Diagnosis and Biopharmaceutics

PNCs are unique in their size, shape, structure, morphology, and physicochemical properties as recommended for many biopharma applications. Apart from nanocomposites, nanofibers, nanofillers, hollow capsules, core-shell capsules, nanorods, nanotubes, nanovacuoles, nanofragments, micelles, liposomes, CNTs, QDs shows greater possibilities for use in medical applications. However, TE, molecular imaging, MRI (Magnetic Resonance Image) scanning, CT (Computed Tomography) scanning, drug/gene/cell delivery, patch therapy by using tissue scaffolds are promising ongoing approaches and methods widely applicable for biomedical therapeutics [31] (see Figure 16.8).

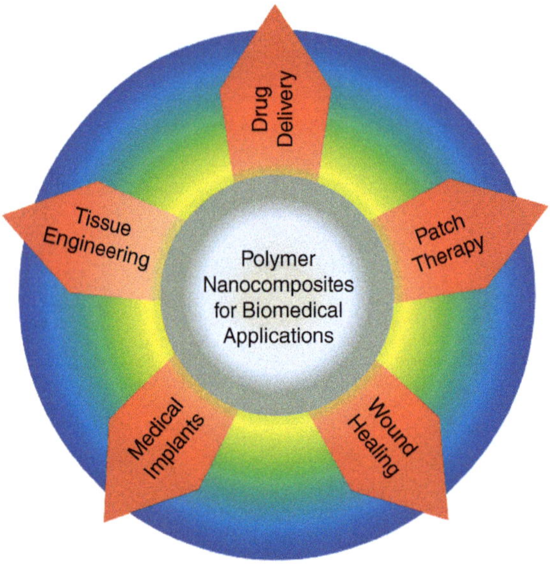

Figure 16.8 Biomedical applications of polymer-based nanocomposites [30–35].

16.7.1 Tissue-Engineering

Scaffold template grafting is the main approach for replacing damaged or destroyed tissue and organs using TE methods. Scaffold grafting TE techniques involve the provision of a scaffold template on the surface of which the cells are replaced under favorable conditions to grow, and thus heal, the damaged portion of tissue [32]. Polymer-based nanocomposites have been reported to provide suitable conditions for the growth of scaffold grafted tissue on the damaged organs of human beings. These PNCs are suitable for growing tissue because of their greater affinity to attach to cells and allow them to grow without any hurdles. However, TE is mainly concerned with tissue implants and replacement of destroyed organs/tissue by means of any unusual activities, for example, damage to cartilage, skin, bone, vessels, and any other parts or organs of the body can be replaced with healthy tissue/organs. Similarly, patch therapy along with TE also has a remarkable impact on organ implantation and replacement without any hazardous issues. Uneven skin has been replaced with even-toned tissue or organs following the applicability of patch therapy and TE [30–33]. Nanoparticles such as ZnO, SiO_2 (silicon dioxide), FeO_3, graphene, reduced graphene, CNTs, QDs, tungsten sulfide, and molybdenum sulfide are referred as strong reinforcing materials in the design of PNCs. These PNCs are mechanistically strong, biosafe, biodegradable, and biocompatible offering good potential for pharmaceutical applications. The addition of filler particles to the polymer matrix (host) with lower concentrations (i.e. 0.2 wt/wt %) leads to enhanced improvement in the properties of PNCs. However, PNCs with comprehensive properties are recommended in the fabrication of lightweight, low density, and foamy materials requiring biocompatible and biodegradable characteristics for bone TE and bone implants. The mechanically reinforced nanocomposites with compressive strength and flexural properties rely on size, shape, morphology, nanostructure, defects, dispersion of nanomaterials which are added to polymer host matrix [32–34].

16.7.2 Patch Therapy

Patch therapy is an advanced nanobiotechnology-based treatment method to treat destroyed or damaged tissue and organs. Polymer-based nanocomposites are loaded with ointments, nanomedicines, or anticancer drugs and the substrate source is fixed on to the damaged tissue or skin parts [35]. PNCs with thermo-sensitive properties show morphological and structural disintegration followed by the release of drug or ointment on the wound or scars requiring healing. Nanocomposites with a thermo-sensitive nature are a boon for the many biomedical potentials and diagnosis methods. These thermos-sensitive nanocomposites show a significant impact on damaged tissue cells and on the malignant portion of the body to promote healing [36]. The usage of thermo-sensitive nanocomposites is an advanced and eminent technique to cure damaged tissue without any pre-treatments and surgery requirements. It is a simple approach to repairing damaged tissues or organs without any side effects and toxicity issues.

16.7.3 Drug-Delivery Systems

Many techniques are applied to treat various cancers but still there are many limitations associated with them. More importantly, biosafety, biocompatibility, biodegradability, side effects, cytotoxic issues, drug-loading efficacy, sustained and dosage-based precise control over the drug release, removal of the drug-carrier tools and nanomaterials, etc. [36, 37]. Over the past couple of decades, scientists and researchers have been working tirelessly to develop efficient nanotechnology-based tools for effective cancer treatments and theranostics. Metal nanoparticles like Fe_2O_3 (iron oxide), AgNPs, AuNPs, SiO2NPs (silica nanoparticles), and ZnONPs have been developed and their potential working efficacy checked through *in vitro* (cell-based examinations) and *in vivo* (animal model) therapeutics, demonstrating that further improvement is needed. Further, polymeric nanoparticles such as protein nanoparticles, solid lipid nanoparticles (SLNPs), liposomal nanoparticles, core-shell nanoparticles, mesosomes nanoparticles, lysosome nanoparticles, micelles, hollow core-shell nanoparticles, QDs, polymer-metal nanoparticles, polymer-inorganic nanoparticles, polymer-organic nanoparticles, ceramic nanoparticles, and polymer-ceramic nanoparticles have also been developed for drug-delivery systems [38]. Polymeric nanoparticles and capsules with surface functionalized properties such as porosity, surface roughness, and a high surface area were also designed for drug-delivery systems. As per the IUPAC (International Union of Pure and Applied Chemistry) nomenclature, pore size has been defined as meso (pore size < 50 nm, in diameter), micro (pore size < 2 nm), nano (pore size < 200 nm), and macro (pore size > 200 nm) with nanoporous materials evidenced as potential drug-carrier systems with the availability of precise and dosage-based drug release [39]. However, the design and development of polymer-based nanocomposites with good reinforcing properties for drug-delivery systems and other cancer theranostics has been explored. The amino-acid-based block copolymers and SiO_2NPs, ZnONPs as filler materials can reinforce the desired properties of nanocomposites and they show remarkable effect as a promising drug-delivery systems for biomedical applications. Further, superparamagnetic nanomaterials, nanotubes, nanorods, nanofibers, and nanocomposites were also used for drug-delivery systems and in the treatment of various cancers such as lung, liver, colon, prostate, ovarian, and skin as well as in the diagnosis of various oncogenic tumors [39–42].

16.7.4 Biosensor and Protein Immobilization

Interestingly, core-shell nanoparticles and fibrous nanoparticles with fluid cores and a solid shell portion can, under favorable conditions, easily entrap nanomedicines and anticancer drugs, and biological entities such as proteins, viruses, and bacteria, without altering their functional activities [41, 42]. Such PNCs can be further recommended for biosensor and biomarker detecting tools. For example, green fluorescent protein is immobilized in nanostructured fibers and nanorods by providing a huge surface area and lower distances for the sensor protein analyzer. PNCs along with nanostructured fibrous materials, core-shell nanomaterials, hollow core-shell nanoparticles, nano-rods, and nanotubes can provide a basic platform for their use as engineering materials as well as life-sciences applications. PNCs with different structural complexity lead to various specific functions with unique characteristics and their usage is not limited to materials science, materials engineering, nanotechnology, biotechnology, pharmaceuticals, or nanobiotechnological applications [43]. They are recommended for many applications irrelevant to the field based on the changing of desired properties. They can then be used for defense systems, transportation systems, aero applications, electric, marine, etc. These PNCs are promising tools for drug delivery, TE, and bone implantation along with biosensor and biomarker applications to detect malignancy or affected portions in specific organs [44].

16.7.5 Drug-release Kinetics

Nowadays, these PNCs are referred as potential payloads for anticancer drugs and nanomedicines for the treatment of various cancers. The spherical nanocomposites with porosity are much more suitable and recommended for drug-delivery systems [45]. However, PNCs shows very controlled and sustained release with precise dosages of drugs. Figure 16.9 (a) and (b) shows the release of drug molecules from a complex structure nanocomposite [46]. The structural disintegration and release of drug molecules with the help of external stimuli factors such as pH, temperature, magnetic, electric, ultrasonication has a key role in release kinetics. These PNCs are considered potential nanocarrier systems for targeted and site-specific drug delivery. The polymer-based nanocomposites do not show any activity on loaded drug molecules while carrying them for assigned release. The combination of nanotechnology and biotechnology can help to minimize the issues associated with drug delivery and cancer theranostics [47]. The biocompatible, biodegradable, and biosafe polymer-based nanocomposites are promising tools for drug-delivery systems.

In addition, Figure 16.10 depicts the concentration-based drug release with respect to time intervals and dose-based therapeutic levels. The meso- and nanoporous carrier composites give a controlled and sustained drug release profile without any harm to body metabolism. Whereas, macro-porous composites give a burst release of drugs that ultimately leads to toxic levels.

However, there are some limitations and issues associated the polymer-based nanocomposites that require addressing. Moreover, there are always risks associated with the filler foreign objects related to their size, shape, morphology, texture, nanostructure, melting point, and crystalline nature as well as their surface properties. Apart from those limitations, there are certain precautions that need to be considered during the design of

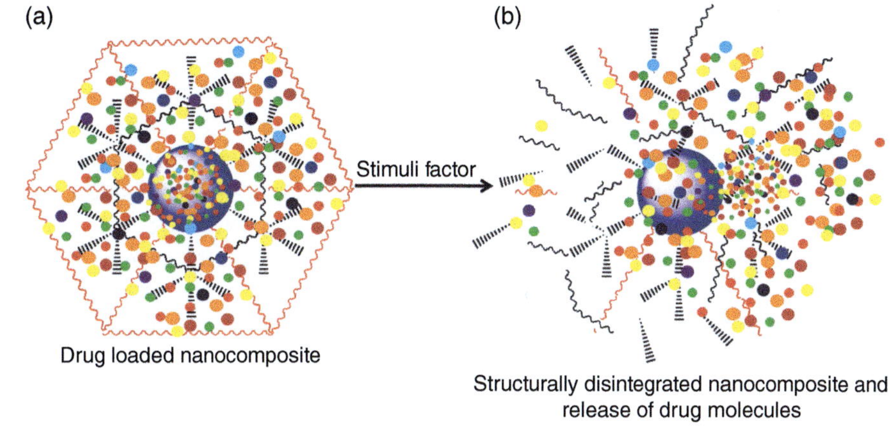

Figure 16.9 The drug-loaded nanocomposite and release upon structural disintegration of nanocomposite with the effect of external stimuli factor. Schematic (a) drug loaded dendritic nanocomposite and (b) drug release upon structural disintegration of dendritic nanocomposite [12, 20, 41–45].

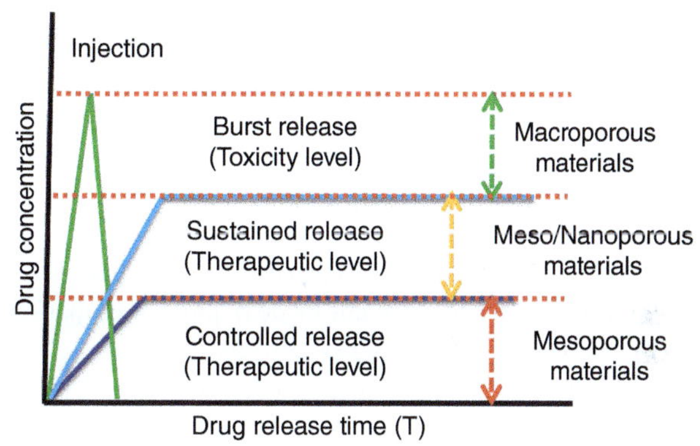

Figure 16.10 The drug release kinetics for different pore sized nanomaterials and their therapeutic levels with respect to drug concentration [6, 12, 20, 44–47].

polymer-based nanocomposites such as nanotoxicology, activity of nanomedicines, risk of assessment, and the detailed study and knowledge of nanomaterials like the host matrix and guest filler foreign objects and particulates [46–48].

16.8 Perspectives of Nanocomposites

PNC materials are considered promising and remarkable tools for biomedical systems. The nanocomposites with improved properties are recommended for many applications such as defense, electromagnetic, electrical and electronics, environmental, transportation, solar energy generation, and as biosensors. Nanocomposites that are biocompatible, biosafe, and biodegradable and have a low density are more suitable for drug-delivery systems. The amino-acid-based host polymer matrix with nanoscale ZnONPs, AuNPs, AgNPs, and SiO_2NPs are certainly promising materials for multipurpose applications.

16.9 Conclusion

Recent advancements in nanocomposites as well polymer-based nanocomposites have been discussed. Different types of polymers and their considerable properties for synthesis of host matrix materials has been described. Further, the definition for nanocomposites and types of nanofiller (foreign objects) materials and their considerable properties were also considered and incorporated in their respective sections. The synthesis of the polymer matrix, the methods, the methodologies, and the development of nanocomposite with improved properties followed by their characterization techniques were also elaborated. Different types of nanocomposites and their medical applications along with other usages for example, in the field of defense, electrical and electronics applications, the plastics industry, environmental, transportation, and solar energy were also identified. Nanocomposites containing surface functionalized properties such as rough and smooth surfaces, specific surface areas, porosity, low density polymer materials and their drug-loading efficacy along with targeted drug delivery and adverse effects were also addressed. Furthermore, biomedical applications of PNCs are predominantly used in TE, bone implantation, as biosensors and biomarkers, in patch therapy and wound healing with precise dose to ensure therapeutic levels were acknowledged.

References

1 Abbasi, R., Mostafavi Amjad, J., Nosrati, H. et al. (2020). Synthesis and characterization of PEGylated iron and graphene oxide magnetic composite for curcumin delivery. *Applied Organometallic Chemistry* 34.

2 Chander, A., Shuai, C., Tirupathi, M., and Guping, T. (2020). Block copolymer [(L-luA-5-BE)-b-(L-AspA-4-BE)]-based nanoflower capsules with thermosensitive morphology and pH-responsive drug release for cancer therapy. *Journal of Materials Chemistry B* 8: 9258–9268.

3 Ahmad, A., Mubharak, N.M., Naseem, K. et al. (2020). Recent advancement and development of chitin and chitosan-based nanocomposite for drug delivery: critical approach to clinical research. *Arabian Journal of Chemistry* 13: 8935–8964.

4 Al Mizan, R. and Islam, M.A. (2020). Synchronization of thermal properties and constituents in nanocomposite: manufacturing, characterization, adjustable properties. *Journal of Applied Polymer Science* (12): 138.

5 Al-Refai, H.H., Ganash, A.A., and Hussein, M.A. (2021). Polythiophene and its derivatives-based nanocomposites in electrochemical sensing: a mini review. *Materials Today Communications* 26: 101935.

6 Amgoth, C., Phan, C., Yumnan, S. et al. (2020). Surface functionalized porous nanomaterials for theranostics. In: *AIP Conference Proceedings*, vol. 2269 (ed. G.N. Rani, J. Anjaiah and P. Raju), 030067.

7 Bahramian, A. (2020). Poly(ethylene terephthalate)-based nanocomposite films as greenhouse covering material: environmental sustainability, mechanical durability, and thermal stability. *Journal of Applied Polymer Science* 138 (10): 49991.

8 Chen, X., Yusuf, A., del Rio, J.S., and Wang, D.-Y. (2021). A facile and robust route to polyvinyl alcohol-based triboelectric nanogenerator containing flame-retardant polyelectrolyte with improved output performance and fire safety. *Nano Energy* 81: 105656.

9 Apostolos, R.C.R., Cipreste, M.F., de Sousa, R.G., and de Sousa, E.M.B. (2020). Multifunctional hybrid nanosystems based on mesoporous silica and hydroxyapatite nanoparticles applied as potential nanocarriers for theranostic applications. *Journal of Nanoparticle Research* 22, 368.

10 Florian, P.E., Icriverzi, M., Ninciuleanu, C.M. et al. (2020). Salecan-clay based polymer nanocomposites for chemotherapeutic drug delivery systems; characterization and in vitro biocompatibility studies. *Materials (Basel)* (23): 13, 5389.

11 Han, L., Shi, J., and Liu, A. (2017). Novel biotemplated MnO_2 1D nanozyme with controllable peroxidase-like activity and unique catalytic mechanism and its application for glucose sensing. *Sensors Actuators B: Chem.* 252: 919–926.

12 Chander, A., Suman, D., Gangappa, D., and Mohan, L. (2018). Self-assembled block copolymer [(BenzA)-b-(PCL)] micelles to orient randomly distributed AuNPs into hollow core-shell morphology and its role as payload for nanomedicines. *Materials Science & Engineering C* 92: 790–799.

13 Hanani, Z., Izanzar, I., Amjoud, M.'b. et al. (2021). Lead-free nanocomposite piezoelectric nanogenerator film for biomechanical energy harvesting. *Nano Energy* 81, 105661.

14 Huang, L., Chen, K., Zhang, W. et al. (2018). ssDNA-tailorable oxidase-mimicking activity of spinel $MnCo_2O_4$ for sensitive biomolecular detection in food sample. *Sensors Actuators B: Chemical* 269: 79–87.

15 Jamei, H.R., Rezaei, B., and Ensafi, A.A. (2020). Ultra-sensitive and selective electrochemical biosensor with aptamer recognition surface based on polymer quantum dots and C_{60}/MWCNTs- polyethylenimine nanocomposites for analysis of thrombin protein. *Bioelectrochemistry* 138: 107701.

16 Junaidi, N.F.D., Othman, N.H., Fuzil, N.S. et al. (2021). Recent development of graphene oxide-based membranes for oil–water separation: a review. *Separation and Purification Technology* 258: 118000.

17 Karthika, V., AlSalhi, M.S., Devanesan, S. et al. (2020). Chitosan overlaid $Fe3O_4$/rGO nanocomposite for targeted drug delivery, imaging, and biomedical applications. *Scientific Reports* 10: 18912.

18 Kausar, A. (2020). Nanocarbon in polymeric nanocomposite hydrogel–design and multi-functional tendencies. *Polymer-Plastics Technology and Materials* 59: 1505–1521.

19 Kubo, M., Okibayashi, K., Kojima, M. et al. (2021). Superhydrophobic polycarbosilane membranes for purification of solar hydrogen. *Separation and Purification Technology* 258: 117998.

20 Chander, A., Gangappa, D., Arunasree, M., and Pradip, P. (2016). Nanoporous capsules of block co-polymers of [(MeO-PEG-NH)-b-(L-GluA)]-PCL for the controlled release of anticancer drugs for therapeutic applications. *Nanotechnology* 27: 125101.

21 Lalegul-Ulker, O. and Elcin, Y.M. (2021). Magnetic and electrically conductive silica-coated iron oxide/polyaniline nanocomposites for biomedical applications. *Materials Science Engineering C* 119: 111600.

22 Li, C.-Q., Zhao, D.-H., Hou, X.-L. et al. (2020). In situ synthesis of multifunctional tellurium nanorods stabilized by polypeptide-engineered for photothermal-sonodynamic combination therapy of tumors. *Chemical Engineering Journal* 417: 127989.

23 Li, F., Chang, Q., Li, N. et al. (2020). Carbon dots-stabilized Cu4O_3 for a multi-responsive nanozyme with exceptionally high activity. *Chemical Engineering Journal* 394, 125045.

24 Li, L., Liu, H., Bian, J. et al. (2020). Ag/Pd bimetal nanozyme with enhanced catalytic and photothermal effects for ROS/hyperthermia/chemotherapy triple-modality antitumor therapy. *Chemical Engineering Journal* 397, 125438.

25 Li, Y., Ma, W., Sun, J. et al. (2020). Electrochemical generation of Fe_3C/N-doped graphitic carbon nanozyme for efficient wound healing in vivo. *Carbon* 159: 149–160.

26 Liu, C., Du, Z., Ma, M. et al. (2020). Carbon monoxide controllable targeted gas therapy for synergistic anti-inflammation. *iScience* 23: 101483.

27 Liu, X., Wang, X., Qi, C. et al. (2019). Sensitive colorimetric detection of ascorbic acid using Pt/CeO_2 nanocomposites as peroxidase mimics. *Applied Surface Science* 479: 532–539.

28 Lu, Y., Ye, W., Yang, Q. et al. (2016). Three-dimensional hierarchical porous PtCu dendrites: a highly efficient peroxidase nanozyme for colorimetric detection of H_2O_2. *Sensors and Actuators B: Chemical* 230: 721–730.

29 Luzi, F., Tortorella, I., Di Michele, A. et al. (2020). Novel nanocomposite PLA films with lignin/zinc oxide hybrids: design, characterization, interaction with mesenchymal stem cells. *Nanomaterials (Basel)* 10 (11): 2176.

30 Muthulakshmi, L., Pavithra, U., Sivaranjani, V. et al. (2020). A novel Ag/carrageenan-gelatin hybrid hydrogel nanocomposite and its biological applications: preparation and characterization. *Journal of the Mechnical Behavior of Biomedcial Materials* 115: 104257.

31 Negahban, Z., Shojaosadati, S.A., and Hamedi, S. (2021). A novel self-assembled micelles based on stearic acid modified schizophyllan for efficient delivery of paclitaxel. *Colloids and Surfaces. B, Biointerfaces* 199: 111524.

32 Niu, X., He, Y., Pan, J. et al. (2016). Uncapped nanobranch-based CuS clews used as an efficient peroxidase mimic enable the visual detection of hydrogen peroxide and glucose with fast response. *Analytica Chimica Acta* 947: 42–49.

33 Ounkaew, A., Kasemsiri, P., Jetsrisuparb, K. et al. (2020). Synthesis of nanocomposite hydrogel based carboxymethyl starch/polyvinyl alcohol/nanosilver for biomedical materials. *Carbohydrate Polymers* 248: 116767.

34 Pirahmadi, P., Kokabi, M., and Alamdarnejad, G. (2020). Polyvinyl alcohol/chitosan/carbon nanotubes electroactive shape memory nanocomposite hydrogels. *Journal of Applied Polymer Science* (11): 138, 49995.

35 Platnieks, O., Gaidukovs, S., Neibolts, N. et al. (2020). Poly(butylene succinate) and graphene nanoplatelet-based sustainable functional nanocomposite materials: structure-properties relationship. *Materials Today Chemistry* 18: 1–17.

36 Rezvova, M.A., Yuzhalin, A.E., Glushkova, T.V. et al. (2020). Biocompatible nanocomposites based on poly(styrene-block-isobutylene-block-styrene) and carbon nanotubes for biomedical application. *Polymers (Basel)* 12 (9): 2158.

37 Seifi, T. and Kamali, A.R. (2020). Anti-pathogenic activity of graphene nanomaterials: a review. *Colloids and Surfaces B: Biointerfaces* 199: 111509.

38 Swapna, V.P., Kaliyathan, A.V., Abhisha, V.S. et al. (2020). Changes in free volume and gas permeation properties of poly(vinyl alcohol) nanocomposite membranes modified using cage-structured polyhedral oligomeric silsesquioxane. *Journal of Applied Polymer Science* 138 (10): 138, 49953.

39 Wang, C., Gao, J., Cao, Y., and Tan, H. (2018). Colorimetric logic gate for alkaline phosphatase based on copper (II)-based metal-organic frameworks with peroxidase-like activity. *Analytica Chimica Acta* 1004: 74–81.
40 Wen, C., Cheng, R., Gong, T. et al. (2020). Beta-cyclodextrin-cholic acid-hyaluronic acid polymer coated Fe_3O_4-graphene oxide nanohybrids as local chemo-photothermal synergistic agents for enhanced liver tumor therapy. *Colloids and Surfaces B: Biointerfaces* 199: 111510.
41 Xu, Z., Sun, P., Zhang, J. et al. (2020). High-efficiency platinum–carbon nanozyme for photodynamic and catalytic synergistic tumor therapy. *Chemical Engineering Journal* 399, 125797.
42 Xu, Z., Xing, W., Hou, Y. et al. (2021). The combustion and pyrolysis process of flame-retardant polystyrene/cobalt-based metal organic frameworks (MOF) nanocomposite. *Combustiona and Flame* 226: 108–116.
43 Yang, Q., Li, L., Zhao, F. et al. (2019). Generation of MnO_2 nanozyme in spherical polyelectrolyte brush for colorimetric detection of glutathione. *Materials Letters* 248: 89–92.
44 Yi, H., Zhang, Y., Zhou, C. et al. (2020). Wonton-like nanoparticles with dual enzyme-mimetic function for the multiple-imaging-guided cancer combined therapy. *Chemical Engineering Journal* 401, 126054.
45 Chander, A., Rompivalasa, S., Tirupathi, M. et al. (2020). Solvent-assisted [(Glycine)-(MP-SiO_2NPs)] aggregate for drug loading and cancer therapy. *ChemistrySelect* 5: 8221–8232.
46 Yuan, W., Li, Z., Xie, X. et al. (2020). Bisphosphonate-based nanocomposite hydrogels for biomedical applications. *Bioactive Materials* 5: 819–831.
47 Zhang, B., Wang, X., Zhao, Y. et al. (2017). Highly photosensitive colorimetric immunoassay for tumor marker detection based on cu(2+) doped ag-AgI nanocomposite. *Talanta* 167: 111–117.
48 Zhang, S., Ming, Y., Wei, Y. et al. (2020). The effect of grafting density on the crystallization behavior of one-dimensional confined polymers. *Journal of Applied Polymer Science* (12): 138, 50064.

17

Toxicity of Nanomaterials

Elham Abohamzeh[1], *M. Sheikholeslami*[2,3], *and Ahmad Shafee*[4]

[1] *Department of Energy, Materials, and Energy Research Center (MERC), Karaj, Iran*
[2] *Department of Mechanical Engineering, Babol Noshirvani University of Technology, Babol, Iran*
[3] *Renewable energy systems and nanofluid applications in heat transfer Laboratory, Babol Noshirvani University of Technology, Babol, Iran*
[4] *Public Authority of Applied Education & Training, College of Technological Studies, Applied Science Department, Shuwaikh, Kuwait*

17.1 Introduction and Recent Advances

An outstanding revolution in science has been observed by nanotechnology. The diverse applications of nanoparticles (NPs) have resulted in the substantial development of nanotechnology in several fields of science in recent decades. The application of nanotechnology knowledge for diagnosis and treatment of diseases has been provided by nanomedicine, leading to improvement of the current techniques and the development of new tools and techniques, and also providing solutions for the early diagnosis of diseases and treating complicated diseases. Due to the unique characteristics of nanomaterials (NMs), they have been reported to be promising materials to be utilized in several biomedical applications including imaging, drug delivery, as well as biosensing [1]. Such NPs possess diverse advantages such as a large surface-to-volume ratio, allowing them to be attached to more units of the drug compared to the particles with the larger size and a similar mass, resulting in a larger loading efficacy. Also, their small size, in the order of biomolecules, allows them to interact with cells, and their unique physicochemical characteristics in comparison with their bulk counterparts offer opportunities for special treatments and their greatly tunable surfaces allow the targeting of cells [2, 3]. Despite various advantages of nano-sized particles for biomedical applications, their utilization is also associated with potential health risks because of their potential toxic impacts and uncontrollable utilization. Higher chemical reactivity in NPs arises from their larger surface-area-to-volume ratio and causes enhanced production of reactive oxygen species (ROS). In fact, a significant parameter in the intrinsic toxicity of NPs is their surface area because their surfaces interact with biological systems [4]. One of the toxicity mechanisms of NPs is the formation of ROS, which results in inflammation, oxidative stress, and damages DNA (deoxyribonucleic acid), cell membranes, as well as proteins. Thus, in biomedical applications including therapeutic

Nanomaterials and Nanotechnology in Medicine, First Edition. Edited by P.M. Visakh.

applications, gene delivery, and drug-delivery systems, it is required to assess the toxicity of NPs [5, 6]. Oxidative stress is induced by prooxidants through inhibiting antioxidants or creating ROS. Through the reaction of cells with NPs, their prooxidant effects are induced via the intracellular generation of ROS [7]. The cellular redox system can be activated by NPs in particular in the lungs where the immune cells such as neutrophils and alveolar macrophages (AMs) serve as direct inducers of ROS. Considerable ROS is induced by AMs and neutrophils [8, 9]. Recently, several *in vivo* and *in vitro* investigations have been conducted with the purpose of understanding the toxicological effects and potential risks of the exposure of various NPs to humans and the environment. The shape-dependency of gold (Au) NPs cytotoxicity has been reported by Steckiewicz et al. [10]. The highest level of cytotoxicity was observed for Au nanostars, which had the greatest anticancer potential, while a lower toxicity level was reported by Au nanospheres, which had low anticancer potential. The impact of the size and surface coating of iron oxide NPs and their biological impacts in *in vivo* and *in vitro* have been investigated by Feng et al. [11]. They noticed the considerable influence of both surface coating and size on the uptake by cells, distribution, excretion, and cytotoxicity of these particles. Gea et al. [12] investigated the genotoxicity, cytotoxicity, and lactate dehydrogenase (LDH) release of three fabricated TiO2 (titanium dioxide) nano-sized particles having dissimilar shapes (platelets, rods, bipyramids) compared to two commercial TiO_2NPs (food grade, P25). They reported low cytotoxicity and also concluded that the light exposure and the shape of NPs affect the cytotoxicity. Moreover, they observed no substantial LDH release. Actually, the aggregation tendency of TiO_2 NPs and also the cellular uptake influence genotoxicity. Mulenos et al. [13] have noticed the low-level toxicity of silver (Ag)NPs considering the observation that the release of just minute quantities of Ag ions was noticed. Kassem et al. [14] reported that the carbon nanotube (CNT) antimicrobial activity depends on different operational factors including temperature, solvent composition, retention time, as well as pH. Based on the study by Mohanta et al. [15], CNTs interaction with heart muscle (myocardium) leads to the toxicity of cardiovascular tissue and induces cell proliferation, muscle damage, bloodstream interruption, as well as atherosclerosis. The main impacts of CNTs are blood clotting and hemolysis. The toxicity properties of few-layer graphene (FLG) have been investigated by Ruiz et al. [16]. They observed a very low toxic profile of FLG, which is highly dispersed in water. It must be mentioned that although quite a few investigations have been performed, the toxicity effects of NP exposures are still poorly understood since there are no important methods for testing NPs for safety and health effects [17]. This chapter discusses the biomedical application of several types of NMs including polymeric NPs, metallic NPs, carbon-based NPs such as fullerenes, graphene, CNTs, quantum dots (QDs), as well as metallic NPs. Moreover, several administration routes for exposure of NMs, cellular uptakes, their mechanism in human body, their long-term fate, and elimination are discussed comprehensively. Additionally, the toxicity of the most important NMs in biomedical applications are discussed. A review of various physicochemical properties of NPs such as surface area, size shape, surface charge, composition, concentration and aggregation, and surface coating that govern the cytotoxicity characteristics is included. The chapter also presents safety regulations for NMs handling.

17.2 Biomedical Applications of Nanomaterials

Nanomedicine, as a medicine branch, has provided the opportunity of applying tools and knowledge of nanotechnology to diagnose and treat diseases with no side effects. With the application of nanotechnology, the available tools can be improved and the development of entirely new techniques is also possible. The exceptional electronic, magnetic, and optical characteristics of NMs can offer excellent potential for a broad range of biomedical applications such as biosensing, imaging, and drug delivery [18–21] as schematically illustrated in Figure 17.1. NMs can be classified based on their special characteristics and materials nature, as metallic NPs, carbon-based NPs such as graphene, CNTs, fullerenes, and QDs, and also polymeric NPs and magnetic NPs. The biomedical applications of these nano-sized particles are discussed in this section.

17.2.1 Metallic Nanoparticles

Metallic NPs are utilized in several biomedical applications including gene targeting, drug delivery, sensing, as well as imaging [22–24]. Some of these NPs also have been observed to possess considerable potential for therapeutic applications [25–28]. The applications of these NPs are consistently increasing since they have been reported to be comparatively less toxic. Over the past few decades, the utilization of metallic NPs and their alloys in different sectors has been investigated in several research works including cancer treatment [29], drug delivery [30], biological labeling [31], catalysis [32], optical devices [33], and sensor technology [34]. Moreover, because of the outstanding surface plasmon resonance (SPR) characteristics of metallic NPs, they are appropriate as markers for biomolecule optical detection. Furthermore, based on results from previous studies, they are promising alternative as antiplatelets [35], antimicrobial agents [27, 36], for drug

Figure 17.1 The utilization of nanomaterials in biomedical applications.

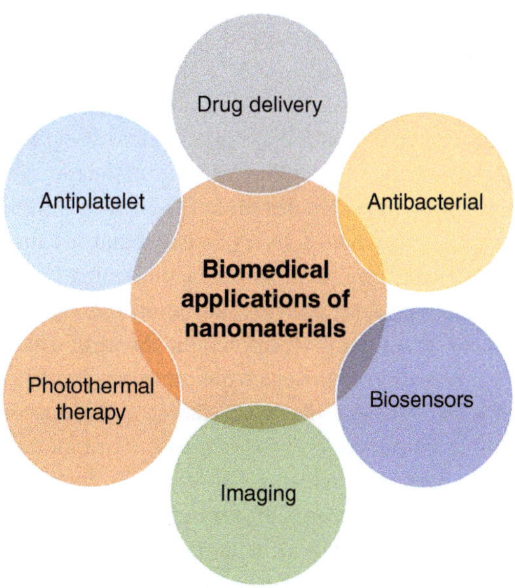

delivery, and the stabilization of proteins. One of the metallic NPs that is widely utilized is nanosilver, which possesses antiparasitic [37] and antimicrobial activity [27]. The support of AgNPs from surface plasmons is advantageous for their extraordinary optical characteristics. The surface plasmons are driven into resonance at the distinct light wavelength and AgNPs have a specific color. Any alteration in the aggregation value of AgNPs can be detected by an apparent change in the particle color in the solvent. AgNPs have been observed to possesses great potential for application in biomedical services including therapeutics and imaging, in particular as antimicrobial agents, considering their exceptional physiochemical characteristics. Moreover, AgNPs have demonstrated antifungal and antiviral characteristics. The antimicrobial activity of AgNPs against yeast has been assessed by Kim et al. [27]. Another application of AgNPs is in biomedical imaging. Ag is also a favorite metallic material for surface-enhanced Raman spectroscopy, due to its easy preparation, excellent surface properties, and excitation wavelengths, which can be easily accessed in the visible range. Aqueous solutions of nanosilver are widely utilized for biolabeling, biodetection, development of microarray type gene probes, detecting DNA, identification of proteins, investigation of different drug interactions with proteins for drug discovery, and realizing the pH impacts on the protein's conformation [38]. The application of AgNPs has also been reported as a biocide to aid the avoidance of burn infection, diabetic ulcers [39], and traumatic wounds [40]. Another application can be for enhancing surface coating for implanted medical devices within or on the body [41].

Another metallic NP with an extensive range of utilization in biomedical devices is Au. The biocompatibility of Au particles is comparatively high because they are biologically inert. Primarily, AuNPs have been applied for labeling applications. Au particles have a great ability for absorbing and scattering visible light. The free electrons in AuNPs are excited with absorbing light to SPR. In general, SPR dominates the optical characteristics of small metallic NPs [42]. The bulk dielectric constant of Au governs the resonance frequency that lies in the visible range of the electromagnetic spectrum [43]. Compared to other metallic NPs, AuNPs have been reported to have high surface reactivity, biocompatible catalytic, electronic, and optical properties. Bio-optical imaging is an important application of AuNPs. It is possible to visualize AuNPs applying several methods, so they can be extremely interesting contrast agents. Moreover, great contrast for transmission electron microscopy imaging can be offered by AuNPs, as well as visible-light imaging applications. Furthermore, AuNPs are utilized for the delivery of the molecule into cells. To achieve this objective, after the absorption of molecules on to the AuNPs' surface, they enter the cells. AuNPs are utilized for anticancer drug delivery, in particular to cancer tissue. Since AuNPs are inert, colloidally stable, small, and comparatively biocompatible and can easily conjugate with ligands, they are used for drug delivery. Based on their photothermal characteristics, they are used for cancer treatment [44]. Upon light absorption excitation of the AuNPs' free electrons occurs. The relaxation of the electrons and transfer of heat energy to the lattice happens with interactions of the AuNPs crystal lattice and electrons. It will be followed by the dissipation of heat from AuNPs into the nearby environment. Therefore, it is possible to heat AuNPs via light absorption, which is subsequently changed into thermal energy. The local destruction of cancerous tissues with no exposure of the entire organism to high temperature via the heat generated by AuNPs is possible [45]. Diagnosis is another biomedical application of AuNPs. For this purpose, AuNPs with several exceptional characteristics including nonlinear optical characteristics as well as colorimetric conductivity, have been investigated for application in biomolecular detection applications. They have the potential to be utilized for detecting nucleic acids and proteins in clinical samples [46–48].

17.2.2 Carbon-based Nanomaterials

All carbon-based NMs possess extremely attractive physicochemical characteristics, including a large surface area, semi-metallic or metallic behavior, high mechanical strength, great electrical and thermal conductivity, ultra-light weight, as well as an ordered structure with great aspect ratio. The successful transformation of carbon nanomaterials (CNMs) into soluble or polar derivatives is possible via chemical transformation, which poses significant challenge to be overcome prior to their application in biological utilities. These materials have been reported to possess an exceptional potential to be used in several biomedical utilities including therapeutics, imaging, and sensing. There are several classifications of carbon-based NPs such as graphene, nanotubes, and fullerenes.

17.2.3 Carbon Nanotubes

CNTs possess exceptional properties such as great thermal and chemical stability, rich electronic characteristics, ultra-light weight, high mechanical strength, and a large surface area. With surface functionalization of CNTs, binding to nucleic acids and proteins is easily possible, and thus CNTs have been considered as promising new components for the delivery of therapeutic molecules [49, 50]. Their tendency for noncovalent interaction with drug molecules, small size, and a great surface-to-volume ratio have made them great tools in drug delivery and they have gained considerable attention recently for their application as drug carriers [51]. Given the electronic characteristics of CNTs they are widely utilized in biosensors. Their charge transport characteristics have made them appropriate for application in different biosensors. CNTs can be integrated with biomolecules and hybrid systems and have the potential to be used as electrochemical biosensors (DNA sensors, immunosensors, and enzyme electrodes). The small size of CNTs is a great advantage for their integration with biomolecules, which allows plugging them into places that are essential for electrochemistry utilization, for instance within proteins [52]. Sensors that are based on CNTs can be utilized for the detection of various biological species like DNA, proteins, toxins, neurotransmitters, as well as glucose [53–55].

17.2.4 Fullerenes

Fullerenes are carbon allotropes with molecules consisting of carbon atoms that are held together by double and single bonds forming a closed mesh. Fullerenes derivatives like C_{60} have exhibited excellent performance as pharmaceutical agents. Several attachment points exist in nanostructures and functionalizing their surfaces it is possible to bind them with tissue. In clinical trials there are already many of these derivatives with great biocompatibility and, even at high doses, relatively low toxicity is observed in them. Based on the previous studies, fullerene derivatives have the potential to be applied as antibacterial agents such as *Mycobacterium tuberculosis*, Streptococcus, *Escherichia coli*, and as antiviral agents, particularly against human immunodeficiency virus [38, 56, 57].

17.2.5 Graphene

In graphene, the carbon atoms are held together by strong sigma bonds in a plate plane with a thickness of one atom. Its exceptional structural characteristics, especially potential biocompatibility, exceptional optical and electronic properties, the propensity for functional modification, and high aspect ratio have made it a very interesting alternative in biomedical applications such as photothermal therapy, bacterial inhibition, imaging, drug delivery, and biosensor development [58, 59]. Biosensors that are based on graphene are applied for sensing different significant biomolecules such as alcohol, dopamine, DNA, and glucose [60, 61]. Moreover, graphene can also act as a great alternative for microbial diagnosis and detection [62]. Based on the exceptional optical characteristics of graphene, they possess the potential to be utilized for cellular imaging [63]. The intrinsic optical characteristics of graphene oxide (GO) and single-walled nanotubes (SWNTs) provide the opportunity for their application as optical probes. SWNTs and GO exhibit optical absorption in the near-infrared and visible ranges. In the near-infrared range, they show great photoluminescence, optical absorption, and resonance Raman scattering, and these properties are used for imaging in biological systems. Moreover, it is also possible to use the great optical absorption of single-wall CNTs in photoacoustic imaging, where local heating via laser light absorption generates sound waves. Compared to traditional ultrasound, the spatial resolution of photoacoustic imaging is higher and penetration in tissue is deeper compared with fluorescence imaging [64].

17.2.6 Quantum Dots

QDs are semiconductor particles with electronic characteristics between discrete molecules and bulk semiconductors. Coating QDs with other materials is necessary when they are used in biological applications in order to avoid toxic metals leakage and to allow aqueous dispersion. QDs have been extensively utilized in photodynamic therapy [65], drug delivery [66, 67], charge transfer-based biosensors [68], quantify fluorescence in *in situ* hybridization signals [69, 70], and in cell labeling and imaging [71, 72].

There are also other significant NPs, besides the already-mentioned NMs, such as magnetic NPs, which are extensively utilized in biomedical devices. The advantageous characteristics of magnetic NPs such as manipulation with externally applied magnetic fields, contrast improvement in magnetic resonance imaging, size optimization upon requirements have made them suitable for several biomedical applications such as sensing, tagging, drug delivery, imaging, as well as protein purification [73, 74].

17.2.7 Polymeric Nanoparticles

Polymeric NPs are colloidal structures consisting of semisynthetic or synthetic polymers. Polymers including chitosan NPs, polycaprolactone, and polysaccharide chitosan-polylactic acid are utilized as drug carriers [75, 76].

17.3 Biodistribution, Mechanism, and Excretion of Nanomaterials

When NMs are used for biomedical applications *in vivo*, their biodistribution, as well as long-term fate, is critical for the effectiveness and applicability of the drug. In particular, when such nanomedicines are used over an extended period of time, repeatedly. The NMs' biodistribution and excretion are dependent on both materials' physicochemical characteristics and the way of applying the material to the body.

17.3.1 Routes of Entry

Drug or active agents are required to be introduced to the considered cell, tissue, and organ to exert their action mode. A significant factor for the efficient employment of NPs is the route of administration. The biodistribution and biokinetics of NMs, which are exposed by intravenous injection are dissimilar to that of those exposed via inhalation [77–79]. Several distribution pathways in the body can result from selecting different routes of administration, which also influence the clearance [80, 81]. The injection is the most direct administration route for NMs. The most common experienced routes are intratumoral, subcutaneous, and intravenous injection. Quick distribution and delivery across the vasculature are advantages of the NMs' direct injection into the bloodstream. Nevertheless, it should be noted that this quick distribution might lead to quick excretion by the reticuloendothelial system (RES), the liver, or the kidneys. Inhalation is another favorable and direct route of NMs' administration, for both systemic delivery and lung disease treatment. Inhaled insulin has been reported to lower blood sugar since the 1920s [82]. The lung possesses extremely attractive structure-functional characteristics including great endothelial permeability and a large internal surface area [83]. Systemic delivery can be facilitated by these features of the lung, making it a favorable organ for the administration of NM [33]. NMs that are inhaled can be cleared by different mechanisms. The most preferable administration route in nanotherapeutics is oral administration, in particular, because of the convenience of application and accordingly great patient compliance. Despite the convenience of oral delivery, this method faces the challenge of poor bioavailability of the drug, which results from poor penetration across intestinal biological membranes, the acidic environment of the stomach, and enzymatic degradation [84]. Nevertheless, based on previous studies, the bioavailability of a drug, which is administered orally can be improved by NM formulations [85]. Dermal absorption is another administration route that delivers the drug directly to the skin. The largest organ in the body is the skin, which offers an efficient barrier to the environment. The possible NM interactions with the skin have been reported in previous studies [86, 87]. Additionally, the skin is a possible target for drug delivery by nanocarriers [88]. Retinal drug delivery is another administration route involving application to the retina of various formulations. The low bioavailability of NMs, which are utilized as drug carriers, can be enhanced partly via optimizing their design [89].

17.3.2 Biological Barriers

NPs have to pass several biological barriers including tissue as well as cellular barriers prior to their entry into the systemic circulation. There are several biological barriers, so drug nanocarriers or NMs should be developed to efficiently influence the diseased barrier or to overcome the barrier to offer an efficient medicinal drug concentration at the diseased site. Internal and external biological barriers could be distinguished based on the administration route and the exposure mode. The primary biological barriers that separate the external environment and the hosts are the mucosa and the skin. The gastrointestinal tract, the retina, the respiratory system, the nasal tract, and the epithelia of the skin are external barriers. The stratum corneum consists of the barrier of the skin, while a mucosal layer covers the other barriers, so NMs should overcome this layer through the previously mentioned exposure routes for reaching the targeted tissue or/and the blood circulation. The RES, the placental barrier, the interstitium and extracellular matrix (ECM), the blood–brain barrier (BBB), the blood–testis barrier (BTB), and the endothelium of the blood vessels are secondary or internal biological barriers. In the body, one of the tightest barriers is the BBB and is responsible for protecting the central nervous system (CNS) against toxicants and restricting the hydrophilic, large molecules diffusion into the cerebrospinal fluid (CSF). After overcoming the tissue barriers and reaching the targeted cells or interacting with cells at the tissue barrier itself, NMs have to face different other remained barriers, such as intracellular membranes and the outer cell membrane. The cell membrane contains a phospholipid bilayer consisting of different membrane proteins, carbohydrates, and lipids, for regulating the entry of large and small molecules into the cell [90]. Following cellular internalization, the transport of NMs to the lysosomes ultimately occurs through endocytic vesicles. The degradation of ingested NPs is possible in the lysosomal compartment because of its acidic environment. As pH-sensitive nanocarriers, CNTs have exhibited an excellent ability in cancer therapy since, in low-pH environments, they have shown improved release of conjugated drugs. The main barriers for targeted NPs to the nucleus are nuclear pores that prevent entry to NPs that are more than 9 nm [91].

17.3.3 Cellular Uptake

The cellular uptake of NMs takes place through the endocytic pathways. There are two kinds of endocytosis namely pinocytosis and phagocytosis. Phagocytosis includes ingesting large particles, like cell debris and microorganisms, at giant vesicle formations named phagosomes (in general $0.25\,\mu m < \text{diameter}$) while pinocytosis includes ingesting molecules with small vesicles (diameter of $0.25\,\mu m$). The NM endocytosis is dependent on the interaction of the cell wall with NM, the type of cell, and the NM (see [92, 93] for further information). Different characteristics of NMs including surface coating, material, shape, and size possess an influence on which kind of uptake mechanism is activated [94–96]. The type of the cell also seems to have a significant effect due to different uptake mechanisms of different cell types and their different reaction with NMs [97]. Following the uptake of NPs by cells, they reach endocytic vesicles that merge with each other for forming early endosomes (EE). There, a complicated system sorts the vesicles in accordance with the membrane-associated proteins and internalized materials type for maturing into degradation or translocation vesicles. Following endocytosis, the travel of formed endosomes to the cell's inner part takes place along the microtubule (MT) network. The required proteins for the maturation of EE are provided by exchanging membrane components and cargo, such as NMs, with the trans-Golgi network (TGN). Eventually, an endolysosome is formed through the fusion of the endosome and a lysosome.

NMs degradation in the cells could be mostly caused by the harsh environment of the lysosomes [98–101]. NPs are exposed to high ionic strength, acidic pH, and up to 50 various degrading enzymes inside the lysosomal compartment. Degradation of biomolecules including polysaccharides, lipids, proteins, RNA (ribonucleic acid), and DNA can occur by these acid hydrolases [102]. The dissolution of AgNPs following their endocytosis into epithelial cells has been reported by Jiang et al. [103]. Based on their results, enzymes caused the largest portion of NP dissolution. The same impacts have been also reported for metallic NPs like Au, iron, and ZnNPs (zinc NP) [104]. On the contrary, the degradation of some other NPs, such as mesoporous silica NPs occurs without degrading enzymes in a phosphate-buffered saline solvent [105]. The characteristics of the NMs including surface oxidation and porosity, and also the composition of the degradation medium such as protein content, flow, and pH influence NP degradation [106].

NP exocytosis has been discussed in a few studies in recent years. Two exocytosis pathways are the regulated and constitutive pathways [77, 107]. Based on previous studies exocytosis could be considerably influenced by different parameters including the characteristics of NM (for instance, shape, surface, and size), their exposure time and concentration, and also cell type [108, 109]. As a result of the possible change of the NPs' surface during the processing of lysosomal and uptake, the exocytosed NPs might show dissimilar surfaces, leading to the recognition of different receptors or targets on the surface of the cell [109, 110].

17.3.4 Biodistribution and Biokinetics of Nanoparticles

The activities of NMs at the level of cellular or tissue depends on the ratio and amount of absorbing, distribution, metabolism, and elimination of the material. These processes in nanoscience are known as biokinetics [187377]. Considering the fact that the ratio of volume-to-surface for NMs is totally different from that of their bulk counterparts, they have also different physicochemical characteristics [111]. The biodegradation parameters like barrier tightness and permeability, and also biodistribution are affected as a result of these differences [112, 113]. Generally, NMs could be detected in local and systemic lymph nodes, CNS, bone marrow, kidneys, spleen, and liver, following intravenous injection, oral exposure, or inhalation [114]. Thus, the investigation of these organs in addition to the targeted organs should be considered in the research studies that are conducted about NMs biokinetics [110]. NMs' physicochemical characteristics, their dissolution/biodurability, and the entry point are parameters that affect NM systemic biokinetics. The definition of biodurability is the tendency to persist dissolving and biotransforming inside the biological medium, resulting in possible NMs bioaccumulation [115]. The biodistribution of NMs can be affected by their properties. The surface chemistry of NMs that influence protein binding specificity and extent plays a significant role in their biodistribution. NPs should have sufficient time for circulation to reach their targeted organs. A mechanism named opsonosis quickly tags naked NPs in circulation for removal. This mechanism is conducted by attaching opsonin, which is a protein, to NPs so they will be flagged to be removed by the macrophages of the immune system through reticuloendothelial cells [116–119]. The NMs that are considered external objects, mostly reach the spleen and liver. A conventional approach for NPs stealth shielding is surface coating of poly(ethylene glycol) (PEG), which results in significantly longer circulation times. It must be noted that although surface coating of NPs with PEG (NP PEGylation) extends the time of circulation it can also have a negative impact on uptake by cells through interference with the interactions of the cell with NPs [120].

Following the entry of NPs into the biological system, proteins quickly surrounded them, resulting in the formation of a corona that controls subsequent interactions with other cells and proteins. The surface charge of the NP influence this process's equilibrium kinetics and should be considered in investigating the behavior of NP *in vivo* [121]. In addition to internalization by cells, the NP biodistribution and eventually its fate could be significantly influenced by surface charge. The impact of surface charge on the ability of Au nanoclusters to overcome the BBB by applying focused ultrasound technology was investigated by Sultan et al. [122]. They noticed that the neutral nanoclusters had the most efficient performance for opening the BBB and entering the brain. In the marginal zone of the spleen white pulp, the accumulation of uncharged NPs is more than negatively and positively charged NPs, while the accumulation of both negatively and positively charged NPs occurs to a greater extent in the spleen red pulp. The accumulation of negatively charged NPs occurs extensively in the liver, while positively charged NPs demonstrated more extensive accumulation in liver hepatocytes. Compared to charged NPs, the immune system might interact with uncharged NPs with the core of 2-nm to a higher extent, showing the importance of determination of NMs sub-organ distribution for imaging and drug-delivery applications [123]. Several studies that have been conducted to study biodistribution of NPs *in vivo* revealed that the effect of composition, size, and also the surface charge of NPs are complicated and these parameters should be taken into account together in the process of design and testing of NPs [81]. The size of a NP is one significant factor in designing a NP and for achieving optimum biodistribution. Based on physiological factors including cellular uptake, molecular targeting, and blood circulation half-life, it can be clearly understood that, in addition to surface composition, the size of the particle is a critical parameter for obtaining therapeutic efficacy [81].

17.3.5 Excretion

The clearance of NPs is another significant concept regarding biodistribution. Following the entry of NPs in the circulatory system, their interaction with cells is possible, however, their distribution in peripheral organs is more probable. The physicochemical characteristics of NPs, especially their surface modifications and structure, and also their size strongly affect their distribution to peripheral organs, and in particular their clearance. There are several pathways through which NMs can be cleared from the body. The most efficient is renal clearance, however, because of the relevant size limitations, it is not possible for several NPs to effectively be excreted by this pathway and undergo biliary clearance where the liver process NPs and they become cleared through the gastrointestinal tract. The main route for the clearance of NPs that could not be excreted by the kidneys is through hepatic clearance. There is a close correlation between mucociliary clearance in the upper airways and gastrointestinal excretion, because of the transport of NPs that are trapped in the mucus passing through the pharynx, and then swallowed. NMs can be cleared through the liver and kidneys in 30 minutes to a couple of days, whereas NMs absorbed via

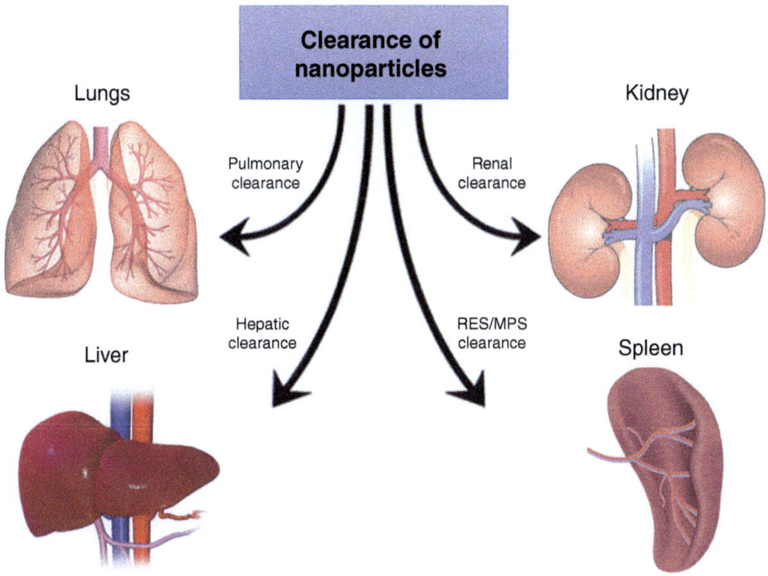

Figure 17.2 Clearance of nanoparticles from different organs.

mononuclear macrophages could remain for much longer, trapped in the RES [89]. Moreover, the RES can also clear NPs. The elimination of NPs from different organs is illustrated in Figure 17.2. The clearance system and the speed of clearance are by the NP size. In the case where NPs are exposed intravenously, the particles with a size of more than 200 nm are rapidly cleared via the mononuclear phagocytic system [124]. When NPs are lower than 5 nm, they would penetrate the kidney capillaries and the renal system will clear them [125]. For NPs with a size between 5 and 200 nm, a longer time of circulation is possible. The lungs also cooperate in the elimination of particles having sizes larger than 100 nm. The distribution and NP translocation have been reported to be dependent on their size, so the NP size also plays a significant role in their toxicity issues [126].

17.4 Toxicity of Nanomaterials

Following the entry of NPs such as Ag, CNTs, graphene, and TiO_2 into the body, they might overcome different cellular barriers and end up in the vulnerable organs including kidney, liver, and lung, and depending on their dose, size, as well as shape, they may induce deleterious impacts such as DNA mutations, mitochondrial damages, and ultimately cell apoptosis or death [126–128]. The toxicity of the most important NMs in biomedical applications is discussed in this section.

17.4.1 Metal-based Nanoparticles

Metal-based NPs have been reported as the most extensively utilized with a variety of applications including environmental remediation, UV blockers in sunscreen, cosmetics, coatings, paints, solar cells, and biomedical fields. Physiochemical characteristics, high reactivity, high stability, and plasmonic and photothermal characteristics provided them with considerable potential to be applied as a therapeutic agent [3]. There is still conflict in the results concerning the toxicity of AuNPs. Based on the reports from some studies, the toxicity of AuNPs is insignificant and these particles are biocompatible [129–132], whereas it is highlighted by others that these particles cause ROS, causing oxidative stress and also due to their physicochemical characteristics, they are very toxic [133–136]. According to the research conducted by Senut et al. [137], the exposure of human embryonic stem cells (hESCs) tp mercaptosuccinic acid-capped AuNPs with a size of 1.5 nm resulted in cohesiveness loss and detachment, leading to ongoing cell death at 0.1 μg/mL concentrations, while approximately no toxic impacts on hESCs were observed by AuNPs with the size of 4 and 14 nm at 10 $\mu g\,ml^{-1}$ concentrations. Hanna et al. [138] reported that the most tested AuNPs, which had a variety of surface coatings (branched polyethylenimine, dendrimers, citrate, polyethylene glycol polyvinylpyrrolidone) and different sizes of 100, 60, and 30 nm had a small or no effect on the reproduction or growth of a nematode, resulting in low toxicity. On the other hand, the positively charged coated AuNPs, like AuNPs coated with branched polyethylenimine caused considerable toxic impacts in *Caenorhabditis elegans*. There are several pathways by which AgNPs could enter the body, which might be followed by the accumulation of these particles inside different organs and tissues, overcoming the BBB and ultimately ending up in the brain [139]. AgNPs have been demonstrated to be more toxic concerning LDH leakage and the increased synthesis of ROS, in comparison with other metallic NPs [140–142]. Based on the previous studies, the toxicity of AgNPs depends on several factors, for instance, contact interactions, environmental conditions, as well as physical and chemical properties of AgNPs [143]. The low level of AgNPs' toxicity has been reported by Mulenos et al. [13] considering the observation that the release of just minute quantities of Ag ions was observed. The amounts were 0.2 μg/mL under anoxic conditions and 6.9 $\mu g\,ml^{-1}$ under oxic conditions, and both were levels considerably less than that of natural conditions.

There are some other metallic NMs besides with biomedical applications. Cobalt-based NPs are utilized in magnetic contrast agents, sensors, catalysts, and pigments [144]. When the human body is exposed to cobalt-based NPs or cobalt, damages to DNA, oxidative stress, inflammatory responses, and morphological transformation in cells are possible [145–147]. The investigation of three different cobalt-based NPs' cellular uptake in lung cells was conducted by Cappellini et al. [144]. The exposure to cobalt (II) oxide NPs and cobalt metal was found to induce oxidative damage and DNA strand breaks.

17.4.2 Carbon Nanomaterials

In recent decades many research works have been performed to investigate carbon, in particular in the CNMs form. The family of carbon has been continually expanded and novel NMs in this family have been developed including graphene, nanodiamonds (NDs), CNTs, carbon nanofibers (CNFs), carbon dots (CDs), and fullerenes, etc. [148]. Some potential toxicity and risks of CNMs are the promotion of free radicals, inflammation, and genotoxicity. Carbon NPs have been reported as great candidates with exceptional properties that has led to their broad employment in a variety of fields such as cosmetology as well as biomedicine. For example, carbon NPs are extensively utilized for cancer therapy and diagnosis, regenerative medicine, and drug delivery [149]. Although these materials have great potential in biomedical applications, as discussed in previous sections, their application may also have negative health impacts. When carbon NPs are introduced to the cell membrane, a sequence of signal transduction cascading events is activated, leading to different cellular responses, including changes in cell metabolism, cell growth, excretion, autophagy, proliferation, differentiation, communication, migration, survival, and cell death. It is probable that carbon NPs damage DNA, resulting in eco genotoxicity contribution, initiation, and promotion of carcinogenesis, or they may affect fertility. So, it seems to crucial to have knowledge about carbon NPs genotoxicity, which provides the opportunity for understanding and minimizing the associated negative impacts in order to protect human health and the environment. Carbon NPs genotoxicity is mainly in relation to their contact with cellular macromolecules, in particular DNA. NPs reach the nucleus by diffusing across nuclear pores or membrane after crossing the cell membrane and interaction with the DNA. Cellular vesicle formation and nuclear shape deformation can be induced by large NP aggregates, and have adverse impacts on the mitosis process, resulting in physical interference with proper chromosome segregation, and the mitotic spindle function. The mechanical damage of chromosomes with such NP aggregates is possible [150]. Several *in vivo* and *in vitro* investigations have been conducted and it has been reported that fullerene and its derivatives exhibited no or a very small level of genotoxicity [151–153]. The usual cause for genotoxicity of fullerene and its derivatives is photo-induced DNA damages by the interaction with NADH (nicotinamide adenine dinucleotide + hydrogen), which is followed by ROS generation [153]. It is important to emphasize that although fullerenes have an antioxidant capacity, providing it with a great potential to be applied as one of the main ingredients in several skin-care products, there is a risk of displaying activities that cause dysfunction or death. FLG, GO, reduced graphene oxide (rGO), and single-layer graphene (G) are members of the graphene family that are used for different applications in different fields. The toxicity of graphene derivatives in the spleen, kidney, liver, and lungs has been investigated in experimental studies [16, 154, 155]. For the investigation of airway exposures, the usual administration pathways have been intravenous administration [155], intra-peritoneal or oral [156],and instillation or inhalation [157]. The deposition of the graphene in the lungs, leading to inflammation and frustrated phagocytosis was reported by one of these studies. The nanoplatelets had a lateral size of up to 25 μm and were inhaled [158]. In another research work, thinner graphene nanoplatelets (10 layers) with a smaller size (lateral size of 2 μm) were instilled intratracheally into the lungs and a three-month observation was performed. In treated mice, slow excretion and disturbing immunological and physiological homeostasis was observed [159]. FLG with a lateral size of 1.25 μm has been reported to persist in the lungs for 28 days after intratracheal inhalation but induce only transient pulmonary impacts [157]. Dose-dependent toxicity has been investigated in recent research conducted by Jaworski et al. [160] where the formation of agglomerates with a size of 4 μm from graphene nanoplatelets caused ROS overgeneration and mitochondrial membrane potential depletion. There was an agreement between the outputs of that research work on skin cells perform by Pelin et al. [161] using the FLG. For CNTs genotoxicity, several mechanisms can be considered such as disrupting mitochondria, interfering with the spindle apparatus and centrosome, indirect or/and direct generation of ROS, and direct interactions with genomic materials. Furthermore, transition metal residues that remained from the synthesis of CNT take part in Fenton-like reaction caused induction of more ROS generation compared to pure CNTs. The results from the study by Donaldson and Poland [162] demonstrated that more generations of ROS or inflammation can be induced by CNTs, resulting in secondary genotoxicity. The assessment of the genotoxicity of graphene and its derivatives in different *in vivo* investigations has shown different results. While no genotoxicity has been exhibited in some outputs, in some others damage in DNA has been reported [162–164]. Many research works highlighted the extremely small levels of CDs cytotoxicity toward normal human breast epithelial cells (MCF-12A), human breast cancer cell line SKBR3 [165], human kidney embryonic 293T cells, and the human hepatocellular carcinoma Hep G2 cells [98, 166]. The greater degree of oxidation, the higher hydrosolubility, and the smaller size of CDs can be possible causes for no toxic impacts in the cells [57]. Nevertheless, it has been revealed that CDs produce ROS response in yeast cells in a dose-dependent manner, and it was subsequently enhanced with exposure to light [99]. For decreasing the adverse toxic impacts and expanding their applications, surface engineering seems to be an effective approach for the fabrication of functionalized CDs with enhanced properties that can fulfill special requirements [100]. Based on *in vivo* investigations, it has been concluded that inducing oxidative stress, which may accompany long exposures to NDs, can be regarded as a cause of NDs toxicity [101, 167]. For instance, Dworak et al. [168] have reported that DNA damage on lymphocytes might be caused by ND-mediated oxidative stress. Generally, NDs have been reported to be nontoxic for different types of cells (skin cells, neurons, immune cells), and are harmless concerning ROS induction [167]. In addition, no important toxicity has been reported in the study performed on six lines of human cells, representing intestine, lung, liver, and kidney, exposed to up to $250\,g\,ml^{-1}$ dose [169]. It is extremely important regarding the applications of NDs in human nanomedicine. Moreover, several studies have emphasized that the advantages of utilizing NDs in various systems are substantially higher than their possible negative impacts [170–172].

17.4.3 Organic-based Nanoparticles

Organic-based NPs including polymers, dendrimers, liposomes, and micelles have been widely applied in therapeutic agents, DNA transfecting agents, modified electrodes, chemical sensors, polymer materials, drug delivery, and different diagnostics/biosensors [173, 174]. Polymeric NPs show promise as anticancer drug carriers because of their great pharmacokinetic characteristics including structure stability, drug release, and drug loading [175]. In addition, the majority of biodegradable polymeric NPs have been reported to be nonimmunologic, noninflammatory, and nontoxic, and also no activation of neutrophils has been observed by them [176]. The results of the study conducted by Voigt et al. [177] revealed the nontoxicity of polybutylcyanoacrylate NPs as drug carriers. In contrast, the toxicity of positively charged NPs has been reported for various types of cells [178].

17.5 Physicochemical Properties and Toxicity of Nanomaterials

With the ever-increasing daily-life applications of NMs, exposure to these materials is more probable, making the assessments of toxicity risks of NMs unavoidable. However, the characteristics of NPs are strongly affected by the chemical and physical properties of NMs. It is necessary to study the physicochemical characteristics of NP-based materials. Generally, it is understood that toxicity stems from the shape and size of NMs, their composition, solubility, surface area, etc. as discussed next.

17.5.1 Nanoparticle Surface Area and Size

The interaction of biological systems and materials depends considerably on the surface area and size of particles. It seems that with a reduction in material size, its surface area increases exponentially with respect to the volume, resulting in increasing the reactivity of the NM surface to its contiguous milieu and on itself [179]. It should be noted that surface area and size possess a significant influence on the ways in which the system eliminates, distributes, and responds to the materials [180]. Moreover, the material size has been reported to affect different biological mechanisms such as particle processing efficiency in the endocytic pathway, cellular uptake, and endocytosis [181, 182]. Generally, the dependency of the toxicity on the size of particles can be attributed to the ability of NPs to penetrate into the biological system [183] and alter different macromolecules structures [116], leading to interference with crucial biological functions. The production of oxidative responses through forming free radicals is among the most important processes for *in vivo* toxicity of the engineered NMs, and size plays a significant role in this mechanism. It has been underlined by several researchers that the ability of particles in forming ROS increases with decreasing the size. Significant hazards to biological systems are reported to be caused by these free radicals mostly through the ensuing inflammatory responses, oxidation of lipids, and DNA damage. The reduction in the NP size comes with an increase in the surface area, resulting in an increase in oxidation, which depends on the dose. The ability of these particles in damaging DNA is significantly more than the particle with a larger size and similar mass dose [184, 185]. Furthermore, the oral toxicity of NPs is also influenced by their size. Generally, the reduction in size leads to an increase in oral toxicity. The oral toxicity of copper was observed to become higher along with reducing the size. Despite the toxicity of smaller particles, larger ones even under greater doses were nontoxic [186].

17.5.2 Crystalline Structure and Composition

Despite the fact that the particle size has a substantial influence on NP toxicity, several studies also have demonstrated that different NP chemistries with similar sizes had comparable toxicities. The results of these studies revealed the role of NP composition on their toxicity issues. Griffitt et al. [185] utilized models of various trophic levels including algal species, daphnids, and zebrafish and noticed in all tested organisms that the toxicity is caused when nanocopper and nanosilver are in their soluble forms, while no toxicity issues were caused by TiO_2 with similar dimensions [187]. The outputs of this research highlighted the effect of compositions in defining NPs toxicities.

17.5.3 Aspect Ratio and Shape of Particles

There are NMs with different shapes such as hexagonal, fibers, spherical, cubic, and rod, as illustrated in Figure 17.3. It is reported for a large number of NPs such as titanium, Au, nickel, allotropies, silica, and CNTs that their toxicity depends on their shape [188–191]. Actually, the processes of membrane wrapping in the living body through phagocytosis and endocytosis are influenced by shape-dependent nanotoxicity [192]. Based on observations, NPs with spherical shapes have easier and faster endocytosis in comparison with NPs with fiber or rod shapes [38, 193]. Moreover, the toxicity of NPs with spherical shapes has been reported to be less regardless of heterogeneity or homogeneity of the particles [194]. Based on studies, spherical nanospheres uptake is faster than that of Au nanorods [190]. In addition, with approaching the aspect ratio of nanorods to unity, its uptake becomes maximum [195]. The results of previous studies revealed that the cytotoxicity of TiO_2 fibers is higher than that of spherical entities [196]. Furthermore, the particle toxicity increases along with augmenting the aspect ratio [197]. Based on the study conducted by Hamilton et al. [191], when the length of TiO_2 fibers is 15 mm, the toxicity is higher when compared to a length of 5 mm and by alveolar macrophages an inflammatory response is initiated in mice.

17.5.4 Surface Charge

The toxicity of NPs is significantly affected by the surface charge as interactions of the biological systems and NPs are defined mostly by it. The surface charge of NPs regulates

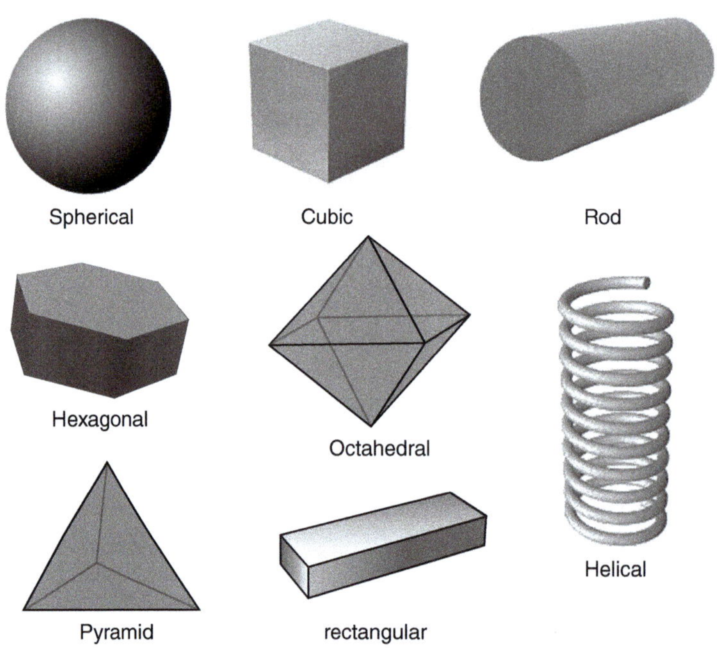

Figure 17.3 Different shapes of nanoparticles.

primarily different aspects of NMs including transmembrane permeability, plasma protein binding, BBB integrity [198], colloidal behavior, and NPs selective adsorption [199, 200]. It must be mentioned that the cellular uptake of neutral and negatively charged NPs is significantly less than that of positively charged NPs because of the improved opsonization of positively charged NPs by the plasma proteins. Furthermore, platelet aggregation and hemolysis are reported to be induced by them [201]. Compared to negatively charged and neutral silicon NPs with negligible cytotoxicity, positively charged silicon NPs have higher cytotoxicity [202]. According to the observations, the permeation of the negatively charged NPs into skins occurs after skin administration when their size is in the range of 50 to 500 nm, while it was not observed for neutral and positively charged particles for regardless of their sizes.

17.5.5 Concentration and Aggregation

The toxicity of NPs can be also affected by their aggregation states. Based on the observations, the accumulation of CNTs occurs mostly in the lungs, spleen, and liver while demonstrating no severe toxicity. However, cytotoxic impacts are induced mainly due to accumulating aggregates over an extended period of time [203]. The pulmonary interstitial fibrosis is enhanced by the adverse impacts of agglomerated CNTs, which is also more than that of well-dispersed CNTs [78].

17.5.6 Influence of Surface Roughness and Surface Coating

The cytotoxic characteristics of NPs can be affected by surface coating through the change of their physicochemical characteristics including chemical reactivity optical, electric, and magnetic properties [17, 59]. Moreover, the surface coating is able to change the toxicity, accumulation, distribution, and pharmacokinetics of NPs. Based on previous studies, it has been reported that the existence of transition metals, oxygen radicals, ozone, and oxygen on the surface of NP results in the generation of ROS and inflammation can be induced by these systems [79, 184, 185], so their associated toxicities matters are definitely affected. For this purpose, in particular, the investigation performed by Fubini et al. [204] revealed that silica-specific cytotoxicity is closely related to the occurrence of ROS and surface radicals on their surfaces. On the other hand, the surface coating can be also applied in order to decrease NPs' toxicity issues. Generally, the adverse impacts of NPs can be mitigated or eliminated by surface coating. Particularly, NPs can be stabilized and toxic ions from NPs can be avoided with appropriate surface coating [205].

17.5.7 Influence of Media/Solvents

The particle agglomeration and dispersion state of NPs is shown to be affected by conditions of solvents/media, which in turn influence the size of particles, resulting in affecting the NP toxicity. The size of carbon black, ZnO (zinc oxide), TiO_2 particles has been reported to be considerably larger in phosphate-buffered saline compared to their size in water. Furthermore, there is a general agreement that in the biological milieu, NPs demonstrate various diameters [206, 207]. Consequently, depending on the composition of media in which NPs is dissolved, the NPs' toxic impacts can be varied, so when the same NPs are suspended in different media, dissimilar toxic demonstrations can be exhibited [208, 209]. In spite of the fact that the solution and physicochemical characteristics of NPs formulations might be improved by the dispersing agent, the NMs toxicity can be also negatively affected by them.

17.6 Regulatory Issues

One critical issue that has caused increasing concern for legislators around the world is the safety of manufactured NMs. To make sure that products that contain NMs are safe, prior to their launch in the marketplace, different international and national regulatory authorities including the Organization for Economic Cooperation and Development (OECD), the International Organization for Standardization (ISO), and European Committee for Standardization (CEN) are concentrating on safety and health concerns and environmental-related aspects, definitions and nomenclature, toxicity protocols, test procedures for characterization and recognizing the NPs toxicity and safety potential, toxicity protocols, and standard test procedures. The responsibility of these standard regulatory bodies is the definition of a framework for setting up nanotechnology working groups, safe utilization of NPs, and the associated tasks to prevent repeated tests and to ensure the completion of their nanotechnology output. The short-term objective is the development of regulations for pharmacokinetics, drug-delivery standards, approaches for the determination of NPs stability in animal models, toxicology screening tests, protocols of particle characterization for nanohealth, the definition of a minimum set of measurements, and NMs exposure routes. In recent years, guidelines have been adapted for different industries and institutes. Although no accurate steps for circumventing problems are provided by these guidelines, they certainly are regulations for best practice for avoiding the potential risks. All these guidelines have a basic framework [210]:

- The identification of uncertainties and assessment of risks in the processes of manufacturing as well as utilization of NMs.
- The elaboration and employment of an efficient method for addressing and controlling the risks.
- Preventing and controlling the required exposure.
- Guarantee to pursue processes that are implemented and performing the required measures in this regard.
- Investigation of the levels of exposure and undertaking correct monitoring.
- Initiation of sufficient health assessment.
- Employment protocols in advance for facing any emergency or accidents.
- Sufficient supervision, informing, and training of the employees or students of industries and institutions, respectively.

Due to the exceptional characteristics of NPs including high surface area, nano-size range, as well as catalytic characteristics, further tests beyond regulations determining toxicity testing are required. With these guidelines, there exists a significant discussion between policymakers and academia for addressing the serious threat raised by NMs [210]. Regulation and legislation for chemical testing are formed by the Toxic Substances Control Act (TSCA). Chemical companies and producers that utilize NMs as end products are obliged to follow the guidelines of the TSCA during the processes of manufacturing. The dissatisfaction of customers and business interruption of a company can be a possible consequence of failing to implement the regulatory requirements. The OECD with accompanied by their business partners (academia, industry, and government) launched the Sponsorship Program in 2007 to test dossiers of 11 produced NMs, including ZnO, TiO_2, SiO_2, nanoclays, CNTs (SWNTs, multiwalled nanotubes [MWNTs]), Ag, Au, fullerenes, dendrimers, and cerium oxide. Proving scientific endpoint data, related to environmental safety and human health is their objective achieved through performing special tests. The aim of the testing program is the assessment of the available NMs guidelines and the comparison of the exceptional characteristics of NMs to their bulk counterparts. Testing is conducted for all physicochemical characteristics of NMs against bulk material to define strategies for evaluating risks and hazards or measures for managing risk for utilization and safety issues of engineered NMs. Two agencies, the Federal Insecticide, Fungicide, and Rodenticide Act (FIFRA) and the Environmental Protection Agency (EPA) are in charge of regulating NMs in the United States. The Food and Drug Administration (FDA) is another regulatory agency in the United States that regulates NMs, in cooperation with EPA, to control NMs' progress in several nanoproduction categories, like food, drugs, medical devices, and cosmetics. Legislations and amendments of the legislations or amendments toward NMs' regulation have been revised by the EPA. The voluntarily sharing of NMs' information by manufacturers has been changed toward obligatory rules for collecting comprehensive information about the set of standards on fabrication, the use of NMs, and their safe disposal, with obligatory preproduction information for new NMs. The rules are strict for collecting both new and available NMs [211]. Some special information on new chemical material is followed by TSCA for revising the risk of NMs on human health and the environment, prior to production of NMs under the Pre-manufacture Notification Rule [211]. Lately, the regulatory bodies have taken measures for controlling the huge NMs utilization, such as decreasing the release and entry of NMs into the ecosystem and environment, the enforcement of the use of personal protective equipment (PPE), and the necessity of conducting tests to provide raw data concerning health and environmental impacts. Some regulations and rules have been set by the European Commission for manufacturers of NMs, prior to their launch in the marketplace and consumption. In the European Union (EU), the Classification, Labelling, and Packaging of Substances and Mixtures (CLP) and the Registration, Evaluation, Authorization, and Restriction of Chemicals (REACH) regulatory bodies intend to offer improved protection to human health and environment and risk management of NMs/substances/products through the evaluation of their intrinsic characteristics. The submission of technical data to the European Chemicals Agency (ECHA) in Helsinki is required for importers and manufacturers. The agency assesses suspicious NMs comprehensively and present safety details to consumers and manufacturers. REACH is a four-tiered process that is followed by them. All information on NMs regulations required by the Commission Communication are contained in "Commission Staff Working Document" and "Second Regulatory Review on Nanomaterials" documents. In Asian countries, compared to the EU and the United States, NMs regulations are still in progress. Nevertheless, still, no regulatory bodies in the United States or the EU have finalized specified regulations for all sizes, shapes, and types of NMs. A majority of NMs are dealt with under the TSCA by the EPA [212].

17.7 Conclusion

Despite the ever-increasing use of NPs in different industrial fields such as their employment in biomedical devices, several research works have reported the adverse effects of nano-sized particles on human health and the environment. Therefore, the potential risks of the exposure of these particles to the human body, which are a serious concern, should be considered in the processes of design, fabrication, and application, which requires better tools for the assessment of NP safety. Defining the toxic characteristics of NPs is possible with the evaluation of the nano-sized characteristics such as crystalline, temperature, density, electronic structure, phase, shape, size, and surface area characteristics like surface functionality and surface electronic structure, and also composition, dimensions, and geometry, exceptional magnetic and quantum characteristics, surface chemistry, surface charge, and reactivity, and nanostructures interactions with biological systems. Nevertheless, biopersistence requires further investigation as it is not yet clear what the undesirable and unintended impacts of NMs are. These are required to ensure that new medical devices and therapies are safe. Nanotoxicity studies are offered and these aim to define how and to what degree the characteristics of materials having nano-size might threaten the living system and environment. Although several studies have found toxicity of these particles, the understanding behind the toxicity is still very limited. The relationship between these particles and their surroundings is not clear – this field is evolving and investigations on new characteristics and their toxicity effects are questionable. There is no correlation between the outputs of previous studies on NP toxicity. Diverse testing methods, testing conditions, and approaches for NMs characterization are observed in these studies. Harmonization of the data from NM toxicity studies can be helpful for further progress. Further investigation on the safety of engineered NPs should be performed and validated models should be developed in future works.

References

1 Sanhai, W.R., Sakamoto, J.H., Canady, R., and Ferrari, M. (2008). Seven challenges for nanomedicine. *Nature Nanotechnology* 3 (5): 242–244.

2 Wang, R., Billone, P.S., and Mullett, W.M. (2013). Nanomedicine in action: an overview of cancer nanomedicine on the market and in clinical trials. *Journal of Nanomaterials* 2013: 629681.

3 Sardoiwala, M.N., Kaundal, B., and Choudhury, S.R. (2018). Development of engineered nanoparticles expediting diagnostic and therapeutic applications across blood–brain barrier. In: *Handbook of Nanomaterials for Industrial Applications* (ed. C.M. Hussain), 696–709. Elsevier.

4 Zoroddu, M.A., Medici, S., Ledda, A. et al. (2014). Toxicity of nanoparticles. *Current Medicinal Chemistry* 21 (33): 3837–3853.

5 Gupta, R. and Xie, H. (2018). Nanoparticles in daily life: applications, toxicity and regulations. *Journal of Environmental Pathology, Toxicology and Oncology* 37 (3).

6 Wang, F., Yu, L., Monopoli, M.P. et al. (2013). The biomolecular corona is retained during nanoparticle uptake and protects the cells from the damage induced by cationic nanoparticles until degraded in the lysosomes. *Nanomedicine: Nanotechnology, Biology and Medicine* 9 (8): 1159–1168.

7 Regoli, F. and Giuliani, M.E. (2014). Oxidative pathways of chemical toxicity and oxidative stress biomarkers in marine organisms. *Marine Environmental Research* 93: 106–117.
8 Jomova, K., Baros, S., and Valko, M. (2012). Redox active metal-induced oxidative stress in biological systems. *Transition Metal Chemistry* 37 (2): 127–134.
9 Fard, J.K., Jafari, S., and Eghbal, M.A. (2015). A review of molecular mechanisms involved in toxicity of nanoparticles. *Advanced Pharmaceutical Bulletin* 5 (4): 447.
10 Steckiewicz, K.P., Barcinska, E., Malankowska, A. et al. (2019). Impact of gold nanoparticles shape on their cytotoxicity against human osteoblast and osteosarcoma in in vitro model. Evaluation of the safety of use and anti-cancer potential. *Journal of Materials Science: Materials in Medicine* 30 (2): 22.
11 Feng, Q., Liu, Y., Huang, J. et al. (2018). Uptake, distribution, clearance, and toxicity of iron oxide nanoparticles with different sizes and coatings. *Scientific Reports* 8 (1): 1–13.
12 Gea, M., Bonetta, S., Iannarelli, L. et al. (2019). Shape-engineered titanium dioxide nanoparticles (TiO_2-NPs): cytotoxicity and genotoxicity in bronchial epithelial cells. *Food and Chemical Toxicology* 127: 89–100.
13 Mulenos, M.R., Liu, J., Lujan, H. et al. (2020). Copper, silver, and titania nanoparticles do not release ions under anoxic conditions and release only minute ion levels under oxic conditions in water: evidence for the low toxicity of nanoparticles. *Environmental Chemistry Letters* 1–10.
14 Kassem, A., Ayoub, G.M., and Malaeb, L. (2019). Antibacterial activity of chitosan nano-composites and carbon nanotubes: a review. *Science of the Total Environment* 668: 566–576.
15 Mohanta, D., Patnaik, S., Sood, S., and Das, N. (2019). Carbon nanotubes: evaluation of toxicity at biointerfaces. *Journal of Pharmaceutical Analysis* 9 (5): 293–300.
16 Ruiz, A., Lucherelli, M.A., Murera, D. et al. (2020). Toxicological evaluation of highly water dispersible few-layer graphene in vivo. *Carbon* 170: 347–360.
17 Shanmugarathinam, A., Madhusudhana, J., Poongothai, M., and Ahamed, I.N. A complete review on toxicity of nanoparticles. In: *Proceedings of the Virtual International Conference on Advancements in Research and Education (VICARE)* (ed. Y. Sarkingobir), 156–168. Association of Global Acadamicians and Researchers & Association of Indian Biologists.
18 Kamat, P.V. (2002). *Photophysical, Photochemical and Photocatalytic Aspects of Metal Nanoparticles*. ACS Publications.
19 Maiti, D., Tong, X., Mou, X., and Yang, K. (2019). Carbon-based nanomaterials for biomedical applications: a recent study. *Frontiers in Pharmacology* 9: 1401.
20 Kalpana, V.N. and Devi Rajeswari, V. (2018). A review on green synthesis, biomedical applications, and toxicity studies of ZnO NPs. *Bioinorganic Chemistry and Applications* 2018: 3569758.
21 Makvandi, P., Wang, C.Y., Zare, E.N. et al. (2020). Metal-based nanomaterials in biomedical applications: antimicrobial activity and cytotoxicity aspects. *Advanced Functional Materials* 30 (22): 1910021.
22 Mandal, A.K. (2017). Copper nanomaterials as drug delivery system against infectious agents and cancerous cells. *Journal of Applied Life Sciences International* 15 (4): 1–8.
23 Zhou, Z., Li, B., Shen, C. et al. (2020). Metallic 1T phase enabling MoS2 nanodots as an efficient agent for photoacoustic imaging guided photothermal therapy in the near-infrared-II window. *Small* 16 (43): 2004173.
24 Si, Y. and Lee, H.J. (2020). Carbon nanomaterials and metallic nanoparticles-incorporated electrochemical sensors for small metabolites: detection methodologies and applications. *Current Opinion in Electrochemistry* 22: 234–243.

25 Fahimirad, S., Ajalloueian, F., and Ghorbanpour, M. (2019). Synthesis and therapeutic potential of silver nanomaterials derived from plant extracts. *Ecotoxicology and Environmental Safety* 168: 260–278.

26 Shrivastava, S., Bera, T., Singh, S.K. et al. (2009). Characterization of antiplatelet properties of silver nanoparticles. *ACS Nano* 3 (6): 1357–1364.

27 Kim, J.S., Kuk, E., Yu, K.N. et al. (2007). Antimicrobial effects of silver nanoparticles. *Nanomedicine: Nanotechnology, Biology and Medicine* 3 (1): 95–101.

28 Li, Y., Leung, P., Yao, L. et al. (2006). Antimicrobial effect of surgical masks coated with nanoparticles. *Journal of Hospital Infection* 62 (1): 58–63.

29 Cordani, M. and Somoza, Á. (2019). Targeting autophagy using metallic nanoparticles: a promising strategy for cancer treatment. *Cellular and Molecular Life Sciences* 76 (7): 1215–1242.

30 Tran, P.H.L., Duan, W., and Tran, T.T.D. (2020). Fucoidan-based nanostructures: a focus on its combination with chitosan and the surface functionalization of metallic nanoparticles for drug delivery. *International Journal of Pharmaceutics* 575: 118956.

31 Nicewarner-Pena, S.R., Freeman, R.G., Reiss, B.D. et al. (2001). Submicrometer metallic barcodes. *Science* 294 (5540): 137–141.

32 Soldevila-Barreda, J.J. and Metzler-Nolte, N. (2019). Intracellular catalysis with selected metal complexes and metallic nanoparticles: advances toward the development of catalytic metallodrugs. *Chemical Reviews* 119 (2): 829–869.

33 Seemala, B., Therrien, A.J., Lou, M. et al. (2019). Plasmon-mediated catalytic O_2 dissociation on Ag nanostructures: hot electrons or near fields? *ACS Energy Letters* 4 (8): 1803–1809.

34 Su, J., Wang, D., Nörbel, L. et al. (2017). Multicolor gold–silver nano-mushrooms as ready-to-use SERS probes for ultrasensitive and multiplex DNA/miRNA detection. *Analytical Chemistry* 89 (4): 2531–2538.

35 Dakshayani, S.S., Marulasiddeshwara, M.B., Kumar, S. et al. (2019). Antimicrobial, anticoagulant and antiplatelet activities of green synthesized silver nanoparticles using Selaginella (Sanjeevini) plant extract. *International Journal of Biological Macromolecules* 131: 787–797.

36 Jain, S. and Mehata, M.S. (2017). Medicinal plant leaf extract and pure flavonoid mediated green synthesis of silver nanoparticles and their enhanced antibacterial property. *Scientific Reports* 7 (1): 1–13.

37 Marimuthu, S., Rahuman, A.A., Rajakumar, G. et al. (2011). Evaluation of green synthesized silver nanoparticles against parasites. *Parasitology Research* 108 (6): 1541–1549.

38 Singh, S.K., Kulkarni, P.P., and Dash, D. Biomedical applications of nanomaterials: an overview. In: *Bio-nanotechnology: A Revolution in Food, Biomedical and Health Sciences* (ed. D. Bagchi, M. Bagchi, H. Moriyama and F. Shahidi), 1–32. Wiley.

39 Almonaci Hernández, C.A., Juarez-Moreno, K., Castañeda-Juarez, M.E. et al. (2017). Silver nanoparticles for the rapid healing of diabetic foot ulcers. *International Journal of Medical Nano Research* 4: 19.

40 Rajendran, N.K., Kumar, S.S.D., Houreld, N.N., and Abrahamse, H. (2018). A review on nanoparticle based treatment for wound healing. *Journal of Drug Delivery Science and Technology* 44: 421–430.

41 Stevens, K.N.J., Crespo-Biel, O., van den Bosch, E.E.M. et al. (2009). The relationship between the antimicrobial effect of catheter coatings containing silver nanoparticles and the coagulation of contacting blood. *Biomaterials* 30 (22): 3682–3690.
42 Daniel, M.-C. and Astruc, D. (2004). Gold nanoparticles: assembly, supramolecular chemistry, quantum-size-related properties, and applications toward biology, catalysis, and nanotechnology. *Chemical Reviews* 104 (1): 293–346.
43 El-Sayed, M.A. (2001). Some interesting properties of metals confined in time and nanometer space of different shapes. *Accounts of Chemical Research* 34 (4): 257–264.
44 Rao, C.V., Manimaran, V., and Damodharan, N. (2020). Review on methods, applications and role of gold nano particles in cancer therapy. *Research Journal of Pharmacy and Technology* 13 (8): 3963–3968.
45 Samad, A., Sultana, Y., and Aqil, M. (2007). Liposomal drug delivery systems: an update review. *Current Drug Delivery* 4 (4): 297–305.
46 Cao, Y.C., Jin, R., and Mirkin, C.A. (2002). Nanoparticles with Raman spectroscopic fingerprints for DNA and RNA detection. *Science* 297 (5586): 1536–1540.
47 McVey, C., Huang, F., Elliott, C., and Cao, C. (2017). Endonuclease controlled aggregation of gold nanoparticles for the ultrasensitive detection of pathogenic bacterial DNA. *Biosensors and Bioelectronics* 92: 502–508.
48 Paul, E.P.R. (2019). Plasmonic Gold Nanostructures for Biodiagnostics and Photothermal Therapeutics. MS Thesis for Graduate School of Vanderbilt University, Nashville, TN.
49 Xue, Y. (2017). Carbon nanotubes for biomedical applications. In: *Industrial Applications of Carbon Nanotubes*, 323–346. Elsevier.
50 Yadav, N., Tyagi, M., Wadhwa, S. et al. (2020). Few biomedical applications of carbon nanotubes. In: *Methods in Enzymology* (ed. C.V. Kumar), 347–363. Elsevier.
51 Saleemi, M.A., Kong, Y.L., Yong, P.V.C., and Wong, E.H. (2020). An overview of recent development in therapeutic drug carrier system using carbon nanotubes. *Journal of Drug Delivery Science and Technology* 59: 101855.
52 Patolsky, F., Weizmann, Y., and Willner, I. (2004). Long-range electrical contacting of redox enzymes by SWCNT connectors. *Angewandte Chemie* 116 (16): 2165–2169.
53 Rasheed, T., Hassan, A.A., Kausar, F. et al. (2020). Carbon nanotubes assisted analytical detection–sensing/delivery cues for environmental and biomedical monitoring. *TrAC Trends in Analytical Chemistry* 132: 116066.
54 Saliev, T. (2019). The advances in biomedical applications of carbon nanotubes. *C—Journal of Carbon Research* 5 (2): 29.
55 Tang, X., Bansaruntip, S., Nakayama, N. et al. (2006). Carbon nanotube DNA sensor and sensing mechanism. *Nano Letters* 6 (8): 1632–1636.
56 Ma, H. and Liang, X.-J. (2010). Fullerenes as unique nanopharmaceuticals for disease treatment. *SCIENCE CHINA Chemistry* 53 (11): 2233–2240.
57 Jović, D., Jaćević, V., Kuča, K. et al. (2020). The puzzling potential of carbon nanomaterials: general properties, application, and toxicity. *Nanomaterials* 10 (8): 1508.
58 Shen, H., Zhang, L., Liu, M., and Zhang, Z. (2012). Biomedical applications of graphene. *Theranostics* 2 (3): 283.
59 Ghosal, K. and Sarkar, K. (2018). Biomedical applications of graphene nanomaterials and beyond. *ACS Biomaterials Science & Engineering* 4 (8): 2653–2703.

60 Liu, Y., Yu, D., Zeng, C. et al. (2010). Biocompatible graphene oxide-based glucose biosensors. *Langmuir* 26 (9): 6158–6160.

61 Wang, W., Su, H., Wu, Y. et al. (2019). Biosensing and biomedical applications of graphene: a review of current progress and future prospect. *Journal of The Electrochemical Society* 166 (6): B505.

62 Mohanty, N. and Berry, V. (2008). Graphene-based single-bacterium resolution biodevice and DNA transistor: interfacing graphene derivatives with nanoscale and microscale biocomponents. *Nano Letters* 8 (12): 4469–4476.

63 Nasrollahi, F., Koh, Y.R., Chen, P. et al. (2019). Targeting graphene quantum dots to epidermal growth factor receptor for delivery of cisplatin and cellular imaging. *Materials Science and Engineering: C* 94: 247–257.

64 Xu, M. and Wang, L.V. (2006). Photoacoustic imaging in biomedicine. *Review of Scientific Instruments* 77 (4): 041101.

65 Sun, J., Guo, Y., Zhu, L. et al. (2017). Photodynamic therapy of human hepatoma using semiconductor quantum dots as sole photosensitizer. *Particle & Particle Systems Characterization* 34 (10): 1600413.

66 Matea, C.T., Mocan, T., Tabaran, F. et al. (2017). Quantum dots in imaging, drug delivery and sensor applications. *International Journal of Nanomedicine* 12: 5421.

67 Su, W., Guo, R., Yuan, F. et al. (2020). Red-emissive carbon quantum dots for nuclear drug delivery in cancer stem cells. *The Journal of Physical Chemistry Letters* 11 (4): 1357–1363.

68 Medintz, I.L., Pons, T., Trammell, S.A. et al. (2008). Interactions between redox complexes and semiconductor quantum dots coupled via a peptide bridge. *Journal of the American Chemical Society* 130 (49): 16745–16756.

69 Bentolila, L.A. and Weiss, S. (2006). Single-step multicolor fluorescence in situ hybridization using semiconductor quantum dot-DNA conjugates. *Cell Biochemistry and Biophysics* 45 (1): 59–70.

70 Medintz, I.L. and Mattoussi, H. (2009). Quantum dot-based resonance energy transfer and its growing application in biology. *Physical Chemistry Chemical Physics* 11 (1): 17–45.

71 Martynenko, I.V., Litvin, A.P., Purcell-Milton, F. et al. (2017). Application of semiconductor quantum dots in bioimaging and biosensing. *Journal of Materials Chemistry B* 5 (33): 6701–6727.

72 Himmelstoß, S.F. and Hirsch, T. (2019). A critical comparison of lanthanide based upconversion nanoparticles to fluorescent proteins, semiconductor quantum dots, and carbon dots for use in optical sensing and imaging. *Methods and Applications in Fluorescence* 7 (2): 022002.

73 Cardoso, V.F., Francesko, A., Ribeiro, C. et al. (2018). Advances in magnetic nanoparticles for biomedical applications. *Advanced Healthcare Materials* 7 (5): 1700845.

74 Mohammed, L., Gomaa, H.G., Ragab, D., and Zhu, J. (2017). Magnetic nanoparticles for environmental and biomedical applications: a review. *Particuology* 30: 1–14.

75 Zununi Vahed, S., Fathi, N., Samiei, M. et al. (2019). Targeted cancer drug delivery with aptamer-functionalized polymeric nanoparticles. *Journal of Drug Targeting* 27 (3): 292–299.

76 Khalid, M. and El-Sawy, H.S. (2017). Polymeric nanoparticles: promising platform for drug delivery. *International Journal of Pharmaceutics* 528 (1–2): 675–691.

77 Burgess, T.L. and Kelly, R.B. (1987). Constitutive and regulated secretion of proteins. *Annual Review of Cell Biology* 3 (1): 243–293.

78 Wick, P., Manser, P., Limbach, L.K. et al. (2007). The degree and kind of agglomeration affect carbon nanotube cytotoxicity. *Toxicology Letters* 168 (2): 121–131.

79 Sayes, C.M., Fortner, J.D., Guo, W. et al. (2004). The differential cytotoxicity of water-soluble fullerenes. *Nano Letters* 4: 1881–1887.

80 Yu, J., Yin, W., Peng, T. et al. (2017). Biodistribution, excretion, and toxicity of polyethyleneimine modified NaYF 4: Yb, Er upconversion nanoparticles in mice via different administration routes. *Nanoscale* 9 (13): 4497–4507.

81 Zhao, Y., Sultan, D., and Liu, Y. (2019). Biodistribution, excretion, and toxicity of nanoparticles. In: *Theranostic Bionanomaterials* (ed. W. Cui and X. Zhao), 27–53. Elsevier.

82 Gänsslen, M. (1925). Uber inhalation von insulin. *Klinische Wochenschrift* 4: 71.

83 Gehr, P., Bachofen, M., and Weibel, E.R. (1978). The normal human lung: ultrastructure and morphometric estimation of diffusion capacity. *Respiration Physiology* 32 (2): 121–140.

84 Yun, Y., Cho, Y.W., and Park, K. (2013). Nanoparticles for oral delivery: targeted nanoparticles with peptidic ligands for oral protein delivery. *Advanced Drug Delivery Reviews* 65 (6): 822–832.

85 Desai, P.P., Date, A.A., and Patravale, V.B. (2012). Overcoming poor oral bioavailability using nanoparticle formulations – opportunities and limitations. *Drug Discovery Today: Technologies* 9 (2): e87–e95.

86 Labouta, H.I. and Schneider, M. (2013). Interaction of inorganic nanoparticles with the skin barrier: current status and critical review. *Nanomedicine: Nanotechnology, Biology and Medicine* 9 (1): 39–54.

87 Alvarez-Román, R., Naik, A., Kalia, Y.N. et al. (2004). Skin penetration and distribution of polymeric nanoparticles. *Journal of Controlled Release* 99 (1): 53–62.

88 Barry, B.W. (2001). Novel mechanisms and devices to enable successful transdermal drug delivery. *European Journal of Pharmaceutical Sciences* 14 (2): 101 114.

89 Bourquin, J., Milosevic, A., Hauser, D. et al. (2018). Biodistribution, clearance, and long-term fate of clinically relevant nanomaterials. *Advanced Materials* 30 (19): 1704307.

90 Warren, G. and Wickner, W. (1996). Organelle inheritance. *Cell* 84 (3): 395–400.

91 Macara, I.G. (2001). Transport into and out of the nucleus. *Microbiology and Molecular Biology Reviews* 65 (4): 570–594.

92 Conner, S.D. and Schmid, S.L. (2003). Regulated portals of entry into the cell. *Nature* 422 (6927): 37–44.

93 Doherty, G.J. and McMahon, H.T. (2009). Mechanisms of endocytosis. *Annual Review of Biochemistry* 78: 857–902.

94 Hillaireau, H. and Couvreur, P. (2009). Nanocarriers' entry into the cell: relevance to drug delivery. *Cellular and Molecular Life Sciences* 66 (17): 2873–2896.

95 Mühlfeld, C. (2008). Translocation and cellular entering mechanisms of nanoparticles in the respiratory tract. *Swiss Medical Weekly* 138 (2728, 27–28): 387–391.

96 Iversen, T.-G., Skotland, T., and Sandvig, K. (2011). Endocytosis and intracellular transport of nanoparticles: present knowledge and need for future studies. *Nano Today* 6 (2): 176–185.

97 Kuhn, D.A., Vanhecke, D., Michen, B. et al. (2014). Different endocytotic uptake mechanisms for nanoparticles in epithelial cells and macrophages. *Beilstein Journal of Nanotechnology* 5 (1): 1625–1636.

98 Tao, H., Yang, K., Ma, Z. et al. (2012). In vivo NIR fluorescence imaging, biodistribution, and toxicology of photoluminescent carbon dots produced from carbon nanotubes and graphite. *Small* 8 (2): 281–290.

99 Bagheri, Z., Ehtesabi, H., Hallaji, Z. et al. (2018). Investigation the cytotoxicity and photo-induced toxicity of carbon dot on yeast cell. *Ecotoxicology and Environmental Safety* 161: 245–250.

100 Liu, W., Li, C., Ren, Y. et al. (2016). Carbon dots: surface engineering and applications. *Journal of Materials Chemistry B* 4 (35): 5772–5788.

101 Totsuka, Y., Higuchi, T., Imai, T. et al. (2009). Genotoxicity of nano/microparticles in in vitro micronuclei, in vivo comet and mutation assay systems. *Particle and Fibre Toxicology* 6 (1): 23.

102 Müller, S., Dennemärker, J., and Reinheckel, T. (2012). Specific functions of lysosomal proteases in endocytic and autophagic pathways. *Biochimica et Biophysica Acta (BBA)-proteins and Proteomics* 1824 (1): 34–43.

103 Jiang, X., Miclăuş, T., Wang, L. et al. (2015). Fast intracellular dissolution and persistent cellular uptake of silver nanoparticles in CHO-K1 cells: implication for cytotoxicity. *Nanotoxicology* 9 (2): 181–189.

104 Sabella, S., Carney, R.P., Brunetti, V. et al. (2014). A general mechanism for intracellular toxicity of metal-containing nanoparticles. *Nanoscale* 6 (12): 7052–7061.

105 Martinez, J.O., Evangelopoulos, M., Chiappini, C. et al. (2014). Degradation and biocompatibility of multistage nanovectors in physiological systems. *Journal of Biomedical Materials Research Part A* 102 (10): 3540–3549.

106 (a) Croissant, J.G., Fatieiev, Y., Almalik, A. et al. (2018). Mesoporous silica and organosilica nanoparticles: physical chemistry, biosafety, delivery strategies, and biomedical applications. *Advanced Healthcare Materials* 7 (4): 1700831. (b) Croissant, J.G., Fatieiev, Y., and Khashab, N.M. (2017). Degradability and clearance of silicon, organosilica, silsesquioxane, silica mixed oxide, and mesoporous silica nanoparticles. *Advanced Materials* 29 (9): 1604634.

107 Alberts, B., Johnson, A., Lewis, J. et al. (2002). *The Molecular Biology of the Cell*. London: Garland Publ, Inc.

108 Chu, Z., Huang, Y., Tao, Q., and Li, Q. (2011). Cellular uptake, evolution, and excretion of silica nanoparticles in human cells. *Nanoscale* 3 (8): 3291–3299.

109 Sakhtianchi, R., Minchin, R.F., Lee, K.-B. et al. (2013). Exocytosis of nanoparticles from cells: role in cellular retention and toxicity. *Advances in Colloid and Interface Science* 201: 18–29.

110 Mahon, E., Salvati, A., Bombelli, F.B. et al. (2012). Designing the nanoparticle–biomolecule interface for "targeting and therapeutic delivery.". *Journal of Controlled Release* 161 (2): 164–174.

111 Krug, H.F. and Wick, P. (2011). Nanotoxicology: an interdisciplinary challenge. *Angewandte Chemie International Edition* 50 (6): 1260–1278.

112 Abra, R.M., Bosworth, M.E., and Hunt, C.A. (1980). Liposome disposition in vivo: effects of pre-dosing with lipsomes. *Research Communications in Chemical Pathology and Pharmacology* 29 (2): 349.

113 Xiao, K., Li, Y., Luo, J. et al. (2011). The effect of surface charge on in vivo biodistribution of PEG-oligocholic acid based micellar nanoparticles. *Biomaterials* 32 (13): 3435–3446.

114 Romberg, B., Hennink, W.E., and Storm, G. (2008), Sheddable coatings for long-circulating nanoparticles, Pharmaceutical research 25 (1): 55–71.

115 Utembe, W., Potgieter, K., Stefaniak, A.B., and Gulumian, M. (2015). Dissolution and biodurability: important parameters needed for risk assessment of nanomaterials. *Particle and Fibre Toxicology* 12 (1): 11.

116 Aggarwal, P., Hall, J.B., McLeland, C.B. et al. (2009). Nanoparticle interaction with plasma proteins as it relates to particle biodistribution, biocompatibility and therapeutic efficacy. *Advanced Drug Delivery Reviews* 61 (6): 428–437.

117 Lee, B.-J. and Nguyen, V.H. (2017). Protein corona: a new approach for nanomedicine design. *International Journal of Nanomedicine* 12: 3137.

118 Nie, S. (2010). Understanding and overcoming major barriers in cancer nanomedicine. *Nanomedicine* 5 (4): 523–528.

119 Frank, M.M. and Fries, L.F. (1991). Complement interactions and functions. *Immunology Today* 12 (9): 322–326.

120 Romberg, B., Hennink, W.E., and Storm, G. (2008). Sheddable coatings for long-circulating nanoparticles. *Pharmaceutical Research* 25 (1): 55–71.

121 Cedervall, T., Lynch, I., Lindman, S. et al. (2007). Understanding the nanoparticle–protein corona using methods to quantify exchange rates and affinities of proteins for nanoparticles. *Proceedings of the National Academy of Sciences* 104 (7): 2050–2055.

122 Sultan, D., Ye, D., Heo, G.S. et al. (2018). Focused ultrasound enabled trans-blood brain barrier delivery of gold nanoclusters: effect of surface charges and quantification using positron emission tomography. *Small* 14 (30): 1703115.

123 Elci, S.G., Jiang, Y., Yan, B. et al. (2016). Surface charge controls the suborgan biodistributions of gold nanoparticles. *ACS Nano* 10 (5): 5536–5542.

124 Moghimi, S.M., Hedeman, H., Muir, I.S. et al. (1993). An investigation of the filtration capacity and the fate of large filtered sterically-stabilized microspheres in rat spleen. *Biochimica et Biophysica Acta (BBA) - General Subjects* 1157 (2): 233–240.

125 Choi, H.S., Liu, W., Misra, P. et al. (2007). Renal clearance of quantum dots. *Nature Biotechnology* 25 (10): 1165–1170.

126 Bahadar, H., Maqbool, F., Niaz, K., and Abdollahi, M. (2016). Toxicity of nanoparticles and an overview of current experimental models. *Iranian Biomedical Journal* 20 (1): 1.

127 Seung, W.S., In, H.S., and Soong, H.U. (2015). Role of physicochemical properties in nanoparticle toxicity. *Nanomaterials* 5: 1351–1365.

128 Tan, K.X., Barhoum, A., Pan, S., and Danquah, M.K. (2018). Risks and toxicity of nanoparticles and nanostructured materials. In: *Emerging Applications of Nanoparticles and Architecture Nanostructures* (ed. A. Barhoum and A.S.H. Makhlouf), 121–139. Elsevier.

129 Mukherjee, S., Sau, S., Madhuri, D. et al. (2016). Green synthesis and characterization of monodispersed gold nanoparticles: toxicity study, delivery of doxorubicin and its bio-distribution in mouse model. *Journal of Biomedical Nanotechnology* 12 (1): 165–181.

130 Negahdary, M., Chelongar, R., Zadeh, S.K., and Ajdary, M. (2015). The antioxidant effects of silver, gold, and zinc oxide nanoparticles on male mice in in vivo condition. *Advanced Biomedical Research* 4: 69.

131 Nelson, B.C., Petersen, E.J., Marquis, B.J. et al. (2013). NIST gold nanoparticle reference materials do not induce oxidative DNA damage. *Nanotoxicology* 7 (1): 21–29.

132 Sung, J.H., Ji, J.H., Park, J.D. et al. (2011). Subchronic inhalation toxicity of gold nanoparticles. *Particle and Fibre Toxicology* 8 (1): 16.

133 Li, J.J., Hartono, D., Ong, C.-N. et al. (2010). Autophagy and oxidative stress associated with gold nanoparticles. *Biomaterials* 31 (23): 5996–6003.

134 Lopez-Chaves, C., Soto-Alvaredo, J., Montes-Bayon, M. et al. (2018). Gold nanoparticles: distribution, bioaccumulation and toxicity. In vitro and in vivo studies. *Nanomedicine: Nanotechnology, Biology and Medicine* 14 (1): 1–12.

135 Thakor, A.S., Paulmurugan, R., Kempen, P. et al. (2011). Oxidative stress mediates the effects of Raman-active gold nanoparticles in human cells. *Small* 7 (1): 126–136.

136 Zhang, X.-D., Wu, H.-Y., Di Wu, Y.-Y.W. et al. (2010). Toxicologic effects of gold nanoparticles in vivo by different administration routes. *International Journal of Nanomedicine* 5 (1): 771–781.

137 Senut, M.C., Zhang, Y., Liu, F. et al. (2016). Size-dependent toxicity of gold nanoparticles on human embryonic stem cells and their neural derivatives. *Small* 12 (5): 631–646.

138 Hanna, S.K., Montoro Bustos, A.R., Peterson, A.W. et al. (2018). Agglomeration of *Escherichia coli* with positively charged nanoparticles can lead to artifacts in a standard *Caenorhabditis elegans* toxicity assay. *Environmental Science & Technology* 52 (10): 5968–5978.

139 Ciucă, A.G., Grecu, C.I., Rotărescu, P. et al. (2017). Nanostructures for drug delivery: pharmacokinetic and toxicological aspects. In: *Nanostructures for Drug Delivery* (ed. E. Andronescu and A.H. Grumezescu), 941–957. Elsevier.

140 Dasgupta, N. and Ramalingam, C. (2016). Silver nanoparticle antimicrobial activity explained by membrane rupture and reactive oxygen generation. *Environmental Chemistry Letters* 14 (4): 477–485.

141 Foldbjerg, R., Dang, D.A., and Autrup, H. (2011). Cytotoxicity and genotoxicity of silver nanoparticles in the human lung cancer cell line, A549. *Archives of Toxicology* 85 (7): 743–750.

142 Haase, A., Tentschert, J., Jungnickel, H. et al. Toxicity of silver nanoparticles in human macrophages: uptake, intracellular distribution and cellular responses. *Journal of Physics: Conference Series* 304: 012030.

143 Guo, Z., Zeng, G., Cui, K., and Chen, A. (2019). Toxicity of environmental nanosilver: mechanism and assessment. *Environmental Chemistry Letters* 17 (1): 319–333.

144 Cappellini, F., Hedberg, Y., McCarrick, S. et al. (2018). Mechanistic insight into reactivity and (geno) toxicity of well-characterized nanoparticles of cobalt metal and oxides. *Nanotoxicology* 12 (6): 602–620.

145 Annangi, B., Bach, J., Vales, G. et al. (2015). Long-term exposures to low doses of cobalt nanoparticles induce cell transformation enhanced by oxidative damage. *Nanotoxicology* 9 (2): 138–147.

146 Wan, R., Mo, Y., Feng, L. et al. (2012). DNA damage caused by metal nanoparticles: involvement of oxidative stress and activation of ATM. *Chemical Research in Toxicology* 25 (7): 1402–1411.

147 Wan, R., Mo, Y., Zhang, Z. et al. (2017). Cobalt nanoparticles induce lung injury, DNA damage and mutations in mice. *Particle and Fibre Toxicology* 14 (1): 38.

148 Patel, K.D., Singh, R.K., and Kim, H.-W. (2019). Carbon-based nanomaterials as an emerging platform for theranostics. *Materials Horizons* 6 (3): 434–469.

149 Madannejad, R., Shoaie, N., Jahanpeyma, F. et al. (2019). Toxicity of carbon-based nanomaterials: reviewing recent reports in medical and biological systems. *Chemico-Biological Interactions* 307: 206–222.

150 Liu, Y., Luo, Y., Wu, J. et al. (2013). Graphene oxide can induce in vitro and in vivo mutagenesis. *Scientific Reports* 3: 3469.

151 Mrđanovic, J., Šolajić, S., Bogdanović, V. et al. (2012). Effects of fullerenol nano particles $C_{60}(OH)_{24}$ on micronuclei and chromosomal aberrations' frequency in peripheral blood lymphocytes. *Journal of Nanomaterials and Biostructures* 7: 673–386.

152 Sumi, N. and Chitra, K.C. (2019). Cytogenotoxic effects of fullerene C60 in the freshwater teleostean fish, Anabas testudineus (Bloch, 1792). *Mutation Research, Genetic Toxicology and Environmental Mutagenesis* 847: 503104.

153 Samadian, H., Salami, M.S., Jaymand, M. et al. (2020). Genotoxicity assessment of carbon-based nanomaterials; have their unique physicochemical properties made them double-edged swords? *Mutation Research, Reviews in Mutation Research* 783: 108296.

154 Rodrigues, A.F., Newman, L., Jasim, D.A. et al. (2018). Immunological impact of graphene oxide sheets in the abdominal cavity is governed by surface reactivity. *Archives of Toxicology* 92 (11): 3359–3379.

155 Jasim, D.A., Murphy, S., Newman, L. et al. (2016). The effects of extensive glomerular filtration of thin graphene oxide sheets on kidney physiology. *ACS Nano* 10 (12): 10753–10767.

156 Amrollahi-Sharifabadi, M., Koohi, M.K., Zayerzadeh, E. et al. (2018). In vivo toxicological evaluation of graphene oxide nanoplatelets for clinical application. *International Journal of Nanomedicine* 13: 4757.

157 Mao, L., Hu, M., Pan, B. et al. (2015). Biodistribution and toxicity of radio-labeled few layer graphene in mice after intratracheal instillation. *Particle and Fibre Toxicology* 13 (1): 7.

158 Schinwald, A., Murphy, F.A., Jones, A. et al. (2012). Graphene-based nanoplatelets: a new risk to the respiratory system as a consequence of their unusual aerodynamic properties. *ACS Nano* 6 (1): 736–746.

159 Park, E.J., Lee, S.J., Lee, K. et al. (2017). Pulmonary persistence of graphene nanoplatelets may disturb physiological and immunological homeostasis. *Journal of Applied Toxicology* 37 (3): 296–309.

160 Jaworski, S., Strojny, B., Sawosz, E. et al. (2019). Degradation of mitochondria and oxidative stress as the main mechanism of toxicity of pristine graphene on U87 glioblastoma cells and tumors and HS-5 cells. *International Journal of Molecular Sciences* 20 (3): 650.

161 Pelin, M., Fusco, L., Martín, C. et al. (2018). Graphene and graphene oxide induce ROS production in human HaCaT skin keratinocytes: the role of xanthine oxidase and NADH dehydrogenase. *Nanoscale* 10 (25): 11820–11830.

162 Donaldson, K. and Poland, C.A. (2012). Inhaled nanoparticles and lung cancer – what we can learn from conventional particle toxicology. *Swiss Medical Weekly* 142 (2526).

163 Kim, J.K., Shin, J.H., Lee, J.S. et al. (2016). 28-day inhalation toxicity of graphene nanoplatelets in Sprague-Dawley rats. *Nanotoxicology* 10 (7): 891–901.

164 Mendonça, M.C.P., Soares, E.S., de Jesus, M.B. et al. (2016). Reduced graphene oxide: nanotoxicological profile in rats. *Journal of Nanobiotechnology* 14 (1): 53.

165 Esfandiari, N., Bagheri, Z., Ehtesabi, H. et al. (2019). Effect of carbonization degree of carbon dots on cytotoxicity and photo-induced toxicity to cells. *Heliyon* 5 (12): e02940.

166 Zhao, Q.-L., Zhang, Z.-L., Huang, B.-H. et al. (2008). Facile preparation of low cytotoxicity fluorescent carbon nanocrystals by electrooxidation of graphite. *Chemical Communications* 41: 5116–5118.

167 Karpeta-Kaczmarek, J., Kędziorski, A., Augustyniak-Jabłokow, M.A. et al. (2018). Chronic toxicity of nanodiamonds can disturb development and reproduction of *Acheta domesticus* L. *Environmental Research* 166: 602–609.

168 Dworak, N., Wnuk, M., Zebrowski, J. et al. (2014). Genotoxic and mutagenic activity of diamond nanoparticles in human peripheral lymphocytes in vitro. *Carbon* 68: 763–776.

169 Paget, V., Sergent, J.A., Grall, R. et al. (2014). Carboxylated nanodiamonds are neither cytotoxic nor genotoxic on liver, kidney, intestine and lung human cell lines. *Nanotoxicology* 8 (Suppl. 1): 46–56.

170 Balek, L., Buchtova, M., Bosakova, M.K. et al. (2018). Nanodiamonds as "artificial proteins": regulation of a cell signalling system using low nanomolar solutions of inorganic nanocrystals. *Biomaterials* 176: 106–121.

171 Ibrahim, M., Xue, Y., Ostermann, M. et al. (2018). In vitro cytotoxicity assessment of nanodiamond particles and their osteogenic potential. *Journal of Biomedical Materials Research Part A* 106 (6): 1697–1707.

172 Alawdi, S.H., El-Denshary, E.S., Safar, M.M. et al. (2017). Neuroprotective effect of nanodiamond in Alzheimer's disease rat model: a pivotal role for modulating NF-κB and STAT3 signaling. *Molecular Neurobiology* 54 (3): 1906–1918.

173 Farré, M. and Barceló, D. (2012). Introduction to the analysis and risk of nanomaterials in environmental and food samples. In: *Comprehensive Analytical Chemistry* (ed. M. Farré and D. Barceló), 1–32. Elsevier.

174 Vashist, S.K. (2013). Nanomaterials-based health care and bioanalytical applications: trend and prospects. *Journal of Nanomaterials & Molecular Nanotechnology* 2 (2): 1–6.

175 Li, B., Li, Q., Mo, J., and Dai, H. (2017). Drug-loaded polymeric nanoparticles for cancer stem cell targeting. *Frontiers in Pharmacology* 8: 51.

176 Kumari, A., Yadav, S.K., and Yadav, S.C. (2010). Biodegradable polymeric nanoparticles based drug delivery systems. *Colloids and Surfaces B: Biointerfaces* 75 (1): 1–18.

177 Voigt, N., Henrich-Noack, P., Kockentiedt, S. et al. (2014). Toxicity of polymeric nanoparticles in vivo and in vitro. *Journal of Nanoparticle Research* 16 (6): 2379.

178 Elliott, J.T., Rösslein, M., Song, N.W. et al. (2017). Toward achieving harmonization in a nanocytotoxicity assay measurement through an interlaboratory comparison study. *ALTEX-Alternatives to Animal Experimentation* 34 (2): 201–218.

179 Gatoo, M.A., Naseem, S., Arfat, M.Y. et al. (2014). Physicochemical properties of nanomaterials: implication in associated toxic manifestations. *BioMed Research International* 2014: 498420.

180 Powers, K.W., Palazuelos, M., Moudgil, B.M., and Roberts, S.M. (2007). Characterization of the size, shape, and state of dispersion of nanoparticles for toxicological studies. *Nanotoxicology* 1 (1): 42–51.

181 Nel, A., Xia, T., Mädler, L., and Li, N. (2006). Toxic potential of materials at the nanolevel. *Science* 311 (5761): 622–627.

182 Aillon, K.L., Xie, Y., El-Gendy, N. et al. (2009). Effects of nanomaterial physicochemical properties on in vivo toxicity. *Advanced Drug Delivery Reviews* 61 (6): 457–466.

183 Lovrić, J., Bazzi, H.S., Cuie, Y. et al. (2005). Differences in subcellular distribution and toxicity of green and red emitting CdTe quantum dots. *Journal of Molecular Medicine* 83 (5): 377–385.

184 Risom, L., Møller, P., and Loft, S. (2005). Oxidative stress-induced DNA damage by particulate air pollution. *Mutation Research, Fundamental and Molecular Mechanisms of Mutagenesis* 592 (1–2): 119–137.

185 Donaldson, K. and Stone, V. (2003). Current hypotheses on the mechanisms of toxicity of ultrafine particles. *Annali dell'Istituto superiore di sanitÃ* 39 (3): 405–410.
186 Chen, Z., Meng, H., Xing, G. et al. (2006). Acute toxicological effects of copper nanoparticles in vivo. *Toxicology Letters* 163 (2): 109–120.
187 Griffitt, R.J., Luo, J., Gao, J. et al. (2008). Effects of particle composition and species on toxicity of metallic nanomaterials in aquatic organisms. *Environmental Toxicology and Chemistry: An International Journal* 27 (9): 1972–1978.
188 Petersen, E.J. and Nelson, B.C. (2010). Mechanisms and measurements of nanomaterial-induced oxidative damage to DNA. *Analytical and Bioanalytical Chemistry* 398 (2): 613–650.
189 Ispas, C., Andreescu, D., Patel, A. et al. (2009). Toxicity and developmental defects of different sizes and shape nickel nanoparticles in zebrafish. *Environmental Science & Technology* 43 (16): 6349–6356.
190 Chithrani, B.D., Ghazani, A.A., and Chan, W.C.W. (2006). Determining the size and shape dependence of gold nanoparticle uptake into mammalian cells. *Nano Letters* 6 (4): 662–668.
191 Hamilton, R.F., Wu, N., Porter, D. et al. (2009). Particle length-dependent titanium dioxide nanomaterials toxicity and bioactivity. *Particle and Fibre Toxicology* 6 (1): 35.
192 Verma, A. and Stellacci, F. (2010). Effect of surface properties on nanoparticle–cell interactions. *Small* 6 (1): 12–21.
193 Champion, J.A. and Mitragotri, S. (2006). Role of target geometry in phagocytosis. *Proceedings of the National Academy of Sciences* 103 (13): 4930–4934.
194 Lee, M.-K., Lim, S.-J., and Kim, C.-K. (2007). Preparation, characterization and in vitro cytotoxicity of paclitaxel-loaded sterically stabilized solid lipid nanoparticles. *Biomaterials* 28 (12): 2137–2146.
195 Chen, Y.-S., Hung, Y.-C., Liau, I., and Huang, G.S. (2009). Assessment of the in vivo toxicity of gold nanoparticles. *Nanoscale Research Letters* 4 (8): 858.
196 Hsiao, I.L. and Huang, Y.-J. (2011). Effects of various physicochemical characteristics on the toxicities of ZnO and TiO_2 nanoparticles toward human lung epithelial cells. *Science of the Total Environment* 409 (7): 1219–1228.
197 Fubini, B., Fenoglio, I., Tomatis, M., and Turci, F. (2011). Effect of chemical composition and state of the surface on the toxic response to high aspect ratio nanomaterials. *Nanomedicine* 6 (5): 899–920.
198 Pietroiusti, A., Massimiani, M., Fenoglio, I. et al. (2011). Low doses of pristine and oxidized single-wall carbon nanotubes affect mammalian embryonic development. *ACS Nano* 5 (6): 4624–4633.
199 Hoshino, A., Fujioka, K., Oku, T. et al. (2004). Physicochemical properties and cellular toxicity of nanocrystal quantum dots depend on their surface modification. *Nano Letters* 4 (11): 2163–2169.
200 Georgieva, J.V., Kalicharan, D., Couraud, P.-O. et al. (2011). Surface characteristics of nanoparticles determine their intracellular fate in and processing by human blood–brain barrier endothelial cells in vitro. *Molecular Therapy* 19 (2): 318–325.
201 Goodman, C.M., McCusker, C.D., Yilmaz, T., and Rotello, V.M. (2004). Toxicity of gold nanoparticles functionalized with cationic and anionic side chains. *Bioconjugate Chemistry* 15 (4): 897–900.

202 Bhattacharjee, S., de Haan, L.H.J., Evers, N.M. et al. (2010). Role of surface charge and oxidative stress in cytotoxicity of organic monolayer-coated silicon nanoparticles towards macrophage NR8383 cells. *Particle and Fibre Toxicology* 7 (1): 25.

203 Yang, S.-T., Wang, X., Jia, G. et al. (2008). Long-term accumulation and low toxicity of single-walled carbon nanotubes in intravenously exposed mice. *Toxicology Letters* 181 (3): 182–189.

204 Fubini, B., Giamello, E., Volante, M., and Bolis, V. (1990). Chemical functionalities at the silica surface determining its reactivity when inhaled. Formation and reactivity of surface radicals. *Toxicology and Industrial Health* 6 (6): 571–598.

205 Kirchner, C., Liedl, T., Kudera, S. et al. (2005). Cytotoxicity of colloidal CdSe and CdSe/ZnS nanoparticles. *Nano Letters* 5 (2): 331–338.

206 Sager, T.M., Porter, D.W., Robinson, V.A. et al. (2007). Improved method to disperse nanoparticles for in vitro and in vivo investigation of toxicity. *Nanotoxicology* 1 (2): 118–129.

207 Jiang, J., Oberdörster, G., and Biswas, P. (2009). Characterization of size, surface charge, and agglomeration state of nanoparticle dispersions for toxicological studies. *Journal of Nanoparticle Research* 11 (1): 77–89.

208 Colvin, V.L. (2003). The potential environmental impact of engineered nanomaterials. *Nature Biotechnology* 21 (10): 1166–1170.

209 Hou, W.-C., Westerhoff, P., and Posner, J.D. (2013). Biological accumulation of engineered nanomaterials: a review of current knowledge. *Environmental Science: Processes & Impacts* 15 (1): 103–122.

210 Ganguly, P., Breen, A., and Pillai, S.C. (2018). Toxicity of nanomaterials: exposure, pathways, assessment, and recent advances. *ACS Biomaterials Science & Engineering* 4 (7): 2237–2275.

211 Duvall, M. and Wyatt, A. (2011). *Regulation of Nanotechnology and Nanomaterials at EPA and around the World: Recent Developments and Context.* Washington, DC: Beveridge & Diamond, PC.

212 Naqvi, S. and Flora, S.J.S. (2020). Nanomaterial's toxicity and its regulation strategies. *Journal of Environmental Biology* 41 (4): 659–671.

18

Multifunctional Nanomaterials for Medical Applications

Jitha S. Jayan[1], Ramya Rajan[1], Saritha Appukuttan[1], and Kuruvilla Joseph[2]

[1] *Department of Chemistry, Amrita Vishwa Vidyapeetham, Kollam, Kerala, India*
[2] *Department of Chemistry, Indian Institute of Space Science and Technology, Valiyamala, Kerala, India*

18.1 Introduction and Recent Advances

Nanomaterials are considered to be functional materials having wide applications in different fields [1]. The advent of nanotechnology has brought tremendous change in the medicinal field due to the advancement in diagnostic, therapeutic, and targeting applications [2–4]. Based on the composition of the nanoparticles (NPs), it can be broadly classified as metallic, carbon-based, organic, and inorganic [5–8]. These NPs can be made suitable for medical application either by simple chemical modification, self-assembly, or by functionalization with other polymers [9–11]. The exceptionally small size and enhanced surface area, surface-to-volume ratio etc. makes them suitable for adsorption, dissolving, and covalent attachment of targeting and diagnostic agents so as to make it appropriate for targeting and diagnostic application by forming diagnostic/therapeutic complexes [12]. The complexes thus formed are highly efficient in sending signals and delivering drugs for diagnosis and targeting activities respectively [13–15]. Targeted drug delivery is considered as smart drug delivery due to the efficiency in selective targeting of the diseased cells especially cancer cells, which can be attained by the conjugation with targeting moieties, genes, fluorescence dyes, or drugs [16, 17]. Anti-LDLR (low-density lipoprotein receptor) antibody modified NPs have been recently employed for targeting hypoxia and they exhibited exceptional characteristics over chemotherapy due to the combined chemo and phototherapeutic characteristics [18]. NPs can be even functionalized with ligands for ensuring specific targeting of cancer cells [19–21]. Nowadays, the imaging, therapeutic, and targeted drug-releasing applications of nanomaterials are gaining wider attention [22–25]. Theranostics is a concept of the combination of diagnostics and therapeutics, which aims to personalize monitoring and treatment [26–30]. Recently, NPs have been explored for diagnostic and treatment due to the unique physical and chemical properties [31]. Hence, theranostics is now giving way to nanotheranostics, which helps in providing imaging, diagnosing of disease,

Nanomaterials and Nanotechnology in Medicine, First Edition. Edited by P.M. Visakh.

and its therapy without affecting healthier cells [32]. The use of nanotheranostics avoids the invasive treatment, improves prognosis, and consequently helps clinicians to make advanced therapeutic decisions. Thus, it ensures early detection and treatment of diseases making possible the real-time monitoring of drug release with high efficiency [33, 34]. Recently, Bai et al. [35] made block copolymer NPs as theranostic agents for the treatment and diagnosis of coronary microthrombosis. This strategy aims to monitor the response of the patient, enhance drug efficacy and treatment, and avoid the chances of incorrect treatment thereby saving time and cost [36]. It can be treated as an emerging science for meeting regulatory and financial constraints [37]. This technique ensures the damage of affected cells at the time of diagnosis itself, which is more beneficial in the case of cancer treatment and hence the growth can be retarded to some extent. Moreover, light-responsive NPs are gaining wide attention due to their phosphorescence, fluorescence, photothermal effects, and plasmonic characteristics, which can be utilized for making wearable and implantable diagnostics as well as therapeutic devices [38]. Light-responsive materials like plasmonic gold (Au) nanostructures, quantum dots (QDs) [39], organic semiconductors [40], transition metal dichalcogenides [41] and upconversion NPs (UNPs) [42] have potential use for diagnostic and therapeutic purposes. Light-based therapies like photodynamic therapy (PDT) [43], photothermal therapy (PTT) [44], optogenetic therapy [45], and photobiomodulation (PBM) [46] have emerged as photonic healthcare applications. Owing to the biocompatibility and exceptional chemical and physical stability and ease of functionalization, metal-based NPs have potential applications in the medical field [47, 48]. These NPs are capable of forming conjugated products with drugs, proteins, enzymes, antibodies, fluorescent dyes, and nucleic acids and thus have enhanced biological activities [49, 50]. In addition to these materials, polymers with functional groups are utilized for making polymeric-NP-based theranostics. Entrapment, conjugation, encapsulation, and absorption of diagnostic agents and drugs in polymer-based particles can be utilized for the combined loading of both therapeutic and diagnostic agents for achieving the theranostic effects at cellular and molecular levels [51–53]. Au nanocages [54], graphene-based materials [55], layered double hydroxides [56] etc. find utility in cancer detection and therapeutics and are having advanced application in the field of medicine. Moreover, NPs are adequate in making scaffolds appropriate for tissue-engineering application [57]. Nanomaterial-based scaffolds are used to fabricate complex tissue or organs and are being clinically used as a standard treatment. Carbon-based nanomaterials are being applied in the field of regenerative medicine and tissue engineering due to their inherent ability to show differentiation and the capacity for attachment to stem cells [58, 59]. Nanomaterials are used for the toughening of polymer-based scaffolds for ensuring bone regeneration and to improve the mechanical and morphological properties [60–62]. General applications of multifunctional nanomaterials are represented in Figure 18.1. Thus, NPs are used in the medicinal field from the diagnosis, therapy, to tissue regeneration. Many new experiments continue to develop materials that have targeting and therapeutic efficiency. Sustained inventions in nanomaterial research in tuning the advanced applications of targeted theranostic materials, stimuli responsive, and regenerative materials are in demand. Hence, the development of functional nanomaterials and tuning of its application is considered as a vibrant research field and studies continue in different parts of the world since healthcare is of prime concern.

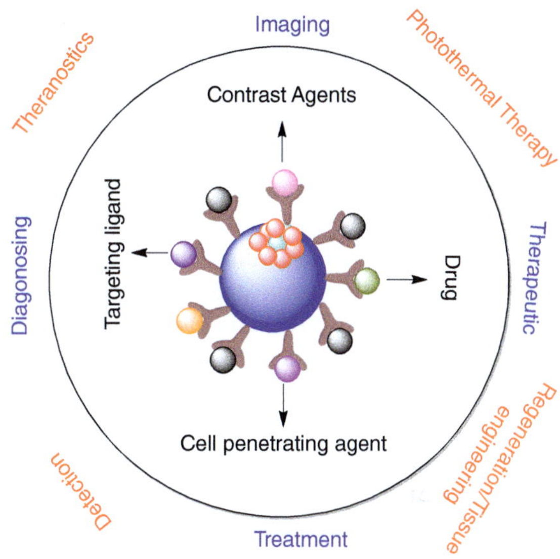

Figure 18.1 General applications of multifunctional nanomaterials.

18.2 Multifunctional Nanomaterials

Multifunctional nanomaterials are considered nanomaterials that have mutual influences of different properties, and are having most promising applications compared to normal NPs. Compared to conventional monofunctional NPs, multifunctional NPs are capable of integrating different functionalities for a highly sophisticated application [63, 64]. It is possible to combine different properties like electronic, optical, magnetic, and catalytic properties together [65]. Considering the medical application, it is possible to perform synergic goals like diagnosing and therapy by enhancing the codelivery of imaging, diagnosing, and therapeutic agents aimed at the targeted application. Thus, it is very effective in ensuring theranostics applications to a greater extent [66, 67]. Specific targetability, biocompatibility, stability, and bioimaging are the general advantages of multifunctional nanomaterials [66]. Among the various NPs, inorganic and polymer-based NPs are attracting more attention due to the enhanced clinical activity. Low toxicity and biocompatible natures are the key conditions essential for multifunctional NPs to be utilized as a therapeutic agent especially against cancer [68, 69]. But toxicity is the one of the major issues associated with inorganic NPs [70], hence the functionalization using polymers is considered as a solution to overcome this issue [71]. Polymer coatings like shells over the inorganic NPs reduce toxicity to a great extent. Moreover, the multifunctionality further helps in enhancing the stability, biocompatibility, and it will aid grafting or loading of drugs. The polymer shells are also capable of enhancing the photothermal activity for attaining better therapeutic characteristics [72, 73]. Due to the advanced magnetic properties and ease of functionalization, superparamagnetic iron (Fe) NPs (SPION: Superparamagnetic iron oxide NPs) are widely used as contrasting agents [74]. Au nanocages, which are capable of showing surface plasmon resonance (SPR) can be functionalized so as to improve imaging and PDT [75]. Moreover, the hollow structures of the Au nanocages can be utilized for ultrasound treatment and upon further modification it can be made more effective for cancer treatment [76]. Thus, the combination of Au nanocages, Fe NPs, and polymers ensure multimodal imaging and enhances the efficiency of ultrasound therapy with biocompatibility and biosafety characteristics [77]. Thus, multifunctional NPs are multifaceted and are found to be effective in imaging, targeting, and drug-delivery applications. UNPs like lanthanide-doped rare earths have photon-emitting efficiency in the near-infrared (IR) region and are good in biomedical application [78, 79]. Due to enhanced biostability and tissue-penetration ability, the absence of auto-fluorescence, and chemical stability, UNPs are efficient when compared to conventional QDs and are have *in vivo* sensitivity [80]. However, the surface engineering of these UNPs is helpful in enhancing the existing functionalities as well as creating additional functionalities [81]. Photosensitizers and drug-loaded UNPs are capable of use in imaging and therapeutic applications [82–84]. Thus, multifunctional NPs are superior to monofunctional NPs in activity and efficiency. Recently, multimodal imaging and PTT modification were made to $Cu_{2-x}Se$ (crystalline copper selenide) NPs by PEGylation (PEG: polyethylene glycol), thus fabricated multifunctional NPs were showing *in vitro* and *in vivo* stability and biocompatibility. Multifunctional groups are coordinated with the NPs thus, the coordination of radioisotopes ensured single photon emission computed tomography imaging and the photothermal conversion ensured the photoacoustic (PA) imaging and PTT [85]. Figure 18.2 represents the synthesis method and the applications of PEGylated $Cu_{2-x}Se$ NPs. The combination of molecular

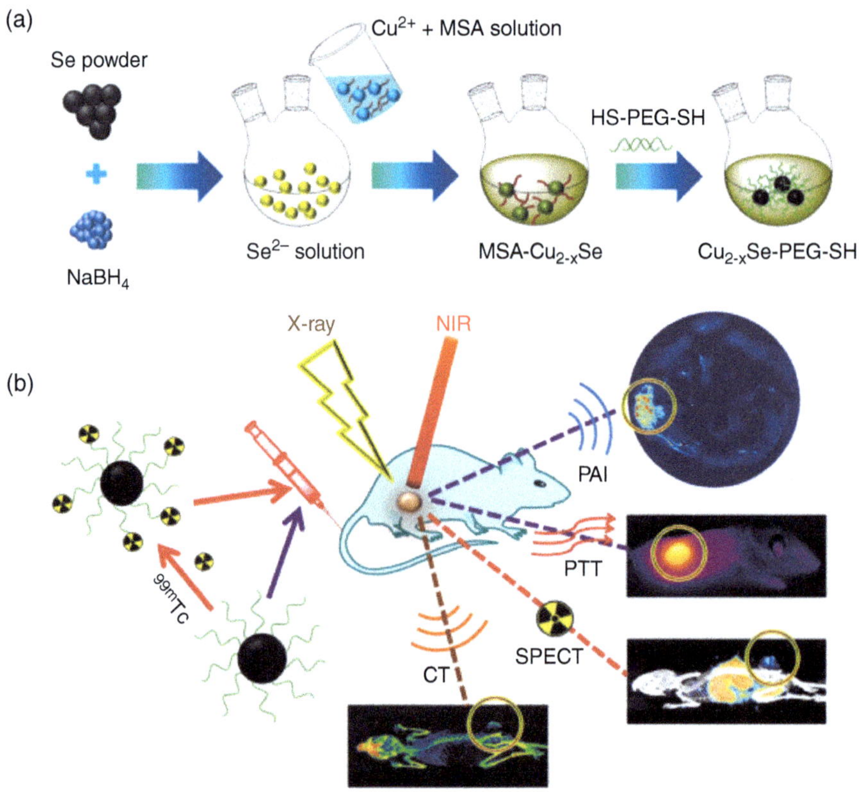

Figure 18.2 (a) Schematic illustration of the synthesis of multifunctional Cu_2-xSe-PEG-SH nanoparticles and (b) their applications in PA imaging, CT imaging, SPECT imaging, and photothermal therapy of cancer [85]. *Source:* Reproduced with permission from Wiley.

imaging and nanotechnology has tremendous effect in cancer theragnosis [86]. Polymeric NPs like chitosan-based NPs with anticancer drug and fluorescent dye can be utilized for multi-functional applications like imaging and targeting [87].

18.3 Diagnostic Application

For better disease management, early diagnosis is critical. There is a lot of research in with the aim of early diagnosis to support prompt arrest and better management of diseases. Medical diagnostics has been defined as the discipline or practice of diagnosis, which involves determining and describing a disease state and its causative factors, using signs and symptoms obtained from patient history, from physical examination, or patients' specimen using an array of diagnostic techniques [88]. The branch of science termed nanodiagnostics is helpful for early disease detection even before the emergence of symptoms and is the application of nanobiotechnology in molecular diagnostics [89]. It provides various options for early diagnosis of diseases. Nanotechnology, deals with NPs and it has been used to develop tools that provide the detection of biomolecules and analytes relevant for diagnosis purposes [90]. Nanodiagnostics allow diagnosis to be carried out on a nanoscale, which has led to the development and use of handy easy-to-use devices [90, 91]. The uniqueness of nanomaterials is that their size is similar to most biological molecules and structures and hence they can be useful for both *in vivo* and *in vitro* biomedical research and applications [92].

The integration of biology with nanomaterials that possess unique properties has led to research and development of diagnostic devices [93]. The unique properties of NPs, particularly with respect to optical and magnetic properties, can be manipulated to achieve a detectable signal. In NP-based diagnostic assays a measurable characteristic signal of the target biomolecule is achieved upon binding of the NP probe to the target biomolecule. Nanoshells, metal NPs, and QDs are promising nanostructural probes used for this type of diagnostic applications [94]. The other types of NPs utilized for the development of nanodiagnostics include nanotubes, nanocrystals, nanorobots, and nanowires. Also, there are various NP-based platforms that are employed in nanodiagnostics such as nanotechnology-based biochips, nanofluidic microarrays, protein nanobiochips, micro-electro-mechanical systems, nanobiosensors, and nanoscale single-cell identifiers [88]. An example of a nanowire-based platform used in nanodiagnostic devices include the powerful general platforms for ultrasensitive direct electrical detection of biological and chemical species [95]. Another example of nanowire-based platform that is employed in nanodiagnostics include the silicon nanowire (SiNW) used in sensors as field effect transistors [96] (FETs-SiNWs) are used in diagnostic detection of several prostate cancer biomarkers, such as PSA (prostate-specific antigen), for diagnosing and monitoring prostate cancer and also predicting the risks. Figure 18.3 summarizes the nanodevices that are used in clinical applications [93].

Early diagnosis of cancer is one of the important research areas in which nanomedicine has emerged to play a vital role and thereby support cancer management. PSA is an important and the most widely used biomarker for the early diagnosis of prostate cancer and its subsequent treatment. Electrical detection of PSA using MoS_2 (molybdenum disulfide) nanosheets is highly sensitive and a label-free immunoassay format is known. This method has double the sensitivity of the classic Si-FET sensors [97]. The MoS_2 nanosheet is a two-dimensional

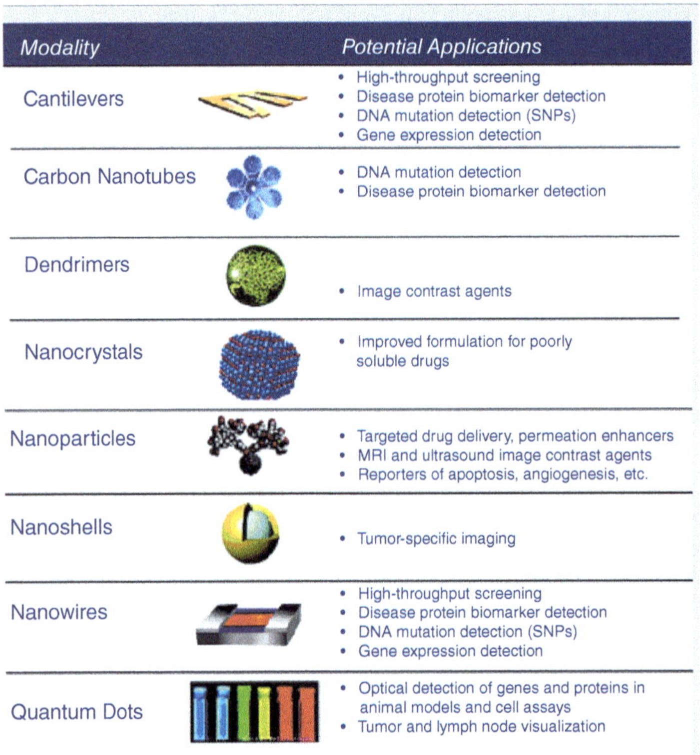

Figure 18.3 Types of nanodevices used in clinical applications [93]. *Source:* Reproduced with permission from Elsevier.

layered nanomaterial analogous to graphene but with a higher fluorescence-quenching ability making it suitable for development as a biomarker. Using this unique feature a novel aptamer-functionalized MoS_2 nanosheet fluorescent biosensor that detects PSA for early diagnosis of prostate cancer was developed by Kong et al. [98]. Nanomaterials with unique intrinsic physical properties such as inorganic NPs including semiconductor QDs, IO NPs, and Au NPs have been developed for diagnosis in biomedical imaging applications [99].

The advantage of NPs in biomolecular imaging is that their properties can be tailored to suit the required needs such as target specificity and high sensitivity. The surfaces of these NPs can be modified with ligands to target specific biomarkers associated with various diseases. Carbon nanotubules (CNTs) have demonstrated great potential in the area of biosensing and bioanalysis of typical biological molecules at early stages of cancer. Novel methods in which CNTs have been modified to have high sensitivity and wide linear and ultralow detection have been achieved. Also, the newly developed system of CNTs with similar selectivity is quick and can save time compared to conventional ELISA (enzyme-linked immunosorbent assay) commercial test kits [100]. Alzheimer's disease (AD) is a common neurodegenerative disorder in which the treatment after the onset of clinical symptoms cannot stop or reverse the disease progression. An early diagnosis is critical to reduce the associated risk by either preventing or delaying the progression of the disease. Recently, synthesis of closely packed and unidirectionally aligned CNT arrays for extremely sensitive and selective detection of multiple AD core biomarkers in blood plasma has been reported. This demonstrated early AD diagnosis using the aligned CNT-based, multiplexed, and electrical sensing platform [101].

18.4 Therapeutic Application

One of the applications of NPs in medicine that is extensively researched and developed is the use of NPs to deliver drugs, heat, light, and other substances to specific types of cells (such as cancer cells). These particles are often designed and synthesized with features and specific physicochemical properties such that they are attracted to diseased cells, which allows direct treatment of those cells. These techniques reduce damage to healthy cells in the body thereby reducing toxicity and allows for earlier detection of disease.

The beginning of the evolutionary journey of polymers with therapeutic applications can be tracked to the work of German scientist Horst Jatzkewitz back in the 1950s that included the synthesis of polyvinylpyrrolidone-mescaline conjugate with different linkers and the study of drug release of mescaline through this polymer–peptide–mescaline conjugate [102]. The USFDA's (US Food and Drug Administration) approval of Doxil, a liposome-based therapeutic nanomaterial containing doxorubicin (DOX), in 1995 for the treatment of Kaposi's sarcoma [103] provided another step forward. Since then, a plethora of nanomaterials have been designed and researched for use as a drug-delivery systems (DDSs). In the past several decades, the research and development interest toward the development of biodegradable NPs, as effective DDSs, has increased, which is also evident from the numbers of related publications and publication citations [104]. Various polymers have been employed in drug-delivery research as they have shown a capability to effectively deliver the drug to a target site and thus increase the therapeutic benefit thereby also minimizing the side effects, particularly for toxic drugs [105]. The steady increase in the number of polypeptide-based compounds reaching preclinical studies and clinical trials confirms the suitability of polypeptide-based materials for drug-delivery applications [106]. There has been a progressive increase in commercially available NP-based therapeutic products [107], which includes predominantly the liposomal drugs and polymer–drug conjugate classes. Table 18.1 list clinically approved NP-based therapeutics.

The most prominent role of nanomaterials in therapeutics in past has been as DDSs with a vast history of evolution. Nevertheless, functional nanomaterials have also shown good progress as direct modalities for the treatment of common diseases such as cancers, abnormal blood-vessel growth, and infectious diseases and tissues [104]. For example, fullerenes are used actively against neurodegenerative diseases due to their neuroprotective effect [109]. Recently, nanomaterials have also attracted scientists interested in cancer therapy due to their versatile physical and chemical properties [110]. Magnetic NPs (MNPs) and CNTs have also found utility directly as therapeutic materials for killing cancerous and precancerous cells by thermal ablation, which refers to the destruction of tissue by extreme hyperthermia or hypothermia that involves the temperature change concentrated on a focal zone in and around the tumor via use of an external stimulus [111]. Recent studies have also shown the potential use of NPs in gene delivery and gene therapy involving stem cells. Researchers at North Carolina State University are developing a method to deliver cardiac stem cells to damaged heart tissue. The various types of NPs have shown improved effects when compared to the traditional treatments. The studies devote the ability of NPs

Table 18.1 Clinically approved nanoparticle-based therapeutics [108].

Composition	Trade name	Company	Indication	Administration
Liposomal platforms				
Liposomal amphotericin B	Abelcet	Enzon	Fungal infections	i.v.
Liposomal amphotericin B	AmBisome	Gilead Sciences	Fungal and protozoal infections	i.v.
Liposomal cytarabine	DepoCyt	SkyePharma	Malignant lymphomatous meningitis	i.t.
Liposomal daunorubicin	Dauno Xome	Gilead Sciences	HIV-related Kaposi's sarcoma	i.v.
Liposomal doxorubicin	Myocet	Zeneus	Combination therapy with cyclophosphamide in metastatic breast cancer	i.v.
Liposomal IRIV vaccine	Epaxal	Berna Biotech	Hepatitis A	i.m.
Liposomal IRIV vaccine.	Inflexal V	Berna Biotech	Influenza	i.m
Liposomal morphine	DepoDur	SkyePharma,	Endo postsurgical analgesia	Epidural
Liposomal verteporfin degeneration,	Visudyne QLT,	Novartis	Age-related macular pathologic myopia, ocular histoplasmosis	i.v.
Liposome-PEG doxorubicin	Doxil/ Caelyx	Ortho Biotech, Schering-Plow	HIV-related Kaposi's sarcoma, metastatic breast cancer, metastatic ovarian cancer	i.m.
Micellular estradiol	Estrasorb	Novavax	Menopausal therapy	Topical
Polymeric platforms				
L-Glutamic acid, L-alanine, L-lysine, and L-tyrosine copolymer	Copaxone	TEVA Pharmaceuticals	Multiple sclerosis	s.c.
Methoxy-PEG-poly(D,L-lactide) taxol	Genexol-PM	Samyang	Metastatic breast cancer	i.v.
PEG–adenosine deaminase	Adagen	Enzon	Severe combined immunodeficiency disease associated with ADA deficiency	i.m.

Table 18.1 (Continued)

Composition	Trade name	Company	Indication	Administration
PEG-anti-VEGF aptamer	Macugen	OSI Pharmaceuticals	Age-related macular degeneration	i.r.
PEG–a-interferon 2a	Pegasys	Nektar, Hoffmann-La Roche	Hepatitis B, hepatitis C	s.c.
PEG–GCSF	Neulasta	Amgen	Neutropenia associated with cancer chemotherapy	s.c.
PEG–HGF	Somavert	Nektar, Pfizer	Acromegaly	s.c.
PEG–L-asparaginase	Oncaspar	Enzon	Acute lymphoblastic leukemia	i.v., i.m.
Poly(allylamine hydrochloride)	Renagel	Genzyme	End-stage renal disease	Oral
Other platforms				
Albumin-bound paclitaxel	Abraxane	AbraxisBio Science, AstraZeneca	Metastatic breast cancer	i.v.
Nanocrystalline aprepitant	Emend	Elan, Merck	Antiemetic	Oral
Nanocrystalline fenofibrate	Tricor	Elan, Abbott	Antihyperlipidemic	Oral
Nanocrystalline sirolimus	Rapamune	Elan, Wyeth Pharmaceuticals	Immunosuppressant	Oral

ADA, adenosine deaminase; GCSF, granulocyte colony-stimulating factor; HGF, hepatocyte growth factor; HIV, human immunodeficiency virus; i.m., intramuscular; i.r., intravitreous; IRIV, immunopotentiating reconstituted influenza virosome; i.t., intrathecal; i.v., intravenous; PEG, polyethyleneglycol; s.c., subcutaneous; VEGF, vascular endothelial growth factor.
Source: Reproduced with permission from Wiley.

in gene therapy to their unique chemical and physical advantages that in turn influences stem cell activity [112]. As gene therapy for cardiovascular disease is becoming increasingly popular, the future research on NPs in this area looks bright.

18.5 External Stimuli-responsive Nanoparticles for Medicinal Applications

Stimuli-responsive nanomaterials propose an alternative way toward the development of another drug delivery based on spatiotemporally controllable properties. External stimuli-responsive nanomaterials have potential application in theranostics due to their unique characteristics. Stimuli-responsive smart carriers are preferred over conventional drug carriers due to the advantage of this stimuli-responsive behavior [113]. Stimuli-responsive materials have been developed over the last few decades, and are considered to be smart materials due to the predictable changes in properties according to the environment. These can be considered as pioneers of the next generation interactive materials [114]. The inherent nature-to-sense behavior is considered a gift, which is more relevant in biological systems with self-response and regulation. Generally, in stimuli-responsive DDS, after injection, nanocarriers will leak and reach the targeted cells by active or passive ways. Then the delivery system is activated by specific triggers and release occurs at the target sites. Specific triggers can be delivered by intrinsic or external stimuli; intrinsic is considered within a diseased cell, like pH, while external stimuli are associated with outside the body like ultrasound, magnetic fields, and light. So, for a stimuli-responsive system, the system should be capable of releasing the drug by a transformation of a physical or compositional structure based on the intrinsic or external stimuli. Multifunctional nanomaterials can be triggered by the chemical modification so as to be active to the stimuli. Thus, different stimuli-responsive ligands can be introduced to NPs generally MNPs so as to make it suitable for a particular stimuli like electric fields, ultrasound, magnetic fields, light, pH, temperature, redox, and enzymes to achieve diverse applications [115–117]. This section discusses new trends in the development of stimuli-responsive multifunctional NPs for medicinal applications [118]. Figure 18.4 gives a schematic representation of MNP-based stimuli-responsive DDSs.

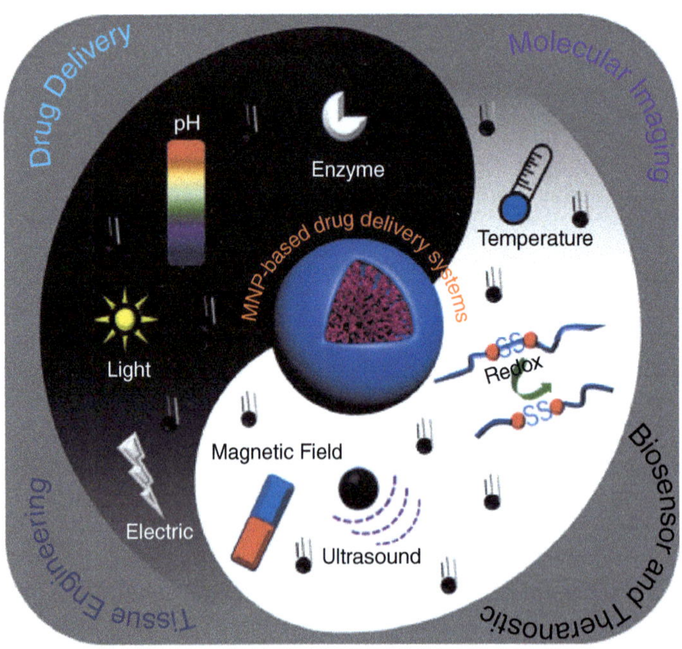

Figure 18.4 Schematic illustration of stimuli-responsive MNP-based DDSs for various biomedical applications [118]. *Source:* Reproduced with permission from Wiley.

18.5.1 Light Responsive

Photo-responsive therapy has made tremendous advances in the field of treatment especially in cancer treatment due to its noninvasiveness and spatiotemporally controllable ability. Light-triggered treatments, such as PDT and PTT and light-triggered drug delivery, are based on photosensitive materials [119]. UNPs and silica-based multifunctional NPs are finding potential applications in the field of PDT for cancer and have magnetic resonance contrast characteristics. These NPs were potentially used in the treatment due to their: capability of excitement in the near IR (NIR); cell penetration ability without cell damage; ability to emit various wavelengths; ability to kill cancer cells; and simultaneous imaging applications. Hence, these materials are highly efficient in diagnosis and PDT [120]. Due to strong SPR characteristics and X-ray attenuating abilities, tungsten NPs and dichalcogenides have potential applications in X-ray computer tomography (CT) and PTT of cancer, with poor biocompatibility and control of the surface chemistry as major drawbacks [121–125]. Hence, multifunctional systems like tungsten-bronze nanorods and polyelectrolyte-coated rods are capable of improving the biocompatibility and higher optical absorption in the region of NIR enhances the PTT application. It was also found efficient as NIR-induced ablation of cancer by a PTT/PDT method with less cytotoxicity [126]. Generally, in PDT reactive oxygen species (ROS) are responsible for the destruction of cancerous cell. These species release gaseous O_2 and increase the intramural O_2 level that effectively destroys the targeted cells. The attachment of species that are responsible for the oxygen release is highly expensive and sometimes the intrinsic pressure acts as a barrier to the penetration of O_2. To overcome, these risks Zhang et al. [127] developed multifunctional Bi_2WO_6 (bismuth tungstate) NPs having enhanced CT imaging and PDT ability without releasing O_2. This helps in the detection of tumor cells with high resolution and the CT-guided tumor ablation is also seen. The *in vivo* antitumor effects were analyzed in mice, by dividing them into four group: indocyanine green (ICG), saline, $W_{18}O_{49}$ (tungsten oxide), and Bi_2WO_6 groups. In saline, the tumor grows rapidly, but in ICG and $W_{18}O_{49,}$ the tumor growth was feeble after the laser irradiation but later grew abruptly. In the case of $Bi_2WO_{6,}$ continuous abatement is observed due to the efficient photodynamic destruction of the tumor as in the histological observation shown in Figure 18.5. The self-assembly of supramolecules can be utilized for making a nanoplatform for the self-assembled co-ordination of an MRI (magnetic resonance imaging) agent, photosensitive drug, and contrast agent. The multidimensional biometal organic NPs thus obtained possess inherent good biocompatibility, drug-loading capability, and robust stability. The co-operative assembly of different components leads to the delivery of photosensitizers that enhance the PDT against cancer cells. Moreover, the MRI agents assembled in it ensures the detection of the tumor cells and hence acts as a nanotheranostic [128].

Hou et al. [129] developed new multifunctional Fe_3O_4@Dex/TPP/PpIX/ss-mPEG NPs capable of producing O_2 by utilizing the excess of H_2O_2 in tumor cells by the Fenton reaction, thus avoiding the short half-life of ROS via mitochondrial targeting. The excellent MRI ability enhances the real-time monitoring of effected cells. The relaxation ratio of 21.7 obtained from Figure 18.6a indicates the imaging ability and the intracellular MRI shows that the signal was reduced to half within four hours indicating the enhanced MRI effect in cancerous cells (Figure 18.6b). Figure 18.6c shows the reduction in signal of the tumor

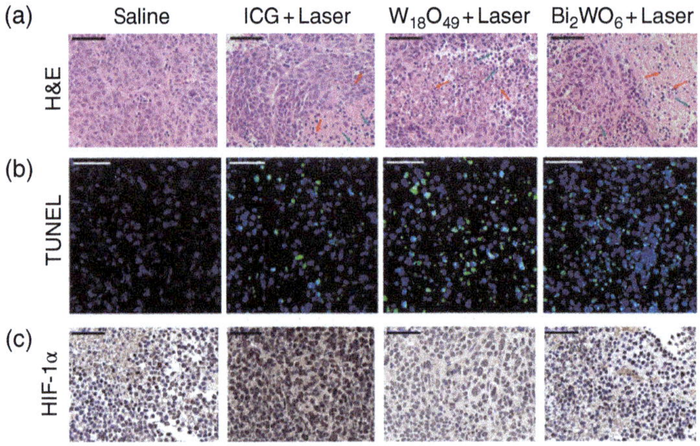

Figure 18.5 (a) Histological observation of the tumor tissues after different treatments. The tumor sections were stained with hematoxylin and eosin (H&E), (red arrow: necrosis; blue arrow: apoptosis). (b) Detection of cell apoptosis in the tumor tissues after the different treatments was detected with immunofluorescence. The tumor sections were stained with deoxynucleotidyl transferase dUTP nick end labeling (TUNEL) and then were observed with an optical microscope. (c) Hypoxia-inducible factor (HIF)-1α staining of the tumor slides from HeLa tumor-bearing mice after different treatments. Scale bars: 50 μm [127]. *Source:* Reproduced with permission from American Chemical Society.

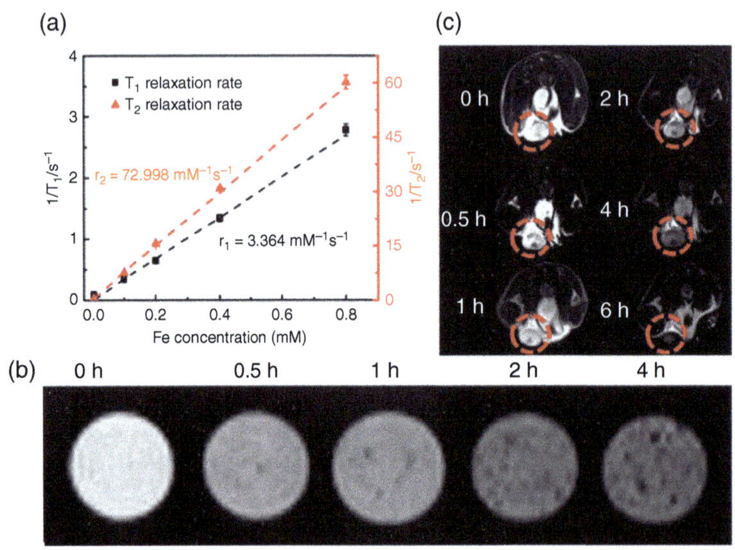

Figure 18.6 *In vitro* and *in vivo* MRI of Fe_3O_4@Dex-TPP/PpIX/ss-mPEG. (a) Transverse and longitudinal relaxation rates of Fe_3O_4@Dex-TPP/PpIX/ss-mPEG. (b) T2-weighted MR images of Fe_3O_4@Dex-TPP/PpIX/ss-mPEG in 4T1 cells after coincubated for 0, 0.5, 1, 2, and 4 hours. (c) *in vivo* T2-weighted MR images of Fe_3O_4@Dex-TPP/PpIX/ss-mPEG in 4T1 tumor-bearing Balb/c mice [129]. *Source:* Reproduced with permission from American Chemical Society.

after intravenous injection indicating the MRI effect of the NPs. Multifunctional silica-based NPs are generally categorized as a single modal therapeutic material, but the photosensitizer and folic-acid-modified silica-based NPs, are capable of showing PDT due to their ability to produce singlet oxygen and loading of anticancer to achieve a chemotherapy effect. Two-photon excitation of silica NP enhances fluorescence imaging [130]. Carbon–silica nanocomposites (CSNs) having PTT and PDT properties with inherent immunoadjuvant characteristics were fabricated by Wang et al. [131] and the photoactivity was confirmed from the absorption band obtained in the visible to NIR region. Usually, carbon-based NPs like GO (graphene oxide) and CNT are capable of showing biodegradation, but nonuniform morphology and size leads to serious side effects in *in vivo* treatment. However, CSN-based NPs are capable of degrading without enzymatic action, which enhances the possibility of clinical translation.

Multifunctional NPs introduced as self-healing materials like hydrogels are recently receiving attention in cancer treatment due to their drug-loading and releasing characteristics [132–134]. Block copolymer-based NPs are capable of loading the drug by emulsification [135] and it is highly active in making multifunctional NPs based on hybrid gels [136]. MNPs are capable of producing heat by the Neel and Brownian relaxations and the generated heat can be utilized for the destruction of tumor cells [137, 138]. Thus Wang et al. [139] made Mn-Zn ferrite-based mesoporous silica nanospheres that are highly efficient in producing hydroxyl radicals against cancer cells by the Fenton reaction. The MRI characteristics and the magnetothermal and chemodynamic therapy can be utilized for diagnostic and therapeutic applications. Utilization of covalent organic frameworks (COFs) as templates for making multifunctional NPs have potential application in cancer therapy. The organic-framework-based hyaluronic-acid-modified Au NPs are capable of producing oxygen by reacting with intratumoral hydrogen peroxide as shown in Figure 18.7 [140]. Multifunctional NPs that have excellent reactive O_2 species would provide a better therapeutic platform [141]. Multifunctional nanocomplexes made of polysaccharides [142], bioreducible dendrimers [143, 144], photosensitizer [145], and oxygen-evolving magnetic NPs [145, 146] are capable of overcoming tumor hypoxia, tumor mediated immunosuppression, suppressing oncogenic signals, and thus enhance PDT efficiency by utilizing the selective turn-on efficiency of the photosensitizer. These complexes are also capable of targeting, photodynamic, and immunotherapy applications [147]. Photosensitizer-encapsulated copolymer NPs are functional materials for PDT application with a higher targeting ability and efficiency [148, 149]. DNA (deoxyribonucleic acid)-polymer-assisted UCNPs are considered better PDT materials due to their recognition and loading capacity. The loading of photosensitizers in the NPs triggers the cancer therapy in the NIR region by generating O_2 ROS and DNAzyme triggered gene silencing [150].

Mesoporous nanosilica-based intelligent composites utilizing the on/off strategy in loading the metal and the contrasting agents are highly efficient in PDT applications. The on/off mechanism helps in the precise loading of diagnosing and therapeutic agents and guarantees a high loading capacity [151]. Yuan et al. [152] made a nanotherapeutic platform of mesoporous polydopamine containing L-arginine (L-Arg) and ICG that was capable of producing heat and reactive oxygen through NIR exposure thus, it can be utilized for PTT. Biofilm elimination was offered by the presence of nitric oxide species released by the NPs capable of enhancing PDT and PTT. Highly hierarchical passion fruit-like exosomes

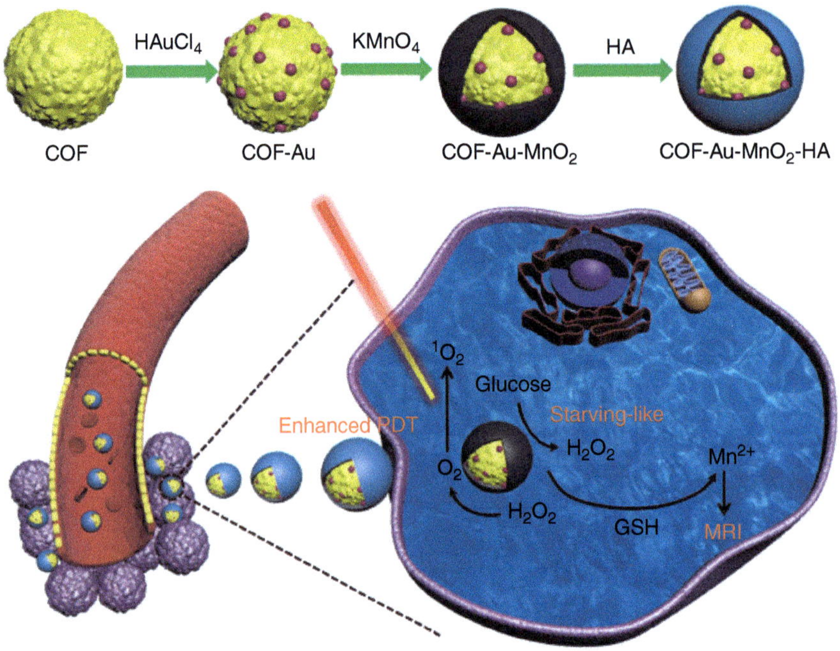

Figure 18.7 Schematic illustration of the synthetic process of COF-Au-MnO_2-HA and the therapeutic mechanism [140]. *Source:* Reproduced with permission from Springer.

and Au-based NPs, which are capable of encapsulating photosensitizers are used as nanovehicles for cancer theranostics. These nanovehicles have good biocompatibility, long-standing retention, and higher penetration behavior and promise noninvasive clinical application [153]. Combined PTT and PDT treatments are attributed by the modified hollow nanostructures that are capable of avoiding the drawbacks of photosensitizing agents like low light-conversion efficiency, a hostile tumor microenvironment, and intrinsic toxicity [154]. Dopamine-enveloped $W_{18}O_{49}$ nanodots emerged as effective for PDT and PTT and are capable of acting together under a single treatment using a laser. The higher biocompatibility and activity against cancer cells without effecting normal cells makes it a highly efficient multifunctional nano-platform [155]. MNP-based core and Au-based shell structured materials having nanostars are utilized for imaging and NIR-triggered drug delivery. The SPR effect of Au causes a temperature rise under NIR exposure; this light-to-heat conversion is responsible for the release of the drug [156]. Multifunctional NPs are integrated into cancer therapy due to advanced PTT applications [157, 158]. In order to avoid the lower therapy efficiency due to lower penetration depth Wang et al. [159] made multifunctional bismuth-based phthalocyanine manganese composites capable of showing combined PDT/PTT efficiency. Excellent efficiency in converting NIR to heat makes them effective against tumor cells. NIR-responsive nanodots, bioactive Na-Cd-Si, and inorganic materials are effectively used for PTT applications, especially for tumor treatment [160–162]. Carboxybetaine- and Fe_3O_4-based biomimicking membranes that have higher blood circulation and higher potential PTT efficiency are being developed as potential therapeutic agents [163]. Thermal biomolecule-mediated bismuth-based NPs having CT and IR thermal-imaging systems with drug-loading efficiency have been utilized for pH responsive antitumor material [164]. Recently, materials like nanoclusters [165], nanocrystal dendrimer composites [166], Bi_2WO_6NPs [127], metal–organic framework-based composites [167], core-shell NPs [168], and hybrid nanospheres [169] etc. are being developed as functional materials for PDT application. Thus, multifunctional NPs for the development of PDT due to the advantages over chemotherapeutic methods are achieving much research interest.

18.5.2 Thermal Responsive

Temperature is considered as the one of the most important stimuli responsible for the DDS by the spatiotemporal control and this ensures treatment without the surgical operation of affected tissue [170]. Temperature-dependent hydrogels are considered one of the most promising materials for controlled drug release due to their ability to remain as sol at room temperature and as gel at higher temperature. Thus, drugs that are loaded can be released in a sustained manner. Other hydrogels that have the ability to show further transformation from gel to sol can be utilized for controlled release. Thus, by making NIR as the switch to trigger the temperature change it is possible to control the drug release. Thus, photothermal treatment can also be considered to be a temperature-responsive method [171–173]. Thermo-responsive polymer-based NPs having core-shell structures are capable of showing thermo-responsive fluorescence and drug release. Higher loading can be obtained by enabling a better interaction of NPs with the drug, they are also capable of releasing higher amount of the drug at a temperature higher than the LCST (lower critical solution temperature) [174]. P(NIPAM-MAm)-coated and Fe_3O_4 NP-embedded hollow mesoporous silica nanocomposite have been utilized for drug-delivery applications due to their enhanced encapsulation efficiency and capacity, showing enhanced release above LCST [175]. Cai et al. [176] utilized Au nanostars and mesoporous silica-based NPs with core-shell structures for thermo-responsive drug delivery by the irradiation of NIR laser. Thermo-responsive DDS are capable of offering on-demand release of a drug according to the tumor environment. MNPs are utilized in the stimuli-responsive drug delivery due to their pH and thermal-responsive behavior [177, 178]. But biocompatibility is a major concern, hence the chitosan-modified MNPs with thermo-responsive poly(N-isopropylacrylamide) (PNIPAAm) could offer advanced efficiency in the thermo-responsive drug delivery [179, 180]. Fathi et al. [181] studied the drug-releasing effect of chitosan-grafted methotrexate-conjugated MNPs, and analyzed the cumulative thermo-responsive drug release as a function of time as shown in Figure 18.8. It was observed that 37°C at pH 5.5, showed enhanced drug release. As temperature increases, the drug release decreases due to the hydrophobic and steric interactions between the MNPs and the polymer. The drug release was higher in acidic conditions than in basic conditions. Generally, polymers with lower LCST show temperature-responsive behavior and thus polymer-based naocomposites can be utilized for the thermo-responsive release as described earlier. Equilibrium swelling studies showed the thermo-responsive behavior of poly(N-vinylcaprolactam) (PNVCL)-based MNPs and biocompatibility of these composites confirm the cell viability. Above the LCST, hydrophobic behavior is higher and thus it reduces the drug-carrier interaction [182]. Recently, polymer- and copolymer-based MNPs have received wider attention in thermo-responsive

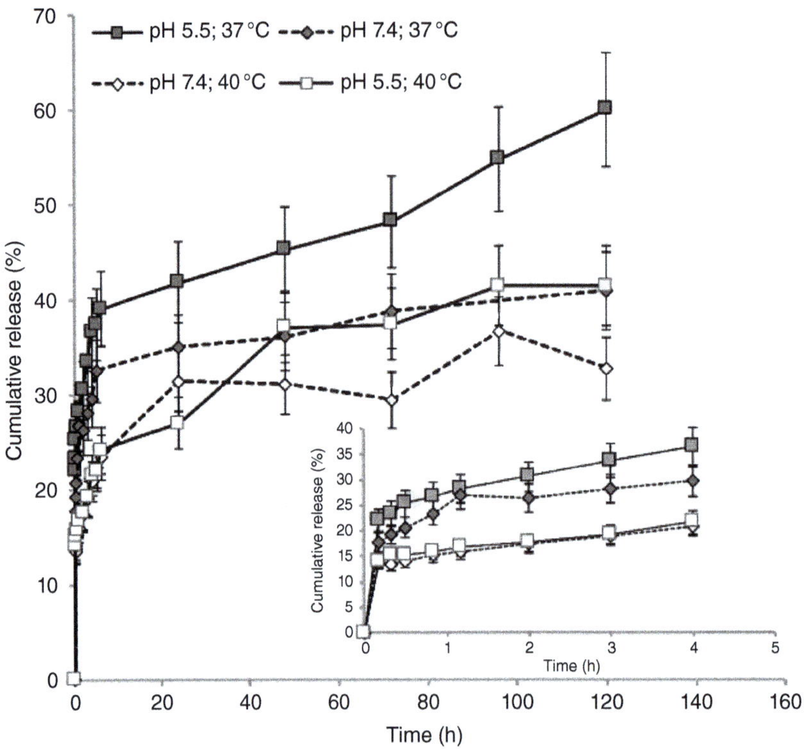

Figure 18.8 Temperature- and pH-dependent cumulative release of erlotinib from MTXCSC@MNPs. Inset: Drug release during the first 4 hours period [181]. *Source:* Reproduced with permission from Elsevier.

drug release [183, 184] in order to avoid direct administration of drugs especially anticancer drugs. Mesoporous nanosilica with self-assembled peptide-based core-shell materials with $ZnO@Fe_3O_4$ cores are used as thermo-responsive DDS due to their thermal conversion and microwave absorbing characteristics. Local heating in the tumor environment due to the microwave irradiation leads to the deassembling of peptides and subsequently the release off drugs [185]. Hierarchical-structured nanocarriers made of flower-shaped hollow nano-ruthenium-based composites were capable of showing temperature-dependent drug delivery [186]. Amphiphilic polymer-based combs are also efficient materials in theranostics due to their multimodal and thermo-responsive properties [187]. Thermo-responsive and self-healing nanocomposite hydrogels are capable of showing a hydrophilic to hydrophobic transition leading to shrinkage causing the drug release [188].

18.5.3 Ultrasonic Responsive

Ultrasonic sound has a tremendous effect in controlled drug delivery due to its noninvasiveness, regulated tissue penetration, and absence of radiation [189]. Moreover, it can penetrate deeper into the body than light and has the ability to reverse drug resistance in tumor cells [190]. By setting high intensity sound waves outside the targeted cells it is possible to generate increased temperature thus, increased localized temperature shows induced tumor necrosis and apoptosis [76]. Thus, ultrasound-targeted technology serves as an encouraging tool for increasing targeted delivery [191, 192]. The ultrasound-targeted microbubble destruction (UTMD) technique is considered a versatile technique in the treatment of cancer and is capable of enhancing targeted drug delivery [191, 192]. UTMD is capable of enhancing the codelivery of a microRNA inhibitor and gemcitabine by a dendrimer-based platform [193]. It also aids in the accumulation of MnO_2 in order to enhance therapeutic efficiency and diagnostic sensitivity. The therapeutic effect was evaluated by monitoring the size change associated with the tumor after the therapy as shown in Figure 18.9. DOX and MnO_2 NPs coloaded within poly (N-vinylcaprolactam) nanogels. eDOX/MnO_2@PVCL NGs are capable of suppressing tumor growth compared to free DOX, especially under UTMD [194].

Ultrasound-responsive imaging has advantages like improved blood circulation, and enhanced cellular interaction and detection with a conventional ultrasound method [195]. It is possible to control the depth of penetration by tuning the power density, frequency, and irradiation time [196, 197]. It is a safe and beneficial method for the *in vivo* circulation of DDS [198]. By adjusting the frequency in the focal site and at weak sites less side effects result. Thus, high intensity focal ultrasound (HIFU) can be considered as an attractive

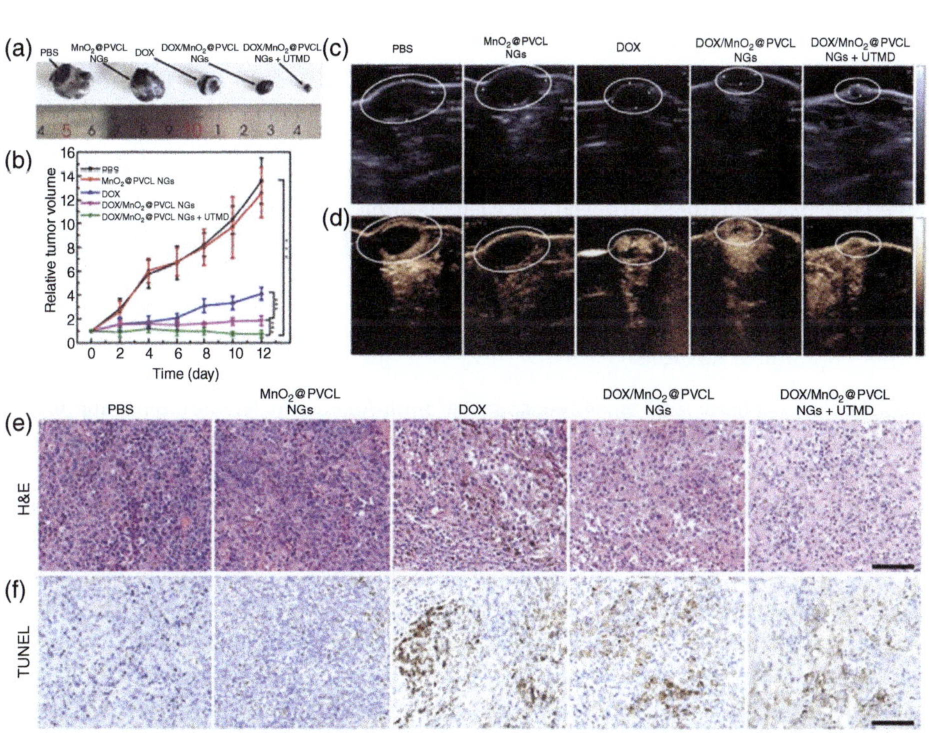

Figure 18.9 (a) Photograph of tumors on day 12 after different treatments. (b) Tumor growth curves after different treatments (n = 5). Tumor volumes (V) were normalized to their initial values (V0). (c) B-mode US images and (d) CEUS images of the tumor on day 12 after different treatments. (e) H&E staining and (f) TUNEL staining of tumor slices taken on day 12 after different treatments. Scale bar for (e) and (f) represents 200 μm for each panel. [194]. *Source:* Reproduced with permission from Ivyspring International Publisher.

stimulus for drug release [199]. Polydopamine-coated mesoporous silica NPs (MSNs) show excellent pH and ultrasound responsiveness for controlled drug release. HIFU is capable of dissipating thermal energy and thus generating high temperatures in the localized region [200]. Ultrasound imaging can also overcome the disadvantages of optical imaging, i.e. the depth limit, and ensures high resolution [201]. Wu et al. [202] utilized a combination of nanotechnology and focused ultrasound (FUS) for the delivery of drugs for cancer treatment, they have used hollow mesoporous organo-silica NPs (HMONs) and integrated $Cu_{2-x}Se$ ultrasmall particles for this purpose. Microbubble-based NPs and complexes, having superparamagnetic properties, are highly active for ultrasound and MR imaging [203–206]. Mn (manganese)-functionalized inorganic NPs are used due to their high stability and nano-size effect. Perfluorocarbon microbubbles with SPIONs and Mn are capable of showing ultrasound and MR imaging due to the combined contribution of all the materials under ultrasound irradiation [206, 207]. ROS-enzyme modified SPIO (superparamagnetic iron oxide) nanogels are capable of interacting with ROS in the tumor environment and result in an enhancement in ultrasound imaging [208, 209]. Dual-enzyme-loaded multifunctional hybrid nanogel probes are capable of detecting the bubbles of oxygen or hydrogen peroxide generated in the tumor environment [210]. Micron-sized hair-derived robots (hairbots) are utilized as an ultrasound contrasting agent [211]. HIFU ablation and chemotherapy are efficient against tumor growth via a synergistic effect. The thermal effect of the HIFU action induces the transition of composites and some changes will be observed in the acoustic environment, which helps imaging; moreover, the transitions are also capable of aiding drug release [212]. Thus, ultrasonic-responsive NPs are highly active in refining and improving imaging and drug-delivery applications [213–217]. Thus, ultrasound NPs provide a new multifunctional theranostic platform for tumor treatment. They are capable of showing enhanced application in both imaging and drug delivery and thus can be considered as an effective invasive method in diagnosis and treatment of tumor.

18.5.4 Biological Responsive

Biomolecule-responsive and enzyme-responsive multifunctional NPs are receiving wider attention in the biomedical filed due to their efficiency in therapeutic and drug-delivery application [218]. Biomolecule-responsive materials are capable of drug delivery by monitoring the difference in the biochemical or physiological conditions in the normal and pathological conditions [219–221]. Glucose-responsive acupuncture needles using multifunctional NPs are fabricated using glucose oxidase and, 4-mercaptobenzoic-acid-modified polystyrene-coated surface-enhanced Raman scattering (SERS) materials are good at detecting the glucose content by converting glucose to gluconic acid and thus changing the pH [222]. SERS-based materials are receiving wider attention due to their ability to detect biomolecules. Surface functionalization and portioned and charged layers are considered ways to enhance the efficiency of SERS-based biomolecular detectors [223, 224]. Molecular traps are used for *in vitro* detection [225], SERS-based microanalysis techniques are used for the *in vivo* detection of biomolecules [226]. Adenosine triphosphate (ATP)-responsive DDS are getting wider attention in monitoring and tracking in the DDS, ATP is considered molecular units of energy transfer [227–230], but increased ATP content indicates pathological process [231–234]. Lai et al. [235] prepared polypeptide-wrapped mesoporous silica-coated multicolor UNPs (UNP@MSN) functionalized with zinc dipicolylamine analog (TDPA-Zn^{2+}) for developing ATP-responsive DDS. Thus, manufactured biocompatible NPs are introduced into the targeted cells that are capable of taking-up the NPs, the folic-acid modification ensures the targeted delivery and monitors the luminescent signals for tracking the drug release. ATP-responsive drug release is schematically shown in Figure 18.10.

ATP aptamer-sandwiched DNA structures grafted to MSN are capable of releasing guest molecules from the system in the presence of ATP due to the interaction of aptamer with ATP, which results in a controlled release. Thus, these materials can be utilized for the fabrication of ATP-responsive devices [232]. DNA and graphene cross-linked ATP aptamer-based aggregates are capable of releasing the drug DOX in response to ATP concentration by the formation of TAP-ATP (tumor-activated prodrug-ATP) aptamer complex formation [236]. Tumor-specific enzymes like prostate-specific membrane antigen (PSMA) and matrixmetaloproteinase-2 (MMP-2) responsive NPs are utilized for the delivery of drugs against cancer [237–240]. Recently, amphiphilic-block copolymer-based micellar systems were used for MMP-2 enzyme-responsive drug delivery. NPs with high biocompatibility and high drug-loading efficiency are capable of forming higher cellular uptake and antitumor activity [241]. Enzyme-responsive stealth peptide-coated NPs are capable of detecting tumors and are very effective in tumor therapy with enhanced cellular uptake [242, 243]. Enzyme-responsive NPs ensure the drug accumulation at the targeted site [189], controlled drug release can be attained by the degradation of NPs upon the enzyme exposure [244, 245]. Enzyme-responsive treatment ensures reduced toxicity, better tissue or membrane penetration, improved drug accumulation, extended circulation time, and controllable release of therapeutic agents. Enzyme-responsive NPs can be prepared easily and have a short response time and hence provides a highly active method of therapeutics [246]. Due to the high complexity of the tumor environment, the development and application of biologically responsive NPs still has challenges. Thus, research continues to achieve high selectivity, *in vivo* efficiency, and biosafety.

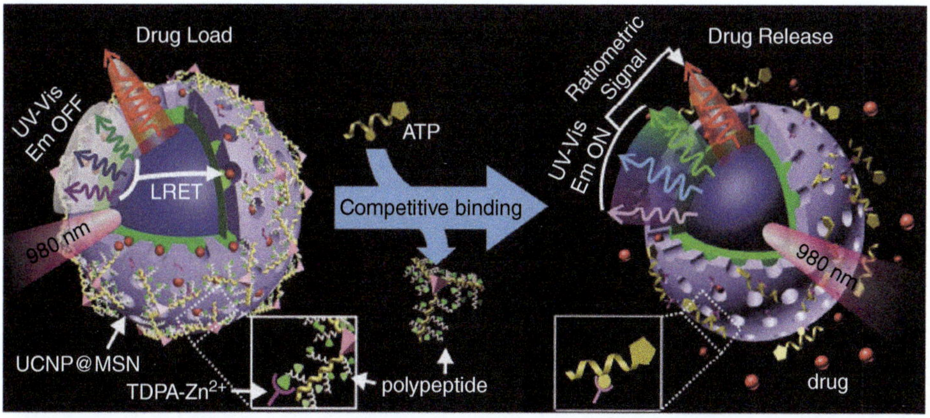

Figure 18.10 Schematic representation of the real-time monitoring of ATP-responsive drug release from polypeptide wrapped TDPA-$Zn2^{+}$–UCNP@MSNs [235]. *Source:* Reproduced with permission from American chemical Society.

18.6 Regulatory Issues

Due to the efficiency of multifunctional NPs diagnostic, therapeutic, and targeting applications, immense effort is taken in the development of functional NPs for nanotherapeutic applications [247, 248]; nanotechnology-based DDS has made tremendous changes in dealing with the problems in the treatment of cancer, infection, diabetes, blood disorders, neurodegenerative diseases, and orthopedic problems [249]. This is attained by improving the characteristics of NPs in showing higher drug absorption capability, proper targeting, enhanced stability, reproducibility, long-term use, and controlled release [250]. Most of the nanotherapeutics that are presently in use are administrated mostly by parenteral route but some are taken orally [251]. Research is being carried out on nonparenteral administration of NPs like ocular, pulmonary, nasal, dermal, and vaginal [252, 253]. Safety data of the clinical studies of nanotherapeutic NPs are considered a challenging task. Control over the manufacturing process for making functional NPs appropriate for therapeutic applications is indeed a serious challenge [254]. Quality-by-Design (QbD) is an appropriate method for the systematic analysis and control of nanomedicine [255]. The properties might vary while carrying out the manufacturing process, hence the identification of quality attributes is very important [256]. Quality testing of the actual medicine is not done but it is built on the basis of its therapeutic purposes [257]. Biocompatibility is one of the important parameters in the design of DDS hence the preclinical studies are highly important [258]. Serious efforts have been taken for the development of risk assessment guidelines [259–262]. NPs generally cause many hazards to health because of their ability to penetrate the body [263, 264]. They are capable of destroying cell membranes, organs, and even the DNA [265–267]. So, importance should be given to the nanotoxicological analysis in the fabrication stage. Technologies like dip-pen nanolithography (DPN), nano-topographical studies, and stiffness have importance in regulating the cell and NP interaction [268, 269]. Combining the clinical, toxicological, and biological analysis is required for the efficient analysis of nanotoxicology [270]. Moreover, *in vitro* models are considered inexpensive, reliable, and rapid methods of analyzing nanotoxicology. Regulatory mechanisms are highly desirable to overcome the associated problems and strengthen the area of nanotherapeutics [271–273]. For commercialization of a nanomedicine, a systematic comparison of safety, bioequivalence, efficacy, and quality with the reference medicine is inevitable [274, 275]. For regulatory purposes, the European Medicines Agency has drawn up some guidelines and a framework for biological nanomedicines [276]. Similar to this agency, different governments, pharmaceutical departments, health ministries, the World Health Organization etc. are developing new rules and regulations to ensure the safe usage of nanomedicines [277–279]. By following the guidelines and rules on nanomedicine, multifunctional NPs can be tuned for outstanding nanotheranostic applications.

18.7 Conclusion and Future Perspectives

Multifunctional NPs have made several breakthrough changes and continue to grow incessantly as a vital part in healthcare that aims at nanotheranostic and stimuli-responsive advanced applications. The advantages of these NPs in diagnostic, therapeutic, as well as targeting applications, is due to the efficiency in targeted drug delivery and controlled release over other materials. Owing to the fast growth and subsequent utility, multifunctional NPs are receiving much attention and hence a separate field called nanotheranostics has recently emerged for simultaneous diagnostic and therapeutic application. Stimuli-responsive devices for real-time monitoring are still in their developmental infancy and research has to be carried out to avoid dual modality limitation and to improve the targeting and drug-loading efficiency. Designing, fabrication, and application of multifunctional NPs with enhanced targeting efficiency, biocompatibility, and activity is still a major research problem for the development of highly sophisticated health devices. Nanodevices with less cytotoxic effects are highly essential in the present scenario. Hence, it is mandatory to follow the guidelines, theories, and promote DPN-like techniques. Hitherto, research in the field of multifunctional NPs is dynamic and vibrant and has some distance to go for the full emergence of techniques like nanorobotics. So there remains a vast diversity of materials and methods for young researchers/scientists to explore to contribute toward the development of advanced devices from multifunctional NPs.

References

1 Yang, Z., Song, J., Tang, W. et al. (2019). Stimuli-responsive nanotheranostics for real-time monitoring drug release by photoacoustic imaging. *Theranostics* 9 (2): 526.

2 Bhaskar, S., Tian, F., Stoeger, T. et al. (2010). Multifunctional nanocarriers for diagnostics, drug delivery and targeted treatment across blood-brain barrier: perspectives on tracking and neuroimaging. *Particle and Fibre Toxicology* 7 (1): 3.

3 Boisseau, P. and Loubaton, B. (2011). Nanomedicine, nanotechnology in medicine. *Comptes Rendus Physique* 12 (7): 620–636.

4 Sahoo, S.K., Parveen, S., and Panda, J.J. (2007). The present and future of nanotechnology in human health care. *Nanomedicine: Nanotechnology, Biology and Medicine* 3 (1): 20–31.

5 Schrand, A.M., Rahman, M.F., Hussain, S.M. et al. (2010). Metal-based nanoparticles and their toxicity assessment. *Wiley Interdisciplinary Reviews. Nanomedicine and Nanobiotechnology* 2 (5): 544–568.

6 Bozano, L.D., Kean, B.W., Beinhoff, M. et al. (2005). Organic materials and thin-film structures for cross-point memory cells based on trapping in metallic nanoparticles. *Advanced Functional Materials* 15 (12): 1933–1939.

7 Sharma, R.K., Sharma, S., Dutta, S. et al. (2015). Silica-nanosphere-based organic–inorganic hybrid nanomaterials: synthesis, functionalization and applications in catalysis. *Green Chemistry* 17 (6): 3207–3230.

8 Khan, I., Saeed, K., and Khan, I. (2019). Nanoparticles: properties, applications and toxicities. *Arabian Journal of Chemistry* 12 (7): 908–931.

9 Afonin, K.A., Grabow, W.W., Walker, F.M. et al. (2011). Design and self-assembly of siRNA-functionalized RNA nanoparticles for use in automated nanomedicine. *Nature Protocols* 6 (12): 2022.

10 Hau, S.K., Cheng, Y.-J., Yip, H.-L. et al. (2010). Effect of chemical modification of fullerene-based self-assembled monolayers on the performance of inverted polymer solar cells. *ACS Applied Materials & Interfaces* 2 (7): 1892–1902.

11 Erathodiyil, N. and Ying, J.Y. (2011). Functionalization of inorganic nanoparticles for bioimaging applications. *Accounts of Chemical Research* 44 (10): 925–935.

12 Jafari, S., Derakhshankhah, H., Alaei, L. et al. (2019). Mesoporous silica nanoparticles for therapeutic/diagnostic applications. *Biomedicine & Pharmacotherapy* 109: 1100–1111.

13 Jo, H. and Ban, C. (2016). Aptamer–nanoparticle complexes as powerful diagnostic and therapeutic tools. *Experimental & Molecular Medicine* 48 (5): e230–e230.

14 Ji, T., Zhao, Y., Ding, Y., and Nie, G. (2013). Using functional nanomaterials to target and regulate the tumor microenvironment: diagnostic and therapeutic applications. *Advanced Materials* 25 (26): 3508–3525.

15 Liu, Q., Jin, C., Wang, Y. et al. (2014). Aptamer-conjugated nanomaterials for specific cancer cell recognition and targeted cancer therapy. *NPG Asia Materials* 6 (4): e95–e95.

16 Lee, S., Park, H., Choi, J. et al. (2011). Multifunctional nanoparticles for targeted chemophotothermal treatment of cancer cells. *Angewandte Chemie* 123 (33): 7723–7728.

17 Yu, M.K., Park, J., and Jon, S. (2012). Targeting strategies for multifunctional nanoparticles in cancer imaging and therapy. *Theranostics* 2 (1): 3.

18 Song, C., Tang, C., Xu, W. et al. (2020). Hypoxia-targeting multifunctional nanoparticles for sensitized chemotherapy and phototherapy in head and neck squamous cell carcinoma. *International Journal of Nanomedicine* 15: 347.

19 Koo, H., Huh, M.S., Sun, I.-C. et al. (2011). In vivo targeted delivery of nanoparticles for theranosis. *Accounts of Chemical Research* 44 (10): 1018–1028.

20 Vannucci, L., Falvo, E., Fornara, M. et al. (2012). Selective targeting of melanoma by PEG-masked protein-based multifunctional nanoparticles. *International Journal of Nanomedicine* 7: 1489.

21 Andriola Silva, A.K., Di Corato, R., Gazeau, F. et al. (2012). Magnetophoresis at the nanoscale: tracking the magnetic targeting efficiency of nanovectors. *Nanomedicine* 7 (11): 1713–1727.

22 Chomoucka, J., Drbohlavova, J., Huska, D. et al. (2010). Magnetic nanoparticles and targeted drug delivering. *Pharmacological Research* 62 (2): 144–149.

23 Parveen, S., Misra, R., and Sahoo, S.K. (2012). Nanoparticles: a boon to drug delivery, therapeutics, diagnostics and imaging. *Nanomedicine: Nanotechnology, Biology and Medicine* 8 (2): 147–166.

24 Cheng, L., Wang, X., Gong, F. et al. (2020). 2D nanomaterials for cancer theranostic applications. *Advanced Materials* 32 (13): 1902333.

25 Yew, Y.P., Shameli, K., Miyake, M. et al. (2020). Green biosynthesis of superparamagnetic magnetite Fe_3O_4 nanoparticles and biomedical applications in targeted anticancer drug delivery system: a review. *Arabian Journal of Chemistry* 13 (1): 2287–2308.

26 Chen, X., Song, J., Chen, X., and Yang, H. (2019). X-ray-activated nanosystems for theranostic applications. *Chemical Society Reviews* 48 (11): 3073–3101.

27 Rizzo, L.Y., Theek, B., Storm, G. et al. (2013). Recent progress in nanomedicine: therapeutic, diagnostic and theranostic applications. *Current Opinion in Biotechnology* 24 (6): 1159–1166.

28 Kojima, R., Aubel, D., and Fussenegger, M. (2015). Novel theranostic agents for next-generation personalized medicine: small molecules, nanoparticles, and engineered mammalian cells. *Current Opinion in Chemical Biology* 28: 29–38.

29 Li, X., Kim, J., Yoon, J., and Chen, X. (2017). Cancer-associated, stimuli-driven, turn on theranostics for multimodality imaging and therapy. *Advanced Materials* 29 (23): 1606857.

30 Vats, S., Singh, M., Siraj, S. et al. (2017). Role of nanotechnology in theranostics and personalized medicines. *Journal of Health Research and Review* 4 (1): 1.

31 Yang, K., Feng, L., Shi, X., and Liu, Z. (2013). Nano-graphene in biomedicine: theranostic applications. *Chemical Society Reviews* 42 (2): 530–547.

32 Wong, X.Y., Sena-Torralba, A., Álvarez-Diduk, R. et al. (2020). Nanomaterials for nanotheranostics: tuning their properties according to disease needs. *ACS Nano* 14 (3): 2585–2627.

33 Kim, T.H., Lee, S., and Chen, X. (2013). Nanotheranostics for personalized medicine. *Expert Review of Molecular Diagnostics* 13 (3): 257–269.

34 Mura, S. and Couvreur, P. (2012). Nanotheranostics for personalized medicine. *Advanced Drug Delivery Reviews* 64 (13): 1394–1416.

35 Bai, S., Liao, J., Zhang, B. et al. (2020). Multimodal and multifunctional nanoparticles with platelet targeting ability and phase transition efficiency for the molecular imaging and thrombolysis of coronary microthrombi. *Biomaterials Science* 8 (18): 5047–5060.

36 Pene, F., Courtine, E., Cariou, A., and Mira, J.-P. (2009). Toward theragnostics. *Critical Care Medicine* 37 (1): S50–S58.

37 Xie, H.-G. and Frueh, F.W. (2005). Pharmacogenomics steps toward personalized medicine. *Personalized Medicine* 2 (4): 325–337.

38 Lee, G.-H., Moon, H., Kim, H. et al. (2020). Multifunctional materials for implantable and wearable photonic healthcare devices. *Nature Reviews Materials* 1–17.

39 Shirasaki, Y., Supran, G.J., Bawendi, M.G., and Bulović, V. (2013). Emergence of colloidal quantum-dot light-emitting technologies. *Nature Photonics* 7 (1): 13.

40 Kim, H., Park, Y., Beack, S. et al. (2017). Dual-color-emitting carbon nanodots for multicolor bioimaging and optogenetic control of ion channels. *Advancement of Science* 4 (11): 1700325.

41 Manzeli, S., Ovchinnikov, D., Pasquier, D. et al. (2017). 2D transition metal dichalcogenides. *Nature Reviews Materials* 2 (8): 17033.

42 Haase, M. and Schäfer, H. (2011). Upconverting nanoparticles. *Angewandte Chemie, International Edition* 50 (26): 5808–5829.

43 Liu, K., Xing, R., Zou, Q. et al. (2016). Simple peptide-tuned self-assembly of photosensitizers towards anticancer photodynamic therapy. *Angewandte Chemie* 128 (9): 3088–3091.

44 Shao, J., Xie, H., Huang, H. et al. (2016). Biodegradable black phosphorus-based nanospheres for in vivo photothermal cancer therapy. *Nature Communications* 7 (1): 1–13.

45 Deisseroth, K. (2011). Optogenetics. *Nature Methods* 8 (1): 26–29.

46 Hamblin, M.R., Huang, Y., and Heiskanen, V. (2019). Non-mammalian hosts and photobiomodulation: do all life-forms respond to light? *Photochemistry and Photobiology* 95 (1): 126–139.

47 Pissuwan, D., Gazzana, C., Mongkolsuk, S., and Cortie, M.B. (2020). Single and multiple detections of foodborne pathogens by gold nanoparticle assays. *Wiley Interdisciplinary Reviews. Nanomedicine and Nanobiotechnology* 12 (1): e1584.

48 Hu, X., Zhang, Y., Ding, T. et al. (2020). Multifunctional gold nanoparticles: a novel nanomaterial for various medical applications and biological activities. *Frontiers in Bioengineering and Biotechnology* 8: 990.

49 Ramalingam, V., Raja, S., Sundaramahalingam, S., and Rajaram, R. (2019). Chemical fabrication of graphene oxide nanosheets attenuates biofilm formation of human clinical pathogens. *Bioorganic Chemistry* 83: 326–335.

50 Slocik, J.M., Stone, M.O., and Naik, R.R. (2005). Synthesis of gold nanoparticles using multifunctional peptides. *Small* 1 (11): 1048–1052.

51 Muthu, M.S. and Singh, S. (2009). Targeted nanomedicines: effective treatment modalities for cancer, AIDS and brain disorders. *Nanomedicine* 4 (1): https://doi.org/10.2217/17435889.4.1.105.

52 Muthu, M.S. and Feng, S.-S. (2013). *Theranostic Liposomes for Cancer Diagnosis and Treatment: Current Development and Pre-Clinical Success*. Taylor & Francis.

53 Peng, H., Liu, X., Wang, G. et al. (2015). Polymeric multifunctional nanomaterials for theranostics. *Journal of Materials Chemistry B* 3 (34): 6856–6870.

54 Xia, X. and Xia, Y. (2014). Gold nanocages as multifunctional materials for nanomedicine. *Frontiers of Physics* 9 (3): 378–384.

55 Gong, W., Hu, Z., Liang, Y. et al. (2018). Graphene-based multifunctional nanomaterials in cancer detection and therapeutics. *Journal of Nanoscience and Nanotechnology* 18 (8): 5155–5170.

56 Zümreoglu-Karan, B. and Ay, A. (2012). Layered double hydroxides – multifunctional nanomaterials. *Chemical Papers* 66 (1): 1–10.

57 Abraham, E., Weber, D.E., Sharon, S. et al. (2017). Multifunctional cellulosic scaffolds from modified cellulose nanocrystals. *ACS Applied Materials & Interfaces* 9 (3): 2010–2015.

58 Menaa, F., Abdelghani, A., and Menaa, B. (2015). Graphene nanomaterials as biocompatible and conductive scaffolds for stem cells: impact for tissue engineering and regenerative medicine. *Journal of Tissue Engineering and Regenerative Medicine* 9 (12): 1321–1338.

59 Reddi, A.H., Becerra, J., and Andrades, J.A. (2011). Nanomaterials and hydrogel scaffolds for articular cartilage regeneration. *Tissue Engineering. Part B, Reviews* 17 (5): 301–305.

60 Hou, Y., Wang, W., and Bártolo, P. (2020). Investigating the effect of carbon nanomaterials reinforcing poly (ε-Caprolactone) printed scaffolds for bone repair applications. *International Journal of Bioprinting* 6 (2): 266.

61 Gandolfi, M.G., Gardin, C., Zamparini, F. et al. (2020). Mineral-doped poly (L-lactide) acid scaffolds enriched with exosomes improve osteogenic commitment of human adipose-derived mesenchymal stem cells. *Nanomaterials* 10 (3): 432.

62 Ye, G., Bao, F., Zhang, X. et al. (2020). Nanomaterial-based scaffolds for bone tissue engineering and regeneration. *Nanomedicine* 15 (20): 1995–2017.

63 Suzdalev, I.P. (2009). Multifunctional nanomaterials. *Russian Chemical Reviews* 78 (3): 249.

64 Srinivasan, M., Rajabi, M., and Mousa, S.A. (2015). Multifunctional nanomaterials and their applications in drug delivery and cancer therapy. *Nanomaterials* 5 (4): 1690–1703.

65 Nam, N.H. and Luong, N.H. (2019). Nanoparticles: synthesis and applications. In: *Materials for Biomedical Engineering* (ed. V. Grumezescu and A. Grumezescu), 211–240. Elsevier.

66 Sharma, R., Agrawal, U., Mody, N. et al. (2016). Engineered nanoparticles as a precise delivery system in cancer therapeutics. In: *Engineering of Nanobiomaterials* (ed. A.M. Grumezescu), 397–427. Elsevier.

67 Vasile, C. (2019). Polymeric nanomaterials: recent developments, properties and medical applications. In: *Polymeric Nanomaterials in Nanotherapeutics* (ed. C. Vasile), 1–66. Elsevier.

68 Ge, R., Lin, M., Li, X. et al. (2017). Cu_2+ - loaded polydopamine nanoparticles for magnetic resonance imaging-guided pH-and near-infrared-light-stimulated thermochemotherapy. *ACS Applied Materials & Interfaces* 9 (23): 19706–19716.

69 Jabeen, F., Najam-ul-Haq, M., Javeed, R. et al. (2014). Au-nanomaterials as a superior choice for near-infrared photothermal therapy. *Molecules* 19 (12): 20580–20593.

70 Sharifi, S., Behzadi, S., Laurent, S. et al. (2012). Toxicity of nanomaterials. *Chemical Society Reviews* 41 (6): 2323–2343.

71 Li, Z., Hu, Y., Howard, K.A. et al. (2016). Multifunctional bismuth selenide nanocomposites for antitumor thermo-chemotherapy and imaging. *ACS Nano* 10 (1): 984–997.

72 Zhang, H., Ren, T., Ji, Y. et al. (2015). Selective modification of halloysite nanotubes with 1-pyrenylboronic acid: a novel fluorescence probe with highly selective and sensitive response to hyperoxide. *ACS Applied Materials & Interfaces* 7 (42): 23805–23811.

73 Tao, K., Liu, S., Wang, L. et al. (2020). Targeted multifunctional nanomaterials with MRI, chemotherapy and photothermal therapy for the diagnosis and treatment of bladder cancer. *Biomaterials Science* 8 (1): 342–352.

74 Zhou, X.-Z. and Zhou, X.-Y. (2012). The killing effect of NY-ESO-1-sensitzed dendritic cells-induced cellular toxic lymphocyte on human retinoblastoma cell. *Chinese Journal of Experimental Ophthalmology* 30 (7): 586–591.

75 Wang, X., Gao, F., and Zhang, X. (2017). Initiator-loaded gold nanocages as a light-induced free-radical generator for cancer therapy. *Angewandte Chemie* 129 (31): 9157–9161.

76 Li, W., Cai, X., Kim, C. et al. (2011). Gold nanocages covered with thermally-responsive polymers for controlled release by high-intensity focused ultrasound. *Nanoscale* 3 (4): 1724–1730.

77 Wang, M., Yang, Q., Li, M. et al. (2020). Multifunctional nanoparticles for multimodal imaging-guided low-intensity focused ultrasound/immunosynergistic retinoblastoma therapy. *ACS Applied Materials & Interfaces* 12 (5): 5642–5657.

78 Chatterjee, D.K., Gnanasammandhan, M.K., and Zhang, Y. (2010). Small upconverting fluorescent nanoparticles for biomedical applications. *Small* 6 (24): 2781–2795.

79 Takekawa, J.Y., Prosser, D.J., Newman, S.H. et al. (2010). Victims and vectors: highly pathogenic avian influenza H5N1 and the ecology of wild birds. *Avian Biology Research* 3 (2): 51–73.

80 Cheng, L., Yang, K., Zhang, S. et al. (2010). Highly-sensitive multiplexed in vivo imaging using PEGylated upconversion nanoparticles. *Nano Research* 3 (10): 722–732.

81 Zhang, H., Li, Y., Ivanov, I.A. et al. (2010). Plasmonic modulation of the upconversion fluorescence in NaYF4: Yb/Tm hexaplate nanocrystals using gold nanoparticles or nanoshells. *Angewandte Chemie, International Edition* 49 (16): 2865–2868.

82 Wang, C., Cheng, L., and Liu, Z. (2011). Drug delivery with upconversion nanoparticles for multi-functional targeted cancer cell imaging and therapy. *Biomaterials* 32 (4): 1110–1120.

83 Qian, H.S., Guo, H.C., Ho, P.C. et al. (2009). Mesoporous-silica-coated up-conversion fluorescent nanoparticles for photodynamic therapy. *Small* 5 (20): 2285–2290.

84 Cheng, L., Yang, K., Li, Y. et al. (2011). Facile preparation of multifunctional upconversion nanoprobes for multimodal imaging and dual-targeted photothermal therapy. *Angewandte Chemie* 123 (32): 7523–7528.

85 Zhang, S., Sun, C., Zeng, J. et al. (2016). Ambient aqueous synthesis of ultrasmall PEGylated $Cu_{2-x}Se$ nanoparticles as a multifunctional theranostic agent for multimodal imaging guided photothermal therapy of cancer. *Advanced Materials* 28 (40): 8927–8936.

86 Ozdemir, V., Williams-Jones, B., Glatt, S.J. et al. (2006). Shifting emphasis from pharmacogenomics to theragnostics. *Nature Biotechnology* 24 (8): 942–946.

87 Kim, K., Kim, J.H., Park, H. et al. (2010). Tumor-homing multifunctional nanoparticles for cancer theragnosis: simultaneous diagnosis, drug delivery, and therapeutic monitoring. *Journal of Controlled Release* 146 (2): 219–227.

88 Jackson, T.C., Patani, B.O., and Ekpa, D.E. (2017). Nanotechnology in diagnosis: a review. *Advances in Nanoparticles.* 6 (03): 93.

89 Moffatt, S. (2016). Nanodiagnostics: a revolution in biomedical nanotechnology. *MedCrave Online Journal of Proteomics & Bioinformatics* 3 (2): 80.

90 Baptista, P.V. (2014). Nanodiagnostics: leaving the research lab to enter the clinics? *Diagnosis* 1 (4): 305–309.

91 Jain, K.K. (2003). Nanodiagnostics: application of nanotechnology in molecular diagnostics. *Expert Review of Molecular Diagnostics* 3 (2): 153–161.

92 Curtis, A. and Wilkinson, C. (2001). Nantotechniques and approaches in biotechnology. *Trends in Biotechnology* 19 (3): 97–101.

93 Alharbi, K.K. and Al-Sheikh, Y.A. (2014). Role and implications of nanodiagnostics in the changing trends of clinical diagnosis. *Saudi Journal of Biological Sciences* 21 (2): 109–117.

94 West, J.L. and Halas, N.J. (2003). Engineered nanomaterials for biophotonics applications: improving sensing, imaging, and therapeutics. *Annual Review of Biomedical Engineering* 5 (1): 285–292.

95 Jabir, N.R., Tabrez, S., Ashraf, G.M. et al. (2012). Nanotechnology-based approaches in anticancer research. *International Journal of Nanomedicine* 7: 4391.

96 Reimhult, E. and Höök, F. (2015). Design of surface modifications for nanoscale sensor applications. *Sensors* 15 (1): 1635–1675.

97 Lee, J., Dak, P., Lee, Y. et al. (2014). Two-dimensional layered MoS_2 biosensors enable highly sensitive detection of biomolecules. *Scientific Reports* 4: 7352.

98 Kong, R.-M., Ding, L., Wang, Z. et al. (2015). A novel aptamer-functionalized MoS_2 nanosheet fluorescent biosensor for sensitive detection of prostate specific antigen. *Analytical and Bioanalytical Chemistry* 407 (2): 369–377.

99 Cho, E.C., Glaus, C., Chen, J. et al. (2010). Inorganic nanoparticle-based contrast agents for molecular imaging. *Trends in Molecular Medicine* 16 (12): 561–573.

100 Cui, D., Tian, F., Coyer, S.R. et al. (2007). Effects of antisense-Myc-conjugated single-walled carbon nanotubes on HL-60Cells. *Journal of Nanoscience and Nanotechnology* 7 (4–5): 1639–1646.

101 Meng, N., Ren, X., Santagiuliana, G. et al. (2019). Ultrahigh β-phase content poly (vinylidene fluoride) with relaxor-like ferroelectricity for high energy density capacitors. *Nature Communications* 10 (1): 1–9.

102 Jatzkewitz, H. (1955). An ein kolloidales Blutplasma-Ersatzmittel (Polyvinylpyrrolidon) gebundenes Peptamin (Glycyl-L-leucyl-mezcalin) als neuartige Depotform für biologisch aktive primäre Amine (Mezcalin). *Zeitschrift für Naturforschung Part B* 10 (1): 27–31.

103 Brem, H., Piantadosi, S., Burger, P.C. et al. (1995). Placebo-controlled trial of safety and efficacy of intraoperative controlled delivery by biodegradable polymers of chemotherapy for recurrent gliomas. *Lancet* 345 (8956): 1008–1012.

104 Xue, X., Wang, F., and Liu, X. (2011). Emerging functional nanomaterials for therapeutics. *Journal of Materials Chemistry* 21 (35): 13107–13127.

105 Kreuter, J. (2014). *Colloidal Drug Delivery Systems*, vol. 66. CRC Press.

106 Duro-Castano, A., Conejos-Sánchez, I., and Vicent, M.J. (2014). Peptide-based polymer therapeutics. *Polymers (Basel)* 6 (2): 515–551.

107 Wagner, V., Dullaart, A., Bock, A.-K., and Zweck, A. (2006). The emerging nanomedicine landscape. *Nature Biotechnology* 24 (10): 1211–1217.

108 Zhang, L., Gu, F.X., Chan, J.M. et al. (2008). Nanoparticles in medicine: therapeutic applications and developments. *Clinical Pharmacology and Therapeutics* 83 (5): 761–769.

109 Dugan, L.L., Turetsky, D.M., Du, C. et al. (1997). Carboxyfullerenes as neuroprotective agents. *Proceedings of the National Academy of Sciences* 94 (17): 9434–9439.

110 Palazzolo, S., Hadla, M., Spena, C.R. et al. (2019). Proof-of-concept multistage biomimetic liposomal DNA origami nanosystem for the remote loading of doxorubicin. *ACS Medicinal Chemistry Letters* 10 (4): 517–521.

111 Hao, R., Xing, R., Xu, Z. et al. (2010). Synthesis, functionalization, and biomedical applications of multifunctional magnetic nanoparticles. *Advanced Materials* 22 (25): 2729–2742.

112 Sun, Y., Lu, Y., Yin, L., and Liu, Z. (2020). The roles of nanoparticles in stem cell-based therapy for cardiovascular disease. *Frontiers in Bioengineering and Biotechnology* 8: 947.

113 Yao, J., Feng, J., and Chen, J. (2016). External-stimuli responsive systems for cancer theranostic. *Asian Journal of Pharmaceutical Sciences* 11 (5): 585–595.

114 McCune, J.A., Mommer, S., Parkins, C.C., and Scherman, O.A. (2020). Design principles for aqueous interactive materials: lessons from small molecules and stimuli-responsive systems. *Advanced Materials* 32 (20): 1906890.

115 Karimi, M., Ghasemi, A., Zangabad, P.S. et al. (2016). Smart micro/nanoparticles in stimulus-responsive drug/gene delivery systems. *Chemical Society Reviews* 45 (5): 1457–1501.

116 Soppimath, K.S., Aminabhavi, T.M., Dave, A.M. et al. (2002). Stimulus-responsive "smart" hydrogels as novel drug delivery systems. *Drug Development and Industrial Pharmacy* 28 (8): 957–974.

117 Hajebi, S., Rabiee, N., Bagherzadeh, M. et al. (2019). Stimulus-responsive polymeric nanogels as smart drug delivery systems. *Acta Biomaterialia* 92: 1–18.

118 Yang, H.Y., Li, Y., and Lee, D.S. (2018). Multifunctional and stimuli-responsive magnetic nanoparticle-based delivery systems for biomedical applications. *Advances in Therapy* 1 (2): 1800011.

119 Dai, J., Wu, X., Ding, S. et al. (2020). Aggregation-induced emission photosensitizers: from molecular design to photodynamic therapy: miniperspective. *Journal of Medicinal Chemistry* 63 (5): 1996–2012.

120 Zhao, Z., Han, Y., Lin, C. et al. (2012). Multifunctional core–shell upconverting nanoparticles for imaging and photodynamic therapy of liver cancer cells. *Chemistry - An Asian Journal* 7 (4): 830–837.

121 Jang, B., Park, J.-Y., Tung, C.-H. et al. (2011). Gold nanorod–photosensitizer complex for near-infrared fluorescence imaging and photodynamic/photothermal therapy in vivo. *ACS Nano* 5 (2): 1086–1094.

122 Lin, J., Wang, S., Huang, P. et al. (2013). Photosensitizer-loaded gold vesicles with strong plasmonic coupling effect for imaging-guided photothermal/photodynamic therapy. *ACS Nano* 7 (6): 5320–5329.

123 Yang, G., Phua, S.Z.F., Bindra, A.K., and Zhao, Y. (2019). Degradability and clearance of inorganic nanoparticles for biomedical applications. *Advanced Materials* 31 (10): 1805730.

124 Wang, S., Zhao, J., Yang, H. et al. (2017). Bottom-up synthesis of WS_2 nanosheets with synchronous surface modification for imaging guided tumor regression. *Acta Biomaterialia* 58: 442–454.

125 Saxena, V., Sadoqi, M., and Shao, J. (2003). Degradation kinetics of indocyanine green in aqueous solution. *Journal of Pharmaceutical Sciences* 92 (10): 2090–2097.

126 Guo, W., Guo, C., Zheng, N. et al. (2017). $CsxWO_3$ nanorods coated with polyelectrolyte multilayers as a multifunctional nanomaterial for bimodal imaging-guided photothermal/photodynamic cancer treatment. *Advanced Materials* 29 (4): 1604157.

127 Zhang, C., Ren, J., Hua, J. et al. (2018). Multifunctional Bi_2WO_6 nanoparticles for CT-guided photothermal and oxygen-free photodynamic therapy. *ACS Applied Materials & Interfaces* 10 (1): 1132–1146.

128 Zhang, H., Liu, K., Li, S. et al. (2018). Self-assembled minimalist multifunctional theranostic nanoplatform for magnetic resonance imaging-guided tumor photodynamic therapy. *ACS Nano* 12 (8): 8266–8276.

129 Hou, H., Huang, X., Wei, G. et al. (2019). Fenton reaction-assisted photodynamic therapy for cancer with multifunctional magnetic nanoparticles. *ACS Applied Materials & Interfaces* 11 (33): 29579–29592.

130 Li, S., Zhang, Y., He, X.-W. et al. (2020). Multifunctional mesoporous silica nanoplatform based on silicon nanoparticles for targeted two-photon-excited fluorescence imaging-guided chemo/photodynamic synergetic therapy in vitro. *Talanta* 209: 120552.

131 Wang, H., Pan, X., Wang, X. et al. (2020). Degradable carbon–silica nanocomposite with immunoadjuvant property for dual-modality photothermal/photodynamic therapy. *ACS Nano* 14 (3): 2847–2859.

132 Bao, C., Liu, B., Li, B. et al. (2020). Enhanced transport of shape and rigidity-tuned α-lactalbumin nanotubes across intestinal mucus and cellular barriers. *Nano Letters* 20 (2): 1352–1361.

133 Song, J., Yuan, C., Jiao, T. et al. (2020). Multifunctional antimicrobial biometallohydrogels based on amino acid coordinated self-assembly. *Small* 16 (8): 1907309.

134 Mao, C., Xiang, Y., Liu, X. et al. (2017). Photo-inspired antibacterial activity and wound healing acceleration by hydrogel embedded with ag/ag@ AgCl/ZnO nanostructures. *ACS Nano* 11 (9): 9010–9021.

135 Vincent, C., Kiran, M.D., Jayan, J.S., and Saritha, A. (2020). Determination of drug loading efficiency of nano capsules formed by the O/W emulsification of block copolymer. *Materials Today: Proceedings* 26 (2): 720–723.

136 Jayan, J.S., Saritha, A., Deeraj, B.D.S., and Joseph, K. (2020). Triblock copolymer grafted graphene oxide as nanofiller for toughening of epoxy resin. *Materials Chemistry and Physics* 248: 122930.

137 Zhang, Q., Wu, J., Wang, J. et al. (2020). A neutrophil-inspired supramolecular nanogel for magnetocaloric–enzymatic tandem therapy. *Angewandte Chemie* 132 (9): 3761–3767.

138 Wu, H., Liu, L., Song, L. et al. (2019). Enhanced tumor synergistic therapy by injectable magnetic hydrogel mediated generation of hyperthermia and highly toxic reactive oxygen species. *ACS Nano* 13 (12): 14013–14023.

139 Wang, C., Zhao, N., Huang, Y. et al. (2020). Coordination of injectable self-healing hydrogel with Mn-Zn ferrite@ mesoporous silica nanospheres for tumor MR imaging and efficient synergistic magnetothermal-chemo-chemodynamic therapy. *Chemical Engineering Journal* 401: 126100.

140 Cai, L., Hu, C., Liu, S. et al. (2020). A covalent organic framework-based multifunctional therapeutic platform for enhanced photodynamic therapy via catalytic cascade reactions. *Science China Materials* 64 (2): 488–497.

141 Liang, J., Jin, X., Chen, B. et al. (2020). Doxorubicin-loaded pH-responsive nanoparticles coated with chlorin e6 for drug delivery and synergetic chemo-photodynamic therapy. *Nanotechnology* 31 (19): 195103.

142 Wang, Y., Yang, M., Qian, J. et al. (2019). Sequentially self-assembled polysaccharide-based nanocomplexes for combined chemotherapy and photodynamic therapy of breast cancer. *Carbohydrate Polymers* 203: 203–213.

143 Li, J., Liang, H., Liu, J., and Wang, Z. (2018). Poly (amidoamine)(PAMAM) dendrimer mediated delivery of drug and pDNA/siRNA for cancer therapy. *International Journal of Pharmaceutics* 546 (1–2): 215–225.

144 Yang, Y.-Y., Wang, X., Hu, Y. et al. (2014). Bioreducible POSS-cored star-shaped polycation for efficient gene delivery. *ACS Applied Materials & Interfaces* 6 (2): 1044–1052.

145 Zeng, L., Xiang, L., Ren, W. et al. (2013). Multifunctional photosensitizer-conjugated core–shell Fe_3O_4@ $NaYF_4$: Yb/Er nanocomplexes and their applications in T_2-weighted magnetic resonance/upconversion luminescence imaging and photodynamic therapy of cancer cells. *RSC Advances* 3 (33): 13915–13925.

146 Cho, N.-H., Cheong, T.-C., Min, J.H. et al. (2011). A multifunctional core–shell nanoparticle for dendritic cell-based cancer immunotherapy. *Nature Nanotechnology* 6 (10): 675–682.

147 Chung, C.-H., Lu, K.-Y., Lee, W.-C. et al. (2020). Fucoidan-based, tumor-activated nanoplatform for overcoming hypoxia and enhancing photodynamic therapy and antitumor immunity. *Biomaterials* 257: 120227.

148 Ye, Z., Hua, D., Rao, Y. et al. (2020). Targeted photodynamic therapy with a novel photosensitizer cercosporin encapsulated multifunctional copolymer. *Colloids and Surfaces A: Physicochemical and Engineering Aspects* 585: 124136.

149 Ha, S.Y.Y., Zhou, Y., Fong, W.-P., and Ng, D.K.P. (2020). Multifunctional molecular therapeutic agent for targeted and controlled dual chemo-and photodynamic therapy. *Journal of Medicinal Chemistry* 63 (15): 8512–8523.

150 Jin, Y., Wang, H., Li, X. et al. (2020). A multifunctional DNA polymer assisted upconversion therapeutic nanoplatform for enhanced photodynamic therapy. *ACS Applied Materials & Interfaces*.

151 Du, W., Liu, T., Xue, F. et al. (2020). Confined nanoparticles growth within hollow mesoporous nanoreactors for highly efficient MRI-guided photodynamic therapy. *Chemical Engineering Journal* 379: 122251.

152 Yuan, Z., Lin, C., He, Y. et al. (2020). Near-infrared light-triggered nitric-oxide-enhanced photodynamic therapy and low-temperature photothermal therapy for biofilm elimination. *ACS Nano* 14 (3): 3546–3562.

153 Pan, S., Pei, L., Zhang, A. et al. (2020). Passion fruit-like exosome-PMA/Au-BSA@ Ce6 nanovehicles for real-time fluorescence imaging and enhanced targeted photodynamic therapy with deep penetration and superior retention behavior in tumor. *Biomaterials* 230: 119606.

154 Zeng, W., Zhang, H., Deng, Y. et al. (2020). Dual-response oxygen-generating MnO_2 nanoparticles with polydopamine modification for combined photothermal-photodynamic therapy. *Chemical Engineering Journal* 389: 124494.

155 Sheng, J., Zhang, L., Deng, L. et al. (2020). Fabrication of dopamine enveloped WO_{3-x} quantum dots as single-NIR laser activated photonic nanodrug for synergistic photothermal/photodynamic therapy against cancer. *Chemical Engineering Journal* 383: 123071.

156 Tomitaka, A., Arami, H., Ahmadivand, A. et al. (2020). Magneto-plasmonic nanostars for image-guided and NIR-triggered drug delivery. *Scientific Reports* 10 (1): 1–10.

157 Yang, J., Dai, D., Lou, X. et al. (2020). Supramolecular nanomaterials based on hollow mesoporous drug carriers and macrocycle-capped CuS nanogates for synergistic chemo-photothermal therapy. *Theranostics* 10 (2): 615.

158 Xu, J., Cheng, X., Chen, F. et al. (2020). Fabrication of multifunctional polydopamine-coated gold nanobones for PA/CT imaging and enhanced synergistic chemo-photothermal therapy. *Journal of Materials Science and Technology* 63: 97–105.

159 Wang, Z., Jia, T., Sun, Q. et al. (2020). Construction of bi/phthalocyanine manganese nanocomposite for trimodal imaging directed photodynamic and photothermal therapy mediated by 808 nm light. *Biomaterials* 228: 119569.

160 Li, F., Li, T., Zhi, D. et al. (2020). Novel ultrasmall multifunctional nanodots for dual-modal MR/NIR-II imaging-guided photothermal therapy. *Biomaterials* 256: 120219.

161 Bao, Z., Liu, X., Liu, Y. et al. (2016). Near-infrared light-responsive inorganic nanomaterials for photothermal therapy. *Asian Journal of Pharmaceutical Sciences* 11 (3): 349–364.

162 Ma, L., Zhou, Y., Zhang, Z. et al. (2020). Multifunctional bioactive Nd-Ca-Si glasses for fluorescence thermometry, photothermal therapy, and burn tissue repair. *Science Advances* 6 (32): eabb1311.

163 Peng, S., Ouyang, B., Men, Y. et al. (2020). Biodegradable zwitterionic polymer membrane coating endowing nanoparticles with ultra-long circulation and enhanced tumor photothermal therapy. *Biomaterials* 231: 119680.

164 Liu, C., Zhang, L., Chen, X. et al. (2020). Biomolecules-assisted synthesis of degradable bismuth nanoparticles for dual-modal imaging-guided chemo-photothermal therapy. *Chemical Engineering Journal* 382: 122720.

165 Cai, Y., Ni, D., Cheng, W. et al. (2020). Enzyme-triggered disassembly of perylene monoimide-based nanoclusters for activatable and deep photodynamic therapy. *Angewandte Chemie* 132 (33): 14118–14122.

166 Liang, S., Sun, C., Yang, P. et al. (2020). Core-shell structured upconversion nanocrystal-dendrimer composite as a carrier for mitochondria targeting and catalase enhanced anti-cancer photodynamic therapy. *Biomaterials* 240: 119850.

167 Pan, Y., Luo, Z., Wang, X. et al. (2020). A versatile and multifunctional metal–organic framework nanocomposite toward chemo-photodynamic therapy. *Dalton Transactions* 49 (16): 5291–5301.

168 Kim, Y., Na, H., Kim, S. et al. (2015). One-pot synthesis of multifunctional Au@ graphene oxide nanocolloid Core@ Shell nanoparticles for Raman bioimaging, photothermal, and photodynamic therapy. *Small* 11 (21): 2527–2535.

169 Shi, L., Hu, F., Duan, Y. et al. (2020). Hybrid nanospheres to overcome hypoxia and intrinsic oxidative resistance for enhanced photodynamic therapy. *ACS Nano* 14 (2): 2183–2190.

170 Ruel-Gariepy, E. and Leroux, J.-C. (2004). In situ-forming hydrogels–review of temperature-sensitive systems. *European Journal of Pharmaceutics and Biopharmaceutics* 58 (2): 409–426.

171 Qu, Y., Chu, B.Y., Peng, J.R. et al. (2015). A biodegradable thermo-responsive hybrid hydrogel: therapeutic applications in preventing the post-operative recurrence of breast cancer. *NPG Asia Materials* 7 (8): e207–e207.

172 Shi, K., Wang, Y.-L., Qu, Y. et al. (2016). Synthesis, characterization, and application of reversible PDLLA-PEG-PDLLA copolymer thermogels in vitro and in vivo. *Scientific Reports* 6: 19077.

173 Jia, Y.P., Shi, K., Yang, F. et al. (2020). Multifunctional nanoparticle loaded injectable thermoresponsive hydrogel as NIR controlled release platform for local photothermal immunotherapy to prevent breast cancer postoperative recurrence and metastases. *Advanced Functional Materials* 30 (25): 2001059.

174 Pandey, N., Menon, J.U., Takahashi, M. et al. (2020). Thermo-responsive fluorescent nanoparticles for multimodal imaging and treatment of cancers. *Nano* 4 (1): 1.

175 Asghar, K., Qasim, M., Dharmapuri, G., and Das, D.D. (2020). Thermoresponsive polymer gated and superparamagnetic nanoparticles embedded hollow mesoporous silica nanoparticles as smart multifunctional nanocarrier for targeted and controlled delivery of doxorubicin. *Nanotechnology* 31 (45): 455604.

176 Cai, Z., Zhang, Y., He, Z. et al. (2020). NIR-triggered chemo-photothermal therapy by thermosensitive gold nanostar@ mesoporous silica@ liposome-composited drug delivery systems. *ACS Applied Bio Materials* 3 (8): 5322–5330.

177 Kang, T., Li, F., Baik, S. et al. (2017). Surface design of magnetic nanoparticles for stimuli-responsive cancer imaging and therapy. *Biomaterials* 136: 98–114.

178 Aghanejad, A., Babamiri, H., Adibkia, K. et al. (2018). Mucin-1 aptamer-armed superparamagnetic iron oxide nanoparticles for targeted delivery of doxorubicin to breast cancer cells. *BioImpacts: BI* 8 (2): 117.

179 Zhao, J., Zhao, X., Guo, B., and Ma, P.X. (2014). Multifunctional interpenetrating polymer network hydrogels based on methacrylated alginate for the delivery of small molecule drugs and sustained release of protein. *Biomacromolecules* 15 (9): 3246–3252.

180 Fathi, M., Majidi, S., Zangabad, P.S. et al. (2018). Chitosan-based multifunctional nanomedicines and theranostics for targeted therapy of cancer. *Medicinal Research Reviews* 38 (6): 2110–2136.

181 Fathi, M., Barar, J., Erfan-Niya, H., and Omidi, Y. (2020). Methotrexate-conjugated chitosan-grafted pH-and thermo-responsive magnetic nanoparticles for targeted therapy of ovarian cancer. *International Journal of Biological Macromolecules* 154: 1175–1184.

182 Anirudhan, T.S. and Christa, J. (2020). Temperature and pH sensitive multi-functional magnetic nanocomposite for the controlled delivery of 5-fluorouracil, an anticancer drug. *Journal of Drug Delivery Science and Technology* 55: 101476.

183 Pourjavadi, A., Kohestanian, M., and Streb, C. (2020). pH and thermal dual-responsive poly (NIPAM-co-GMA)-coated magnetic nanoparticles via surface-initiated RAFT polymerization for controlled drug delivery. *Materials Science and Engineering: C* 108: 110418.

184 Chung, C.K., Fransen, M.F., van der Maaden, K. et al. (2020). Thermosensitive hydrogels as sustained drug delivery system for CTLA-4 checkpoint blocking antibodies. *Journal of Controlled Release* 323: 1–11.

185 Shi, Z., Yang, C., Li, R., and Ruan, L. (2020). Microwave thermal-triggered drug delivery using thermosensitive peptide-coated core–shell mesoporous silica nanoparticles. *Journal of Materials Science* 55 (14): 6118–6129.

186 Zhou, H., Gong, Y., Liu, Y. et al. (2020). Intelligently thermoresponsive flower-like hollow nano-ruthenium system for sustained release of nerve growth factor to inhibit hyperphosphorylation of tau and neuronal damage for the treatment of Alzheimer's disease. *Biomaterials* 237: 119822.

187 Zhu, J., Zhang, H., Chen, K. et al. (2020). Peptidic monodisperse PEG "comb" as multifunctional "add-on" module for imaging-traceable and thermo-responsive theranostics. *Advanced Healthcare Materials* 9 (3): 1901331.

188 Wang, C., Zhao, N., and Yuan, W. (2020). NIR/thermoresponsive injectable self-healing hydrogels containing Polydopamine nanoparticles for efficient synergistic cancer Thermochemotherapy. *ACS Applied Materials & Interfaces* 12 (8): 9118–9131.

189 Mura, S., Nicolas, J., and Couvreur, P. (2013). Stimuli-responsive nanocarriers for drug delivery. *Nature Materials* 12 (11): 991–1003.

190 Deelman, L.E., Declèves, A.-E., Rychak, J.J., and Sharma, K. (2010). Targeted renal therapies through microbubbles and ultrasound. *Advanced Drug Delivery Reviews* 62 (14): 1369–1377.

191 Xing, L., Shi, Q., Zheng, K. et al. (2016). Ultrasound-mediated microbubble destruction (UMMD) facilitates the delivery of CA19-9 targeted and paclitaxel loaded mPEG-PLGA-PLL nanoparticles in pancreatic cancer. *Theranostics* 6 (10): 1573.

192 Lin, L., Fan, Y., Gao, F. et al. (2018). UTMD-promoted co-delivery of gemcitabine and miR-21 inhibitor by dendrimer-entrapped gold nanoparticles for pancreatic cancer therapy. *Theranostics* 8 (7): 1923.

193 Lentacker, I., De Cock, I., Deckers, R. et al. (2014). Understanding ultrasound induced sonoporation: definitions and underlying mechanisms. *Advanced Drug Delivery Reviews* 72: 49–64.

194 Xu, F., Zhu, J., Lin, L. et al. (2020). Multifunctional PVCL nanogels with redox-responsiveness enable enhanced MR imaging and ultrasound-promoted tumor chemotherapy. *Theranostics* 10 (10): 4349.

195 Meng, D., Guo, L., Shi, D. et al. (2019). Charge-conversion and ultrasound-responsive O-carboxymethyl chitosan nanodroplets for controlled drug delivery. *Nanomedicine* 14 (19): 2549–2565.

196 Xuan, J., Pelletier, M., Xia, H., and Zhao, Y. (2011). Ultrasound-induced disruption of amphiphilic block copolymer micelles. *Macromolecular Chemistry and Physics* 212 (5): 498–506.

197 Kim, H., Matsuda, H., Zhou, H., and Honma, I. (2006). Ultrasound-triggered smart drug release from a poly (dimethylsiloxane)–mesoporous silica composite. *Advanced Materials* 18 (23): 3083–3088.

198 Böhmer, M.R., Klibanov, A.L., Tiemann, K. et al. (2009). Ultrasound triggered image-guided drug delivery. *European Journal of Radiology* 70 (2): 242–253.

199 Wang, J., Pelletier, M., Zhang, H. et al. (2009). High-frequency ultrasound-responsive block copolymer micelle. *Langmuir* 25 (22): 13201–13205.

200 Li, X., Xie, C., Xia, H., and Wang, Z. (2018). pH and ultrasound dual-responsive polydopamine-coated mesoporous silica nanoparticles for controlled drug delivery. *Langmuir* 34 (34): 9974–9981.

201 Yao, J., Kaberniuk, A.A., Li, L. et al. (2016). Multiscale photoacoustic tomography using reversibly switchable bacterial phytochrome as a near-infrared photochromic probe. *Nature Methods* 13 (1): 67–73.

202 Wu, M., Chen, W., Chen, Y. et al. (2018). Focused ultrasound-augmented delivery of biodegradable multifunctional nanoplatforms for imaging-guided brain tumor treatment. *Advancement of Science* 5 (4): 1700474.

203 Yang, F., Li, L., Li, Y. et al. (2008). Superparamagnetic nanoparticle-inclusion microbubbles for ultrasound contrast agents. *Physics in Medicine and Biology* 53 (21): 6129.

204 Yang, F., Li, Y., Chen, Z. et al. (2009). Superparamagnetic iron oxide nanoparticle-embedded encapsulated microbubbles as dual contrast agents of magnetic resonance and ultrasound imaging. *Biomaterials* 30 (23–24): 3882–3890.

205 Sun, Y., Zheng, Y., Ran, H. et al. (2012). Superparamagnetic PLGA-iron oxide microcapsules for dual-modality US/MR imaging and high intensity focused US breast cancer ablation. *Biomaterials* 33 (24): 5854–5864.

206 Ao, M., Wang, Z., Ran, H. et al. (2010). Gd-DTPA-loaded PLGA microbubbles as both ultrasound contrast agent and MRI contrast agent–a feasibility research. *Journal of Biomedical Materials Research Part B: Applied Biomaterials* 93 (2): 551–556.

207 Yang, F., Hu, S., Zhang, Y. et al. (2012). A hydrogen peroxide-responsive O_2 Nanogenerator for ultrasound and magnetic-resonance dual modality imaging. *Advanced Materials* 24 (38): 5205–5211.

208 Nayak, S. and Lyon, L.A. (2005). Soft nanotechnology with soft nanoparticles. *Angewandte Chemie, International Edition* 44 (47): 7686–7708.

209 Sasaki, Y. and Akiyoshi, K. (2010). Nanogel engineering for new nanobiomaterials: from chaperoning engineering to biomedical applications. *Chemical Record* 10 (6): 366–376.

210 Wang, X., Niu, D., Li, P. et al. (2015). Dual-enzyme-loaded multifunctional hybrid nanogel system for pathological responsive ultrasound imaging and T_2-weighted magnetic resonance imaging. *ACS Nano* 9 (6): 5646–5656.

211 Singh, A.V., Ansari, M.H.D., Dayan, C.B. et al. (2019). Multifunctional magnetic hairbot for untethered osteogenesis, ultrasound contrast imaging and drug delivery. *Biomaterials* 219: 119394.

212 Tang, H., Guo, Y., Peng, L. et al. (2018). In vivo targeted, responsive, and synergistic cancer nanotheranostics by magnetic resonance imaging-guided synergistic high-intensity focused ultrasound ablation and chemotherapy. *ACS Applied Materials & Interfaces* 10 (18): 15428–15441.

213 Guo, C., Jin, Y., and Dai, Z. (2014). Multifunctional ultrasound contrast agents for imaging guided photothermal therapy. *Bioconjugate Chemistry* 25 (5): 840–854.

214 Chen, J., Ratnayaka, S., Alford, A. et al. (2017). Theranostic multilayer capsules for ultrasound imaging and guided drug delivery. *ACS Nano* 11 (3): 3135–3146.

215 Meng, Z., Zhou, X., She, J. et al. (2019). Ultrasound-responsive conversion of microbubbles to nanoparticles to enable background-free in vivo photoacoustic imaging. *Nano Letters* 19 (11): 8109–8117.

216 Xia, H., Zhao, Y., and Tong, R. (2016). Ultrasound-mediated polymeric micelle drug delivery. In: *Therapeutic Ultrasound* (ed. A. Bouakaz and J.-M. Escoffre), 365–384. Springer.

217 Husseini, G.A. and Pitt, W.G. (2008). Micelles and nanoparticles for ultrasonic drug and gene delivery. *Advanced Drug Delivery Reviews* 60 (10): 1137–1152.

218 Cai, W., Luo, T., Mao, L., and Wang, M. (2020). Spatiotemporal delivery of CRISPR/Cas9 genome editing machinery using stimuli-responsive vehicles. *Angewandte Chemie*.

219 Zhang, Z., Balogh, D., Wang, F. et al. (2013). Biocatalytic release of an anticancer drug from nucleic-acids-capped mesoporous SiO_2 using DNA or molecular biomarkers as triggering stimuli. *ACS Nano* 7 (10): 8455–8468.

220 de la Torre, C., Mondragón, L., Coll, C. et al. (2014). Cathepsin-B induced controlled release from peptide-capped mesoporous silica nanoparticles. *Chemistry - A European Journal* 20 (47): 15309–15314.

221 Li, L. and Wang, H. (2013). Enzyme-coated mesoporous silica nanoparticles as efficient antibacterial agents in vivo. *Advanced Healthcare Materials* 2 (10): 1351–1360.

222 Dong, J., Tao, Q., Guo, M. et al. (2012). Glucose-responsive multifunctional acupuncture needle: a universal SERS detection strategy of small biomolecules in vivo. *Analytical Methods* 4 (11): 3879–3883.

223 Stuart, D.A., Yuen, J.M., Shah, N. et al. (2006). In vivo glucose measurement by surface-enhanced Raman spectroscopy. *Analytical Chemistry* 78 (20): 7211–7215.

224 Alvarez-Puebla, R.A., Arceo, E., Goulet, P.J.G. et al. (2005). Role of nanoparticle surface charge in surface-enhanced Raman scattering. *The Journal of Physical Chemistry B* 109 (9): 3787–3792.

225 Álvarez-Puebla, R.A., Contreras-Cáceres, R., Pastoriza-Santos, I. et al. (2009). Au@ pNIPAM colloids as molecular traps for surface-enhanced, spectroscopic, ultra-sensitive analysis. *Angewandte Chemie* 121 (1): 144–149.

226 Dong, J., Chen, Q., Rong, C. et al. (2011). Minimally invasive surface-enhanced Raman scattering detection with depth profiles based on a surface-enhanced Raman scattering-active acupuncture needle. *Analytical Chemistry* 83 (16): 6191–6195.

227 Sperlágh, B. and Vizi, S.E. (1996). Neuronal synthesis, storage and release of ATP. *Seminars in Neuroscience* 8 (4): 175–186; Elsevier.

228 Beis, I. and Newsholme, E.A. (1975). The contents of adenine nucleotides, phosphagens and some glycolytic intermediates in resting muscles from vertebrates and invertebrates. *Biochemical Journal* 152 (1): 23–32.

229 Zhou, Y., Tozzi, F., Chen, J. et al. (2012). Intracellular ATP levels are a pivotal determinant of chemoresistance in colon cancer cells. *Cancer Research* 72 (1): 304–314.

230 Fan, Y.-Y., Deng, X., Wang, M. et al. (2020). A dual-function oligonucleotide-based ratiometric fluorescence sensor for ATP detection. *Talanta* 219: 121349.

231 Mo, R., Jiang, T., and Gu, Z. (2014). Enhanced anticancer efficacy by ATP-mediated liposomal drug delivery. *Angewandte Chemie* 126 (23): 5925–5930.

232 He, X., Zhao, Y., He, D. et al. (2012). ATP-responsive controlled release system using aptamer-functionalized mesoporous silica nanoparticles. *Langmuir* 28 (35): 12909–12915.

233 Mo, R., Jiang, T., DiSanto, R. et al. (2014). ATP-triggered anticancer drug delivery. *Nature Communications* 5 (1): 1–10.

234 Deng, J. and Walther, A. (2020). ATP-responsive and ATP-fueled self-assembling systems and materials. *Advanced Materials* 32 (42): 2002629.

235 Lai, J., Shah, B.P., Zhang, Y. et al. (2015). Real-time monitoring of ATP-responsive drug release using mesoporous-silica-coated multicolor upconversion nanoparticles. *ACS Nano* 9 (5): 5234–5245.

236 Mo, R., Jiang, T., Sun, W., and Gu, Z. (2015). ATP-responsive DNA-graphene hybrid nanoaggregates for anticancer drug delivery. *Biomaterials* 50: 67–74.

237 Tai, W., Shukla, R.S., Qin, B. et al. (2011). Development of a peptide–drug conjugate for prostate cancer therapy. *Molecular Pharmaceutics* 8 (3): 901–912.

238 Ke, W., Zha, Z., Mukerabigwi, J.F. et al. (2017). Matrix metalloproteinase-responsive multifunctional peptide-linked amphiphilic block copolymers for intelligent systemic anticancer drug delivery. *Bioconjugate Chemistry* 28 (8): 2190–2198.

239 Nguyen, M.M., Carlini, A.S., Chien, M. et al. (2015). Enzyme-responsive nanoparticles for targeted accumulation and prolonged retention in heart tissue after myocardial infarction. *Advanced Materials* 27 (37): 5547–5552.

240 Li, Y., Zhao, Z., Liu, H. et al. (2019). Development of a tumor-responsive nanopolyplex targeting pancreatic cancer cells and stroma. *ACS Applied Materials & Interfaces* 11 (49): 45390–45403.

241 Barve, A., Jain, A., Liu, H. et al. (2020). Enzyme-responsive polymeric micelles of cabazitaxel for prostate cancer targeted therapy. *Acta Biomaterialia* 113: 501–511.

242 Wu, L., Lin, B., Yang, H. et al. (2019). Enzyme-responsive multifunctional peptide coating of gold nanorods improves tumor targeting and photothermal therapy efficacy. *Acta Biomaterialia* 86: 363–372.

243 Ding, Y., Sun, Z., Tong, Z. et al. (2020). Tumor microenvironment-responsive multifunctional peptide coated ultrasmall gold nanoparticles and their application in cancer radiotherapy. *Theranostics* 10 (12): 5195.

244 Wang, Z., Wang, Y., Jia, X. et al. (2019). MMP-2-controlled transforming micelles for heterogeneic targeting and programmable cancer therapy. *Theranostics* 9 (6): 1728.

245 Yavuz, B., Chambre, L., and Kaplan, D.L. (2019). Extended release formulations using silk proteins for controlled delivery of therapeutics. *Expert Opinion on Drug Delivery* 16 (7): 741–756.

246 Li, M., Zhao, G., Su, W.-K., and Shuai, Q. (2020). Enzyme-responsive nanoparticles for anti-tumor drug delivery. *Frontiers in Chemistry* 8: 00647.

247 Tinkle, S., McNeil, S.E., Mühlebach, S. et al. (2014). Nanomedicines: addressing the scientific and regulatory gap. *Annals of the New York Academy of Sciences* 1313 (1): 35–56.

248 Bawa, R., Melethil, S., Simmons, W.J., and Harris, D. (2008). Nanopharmaceuticals: patenting issues and FDA regulatory challenges. *The SciTech Lawyer* 5 (2): 10–15.

249 Petros, R.A. and DeSimone, J.M. (2010). Strategies in the design of nanoparticles for therapeutic applications. *Nature Reviews. Drug Discovery* 9 (8): 615–627.

250 Shrestha, M. (2012). Nanotechnology to revolutionize medicine. *Journal of Drug Delivery and Therapeutics* 2 (5): 156–161.

251 Hafner, A., Lovrić, J., Lakoš, G.P., and Pepić, I. (2014). Nanotherapeutics in the EU: an overview on current state and future directions. *International Journal of Nanomedicine* 9: 1005.

252 Prasad, M., Lambe, U.P., Brar, B. et al. (2018). Nanotherapeutics: an insight into healthcare and multi-dimensional applications in medical sector of the modern world. *Biomedicine & Pharmacotherapy* 97: 1521–1537.

253 Salatin, S., Ahmadian, E., Mokhtarpour, M. et al. Introduction: an overview of the non-parenteral delivery of nanomedicine. In: *Theory and Applications of Nonparenteral Nanomedicines* (ed. P. Kesharwani, S. Taurin and K. Greish), 1–25.

254 Bowman, D., Maynard, A., and Gaspar, R. (2010). Therapeutic products: regulating drugs and medical devices. In: *International Handbook on Regulating Nanotechnologies* (ed. G.A. Hodge, D.M. Bowman and A.D. Maynard). Northampton, MA: Edward Elgar Pub.

255 Sainz, V., Conniot, J., Matos, A.I. et al. (2015). Regulatory aspects on nanomedicines. *Biochemical and Biophysical Research Communications* 468 (3): 504–510.

256 Soares, S., Sousa, J., Pais, A., and Vitorino, C. (2018). Nanomedicine: principles, properties, and regulatory issues. *Frontiers in Chemistry* 6: 360.

257 US Food and Drug Administration (2011). Guidance for industry considering whether an FDA-regulated product involves the application of nanotechnology. *Biotechnology Law Report* 30 (5): 613–616.

258 Keck, C.M. and Müller, R.H. (2013). Nanotoxicological classification system (NCS) – a guide for the risk-benefit assessment of nanoparticulate drug delivery systems. *European Journal of Pharmaceutics and Biopharmaceutics* 84 (3): 445–448.

259 Hussain, S.M., Warheit, D.B., Ng, S.P. et al. (2015). At the crossroads of nanotoxicology in vitro: past achievements and current challenges. *Toxicological Sciences* 147 (1): 5–16.

260 Zolnik, B.S. and Sadrieh, N. (2009). Regulatory perspective on the importance of ADME assessment of nanoscale material containing drugs. *Advanced Drug Delivery Reviews* 61 (6): 422–427.

261 Ragelle, H., Danhier, F., Préat, V. et al. (2017). Nanoparticle-based drug delivery systems: a commercial and regulatory outlook as the field matures. *Expert Opinion on Drug Delivery* 14 (7): 851–864.

262 Emdad, L., Bhoopathi, P., Talukdar, S. et al. (2019). Recent insights into apoptosis and toxic autophagy: the roles of MDA-7/IL-24, a multidimensional anti-cancer therapeutic. In: *Seminars in Cancer Biology*, vol. 66, 140–154. Elsevier.

263 Ryman-Rasmussen, J.P., Riviere, J.E., and Monteiro-Riviere, N.A. (2006). Penetration of intact skin by quantum dots with diverse physicochemical properties. *Toxicological Sciences* 91 (1): 159–165.

264 Maynard, A.D. (2006). Nanotechnology: assessing the risks. *Nano Today* 1 (2): 22–33.

265 Drobne, D. (2007). Nanotoxicology for safe and sustainable nanotechnology. *Archives of Industrial Hygiene and Toxicology* 58 (4): 471–478.

266 Xia, T., Kovochich, M., Brant, J. et al. (2006). Comparison of the abilities of ambient and manufactured nanoparticles to induce cellular toxicity according to an oxidative stress paradigm. *Nano Letters* 6 (8): 1794–1807.

267 Penn, A., Murphy, G., Barker, S. et al. (2005). Combustion-derived ultrafine particles transport organic toxicants to target respiratory cells. *Environmental Health Perspectives* 113 (8): 956–963.

268 Wang, K., He, X., Linthicum, W. et al. (2017). Carbon nanotubes induced fibrogenesis on nanostructured substrates. *Environmental Science. Nano* 4 (3): 689–699.

269 Yang, Y., Wang, K., Gu, X., and Leong, K.W. (2017). Biophysical regulation of cell behavior – cross talk between substrate stiffness and nanotopography. *Engineering* 3 (1): 36–54.

270 Song, L., Wang, K., Li, Y., and Yang, Y. (2016). Nanotopography promoted neuronal differentiation of human induced pluripotent stem cells. *Colloids and Surfaces B: Biointerfaces* 148: 49–58.

271 Gaspar, R. and Duncan, R. (2009). Polymeric carriers: preclinical safety and the regulatory implications for design and development of polymer therapeutics. *Advanced Drug Delivery Reviews* 61 (13): 1220–1231.

272 Dorbeck-Jung, B.R. and Chowdhury, N. (2011). Is the European medical products authorisation regulation equipped to cope with the challenges of nanomedicines? *Law Policy* 33 (2): 276–303.

273 Dobrovolskaia, M.A. and McNeil, S.E. (2013). Understanding the correlation between in vitro and in vivo immunotoxicity tests for nanomedicines. *Journal of Controlled Release* 172 (2): 456–466.

274 Mühlebach, S. (2018). Regulatory challenges of nanomedicines and their follow-on versions: a generic or similar approach? *Advanced Drug Delivery Reviews* 131: 122–131.

275 Astier, A., Barton Pai, A., Bissig, M. et al. (2017). How to select a nanosimilar. *Annals of the New York Academy of Sciences* 1407 (1): 50–62.

276 Mühlebach, S., Borchard, G., and Yildiz, S. (2015). Regulatory challenges and approaches to characterize nanomedicines and their follow-on similars. *Nanomedicine* 10 (4): 659–674.

277 Bhatia, P. and Chugh, A. A multilevel governance framework for regulation of nanomedicine in India. *Nanotechnology Reviews [Internet]* 6 (4): 373–382. https://www.degruyter.com/view/journals/ntrev/6/4/article-p373.xml (accessed 23 February 2022).

278 Ventola, C.L. (2012). The nanomedicine revolution: part 3: regulatory and safety challenges. *Pharmacology and Therapeutics* 37 (11): 631.

279 Paradise, J. (2019). Regulating nanomedicine at the Food and Drug Administration. *AMA Journal of Ethics* 21 (4): 347–355.

19

Antimicrobial Applications of Nanoparticles

Moataz A. Elsawy and Mohamed H. Mostafa

Polymer Laboratory, Petrochemical Department, Egyptian Petroleum Research Institute, Nasr City, Cairo, Egypt

19.1 Introduction

Increased rates of infectious diseases and illnesses arising from the widespread pathogens like bacteria, fungi, and viruses are and will always be of great concern. The outbreaks of these diseases disrupt daily activities, exhibit high costs of resolution, and result in both economic recessions of countries and, in severe cases, death [1]. Therefore, it is of the utmost importance to inhibit, if not eliminate the growth and spread of pathogenic microbes on goods, products, hospital equipment, consumable clinical materials, devices, and especially packaged foods. Thus, researchers and pharmaceutical companies across the world have been driven to investigate new and effective potent antimicrobial agents in order to curb the spread of diseases [2].

Recently, controlling microorganisms through using antimicrobial agents like nanoparticles (NPs) has received great interest across the globe. The interest in using NPs come from the morphological and physicochemical properties including the surface-area-to-volume ratio and the potency toward a wide spectrum of microorganisms. Additionally, the presence of surface charges on these NPs could facilitate binding to the opposite surface charges of the microorganisms, resulting in effective antimicrobial activities. Also, the effective durability and longevity of these NPs, when used in antimicrobial applications, are promising owing to their insolubility as well as the close interaction with microbial membranes [3].

In general, NPs show distinct properties compared to the same material in the bulk state, which can be attributed to the fact that the decrease in the particle size results in a significant increase in the surface-to-volume ratio of NPs. Indeed, the proportion of the surface molecule is considerably increased in nanometer dimensions, which in turn enhances some properties of the particles such as mass transfer, dissolution rate, heat treatment, and catalytic activity [4]. The antimicrobial activity of metal NPs (MeNPs) such as silver NPs (AgNPs), titanium NPs (TiNPs), gold NPs (AuNPs), zinc NPs (ZnNPs), and copper NPs (CuNPs) has been known and applied for centuries [5]. Metal-oxide NPs (MeONPs) such as titanium dioxide NPs (TiO2NPs), silver oxide NPs (Ag2ONPs), copper oxide NPs (CuONPs), zinc oxide NPs (ZnONPs), magnesium oxide NPs (MgONPs), and calcium oxide NPs (CaONPs) were

Nanomaterials and Nanotechnology in Medicine, First Edition. Edited by P.M. Visakh.

recognized to possess antimicrobial activity [6]. Several studies showed that polymeric NPs inhibit several microbial species [7]. Also, polymer nanocomposites have been reported as key materials in many applications because of their antimicrobial activity [8].

In this chapter, we focus on the latest findings on antimicrobial activity of NPs including MeNPs, MeONPs, polymeric NPs, and polymer nanocomposites. Additionally, we discuss their application as antimicrobial agents in various fields including biomedical applications, food industry, water disinfection, and textiles.

19.2 Antimicrobial Properties of Nanoparticles

19.2.1 Metal-oxide Nanoparticles

MeONPs have attracted considerable interest owing to their unique physical and chemical properties. MeONPs include various types of oxides like ZnONPs, CuONPs, TiO_2NPs, and Fe_3O_4NPs. MeONPs could be synthesized by a wide variety of methods, which can be categorized into bottom-up or top-down approaches [9]. The bottom-up approach uses atoms and molecules to create NPs via chemical or biological methods. Conversely, the top-down approach refers to the cutting or slicing of bulk material to obtain NPs via physical methods such as (mechanical milling, physical vapor deposition, thermal evaporation, ultrasonic irradiation, or laser ablation) or chemical methods like chemical etching [10]

With their outstanding physical, electrical, chemical, optical, magnetic, and biological properties, MeONPs are of immense importance to scientists as antimicrobial agents. Table 19.1 lists the antimicrobial activity of MeONPs against various types of microorganisms. The availability of many MeONPs with diverse functional and physicochemical

Table 19.1 Antimicrobial activity of MeONPs toward various microorganisms.

MeONPs	Antimicrobial property	Target microorganism	Ref.
	Antibacterial activity	*E. coli 0157:H7*	[11]
		B. subtilis	[12]
		P. fluorescens	[13]
		Listeria monocytogenes	[11]
		Salmonella enteritidis	[14]
		S. aureus	[15]
		Salmonella typhimurium	[16]
ZnONPs	Antifungal activity	*Botrytis cinerea*	[17]
		P. expansum	[17]
		A. flavus	[18]
		A. niger	[18]
		C. albicans	[19]
		Rhizopus stolonifera	[20]
	Antiviral activity	*Herpes simplex virus type-1 & 2*	[21]
		Transmissible gastroenteritis virus (TGEV)	[22]
	Antiparasitic activity	*Leishmania major*	[23]
		Plasmodium	[24]

Table 19.1 (Continued)

MeONPs	Antimicrobial property	Target microorganism	Ref.
CuONPs	Antibacterial activity	*B. subtilis*	[25]
		L. monocytogenes	[26]
		S. aureus	[25]
		E. coli	[27]
	Antiparasitic activity	*Entamoeba histolytica*	[28]
		Cryptosporidium parvum	[28]
		Plasmodium falciparum	[29]
TiO_2NPs	Antibacterial activity	*E. coli*	[30]
		S. aureus	[30]
		L. monocytogenes	[31]
		S. enteritidis	[32]
		P. fluorescens	[33]
	Antifungal activity	*P. expansum*	[34]
		A. niger	[35]
		Fusarium oxysporum	[36]
	Antiviral activity	*Influenza virus*	[37]
		Qβ and T4 bacteriophages	[38]
	Antiparasitic activity	*Leishmania tropica*	[39]
		Leishmania infantum	[40]
		Leishmania major	[41]
		Anopheles stephensi	[42]
		Aedes aegypti	[42]
MgONPs	Antibacterial activity	*B. subtilis*	[43]
		E. coli	[44]
		S. aureus	[45]
		Pseudomonas aeruginosa	[43]
	Antifungal activity	*C. albicans*	[46]
		A. niger	[47]
		P. oxalicum	[47]
	Antiparasitic activity	*Leishmania major*	[48]
Al_2O_3NPs	Antibacterial activity	*E. coli*	[49]
		P. aeruginosa	[50]
		S. aureus	[51]
	Antifungal activity	*Candida spp*	[52]

properties makes them good antimicrobial agents for many applications. However, information about the molecular mechanisms of antimicrobial activities of MeONPs is still in its infancy, and elucidation of the mechanisms of MeONPs is considered the subject of rigorous research. The possible mechanisms of MeONPs antimicrobial activity are discussed according to the previous studies and presented in Figure 19.1 [53]

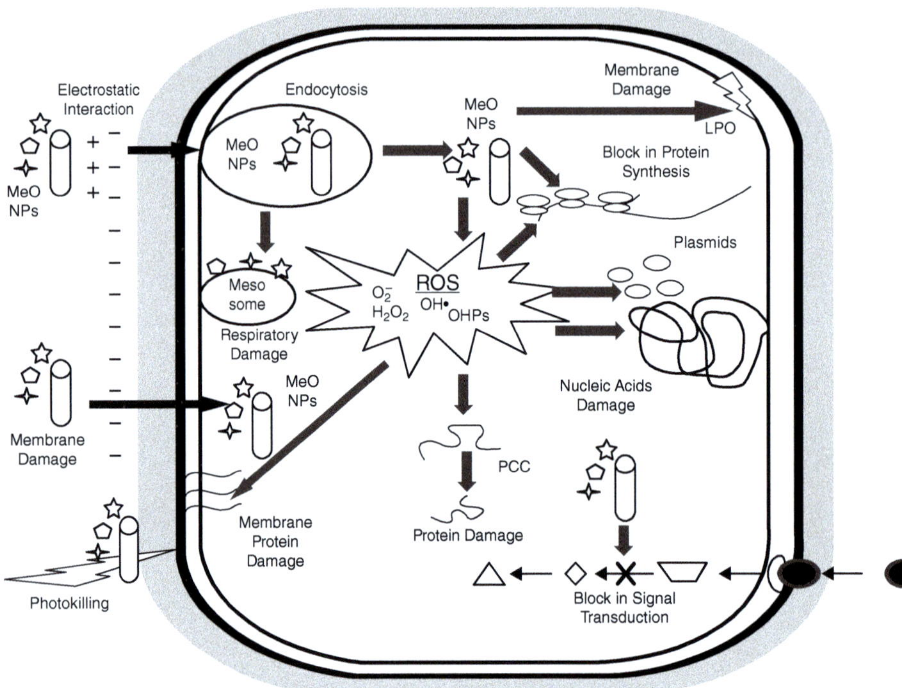

Figure 19.1 Overview of antimicrobial mechanisms by MeONPs [53].

19.2.1.1 Cell Membrane Damage by Electrostatic Interaction

The presence of negatively charged chemical groups of polymers on the bacterial membranes makes them sites of metal cation attraction. The negative surface charges of bacteria and spores at biological pH are due to the carboxylic acid groups in the proteins [54]. The electrostatic interactions between the opposite charges of bacterial membranes and MeONPs results in the accumulation of MeONPs on the cell surface and allow entry into the bacteria. Thus, MeONPs can exert toxicity to microorganisms. Gram-negative bacteria exhibit a greater negative charge compared with the gram-positive bacteria and hence the electrostatic interaction will be stronger in gram-negative strains [55]. The electrostatic attraction force is dependent on the surface area available for interaction, which is extensively offered by MeONPs compared to their native large-sized particles, thus imparting cytotoxicity. The attachment of MeONPs alters the structure and permeability of the cell membrane, leading to membrane disruption and loss of intracellular contents. Consequently, growth is inhibited as an after effect of the electrostatic interaction between MeONPs and the cell surface [56].

19.2.1.2 Disturbance in Metal/Metal Ion Homeostasis

Metal ion homeostasis is important for microbial survival as it regulates metabolic functions via assisting coenzymes, cofactors, and catalysts. When bacteria exhibit excess metal or metal ions, there will be disorder in metabolic functions. Metal ions bind with DNA (deoxyribonucleic acid) and disrupt the helical nature via cross-linking between and within DNA strands. Metal ions released from the MeONPs possess positive charges, which neutralize the charges present on lipopolysaccharides and thus increase the permeabilization of the outer membrane. Therefore, bacterial growth is slowed down as the membranes become disorganized, with the increased permeability attributed to the accumulation of MeONPs in the cells [57]. The strong binding between MeONPs and the outer membrane inhibits active transport and the activities of dehydrogenase and periplasmic enzymes. Also, the released metal ions from MeONPs in cells are reduced to metal atoms by thiol groups (–SH) in enzymes and proteins, thus inactivating the essential metabolic proteins blocking respiration and leading to cell death. Moreover, the interaction between metal ions and –SH groups in the peptidoglycan layer leads to cell wall destruction. MeONPs release metal ions via adsorption, dissolution, and hydrolysis, which are toxic and abrasive to bacteria and hence lyse the cells. The abrasive nature of MeONPs generates surface defects, thus causing physical damages to the cell wall [58].

19.2.1.3 Production of Reactive Oxygen Species and Oxidative Stress

Once MeONPs enter bacteria, they induce the production of reactive oxygen species (ROS) (hydroxyl radicals (OH•), superoxide anion (O_2•−), hydrogen peroxide (H_2O_2), and organic hydroperoxides), which are pernicious to bacteria, causing damage to almost all organic biomolecules (carbohydrates, amino acids, lipids, proteins, and nucleic acids) and eventually causing microbial death [59]. The physicochemical properties of MeONPs like surface area, electrophilic nature, and diffusibility determine the amount of ROS produced in the bacteria. Generation of ROS eventually results in oxidative stress, which is one of the mechanisms mediated by MeONPs against microorganisms to kill them [60].

19.2.1.4 Protein and Enzyme Dysfunction

Another mode of antibacterial activity possessed by MeONPs is protein dysfunction. Metal ions released from MeONPs catalyze the oxidation of amino acid side chains, leading to protein-bound carbonyls. The carbonylation level within the protein molecules acts as a marker for oxidative protein damage. This protein carbonylation will result in loss of the catalytic activity in the case of enzymes, ultimately triggering protein degradation. Additionally, these metal ions react with the –SH groups of various enzymes and proteins, rendering them inactive [61]. Some metals such as Ag may act as a weak acid and react with soft bases such as sulfur and phosphorus, which are the main constituents of proteins and DNA, respectively. The released metals could interact with these soft bases and damage DNA, thus leading to cell death [62].

19.2.1.5 Genotoxicity and Signal Transduction Inhibition

Regarding their electrical properties, MeONPs interact with nucleic acids, especially genomic and plasmid DNA and suppress cell division of microbes through disrupting the replication processes of chromosomal and plasmid DNA. In rare cases, metal ions bind to the 30S ribosomal subunit, thus stalling movement of the ribosome complex and blocking the translation of proteins from mRNA [53]. Signal transduction in bacteria is also affected by MeONPs. MeONPs dephosphorylate the phosphotyrosine, which is an important component of the signal transduction mechanism in bacteria, leading inhibition of signal transduction and ultimately obstruction of the growth of bacteria [63].

19.2.1.6 Photokilling

Photokilling takes place when MeONPs are exposed to microorganisms in the presence of light. Only transition metal oxides can be photosensitized, not all metal oxides. The light effect results in photochemical alteration of the cell membrane, diminution in superoxide dismutase (SOD) activity, Ca^{2+} permeability, damage to proteins/DNA, and abnormal cell division. The generation of $O^{2\bullet-}$ is the primary reason wherein the generation and release of electrons occur via light. These electrons are grabbed by MeONPs, thus producing more ROS [64].

19.2.2 Metal Nanoparticles

MeNPs have been widely used as antimicrobial agents for centuries in several countries thanks to their desired antimicrobial properties. Some examples of these NPs are AgNPs, AuNPs, and CuNPs. MeNPs can be prepared by various physical, chemical, and biological methods. MeNPs exhibit several advantages like their small size and larger surface area, which make them good candidates for many antimicrobial applications [65]. Table 19.2 shows the antimicrobial activity of MeNPs against various types of microorganisms.

The mechanisms for the antimicrobial activity of MeNPs are still poorly understood. The explanation of this could be due to the multifactorial nature of the activity, which makes it difficult to decouple the individual mechanisms. Several proposed mechanisms have been derived from the physical interactions and chemical interactions like production of ROS and the increased dissolution of metal cations. The microbial target sites involved in these mechanisms are the cell membrane, nucleic acids, membrane-bound proteins, and the inhibition of enzyme activity. Thus, it is proposed that it is difficult for the microbes to resist these NPs to protect all of these cellular components [103]. Theses mechanisms will be discussed.

19.2.2.1 Physical Interactions

Physical interactions involve both interactions with the cell wall and intracellular activity. The cell wall represents the barrier between the cell and its external environment. It also facilitates important metabolic pathways such as the electron transport chain, so disruption to the cell wall can be fatal for the cell. MeNPs exhibit surface charges, which could bind to the oppositely charged cell wall components via electrostatic interactions. These interactions alter the charge of the membrane, leading to the disturbance of the membrane and finally its damage. The damage of the membrane generates leakage of fluid from the cytosol and irreparable damage to the cell wall and ATP (adenosine triphosphate) production, effectively inactivating the cells [65, 104].

19.2.2.2 Ion Dissolution

MeNPs dissolve the metal ions at higher rate compared to the bulk material as a result of the larger surface area of NPs compared to the bulk state. The metal ions could interact with many cellular components like proteins, DNA, and the cellular membrane. These ions could be taken up by the bacteria and fungi via a suite of transport proteins, which control the uptake of metal species [105]. Metal ions could inhibit the enzyme activity through metal-catalyzed oxidation of amino acid residues in proteins [106]. They also are able to produce ROS, either

Table 19.2 Antimicrobial activity of some MeNPs toward microorganisms.

MeNPs	Antimicrobial property	Target microorganism	Ref.
AgNPs	Antibacterial activity	*E. coli*	[66]
		B. subtilis	[67]
		S. aureus	[68]
			[69]
		Enterococcus faecalis	[70]
		Klebsiella pneumoniae	[71]
		Salmonella typhi	[72]
	Antifungal activity	*Trichophyton mentagrophytes*	[73]
		Bipolaris sorokiniana	[74]
		Magnaporthe grisea	[75]
		Microsporum gypseum	[73]
		Microsporum canis	[73]
	Antiviral activity	*Influenza virus*	[76]
		Herpes Simplex virus	[77]
		parainfluenza virus type 3	[77]
		Respiratory syncytial virus	[78]
		Monkey pox virus	[79]
	Antiparasitic activity	*Leishmania tropica*	[80]
		Leishmania infantum	[81]
		Entamoeba histolytica	[82]
		Cryptosporidium parvum	[83]
		Giardia lamblia	[84]
		Toxoplasma gondii	[85]
		Leishmania major	[86]
		Plasmodium falciparum	[87]
		malaria vector	[88]
AuNPs	Antibacterial activity	*S. aureus*	[89]
		Enterococcus faecium	[90]
		E. coli	[91]
		Bacillus subtilis	[91]
		K. pneumoniae	[90]
	Antifungal activity	*Puccinia graminis tritci*	[92]
		A. flavus	[92]
		C. albicans	[93]
	Antiviral activity	*HIV virus*	[94]
		Influenza virus	[95]
		Measles virus	[96]
	Antiparasitic activity	*G. lamblia*	[97]
		Plasmodium falciparum	[98]
		Toxoplasma gondii	[99]
		Leishmania donovani	[100]
		Leishmania major	[101]
		C. parvum	[102]

directly in the case of redox-active metals, or via damage to the Fe—S clusters within proteins, which liberate redox-active Fe (iron) ions. Moreover, metal ions inhibit the nutrient uptake, for example gallium ions (Ga^{3+}) have been shown to kill bacteria using a "Trojan horse" mechanism, whereby the cells mistake Ga^{3+} for Fe^{3+} ions, owing to the similar chemical properties. The cell becomes inactivated via the inhibition of metabolic activity since the bacteria are incapable of reducing Ga^{3+}. Additionally, the positive metal ions could cause damage to the membrane via electrostatic attractions with the electronegative membrane and some integral proteins like those involved in the electron transport chain [107]. Furthermore, metal ions possess genotoxic activity, which causes damage to DNA, resulting in the cell death [108].

19.2.2.3 Production of Reactive Oxygen Species

The ROS like singlet oxygen ($^1O_2$), hydroxyl radicals (•OH), superoxide anion radicals ($^{\bullet}O_2^-$), and hydrogen peroxide (H_2O_2) are generated endogenously in the cell via natural processes. In higher concentrations, ROS cause oxidative stress to microbial cells, causing significant damage to the cell membrane and degrading important proteins and nucleic acids, ultimately leading to the cell death. MeNPs could induce cells to increase the production of ROS via metabolic responses by the promotion or suppression of ROS-related enzymes. The amount of ROS generated from MeNPs primarily depends on the size and chemistry of the NPs [59, 109].

19.2.3 Polymer-based Nanoparticles

Recently, there has been great interest for the development of polymeric NPs as antimicrobial agents. This is attributed to fact that polymers can be formulated into NPs (e.g. vesicles, star polymers, micelles, and inorganic-polymer hybrids) of various sizes and shapes, enabling their use in wide variety of applications [110, 111]. The main advantage of polymeric NPs is the multivalency, where the cluster of (multiple) functional groups from NP constructs allows for higher cell recognition and binding capabilities compared to linear polymers. Additionally, polymer NPs like micelles, vesicles, or star polymers enable the efficient encapsulation of cargo molecules, which can be released at targeted sites. In addition, the inorganic-polymer hybrid NPs provide new avenues for synergistic therapy (e.g. photodynamic therapy) and/or diagnostic purposes (e.g. biosensing) [112–114].

Polymers could have favorable antimicrobial properties via incorporating cationic and hydrophobic moieties into their chains. The overall cationic charge of the polymer allows for interaction with the negatively charged bacterial cell walls, while the hydrophobic counterparts make the microbial membrane penetration easy. Also, it should be noted that antimicrobial polymers with differing chemical structures such as quaternary ammonium-containing polymers without any hydrophobic components have been reported [115]. Polymer-based NPs that exhibit antimicrobial activity could be classified into various classes based on their complex macromolecular architecture. These classes involve self-assembled polymer NPs, star polymer NPs, and inorganic-polymer hybrid NPs [7].

19.2.3.1 Self-assembled Polymer Nanoparticles

Self-assembly has been reported as one of the most used ways for the construction of nano-sized biomaterials owing to the relative ease, versatility, and precision of the method, enabling the incorporation of functions such as stimuli responsiveness, recognition, and targeting. The formation of complex molecular structures in the most thermodynamically stable form like micelles and vesicles is made possible by multiple weak and noncovalent interactions between the chemical building blocks [116].

19.2.3.2 Star Polymer Nanoparticles

In spite of the versatility of self-assembly and the potential presented by self-assembled polymer NPs in the antimicrobial field, the reversible and dynamic nature of self-assembly renders the nanostructures prone to disassembly, providing that they are not sufficiently robust for practical applications. In contrast to self-assembled polymer NPs, star polymer NPs are covalently bonded unimolecular nanostructures with a branched architecture, consisting of several linear chains linked to a central core, thereby forming three-dimensional (3D) globular structures. These NPs have gained much interest owing to their unique structural properties. It is possible to synthesize high molecular weight star-shaped polymer NPs while still retaining a solubility and viscosity similar to that of low molecular weight linear or branched polymers. Other attractive advantages of star polymer NPs involve their encapsulation capabilities and wide range of internal and peripheral functionalities. Recently, star polymers have been widely reported in the field of nanomedicine with applications in drug delivery, gene therapy, bioimaging, and tissue engineering [117, 118].

19.2.3.3 Inorganic-polymer Hybrid Nanoparticles

Inorganic-polymer core-shell NPs consist of a metal- or silica-based core with a polymer shell. In order to increase the applicability of inorganic NPs like metal or MeONPs in various fields, they are coated with polymer to improve the oxidation stability of the core, reducing the aggregation of particles in solution, and improving mammalian cell compatibility for bio-applications. Recently, there has been great interest in developing antimicrobial silica-polymer core-shell NPs and polymer-coated metal-based NPs [119].

19.3 Antimicrobial Applications of Nanoparticles

NPs have been extensively studied as antimicrobial agents in various applications like biomedical applications, the food industry, water disinfection, and textiles. In the following, the applications of some NPs are discussed.

19.3.1 Biomedical Applications

Regarding the biomedical sector, there is an increase in the rates of diseases coming from pathogenic microorganisms like bacteria, fungi, and viruses, which are of great concern. The pathogenic contamination could be difficult to limit and could become fatal in some cases by the time it is realized. Concerning the medical facilities like laboratories, hospitals, or at pharmaceutical industrial sites, there is always a possibility of contamination or disease transmission via various routes owing to the disease-causing microorganisms. Hospitals and laboratories represent the highest risk of contamination as a result of the nature of the environment. Hence, with a high traffic of patients, the potential of cross contamination by many disease-causing microorganisms is high. Transmission of microorganisms via liquids, vapors, or mists in the air and contact through persons, equipment, and materials are inevitable. Therefore, it is important to develop antimicrobial technologies in order to limit the growth and spread of diseases directly at the source in the medical sector [1]. NPs with antimicrobial properties that can disrupt the growth and destroy the pathogenic microorganisms have been widely investigated as potent materials against various pathogens. Some NPs that have been investigated are ZnONPs, TiO_2NPs, CuONPs, AgNPs, AuNPs, and chitosan NPs. Compared to the bulk materials, nanomaterials are able to cross cell membranes easily, leading to a destructive effect on the bacterial cell [120].

19.3.1.1 Applications in Endodontic Irrigants

The mouth is considered as a host to various microorganisms and could act as incubation sites when not properly cared for. The exposed underlying tissue is always infected especially after a dental procedure [121]. Antibiotics are utilized to combat and prevent the infection of the area. However, the antibiotics could exhibit some adverse effects on the liver and in some cases could cause the patient to react. During dental surgery and reconstruction, the use of irrigants like hypochlorite and chlorhexidine is common to clean, flush, and disinfect the areas that are susceptible to infection. However, these irrigants are temporary fixes, as they are eventually washed away, dried up, or soaked up by gauze, leaving no protection against a host of microorganisms. Also, they could be harmful to the host as they can lead to tissue destruction on a cellular level [122].

The use of antimicrobial NPs in irrigants or aqueous solutions could greatly enhance, prolong, and prevent the microbial activities. The large surface area of these particles leads to increased contact efficiency with microbes, hence increasing antimicrobial activities against pathogens. Also, they are able to penetrate nooks and crannies where microorganisms can easily grow and remain embedded after the irrigant is removed [123]. Recently, AgNPs in an aqueous solution has been used as a potential irrigant due to their ability to limit the activities of an array of both gram-negative and gram-positive bacteria via binding to the cell walls and releasing silver ions. Other studies have revealed that the longer contact time of antimicrobial NPs like AgNPs with well-developed biofilms is required to produce significant antimicrobial activities especially at very low concentrations [124]. Similar reports show that using AgNPs and ZnONPs in aqueous solution at concentrations of 1 and 26% respectively, gives comparable results to chlorhexidine and hypochlorite irrigants [125]. Although AgNPs are expensive, they can be used at low concentrations in aqueous solution without adding any surfactants and produce remarkable effects on the prevention of bacterial growth. Additionally, they can remain within the cavities to maintaining antimicrobial activities, resulting in rapid recovery and healing of the surrounding tissue [126].

19.3.1.2 Applications in Medical Devices and Consumables

Medical personal protective equipment (PPE) like overalls, gowns, masks, and hair covers are consumed globally on a large scale and disposed of after a very short period of use. This is carried out to limit the exposure and transport of microorganisms from one individual to another. It is apparent that PPE must be kept in pristine and clean conditions and limit their exposure to potential hazards to prevent infection by microbes [127]. The use of nanotechnology could be a suitable method to increase the life time of PPE and their reuse without the major concern of microbial contamination. This can be done by incorporating potent antimicrobial NPs like AgNPs and AuNPs and their composites into these materials where some could be used for increased periods of time and reused when laundered, while providing added protection to the wearer as microbes could be eliminated on contact [128].

The use of antimicrobial NPs for face masks has been investigated. Face masks function to protect the user from airborne bacteria and bacteria in liquids like bodily fluids. The outer layers of face masks are hydrophobic to prevent absorption of liquids and the wetting of the inner layer. However, a considerable amount of fluid such as blood, sweat, and other body fluids could make the mask fail, resulting in transmission of bacteria from the outer layer to the inner one. Introducing antimicrobial NPs into the mask layers could limit the microbial activities to prevent potential infection of the wearer [129]. The combination of $AgNO_3$ and TiO_2 as antimicrobial agents shows 100% reduction in the amount of *Escherichia coli* and *Staphylococcus aureus* after 48 hours of incubation for masks coated with antimicrobial particles, whereas the uncoated masks revealed a 25% and 50% increase in the viable count of *E. coli* and *S. aureus*, respectively [130]. In another study, Ag solution was made from $AgNO_3$ and starch at concentrations of 50ppm and 100 ppm of AgNPs. Three layered face masks were soaked in this solution for a time period and allowed to air dry. Tests against gram-positive and negative bacteria were carried out to determine the antimicrobial efficacy of the treated masks. The obtained results indicated that solutions with higher amounts of AgNPs were more effective in limiting bacterial growth [131].

Medical consumable tools including surgical needles and blades are used interchangeably and placed on a sterile tray when not in use during surgery. Although sterilization is performed to keep the surroundings clean, it is impossible to completely prevent microbial contact during surgery. Creating a biocidal surface on these tools could be a suitable way to prevent contamination where the applied surface could destroy any microorganisms when contact is made, thus reducing the chances of transmitting infection [132].

19.3.1.3 Applications in Dental Composites

Dental composites that are composed of polymer resin and inorganic fillers are widely utilized in dentistry as fillers for reconstruction of artificial teeth, cavities, and other structures within the mouth. These materials have to be resilient, durable, and able to withstand the vigorous activities and environment in the mouth such as varying pH and mechanical abrasion during chewing and contact with food. Therefore, coupling agents are used to chemically modify the surface of the filler, thereby allowing the formation of chemical bonds with the cured polymer resin. Without this, leaching can occur over time. This modification would ensure that there was no leaching of any component of the dental material while also improving its physical properties and durability. However, the growth of bacteria can form biofilms on these surfaces and result in more plaque accumulation in comparison to other surfaces such as enamel [133]. Antimicrobial NPs such as ZnONPs embedded within dental composite materials have demonstrated antibacterial activities and led to a reduction in the growth of bacterial biofilms against *Streptococcus sobrinus* by as much as 80% [134]. Potent and well-known antimicrobial agents such as AgNPs have been shown to be effective when incorporated into acrylic polymer material for denture applications with minimal toxicity toward humans [135].

19.3.2 Food Industry

The food industry is one of the industrial sectors where antimicrobial NPs play an important role. Recently, there has been a growing interest about not only food preservation and quality maintenance but also food safety for consumers. Foodborne diseases are caused through consuming foods or beverages that are contaminated by pathogenic microorganisms. Some of the most common pathogens are *Salmonella*, norovirus, *Clostridium perfringens*, *S. aureus*, *Listeria*, and *Campylobacter*, which can also lead to eventual death. Currently, there is much effort used to investigate how to prevent foodborne diseases, particularly to protect food from the growth of microorganisms during storage [136]. Antimicrobial NPs could be used for various purposes in the food industry including food packaging and edible films as well as food additives [137].

19.3.2.1 Food Packaging

Food packaging is an important item in the food industry as it plays a vital role in modern society. The functions of food packaging include marketing, protection, and preservation. Microbial food spoilage, a major global concern, minimizes the shelf life of food and increases the risk of foodborne illness. To improve both the safety and shelf life of food, many food preservation strategies have been employed such as pasteurization, chemical preservation, thermal sterilization, and freezing. Antimicrobial packaging offers a promising alternative method for minimizing microbial activity in packaged foods. Antimicrobial packaging could interact with the surfaces of packaged food releasing biocide to inhibit the growth of pathogens such as *E. coli, Campylobacter jejuni, Listeria monocytogenes,* and *Salmonella typhi* [138].

Various NPs have been developed and explored for food-packaging applications. The role of these NPs in food packaging is to enhance the mechanical, thermal, barrier, and antimicrobial properties of the produced materials. Based on the obtained benefits from the incorporation of NPs in active food packaging, nanotechnology plays a remarkable role in developing food-packaging systems to satisfy the requirements of producers and consumers [139].

AgNPs exhibit high thermal stability and antimicrobial properties against several types of microorganisms, making them appealing antimicrobial NPs in food-packaging systems. The antimicrobial activity of AgNPs is due to several factors such as adhesion and rupture of cell surfaces, increase in permeability, degradation of lipopolysaccharides, and binding of Ag to electron donor groups in biological molecules containing sulfur, oxygen, or nitrogen. AgNPs possess effective antimicrobial activity against *S. aureus*, appearing from the lysis of the *S. aureus* cells when exposed to AgNPs for 12 hours. The breakdown of cell walls leads to the release of cellular contents into the surrounding environment and the emptying of the cytoplasm [140].

Many reports have studied the antimicrobial properties of AgNPs incorporated in polymers for packaging applications. It has been reported that *in situ* AgNPs were synthesized with linear polyurethane and hyperbranched polyurethane to produce nanocomposites with antimicrobial activities [141]. Spherical-shaped AgNPs were incorporated in poly (butylene adipate-co-terephthalate) and the final product was used as antimicrobial packaging material [142]. In another study, the shelf life of chicken breast fillets wrapped with

Ag NP low-density polyethylene (LDPE) film was shown to be extended substantially in comparison to the control LDPE film [143]. Ternary nanocomposite films produced from AgNPs, cellulose nanocrystals, and polylactic acid exhibited good antimicrobial effect against *S. aureus* and *E. coli* [144]. In another report, the shelf life of Fior di latte cheese was increased via using antimicrobial packaging composed of alginate with Ag-montmorillonite NPs [145]. The biosynthesized AgNPs by *Bacillus subtilis* were incorporated into chitosan matrix to produce nanocomposite films for packaging applications [146].

CuNPs and CuONPs have also been known for their antimicrobial properties against foodborne pathogens. It has been reported that LDPE/CuNPs nanocomposite (LDPE/Cu) films could be beneficial in food packaging where the shelf life is short. Products such as Peda cheese that has a shelf life of only 24 hours can benefit substantially from such antimicrobial packaging materials. Sweet Peda wrapped with LDPE/CuNPs film did not show degradation up to eight days and no further increase in the number of viable cells was noticed after the eighth day [147]. CuNPs were combined with other NPs in order to enhance their antimicrobial effect compared with CuNPs alone. For instance, Ag/CuNPs were incorporated into guar gum in order to produce antimicrobial food packaging film. The film showed antimicrobial activity against both gram-positive and gram-negative bacteria [148]. Liners made from LDPE-based composite films were prepared with Ag/CuNPs and cinnamon essential oil for chicken meat packaging. The developed composite films exhibited successful inhibition of *L. monocytogenes*, *Salmonella typhimurium*, and *C. jejuni* growth in the chicken sample over 21 days storage at 4 °C [149]. Bimetallic NPs hybrid (Ag/CuNPs) incorporated with fish skin gelatin nanocomposites were prepared resulting in improved antimicrobial properties. These nanocomposite films showed remarkable bactericidal effect against gram-negative as well as gram-positive bacteria, giving them the potential to be used as packaging material for extending the shelf life of food [150]. CaONPs show antimicrobial activity, which could be related to its alkaline effect as well as the activated oxygen generated from CaONPs. CaONPs possess antimicrobial properties against *Pseudomonas aeruginosa* growth and biofilm formation suggesting that CaONPs have great potential in a wide variety of applications such as food packaging. Aluminum oxide NPs (Al2O3NPs) can also provide inhibitory effect to the bacteria via electrostatic interactions, leading to cell death [151].

ZnONPs, one of the most widely applied MeONPs, can be applied in distinct fields including chemical sensors, photocatalysis, and biomedical fields attributed to its photo activity, low production cost, abundant availability, biocompatibility, and antimicrobial activity [152]. ZnONPs have been incorporated with various polymers to prepare nanocomposites with an effective antimicrobial activity for food packaging. For instance, poly (hydroxybutyrate-co-hydroxyvalerate) nanocomposites containing ZnONPs were produced for antimicrobial packaging of foods like cheese, baked goods, cereals, and meat products [153]. Poly lactic acid (PLA) is another biodegradable polymer that can be used to prepare packaging materials with ZnONPs. PLA/ZnONPs composite films were prepared with excellent antibacterial effect against *E. coli* and improved barrier properties for packaging applications [154]. Polyvinyl chloride film coated with ZnONPs showed effective antimicrobial properties and was used as a packaging material. The film possessed good inhibition effect on the growth of *E. coli* and *S. aureus* and its ability was due to the presence of ZnONPs. The film exhibited more effective antibacterial activity for *S. aureus*.

However, the antifungal activity of the film against *Aspergillus flavus* and *Penicillium citrinum* was not observed, attributed to the complexity of the fungal cell wall and ZnONPs without irradiated UV (ultraviolet) light or an insufficient number of NPs. [155]. The high bactericidal activity against *E. coli* and *S. aureus* was noticed in ZnO/TiO_2NPs incorporated into linear low-density polyethylene (LLDPE) packaging films [156]. Solution-casted nanocomposite films were developed from carboxymethyl cellulose, chitosan, and ZnONPs for potential soft white cheese packaging. The shelf life of white soft cheese packaged with nanocomposite film was extended up to 30 days. The yeast and mold count were not detectable in the cheese after 15 days of storage at 7 °C [157]. MgONPs exhibit antimicrobial activity against various microorganisms. This was related to the ability of MgONPs to produce high concentrations of superoxide anions (—O_2) on MgO surfaces that could react with the carbonyl groups of peptide linkages in bacterial cell walls to destroy them [158].

TiO_2NPS have been extensively utilized in many sectors thanks to their low production cost, availability, biocompatibility, photocatalytic behavior, and antimicrobial performance [159]. TiO_2NPs were reported as good antimicrobial agents against many microorganisms. TiO_2NPs can damage the bacterial cell membrane via oxidizing their polyunsaturated phospholipids making it an appealing material for antimicrobial packaging applications [138]. TiO_2NPs and ZnONPs incorporated into poly (ethylene terephthalate)/poly (butylene succinate) binary blend thin films exhibited antimicrobial activity against *S. aureus* and *E. coli* [160]. As a result of the remarkable antimicrobial activity, novel active packaging based on carboxymethyl cellulose, chitosan, and ZnONPs has been used for extending the shelf life of bread [161]. Doped and undoped TiO_2NPs with AgNPs were mixed with polystyrene solution to coat rice-straw-derived paper for packaging application [162].

19.3.2.2 Food Additives

In addition to using antimicrobial NPs in food packaging, they could be directly utilized as additives in beverages and foods. The direct addition of antimicrobial agents has been explored in beverages such as milk and juices. The inclusion of these antimicrobial agents in foods improves the safety of the products and inhibits the growth of microbes and hence prevents from foodborne diseases. Similarly, in edible films, natural antimicrobial NPs have recently gained considerable attention owing to the increasing consumer demand on food safety. However, both synthetic and natural agents can result in undesired changes when embedded in the food system like active loss of nutrients or modification of texture. Thus, a low concentration so that the quality of the food products is not altered but is high enough to inhibit microbial growth. In this regard, nanoencapsulation of antimicrobial agents represents an efficient way to improve their activity to protect from the results of an interaction with food ingredients, augmenting its stability and dispersability. Selectively locating them in areas where microorganisms preferentially proliferate, for instance in water-rich phases aids in reducing interaction with the food [163].

Nisin is regarded as one of the most employed food natural preservatives, related to the "Generally Recognized as Safe (GRAS)" category and is permitted by the US FDA and by the European Union (E234). Various methods have been developed for including nisin into NPs for its application as a food additive with enhanced capacities [164]. Some of these studies have compared the antimicrobial activities of free nisin and that obtained when polymer NPs are incorporated with nisin. For instance, nisin-loaded Ch/AlgNPs (~200 nm of diameter) were evaluated as a nanoadditive in pasteurized and raw milks. It was indicated that nisin-loaded NPs could inhibit the growth of *S. aureus* bacteria more effectively during longer incubation periods than free nisin. The evaluation, based on the total counts, demonstrates a similar reduction of *S. aureus* bacteria population in pasteurized milk for free nisin and for nisin-loaded NPs for up to 24 hours. However, the most significant reduction was for the nisin-loaded NPs at 48 hours of incubation when compared to the free nisin. Also, a more significant reduction was observed at the longer incubation times for raw milk samples containing NPs. This could be attributed to the fact that free nisin interacts with the ingredients present in the milk and thus the antimicrobial activity decreases with time, while Ch/AlgNPs protect nisin and extend its activity [165].

Other studies showed that nisin was encapsulated into NPs composed of three biocompatible and food permissible polymers: chitosan, alginate, and Pluronic F68. The efficiency of the loaded NPs was determined in tomato juice and compared with the obtained results from adding free nisin and unloaded NPs. In a control experiment without additives, the growth of microorganisms (*Micrococcus luteus* [MTCC 1809], *Salmonella enterica* [MTCC 1253], *P. aeruginosa* [MTCC 424], and *Enterobacter aerogenes* [MTCC 2823]) was noticed within two days. In the case of unloaded NPs, no growth was observed until two months. Regarding free nisin, the growth inhibition was prolonged up to five months but the seed color changed to black. The best results were obtained with nisin-loaded NPs, which exhibited an antimicrobial effect on tomato juice over six months. Therefore, it was concluded that the antimicrobial properties of nisin were prolonged with encapsulation in polymeric NPs [166].

19.3.2.3 Edible Films

Recently, edible packaging utilizing edible and biodegradable biopolymers has shown itself to be a good alternative to synthetic film, though this does not mean a total replacement synthetic film. The aim of using edible films as packaging materials is to improve food safety, food quality, and extended the shelf life of foods. These biopolymers include polysaccharides (starch alginate, pectin, chitosan, and cellulose) and proteins (whey protein, zein, wheat gluten, soy, collagen, and casein), which could act as a barrier to control the diffusion of water and gases and could also serve as carriers of various food additives such as vitamins, antioxidants, and antimicrobials. Particularly, utilizing edible films containing antimicrobial agents has shown them to efficiently eliminate microbial growth and extend the shelf life of various foodstuffs. The incorporation of NPs as antimicrobials in the design of edible films leads to films with improved properties not only antimicrobial ones but also thermal and mechanical characteristics [167].

AgNPs were used to prepare antimicrobial nanocomposites owing to their antimicrobial properties. Incorporation of AgNPs in the formulation of edible films has also been evaluated. For example, the binary blends of both agar and banana were reinforced with AgNPs to prepare edible films. These films showed improved antimicrobial activity against the foodborne pathogenic bacteria *L. monocytogenes* (ATCC 15313) and *E. coli* (ATCC 43895) [168]. Also, edible films were prepared from ZnONPs and essential oils from oregano (*Origanum minutiflorum*) and rosemary (*Rosmarinus officinalis*) as antimicrobial agents. These films were able to inhibit the growth of *L. monocytogenes* and *S. aureus*, extend product shelf life, and reduce the risk of pathogens [169]. Nanoemulsions of clove bud (*Syzygium aromaticum*) and oregano (*Origanum vulgare*) were prepared via ultrasonication and were dispersed in methylcellulose to produce antimicrobial edible films [170].

19.3.3 Water Disinfection

Water is regarded as the fundamental resource for society both for human consumption and industrial processes. Recently, water resources have exhibited leakage with prolonged droughts becoming a common phenomenon around the globe, which constitutes a serious threat to the development of nations. Though our planet reveals great amounts of water, only a limited amount of this resource is available for human consumption. This is because most of the water is not suitable for direct use owing to its higher salinity or presence of chemical or biological contaminants. Additionally, after water is used in cities or for industrial purposes, various treatments are required to eliminate the contaminants before its deposition or recirculation. Hence, the water purification technologies represent a key step in coping with society's water demands. Enabling water purification for human consumption from alternative sources such as seawater is a necessity of high priority [171].

Nanotechnology has acted as a novel alternative for improving existing technologies in water purification. Water can be obtained from: fresh water sources like lakes and rivers; saline water from sources like oceans; and used-water collection systems such as sewage. However, various chemical and biological contaminants are present in these water sources. Thus, there are many areas of application of NPs in water purification. With regard to antimicrobial applications, there are three major areas where NPs have exhibited promising results at a laboratory scale [172].

i) Addition of ion leaching NPs to water for direct biocidal activity.
ii) Addition of NPs to water as disinfectants triggered by light sources.
iii) Application of antimicrobial NPs to filtration membranes to avoid biofouling.

19.3.3.1 Addition of Ion Leaching Nanoparticles to Water for Direct Biocidal Activity

NPs with the ability to produce water-soluble ions were employed for water disinfection. In this regard, MeNPs are directly added to the contaminated water and a constant release of the metal ions into the water over time occurs on contact with the water. The NPs and the dissolved metallic ions interact with microbes in several ways, resulting in their death. Some of the investigated NPs for this strategy are AgNPs, CuNPs, FeNPs, MgNPs, and ZnNPs. Regarding these cases, the biocidal mechanisms of NPs and ions in antimicrobial water treatment are not completely understood yet, but various mechanisms were proposed like absorption, membrane attachment, interaction with biomolecules, and oxidative damage [173]. In recent years, it has been indicated that bimetallic Fe—Cu NPs release two types of Cu ions, Cu (I) and Cu (II), owing to different ion-generation methods. These ions have shown selective biocidal activities toward bacteria and virus, respectively. This report [174] revealed that the NPs able to produce several ions simultaneously are systems with a broader action spectrum. In a real water disinfection operation on a large scale many types of NPs can be utilized together to achieve a synergistic effect against microorganisms.

19.3.3.2 Addition of Light-triggered Nanoparticles to Water as Disinfectants

This strategy could be applied to fresh and saline water and involves the direct addition of NPs to the contaminated water under the exposure of a light emitting source. This activates the NPs through catalyzing oxidation or reduction reactions directly onto the biomolecules or generating ROS. These ROS have the ability to cause oxidative damage to the biological macromolecules like DNA and cellular membranes, consequently leading to cellular death. The photoactivation of NPs takes place via the absorption of photons on an electronic band structure that excites electrons on the NPs surface, making them available for inducing reduction and oxidation reactions in the surrounding molecules. Metals can be utilized as photocatalysts with biocidal properties in water treatment. MeONPs such as ZnONPs, TiO2NPs, and CuONPs could be used as photocatalysts with antimicrobial properties [131].

19.3.3.3 Water Cleansing using Antimicrobial Nanoparticles on Filtration Membranes

Filtration membranes enable the purification of water with high suspended solid contents. Nanofiltration is a proven alternative for the purification of seawater. Unfortunately, filtration operations are prone to the formation of biofilms on the membrane surface over time that compromises the activity of the membranes. Additionally, when polymeric membranes, which are polyamide-based membranes, are utilized they are sensitive to the use of biocidal chemicals such as chloride. A method to avoid the proliferation of microbes over the filtration membranes is use the NPs on the membrane surfaces so that they can exert antimicrobial activity. This method provides the opportunity to accomplish both the disinfection and filtration of water simultaneously. A first alternative for the control of biofouling over filtration membranes is to add ion-leaching NPs to the surface of membrane. The ions leached from the NPs exhibit antimicrobial activities against microorganisms in the membrane surroundings [175].

NPs could be added to the membrane via either synthesizing them first or binding the synthesized NPs to membranes as a second step, or they could be formed *in situ* over the filtration membranes. Some of the ways for the incorporation of NPs onto polymeric filtration membranes are dip coating, blending, interfacial polymerization, and grafting polymerization. Graft polymerization is advantageous as it is able to bind NPs to filtration membranes via covalent bonds, providing the desired antimicrobial activity. However, there is difficulty in large-scale commercial modification of membranes and the lack of process development persists [176].

MeONPs are regarded as the most studied NPs with antimicrobial properties that could be embedded into polymeric filtration membranes. Among several MeONPs added to polymeric membranes, the use of ZnONPs, AgONPs, and TiO_2NPs has been reported. Reapplication to the membrane is inevitable as the sustained release of metal ions over time eventually leads to the full depletion of the NPs [177].

19.3.3.4 Environmental and Techno-economical Aspects of Water Disinfection using Nanoparticles

When using NPs for water disinfection, the economic analysis and life cycle of NPs should considered particularly for higher volume water-treatment systems. This is to avoid releasing post-water-treatment NPs intothe environment and to eliminate the costs related to using high-cost NPs in treatment process. In recent years, concerns have been raised regarding the toxicity for aquatic organisms and the development of microbial resistance to metallic NPs when released from wastewater treatment facilities into aquatic ecosystems. The two technological ways to the controlled use of NPs are: the recovery after disinfection via direct dissolution in polluted water; and/or the fixation on a support that then comes in contact with polluted water. When NPs are used directly by dissolution in water, they must be separated from the water after treatment. Current water-treatment plants are able to remove about 80% of NPs from the treated water [178]. Recent studies showed that there is a very low potential for NPs to be left in the wastewater from water-treatment plants owing to the high removal rate efficiency. Hence, even though NPs are released into the environment after water treatment, their concentration could be so low that their effects could be comparable to those observed by the naturally occurring NPs in the environment. Nevertheless, the important technical challenge is total recovery and reuse of NPs over the continuous generation of freshly synthesized NPs. This is beneficial for sustainability and for cost management especially for large quantity use like in water-treatment plants. The challenge of NP recovery and reuse has been poorly studied and reported in the literature [179].

Several studies showed that metallic NPs like AgNPs can be removed from water after treatment in wastewater plants via efficient clarification process design and control, with projected efficiencies of the recovery of NPs of about 80% [180]. Recent trends in the research toward the recovery of NPs have emerged that involve the use of superparamagnetic particles decorated with NPs that have the ability to promote disinfection and magnetic separation thereafter. For instance, magnetic iron particles are utilized as a support for exploiting NPs' antimicrobial action with an easy recovery of the NP-support system [181]. Magnetic barium phosphate flakes have been used as support for AgNPs and FeNPs in water-treatment. This method seems promising, however further research is required to demonstrate the feasibility of this process on a large scale, both from an economic and technical stand point [182]. The second method for the use of NPs is the immobilization on a support to avoid a subsequent recovery process. This is a technological option that has attracted remarkable attention owing to its efficiency, reduced complexity, and lower cost. For instance, a patented design that involves a manufacturing process for filtration membranes with incorporated AgNPs to achieve the purification and disinfection of water has been developed recently. This strategy is considered as advantageous in terms of achieving water treatment in a single step but it usually requires a high energy input to achieve filtration since the pore size of the membrane decreases [183].

19.3.4 Textiles

Textiles are another vector for transmission of microorganisms and infectious diseases, particularly in the healthcare sectors such as hospitals where clinicians and patients are exposed to bodily fluids that contain microscopic deposits, which are sometimes inevitable and go unnoticed. Lab coats of doctors and technicians, bed linen, patient robes, and curtains are potential carriers of microorganisms. Thus, these fabrics provide an excellent environment for harboring and growing pathogenic microorganisms since they present large surface areas that can hold oxygen and moisture. Thus, great attention has been paid toward the development of textiles involving decorative, protective, and technical textile products with antimicrobial properties owing to the need to protect users from pathogenic microorganisms that lead to health problems. In addition, installing antimicrobial actions saves fabrics from unfavorable discoloration or damage caused by the proliferation of bacteria resulting in rotting and low functionality. Thus, an antimicrobial coating could be used to minimize the spread of microorganisms and to protect textiles [184]. The stability and durability of metal NPs such as AgNPs and MeONPs like ZnONPs, CuONPs, and TiO_2NPs make them effective and attractive antimicrobial agents in textiles. The improved antimicrobial actions that arise from these NPs stem from their very small size. Hence, the combination of the actions of (i) an increased surface area that facilities interaction with microorganisms and (ii) an increased solubility caused by the increased surface area contributes to the improved antimicrobial behavior of NPs. The antimicrobial activity of biocidal NPs is dependent on the agent and its interaction with the microorganisms [185].

AgNPs possess antimicrobial activity against a wide variety of microorganisms including bacteria, viruses, fungi, and yeasts. The antimicrobial mechanism of AgNPs is not fully understood. The most common mechanisms proposed in the literature are: (i) The accumulation of AgNPs and passage via the cell membrane and interruption of its proper function; (ii) the damage of cell wall that leads to high membrane permeability and interference with the cell respiratory system; (iii) the gradual leaching of Ag ions into the microbial cells, which in the presence of dissolved oxygen causes the production of ROS; and (iv) the interaction between absorbed Ag ions with sulfur-containing proteins in the cell and phosphorous-containing compounds such as DNA, resulting in cell death [140]. Sodium-citrate stabilized AgNPs were prepared and chemically bound to cotton fibers using citric acid to impart antimicrobial activity against *S. aureus* and *E. coli*. It was found that the durability and efficacy of the treated fiber was showcased by a 100% bacterial reduction against both types of bacteria prior to washing and 92% and 99% reduction for *S. aureus* and for *E. coli* after 50 washing cycles, respectively. The absorption of AgNPs on the cytoderm of bacteria, followed by the subsequent penetration of the cytomembrane to interfere with the normal functions of cells was the mechanism of cell death [186].

MeONPs could be utilized in textiles as antimicrobial agents. This could be as a result of their photocatalytic nature and ability to produce electrons and holes under UV/irradiation followed by the generation of ROS that activates various toxic pathways like membrane leakage and intracellular damage. This leads to the interruption of the normal metabolism of bacteria and the oxidation of organic components of cells resulting in the cell death [185]. In a report by Zahid et al., self-cleaning cotton fabrics were prepared using manganese-doped TiO_2 NPs immobilized into a polydimethylsiloxane matrix and the antimicrobial properties of the product were tested against *S. aureus* and *Klebsiells pneumoniae*. Results showed 100% reduction in the bacteria within 120 minutes of incubation under sunlight [187].

Other MeONPs like ZnO, CaO, MgO, and CuO also show antimicrobial activity and have the potential to be used in antimicrobial applications. Antibacterial textiles based on ZnONPs and TiO_2NPs were fabricated and applied to the cotton fabrics in the pad-cure-dry method. The obtained results from antibacterial tests using the agar diffusion method against *B. subtilis* and *E. coli* indicated that the inhibition zones around the treated fibers prior to washing had increased and by increasing the washing cycles, the inhibition zone increased [188]. ZnONPs were incorporated into cotton fabrics using various concentrations of hexamethyltriethylene tetramine. The results of antimicrobial activity of the treated cotton fabrics showed improved antimicrobial behavior against *E. coli* and *S. aureus*. Also, there was marginal increase in antimicrobial activity with an increase in the concentration of the antimicrobial agent [189]. Ghayempour and Montazer synthesized ZnONPs that were stabilized by tragacanth gum on cotton surfaces. The antimicrobial behavior of this treated cotton was tested against *Candiba albicans, S. aereus*, and *E. coli* exhibiting excellent efficiency. This efficiency was due to the production of ROS species by ZnONPs that disrupted the microbial membrane [190]. El-Naggar et al. coated cotton fabric with ZnONPs capped with date seed extract and compared the effects of capping on the antimicrobial activity against *P. aeruginosa, B. subtilis*, and *E. coli*. The results revealed that the capped and uncapped treated fabric exhibited excellent antimicrobial activity against all the tested microorganisms. However, the durability of the uncapped fabric was remarkably affected via washing since inhibition zones completely disappeared after 15 wash cycles [191].

The antimicrobial properties of CuONPs coated over cotton fabrics were tested against *E. coli, Proteus vulgaris*, and *S. aureus* by using the agar diffusion method as studied by Sathiyavimal et al. The results showed clear zones of bacterial inhibition around the cotton fabrics [192]. The environmentally friendly and bioactive chitosan NP not only exhibits antimicrobial activity in textiles but also improves the strength and durability of fabrics. Lu et al. prepared chitosan NPs and applied the NPs to wool fabrics using a dipping process. The antibacterial properties indicated excellent results, showing the ability to kill bacteria. The bacterial activity, depending on the size and concentration of chitosan NPs, was due to the production of -NH^{3+} anions that interact with the cell wall of bacteria and limits growth [193]. Using nanocomposites of antimicrobial agents in textiles has been shown to exhibit positive synergistic antimicrobial effect compared to the single nanomaterial. Gaminian and Montazer improved the antibacterial activity of polyester fabric using Cu_2O/TiO_2 nanocomposites [194]. In all of these studies, the textile exhibited superior antimicrobial properties compared to the single nanomaterial as a result of the synergistic effect of the NPs.

In summary, using antimicrobial NPs in textiles has proven to be effective against microorganisms, depending on which NPs were used. Also, the durability and longevity of these antimicrobial fabrics/textiles via simulated use has shown that the antimicrobial efficacy remains fairly constant. Research is required to develop ways of lowering the cost and increasing the volume involved in the production of NPs to make applications viable and affordable, thereby, improving antimicrobial activities and minimizing the spread of diseases.

19.3.5 Other Applications

19.3.5.1 Antimicrobial Nanoparticles in Coating

Antimicrobial agents are incorporated into most cleaning agents and household products since they inhibit the growth of microorganism upon application. However, antimicrobial actions have a short effective lifespan since the microorganisms find their way back on to the treated surfaces. Thus, for surfaces that require the constant inhibition of microorganisms like medical devices and food preparation surfaces, there is an urgent need for permanent treatment to inhibit the microbial activity. An antimicrobial coating is important for inhibiting the growth of microbes' aggregates (biofilms) on the exposed surfaces, thereby, limiting biodeterioration, health, and the environmental problems. Biofilms are defined as a community of microbes that possess high resistance against antibiotics that irreversibly adhere to surfaces like suture materials and therapeutic devices resulting in the spread of infectious diseases [131].

One of the strategies employed to combat the problems created by microorganisms is to coat the surfaces or substrates like cosmetics, textiles, clothes, or personal care products that are exposed to microbial species with antimicrobial agents. The demands for self-disinfection and antimicrobial surfaces have been on the rise owing to the health hazards created by harmful bacteria. Antimicrobial coatings could be applied to surfaces (counters, walls, medical devices, door handles), water purifiers, buried pipelines, and textiles. Moreover, antimicrobial coatings could protect against premature biodegradation, corrosion, and the loss of paint's integrity, arising from the discoloration from microbial activities [195].

The approach of inhibiting the bacterial growth via coating could be classified based on the mechanism of action. One approach involves the incorporation of antimicrobial compounds by soaking, mixing, or coating surfaces/materials with an antimicrobial agent. The problem of this method is the depletion in the long term since the biocidal agent will drop below the minimum inhibitory concentration as it leaches away, leading to increased microbial resistance and accumulation of dead microbes on the surface as well as biofilm formation. The other approach is to create antifouling surfaces to prevent the adhesion of bacteria and accumulation of living and dead bacteria on surfaces, thereby preventing biofilm formation via coating with hydrophilic polymers or zwitterionic polymers. Although, this antifouling coating could inhibit the bacterial adhesion to surfaces, they do not kill bacteria. Thus, the potential for microbial colonization and biofouling exists. Some researchers integrated both antimicrobial and antifouling agents in coatings to simultaneously achieve antimicrobial activities and antiadhesion, hence, extending the antimicrobial activity service life [196, 197].

With the advances in nanotechnology, there is growing interest in developing antimicrobial NPs that could be used for coating applications. These antimicrobial NPs could make changes to the physical, chemical, and optical properties of the coatings. AgNPs with a high surface area are known to exhibit increased antimicrobial activities against a broad spectrum of bacteria, virus, fungi, and drug-resistant bacteria. Naik and Ratna synthesized a nanocomposite coating of hyperbranched urethane alkyd resin containing AgNPs, in which AgNPs were obtained from silver benzoate. The nanocomposite coating possessed complete inhibition against *Serratia marcescens* at 0.5 wt.% concentration of AgNPs [198]. Chen et al. prepared AgNPs via reducing silver ions in aqueous solutions and incorporated them evenly on double-layered hydroxide (LDH). The Ag-LDH coatings were deposited on a glass substrate and were tested against *E. coli, P. aeruginosa, B. subtilis,* and *S. aureus*. The results indicated that 100% of bacteria were inhibited in three hours. The recycle runs were carried out in order to elucidate the stability of the Ag-LDH coating. After four consecutive recycle runs, the antimicrobial activity against all of the tested bacteria was at a minimum of 97% death rate, with 74% of Ag remaining in the coating. This indicates that not only are AgNPs excellent for antimicrobial coating, but the antimicrobial effect is also retained for a long period of time [199]. MeONPs such as oxides of Zn and Ti have been investigated as antimicrobial agents in coating applications. The antimicrobial activity of these NPs is associated with their photocatalytic and hydrophilic nature and their ability to produce electrons and holes under UV light illumination. TiO_2NPs is regarded as the most common metal oxide utilized in antimicrobial and antifouling coating applications as a result of the combination between its chemical stability, relatively low density compared to other metals, corrosion resistance, good antimicrobial activity, and high strength-to-weight ratio [200].

Marine biofouling is a common problem that affects maritime industries across the world. In the past, an antifouling coating based on tributyltin has been used to address this concern. However, as a result of the high toxicity of this agent toward nontargeted organisms, there is great interest for developing less-toxic antifouling coatings [201]. A nanocomposite composed of chitosan and low-toxic ZnO coatings was investigated for antidiatom, antibacterial, and antibiofilm activity using *Pseudoalteromonas nigrifaciens* as test bacteria. Results showed that nanocomposites exhibited better antibiofilm properties compared to chitosan alone. Chitosan deactivates bacteria via irreversibly binding to the cells making it unavailable to bind to other cells. However, this consumes chitosan and decreases the efficacy as an antimicrobial agent within a short period of time. In the absence of light, the chitosan/ZnO nanocomposite had no antibacterial activity owing to the lack of photocatalysis of ZnONPs [202]. Therefore, it is important to note that coating with this technology requires exposure to a light source, which is a limiting factor in its potential applications. Due to the toxicity, to marine life, of organotin-based antifouling agents the use of Cu_2O as a replacement biofilm agent has been investigated. In a report by Saeed et al., the antimicrobial and antibiofilm properties of epoxy coatings reinforced with Cu_2O NPs were studied. Results showed excellent antibacterial properties against gram-positive and gram-negative bacteria at 0.3 wt. % Cu_2O concentration. The increase or the decrease in the stated concentration indicated a decrease in antimicrobial activity owing to the dependence of antimicrobial activity on the oxidation of Cu_2ONPs. In terms of biofilm formation, all treated-surface-based biofilm cells died within four days [203].

19.3.5.2 Antimicrobial Nanoparticles in Paint Formulations

Paints are highly susceptible to attack and colonization by microorganisms owing to the presence of cellulosic compounds utilized as thickeners. These serve as a carbon source for consumption and the growth of microorganisms. Consequently, this results in color changes, staining, loosening of paints, and the potential spread of infectious diseases. Therefore, adding antimicrobial agents to paint formulations in small quantities is necessary in order to prevent the growth of bacteria and molds, as well as to preserve the functionality and appearance. The common materials that are utilized in paint formulations are Ag and metal oxides or a combination of both to obtain increased antimicrobial activity efficiency. With the advent of nanotechnology some researchers have studied both the incorporation of antimicrobial NPs in paints and their antimicrobial properties [204].

A report carried out by Lateef et al. showed that using AgNPs synthesized from bean extract as an additive to emulsion paint resulted in improved antimicrobial activity against a multidrug-resistant bacteria and fungal pathogens [205]. In a study by Tornero et al., ZnO partially coated with AgNPs incorporated in waterborne paints showed excellent antimicrobial activity against bacteria and mold growth. The study outcomes demonstrated 100% inhibition against *S. aereus*, *E. coli*, and *Aspergillus niger* at a concentration of 0.15%. The long-term (six months) antifungal property of these synergetic materials against molds was retained for the most part, the antimicrobial activity decreased owing to the leaching out of biocide in the treated samples when tested against bacteria. This observation revealed that these materials are effective over a prolonged time against molds. The researchers attributed this observation to the oxidation of Ag ions, which can form soluble complexes with the free organic molecules. These complexes can either mask the Ag or leach out when in contact with water, hence decreasing the antimicrobial activity of the AgNP-based paints. On the other hand, ZnO maintained its antifungal activity, with this attributed to its poor solubility in water [204].

The antifungal activity of AgNPs and ZnONPs on indoor waterborne paints against *Chaetomium globosum* and *A. alternate* was investigated in order to evaluate the bioresistance of acrylic paints with incorporated NPs. The results revealed excellent fungal growth inhibition at 5.8 mg/100 g concentration of AgNPs, with a diminishing effect as the concentration decreased. ZnONPs exhibited unfavorable inhibition demonstrated by the abundant growth of fungi after four days [206]. The utilization of NPs as an additive or a functional filler in materials has been questioned in terms of the cytotoxicity to mammalian cells. Some studies show that the product is regarded as nontoxic below certain limits while others argue about its capability of penetrating the human skin, thereby resulting in cytotoxicity and genotoxicity to human cells. In this area, Smulders et al. carried out an *in vitro* investigation to compare the toxicities on human bronchial and monocytic cells of TiO_2NPs, AgNPs and SiO_2NPs in: the particulate form; paints containing these NPs; and control paints. The results indicated that AgNPs and TiO_2NPs exhibited some cytotoxic effects at higher doses starting at 27 g ml^{-1} and 81 g ml^{-1} as a particle, respectively; whereas SiO2 (silica dioxide) and all aged paints did not [207].

19.4 Conclusions

NPs have exhibited great antimicrobial activities, enabling their potential use for a wide variety of antimicrobial applications. The reason behind their outstanding antimicrobial properties is their high surface area and size, resulting in increased contact with microorganisms. The use of nanotechnology in targeting the NPs in antimicrobial applications across many sectors has led to the translation of fundamental research from literature to practical and commercially available materials and applications such as in the medical sector, food industry, water purification, textiles, coatings, and paints.

With regard to the medical sector, using antimicrobial NPs has significantly reduced the potential spread and growth of infection via their incorporation in disposable items, medical devices, and application in surgical procedures. Despite the relative high cost of many antimicrobial NPs such as gold, silver, titanium oxides, copper, and their compounds, it has been indicated that their activities are high. At very low concentrations and size spectrum they could be used to effectively prevent the growth of microorganisms, thus decreasing cost via minimal material consumption. Moreover, compared to the cost of healthcare arising from microbial infection treatment and other personal losses, using antimicrobial NPs is feasible. An average antimicrobial biocide utilized in applications like surface treatment or oral decontamination during surgery is limited in their effectiveness since they wash away or wear out after a short period of time and have to be replaced. However, using NPs could provide long-lasting effects as they are embedded in the body and slowly release ions to inhibit the microbial growth. As result of their large surface area, NPs could be easily incorporated in environments where microorganisms grow, destroy them, and remain active thereafter.

Regarding the food industry, NPs have been used as antimicrobial agents where they are able to inhibit foodborne microbial growth, extend the shelf life, and decrease the chemical preservatives used in packaged products. However, safety concerns related to the leaching out of these NPs into the product is a major problem for industrial-scale production and in the use of such food packaging materials. The chemical bonding and other physical means of attaching the NPs to the packaging matrix systems can mitigate the leaching concerns. However, more research into the conjugation platforms and their effects on the antimicrobial activity of the NPs need to be conducted. Also, more clinical studies are required to completely understand the resulting effects of migrated metal ions from food packaging materials into food on the long-term health of consumers. Using preservation ways like thermal processes or modified atmospheres combining with antimicrobial packaging can lead to synergistic actions that can increase safety by using low amounts of antimicrobial agents.

Concerning water disinfection, the use of NPs has been widely explored as a promising alternative for securing water supply. However, there are some challenges to be considered in pursuing the large-scale implementation of these technologies in the replacement of existing ones. The direct adding of ion-leaching NPs to contaminated water requires continuous replenishing to compensate for the spent NPs leaching from their surface. This presents an economical challenge to the viability of this alternative method, given the difficulties and high cost in synthesizing NPs on a large scale. This can be alleviated to a certain extent if the recovery of NPs is further developed. For example, utilizing supported NP systems that can be separated via magnetic separation is a promising alternative, which is expected to grow in the near future. The use of light-triggered NPs for water purification and NP-decorated filtration membranes has a major constrain in the consumption of large amounts of energy during the process.

Using antimicrobial NPs in textiles has been widely investigated as these NPs could impart antimicrobial properties to the textiles. Consequently, this results in inhibition of the microorganisms and prevention of formation of biofilms on the surface of textiles such as cloths. Also, the incorporation of NPs in paint formulations could be a suitable way to stop biofilm formation on surfaces.

References

1 Sarmah, P., Dan, M.M., Adapa, D., and Sarangi, T.K. (2018). A review on common pathogenic microorganisms and their impact on human health. *Electronic Journal of Biology* 14: 50–58.

2 Rajapaksha, P., Elbourne, A., Gangadoo, S. et al. (2019). A review of methods for the detection of pathogenic microorganisms. *Analyst* 144: 396–411.

3 Khezerlou, A., Alizadeh-Sani, M., Azizi-Lalabadi, M., and Ehsani, A. (2018). Nanoparticles and their antimicrobial properties against pathogens including bacteria, fungi, parasites and viruses. *Microbial Pathogenesis* 123: 505–526.

4 Khan, I., Saeed, K., and Khan, I. (2019). Nanoparticles: properties, applications and toxicities. *Arabian Journal of Chemistry* 12: 908–931.

5 Sánchez-López, E., Gomes, D., Esteruelas, G. et al. (2020). Metal-based nanoparticles as antimicrobial agents: an overview. *Nanomaterials* 10: 1–39.

6 Azam, A., Ahmed, A.S., Oves, M. et al. (2012). Antimicrobial activity of metal oxide nanoparticles against gram-positive and gram-negative bacteria: a comparative study. *International Journal of Nanomedicine* 7: 6003.

7 Lam, S.J., Wong, E.H.H., Boyer, C., and Qiao, G.G. (2018). Antimicrobial polymeric nanoparticles. *Progress in Polymer Science* 76: 40–64.

8 Zahran, M. and Marei, A.H. (2019). Innovative natural polymer metal nanocomposites and their antimicrobial activity. *International Journal of Biological Macromolecules* 136: 586–596.

9 Oskam, G. (2006). Metal oxide nanoparticles: synthesis, characterization and application. *Journal of Sol-Gel Science and Technology* 37: 161–164.

10 Nikam, A.V., Prasad, B.L.V., and Kulkarni, A.A. (2018). Wet chemical synthesis of metal oxide nanoparticles: a review. *CrystEngComm* 20: 5091–5107.

11 Jin, T., Sun, D., Su, J.Y. et al. (2009). Antimicrobial efficacy of zinc oxide quantum dots against l*isteria monocytogenes*, *Salmonella enteritidis*, and *Escherichia coli* O157: H7. *Journal of Food Science* 74: M46–M52.

12 Hsueh, Y.-H., Ke, W.-J., Hsieh, C.-T. et al. (2015). ZnO nanoparticles affect *Bacillus subtilis* cell growth and biofilm formation. *PLoS One* 10: e0128457.

13 Ahmadi, A., Ahmadi, P., and Ehsani, A. (2020). Development of an active packaging system containing zinc oxide nanoparticles for the extension of chicken fillet shelf life. *Food Science & Nutrition* 8: 5461–5473.

14 Vidovic, S., Elder, J., Medihala, P. et al. (2015). ZnO nanoparticles impose a panmetabolic toxic effect along with strong necrosis, inducing activation of the envelope stress response in *Salmonella enterica* serovar enteritidis. *Antimicrobial Agents and Chemotherapy* 59: 3317–3328.

15 Banoee, M., Seif, S., Nazari, Z.E. et al. (2010). ZnO nanoparticles enhanced antibacterial activity of ciprofloxacin against *Staphylococcus aureus* and *Escherichia coli*. *Journal of Biomedical Materials Research. Part B, Applied Biomaterials* 93: 557–561.

16 Akbar, A. and Anal, A.K. (2014). Zinc oxide nanoparticles loaded active packaging, a challenge study against *Salmonella typhimurium* and *Staphylococcus aureus* in ready-to-eat poultry meat. *Food Control* 38: 88–95.

17 He, L., Liu, Y., Mustapha, A., and Lin, M. (2011). Antifungal activity of zinc oxide nanoparticles against Botrytis cinerea and *Penicillium expansum*. *Microbiological Research* 166: 207–215.

18 Rajiv, P., Rajeshwari, S., and Venckatesh, R. (2013). Bio-fabrication of zinc oxide nanoparticles using leaf extract of *Parthenium hysterophorus* L. and its size-dependent antifungal activity against plant fungal pathogens. *Spectrochimica Acta. Part A, Molecular and Biomolecular Spectroscopy* 112: 384–387.

19 Lipovsky, A., Nitzan, Y., Gedanken, A., and Lubart, R. (2011). Antifungal activity of ZnO nanoparticles – the role of ROS mediated cell injury. *Nanotechnology* 22: 105101.

20 Sardella, D., Gatt, R., and Valdramidis, V.P. (2017). Physiological effects and mode of action of ZnO nanoparticles against postharvest fungal contaminants. *Food Research International* 101: 274–279.

21 Antoine, T.E., Mishra, Y.K., Trigilio, J. et al. (2012). Prophylactic, therapeutic and neutralizing effects of zinc oxide tetrapod structures against herpes simplex virus type-2 infection. *Antiviral Research* 96: 363–375.

22 Chai, W., Zakrzewski, S.S., Günzel, D. et al. (2014). High-dose dietary zinc oxide mitigates infection with transmissible gastroenteritis virus in piglets. *BMC Veterinary Research* 10: 75.

23 Delavari, M., Dalimi, A., Ghaffarifar, F., and Sadraei, J. (2014). In vitro study on cytotoxic effects of ZnO nanoparticles on promastigote and amastigote forms of *Leishmania major* (MRHO/IR/75/ER). *Iranian Journal of Parasitology* 9: 6.

24 Paul, K.B., Kumar, S., Tripathy, S. et al. (2016). A highly sensitive self assembled monolayer modified copper doped zinc oxide nanofiber interface for detection of *Plasmodium falciparum* histidine-rich protein-2: targeted towards rapid, early diagnosis of malaria. *Biosensors and Bioelectronics* 80: 39–46.

25 Pandiyarajan, T., Udayabhaskar, R., Vignesh, S. et al. (2013). Synthesis and concentration dependent antibacterial activities of CuO nanoflakes. *Materials Science & Engineering, C: Materials for Biological Applications* 33: 2020–2024.

26 Shankar, S., Wang, L.-F., and Rhim, J.-W. (2017). Preparation and properties of carbohydrate-based composite films incorporated with CuO nanoparticles. *Carbohydrate Polymers* 169: 264–271.

27 Ren, G., Hu, D., Cheng, E.W.C. et al. (2009). Characterisation of copper oxide nanoparticles for antimicrobial applications. *International Journal of Antimicrobial Agents* 33: 587–590.

28 Saad, A.H.A., Soliman, M.I., Azzam, A.M., and Mostafa, A.B. (2015). Antiparasitic activity of silver and copper oxide nanoparticles against *Entamoeba histolytica* and *Cryptosporidium parvum* cysts. *Journal of the Egyptian Society of Parasitology* 240: 1–10.

29 Pansambal, S., Deshmukh, K., Savale, A. et al. (2017). Phytosynthesis and biological activities of fluorescent CuO nanoparticles using *Acanthospermum hispidum* L. extract. *Journal of Nanostructures* 7: 165–174.

30 Santhoshkumar, T., Rahuman, A.A., Jayaseelan, C. et al. (2014). Green synthesis of titanium dioxide nanoparticles using *Psidium guajava* extract and its antibacterial and antioxidant properties. *Asian Pacific Journal of Tropical Medicine* 7: 968–976.

31 Ammendolia, M.G., Iosi, F., De Berardis, B. et al. (2014). *Listeria monocytogenes* behaviour in presence of non-UV-irradiated titanium dioxide nanoparticles. *PLoS One* 9: e84986.

32 Alizadeh-Sani, M., Hamishehkar, H., Khezerlou, A. et al. (2020). Kinetics analysis and susceptibility coefficient of the pathogenic bacteria by titanium dioxide and zinc oxide nanoparticles. *Advanced Pharmaceutical Bulletin* 10: 56.

33 Karunakaran, G., Suriyaprabha, R., Manivasakan, P. et al. (2013). Impact of nano and bulk ZrO_2, TiO_2 particles on soil nutrient contents and PGPR. *Journal of Nanoscience and Nanotechnology* 13: 678–685.

34 Su, S., Ye, L., Tian, Q. et al. (2020). Photoelectrocatalytic inactivation of *Penicillium expansum* spores on a Pt decorated TiO_2/activated carbon fiber photoelectrode in an all-solid-state photoelectrochemical cell. *Applied Surface Science* 515: 145964.

35 Akinola, P.O., Lateef, A., Asafa, T.B. et al. (2020). Multifunctional titanium dioxide nanoparticles biofabricated via phytosynthetic route using extracts of *Cola nitida*: antimicrobial, dye degradation, antioxidant and anticoagulant activities. *Heliyon* 6: e04610.

36 Mukherjee, K., Acharya, K., Biswas, A., and Jana, N.R. (2020). TiO_2 nanoparticles co-doped with nitrogen and fluorine as visible-light-activated antifungal agents. *ACS Applied Nano Materials* 3: 2016–2025.

37 Mazurkova, N.A., Spitsyna, Y.E., Shikina, N.V. et al. (2010). Interaction of titanium dioxide nanoparticles with influenza virus. *Nanotechnologies in Russia* 5: 417–420.

38 Ishiguro, H., Nakano, R., Yao, Y. et al. (2011). Photocatalytic inactivation of bacteriophages by TiO_2-coated glass plates under low-intensity, long-wavelength UV irradiation. *Photochemical and Photobiological Sciences* 10: 1825–1829.

39 Saleh, A.H. (2019). Potential role of titanium dioxide (TiO_2) nanoparticles against the toxicity of *Leishmania Tropica* in adult albino male rats. *JGPT* 11: 453–457.

40 Allahverdiyev, A.M., Abamor, E.S., Bagirova, M. et al. (2013). Investigation of antileishmanial activities of TiO_2@ Ag nanoparticles on biological properties of *L. tropica* and *L. infantum* parasites, in vitro. *Experimental Parasitology* 135: 55–63.

41 Dolat, E., Salarabadi, S.S., Layegh, P. et al. (2020). The effect of UV radiation in the presence of TiO_2-NPs on Leishmania major promastigotes. *Biochimica et Biophysica Acta (BBA) - General Subjects* 1864: 129558.

42 Suman, T.Y., Ravindranath, R.R.S., Elumalai, D. et al. (2015). Larvicidal activity of titanium dioxide nanoparticles synthesized using *Morinda citrifolia* root extract against *Anopheles stephensi, Aedes aegypti* and *Culex quinquefasciatus* and its other effect on non-target fish. *Asian Pacific Journal of Tropical Disease* 5: 224–230.

43 Palanisamy, G. and Pazhanivel, T. (2017). Green synthesis of MgO nanoparticles for antibacterial activity. *International Research Journal of Engineering and Technology* 4: 137–141.

44 Jin, T. and He, Y. (2011). Antibacterial activities of magnesium oxide (MgO) nanoparticles against foodborne pathogens. *Journal of Nanoparticle Research* 13: 6877–6885.

45 Krishnamoorthy, K., Manivannan, G., Kim, S.J. et al. (2012). Antibacterial activity of MgO nanoparticles based on lipid peroxidation by oxygen vacancy. *Journal of Nanoparticle Research* 14: 1063.

46 Karimiyan, A., Najafzadeh, H., Ghorbanpour, M., and Hekmati-Moghaddam, S.H. (2015). Antifungal effect of magnesium oxide, zinc oxide, silicon oxide and copper oxide nanoparticles against *Candida albicans. Zahedan Journal of Research in Medical Sciences* 17: 19–23.

47 Sierra-Fernandez, A., De la Rosa-García, S.C., Gomez-Villalba, L.S. et al. (2017). Synthesis, photocatalytic, and antifungal properties of MgO, ZnO and Zn/Mg oxide nanoparticles for the protection of calcareous stone heritage. *ACS Applied Materials and Interfaces* 9: 24873–24886.

48 Bafghi, A.F., Daghighi, M., Daliri, K., and Jebali, A. (2015). Magnesium oxide nanoparticles coated with glucose can silence important genes of *Leishmania major* at sub-toxic concentrations. *Colloids and Surfaces. B, Biointerfaces* 136: 300–304.

49 Borthakur, P., Hussain, N., Darabdhara, G. et al. (2018). Adhesion of gram-negative bacteria onto α-Al_2O_3 nanoparticles: a study of surface behaviour and interaction mechanism. *Journal of Environmental Chemical Engineering* 6: 3933–3941.

50 Baysal, A. and Saygin, H. (2019). Physico-chemical and toxicological behaviour of Al_2O_3 nanoparticles in fine particulate matter. *Environmental Engineering and Management Journal* 18.

51 Manikandan, V., Jayanthi, P., Priyadharsan, A. et al. (2019). Green synthesis of pH-responsive Al2O3 nanoparticles: application to rapid removal of nitrate ions with enhanced antibacterial activity. *Journal of Photochemistry and Photobiology A: Chemistry* 371: 205–215.

52 Jalal, M., Ansari, M.A., Shukla, A.K. et al. (2016). Green synthesis and antifungal activity of Al_2O_3 NPs against fluconazole-resistant Candida spp isolated from a tertiary care hospital. *RSC Advances* 6: 107577–107590.

53 Raghunath, A. and Perumal, E. (2017). Metal oxide nanoparticles as antimicrobial agents: a promise for the future. *International Journal of Antimicrobial Agents* 49: 137–152.

54 Alfei, S. and Schito, A.M. (2020). Positively charged polymers as promising devices against multidrug resistant gram-negative bacteria: a review. *Polymers (Basel)* 12: 1195.

55 Beveridge, T.J. (1999). Structures of gram-negative cell walls and their derived membrane vesicles. *Journal of Bacteriology* 181: 4725–4733.

56 Ren, E., Zhang, C., Li, D. et al. (2020). Leveraging metal oxide nanoparticles for bacteria tracing and eradicating. *View* 1: 20200052.

57 Chandrangsu, P., Rensing, C., and Helmann, J.D. (2017). Metal homeostasis and resistance in bacteria. *Nature Reviews Microbiology* 15: 338.

58 Gold, K., Slay, B., Knackstedt, M., and Gaharwar, A.K. (2018). Antimicrobial activity of metal and metal-oxide based nanoparticles. *Advances in Therapy* 1: 1700033.

59 Fu, P.P., Xia, Q., Hwang, H.-M. et al. (2014). Mechanisms of nanotoxicity: generation of reactive oxygen species. *Journal of Food and Drug Analysis* 22: 64–75.

60 Karakoti, A.S., Hench, L.L., and Seal, S. (2006). The potential toxicity of nanomaterials – the role of surfaces. *JOM* 58: 77–82.

61 Lynch, I. and Dawson, K.A. (2008). Protein-nanoparticle interactions. *Nano Today* 3: 40–47.

62 Aggarwal, P., Hall, J.B., McLeland, C.B. et al. (2009). Nanoparticle interaction with plasma proteins as it relates to particle biodistribution, biocompatibility and therapeutic efficacy. *Advanced Drug Delivery Reviews* 61: 428–437.

63 Vega-Jiménez, A.L., Vázquez-Olmos, A.R., Acosta-Gío, E., and Álvarez-Pérez, M.A. (2019). In vitro antimicrobial activity evaluation of metal oxide nanoparticles. In: *Nanoemulsions* (ed. K.S. Koh and V.L. Wong). IntechOpen.

64 Chen, H. and Wang, L. (2014). Nanostructure sensitization of transition metal oxides for visible-light photocatalysis. *Beilstein Journal of Nanotechnology* 5: 696–710.

65 Slavin, Y.N., Asnis, J., Häfeli, U.O., and Bach, H. (2017). Metal nanoparticles: understanding the mechanisms behind antibacterial activity. *Journal of Nanobiotechnology* 15: 1–20.

66 Krce, L., Šprung, M., Maravić, A. et al. (2020). Bacteria exposed to silver nanoparticles synthesized by laser ablation in water: modelling *E. coli* growth and inactivation. *Materials (Basel)* 13: 653.

67 Hsueh, Y.-H., Lin, K.-S., Ke, W.-J. et al. (2015). The antimicrobial properties of silver nanoparticles in *Bacillus subtilis* are mediated by released Ag+ ions. *PLoS One* 10: e0144306.

68 Vu, X.H., TTT, D., TTH, P. et al. (2018). Synthesis and study of silver nanoparticles for antibacterial activity against *Escherichia coli* and *Staphylococcus aureus*. *Advances in Natural Sciences: Nanoscience and Nanotechnology* 9: 25019.

69 Swolana, D., Kępa, M., Idzik, D. et al. (2020). The antibacterial effect of silver nanoparticles on *Staphylococcus epidermidis* strains with different biofilm-forming ability. *Nanomaterials* 10: 1010.

70 Halkai, K.R., Mudda, J.A., Shivanna, V. et al. (2018). Antibacterial efficacy of biosynthesized silver nanoparticles against *Enterococcus faecalis* biofilm: an in vitro study. *Contemporary Clinical Dentistry* 9: 237.

71 Hamida, R.S., Ali, M.A., Goda, D.A. et al. (2020). Cytotoxic effect of green silver nanoparticles against ampicillin-resistant *Klebsiella pneumoniae*. *RSC Advances* 10: 21136–21146.

72 Abd-Elhakeem, M.A., Badawy, I., and Raafat, A. (2016). Efficacy of silver nanoparticles as antimicrobial agent against salmonella infection and accompanied biochemical, immunological and histopathological changes in rats. *Egyptian Journal of Pure and Applied Science* 54 (2): 13–19.

73 Mousavi, S.A.A., Salari, S., and Hadizadeh, S. (2015). Evaluation of antifungal effect of silver nanoparticles against *Microsporum canis, Trichophyton mentagrophytes* and *Microsporum gypseum*. *Iranian Journal of Biotechnology* 13: 38.

74 Mishra, S., Singh, B.R., Singh, A. et al. (2014). Biofabricated silver nanoparticles act as a strong fungicide against *Bipolaris sorokiniana* causing spot blotch disease in wheat. *PLoS One* 9: e97881.

75 Jo, Y.-K., Kim, B.H., and Jung, G. (2009). Antifungal activity of silver ions and nanoparticles on phytopathogenic fungi. *Plant Disease* 93: 1037–1043.

76 Park, S., Ko, Y.-S., Lee, S.J. et al. (2018). Inactivation of influenza a virus via exposure to silver nanoparticle-decorated silica hybrid composites. *Environmental Science and Pollution Research International* 25: 27021–27030.

77 Gaikwad, S., Ingle, A., Gade, A. et al. (2013). Antiviral activity of mycosynthesized silver nanoparticles against herpes simplex virus and human parainfluenza virus type 3. *International Journal of Nanomedicine* 8: 4303.

78 Morris, D., Ansar, M., Speshock, J. et al. (2019). Antiviral and immunomodulatory activity of silver nanoparticles in experimental RSV infection. *Viruses* 11: 732.

79 Rogers, J.V., Parkinson, C.V., Choi, Y.W. et al. (2008). A preliminary assessment of silver nanoparticle inhibition of monkeypox virus plaque formation. *Nanoscale Research Letters* 3: 129.

80 Mohammed, O.T., Abdulkhaliq, R.J., and Mohammed, S.T. (2019). The effects of *Fusarium graminarum* silver nanoparticles on *Leishmania tropica*. *Journal of Physics Conference Series* 1294, IOP Publishing: 62075.

81 Baiocco, P., Ilari, A., Ceci, P. et al. (2011). Inhibitory effect of silver nanoparticles on trypanothione reductase activity and *Leishmania infantum* proliferation. *ACS Medicinal Chemistry Letters* 2: 230–233.

82 Zahra'a, A.A., Mustafa, T.A., Ardalan, N.M., and Idan, E.M. (2017). In vitro toxicity evaluation of silver nanoparticles on Entamoeba histolytica trophozoite. *Baghdad Science Journal* 14: 509–515.

83 Cameron, P., Gaiser, B.K., Bhandari, B. et al. (2016). Silver nanoparticles decrease the viability of *Cryptosporidium parvum* oocysts. *Applied and Environmental Microbiology* 82: 431–437.

84 Said, D.E., Elsamad, L.M., and Gohar, Y.M. (2012). Validity of silver, chitosan, and curcumin nanoparticles as anti-giardia agents. *Parasitology Research* 111: 545–554.

85 Vergara-Duque, D., Cifuentes-Yepes, L., Hincapie-Riaño, T. et al. (2020). Effect of silver nanoparticles on the morphology of toxoplasma gondii and salmonella braenderup. *Journal of Nanotechnology* 2020 (2): 1–11.

86 Dolat, E., Rajabi, O., Salarabadi, S.S. et al. (2015). Silver nanoparticles and electroporation: their combinational effect on *Leishmania major*. *Bioelectromagnetics* 36: 586–596.

87 Avitabile, E., Senes, N., D'Avino, C. et al. (2020). The potential antimalarial efficacy of hemocompatible silver nanoparticles from Artemisia species against *P. falciparum* parasite. *PLoS One* 15: e0238532.

88 Nalini, M., Lena, M., Sumathi, P., and Sundaravadivelan, C. (2017). Effect of phyto-synthesized silver nanoparticles on developmental stages of malaria vector, *Anopheles stephensi* and dengue vector, *Aedes aegypti*. *Egyptian Journal of Basic Applied Sciences* 4: 212–218.

89 Elbehiry, A., Al-Dubaib, M., Marzouk, E., and Moussa, I. (2019). Antibacterial effects and resistance induction of silver and gold nanoparticles against *Staphylococcus aureus*-induced mastitis and the potential toxicity in rats. *Microbiologyopen* 8: e00698.

90 Vaidya, M.Y., McBain, A.J., Butler, J.A. et al. (2017). Antimicrobial efficacy and synergy of metal ions against *Enterococcus faecium*, *Klebsiella pneumoniae* and *Acinetobacter baumannii* in planktonic and biofilm phenotypes. *Scientific Reports* 7: 1–9.

91 Shamaila, S., Zafar, N., Riaz, S. et al. (2016). Gold nanoparticles: an efficient antimicrobial agent against enteric bacterial human pathogen. *Nanomaterials* 6: 71.

92 Jayaseelan, C., Ramkumar, R., Rahuman, A.A., and Perumal, P. (2013). Green synthesis of gold nanoparticles using seed aqueous extract of *Abelmoschus esculentus* and its antifungal activity. *Industrial Crops and Products* 45: 423–429.

93 Rahimi, H., Roudbarmohammadi, S., Delavari, H., and Roudbary, M. (2019). Antifungal effects of indolicidin-conjugated gold nanoparticles against fluconazole-resistant strains of *Candida albicans* isolated from patients with burn infection. *International Journal of Nanomedicine* 14: 5323.

94 Vijayakumar, S. and Ganesan, S. (2012). Gold nanoparticles as an HIV entry inhibitor. *Current HIV Research* 10: 643–646.

95 Sametband, M., Shukla, S., Meningher, T. et al. (2011). Effective multi-strain inhibition of influenza virus by anionic gold nanoparticles. *MedChemComm* 2: 421–423.

96 Meléndez-Villanueva, M.A., Morán-Santibañez, K., Martínez-Sanmiguel, J.J. et al. (2019). Virucidal activity of gold nanoparticles synthesized by green chemistry using garlic extract. *Viruses* 11: 1111.

97 Al-Ardi, M.H. (2020). The uses of gold nanoparticles and *Citrullus colocynthis* L. nanoparticles against of *Giardia lamblia* in vivo. *Clinical Epidemiology and Global Health* 8 (4): 1282–1286.

98 Varela-Aramburu, S., Ghosh, C., Goerdeler, F. et al. (2020). Targeting and inhibiting *Plasmodium falciparum* using ultra-small gold nanoparticles. *ACS Applied Materials and Interfaces* 12: 43380–43387.

99 Adeyemi, O.S., Murata, Y., Sugi, T., and Kato, K. (2017). Inorganic nanoparticles kill *Toxoplasma gondii* via changes in redox status and mitochondrial membrane potential. *International Journal of Nanomedicine* 12: 1647.

100 Alti, D., Veeramohan Rao, M., Rao, D.N. et al. (2020). Gold–silver bimetallic nanoparticles reduced with herbal leaf extracts induce ROS-mediated death in both promastigote and amastigote stages of *Leishmania donovani*. *ACS Omega* 5: 16238–16245.

101 Sazgarnia, A., Taheri, A.R., Soudmand, S. et al. (2013). Antiparasitic effects of gold nanoparticles with microwave radiation on promastigotes and amastigotes of *Leishmania major*. *International Journal of Hyperthermia* 29: 79–86.

102 Jain, S., Huang, Z., Dixon, B.R. et al. (2019). *Cryptosporidium parvu*m oocyst directed assembly of gold nanoparticles and graphene oxide. *Frontiers of Chemical Science and Engineering* 13: 608–615.

103 Lemire, J.A., Harrison, J.J., and Turner, R.J. (2013). Antimicrobial activity of metals: mechanisms, molecular targets and applications. *Nature Reviews Microbiology* 11: 371–384.

104 Singh, A., Gautam, P.K., Verma, A. et al. (2020). Green synthesis of metallic nanoparticles as effective alternatives to treat antibiotics resistant bacterial infections: a review. *Biotechnology Reports* 25: e00427.

105 Turel, I. and Kljun, J. (2011). Interactions of metal ions with DNA, its constituents and derivatives, which may be relevant for anticancer research. *Current Topics in Medicinal Chemistry* 11: 2661–2687.

106 Jomova, K., Baros, S., and Valko, M. (2012). Redox active metal-induced oxidative stress in biological systems. *Transition Metal Chemistry* 37: 127–134.

107 Goss, C.H., Kaneko, Y., Khuu, L. et al. (2018). Gallium disrupts bacterial iron metabolism and has therapeutic effects in mice and humans with lung infections. *Science Translational Medicine* 10 (460): eaat7520.

108 Dutta, S., Mitra, M., Agarwal, P. et al. (2018). Oxidative and genotoxic damages in plants in response to heavy metal stress and maintenance of genome stability. *Plant Signaling & Behavior* 13: e1460048.

109 Abdal Dayem, A., Hossain, M.K., Bin, L.S. et al. (2017). The role of reactive oxygen species (ROS) in the biological activities of metallic nanoparticles. *International Journal of Molecular Sciences* 18: 120.

110 Zhu, Y., Yang, B., Chen, S., and Du, J. (2017). Polymer vesicles: mechanism, preparation, application, and responsive behavior. *Progress in Polymer Science* 64: 1–22.

111 Ren, J.M., McKenzie, T.G., Fu, Q. et al. (2016). Star polymers. *Chemical Reviews* 116: 6743–6836.

112 Elsabahy, M. and Wooley, K.L. (2012). Design of polymeric nanoparticles for biomedical delivery applications. *Chemical Society Reviews* 41: 2545–2561.

113 Rai, R., Alwani, S., and Badea, I. (2019). Polymeric nanoparticles in gene therapy: new avenues of design and optimization for delivery applications. *Polymers (Basel)* 11: 745.

114 Hadinoto, K., Sundaresan, A., and Cheow, W.S. (2013). Lipid–polymer hybrid nanoparticles as a new generation therapeutic delivery platform: a review. *European Journal of Pharmaceutics and Biopharmaceurtics* 85: 427–443.

115 Carmona-Ribeiro, A.M. and de Melo Carrasco, L.D. (2013). Cationic antimicrobial polymers and their assemblies. *International Journal of Molecular Sciences* 14: 9906–9946.

116 Mouffouk, F., Simão, T., Dornelles, D.F. et al. (2015). Self-assembled polymeric nanoparticles as new, smart contrast agents for cancer early detection using magnetic resonance imaging. *International Journal of Nanomedicine* 10: 63.

117 Lam, S.J., Wong, E.H.H., O'Brien-Simpson, N.M. et al. (2016). Bionano interaction study on antimicrobial star-shaped peptide polymer nanoparticles. *ACS Applied Materials and Interfaces* 8: 33446–33456.

118 JA, M.C., Sharbeen, G., Kavallaris, M., and Phillips, P.A. (2019). The use of star polymer nanoparticles for the delivery of siRNA to mouse orthotopic pancreatic tumor models. *RNA Interference and Cancer Therapy*, Springer 329–353.

119 Soares, D.C.F., Domingues, S.C., Viana, D.B., and Tebaldi, M.L. (2020). Polymer-hybrid nanoparticles: current advances in biomedical applications. *Biomedicine and Pharmacotherapie* 131: 110695.

120 Wang, L., Hu, C., and Shao, L. (2017). The antimicrobial activity of nanoparticles: present situation and prospects for the future. *International Journal of Nanomedicine* 12: 1227.

121 Li, X., Kolltveit, K.M., Tronstad, L., and Olsen, I. (2000). Systemic diseases caused by oral infection. *Clinical Microbiology Reviews* 13: 547–558.

122 Zehnder, M. (2006). Root canal irrigants. *Journal of Endodontics* 32: 389–398.

123 Ibrahim, A.I.O., Moodley, D.S., Petrik, L., and Patel, N. (2017). Use of antibacterial nanoparticles in endodontics. *South African Dental Journal* 72: 105–112.

124 Rodrigues, C.T., De Andrade, F.B., De Vasconcelos, L. et al. (2018). Antibacterial properties of silver nanoparticles as a root canal irrigant against *Enterococcus faecalis* biofilm and infected dentinal tubules. *International Endodontic Journal* 51: 901–911.

125 De Almeida, J., Cechella, B.C., Bernardi, A.V. et al. (2018). Effectiveness of nanoparticles solutions and conventional endodontic irrigants against *Enterococcus faecalis* biofilm. *Indian Journal of Dental Research* 29: 347.

126 Burduşel, A.-C., Gherasim, O., Grumezescu, A.M. et al. (2018). Biomedical applications of silver nanoparticles: An up-to-date overview. *Nanomaterials* 8: 681.

127 Singh, N., Tang, Y., and Ogunseitan, O.A. (2020). Environmentally sustainable management of used personal protective equipment. *Environmental Science and Technology* 54: 8500–8502.

128 Mansi, A., Boccuni, F., and Iavicoli, S. (2019). Nanomaterials as a new opportunity for protecting workers from biological risk. *Industrial Health* 197–2018.

129 Hiragond, C.B., Kshirsagar, A.S., Dhapte, V.V. et al. (2018). Enhanced anti-microbial response of commercial face mask using colloidal silver nanoparticles. *Vacuum* 156: 475–482.

130 Li, Y., Leung, P., Yao, L. et al. (2006). Antimicrobial effect of surgical masks coated with nanoparticles. *Journal of Hospital Infection* 62: 58–63.

131 Ogunsona, E.O., Muthuraj, R., Ojogbo, E. et al. (2020). Engineered nanomaterials for antimicrobial applications: a review. *Applied Materials Today* 18: 100473.

132 Mariappan, N. (2019). Recent trends in nanotechnology applications in surgical specialties and orthopedic surgery. *Biomedical and Pharmacology Journal* 12: 1095–1127.

133 Habib, E., Wang, R., Wang, Y. et al. (2016). Inorganic fillers for dental resin composites: present and future. *ACS Biomaterials Science & Engineering* 2: 1–11.

134 Aydin Sevinç, B. and Hanley, L. (2010). Antibacterial activity of dental composites containing zinc oxide nanoparticles. *Journal of Biomedical Materials Research. Part B Applied Biomaterials* 94: 22–31.

135 Kassaee, M.Z., Akhavan, A., Sheikh, N., and Sodagar, A. (2008). Antibacterial effects of a new dental acrylic resin containing silver nanoparticles. *Journal of Applied Polymer Science* 110: 1699–1703.

136 Newell, D.G., Koopmans, M., Verhoef, L. et al. (2010). Food-borne diseases – the challenges of 20 years ago still persist while new ones continue to emerge. *International Journal of Food Microbiology* 139: S3–S15.

137 Chudobová, D., Cihalova, K., Kopel, P. et al. (2015). Antimicrobial nanomaterials in the food industry. *Kvasny Prumysl* 61: 51–56.

138 Nakazato, G., Kobayashi, R.K.T., Seabra, A.B., and Duran, N. (2017). Use of nanoparticles as a potential antimicrobial for food packaging. *Food Preservation*, Elsevier 413–447.

139 Huang, Y., Mei, L., Chen, X., and Wang, Q. (2018). Recent developments in food packaging based on nanomaterials. *Nanomaterials* 8: 1–29.

140 Das, C.G.A., Kumar, V.G., Dhas, T.S. et al. (2020). Antibacterial activity of silver nanoparticles (biosynthesis): a short review on recent advances. *Biocatalysis and Agricultural Biotechnology* 27: 101593.

141 Deka, H., Karak, N., Kalita, R.D., and Buragohain, A.K. (2010). Bio-based thermostable, biodegradable and biocompatible hyperbranched polyurethane/Ag nanocomposites with antimicrobial activity. *Polymer Degradation and Stability* 95: 1509–1517.

142 Shankar, S. and Rhim, J.-W. (2016). Tocopherol-mediated synthesis of silver nanoparticles and preparation of antimicrobial PBAT/silver nanoparticles composite films. *LWT- Food Science and Technology* 72: 149–156.

143 Azlin-Hasim, S., Cruz-Romero, M.C., Morris, M.A. et al. (2015). Effects of a combination of antimicrobial silver low density polyethylene nanocomposite films and modified atmosphere packaging on the shelf life of chicken breast fillets. *Food Packaging and Shelf Life* 4: 26–35.

144 Yu, H.-Y., Yang, X.-Y., Lu, F.-F. et al. (2016). Fabrication of multifunctional cellulose nanocrystals/poly (lactic acid) nanocomposites with silver nanoparticles by spraying method. *Carbohydrate Polymers* 140: 209–219.

145 Mastromatteo, M., Conte, A., Lucera, A. et al. (2015). Packaging solutions to prolong the shelf life of Fiordilatte cheese: bio-based nanocomposite coating and modified atmosphere packaging. *LWT- Food Science and Technology* 60: 230–237.

146 Youssef, A.M., Abdel-Aziz, M.S., and El-Sayed, S.M. (2014). Chitosan nanocomposite films based on Ag-NP and Au-NP biosynthesis by *Bacillus subtilis* as packaging materials. *International Journal of Biological Macromolecules* 69: 185–191.

147 Lomate, G.B., Dandi, B., and Mishra, S. (2018). Development of antimicrobial LDPE/Cu nanocomposite food packaging film for extended shelf life of peda. *Food Packaging and Shelf Life* 16: 211–219.

148 Arfat, Y.A., Ejaz, M., Jacob, H., and Ahmed, J. (2017). Deciphering the potential of guar gum/Ag-Cu nanocomposite films as an active food packaging material. *Carbohydrate Polymers* 157: 65–71.

149 Ahmed, J., Mulla, M., Arfat, Y.A. et al. (2018). Compression molded LLDPE films loaded with bimetallic (Ag-Cu) nanoparticles and cinnamon essential oil for chicken meat packaging applications. *LWT* 93: 329–338.

150 Arfat, Y.A., Ahmed, J., Hiremath, N. et al. (2017). Thermo-mechanical, rheological, structural and antimicrobial properties of bionanocomposite films based on fish skin gelatin and silver-copper nanoparticles. *Food Hydrocolloids* 62: 191–202.

151 Shkodenko, L., Kassirov, I., and Koshel, E. (2020). Metal oxide nanoparticles against bacterial biofilms: perspectives and limitations. *Microorganisms* 8: 1545.

152 Mostafa, M.H., Elsawy, M.A., Darwish, M.S.A. et al. (2020). Microwave-assisted preparation of chitosan/ZnO nanocomposite and its application in dye removal. *Materials Chemistry and Physics* 248: 122914.

153 Díez-Pascual, A.M. and Díez-Vicente, A.L. (2014). ZnO-reinforced poly(3-hydroxybutyrate-co-3-hydroxyvalerate) bionanocomposites with antimicrobial function for food packaging. *ACS Applied Materials and Interfaces* 6: 9822–9834.

154 Kim, I., Viswanatahn, K., Kasi, G. et al. (2019). Poly (lactic acid)/ZnO bionanocomposite films with positively charged ZnO as potential antimicrobial food packaging materials. *Polymers* 11 (19): -1427.

155 Li, X., Xing, Y., Jiang, Y. et al. (2009). Antimicrobial activities of ZnO powder-coated PVC film to inactivate food pathogens. *International Journal of Food Science and Technology* 44: 2161–2168.

156 Saharudin, K.A., Sreekantan, S., Basiron, N. et al. (2018). Bacteriostatic activity of LLDPE nanocomposite embedded with sol–gel synthesized TiO_2/ZnO coupled oxides at various ratios. *Polymers (Basel)* 10: 878.

157 Youssef, A.M., El-Sayed, S.M., El-Sayed, H.S. et al. (2016). Enhancement of Egyptian soft white cheese shelf life using a novel chitosan/carboxymethyl cellulose/zinc oxide bionanocomposite film. *Carbohydrate Polymers* 151: 9–19.

158 Nguyen, N.-Y.T., Grelling, N., Wetteland, C.L. et al. (2018). Antimicrobial activities and mechanisms of magnesium oxide nanoparticles (nMgO) against pathogenic bacteria, yeasts, and biofilms. *Scientific Reports* 8: 1–23.

159 Hussein, L.I., Abdaleem, A.H., Darwish, M.S.A. et al. (2020). Chitosan/TiO_2 nanocomposites: effect of microwave heating and solution mixing techniques on physical properties. *Egyptian Jorunal of Chemistry* 63 (2): 449–460.

160 Threepopnatkul, P., Wongnarat, C., Intolo, W. et al. (2014). Effect of TiO_2 and ZnO on thin film properties of PET/PBS blend for food packaging applications. *Energy Procedia* 56: 102–111.

161 Noshirvani, N., Ghanbarzadeh, B., Mokarram, R.R., and Hashemi, M. (2017). Novel active packaging based on carboxymethyl cellulose-chitosan-ZnONPs nanocomposite for increasing the shelf life of bread. *Food Packaging and Shelf Life* 11: 106–114.

162 Youssef, A.M., Kamel, S., and El-Samahy, M.A. (2013). Morphological and antibacterial properties of modified paper by PS nanocomposites for packaging applications. *Carbohydrate Polymers* 98: 1166–1172.

163 Ameta, S.K., Rai, A.K., Hiran, D. et al. (2020). Use of nanomaterials in food science. In: *Biogenic Nano-Particles and Their Use in Agro-Ecosystems* (ed. M. Ghorbanpour, P.B.A. Varma and D.K. Choudhary), 457–488. Springer.

164 Khan, I. and Oh, D.-H. (2016). Integration of nisin into nanoparticles for application in foods. *Innovative Food Science and Emerging Technologies* 34: 376–384.

165 Zohri, M., Alavidjeh, M.S., Haririan, I. et al. (2010). A comparative study between the antibacterial effect of nisin and nisin-loaded chitosan/alginate nanoparticles on the growth of *Staphylococcus aureus* in raw and pasteurized milk samples. *Probiotics and Antimicrobial Proteins* 2: 258–266.

166 Bernela, M., Kaur, P., Chopra, M., and Thakur, R. (2014). Synthesis, characterization of nisin loaded alginate–chitosan–pluronic composite nanoparticles and evaluation against microbes. *LWT- Food Science and Technology* 59: 1093–1099.

167 Cagri, A., Ustunol, Z., and Ryser, E.T. (2004). Antimicrobial edible films and coatings. *Journal of Food Protection* 67: 833–848.

168 Orsuwan, A., Shankar, S., Wang, L.-F. et al. (2016). Preparation of antimicrobial agar/banana powder blend films reinforced with silver nanoparticles. *Food Hydrocolloids* 60: 476–485.

169 Khalaf, H.H., Sharoba, A.M., El-Tanahi, H.H., and Morsy, M.K. (2013). Stability of antimicrobial activity of pullulan edible films incorporated with nanoparticles and essential oils and their impact on Turkey deli meat quality. *Journal of Food and Dairy Sciences* 4: 557–573.

170 Otoni, C.G., Pontes, S.F.O., Medeiros, E.A.A., and de FF, S.N. (2014). Edible films from methylcellulose and nanoemulsions of clove bud (*Syzygium aromaticum*) and oregano (*Origanum vulgare*) essential oils as shelf life extenders for sliced bread. *Journal of Agricultural and Food Chemistry* 62: 5214–5219.

171 Pichel, N., Vivar, M., and Fuentes, M. (2019). The problem of drinking water access: a review of disinfection technologies with an emphasis on solar treatment methods. *Chemosphere* 218: 1014–1030.

172 Li, Q., Mahendra, S., Lyon, D.Y. et al. (2008). Antimicrobial nanomaterials for water disinfection and microbial control: potential applications and implications. *Water Research* 42: 4591–4602.

173 Pradeep, T. (2009). Noble metal nanoparticles for water purification: a critical review. *Thin Solid Films* 517: 6441–6478.

174 Amin, M.T., Alazba, A.A., and Manzoor, U. (2014). A review of removal of pollutants from water/wastewater using different types of nanomaterials. *Advances in Materials Science and Engineering* 1: 1–24.

175 Bagheri, M. and Mirbagheri, S.A. (2018). Critical review of fouling mitigation strategies in membrane bioreactors treating water and wastewater. *Bioresource Technology* 258: 318–334.

176 Zhu, J., Hou, J., Zhang, Y. et al. (2018). Polymeric antimicrobial membranes enabled by nanomaterials for water treatment. *Journal of Membrane Science* 550: 173–197.

177 Ng, L.Y., Mohammad, A.W., Leo, C.P., and Hilal, N. (2013). Polymeric membranes incorporated with metal/metal oxide nanoparticles: a comprehensive review. *Desalination* 308: 15–33.

178 Good, K.D., Bergman, L.E., Klara, S.S. et al. (2016). Implications of engineered nanomaterials in drinking water sources. *Journal of American Water Works Association* 108: 1–17.

179 Westerhoff, P., Atkinson, A., Fortner, J. et al. (2018). Low risk posed by engineered and incidental nanoparticles in drinking water. *Nature Nanotechnology* 13: 661–669.

180 Vilela, P., Liu, H., Lee, S. et al. (2018). A systematic approach of removal mechanisms, control and optimization of silver nanoparticle in wastewater treatment plants. *Science of the Total Environment* 633: 989–998.

181 Sharifi, R., Hassani, A.H., Panahi, H.A., and Borghei, M. (2018). Performance of silver nanoparticle fixed on magnetic iron nanoparticles (Fe_3O_4–Ag) in water disinfection. *Micro and Nano Letters* 13: 436–441.

182 Zhang, F., Lee, M.H., Huang, Y. et al. (2019). Effective water disinfection using magnetic barium phosphate nanoflakes loaded with ag nanoparticles. *Journal of Cleaner Production* 218: 173–182.

183 Linhares, A.M.F., Grando, R.L., Borges, C.P., and da Fonseca, F.V. (2018). Technological prospection on membranes containing silver nanoparticles for water disinfection. *Recent Patents on Nanotechnology* 12: 3–12.

184 Mejía, M.L., Zapata, J., Cuesta, D.P. et al. (2017). Properties of antibacterial nano textile for use in hospital environments. *Revista Ingenieria Biomédica* 11: 13–19.

185 Joshi, M. and Roy, A. (2018). Antimicrobial textiles based on metal and metal oxide nano-particles. In: *Nanomaterials in the Wet Processing of Textiles* (ed. S. Ul-Islam and B.S. Butola), 71–111. Beverly, MA: Scrivener Publishing LLC.

186 Wu, Y., Yang, Y., Zhang, Z. et al. (2019). Fabrication of cotton fabrics with durable antibacterial activities finishing by Ag nanoparticles. *Textile Research Journal* 89: 867–880.

187 Zahid, M., Papadopoulou, E.L., Suarato, G. et al. (2018). Fabrication of visible light-induced antibacterial and self-cleaning cotton fabrics using manganese doped TiO_2 nanoparticles. *ACS Applied Bio Materials* 1: 1154–1164.

188 Abou Elmaaty, T.M. and Mandour, B.A. (2018). ZnO and TiO_2 nanoparticles as textile protecting agents against UV radiation: a review. *Asian Journal of Chemical Sciences* 1–14.

189 Shaheen, T.I., El-Naggar, M.E., Abdelgawad, A.M., and Hebeish, A. (2016). Durable antibacterial and UV protections of in situ synthesized zinc oxide nanoparticles onto cotton fabrics. *International Journal of Biological Macromolecules* 83: 426–432.

190 Ghayempour, S. and Montazer, M. (2017). Ultrasound irradiation based in-situ synthesis of star-like Tragacanth gum/zinc oxide nanoparticles on cotton fabric. *Ultrasonics Sonochemistry* 34: 458–465.

191 El-Naggar, M.E., Shaarawy, S., and Hebeish, A.A. (2018). Multifunctional properties of cotton fabrics coated with in situ synthesis of zinc oxide nanoparticles capped with date seed extract. *Carbohydrate Polymers* 181: 307–316.

192 Sathiyavimal, S., Vasantharaj, S., Bharathi, D. et al. (2018). Biogenesis of copper oxide nanoparticles (CuONPs) using Sida acuta and their incorporation over cotton fabrics to prevent the pathogenicity of gram negative and gram positive bacteria. *Journal of Photochemistry and Photobiology, B: Biology* 188: 126–134.

193 Lu, Y., Chen, Y., Lin, H. et al. (2010). Preparation of chitosan nanoparticles and their application to *Antheraea pernyi* silk. *Journal of Applied Polymer Science* 117: 3362–3369.

194 Gaminian, H. and Montazer, M. (2015). Enhanced self-cleaning properties on polyester fabric under visible light through single-step synthesis of cuprous oxide doped nano-TiO_2. *Photochemistry and Photobiology* 91: 1078–1087.

195 Ahonen, M., Kahru, A., Ivask, A. et al. (2017). Proactive approach for safe use of antimicrobial coatings in healthcare settings: opinion of the COST action network AMiCI. *International Jorunal of Environmental Research and Public Health* 14: 366.

196 Fu, Y., Yang, Y., Xiao, S. et al. (2019). Mixed polymer brushes with integrated antibacterial and antifouling properties. *Progress in Organic Coatings* 130: 75–82.

197 Knetsch, M.L.W. and Koole, L.H. (2011). New strategies in the development of antimicrobial coatings: the example of increasing usage of silver and silver nanoparticles. *Polymers (Basel)* 3: 340–366.

198 Naik, R.B. and Ratna, D. (2015). Synthesis of silver nanoparticles embedded novel hyperbranched urethane alkyd-based nanocomposite for high solid antimicrobial coating application. *Journal of Coatings Technology and Research* 12: 1073–1083.

199 Chen, C., Gunawan, P., Lou, X.W., and Xu, R. (2012). Silver nanoparticles deposited layered double hydroxide nanoporous coatings with excellent antimicrobial activities. *Advanced Functional Materials* 22: 780–787.

200 Cartwright, A., Jackson, K., Morgan, C. et al. (2020). A review of metal and metal-oxide nanoparticle coating technologies to inhibit agglomeration and increase bioactivity for agricultural applications. *Agronomy* 10: 1018.

201 Bannister, J., Sievers, M., Bush, F., and Bloecher, N. (2019). Biofouling in marine aquaculture: a review of recent research and developments. *Biofouling* 35: 631–648.

202 Al-Naamani, L., Dobretsov, S., Dutta, J., and Burgess, J.G. (2017). Chitosan-zinc oxide nanocomposite coatings for the prevention of marine biofouling. *Chemosphere* 168: 408–417.

203 El Saeed, A.M., Abd El-Fattah, M., Azzam, A.M. et al. (2016). Synthesis of cuprous oxide epoxy nanocomposite as an environmentally antimicrobial coating. *International Journal of Biological Macromolecules* 89: 190–197.

204 Tornero, A.C.F., Blasco, M.G., Azqueta, M.C. et al. (2018). Antimicrobial ecological waterborne paint based on novel hybrid nanoparticles of zinc oxide partially coated with silver. *Progress in Organic Coatings* 121: 130–141.

205 Lateef, A., Azeez, M.A., Asafa, T.B. et al. (2016). Biogenic synthesis of silver nanoparticles using a pod extract of *Cola nitida*: antibacterial and antioxidant activities and application as a paint additive. *Journal of Taibah University for Science* 10: 551–562.

206 Bellotti, N., Romagnoli, R., Quintero, C. et al. (2015). Nanoparticles as antifungal additives for indoor water borne paints. *Progress in Organic Coatings* 86: 33–40.

207 Smulders, S., Luyts, K., Brabants, G. et al. (2015). Toxicity of nanoparticles embedded in paints compared to pristine nanoparticles, in vitro study. *Toxicology Letters* 232: 333–339.

Index